T0401824

The Black Sea Encyclopedia

Sergei R. Grinevetsky • Igor S. Zonn •
Sergei S. Zhiltsov • Aleksey N. Kosarev •
Andrey G. Kostianoy

The Black Sea Encyclopedia

 Springer

Sergei R. Grinevetsky
Parliament of Ukraine
Kiev
Ukraine

Igor S. Zonn
S.Yu. Vitte Moscow University
Moscow, Russia
and Engineering Scientific-Production Center
    for Water Economy Reclamation & Ecology
Moscow, Russia

Sergei S. Zhiltsov
Peoples' Friendship University of Russia
Moscow, Russia
and
Diplomatic Academy of the Russian
Ministry of Foreign Affairs
Moscow, Russia

Aleksey N. Kosarev
Dept. Oceanology
Lomonosov Moscow State University
Faculty of Geography
Vorobjovy Gory
Moscow, Russia

Andrey G. Kostianoy
P.P. Shirshov Inst. Oceanology
Russian Academy of Sciences
Moscow, Russia
and
S.Yu. Vitte Moscow University
Moscow, Russia

ISBN 978-3-642-55226-7   ISBN 978-3-642-55227-4 (eBook)
DOI 10.1007/978-3-642-55227-4
Springer Heidelberg New York Dordrecht London

Library of Congress Control Number: 2014949201

Printed on acid-free paper

Springer is part of Springer Science+Business Media (www.springer.com)

Contents

Introduction

"The Black Sea Encyclopedia" is the third one in the new series of encyclopedias about the seas of the former Soviet Union published by Springer-Verlag. The first volume—"The Aral Sea Encyclopedia" was published in 2009, the second volume—"The Caspian Sea Encyclopedia" appeared in 2010. This new volume includes the information about the Black Sea as well as for the Sea of Azov. It is planned that the series will be continued by the Far Eastern seas—the Japan, Okhotsk and Bering seas, Russian Arctic seas, and the Baltic Sea.

For several thousands years the Black and Azov Seas Region was shared by different civilizations from Ancient Greece to Mogol Khans and the Osman Empire. The region was an arena of continuous wars for the Black Sea coasts and control over the Black Sea between Russia and Turkey in eighteenth to nineteenth centuries, that continued by World War I and World War II in twentieth century. The disintegration of the Socialist block in 1989 (Bulgaria and Romania were socialist countries), disintegration of the USSR in 1991 when the new independent states of Ukraine and Georgia appeared, expansion of NATO over Turkey in 1952, and Bulgaria and Romania in 2004, independence of Abkhazia which left Georgia in 2008, all these political events have radically changed the political and economic situation in the Black Sea Region. This increase in the number of the Black Sea legal entities from four to seven has given rise to a whole tangle of geopolitical, economic, international legal, ethnic and environmental problems, each of which demands its own approach and settlement mechanism.

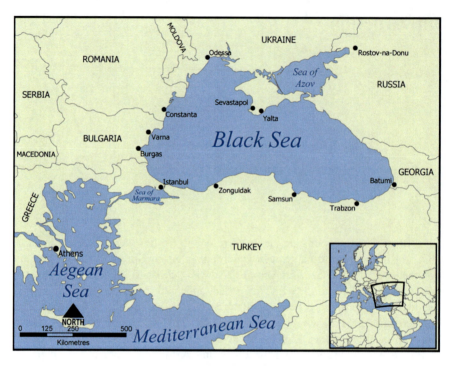

The Black Sea Region (http://climatelab.org/@api/deki/files/222/=Black_Sea_map.png)

   The Black Sea region has always attracted attention, which was caused by the unique features of the nature of the sea, by the diversity of natural resources, and by the great economic and geopolitical importance of the basin of the Black Sea. The Black Sea Region plays a significant role in the European and world economic system. The significance of the Black Sea Region is related, to a great extent, to its multiple sea and land transport routes and oil and gas pipelines, which play a significant role in export of oil, gas, wood, grain and other resources and products from Russia, Kazakhstan, Azerbaijan, Turkey and other countries. The Black Sea region is well known as a recreation area. Fishery is also retained in the region's economy, although, today its significance has strongly decreased. The role of hydrocarbon resources, to date, has been restricted to oil and gas prospecting in the shelf areas of the sea.

   The high degree of urbanization of the coastal zone of the Black Sea represents a permanent menace of marine environment pollution. The principal pollution sources include industrial wastes of large cities and their sanitary condition. Rivers also influence the pollution of marine environment. This is especially manifested in shelf regions such as the northwestern part of the Black Sea.

   One of the most important economic trends of the region is the transport development. Here, intensive international navigation takes place and a significant number of large ports are located. In so doing, a large part of the traffic is related to the transport of oil and oil products, which represents a permanent source of pollution of the sea. Oil enters the environment as a result of illegal, accidental, and operational discharges from vessels and oil terminals, as well as from land-based sources. Almost a half of the inputs of oil from land-based activities are brought to the Black Sea via the Danube River. The most serious aftereffects are related to accidents with ships carrying dangerous cargo, first of all, with tankers. Among the victims of such accidents, there are fishery, mariculture (mussel and shrimp industry), and recreation zones. A special problem is the navigation in the Turkish straits, where the transport, legal, and ecological issues are closely interwoven.

The navigation also affects the migration of organisms, which is often undesirable. In the autumn of 1988, the predator ctenophore *Mnemiopsis leidyi* started its mass development in the Black Sea. Because of the activity of the mnemiopsis population, the planktonic fodder base of planktivorous fishes such as Black Sea anchovy has reduced by 3–5 times; the biomass of these fishes proper, whose juveniles may be directly consumed by this ctenophore, has also decreased.

The particularity of the natural conditions of the Black Sea lies in the fact that it the largest basin in the world with a permanent halocline and a two-layered structure of the waters. The intensive pycno- and halocline prevents the waters from vertical mixing and oxygen penetration to deeper layers even in the period of the development of the wintertime vertical convection. Therefore, the entire water column below a depth of 100–200 m represents an inanimate hydrogen sulphide zone, in which only anoxic processes take place. About 90 % of the water volume does not participate in the processes of self-purification of the sea.

Owing to the isolated inland position of the Black Sea, the formation of its hydrological regime proceeds under the control of external factors such as heat and moisture fluxes, momentum across the sea surface, and riverine runoff. Therefore, the sea is distinguished by a high degree of variability in the hydrological and hydrochemical conditions, especially in shallow-water shelf areas. This, in turn, affects the biocenoses and, finally, leads to general changes in the ecosystem of the sea. Its pelagic ecosystem features a low resistance and is very sensitive to climatic changes and anthropogenic impacts, which hides manifestations of natural factors.

The natural regime of the Black Sea is relatively well studied. The peak of intensity of the studies occurred in the 1960–1980s, when a great number of expeditions was performed. Starting from the 1990s, after the disintegration of the Soviet Union, who carried out the bulk of the regular observations, their number has sharply decreased. Meanwhile, during the last two decades, international activity at sea has intensified and a significant number of expedition have been performed under the auspices of special programs (CoMSBlack, NATO TU–Black Sea, Black Sea Environmental Programme, ARENA, IASON, ASCABOS, SESAME, ALTICORE, MOPED, a series of other projects of the INTAS, INCO-Copernicus, NATO, International Atomic Energy Agency, UNESCO IOC, as well as those of the Russian Academy of Sciences, the Ministry of Education and Science of the Russian Federation, and the Russian Foundation for Basic Research) with the use of modern instruments and techniques for research.

Intensification of the Black Sea studies was mainly defined by three principal reasons: the influence of the regional climate changes during the last decade of the past century on the entire Black Sea ecosystem; the strongest impact of species–invaders on the pelagic and bottom biocenoses of the basin and catastrophic reduction in their commercial potential; and, finally, the large-scale construction and plans of construction of oil terminals and underwater gas pipelines such as the oil terminal of the Caspian Pipeline Consortium (CPC) on the Russian shelf (2001), the underwater gas pipelines "Blue Flow" (2003), "Dzhubga-Lazarevskoe-Sochi" (2010), "Blue Flow – 2" and "South Stream" (nearest future).

On March 15, 2007, in Athens, an agreement was signed between the governments of Russian Federation, Republic of Bulgaria and Greek Republic about the cooperation for construction and use of the Burgas—Alexandroupolis oil pipeline. This project implied an extension of the CPC system and a significant increase in the volumes of the oil transported by tankers between the CPC terminal near Novorossiisk and Burgas, since the throughput capacity of the pipeline should range from 35 to 50 Mt per year. In 2011 Bulgaria refused to participate in the project, which was automatically stopped.

On June 23, 2007, Russian "Gazprom" Company and Italian "Eni" unveiled a plan for a big new pipeline to take Russian gas under the Black Sea to Europe. The 900-km "South Stream" pipeline would submerge on the Russian coast, come ashore in Bulgaria, and then branch to Austria and Slovenia in one spur and to South Italy in another. All these projects required an adequate estimation of the aftereffects of the increasing anthropogenic stress on the sea environment.

As to the international legislation, there are selected unsolved issues such as those about the delimitation of near-shore aquatic areas between the Black Sea countries and the demarcation of the boundaries in the Black Sea and the Sea of Azov, for example in the regions of Tuzla Island in the Kerch Strait and Zmeinyi (Snake) Island in the western part of the sea. On February 3, 2009, the International Court of Justice delivered its judgment, which divided the sea area of the Black Sea along a line which was between the claims of the Ukraine and Romania, and Zmeinyi Island became a territory of the Ukraine.

As most of the publications on the Black and Azov Seas was published in Russian with a limited access to foreign readers (see References), we hope that "The Black Sea Encyclopedia" will help to understand specific Russian, Ukrainian, Georgian, Turkish, Bulgarian, and Romanian terms, terminology, and names of geographical features of the Black and Azov seas much better.

The last pages of Encyclopedia includes the Chronology of historical events having relation to the Black and Azov Seas development, exploration and study since eighteenth century to the present.

This publication is intended for a wide public—from decision-makers to school pupils, for all those who are interested in the problems of this region, its geography, history, ethnography, economics and ecology. This book is intended for specialists working in various fields of physical oceanography, marine chemistry, pollution studies, and biology. It may also be useful to undergraduate and graduate students of oceanography. The authors hope that this monograph will complement knowledge of the nature of the Black Sea and Sea of Azov. Detailed information to this handbook on particular issues may be obtained from the reference list.

We would like to thank the editors at Springer-Verlag for their interest in the Black Sea and Sea of Azov and their support for the present publication. We would like to thank Springer-Verlag for the steady and long-term interest to the Black Sea, which resulted in a significant list of publications (see References).

We are thankful to our good assistants in Moscow—Tatyana Abakumova for her preparation of the manuscript and Elena Kostianaya (Scientific and Coordination Oceanological Center at the P.P. Shirshov Institute of Oceanology, Russian Academy of Sciences) for her painstaking work on search and selection of information materials.

We are very thankful to our colleague Dmytro Solovyov (Marine Hydrophysical Institute, Sevastopol) and Evgenia Kostyanaya who provided sets of photos from their personal archives.

We are thankful to Dr. S.M. Shapovalov, head of the Scientific and Coordination Oceanological Center at the P.P. Shirshov Institute of Oceanology, Russian Academy of Sciences, for the support of our activities in preparation and publication of Encyclopedias of the Russian seas in Russian and English editions.

Moscow, 1 February 2014

P.S.: Dramatic political events in the Ukraine, the power overturn, and destabilization of the internal situation in the country in February and March 2014 led to a change of political map of the Black Sea Region. On 11th March 2014 Parliament of Autonomous Republic of Crimea adopted a Declaration of independence. On 16th March a referendum on the status of the Crimea took place. The official result from the Autonomous Republic of Crimea was a 96.77 % vote for integration of the region into the Russian Federation with an 83.1 % voter turnout. Based on these results, on 18th March 2014 the Crimea became a part of the Russian Federation, including the appropriate sea waters in the Black and Azov seas. On 12th May 2014 Donetsk and Lugansk People's Republics (former regions of the Ukraine) also declared independence from the Ukraine. We, the authors of the book, have had no time to correct the terms and maps related to the Crimea as a former territory of the Ukraine, as well as other political, administrative and territorial changes which occurred in the country, because the book was written before the end of 2013.

A

Abana – a village on the coast of the Black Sea, Turkey.

Abaza – strong north-easterly, easterly wind or one near the western shores of the Black Sea and in the low reaches of the Danube. Sometimes, its force is that of a severe storm. Blows from the Caucasus coast. In winter is accompanied by drifts and hard frosts. A hazard to fishing boats. Presumably, has to do with "Abazinets", in Crimean Tatar "Abaza"—"an unorthodox, stupid person". Another interpretation is "wind blowing from Abkhazia".

Abazins – (selfname "abaza") people in the North Caucasus. The Abazin Language belongs to the Abkhaz-Adyg group of the North-Caucasian language family. Ancestors of A. were the neighbors of closely-related Abkhazians and apparently by the end of the first millennium were partly assimilated by them. In the fourteenth to sixteenth centuries, part of A. moved from the Black Sea coast to the North Caucasus (the upper reaches of the rivers Urup, Laba, Kuban, Kuma and others), where they settled side by side with Adygs. During the fifteenth to sixteenth centuries, Abazin princes played an active role in political life of the region, but in the seventeenth century they became dependent on the rulers of Kabarda and West Adygea, eventually—on the Crimean Khans. Early in the seventeenth century they adopted Suniite Islam. From the first half of the nineteenth century, together with Karachaevs, Cherkes and other peoples, they constituted part of the Russian Empire (mainly, the territory of Kuban region). After the Caucasus war of 1817–1864, many A. moved to the Middle East. Main occupations of A.: agriculture and cattle-raising (including distant-range cattle-raising).

Abich Wilhelm Herman Wilhelmovich (sometimes Vassilyevich) (1806–1886) – geologist who earned during his lifetime the title "Founding father of the Caucasus and Caspian Sea Geology". Was born in Berlin, a German by birth. In 1831–1833, participated in expeditions aimed at exploring Italy's volcanic areas. At the recommendation of the German natural scientist and geographer A.Humboldt, was invited to join civil service in Russia. From 1841—Professor of the Natural History Chair at Derpt (Tallinn) University; in 1847 assigned to the Mining Engineers Corps.

S.R. Grinevetsky et al., *The Black Sea Encyclopedia*,
DOI 10.1007/978-3-642-55227-4_2, © Springer-Verlag Berlin Heidelberg 2015

Took part in a number of expeditions to explore the Caucasus and Transcaucasia (1844–1876), where the conditions, as he believed, might be similar to the areas of Italy and South Europe familiar to him. A. studied the causes and aftermath of the 1840 earthquake in the area of Grand Ararat, examined seismicity around Shemaha Town and established relationship between earthquakes and tectonic structure of the area. A. explored the shores of the Black and Caspian Seas, studied mineral deposits, mineral springs, glaciers, etc. In 1844, A. took up the study of the oil fields on Apsheron Peninsula and was the first to identify a number of regularities governing their formation. He noted a certain genetic correlation between commercial accumulations of oil and anticline structures of the earth's crust. A. laid scientific foundations of geothermal investigations into oil fields. He formulated the theory of secondariness, i.e. of oil migration. A. established genetic relationship between mud volcanicity and the presence of oil and gas in the subsoil, and was the first to hypothesize about the role of tectonic ruptures in the formation and destruction of oil and gas deposits. He was the first to try and make a chemical classification of igneous rocks. In 1865, A. explored geological conditions of the Kerch ore occurrence. In 1878–1887, A. published "Geological Explorations of the Caucasus Countries" (v. 1–3). In 1853, he was elected an academician, and in 1866— honorable member of Saint Petersburg Academy of Sciences in Russia.

"Abkhazia" – passenger motor ship that gave the name to a series of ships. Built at the Baltic shipyard in 1930. Used to cruise along the Crimea-Caucasus line in the USSR. When the World War II broke out, the ship was mobilized and handed over to medical service of the USSR Black Sea Navy, and was used as a sanitation carrier. The ship was reequipped in Odessa. Before it expired finally, the ship had made 36 evacuation runs, transporting 32,000 wounded and evacuees. On June 10, 1942, as a result of an air raid the ship ran aground and the ensuing fire destroyed its superstructures. In 1951, when the Sevastopol harbor was being cleared, "A." was retrieved and in 1956 it was scrapped.

Abkhazia, Republic of Abkhazia – an independent state since August 2008. Before August 2008 it was an Autonomous Republic, part of Georgia. Sometimes it is called "orange republic" with a clear reference to the color of ebony, oranges and tangerines that grow here in abundance. Situated on the eastern Black Sea coast. Area—8.6 thousand km^2, population 240,700 people (2011). Prior to the 1992–1993 war with Georgia A. inhabited by Abkhazians (18 %), Georgians (45 %), Russians, Armenians and other nationalities. In 1992–1993 the Georgian-Abkhazian conflict broke out. The result was that nearly 300,000 Georgians fled from A. (around 60,000 came back). Prior to those events, there were 125 hotels and health resorts in A. In 2005, 32 such facilities were operational, among them the "Pitsunda" complex, the "Aitar" hotel, the "Moskva" sanatorium. There were two hydro-power plants in A.—Perepadnaya and Ingurskaya. There used to operate eight coal mines in Tkvarcheli District; at present, coal is mined in two mining areas by a Turkish firm and is exported to Turkey by sea. The bulk of the products of two tobacco-processing factories is also exported to Turkey. The transportation flows— coal, tobacco, timber—are channeled via the seaports of which there are three in

A.—Sukhumi, Pitsunda and Ochamchira. In all, there were 25 factories and plants in the postwar A. These are, for the most part, food-processing industry facilities, among them the Joint Venture JSC "Wines and Soft Drinks of Abkhazia". Novoafon natural karstic cave has a unique suite of underground halls.

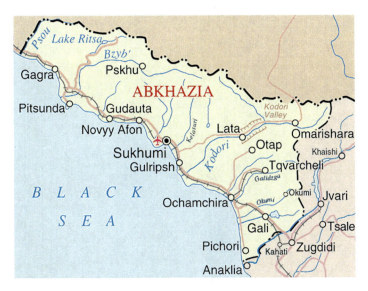

Map of Abkhazia (http://thelangarhall.com/wp-content/uploads/2008/08/abkhazia.jpg)

Abkhazians – (selfname–"Apsua") people living in the Republic of Abkhazia on the southern slopes of the Greater Caucasus Mountain Range, a small part of it settled in the neighboring districts of the Republic of Ajara. Abkhaz population was 93,000 people (1992). The Abkhaz language is part of the Adygo-Abkhazian group of the North-Caucasus language family. It falls into two main dialects: Abju and Bzyb (Gudaut). By language and culture, A. are close to the Adyg peoples of the North Caucasus; by some cultural and lifestyle traits—to Georgians. Religion: in the south, dominant are adherents to the Russian Orthodox Church, in the north— Suniite Moslems. Towards the end of the first millennium B.C., there came into being two large tribal associations: Abazgs and Apsils who subsequently united and lay the foundations for the Abkhazian nationality. From the 1980s, A. were part of the Georgian Kingdom. At the turn of the sixteenth to seventeenth centuries—under the rule of Turkey. From 1810—part of Russia, from 1921—part of Georgia SSR, from 1991—part of Georgia, from July 1992—in Republic of Abkhazia.

Abrau, Lake – (Turk. *"precipice, caving"*) situated 14 km west of Novorossiisk at the height of 84 m above the MSL. Occupies a steep basin and is relatively deep (13 m), its length 3.1 km and width around 0.6 km. Area 1.6 km^2—the largest lake of the north-western Caucasus. A fresh-water lake, sources of alimentation—the water of the small Abrau R. flowing into it, mountain springs, creeks and precipitation. Enclosed lake. Its water percolates through limestones toward the seashore.

Seasonal variations of the level reach 2–3 m. The water is usually turbid, as it contains mud suspensions. Ice is only formed during very cold winters. Fish fauna is diverse. There is the Abrau-Durso village on the shore of the lake and a small park with rare plants: Pitsunda pine, Himalayan cedar, chestnut, cedar.

Abrau Lake (Photo by Dmytro Solovyov)

Abrau-Durso – urban village (since 1948), situated 14 km west of Novorossiisk, Krasnodar Territory, Russia. Initially, consisted of two neighboring settlements: Abrau—on the shore of Abrau Lake (Abkhaz—*"caving in the mountains"*) and Durso—on the bank of the Durso River (from Abkhaz "durdsu"—*"four springs"*). Former appanage estate of the Emperor Alexander II on the Caucasian Black Sea coast, in which there was a viticulture and wine-making farm, and wines and champagne of repute well-known in Russia were fabricated. The farm was established by the Ministry for Imperial Court and Appanages in 1870. Grape-breeding experiments were commenced under the guidance of the Black Sea District agronomist F.I.Geiduk, who in 1872 purchased abroad the selection Rhine grapevines of the "Riesling" and "Portugeser" varieties for the purposes of breeding at A.-D. In 1873–1874, part (about 8,000) bushes was planted at A.-D. in a plantation of over 3.2 thousand square sazhen (one sazhen equaled to 7 feet or 2.13 m). Later, "Sauvignon", "Cabernet", "Pinaut-Fran", "Pinaut-Gri", "Traminer" and other varieties were also cultivated. The first crop of grapes was harvested at A.-D. in 1877. In 1882, a special cellar equipped for wine-making was built (in 1886—the second cellar for 13.6 thousand buckets). By 1882, the wine-grower E.Wedel developed the table wines of repute "Riesling" and "Cabernet-Abrau". By the mid-1890s, the types of A.-D. wines of repute were developed finally. These included: "Sautern", "Laphite", "Bordeau" and "Burgundy". The A.-D. wines were prize-winners of the All-Russian Exhibition of Industries and Arts (Moscow, 1882), the Yalta Exhibition of Horticulture, Wine-Making and Tobacco-Growing (1884), All-Russian Agricultural Exhibition in Kharkov (1887). In 1888–1889, there began most intensive development of the A.-D. farm which was associated with the purchase of the estates that used to belong to Prince D.Chavchavadze (in Kakhetia), Prince S.M.Vorontsov ("Massandra", "Ai-Danil" in the Crimea) as well as of Prince Vorontsov's wine-house. Most of the A.-D. wines were marketed through the outlets of this winehouse in Warsaw, Moscow, Odessa, Petersburg, Kharkov and Yalta. Thanks to the application and propaganda of contemporary science achievements A.-D. was made one of the model farms in the circum-Black Sea area.

In 1891, prince L.S.Golitsyn, who was appointed the chief wine-maker of the Appanage Department, analyzed viticulture and wine-making at A.-D. On his recommendations, preparations were commenced for making sparkling wine by a champagne technique and for the construction of more advanced cellars. Champagne varieties of grapes were cultivated from 1893, and before long nearly 1/9 of the entire area under vineyards was occupied by such varieties. During 1894–1900, the winery building and five tunnel-type cellars were set up at A.-D., and a road to Novoriossiisk was laid. The French wine specialist V.Tiebaut was invited to organize production of champagne in 1895 (he worked until 1899). In 1896, first bottling of champagne with subsequent fermentation was carried out—13,000 bottles. The French wine-grower E.Robinet was invited to supervise the fabrication of sparkling wine at appanage farms. The first lot of "Abrau" brand champagne was released in 1898. As sparkling wine production gained momentum, its significance to A.-D. increased. After the 1917 October Revolution in Russia, the farm was nationalized and became an agro-industrial complex "A.-D".

During the Soviet time, it was a state farm, then an integrated plant of the same specialization. Since 2011 SVL Group (London) is a single owner of the "A.-D" complex. In 2010, two sparkling wines from "A.-D" "Abrau Durso Imperial Cuvee L'art Nouveau brut" (year 2008) and "Abrau Durso Premium Red semi-sweet" (year 2007) won two bronze medals at a major international wine competition International Wine & Spirit Competition—2010 (IWSC).

Abrau-Durso Winery (Photo by Andrey Kostianoy)

Accidental oil spill – a type of marine environment pollution inflicting consider-able damage on flora and fauna as well as on the economy of the coastal countries in the polluted area. By nature and size, they distinguish between two types of AOS. The first type includes spills associated with leaks of oil and oil products from tankers, onshore and offshore storages of oil and gas resulting from accidents when the total volume of spilt oil may reach 400,000 tons (a tanker going aground, collisions, explosion, fire, etc.) as well as emergency oil blowouts during drilling on the continental shelf. The second type of AOS includes leaks occurring as a result of faulty or damaged process equipment, subsea or surface pipelines and connection hoses on the tankers, marine oil storages and drilling platforms or caused by errors in the operation of cargo-pumping systems.

Achilles – (Lat. Achilleus, Achilles) son of Peleus and Thetis one of the bravest Greek heroes who besieged Troy. A.'s mother willing to make her son immortal, dipped him in holy waters of the Styx River. Only the heel by which Thetis held A., did not touch the water and remained vulnerable (hence, the expression Achilles's heel, i.e. weak point). A. was given away to the centaur Chiron for upbringing. According to a prediction popular at that time, Troy could not be conquered without the participation of A., and the man was either to die at Troy's walls or to live a long life. A.'s mother disguised her son in woman's outfit and hid him on Skiros Island. By playing a trick, Odyssey found A., and they went to war together, where A. performed numerous feats. A. died of an arrow which Paris shot straight at A.'s heel.

In the poem "Aithiopis" by Arctinus of Miletus the Greek poet of the eighth century B.C., talking about the Troyan war, when A. died, says that after A.'s death Thetis stole her sons corpse from the fire and took it to White Island (Achilles's Island, Snake Island). Flavius Arrianus in his "Periplus of Pontus Euxinus" tells of a legend, whereby Thetis lifted her son from the seabed and placed him on the island. The name of A. is related to the ruins of a temple that existed on the Northern Spit or Chushka Spit, in Taman Bay. In 1823, they even tried to lift the columns of that temple. The temple's ruins could be observed as late as in the first half of the twentieth century. From the messages of ancient authors it is well known that a cult of the fearless Helennic hero A. was widespread in the towns and settlements on the coasts of Pontus Euxinus (Black Sea). Even now one of the capes in the north-east of Kerch Strait is named after him. It was believed that A. himself used to visit the shores of Bosporus Cimmerian on more than one occasion.

Achuevo – urban village in Slavyansky District of Krasnodar Territory, Russia. Situated on the Azov Sea shore, in the mouth of the Protoka River, 37 km north-east of Temryuk City. Fish plant, fish farm. Achuev caviar used to be a particularly valuable product in Russia at the end of the nineteenth- early in the twentieth century.

Activity Center on Matters of Integrated Coastal Reserves Management of the Black and Azov Seas (ICAM) – established in 1994 within the framework of the Black Sea Environmental Program of the Global Ecological Fund. The primary goal of the Center is the development of regional and national strategy and policy in the field of ICAM. Originally, the Center was based in Novorossiisk, subsequently, it was transferred to Krasnodar, Russia.

Ada – (Turk. *"island"*) one of the medieval names of the Crimean Peninsula.

Adalary Rocks – (Turk. *"rocky islands"*)—rocks-islands in Gurzuf bight (Crimea), rising from the Black Sea as high as 38 and 48 m, viewed from afar look like small gothic castles. Natural monuments. Permanent inhabitants—sea birds.

Adalary rocks http://yalta.ru.russian-women.net/images/photoalbum/yalta/wallpapers_1024x600/Ni784865.jpg

Adashev Daniil Fedorovich (? –1563) – Russian military figure. Military adviser of Ivan IV for the siege and storming of Kazan in 1552. In 1553–1554, commanded troops when crushing the revolt in the Volga Region. During the Livonian War of 1558–1583—voevode (leader) of the Advance Regiment that participated in storming Narva and other cities. In February-September of 1559—first voevode of troops (8,000 people) who took part in the Crimean campaign. The force led by A. went by ships down the Dnieper from Kremenchug to the Black Sea, where in the course of the sea battle two Turkish ships were seized. Having landed on the western coast of the Crimean Peninsula, Russian troops defeated several Tatar detachments and set free from captivity a large number of Russian and Lithuanians. From 1560, commanded a detail (artillery) in Livonia. Subsequently, fell out of favor with the Tzar and was executed.

Adjaria (Adjara), Republic of – officially the Autonomous Republic of Adjaria of Georgia. Adjaria is located in the southwestern corner of Georgia, bordered by Turkey to the south and the eastern end of the Black Sea. Adjaria is also known as Adjara, Ajara, Adzhara, Ajaria, Adzharia, Achara, Acharia and Ajaristan. Under the Soviet Union, it was known as the Adjarian Autonomous Soviet Socialist Republic (Adjar ASSR).

Adjaria has been part of Colchis and Caucasian Iberia since ancient times. It was colonized by Greeks in the fifth century B.C., then it fell under Rome in the second century B.C. It became part of the region of Egrisi before being incorporated into the unified Georgian Kingdom in the ninth century A.D. The Ottomans conquered Adjaria in 1614 and local people were converted to Islam in this period. The Ottomans were forced to cede Adjaria to the expanding Russian Empire in 1878. After a temporary occupation by Turkish and British troops in 1918–1920, Adjaria became part of the Democratic Republic of Georgia in 1920. After a brief military conflict in March 1921, Ankara's government ceded the territory to Georgia due to Article VI of Treaty of Kars on condition that autonomy is provided for the muslim population. The Soviet Union established the Adjaria Autonomous Soviet Socialist Republic in 1921 in accord with this clause. Thus, Adjaria was still a component part of Georgia, but with considerable local autonomy. After the dissolution of the Soviet Union in 1991, Adjaria became part of a newly independent Republic of Georgia. The local legislative body is the Parliament. The head of the region's government is the Council of Ministers of Adjaria which is nominated by the President of Georgia. Adjaria is subdivided into six administrative units: City of Batumi (a capital of Adjaria, 124,000 population) and five districts.

Adjaria is well known for its humid climate (especially along the coastal regions) and prolonged rainy weather, although there is plentiful sunshine during the spring and summer months. Adjaria receives the highest amounts of precipitation in Georgia, Caucasus and the Black Sea. It is also one of the wettest temperate regions in the northern hemisphere. The west-facing slopes of the Meskheti Range receive up to 4,500 mm of precipitation per year. The coastal lowlands receive most of the precipitation in the form of rain (due to the area's subtropical climate). September and October are usually the wettest months. The interior parts of Adjaria are considerably drier than the coastal mountains and lowlands. Winter usually brings significant snowfall to the higher regions of Adjaria, where snowfall often reaches several meters. Average summer temperatures are between 22–24 °C in the lowland areas and 17–21 °C in the highlands. Average winter temperatures are between 4–6 °C along the coast while the interior areas and mountains average around −3–2 °C. Some of the highest mountains of Adjaria have average winter temperatures of −8 to −7 °C.

Adjara has good land for growing tea, citrus fruits and tobacco. Mountainous and forested, the region has a subtropical climate, and there are many health resorts. Tobacco, tea, citrus fruits, and avocados are leading crops; livestock raising is also important. Industries include tea packing, tobacco processing, fruit and fish canning, oil refining, and shipbuilding. Batumi is an important gateway for the shipment of goods heading into Georgia, Azerbaijan and Armenia. The port of Batumi is used for the shipment of oil from Kazakhstan and Turkmenistan. Its oil refinery handles the Caspian oil pumped from Azerbaijan via pipeline to Supsa port and is transported from there to Batumi by rail. The Adjarian capital is a centre for shipbuilding and manufacturing. Adjaria is the main center of Georgia's coastal tourism industry.

Most of Adjara's territory either consists of hills or mountains. The highest mountains rise more than 3,000 m. Around 60 % of Adjaria is covered by forests. Many parts of the Meskheti Range are covered by temperate rain forests. . In the mountains, the so-called lower belt, under 500–600 m, in the woods of the so-called Kolkhida type, chestnut is prevalent, although encountered are oak, beech, hornbeam, alder and other tree species with an evergreen shrub layer of vines, common laurel cherry, pontic rhododendron (common box tree occurs in gorges), from 500 to 1,500 m—mixed forests with prevalence of beech, superseded by coniferous woods (pine and silver fir), from 2,000 m—subalpine, then alpine grasslands. In the maritime strip—cultivated subtropical vegetation (palms, magnolias, Grecian laurel, and other plants). The rivers belong to the Black Sea basin, the main river— Ajaris-Tskali (trubutary of the plentiful Chorokh River); Chokoli, Kintrishi, Chakvis-Tskali and other rivers flow into the Black Sea direct. In the maritime belt, dominant are alluvial and, to some extent, boggy soils, in submontane areas— mainly, red soils. Hinterland Adjaria is dominated by brown soils, more elevated areas—by podzolized light-brown mountain forest soils, while in alpine areas predominant are mountain meadow soils.

The population of Adjaria is 382,400 (2009). The Adjarians (Ajars) are an ethnographic group of the Georgian people who speak a group of local dialects known collectively as Adjarian. The written language is Georgian. The Georgian population of Adjaria had been generally known as "Muslim Georgians" until the 1926 Soviet census which listed them as "Ajars" and counted 71,000 of them. Today, calling them "Muslim Georgians" would be a misnomer in any case as Adjarians are now about half Christians. Ethnic minorities include Russians, Armenians, Greeks, Abkhaz, etc. The re-establishment of Georgia's independence accelerated re-Christianisation, especially among the young people. However, there are still remaining Sunni Muslims. According to the 2006 estimates by the Department of Statistics of Adjaria, 63 % are Georgian Orthodox Christians, and 30 % Muslim. The remaining are Armenian Christians (2.3 %), Roman Catholics (0.2 %), and others (6 %). Most dwellers live in the maritime strip. The cities: Batumi, Kobuleti. Health resorts: Batumi, Kobuleti, Tsikhisdziri, Makhindjauri, Green Cape.

Map of Adjaria (http://upload.wikimedia.org/wikipedia/commons/2/25/Adjara_map.png)

"Adjaria" – (1) (formerly "Adjaristan") the USSR passenger ship of the "Crymchak-ship" series. Built in Leningrad (USSR) at the Baltic shipyard in 1927. Used to cruise along the Crimea-Caucasus line. Joined military shipments, while remaining part of the Black Sea Shipping Line. During the Great Patriotic War (WWII), on July 23, 1941, in the vicinity of Dofinovka (not far from Odessa) was subjected to an air raid of German aircrafts. After being hit by several bombs, the ship ran aground at the depth of 7.5 m, the upper deck and superstructures remaining above the water. All attempts to salvage "A." proved futile. After the war, the hull was recovered and scrapped.

(2) passenger motor ship which was used to cruise along the Crimea-Caucasus line in the USSR. Place built—Mathias Thesen, Wismar (GDR). Year of construction 1964. Length—122 m, width—16 m, depth—7.6 m, displacement—5,640, speed—17.4 knots, range—6,500 miles, endurance—16 days, passengers—315, crew—134. "Adjaria"—was sold several times, ending up in 1996, was cut on scrap.

"Adjaria" (1964) (http://ships.at.ua/chmp/Adzaria-1.jpg)

Adler – (Turk.—*"adilar"*—*"islands"*, Abkhaz.—*"artlar"*—*"jetty"*) a health resort village, part of the Greater Sochi administrative center of Adler (health resort) area; railway station airport—Sochi Air Gate, the southern-most health resort of Sochi health resort tourist area. Is at 22 km south-east of Sochi Railway Station and 8 km from Khosta Railway Station. Situated on the Black Sea coast of the Caucasus, in the Mzymta River Valley. There is a landscaped pebble beach. In the past, A. used to be a small clannish mountain village. In 1837, a Russian naval force landed on Adler Cape. The Decembrist and writer A.A.Bestuzhev-Marlinsky was killed in action. Not long after that, Russians built the Saint Spirit Fort here. During the 1854 Crimean war, the fort was blasted as the Russian garrison was leaving the place. From 1864 to 1874, there existed the military post Adler, where Kuban Cossacks served. At the end of the 1880s, people began to settle in these areas, and early in the twentieth century A. already was a settlement. Before the 1917 October revolution, A. was known as a place of confinement as it was in a swampy and malaria-infested area, then the swamps were drained.

Adler (health resort) District – established in 1934 г. In 1961, under a Decree of the Presidium of the RSFSR Supreme Soviet (together with Lazarevskoe) was included in the constituent parts of Sochi Health Resort. A.D. is in between the rivers Psou (in the south) and Kudepsta (in the north). The district includes Adler city, the settlements of Moldovka, Kazachii Brod, Galitsino, Kepsha, Veseloe, Krasnaya Polyana and other locations. The administrative center is Adler. Its climate is somewhat different from the other areas of Sochi. During the winter months, it is cool here; in summer the temperature is approximately 1 °C lower than in Khosta District. Precipitation is within 1,200 mm per annum. Besides climate, the natural curative factors include mineral springs and peloids that are widely used in numerous sanatoria. At the distance of 39 km from A., in the Chvizhepse R. valley there is a spring of carbon dioxide arsenic-containing water "Sochi Narzan". Imeretinskaya Blight is rich in clayey muds that are used for mud treatment in many sanatoria of the health resort. There are sanatoria, rest houses, holiday centers and campings in A.D. There is a park of subtropical plants "Southern Crops" (37 km from Sochi), where more than 500 species of the southern flora are collected, the Caucasus State Biosphere Reserve; the highway Adler-Krasnaya Polyana passing amid precipitous rocks, is absolutely beautiful. In the middle of that distance there is Glubokii Yar—subterranean river, a small lake and a 20-m waterfall. Along the same road, there is a pedigree stock trout farm at Kepsha Settlement on the Mzymta R. Here, they propagate and farm rainbow trout, Donaldson and Camloops trouts, steel-head salmon. The farm neighbors Akhshtyr Cave, a most ancient dwelling site of primitive man of the early paleolith time.

Adler health resort (Photo by Andrey Kostianoy)

"Admiral Nakhimov" – the largest Soviet passenger liner on the Black Sea. "A. N." was the former famous vessel "Berlin", one of the first top-notch passenger liners, launched from the shipyards in Germany in 1925. "Berlin" served the Bremerhaven -New-York passenger line till summer 1938. During the Great Patriotic War (WWII) it was an floating hospital. In January 1945 it was torpedoed in the Baltic Sea and sunk offshore of Swinemünde (today Świnoujście in Poland). It was in this condition, i.e. lying on the bottom, that the Soviet Union received the liner after the Nazi Germany was defeated. This notwithstanding, the USSR Register confirmed that the ship was in good repair. Since 1957 it was a cruise liner in the Black Sea. On August 31, 1986 г. At 22:00 "A.N.", with 897 passengers and 346 crew on board, left Novorossiisk for Sochi. At this time, the bulk-carrier "Pyotr Vasev" (built in Japan in 1981) with a freight of 30,000 tons of barley from Canada was entering Novorossiisk Port. Due to violation of the maneuvering regulations and because the ships' courses were intersecting, there occurred a collision of the ships at 15 km from Novorossisk Port and 3.5 km from the shore. The bow of "Pyotr Vasev" ran into the starboard of "A.N.". The powerful bulbous forefoot of "Pyotr Vasev" (a bulbous forefoot is a special cylindrical tip that helps to overcome water resistance when the ship is moving) punched a huge 90 m^2 hole in the hull of "A.N." and the ship sank in 7–8 min. 820 people were rescued, while 423 people died.

Passenger liner "Admiral Nakhimov" (http://www.liveinternet.ru/tags/)

"Admiral Vladimirsky" – Soviet oceanographic research vessel (built in 1975) that made a global cruise in 1982–1983. The ship was named after Admiral L.A. Vladimirsky. "A.V." represented a series of six largest research vessels of the USSR Naval Forces hydrographic fleet. The ship took an active part in the exploration of the World Ocean, performing the depth, geophysical and hydrometeorological measurements. In 1975–1982, "A.V." made several cruises to the Atlantic and Indian Oceans, worked in the Mediterranean and Black seas. In 1982–1983—flagship of the Antarctic expedition sponsored by the Black Sea Navy Hydrographic Service to commemorate the 200th anniversary of the hero-city of Sevastopol. The other ship of the expedition was "Faddey Bellingshausen". "A.V." set out from Sevastopol in December 1982 to repeat the route of the Antarctic pathfinders Fabian Gottlieb Thaddeus von Bellingshausen and M.P.Lazarev. "A.V." crossed the South Polar Circle four times and reached 77°05'S in the vicinity of Ross Island (MacMurdo Strait)—no other Soviet ship had ever traveled that far south hitherto.

Having circled the ice continent, the ship called at the Antarctic Research Stations Molodezhnaya, Mirny, Leningradskaya, Bellingshausen, visited New Zealand. From the ship board researchers determined the current position of the South magnetic pole that had shifted towards the D'Urville Sea. The position of the islands of Bouvet, Franklin, Beaufort, Scott, Bounty, Antipodes, Peter I was determined more accurately, a number of sea mountains was discovered, data on the currents, temperature and salinity of Antarctic waters were obtained, electric and magnetic fields were measured throughout the length of the global cruise. In April of 1983, having completed 147 days of successful cruising, having covered 36,000 miles, "A.V." returned to Sevastopol.

Admiralty – state enterprise for the building, repair and outfitting of the ships. Admiralty was located in convenient places on the seashore or in river mouthes. A. had as constituent units shipyards, docks, workshops, shops and warehouses for storing all kinds of materials, munition and equipment for the ships. During the rule of Peter I, there were about 30 A. The first A. were established in Arkhangelsk (1693), Voronezh (1694): these made provisions for building ships for the Azov flotilla. At the end of the eighteenth century, A. were founded on the Black Sea (in Kherson, Sevastopol and Nikolaev).

In the south of Russia, as it got access to the Black Sea, the first A. was established in 1778 near Glubokaya ('Deep') Jetty in the Dnieper River brackish lagoon. The person in charge of A., fortress and military port construction in Kherson was General—Works Manager I.A.Gannibal. Ship-building which was commenced by laying down in 1799 of the 66-cannon battle ship "Catherine's Glory" was under way until 1826. A the same time, civil ships were also being built: these began merchant shipping runs to the ports of the Black and Mediterranean seas. After the establishment of ports more accessible to ships and of A. in Nikolaev and Sevastopol, A. in Kherson was closed down in 1829.

Sevastopol A. was established in 1783 simultaneously with the laying down of the city and port in Akhtiar Bay. A. was situated on the shore of the Southern Bay and following the construction of a new A. which was called Lazarevskoe, it began to be referred to as Old, subsequently after the port was used for the berthing and repair of mine-laying ships, it began to be called Mine A. The new A. in Sevastopol was built in 1835–1849 at the initiative of M.P.Lazarev, Commander-in-Chief of the Black Sea Navy and Ports, on the northern cape formed by the Yuzhnaya (Southern) and Korabelnaya (Ship) Bays. Prominent in A. structures were ship-building ways, slip-docks of the newest type (1845) and a unique complex of dry docks completed in 1850. Five dock chambers were linked by a large central pool from which a lock canal led to the Korabelnaya Bay. During the Crimean War of 1853–1856 the entire docks complex was demolished by the British. After the Russian Navy on the Black Sea was dismantled, the territory of A. with slip-ways and slip-docks was handed over to the Russian Society for Shipping and Trade (ROPiT) to set up an up-to-date ship-building and ship-repair facility. After the London Convention on Black Sea straits (1871) was adopted, there began construc-tion of the first armor-plated ships in Nikolaev, and in Sevastopol A. ROPiT built the first sea-going armor-plated battle ships of the Black Sea Navy "Chesma", "Sinop" (1883–1887), "Sv. Georgy Pobedonosets" ('St. George Victorious') (1892). In 1882–1886 and in 1893–1897 in lieu of the lock-way of the former Sevastopol Docks, there were built government-issue Alekseevsky (Western) and Aleksandrovsky (Eastern) dry docks that virtually constituted a self-sustained (as part of the military port) Dock A. When Sevastopol became the primary naval base and its authority was transferred from Nikolaev to Sevastopol (in 1897), the ROPiT A. was placed under the purview of the Navy, too and regained its original name—Lazarevskoe A. The last ships built by the A. were the Cruiser "Ochakov" (1902) and fleet destroyer "Ioann Zlatoust" ('John the Chrysostom') (1906). During the years that followed, Lazarevskoe A. with a work force of 3–5,000 workers, was that main repair facility of the fleet. In 1914, the third dry dock was commissioned. Subsequently, Lazarevskoe and Dock A. became a single enterprise—Sevastopol Sea Works.

Nikolaev A. was founded on the Ingul River in 1788 as a shipyard (from 1789—A.) at the initiative of G.A.Potemkin. The advantages of geographical position—a more convenient ship fairway and overland communication with Central Russia—made Nikolaev A. by early nineteenth century a chief ship-building center in the south of Russia. Towards the end of the Crimean War, the A. had at its disposal three large and four small shipbuilding ways, a slipping dock, workshops and storages. Ignoring small ships and vessels on the shipbuilding ways of the A. during the period from 1791 to 1858, 34 battle sailing ships were built, along with 16 frigates, 32 troop carriers, 13 brigs, 12 schooners and 11 corvettes—in all, more than half of the ships on the list of the Black Sea Navy from the time of its establishment until the Crimean War. Following the completion of construction of Russia's largest naval screw-propelled battle ships "Tsesarevich" ('Crown Prince') and "Sinop" and their transfer to Kronstadt, A. functioned as a repair base for the ships of the small Black Sea flotilla. The rebirth and technical retooling of A. a center of armor-plated shipbuilding in the south of Russia commenced in 1872. In 1883, the armor-plated battle ship "Catherine II" was laid which set off the building of ships under the 20-year program of building armor-plated battle ships (1882–1902). Nikolaev A. built 5 out of 8 armor-plated battle ships planned by the program (including the last one—"Knyaz Potemkin-Tavrichesky"). Under the additional program based on the improved design of "Potemkin", a cruiser, three battle-boats, a number of destroyers as well as of ships and vessels of other classes were built. In 1911, the territory of A. was handed over to the Russian Shipbuilding Company (Russud).

Russian Admiralties on the Black Sea retained their importance until mid-nineteenth century, then they were dissolved and were replaced by a system of naval bases, state-run and private shipbuilding and ship-repair plants.

Adrianopole Peace, Adrianopole Peace Treaty (1829) – treaty between Russia and Turkey terminated the Russian-Turkish War of 1828—1829. Signed on 2 (14) September in Adrianopole (Edirne) in the European part of Turkey. The main terms of the would-be peace had been forwarded by Russia to High Porta and at the same time war was declared on April 14 (26), 1828. The terms included: recognition by Porta of the 1827 London Convention, annexation of Anapa and Poti to the Russian Empire, confirmation of the autonomous rights of Serbia and Danube Principalities, pulling down of a number of Turkish fortresses on the Danube, restoring the rights of Russian commercial navigation in the Black Sea straits and opening thereof for unhindered commercial navigation, moderate contribution. In November of 1828, the Turkish side put forward its own terms of peace: recognition by Turkey of the Russian-Turkish Akkerman Convention of 1826 in exchange for Russia's refusal to take part in the destiny of Greece; mediation of great powers guaranteeing territorial integrity of Turkey, etc. In August of 1829 Porta introduced modifications in its terms: consent to recognition in principle with the 1827 London Convention in exchange for multilateral talks on the future of Greece; confirmation of all prior treaties with Russia in exchange for guarantees of territorial integrity. On August 10 (22), 1829, Sultan Mahmud II decided to start immediately peace talks on the Russian terms. Mediation of the envoys of Great Britain, France and other powers that had earlier been sought by these powers and Turkey was ruled out. Russia was guided by the principle "The advantages of retaining the Osman Empire in Europe are greater than the disadvantages" and was doing its best to achieve the conclusion of a peace treaty at the earliest, having made some correctives in the original terms of peace (surrender of Kars, Akhaltsikhe, Akhalkalaki occupied by Russian troops in the Caucasus). Both the parties made mutual concessions: Russia said it wanted none of the part of the Danube principalities offered by the Turkish side during the talks. The Osman Empire agreed to grant freedom to Greece and gave up further attempts to secure the holding of an international conference on the Greek issue.

The A.P. comprised 16 articles of the Main body of the document, the Ancillary Act on the Danube Principalities of Moldavia and Valakhia and the Explanatory Act on the contribution. Turkey was regaining all territories occupied by Russian troops on the European theater of military operations except the Danube mouth with islands (Art. 2–3). The border, as before, passed along the Prut River. Russia was gaining the eastern Black Sea coast from the Kuban River mouth to St. Nickolay jetty on the northern frontier of Ajaria, including Anapa and Poti as well as Akhaltsikhe and Akhalkalaki Fortresses with the adjoining areas. Turkey was giving up claims on the Transcaucasian territories that had been gained by Russia earlier under Turkmanchay Peace of 1828 (Art. 4)—Georgia, Imeretia, Mingrelia, Guria as well as Erivan and Nakhichevan Khanates. The rights of the Danube Principalities in the sphere of self-government were confirmed and broadened, Turkish garrisons were removed from these localities, the national Moldavo-Valakh army was to be instituted (Ancillary Act). Turkey was bound to abide by the decrees of the Akkerman Convention on the Serbian autonomy, give back to it the six alienated districts (Art.6). Russian merchants were entitled to engage in commerce on an extraterritorial and free basis on the entire territory of the Osman Empire. The Black Sea straits of Bosporus and Dardanelles were declared open to commercial shipping (Art. 7). Under Art. 10, Greece was given wide autonomy. The Explanatory Act contained a special provision where the size and contribution payment procedure were set out (in all, 11.5 mln Dutch Ducats). On April 14 (26), 1830 in St. Petersburg, the Explanatory Convention to the A.P. was signed. Bringing together in a single act all provisions relating to the contribution (by 1834, its size was halved). The Convention remained in force until 1836. A.P. by totality of its articles had been in effect until the 1856 Paris Peace Treaty was concluded. Some articles of the A.P. were further developed in the Treaty of Hünkâr İskelesi of 1833, London Conventions of 1840 and 1841 on the international status of the Black Sea straits. The treaty bolstered Russia's positions on the Balkans and in the Black Sea considerably, was conducive to the development of Black Sea trade and had a great impact on the struggle of the Balkan peoples against the domination of the Osman Empire.

Adygea (Republic of Adygea) – is a federal subject of Russia (a republic) enclaved within Krasnodar Krai. Its area is 7,600 km^2 with a population of 440,388 (2010). Its capital is Maykop. The Adyghe people, sometimes known as Circassians, were the ancient dwellers of the North-West Caucasus since the thirteenth century. Cherkess (Adyghe) Autonomous Oblast was established within the Russian SFSR on July 27, 1922, on the territories of Kuban-Black Sea Oblast, primarily settled by the Adyghe people. At that time, Krasnodar was the administrative center. It was renamed Adyghe (Cherkess) Autonomous Oblast on August 24, 1922, soon after its creation. In the first two years of its existence the autonomous oblast was a part of the Russian SFSR, but on October 17, 1924, it was transferred to the jurisdiction of the newly created North Caucasus Krai within the RSFSR. It was renamed Adyghe Autonomous Oblast (AO) in July 1928. On January 10, 1934, Adyghe AO became part of new Azov-Black Sea Krai, which was removed from North Caucasus Krai. Maykop was made the administrative center of the autonomous oblast in 1936. Adyghe AO became part of Krasnodar Krai when it was established on September 13, 1937. On July 3, 1991, the oblast was elevated to the status of a republic under the jurisdiction of the Russian Federation. The Adyghe language (Adyghabze) is a member of the Northwest Caucasian group of Caucasian languages. Along with Russian, Adyghe is the official language of the republic.

The republic is rich in oil and natural gas. Other natural resources include gold, silver, tungsten, and iron. The republic has abundant forests and rich soil. The region is famous for producing grain, sunflowers, tea, tobacco, and other produce. Hog and sheep breeding are also developed. Food, timber, woodworking, pulp and paper, heavy engineering, and metal-working are the most developed industries. Tourism is developing in the republic. Adyghe State University and Maykop State Technological University, both in the capital Maykop, are the two major higher education facilities in Adygea.

Afrotherapy – treatment with sea-foam, sea-foam treatment, sea-foam cure. Ancient Cypriots believed that bathing in sea-foam that is produced in abundance near the Cyprus shores, where, as the legend has it, the goddess Aphrodite came out from the sea-foam, heals. On the Black Sea shore, old fishermen used to treat their rheumatism by rubbing the body with a powder that was obtained from films of the sea-foam dried on the beach. Such films are particularly abundant in autumn on low-lying sandy shores in the north-western part of the sea.

Agoy – a small village on the coast of the Northeastern Black Sea, 6 km north-westward of Tuapse, Russia. Recreation area, beach.

Agreement between Britain, France and Russia (1915) – secret agreement between the Triple Entente participants. The possibility of entering into such agreement called upon, according to the Minister of Foreign Affairs S.D.Sazonov, to settle the issue of Black Sea straits, presented itself when Turkey took part in WWI as an ally of Germany. Great Britain and France in the conditions when they were losing all battles and had their armies retreating in the west of France in 1914—early 1915, were interested to have Russia participate in the war until Germany was defeated finally. At the same time, British and French diplomats were delaying the start of actual negotiations regarding the keeping of their promises to transfer the straits to Russia. At the end of 1914, the British and French troops began the operation of their own with a view to seizing the straits and Constantinople. The Russian government was insisting on reviewing the matter of agreement. This made provisions for the transfer to the Russian Empire of Constantinople (Istanbul), Bosporus and Dardanelles Straits, the entire European coast of the Sea of Marmara along the Enos-Midia line, and part of the Asian coast between Bosporus, the Sakarya River and "the point to be indicated in Izmir Bay", of the Sea of Marmara Islands and Tenedos (Bozcaada) Island and Imbros (Gokceada) Island. The Agreement was the first treaty between the Triple Entente participants dealing with the division of Turkey (subsequently developed further in the Sykes-Picot Agreement in May 1916 and Agreement of Saint-Jean-de-Maurienne in April 1917). As far as the consent of the allies to annex to Russia Constantinoples and the straits is concerned, the Russian Government made an appropriate declaration in the State Duma in autumn of 1916, although the agreement as such remained secret until it was promulgated by the Soviet of Peoples' Commissars at the end of 1917. The Agreement was subsequently canceled by the Peace Decree of December 26 (November 8), 1917, passed by the 2nd Congress of Soviets.

Agreement on parameters of the Black Sea Fleet division – it confirmed that the main base of the B.S. Fleet of the Russian Federation was in Sevastopol. It stated the facilities deployed on the main base as well as the basing points and places of dislocation used by the fleet. The appendices listed the military ships and vessels from the B.S. Fleet belonging to each side, outlined the procedure of division between the sides of the armaments, military equipment and support facilities for the forces of coastal defense, marines and land-based naval aircraft of the BSF. It was signed on 28 May 1997 in Kiev, Ukraine.

Agreement on the status and conditions of deployment of the Black Sea Fleet of the Russian Federation on the territory of Ukraine – it was concluded in Kiev on 28 May 1997 for a term of 20 years with automatic prolongation for next 5-year periods if both sides agree.

Agreement on urgent measures on formation of the Naval Forces of Ukraine and the Naval Fleet of Russia on the basis of the Black Sea Fleet – it was signed by Ukraine and the Russian Federation on 17 June 1993 in Moscow. It envisaged initiation in September 1993 of practical actions by the parties on formation of the Ukrainian naval forces and the Russian naval fleet on the basis of the B.S. Fleet division and defining the conditions for basing of the Russian Navy in Ukraine. It stated that the B.S. Navy shall be divided on the following principles:

detachments and units of all kinds of the fleet of surface craft, submarines, air forces, ground forces, detachments, units, establishments, objects of operational, combat, technical and logistics support, the buildings and structures being on their balance; armaments, ammunition, military equipment and other movable property of B.S. Fleet as of the signatory date of a separate agreement on division and conditions of basing of the B.S. Fleet of Russia on the territory of Ukraine shall be divided between the Russian Federation and Ukraine on the 50/50 % principle;

before actual division of the B.S. Fleet the existing system of the B.S. Fleet basing shall remain in force as stated in the Yalta Agreement between RF and Ukraine of 3 August 1992;

the Russian side will participate in development of the socio-economic infrastructure of the city of Sevastopol and other settlements of Ukraine where the military units, objects and establishments of the Russian Navy will be dislocated; the B.S. Fleet of Russia and Naval Forces of Ukraine consider Crimea facilities their technical base;

financing of the B.S. Fleet prior to its division is realized by the sides in equal shares (50/50 %).

Ainali-Kavak convention – signed between Russia and Turkey on March 10, 1779 in the sultan palace of Ainali-Kavak. The Convention specified some provisions of the Kutchuk-Kainardji Treaty, among them provisions on the navigation of Russian merchant ships through the Black Sea straits. C. set limits on the size and tonnage of the ships. Russian merchant ships were allowed to pass through the straits, provided they of the same parameters/types and had the same number of cannons as British and French merchant ships, weighed 26.4 thousand puds. In case of need, Russia was allowed to hire "ship hands from among the Turkish nationals". These amendment unchanged were included in the first Russian-Turkish Trade Agreement of 1783.

Ai-Petri – mountain, one of the most popular symbols of the Southern Coast of the Crimea. One of the first researchers of Crimean archeology P.I.Keppen explained the name by the linking of the ancient Greek words *"Agnos"*—*"Saint"* and *"Petri"*—the name from which *"Peter"* originated. A.P. was associated with the ancient Greek convent whose ruins are at this location. The height of A.-P.— 1,233 m above the sea level. The mountain is structured by massive limestones and is a fossil reef. Its top culminates in a stone crown in the form of castellations.

Four of these are unusually large, of 12–15 m in height, and there are many smaller ones. These were formed during the weathering of the reef mass dissected by faults. A 3 km-long ropeway commissioned in 1987 leads to the A.-P. castellations. The Ai-Petri meridian is also here: a globe mounted on a granite foundation, with accurate geodetic data.

Ai-Petri Mountain (Photo by Evgenia Kostyanaya)

Ai-Petri Yalla – a peculiar, 1,200–1,300 m (above the sea level) high mountain plateau with a steep, precipitous southern and gently-sloping northern slopes, structured by a thick mass of limestones, which conditions the high rate of karst occurrence. Second from the west Yalla of the Main (southern) ridge of the Crimean mountains overlooking the Black Sea above Alupka and Yalta. Gently-sloping hills alternate with hollows and limestone ridges stepped slopes. Virtually woodless. The amount of precipitation exceeds 1,000 mm a year.

Ai-Petri Yalla (Photo by Evgenia Kostyanaya)

Ai-Todor (Aitodor) Cape – situated between Miskhor and Gaspra, termination of
Mt.Mogabi south-eastern slope (Crimea). Juts out into the Black Sea by a Neptune
trident. Its southernmost tooth—Ai-Todor spur proper—the highest part of the
cape. From the time immemorial, this has been a reference point to the seafarers.
In the fifteenth century, Athanasius Nikitin, coming back from his "travel beyond
three seas", on the way from Balaklava to Feodosyia, skirted this cape. Nowadays,
there is light-house on the summit of the spur, its light can be seen from the distance
of 70 km. The cape is a nature reserve. There still exists a restricted grove of
arbuscle juniper. In the grove—ruins of the ancient Kharaks fortress built by the
Roman legionnaires in the first century. A giant 1,000 year old pistachio-tree grows
near the lighthouse. The tree is included in the Red Data Book. This is one of the
oldest trees of the Crimea. At the summit of another ledge of A.-T.—Liman-
Burun—there is a viewing ground, also referred to as "command bridge", whence
a spectacular view opens up. Near the cape, there is picturesque rocky islet Parus
('Sail'). The castle-cum-palace "Swallow's nest" is built on the Aurora's rock of
A.-T. Cape.

Ai-Todor Cape (Photo by Evgenia Kostyanaya)

Aivazovsky (Gaivazovsky) Ivan (Ovanes) Konstantinovich (1817—1900) –
Russian marine painter, outstanding world-acclaimed master of painting the sea
and sea landscape, academician (1844), professor (1847) of the Fine Arts Academy.
Graduated from the Petersburg Fine Arts Academy in 1837, where he studied in the
landscape studio of Prof. M.N.Vorobyov and French marine painter F.Tanner.
From 1846 lived in Feodosiya (Crimea) permanently. Produced around 4,000
(other sources insist—6,000) paintings, hundreds of drawings and water colors,
most of which are kept in the Feodosiya Aivazovsky Art Gallery (which he built
and presented to the city in 1880), in the Russian Museum, Tretyakov Gallery, the
Hermitage and other museums. The main subject of A.'a oeuvre—romantic por-
trayal of the beauty of the sea, especially of the Black Sea he loved so much, the
play of light and shade on the water, endless change of tinges, bursts of splashes in
the sun rays, the vast expanse of the sea and a mighty rhythm of the waves, the
purple sunsets and moonlight. A. painted diverse types of commercial and naval
sailing ships, the hard work of fishermen and sailors, the confrontation of man and
nature, man's courage. His best known paintings include "Black Sea" (1830),
"Moonlight Night in the Crimea. Gurzuf" (1839), "Shipwreck" (1843), "Tenth
Wave" (1850), "Ships in the Roads" (1859), "Sunrise. Black Sea Shore" (1864),
"Windstorm" (1872) and others. From 1844—painter of the General Naval Staff.
During the Caucasus war (1817–1864) took part in marine amphibian operations
(1839), came to know closely the outstanding Russian admirals M.P.Lazarev, P.S.
Nakhimov, V.A.Kornilov, V.I.Istomin. A. always took interest in the history of
Russian Fleet, and its numerous brilliant victories he praised in his paintings:
"Chesmena Battle" (1848), "Battle of Navarin" (1848), "Battle in Khios Strait",
"Brigantine "Mercury", attacked by two Turkish ships" (1892) and others. In 1869,
A. travels to Egypt to take part in the commissioning of Suez Canal. He paints a
general panorama of the canal and a series of painting with Egyptian landscapes.

To mark the 50th anniversary of discovery of the Antarctic Continent by the Russian seafarers M.P.Lazarev and F.F.Bellingshausen, A. painted one of the first pictures in the history of pictorial art, showing polar ice—"Mountains of ice" (1870). At the age of 81, in 1898, he paints his masterpiece "Amidst the waves". A.'s merits are recognized in many countries worldwide: he was elected member of the Roman, Florence, Stuttgart, Amsterdam Fine Arts Academies, Honorable Member of Petersburg Fine Arts Academy. The renowned British scholar of the Caucasus David Lang wrote: "The fortunate paradox: the son of the mountain-enclosed country Armenia has attained crests that require absolute creative immersion in the sea depths and in the depths of painting skill in depicting the ever-changing nature of the waves".

A. did a lot for Feodosiya town development. He built a public water supply system and a new building of Feodosiya Museum. The laying of a railway to Feodosiya owes to him, too.

I.K. Aivazovsky (http://stat17.privet.ru/lr/0a35f634320abf71fa6a520b34fb55d0)

Aiya, Cape – (Greek "Holy Water") shore ledge 13 km to the south-east of Balaklava (Crimea), delimiting the Southern Coast of Crimea on the west. A.C. is a precipitous edge of the Main Ridge of the Crimean Mountains. Its height is 556 m. There are unique relict communities of flora (28 species of plants are entered in the Red Data Book of Ukraine), including the largest in the Crimea area of relict Stankevich pine. The cape proper, its surrounding territory and sea water area are a reserve: from 1982, the territory of 1,340 ha has been part of the landscape sanctuary of republican significance "Aiya Cape", while the 300-m strip of the sea water area adjoining the cape has, from 1972, been within the boundaries of the "Laspi-Sarych" hydrological aquatic complex. A.C. is portrayed by I.K. Aivazovsky in its painting "Windstorm at Aiya Cape" (1875).

Aiya Cape (Photo by Andrey Kostianoy)

Aiya-Burun, Cape – a small rocky ledge of the Black Sea shore, 2 km to the north-north-west of Fiolent Cape (Crimea).

Ajaro-Imeretinsky Mountain Range, Ajaro-Akhaltsiksky Range – mountain range of Minor Caucasus within the Georgian Republic. Stretches from the Black Sea shore to the Kura River Height up to 2,600 m (the summit Mapistskaro). Structured by marl shales, marls, tuffs and andesites. The slopes are covered with broad-leaved (beech, hornbeam and other species) and coniferous (fir-tree) woods.

Ajigol Lakes – salt lakes situated north-east of Feodosiya on the shore of Feodosiya Bay in Crimea. Separated from the bay by narrow sand-coquina tombolos. Black muds of these lakes possess healing balneal properties. In summer, the lakes almost go dry, leaving some soil sediment, their shores covered with reddish glasswort vegetation.

Ajimushkay quarry – situated to the north-east of Kerch on Kerch Peninsula of the Crimea. Was formed as a result of quarrying building stone—shell rock. During the Civil war in Russia and the Great Patriotic War (WWII), underground galleries of A.Q. were partisan territory, a shelter. When Nazi troops seized Kerch in 1942, some Soviet Army units left for A.Q. Storages, hospitals, military units, thousands of civilians were deployed in the quarry. Besides A.Q. there are Starokarantinskie and Mamaiskie Quarries on Kerch Peninsula.

Akçakoca – a village on the coast of the Black Sea, Turkey.

Akchaabat (Akçaabat) – a village on the coast of the Black Sea, Turkey.

Akgel, Lake – fresh-water lake. Situated in the circum-Black Sea area of Turkey on the left bank of the Sakarya River and is almost permanently associated with its runoff.

Akhali-Afoni – see Novy Afon.

Akhmatov Fedor Antipievich – Captain of a major-general rank. On service— from 1776. Participant of the Russian-Turkish war of 1787–1791. Distinguished himself in battles of June 7 and 17, 1788 at Ochakov that ended in the defeat of the Turkish fleet. Also, displayed courage in the 1790 campaign, especially on November 6, when part of the rowing flotilla led by Major-General de-Ribas (Don Giuseppe de-Ribas-y-Boyons) under the command of A. was moved forward as an advance party to Tulcha Fortress. As a result of courageous actions— involving the landing of amphibious parties from his ships, A. seized the fortress without waiting for the approach of the flotilla main forces. Commanded the Caspian flotilla (1795–1796). In 1797, retired from service.

Akhtaniyazovsky brackish liman – lake in the Kuban River mouth, of 100 km^2 in area. The Cossack Yerek River (Kuban River arm) flows into A.L. Linked with the Sea of Azov.

Akhtiar Bight – old name of Sevastopol Bight after the name of small Tatar settlement Ak-Yar (Akht-Yar) on the northern shore of the bight, in the mouth of the balka currently called Sukharnaya Bight.

Akhtiar, Akyar, Ak-Yar – (Tatar *"ak"*—*"white"*, *"yar"*—*"steep shore, preci- pice near the sea"*) this was the official name of Sevastopol in 1796–1804), the name was given after the Tatar village situated in Akhtiar Bight by order of the Emperor Pavel I. After his death, the city regained its former name of "Sevastopol".

Akhtopol – (formerly, Greek city Agatopolis—"good", "kind", "beautiful") mar- itime climatic health resort in Bulgaria (Burgas Region), to the south-east of Burgas City. Situated on the north-western shore of Akhtopol Bight, on a small peninsula. The climate is mild, marine, with numerous sunny days. The winter is mild, the warmest in the country (6–8 days with snow); mean temperature of January 2.4 °C. The summer is warm; mean temperature of July 23 °C. The typical local features are early spring and long warm autumn. Precipitation is around 700 mm per year. A. is protected from the winds by a rocky peninsula (which is also a natural breakwater) and by Bosna Hill adjoining the area on the west (height under 500 m, northern continuation of the Strandja Mountains). The mild climate, warm sea and wide beach of fine sand (more than 2 km in length), protected by dunes constitute the basis of A. health resort resources and are conducive to climotherapy and thalassotherapy. There are numerous holiday camps, campings in A. and its environs. A.—Bulgaria's smallest (along with Melnik) town in terms of its popu- lation and is the southernmost maritime health resort. In A. which is mentioned in the writings of the ancient Roman geographer Ptolemy, there are still remains of the classical aqueduct, part of medieval fortress walls as well as the Church of Christ Ascension (1776). During the Middle Ages, A. was the residence of the Metropol- itan. As a sea port, A. flourished until the mid-nineteenth century, then gave way to Burgas.

Akhun Greater – mount in the Caucasus, near Sochi, height 663 m. On top of the mountain, there is a 30-m high observation tower with a splendid view of the Greater Caucasus Mountain Range and the Black Sea coast.

Akhun Mount (Photo by Evgenia Kostyanaya)

Akkerman – name of Belgorod-Dnestrovsky town until 1941, Odessa Region, Ukraine.

Akkerman Convention – concluded between Russia and Turkey (September 7, 1826, Akkerman). Supplemented and confirmed the Bucharest Peace of 1812. Turkey was undertaking to honor the privileges of Moldavia, Valakhia and Serbia. Russia was granted freedom of trade on the territory of Turkey and the "right of free navigation along the Constantinople Canal of all Russian merchant ships". Porta never observed the terms of C.

Akkose levee – often referred to as "Uzunlar levee" and "moat". A mighty defense fortification built to protect Bosporus Kingdom against nomads. The levee begins at the northern shore of Uzunlar Lake and, and transversing Kerch Peninsula from the north to the south, from A. to the Black Sea, terminates on the shore of Kazantip Bay near Novootradnoe Village. Excavations indicate that the levee was as high as 2.1 m, its width at the foundation reached 27 m. The moat was 2.8 m deep and 15 m wide. The levee on the inside was provided with crepidoma made of large stone blocks.

Akliman-Konu Bight – picturesque bight with several islets 15 km west of Sinop town, Turkey. The bight depths vary from 2 to 8 m. The excellent sand beach borders on broad-leaved forest.

Ak-Monai Isthmus – the narrowest part (17.5 km) of Kerch Peninsula, by convention, regarded its western boundary.

Ak-Monai levee – ancient defense levee situates in the narrowest part of Kerch Peninsula on Ak-Monai Isthmus, stretching from Kamenka Village, formerly Ak-Monai, to Primorsky Settlement.

Ak-Mosque – see Simferopol.

Alasheev Dmitri Aleksandrovich (1908–1953) – Captain I rank, chief of the Hydrographic Service (HS) of the USSR Pacific Navy. In 1926, entered M.Frunze Higher Nautical School from which he graduated in 1930. In 1933, completed Special courses of the Red Army Naval Commanders (hydrographic division). From April of 1936,—Senior Site supervisor of the Northern Hydrographic Expedition (Arkhangelsk), team leader of the expedition. From 1939 to 1942—attendee of the Naval Academy Hydrographic Department. In 1942–1944—Commander of the Manoeuvering Hydrographic Squad of Black Sea Navy Navigation and Hydrographic Support (NHS). As the chief of the Black Sea Navy HS task force, participated in NHS of amphibious operations near Yuzhnaya Ozereika (1943), Taganrog, Mariupol, Temryuk (1943) and in the Kerch-Eltingen Operation (1944). In 1944, in the same capacity took part in NHS for the seizure of Tulcha, Sulin, Constanţa and Varna ports. From 1944 to 1947, commanded the separate hydrographic squad on the western shore of the Black Sea, where he was exploring Romanian and Bulgarian shores. At the end of 1947, was appointed the Chief of the Baltic Hydrographic Expedition. In 1952, became the head of the Hydrohraphic Service of the Pacific Navy. Decorated with orders and medals. A bay of the Antarctic continent is named after him.

Albena – major maritime health resort in the northern part of the Bulgarian Black Sea coast, 30 km south-east of Dobrichka town and 20 km north-east of Varna City in a picturesque bay as the Batova River flows into the sea. The climate is marine, with very warm summer, abundance of sunny days, insignificant cloud cover and low amount of atmospheric precipitation (around 450 mm per year). The formation of positive climatic conditions is favored by the adjoining plateau Dobrudja on the north-west with the hilly tract Baltata or Mt. Batova covered with dense broad-leaved forest (ash, elm, alder, poplar, maple and other tree species) with several species of lianes and reed thicket. The basis of A. health resort resources are mild climate, warm sea with a gently sloping seabed (the depth at a distance of 100–200 m does not exceed 1.6 m) and the widest in Bulgaria fine-sand beach (length over 6 km, width from 50 to 500 m). All these are conducive to climotherapy and thalassotherapy. A. offers hotels, campaigns, swimming-pools with mineral water; yacht clubs, sporting and children's playgrounds, a water-skiing pier, cultural center. The health resort has been built since 1968. The interesting architecture of health-resort and other structures that are happily combined with the landscape features won the Dimitrov Prize in 1971. A. health resort is the "Blue Flag" prize winner, which is awarded for ecological cleanness of a health resort. A. is one of the most popular maritime health resorts in Bulgaria and a center of rest and tourism of international significance. Today A. has more than 40 hotels for about 15,000 guests.

Albena (http://www.avialine.com/img/photoreports/photoreport_51_7219.jpg)

Alchak, Alchak-Kaya, Cape – (Crimean Tatar *"alchak"—"low"*) closes on the east the large Sudak Bight and separates Sudak Valley from the neighboring Kapsel Valley. A. is an ancient coral reef. It is a low rocky mountain strung in a submeridional direction to the distance of 1,000 m, its height is 152 m above sea level. The northeastern hillside is gently sloping, the western and south-western slopes overlooking Sudak are rather steep. In many locations structured by lime-stones there formed niches, grottos as a result of weathering, and one of the sightseeing attractions is a large ring hole in the western part of "Aeolian Harp" Cape. In 1988, A. of 55 ha in area was declared a reserve.

Alchak Cape (http://stat17.privet.ru/lr/09325529ba3d1bcc63ca5e52187431d7)

Alepu – maritime health resort on the Bulgarian coast of the Black Sea. Situated to the south of Sozopol. A. features one of the largest beaches in the area (45 ha). Includes the coastal strip and a lake-marsh of the same name. There are numerous dunes in the north of the long beach. The lake is characterized by low salinity due to permanent connection with the sea.

Alepu (fox's marsh) – brackish lagoon north of Arkutino in the direction of Agalina Cape on the Bulgarian coast of the Black Sea. The lagoon is shrinking all the time, being drifted with sand. Nearly the whole year it has been two small water bodies that are linked in spring, when there is a fall of precipitation. Previously, when the marsh was connected with the sea, the marsh used to be frequented by gray mullet for feeding. In summer, salinity of A. increased to 14‰. At present, seasonal fluctuations of salinity range from 1 to 5‰. The impact of sea water is only possible in case of a violent storm. The lagoon is hemmed with a 5-km beach with sand dunes. There is an aquatic fauna reserve around the lagoon.

Alexander III (1845–1894) – Russian Emperor, son of Alexander II. After the death of his elder brother Nikolay in 1865, Alexander heir to the throne. As the crown prince of Russian Empire, Alexander became member of the State Council, commander of the guard units and chieftain of all Cossack forces. During the Russian-Turkish war of 1877–1878, he was with in the field army. Commaned the Separate Rushchuk Force in Bulgaria that comprised the 12th and 13th army corps and covered the rear of the Russian troops besieging Plevna. On November 14, 1887 at Trestenka and Mechka the force defeated the Turkish troops that were heading for the besieged garrison as assistance; subsequently, the force acted successfully in seizing the settlements Razgrad, Rushchuk, Osman-Bazar and Sylistria.

A.III was instrumental in establishing the Voluntary Fleet of Russia (from 1878) that became the nucleus of the country's merchant fleet and reserve of its navy. Having taken the crown after the assassination of Alexander II, he repealed the draft of the constitutional reform signed by his father immediately before his death. A. III's manifesto of April 29 (May 11), 1881 set out a program of domestic and foreign policy: support of the order and power, consolidation of church devoutness, and ensuring Russia's national interests.

A.III carried out a number of reforms in the military sphere. The armed forces were grouped in structurally uniform corps and most of these were deployed in the near-border districts. The district commands were so arranged that in case of war these were transformed into field armies. The reserve military personnel were upgraded every year in monthly training camps. New manuals were regularly published and were adopted, military service obligation was introduced in the Caucasus, hinternaland territorial changes were made in Cossack troops (new service districts were instituted, the Don cadette corps was opened, along with Novocherkassk Military School, etc.). The existing fortresses were fortified, and new ones were built, railways were adapted to attaining military goals. In the Navy, new high-speed ships were built, the Black Sea Navy was revived completely.

Foreign policy of Alexander III was characterized by resolute reluctance to interfere in the affairs of other states and by willingness not to get involved in military conflicts; during the years of his reign, Russia did not wage any wars. For his consistent efforts in maintaining peace in Europe, Alexander III was dubbed 'the Peacemaker'.

Aleksandr III (http://content2-foto.inbox.lv/albums109484620/ramtairidi/Sv-Simeona-i-Anny/Alexander-III.jpg)

Alexiano Anton Pavlovich (1749–1810) – Vice-admiral, descendant of Greek sailors. In 1770, began service in the Russian fleet as a volunteer. Took part in the Battle of Navarin (seizure of Navarin fortress), in burning down the Turkish fleet at Chesma. In 1771–1772, cruised within the Archipelago (group of islands in the Aegean Sea). In 1773, promoted to the rank of midshipman. In 1774, cruised within the Archipelago again. In 1775, on board the frigate "Natalia" sailed from Livorno to Kronstadt. In 1776–1779, participated in an expedition to the Mediterranean Sea. In 1780, promoted to the rank of lieutenant. In 1782, traveled from Kronstadt to La Manche and back. During 1783–1786, every year participated in the Baltic campaign. In 1785, promoted to a lieutenant commander. Posted to Kherson. In 1787 was sailing near the Crimean coasts. In 1788, promoted to captain second rank, and in 1790—to captain first rank. In 1788–1790, participated in military cruising and took part in the battles of Fidonisi (1788), when attacking Sinop, in the Kerch Strait and at Hadjibei (1790). In 1791–1798, commanded the ship "Grigory of Great Armenia" of the Black Sea Fleet. In 1798, went to Archipelago with the squadron of the Vice-Admiral F.F.Ushakov. Participated in seizing the islands of Cerigo, Zante, blocked Corfu and fought with a French ship. In 1799, took part in the seizure of Corfu; promoted to commodore captain. In 1880–1881, on board the ship "Russia's Glory" sailed from Konstadt to the Mediterranean Sea. In 1801, promoted to rear-admiral. In 1801–1809, was ashore in Sevastopol, running special errands. In 1805, promoted to blue-flag vice-admiral of the Black Sea shipboard fleet.

Alexiano Panaiotti (?–1788) – Rear-admiral, Greek. In April of 1769 was admitted as a volunteer in Livorno to serve for Russia. Took part in the Battle of Chesma and was promoted to Lieutenant. In 1770–1771, was cruising within the Archipelago (a group of islands in the Aegean Sea) and near the shores of Anatolia, seized many ships, landed an amphibian and occupied Kefalonia Island. In 1772, seized Rodos Fortress by three armed ships, captured in battle six merchant ships with Turkish supplies and delivered these to Naoussa in Paros Island. Then A. was posted with a frigate to Dardanelles Strait. In 1773, took part in seizure of Beirut, in 1774, was sailing near the shores of Rumelia and captured the fortress on Kariboda Island. Promoted to Lieutenant-commander. In 1775 was cruising within the Archipelago. In 1778 was sailing between Revel and Kronstadt, in 1780–1781, sailed to the Mediterranean Sea and back; promoted to Captain second rank (1781). In 1782, as the commander of the ship "Constantine" was sailing in the Baltic Sea. In 1783, promoted to Captain 1st rank and posted to the Black Sea. In 1787, commanded the fregate "St. Andrey" of the Black Sea Fleet. In 1788, stayed with the Liman Flotilla as a Rear-admiral. Participated in successful battles at Ochakov against the Turkish Fleet.

Alkadar – (Arab. *"alk dar"*—*"godly dictate"*) farm "Sophia Perovskaya", former isolated farmstead, then estate "Alkadar" (Lyubimovka settlement), Crimea, Ukraine. In 1834, N.I.Perovsky planted vineyards on the lands of Belbek Valley on a large scale, and already in 1846, Perovsky's wines "reminiscent of the Hungarian" pinned everybody's attention at the All-Russian Exhibition. In the second half of the 1880s, the estate, known at that time as "Primorskoe" ('Maritime') came to be owned by Varvara Stepanovna Perovskaya—mother of the notorious activist of Russia's revolutionary and terrorist organization "People's

Will" Sophia Perovskaya (great-granddaughter of the Hetman Cyrill Razumovsky), commander of the group of bombthrowers on Ekaterininsky Canal on March 1, 1881. V.S.Perovskaya built a new house in the estate which is alive to this day. In 1899, the estate was sold to the dealer in southern-shore wines in Petersburg Fedor Shtal who established new vineyards, built a house with large wine cellars and renamed "Primorskoe" into "Alkadar". After the 1917 October Revolution, in 1920, there was established the state farm "Alkadar" in the valley, one of the first in the Ukraine, and in 1927 it was named "Sophia Perovskaya". There still exists Shtal's wine cellar built in 1890. At present, this is the territory of Lyubimovka Village, Nakhimov District, Sevastopol.

Alma – second largest river of the Crimea. It originates on the northern slopes of Babugan-Yaila. Produced by confluence of the small rivers Sary-Su, Savlykh-Su and Babuganka. Length 84 km. Mean annual runoff 38 million m^3. Flows mainly within the Crimean Mountains to the Black Sea. Alimentation—mixed. Freezes for a short time and not each year. There are two storage reservoirs on the river: Alminskoe and Partizanskoe.

Alupka – a village on the southern coast of Crimea, one of the best maritime climatic health resorts in Crimea, Ukraine. The city is part of Greater Yalta. Situated on the Southern Coast of Crimea, 17 km to the south-west of Yalta, stretching by 4 km. For the first time was mentioned in the tenth century under the name of Alubika. The climate is subtropical of Mediterranean type. On the north, the health resort is protected from cold winds by the slopes of the mountain plateau with the Mt. Ai-Petri. The winter is mild; mean temperature of February is 4 °C. The summer is very warm; mean temperature of August is 24 °C. The summer heat is moderated by sea breezes, swimming in the sea—from June to October. Atmospheric precipitation—350 mm per year. The number of sunny days in a year reaches 263. The mild climate of A., clean, mountain and-sea-air saturated with gummy aroma of the pine, beautiful landscape are all conducive to climotherapy. In 1901 Prof. A.A.Bobrov built the first sanatorium for children suffering from tuberculosis.

During the 1820s, there was established a large manor's farm of the count M.S. Vorontsov (in neo-gothic style, with Mauritanian portal, currently—a museum with collections of furniture, paintings, porcelain) and a luxurious park (today, it is the State Palace-Park Museum-Reserve "Vorontsov's Palace"). The palace was built during 1826–1846, the designer—British architect Edward Blore (1787–1879), a court architect of the King William IV and Queen Victoria. There are the lower and upper landscape parks as part of the palace ensemble; near the palace are magnolia and cypress groves. The upper park features a natural agglomerate of diabasic blocks, the so-called Alupka chaos. The black soil for the block chaoses would be brought by barges from the Ukraine over the Dnieper, then by the Black Sea, and subsequently would be scattered by hand over the slopes. After the Revolution of 1917, the palace was nationalized, in 1921 a museum in the palace was opened and it functions until now. The museum was saved from destruction during the Great

Patriotic War (WWII) of 1941–1945. In February of 1945, the British delegation that participated in the Yalta Conference was accommodated at the palace.

The winery "Alupka" was established in the 1850s.

Alushta – a village on the southern coast of Crimea, maritime climatic health resort, administrative center of Greater Alushta District, Crimea, Ukraine. The village has inherited the name of the Byzantine fortress Aluston (Alusta) which was built here by order of the Byzantine Emperor Justinian I back in the sixth century. One of the hills in the village center until now features a defense tower—remains of the medieval fortress. In the fourteenth to fifteenth centuries Aluston was seized by the Genoese who made it a fortified port; then, in 1475, the city was captured by the Turks who were fought by Russian troops commended by A.V. Suvorov and M.I.Kutuzov in the second half of eighteenth century.

A. is situated 45 km south-east of Simferopol and 33 km north-east of Yalta over which mountain massifs of Babugan-Yaily tower with the highest point in the Crimea Mt. Roman-Kosh, Chatyrdag, Demerdji (from Turk "blacksmith"). The slopes are covered with thick beech and pine woods. There are evergreen trees and shrubs on the territory of the health resort; in the environs, there are vineyards. The sand-and-pebble beach of A. is one of the best on the southern coast of the Crimea. The climate is subtropical of Mediterranean type. The winter is mild, rainy; mean temperature of February is 3 °C. The summer is very warm with prevalence of dry weather and few clouds; mean temperature of August is 23 °C. The autumn is warm and lengthy. Atmospheric precipitation around 400 mm per year, mainly in autumn and winter. Number of sunshine hours around 2,260 per year. Bathing season— from June to October (temperature of water 17–23 °C).

The health resort has been growing since the end of the nineteenth century. Most of the sanatorium facilities are in the health resort suburb of A.—in the Professor's Corner (formerly, Worker's Corner). Here, there is vast man-made beach with the longest in the Crimea concrete quay (7 km). Operating are sanatoria, boarding rooms, water-treatment facility (sea-water, pearl and other baths, curative showers), medicinal beach, rest homes, tourist lodges. Conditions are provided for the treatment of persons with diseases of respiratory organs (other than tuberculosis) and of the nervous system. On the shore of the picturesque bight (delimited from southwest by Mt. Ayudag, from north-east—by Plak Cape), there is an old park—one of the best on the Southern Coast of Crimea (former estate of the Raevskys—model of garden and park architecture with an estate in the Mauritanian style of the end of the nineteenth century; cypresses, magnolias, cedar-trees, a grove of Italian pines, etc.).

There are many natural and architectural-historic monuments in the environs of A. These include the caves of Chatyrdag, weird weathering forms of Mt. Demerdji ("Ghost Valley"). A. adjoins the territory of the Crimean State Nature Reserve (relict species of flora and fauna). In A.—museum of local natural history. Medieval monuments are still alive: remains of Byzantine and Genoese fortresses, fortifications in the vicinity of Mt. Demerdji and settlements on Mt. Kastel; house museum of the outstanding Ukrainian architect academician A.N.Beketov, museum of the writer I.S.Shmelev; literary-and-memorial museum of the writer S.N. Sergeev-Tsensky.

Alushta (http://images.netbynet.ru/direct/38f7d1ad9ed79589eb4704fb59fe1270.JPG)

Amasra – regarded one of the best sea health resorts on the Black Sea coast of Turkey. Situated in the north-western part of the country, 60 km to the north-east of Zonguldak. Population—7000. Founded in the sixth century B.C. Classical name was Sesamus. Attracts thousands of holiday-makers by beautiful nature and two beaches: large beach—with good fine sand, and smaller beach—with pebble. Byzantine fortress, remains of ancient Sesamus. Amasra has two islands: the bigger one is called Büyük ada (Great Island) while the smaller one is called Tavşan adası (Rabbit Island).

Amasra (http://upload.wikimedia.org/wikipedia/commons/0/02/Amasra,_Turkey,_Castle,_view_from_the_island.jpg)

Amasra Bay – cuts into the shore between Çakraz Cape and Amasra Peninsula. The shores of the bay are largely precipitous and steep. The eastern shore is cut by the valleys of several rivers. Its depth at the entrance to the bay is 25 m; eastward of the entrance, the depths decrease and at the eastern shore reach 10 m. The southern part of the bay is called Buyuk Liman.

Amasra Peninsula – situated 6 km to the west-south-west of Çakraz Cape and has two oblong ledges: north-western and south-eastern. The peninsula is connected with the continent by a wide and low isthmus. The north-western ledge is connected with the peninsula by a narrow bay-bar. Waves roll over the bay-bar in stormy weather, hence a bridge has been built here. Lighthouse Amasra is mounted at the top of the peninsula. There is Amasra town on the peninsula, the city is an administrative center. It is linked by shipping routes with the Black Sea ports of Turkey.

Amazon squadron – paramilitary women's cavalry unit made up of the wives and daughters of Balaklava Greeks, formed in the spring of 1787 on order of Prince G. A.Potemkin in Balaklava on the occasion of the arrival in the Crimea of the Empress Catherine II. The squadron (100 women) was commanded by the wife of one of commanders—E.I.Sarandov. The amazons wore a special uniform, were trained in the skill to keep in cavalry formation, to change formations, to handle arms and to shoot in salvo. The welcoming of Catherine II was a success, she was delighted: Mrs. Sarandov was promoted to captain, all amazons were decorated. After the departure of the Empress, the squadron was disbanded.

Amazons – the names of A. came to be used, following the custom of burning out the girls' right breast lest it should be in the way, when the girls were handling arms. According to ancient authors, there used to exist A. on the Don River, on the shores of the Sea of Azov and the Black Sea. The earliest reference to A. was made in "Iliade" by Homer, later in the writings of Hekateus, Eshil, Pindar, Skilax of Karyanda and many others. Skilax say: "Asia begins with the Tanais River and the first Asian people at Pont are Sabromates. This people is ruled by Amazon women".

Amisos – ancient town on the southern coast of the Black Sea founded approximately in mid-sixth century B.C.; subsequently, it became King Mitridate's residence.

Amsara – see **Pitsunda**.

Anadolu or Anatolia – name used for the Asian part of Turkey within Asia Minor Peninsula.

Anadolu Cape – eastern cape of the northern entrance to Bosporus Strait; steep. There is a lighthouse on the cape.

Anaklia – a village on the coast of the Black Sea, maritime climatic health resort in Zugdidi Region, Georgia. It is located 25 km north of Poti City and 25 km south-west of Zugdidi town, near the Inguri River mouth. There is a wide strip of sandy beach stretching 12 km along the sea. The climate is subtropical, humid. The winter is very mild, with no snow; mean temperature of January is 5 °C. The summer is very warm; mean temperature of August is 23 °C. Precipitation is around 1,500 mm per year; heavy rains are not uncommon. Sea and shore breezes reducing the summer heat are well pronounced in A. area. The mild climate is used as a natural curative factor in climotherapy (mostly of the children).

Anaklia (http://www.panoramio.com/photo/2093704)

Anapa – a town on the coast of the Northeastern Black Sea, center of Anapa Region, Krasnodar Territory, Russia. One of the largest recreation and resort area on the Black Sea, numerous hotels and long sand beaches. Situated 52 km north-west of Novorossiisk. In ancient time, there used to stand, in place of Anapa, Gorgippia town of Bosporan Kingdom. Gorgippia (fourth century B.C.—third century) was so called after the name of its ruler Gorgipp. More recent settlements belonged to Byzantinians, dwellers of Genoa, Turks. By the time the first Russians came there in the eighteenth century, the settlement had an Adyg name "Anapa" (from Adyg. *"Ana"* + *"pa"*, where "Ana"—name of a small river flowing through the town, and "pa, pe"—"mouth", i.e. *"town in the mouth of the Ana River"*. Maritime climatic and balneal health resort, largely intended for the children, a federal health resort. Situated on the shore of Anapa Bight surrounded by the spurs of the Greater Caucasus Mountain Range. Population of A.—57,000 people (2010). Man-made plantings prevail; vineyards. Fine sand beach (length 20 km, width 100–150 m), the pride of A., gently-sloping sea-bottom particularly suitable for the bathing of children. The climate is of Mediterranean type. The winter is mild, with an unstable snow cover; mean temperature of January is 1 °C. The spring is short; from mid-April, warm weather is established. The summer is very warm, sunny, of long duration; mean temperature of July is 23 °C. The autumn is warm. Precipitation is around 400 mm per year (mainly, in the form of rainfall), maximum—in December. Prevalent are north-easterly winds. A.—the sunniest health resort on the Black Sea coast of the Caucasus; the number of days without sunshine—48. Clean sea air with high content of oxygen and ozone is rich in ions of sodium chloride, iodide, bromine. The bathing season—from May 15 to October 15. Temperature of sea water in August 24 °C, in June and September 19–20 °C. All this is conducive to aero-, helio- and thalasso-therapy. Alongside the climate, the main natural curative factors of the health resort are peloids and mineral springs. Sulfide silty muds of several deposits are used: Chembur Lake 2 km from A.; Vityazev Liman and Solenoe Lake—88 km from A. and 12 km from Taman, respectively as well as bald mountain peloids (bald mountains Shugo, Azovskaya, Akhtinizovskaya, Gnilaya). These are used for peloid-assisted treatment of the diseases of the locomotor system, nervous system, gynecopathy. Mineral springs of the health resort are used for treatment through drinking in case of digestive apparatus diseases. Treatment is arranged of children and adults with diseases of respiratory organs (except tuberculosis), of blood circulation organs, of the nervous system, skin, gynecopathy.

A. has had an eventful history—2,500 years. In the fourth century B.C., there used to be a Sind settlement—Sind Harbor (Sindika) in place of A. which, following the accession to Bosporus Kingdom in the fourth century B.C., began to be called Gorgippia, too. In the fourteenth century, it became Mapa, a colony of Genoa. In 1475, it was seized by Turkey, transformed into a fortress; annexed to Russia in 1829; from 1846—a town. Health resort development began at the end of the nineteenth century. After the Crimean war, the town was deserted, but in 1856 it was re-annexed to the Russian Empire, this time forever. In 1898, the first sanatorium of Doctor V.A. Budzinsky was opened, in 1902, an institution for mud treatment was opened as part of that sanatorium; in 1905—sanatorium "Bimlyuk" for tuberculosis of bone. In 1921, A. was declared a health resort of state significance, and its rapid growth began. The health resort devastated during the Great Patriotic War (WWII) was mainly restored by the early 1950s, then it was enlarged and remodeled. There function over 200 sanatoria, boarding-houses, rest homes, fitness centers and tourist institutions. The economy of A. is based on the health-resort infrastructure. Wine-making and processing industry are at an advanced stage of development. Railway station and airport are located several km from the town center.

Anapa (http://kartoman.ru/anapa-na-karte-rossii/)

Anatolia – historic area of Turkey. The name of *"Anatolia"* (turk. *"Anadolu"*) from Gr. *"Anatole"*—"East". Byzanty used the name to denote but a small part of Asia Minor, in Osman Empire—name of province, from the 1920s applied to the entire Asian part of Turkey.

Anatolian coast – southern coast of the Black Sea, northern part of Anatolia Peninsula. Stretches to the west of Kalender Cape as far as Bosporus Strait to the distance of over 1,000 km. A.C. has distinctly outlined areas: coastal and hinterland. The coastal area, especially its eastern part, is characterized by the steeply rising rocky mountains overgrown with thick woods; there often occurs precipitation fall here. The hinterland area is separated by a coastal ridge. The difference between the hinterland and coastal areas is particularly significant in the east, where the height of the mountains is maximum. In the west, where the mountains are lower, the difference between the areas is less striking. A high mountain chain stretches from Kalender Cape to Isikli Cape along the eastern part of A.C.; the summits of the chain are snow-capped for most of the year. Mt.Kachkar (3,937 m) towers above this mountain chain. The rocky shore of the Pontic mountains, its height reaching 3,000 m with snow capped summits stretches parallel to the shore and its spurs approach the sea. In the area from the Chorokh River to port Trabzon, the coastal mountains are covered with deciduous forests. There are corn fields, fruit orchards and tobacco plantations near the settlements. The middle part of A.C. from Isikli Cape to Boztepe Cape is also mountainous, but the mountains here are lower; their slopes are cultivated. The mountains are dissected by the valleys of the Terme River and Kizilirmak River. The western part of Anatolian coast from Boztepe Cape to Bosporus Strait is mountainous. Closer to the strait, the mountains grow lower and near the strait proper the shore is rather low, but precipitous. The mountains of the western part that are at a significant distance from the coast are much lower than elsewhere on the coast. The coastal strip between the cities of Inebolu and Eregli differs from the rest of the coast by parallel arrangement of valleys and hills overgrown with woods. Small bays and bights open to northerly winds cut into A.C. The most crucial of these are Sinop Bay and Rize, Pulathane, Samsun and Eregli bights. Less important are Persembe Bight and Bulen, Espiye, Ordu, Fatsa and Unye Bays. The deltas of some rivers on A.C. are increasingly jutting out into the sea due to permanent input of sediments. This is particularly obvious in the case of the rivers Kizilirmak, Esilirmak and Sakarya, in front of the deltas of which a gradual change of depths and formation of shoal banks are noted. In early spring and in autumn, turbid waters of these rivers sometimes are spread into the sea to the distance of more than 8 km from the shore. There are several islets near A.C.; of these, only Kefken reaches 2 km in diameter, the rest are rocks, stones or sand deposits, lying in front of the river mouths, maximum 3.7 km from the shore. There are considerable depths along A.C. The soil near A.C. is largely silt or sand, sometimes shell rock and stone. Magnetic anomaly in the vicinity of A.C. is discovered around Rize Bight and Sinop Bay. It should be remembered that this coast is subject to frequent earthquakes. The main ports on A.C. are these: Trabzon, Tirebolu, Giresun, Samsun, Inebolu, Amasra, Zonguldak and Eregli. The cities on

A.S. are connected by highways or dirt roads. There exists regular shipping traffic between the ports of Trabzon, Samsun, Sinop, Zonguldak and other ports. The cities Samsun, Zonguldak and Eregli are linked with the capital city of Turkey Ankara by railway, and with Batumi, Georgia—by a ferry service.

Anchovy (*Engraulis encrasicholus ponticus*) – see Black Sea anchovy.

Ancient Chersonesus – one of the names of Ancient or Strabon Chersonesus, a town that located on the Mayachny Cape that was destroyed in the second century B.C. during the wars of Chersonesus Taurica with the Scythians. This name came to our times thanks to the treatises of Ancient Greek historian and geographer Strabon (64/63 B.C.—23/24 A.D.).

Ancient Greece – in this way the territory of the Ancient Greek state was referred to. In the eighth to sixth centuries B.C. it extended over the south of the Balkan Peninsula, the islands of the Aegean Sea, the coast of Thrace and the western coast of Asia Minor. A.G. also controlled the northern coast of Africa, the straits and coast of the Black and Azov seas where Greek colonies were established. Chersonesus Taurica (today Sevastopol, Ukraine) was one of such Ancient Greek colonies.

Andreev Alexander Petrovich (1820–1882) – Lieutenant-General of the Russian Fleet. Finished a navigation school. From 1842 to 1853, every year took part in hydrographic surveys of Finnish skerries fairways, at first as a warrant officer, then as a sub-lieutenant of the Board of fleet navigators. Participated in the Crimean War of 1853—1856, (from 1854—Lieutenant). In 1871—1872, was the chief of Separate Survey of the Black Sea Caucasian shore in the hydrographic expedition to the Black Sea and the Sea of Azov. In 1873—1880, as a Lieutenant-Colonel, then Colonel (1875) and Major-General (1880) was in charge of hydrographic operations on Onega Lake. In 1881 promoted to a Lieutenant-General and retired. Decorated with Russian orders and one foreign order, F.P.Litke Gold Medal (1875). Three shoal banks in the Baltic Sea and one shoal bank in the Black Sea are named after him.

Andrusov Nikolay Ivanovich (1861–1924) – Russian geologist and paleontologist, Academician of Saint Petersburg Academy of Sciences (from 1914), Academician of the Academy of Sciences of Ukraine (from 1920), Professor of Yuriev (from 1896) and Kiev (from 1905) Universities, of the Higher Women's Courses in St.-Petersburg (from 1912), staff member of the Geological Committee of Russia (from 1913). His scientific research deals with dynamic and regional geology, stratigraphy, paleontology and oceanology. The founding father of Russian palaeoecology. A. was the first to give a detailed characterization of tectonics, stratigraphy and geomorphology of Kerch Peninsula, was the first, using the materials of his studies of Kerch low-mountain topography, to distinguish Meotic, Cimmerian as well as Tarkhan, Chokarak, Karagan and Chaudin stages of geological deposits. His stratigraphic and paleontologic works deal with the study of Neogene and anthropogenic deposits. The stratigraphic scheme of Neogene marine deposits developed by A. has not lost its significance even now. At the end of the nineteenth century, at the initiative of A. a systematic and integrated study of the Black Sea commenced. As from 1884, A. explored natural terraces around Sudak, discriminating between offshore and continental ones. Concurrent with this, A. explored geological conditions of Kerch ores occurrence. In 1890, the first deep-water expedition on board the battleboat "Chernomorets" was performed. The expedition was led by I.B.Spindler, the hydrographer was F.F.Wrangel. A. was commissioned to be in charge of geological and biological observations. The expedition found there was hydrogen sulfide at the depth of 100 sazhen (around 200 m). For a long time this depth was believed to be the upper boundary of the hydrogen sulfide layer. It was later established that the boundary was dome-shaped. A. participated in oceanographic expeditions to the Sea of Marmara (1824), explored Kara-Bogaz-Gol (1897). In 1890, during the expedition on board the "Chernomorets" A. was the first to find the largest field of manganese concretions in Kalamite Bay of the Black Sea. He discovered on the seabed remains of the post-tertiary fauna of Caspian-looking mollusks. Lomonosov prize-winner of Saint Petersburg Academy of Sciences. In 1912, his paper published and based on Sudak terraces finds served the basis for further geomorphological investigations. The author of 90 scientific papers on geology and geomorphology of the Crimea, among them the unique monograph "Geotectonics of Kerch Peninsula" (1893). During the Russian Civil War, lectured at Tavria University (Simferopol), being actually the founding father of the geographic department.

Ankhial Lake – see Pomorie Lake.

Annular bream or gilthead (*Diplodus annularis*) – fish of the *Sparidae* family. It has the elongated, not high body, compressed on the sides, to 30 cm long. The fish back is olive-green, the sides below the lateral line are silver-gray. On the tail there is a large black spot. In the front row of teeth it has eight wide incisive teeth with a smooth cutting edge. It is a permanent inhabitant of the Black Sea.

Antipa Grigore (1867–1944) – natural scientist, professor, academician, hydrobi-ologist, zoologist, fishery biologist, oceanologist, limnologist, the founder and director of the Natural Science Museum in Bucharest (1894–1944), currently named after the man. The founder of the Biooceanographic Institute in Constanţa. Immortalized on a "200 leys" banknote of National Bank of Romania (1992). Member of the Romanian Academy of Sciences (from 1910), British Veterinary Zoological Society, French Institute of Oceanography, Geographic Societies of Austria and Germany, Agronomy Societies of France and USA. A.'s research laid the foundations of ichthyological science in Romania. For years, he studied crucial commercial fish species of the Black Sea with a special emphasis on sturgeons; he described seven taxons of these (white sturgeon, long-snout st., short-snout st., red st., pointed-snout st., rough st., golis st.). National Institute for Marine Research and Development in Constanţa is named after A.

Aquarium-museum, Sevastopol – situated in Sevastopol, Ukraine, in Maritime Park. Established in 1897 as part of Sevastopol Marine Biological Research Station. One of the initiators of establishing the biological station (commissioned in 1871) was the young anthropologist and ethnographer, subsequently outstanding Russian scholar and traveler N.N. Miklukho-Maklay. The first director of the station Academician A.O. Kovalevsky insisted on and pushed through the construction of a special building for the biological station with a marine aquarium on the shore of Sevastopol Bay, where there used to be the former Nikolaev battery. The building of A. was financed by the owner of Foros, 1st Guild Merchant A.G. Kuznetsov, who was also Russia's "tea king". Wrote the newspapers of that time: "The aquarium is so exquisite that its beauty can be compared with that of aquariums in Berlin, Naples and other European cities". In 1994, thanks to the adoption of up-to-date methods of water treatment and systems of hydrobionts' life support, the exhibits were enlarged with addition of tropical fish and invertebrate species. In 1999, the museum's oldest hall was restored (built in 1897), in which now there are aquariums with fishes of fresh-water regions of Africa, South America as well as terrariums and aqua-terrariums with amphibians and reptiles. In 2003, a new exhibit was added: a living coral reef. Exhibited here are over 15 species of soft and skeleton-forming corals, 2 species of sitting polychaetes, 4 species of echnuses, 5 species of tropical actinias, and 3 species of starfishes. At present, the display is arranged in 4 halls, where 150 species of invertebrates, fishes and creepers are exhibited. At the central entrance of the Aquarium, there stand N.N. Miklukho-Maklay and A.O. Kovalevsky busts.

Aquatory (water area) – from Lat. "aqua"—"water") an area of the water (unlike land surface—"territory") within established boundaries of the sea or port. A. of a port makes provisions, in its navigable space, for maneuvering and berthing of ships.

Arabat Arrow – narrow spit of golden shell sand separating Sivash Lake from the Sea of Azov, the eastern coast of the Crimea, Ukraine. Length is around 80 km, width of 270 m—8 km. In the north, A.A. is separated from the continent by a Thin (Genicheskii) Strait. In the name, the word *"Arrow"—long spit, bay-bar, accompanying the seashore in shallow water*. The attribute "Arabat" is used as it means an ancient Turkish fortification Arabat (from Turk. *"rabat, arabat"—"fortification, fortress"* as well as *"coaching inn, caravan-sarai"*), built in the second half of the seventeenth century on the southern part of the spit. The fortress was seized in 1779 when it was stormed by the Russian army. A.A. began to be developed early in the nineteenth century, when the first salt-works were set up, and at the end of the century development of the spit continued by Ukrainian families that were resettled here.

Arabat arrow (http://www.4turista.ru/files/imagecache/max_s/files/ar10129.JPG)

Arabat Bay – bay on the south-western coast of the Sea of Azov, between Kerch Peninsula and Arabat Arrow. Cuts into the southern shore of the Sea of Azov. The bay is 70 km wide and 38 km long. The width between Kazantip Cape and Arabat Arrow—44 km, depth—9 m. The shores of A.B. are fringed with a shoal, its depths less than 5 m, on which above-water and subsea stones are scattered. The depths in the middle part of A.B. are 8–10 m; closer to the apex of the bay they decrease gradually. At the depths of over 8 m, the soil is silt, closer to the shore—sand. The eastern shores are rocky, the southern and western—low-lying. The south-eastern shore of the bay is high, rocky and is dissected by gullies. 13 and 17 km to the north-east of Kazantip Bay, there extend from this shore precipitous capes Kiten and Krasnyi Kut. The southern shore of the bay is precipitous.

"Arabat" – landscape-botanical reserve of republican significance, Ukraine. A maritime steppe area in the southern termination of Arabat Arrow (600 ha in area). The reserve features the conserved natural steppe plant and animal communities. Of great interest are steppe stows with the participation of virgin tulip steppe. Under protection since 1964, reserve—from 1974.

Arakli – a village on the coast of the Black Sea, Turkey.

Ardesen – a town on the coast of the Black Sea, Turkey.

"Argo" – (1) legendary ship of argonauts; (2) sail-and-rowing vessel, 16.4 m long and 3 m wide, a copy of argonauts' ship built in 1984 by the famous Irish traveler and writer Tim Severin. The "Heroes' Travels" expedition (also known as "The Jason Voyage") set out from Volos in North Greece with an international crew on board and, having covered over 1,000 nautical miles, reached Poti at the eastern coast of the Black Sea in three months.

Argonauts – in Greek mythology, 56 participants of the Colchis (Black Sea) campaign. Their ship named "Argo" was built with the assistance of the Goddess Athena who inserted a piece of the sacred century-old Dodon oak, transmitting by the whisper of its leaves the will of the gods. The A. led by Jason, among whom were Admetus, Dioascura's twins Castor and Polydeuces, Heracles, Meleager, Orpheus, Peleus and Telamon, were to bring back to Greece the golden fleece of the miraculous lamb taken away to Greece by Friks. On the way, A. visited Hypsipyla on Lemnos, had the upper hand over Amik, the king of Bebriks, rescued Finey from harpies, overcame Simplegades (floating rocks). The daughter of the local king, wonder-worker Medea helped Jason to get hold of the golden fleece. The way back was long and tough to A., yet they managed to reach the shores of Greece in the long run. The historic basis of the myth about A. are, obviously, the sea voyages of Greeks, above all, Miletians to the Black Sea shores. The campaign of A. is a frequent subject in the works of classical and post-classical art (relief in Delphis; crater with a pictorial image of A. dating to the mid-fifth century B.C.; epics of Apollonius of Rhodes and Valery Flakk, dramatic trilogy "The Golden Fleece" (1821) by Franz Grillparzer, opera "Medee" (1797) by Luigi Cherubini).

Arhavi – a village on the coast of the Black Sea, Turkey.

Arkadia – maritime climatic and balneal-peloid health resort in Odessa (Ukraine), in the maritime zone (on the south, adjoins Otrada), 6 km from the Odessa Railway Station. Is part of the Odessa group of health resorts. There stretch small but very convenient sand beaches along the shore with rocks jutting out into the sea. There is a picturesque park on the territory of the health resort. The major natural curative factors—climate and peloid of Kuyalnitsky brackish lagoon used in climotherapeutic and peloid treatment procedures. Also widely used is sea water on the basis of which man-made mineral-spring, gaseous, radon and other baths are prepared. The health resort is used mainly for the treatment of patients with diseases of the locomotor system, nervous system and cardio-vascular system and having gynecopathy. The health resort has been in operation since 1921. There function sanatoria, boarding-room facilities, rest homes and tourist camps.

Arkadia (http://fotoduet.ru/vika/ua/odessa/107.jpg)

Arkas Nikolay Andreevich (1816–1881) – Russian Admiral, Chief commander of the Black Sea fleet and ports. Son of a Greek settler. From the age of 13, served in the Black Sea fleet, took part in the Russo-Turkish war of 1828–1829 and Bosporus expedition of 1833, in voyages on the Black and Mediterranean seas and in combat operations near the Caucasus shores. From 1842 served as a lieutenant on the ship "Twelve Apostles". Upon return to the Black Sea, commanded the ship "Bessarabia", then by the newest steam-driven fregate "Vladimir". Was progressively promoted to Lieutenant commander, Captain II rank for distinction. In 1852, appointed aide-de-camp to Nicholas I. During the Crimean War of 1853–1856, supervised the building of rowing battle-boats in Finland and Riga. During the 1855 campaign, besides doing special errands, was building to his own designs battery pontoons for Kronstadt defense in the Gulf of Finland, participated in the work of fleet improvement commissions. In 1855, promoted to Captain I rank and as a promoter of the Russian Shipping and Trade Society (ROPiT), one year later was appointed its director. In 1860, promoted to Rear-admiral, in 1866—to Vice-admiral. In 1871, appointed Chief commander of Nikolaev Port and Military Governor of Nikolaev. When the Black Sea flotilla regained its status that same year, he became Chief Commander of the Black Sea and Ports. He spared no efforts to restore the economy ruined in the course of the Crimean War: ports, shipyards, lighthouses, etc. Under his guidance, the fleet established squadrons of active defense ships, which in addition to Popovkas (round-shaped ships called so after Popov, an admiral that designed them), included high-speed steamers leased by ROPiT and converted to auxiliary cruisers. During war year, these, using artillery and mines, attacked the enemy squadrons in ports and did not allow the Turkish armor-plated squadron to win superiority at high seas. In 1880, the city duma of Nikolaev conferred on him the title of an honored citizen of the city. Early in 1881, retired from the position of the Chief Commander of Black Sea and Ports for health reasons.

Arkhangelsky Andrey Dmitrievich (1879–1940) – Soviet geologist, Academician (from 1926). From 1913 to 1924, worked at the Geological Committee. Professor of Moscow State University (1920–1932) and Moscow Academy of Mining (1924–1932). Director of Geological Institute of the USSR Academy of Sciences (from 1934). In association with I.M. Gubkin, carried out projects exploring the Kursk Magnetic Anomaly. Led expeditions for integrated study of Kazakhstan's geological structure (1936–1938) and geology of the European part of the USSR (1939–1940). V.I. Lenin Prize winner (1928). A.'s works encompass diverse problems of regional geology, stratigraphy, tectonics, petrography of sedimentary rocks, palaeography, etc. His early studies deal with Volga Basin geology, stratigraphy of paleogenic and Cretaceous deposits. The monograph "Upper Cretaceous Deposits of European Russia's East" (1912) laid the foundation of the method of comparative-lithological investigations in the USSR. In 1927, A. and N.M. Strakhov on board the ship "Pervoe Maya" ('May Day') studied the geology and morphology of the Black Sea Depression. The findings of the study are discussed in the monograph "Geological Structure and History of the Black Sea Development" (1938), which contains an analysis of the materials of observations of the Black Sea sea-bed deposits and a map of deposits distribution is presented. Besides, in some works A. focused on the conditions of useful minerals formation: "Conditions Conducive to Petroleum Formation in the North Caucasus" (1927), "On Genesis of Beauxites and the Search for Their New Deposits" (1937). In the works "On the Relationship between Anomalies of the Gravity Force, Magnetic Anomalies and Geological Structure in East Europe" (1924), "Geological Significance of Gravity Force Anomalies in USSR" (1937, co-author) and others, A. introduced geophysical methods in regional geotectonic investigations. In the works "Geological Structure of USSR. European and Middle-Asian Parts" (1932, 4th ed.—"Geological Structure and Geological History of USSR", v. 1–2, 1947–1948) A. established some common regularities of the earth's crust development and made a contemporary presentation of the theory of geosynclines and platforms.

Arkhipo-Osipovka – maritime climatic health resort on the coast of the North-eastern Black Sea, Krasnodar Territory, Russia. Large urban village from 1960, 45 km to the south-east of Gelendzhik on the shore of Vulan Bay, in a basin encircled by the spurs of the Greater Caucasus (Mikhailovsky and Pshadsky Mountain Ranges), baring the way to north-easterly winds. A.O is part of Gelendzhik Health Resort Zone. Village of Vulan fortification (named so for it is located in the Vulan River Mouth) emerged side by side with Mikhailovskoe fortification. In 1839, the small garrison of the fortification was attacked by a force of Cherkess highlanders (Ubykhs) of over 12,000 men. When repelling the attack, the private of Terngin Regiment Arkhip Osipov blasted a powder cellar, killing himself and up to 2,000 attackers. In 1840, Arkhip Osipov became the first Russian soldier forever included in the regiment register. In 1876, a 120-pud cast-iron cross in honor of Osipov was installed for private donations, and to commemorate this feat, Vulanskoe Village was renamed A.-O. in 1889. The pebble-and-sand beach as well as the gently-sloping sandy seabed are convenient for bathing. A.-O. sanatoria specialize in the treatment of patients with functional disturbances of the nervous system, there is a water-treatment facility: sea, man-made carbon dioxide baths, radon baths, healing showers, a swimming pool; climatic pavilion; holiday camps, rest home.

Arkhipo-Osipovka (http://img-fotki.yandex.ru/get/0/milumil.0/0_36_2028d688_L)

Arkutino, Lake (Mechesh swamp) – small fresh-water lake situated in the dunes of the left bank of the Ropotamo River on the Bulgarian coast of the Black Sea. Alimentation predominantly is pluvial. The shores of the lake and in places the lake itself are overgrown with cane. A motel and campaigns have been established in the vicinity of the lake since 1964.

"Armed forces of Southern Russia" (AFSR) – White Guard troops during the Civil War in Russia. They were created in January 1919 by uniting the Volunteer Army and the White Cossack Don Army under command of A.I. Denikin. AFSR also comprised the Crimean-Azov Volunteer Army (January-February 1919), the Turkestan Army (January 1919—February 1920), the Third Separate Army Corps (March-December 1919), the troops of the Northern Caucasus (from March 1919), the Caucasus Army (from May 1919) and the Black Sea fleet. In late July 1919 AFSR numbered over 160 thou people. In 1919–1920 the Armed Forces of Southern Russia suffered a crushing defeat and its remnants retreated to the Crimea where P.N. Vrangel formed on their basis the so-called "Russian Army".

"Armenia" – passenger motorship of the "Abkhazia" type, one of the first passenger liners (Crimchaks), built at the Baltic shipyard in 1930, used to cruise on the Crimea-Caucasus Express Shipping line "Odessa-Batumi-Odessa". Had the cruising range of 4,600 miles, was capable of carrying 518 passengers in class suites, of accommodating 125 sitting passengers and 317 deck passengers, plus nearly 100 tons of cargo, developing the speed of 14.5 knots (around 27 km/h). When the Great Patriotic War (WWII) broke out, "A." was converted into a sanitation transport. Yet, the vessel was also used for delivering military elements in Odessa and Sevastopol. A. made its last passage on the night from November 6 to November 7, 1941. A. left Sevastopol, stopped over in Balaklava, then in Yalta, having taken on board around 5,000 wounded, hospital staff, military personnel and civilians. In the morning of November 7, A. was attacked by a German torpedo bomber "Heinkel-111", and one of the two launched torpedoes hit the target. The liner sank in 4 min. The boats were only able to rescue 8 persons. The sinking of A. is the largest naval catastrophe of the Soviet fleet after the wreck of the "Lenin" motorship.

"Armor-clad ship Potemkin" – see "POTEMKIN", "PRINCE POTEMKIN OF TAVRIA".

Arnoglossus Kesleri – small fish of Bothidae family, has a strongly compressed oval-shaped body, covered with large loosely-attached scale. The eyes are on the left side. The upper side of the body is gray-yellow. Length up to 9 cm. A.K. is omnipresent in the Black Sea.

Arnoldi Lev Vladimirovich (1903–1980) – Soviet hydrobiologist and entomologist. In 1927, graduated from the Hydrobiology Chair of Moscow University, in 1929—post-graduate course of the same chair. From 1921, as a student, worked in a number of expeditions led by N.M. Knipovich and A.S. Zernov. Upon completion of the post-graduate studies, A. worked at the hydrobiological station on Sevan

Lake in Armenia, then in Batumi, from 1932—in the Caspian Fisheries Expedition. In 1934, began to work at Sevastopol Biological Station. In 1936, considering all works authored by him cumulatively, A. was awarded the scientific degree of a Candidate of Biology in 1940—promoted to a senior researcher. By 1941, A. prepared a doctoral thesis on hydrobiology. He was unable to defend it as he was forced to evacuate from Sevastopol. In 1941–1942, worked in South Kazakhstan as an entomologist, then returned to his team at the Sevastopol Biological Station that had moved to Dushambe. In February of 1944 A. was admitted to the doctorate of the Zoological Institute of the Russian Academy of Sciences to deal with coleopterans, later was appointed a Senior Consulting Researcher. In 1954, A. was promoted to Professor. Published around 120 works in the field of hydrobiology and entomology.

Arnoldi Vladimir Mitrofanovich (1871–1924) – Russian biologist, main specialization—morphology and embryology of gymnosperms. Worked abroad a lot, spent 2 years on Java, in the famous Buitenzorg Botanical Gardens. Wrote the book "The Islands of Malay Archipelago Visited". Lay the foundation of "Kharkov School of Algologists" by organizing a team of young scientists dealing with major systems groups of algae and setting up a biological station near Kharkov. In 1901, published the first Russian manual on algology "Introduction in the Study of Lower Organisms". In 1923, elected a Corresponding Member of the Russian Academy of Sciences. In 1920–1921, worked as director of the Novorossiisk Biological Station. Carried out a number of crucial hydrobiological projects in Sudzhuk Lagoon, on Abrau Lake and in circum-Azov brackish lagoons. Corresponding Member of the Russian Academy of Sciences (1923). Authored works on the morphology and taxonomy of green algae and gymnosperms.

Aryan, Quint Appii Flavius (circa 90–95, A.D.–?) – was born in Asia Minor, in the rich Roman province of Vifania, Nicomedia city. Got excellent education, had a command of several languages, dealt with diplomacy, rhetoric, military schooling. Senator. Around 121–124 was awarded the rank of a consul. During 131–137, ruled Kappakodia, one of the crucial Roman provinces in Asia Minor, on the Black Sea coast. It was during that period that he traveled from Trapezund (Trabzon) to Dioscurida—Sebastopolis. A. submitted a report about his voyage to the Emperor Adrian in the form of a periple written both on the basis of his personal impressions, and using other sources—"Periple of Pontius Eucsinian". This writing has reached us in the only manuscript of the tenth century (Palatine Manuscript and its London Copy of the fourteenth to fifteenth centuries). In 147 A. was elected as an arhonteponim (one of the highest office bearers) in Athens. Later, A. distanced himself from the state affairs and devoted all his time to literature. As a writer, he is known by his major work "Alexander's Campaign" (Indian campaign of Alexander the Great).

Artek – maritime climatic health resort, complex of health camps on the southern coast of Crimea and stretching 7 km along the shore from the rock of Genoa Fortress to the south-western foot of the dome-shaped Mountain Ayu-Dag

(Mt. Bear, height 572 m). Adjoins Gurzuf village on the east, 70 km to the south of Simferopol. In USSR time,—All-Union Order of the Red Banner of Labor and Order of Peoples' Friendship V.I. Lenin pioneer camp, at present—International Children's Center Artek. In the sixth century A.D., the Byzanthians set up the first fortress in this place. The climate is subtropical of Mediterranean type, similar to that of Yalta. Maximum humidity falls on December, minimum—on August. A. is protected from the winds by the ridge of the Crimean Mountains. The major natural curative factor—climate, favoring the procedures of climatic prophylaxis, thalassotherapy. First mention of A. is dated 1872. On June 16, 1925, the Central Committee of the Young Communist League and the Russian Red Cross Society at the initiative of its Chairman and Deputy People's Commissar of Public Health Z. P. Soloviov established here a children's medical treatment-and-health facility. The pioneer camp initially comprised 4 tarpaulin tents and accommodated 80 pioneers per shift, in 1926, the tents gave way to timber houses and in 1928 the number of children arriving to the camp to spend holidays was over 1,000. A.—is a complex of children's camps on whose territory (over 300 ha in area) in the parks made up of subtropical plants (total area of 5 parks is 300 ha) there are more than 150 up-to-date buildings and structures, a sports center with a winter swimming pool, a village of science, technology and arts with a Knowledge Center. A. unites 10 camps, 5 of which function all year round. Each camp has an access to the sea with a civil and a medical facility. A. prides itself on its part of cypress trees, a monument of garden-and-park art. There are many picturesque and notable corners on the territory of A., one of these of Suuk-Su Cape.

Artek (http://www.aif.ru/application/public/news/034/45acc9fafb2f7a570fb84e44ab6080be_big.jpg)

Artseulov Konstantin Konstantinovich (1891–1980) – one of the first pilots in Russia—founding fathers of aerobatic flying, grandson of the outstanding marine painter I.K. Aivazovsky. In military service from 1912 (on and off). Used to study at the Sea Cadet Corps (1906–1908), then worked at the St. Petersburg aircraft works, combining work with studies at a flying school, and went in for glider flying. Was flying gliders of his own design. In 1911, obtained the diploma of a pilot-aviator. In 1912, flying instructor at the Sevastopol flying club. Participant of WWI: there, first he served in cavalry, where he commanded a platoon, promoted to a warrant officer. In 1915, passed a military pilot exam. Made around 200 reconnaissance flights. From 1916, flier of a fighter squadron, made 18 successful sorties. That same year was appointed chief of air fighter training at the Sevastopol flying school. Here, in autumn of 1916, A. was the first in the history of Russian aviation to put the plane into the spin and pull it out of the spin. Decorated with orders.

Artvin – one of the Black Sea provinces of Turkey. Situated in the eastern part of the country, borders on Georgia and Armenia. Area—7,436 km^2, population—165,000 people (2011). Center—Artvin.

Asia Minor – a peninsula in the west of Asia. It takes the greater part of the Turkish territory. This name was first used in the fifth—early sixth centuries to distinguish it from Great Asia that enclosed the rest territory of this part of the world. It is washed by the Black, Marmara, Aegean and Mediterranean seas. Its length is 1,000 km (from east to west) and width 400–600 km (from north to south). The whole peninsula, except some small coastal lowlands, is covered by mountains and highlands forming the Asia Minor Plateau. The northern and southern coasts are practically unbroken, there are a few bays here, while the western coast is heavily rugged and abounds in natural harbors. The central part of the peninsula is occupied by the vast semiarid Anatolian Plateau rising to 800–1,500 m with some volcanic cones (Mount Erciyes, 3,916 m) and rare not high "spots" of ridges. The highlands are surrounded by high marginal mountains with steep slopes. The coastal mountains are overgrown with forests or shrubs. The northern marginal chains are called the Pontic Mountains, the southern—the Taurus Mountains. The climate here is subtropical with different mountain-specific features. The mean temperature in January on the coast is +5 °C to +10 °C, in the central parts of the highlands—below 0 °C, in the east—to −15 °C. The mean temperature in July varies from +20 °C in the north to +30 °C in the south dropping to +15 °C on the interior highlands. The marginal mountains receive 1,000–2,000 mm of precipitations a year, while the interior highlands locating in the "wind shadow" only 200–500 mm. The southern and southwestern parts of A.M. have the Mediterranean precipitation regime (dry summer and rainy winter). On the Black Sea slope of the Pontic Mountains the precipitations fall year round (cyclonic rainfalls in winter and rainfalls brought by the northwestern winds in summer), thus, the rainfall maximum occurs here in spring. The mountain rivers are low-water and non-navigable; most of them have the Mediterranean regime (low water in summer and floods in winter), while the regimes of rivers on the Black Sea slope is more even. A group of large tectonic and partially karst Pisidium lakes locates in the Western Taurus. Large tectonic lakes are also found in northwestern Anatolia. The saline Tuz Lake on the Anatolian Plateau is drainless and in summer it turns into dry solonchak. The seaside slopes of marginal mountains are covered with rich broad-leaved forests of the subtropical Cholchis type growing on red, yellow and forest brown soils (Lazistan) and with the coarse-leaved evergreen forests of the Mediterranean type (the south of Taurus) passing into maquis and light pine forests on brown soils. For the inland mountain slopes and plateaus typical are semideserts on the brown steppe and solonchak soils with the thick thorn pillow-like shrubs, deciduous shrubs of the shibliac type and upland steppes on the light chestnut soils. Above altitudes of 600–700 m the mountains are mostly overgrown with coniferous forests (mostly fir trees) on the mountain podzolic soils, still higher at altitudes of 1,800–2,000 m the meadows on the mountain meadow soils are developed. Mountain and desert-steppe forms of hoofed, rodents, birds, reptiles and insects are most characteristic of A.M. The mountain forest fauna is developed in the margins.

Aslanbegov Avraamii Bogdanovich (1822–1900) – Russian Vice-admiral, flagman, sea biographer. Enrolled in the Sea Corps as a cadet (1835). Promoted to midshipman (1837). In 1837–1842, cruised on various ships in the Baltic Sea. In 1842, promoted to Lieutenant and transferred to the Black Sea fleet. In 1843–1845, cruised in the Black Sea. In 1845–1846, sailed from Odessa to Archipelago and back. In 1847–1853, cruised near the eastern coasts of the Black Sea, took part in landing an amphibious party and in minor actions with Cherkesses. Stayed in the garrison of Sevastopol during city defense (1854–1855), was contused in the course of city storming. In 1854, promoted to Captain lieutenant for distinction. In 1855–1858, stayed at Nikolaev port. In 1860, after the formation of Black Sea crews, joined the 2nd composite Black Sea crew. In 1860–1861, moved to the Baltic Sea. In 1862, promoted to Captain II rank. From 1864—in Kronstadt. Promoted to Captain I rank (1866). In 1868, joined the 4th fleet crew. In 1878, promoted to Rear-admiral. In 1879–1882, commanded a squadron of ships on the Pacific. In 1884, appointed Junior flagman of the Baltic fleet. In 1887, promoted to Vice-admiral. Was decorated with numerous orders. Published a number of articles in the "Marine Digest". Two geographic locations in the Sea of Okhotsk are named after him.

"Astaninskie plavni" (plavni–flooded areas) – landscape-bird sanctuary. Situated to the south-west of Kazantip Bay of the Sea of Azov, Crimean Peninsula, Ukraine. Area—50 ha. The lake-plavni features the abundance of migratory and nesting waterfowl. The local fauna includes ruddy shelducks, also encountered are common crane, demoiselle crane, mute swan. Under protection from 1947, sanctuary from 1974.

Atanasovskoe Lake – shallow, salt lake in the offshore part of Burgas Bay northeast of Burgas in the Bulgarian Black Sea zone. The northern part is of a brackish lagoon type, the southern part has all features of a lagoon. The lake is divided by a fill in two parts. Previously, sea water used to reach the lake by seeping through the sand spit, at present the water is supplied by pipes. To reduce the input of fresh water, the three small creeks that used to flow into the northern part of the bay are now enclosed in a diversion canal located in the western part of the bay. In 1912, experiments with salt production were commenced here. Before long, they began building saltworks. A license to operate was issued to the Swiss company "Glarus". In 1928, the saltwork went productive. The Fochan salt-production process is used here. The salt of A.L. is clean.

Atatürk Bridge (Unkapanı Bridge) – is a highway bridge on the Golden Horn in Istanbul, Turkey. It is named after Mustafa Kemal Atatürk, the founder and first President of the Republic of Turkey. It was originally completed in 1836 with the name Hayratiye Bridge. The construction of the Hayratiye Bridge was ordered by the Ottoman Sultan Mahmud II and supervised by Ahmed Fevzi Pasha, the Deputy Admiral of the Ottoman Fleet, at the Imperial Naval Arsenal on the Golden Horn. This original bridge was 400 m long and 10 m wide. In 1875 it was replaced by a second bridge 480 m long and 18 m wide, made of iron and constructed by a French company, and remained in service till 1912, when it was demolished due to reaching the end of its service life. In 1912, the nearby Third Galata Bridge was disassembled in pieces and was re-erected at the site of the demolished Hayratiye Bridge, becoming the third bridge on this site. This bridge was used until 1936, when it was damaged by a storm. The current (fourth) bridge on this site was constructed between 1936 and 1940, and entered service in 1940 with the name Atatürk Bridge. It has a length of 477 m and a width of 25 m.

Atatürk Bridge (Photo by Andrey Kostianoy)

Atherine or Sea smelt (*Atherina hepsetus*) – pelagic schooling fish of *Atherinidae* family. Lives mostly in open waters. Hardly ever 10 cm long. The color of back is sandy, white belly and two silver stripes along the body sides. Pectoral fins directed backward do not reach pelvic fins. Approaches the shore for spawning, trying to avoid fresh-water areas. Lives 4–5 years. Sexual maturity is reached at age 1 year. Propagates in May-June. At the end of autumn, leaves for the southern areas of the

Black Sea to winter. The population is not numerous. A. is used for fish flour preparation.

Athos, Mount – see **New Athos**.

"ATLANTIS-II" – research vessel of the Woods Hole Oceanographic Institution, participating in integrated explorations of the Black Sea in April-May of 1969. As a result of the explorations, the age of three upper sedimentary layers of the Black Sea basin was determined: 3, 7, 25,000 years, respectively. It was established that the current processes of sedimentation on the bottom of the basin are 120 times more intensive than on the seabed of the Atlantic Ocean. It was found that salinity of fossil waters in the bottom sediments at the depth of up to 2 km was equal to 7–8 ‰, i.e. it may be assumed that the deposits were formed in nearly fresh-water conditions.

Atlas of the Black Sea Coasts Morphology and Dynamics – produced in the USSR in 1954 manually. Its authors are A.T. Vladimirov (map delineator) and the artist I. Bukhglender (subsequently, I. Ogurtsov). This is a scientific work in which the entire Black Sea cadaster is depicted using cartographic methods. The Atlas is available for public use and is at P.P. Shirshov Institute of Oceanology, Russian Academy of Sciences, Moscow.

Atlas of the Black and Azov Seas Coasts (1842) – complete atlas of the Black and Azov Seas was compiled by Russian Major-General Ye.P. Manganari. In 1826–1836, the hydrographic expedition led by Captain I rank Ye.P. Manganari accommodated on board the brig "Nikolay" and Yacht "Golubka" ('Dove') worked near the Russian and Turkish coasts of the Black Sea. The expedition culminated in the publication in 1842 of the first "Complete Atlas of the Black and Azov Seas". In 1845, Manganari completed the compilation of the hand-written sailing directions for the Black and Azov Seas. This systematic description of both the seas was a valuable contribution to the history of hydrography. The author was presented a diamond ring for the Atlas.

Atlas of the Don River, Black and Azov Seas (1703) – compiled by Russian Admiral Cornelius Cruys based on the survey made during the Azov campaign. Together with him, the measurements were taken by Peter I, the survey of the Black and Azov seas having been completed. Published in Amsterdam in 1703. That signified transition from an ancient Russian drawing to a geographic map. The Atlas of the Black and Azov seas was provided with a compass grid, the meridians and parallels not indicated. Only the vertical side of map frames show the marks of geographic latitudes spaced at 10 min; the Black Sea map features the marks on the horizontal sides of the frames every 30 min.

Atlesh, Rocks – situated in the extreme west of the Crimean Peninsula in the vicinity of Tarkhankut Cape. The reserve stow is a picturesque shore formed by limestone rocks with numerous caves, grottos, niches, stone arcs, piles of blocks.

Aurelia – large (true) jelly-fish, has an umbrella on top, and lobes under it. The umbrella has a diameter of 10–20 cm, in some cases—over 30 cm. Round the periphery of the umbrella, there are numerous filiform feelers and eight marginal bodies and rhopaliums—medusa's sensory organs. In the center of the umbrella's lower surface is a cross-shaped mouth surrounded by 4 long lobes. Hence the name of A.—"eared medusa". 98 % of A.'s body is water. A. kills or paralyzes its prey with a potent poison—hypotoxin. It is dangerous to small organisms of plankton only, which serve as food for A. On human skin, there may be slight irritations only—"burns", yet these are dangerous for the eyes. The population of A. changes on a regular basis. In the 1980s, its mass in the upper 20-m thick water layer alone totaled 40 million tons.

Aurelia (http://kakadu.509.com1.ru:8018/WWW/fotogal/oboi/bezp/7.jpg)

Authority for providing safe navigation in the Black and Azov seas (UBEKOCHERNAZ) – established in Russia in 1920. In 1918, an order was issued for the Russian Navy and Marine Department announcing the Statute of institution authorities for providing safe navigation—UBEKO. This statute regulated in the Soviet Navy the performance of hydrographic bodies in the fleets and flotillas. Administratively, technically and economically the authorities were under the Main Hydrography Department. In April of 1921, UBEKOCHERNAZ was transferred from Nikolaev to Sevastopol.

Autonomous Republic Crimea (ARC) – established by Decree of the Ukraine's Supreme Rada on February 12, 1991 within the boundaries of the former Crimean Region of the USSR. In compliance with the Ukrainian Constitution, ARC is the basis administrative-and-territorial constituent entity of the Ukraine along with 24 regions, cities of Kiev and Sevastopol. Area 26.1 thousand km^2. Population— about 1.96 million people (2012). Urban population—1.2 million people. Rural population—746,000 people. Average density of population—77 persons/km^2. Center—Simferopol city. ARC is situated on the Crimean Peninsula (CP) washed on the west and south by the Black Sea, on the east—by the Sea of Azov. The northern—greater part, of the Crimean Peninsula is a flat or slightly undulating plain, the southern part is occupied by the Crimean Mountains. Along the southern foot of the mountains, there stretches a coastal strip (2–8 km wide) called the Southern Coast of the Crimea (SCC). Climate of the Crimea is moderately warm, on the southern coast of the Crimea—subtropical. The greater part of CP is in the steppe zone, but there is almost no steppe vegetation left. Mountain slopes are overgrown with shrub thicket and woods (oak-trees, beech and pine). The SCC features Mediterranean type vegetation. The rivers: Salgir, Indol, Biyuk-Karasu, Chernaya, Berbek, Kacha, Alma. The longest is the Salgir River (220 km).

The Crimea is mainly inhabited by Russians, Ukrainians, Tatars. Russians—1180.4 thousand people, Ukrainians—492,000 people, Crimean Tatars—243,000 people, 56 urban villages. The largest city is Simferopol; population 336,000 people (2011). Major cities—Sevastopol, Kerch, Evpatoria, Feodosiya, Yalta and others. The region is agricultural. The largest area of horticulture and viticulture development. Branches of industry: iron-ore, machine-building, chemical, building materials, fishing, canning, wine-making, tobacco-processing and essential oils fabrication. Industry is largely geared to supporting agriculture and fishing. The energy base of industry is Simferopol GRES (state district power plant). Iron-ore industry is based on iron-ore deposits around Kerch. Its production and benefication are carried out by Kamyshburun Integrated Plant. Bromium plants at Saki and Krasnyi Perekop are based on the vast reserves of various salts in the waters of Sivash Lake and Crimean lakes. Under way is the development of flux limestone (Balaklava District of Sevastopol), gypsum, limestone. The fisheries centers are Kerch, Feodosyia, Sevastopol and other cities. In the Black Sea and the Sea of Azov, they catch gray mullet, gobies, flatfish, mackerel, horse mackerel, Kerch herring; a special flotilla is busy catching pilchard in the Atlantic. The fish is processed at fish-canning factories in Kerch and Yalta. Wine-making is common in the vicinity of Yalta and Sudak, in Feodosiya and Simferopol; fruit and vegetable canning—in Simferopol, Djankoy, tobacco processing—in Yalta, Simferopol, Feodosiya. There are good prospects for the development of meat processing, dairy and butter industry as well as macaroni products industry (in Simferopol, Sevastopol, Kerch, Feodosiya, Yalta, Evpatoria, Alushta). There are enterprises for garment-making, knitting factories and leather footwear factories (Simferopol, Kerch, Feodosiya, Sevastopol) and wood-working plants (Simferopol, Servastopol, and other cities).

The leading branches of agriculture are horticulture and viticulture. The main areas of perennial fruit plantations are in the steppe zone of the Crimea. Essential oil crops are cultivated (clary, lavender, Kazanlik rose). A 5,000 ha irrigation system has been built near Simferopol Storage Reservoir. The problem of steppe Crimea irrigation has been resolved by using the Dnieper water filling the North-Crimean Canal. The Crimea is one of Ukraine's major areas of sheep-raising (fine-fleece sheep and semi-fine wool sheep raising).

Major railway line: Melitopol—Sevastopol, Kerch—Kherson. The total length of railway lines is over 600 km. Railway communication between the Crimea and

Caucasus is facilitated by the ferry service across Kerch Strait (re-opened in 2004). Sea transport is at an advanced stage of development. The main seaports are: Sevastopol, Kerch, Yalta, Feodosyia. As far as highways are concerned, of crucial importance is the terminal segment of the highway Moscow-Simferopol as well as of the highway Simferopol—Alushta—Yalta—Sevastopol.

Health-resort and tourist business is an important area of the economy. The abundance of sunny days, warm sea, dry air, availability of woods and parks make the Crimean Peninsula one of the best health resorts on the Black Sea. Most health resorts are situated on the SCC (in Yalta, Gurzuf, Simeiz, Alupka, Alushta, etc.), on the western coast—Evpatoria and Saki.

Autonomous Republic Crimea (http://en.wikipedia.org/wiki/File:Crimeamap.png)

Avinov Alexander Pavlovich (1786—1854) – Russian seafarer, Admiral. In 1804–1807, served as a midshipman on British naval ships, took part in the Battle of Trafalgar in 1805. In 1808–1817 served in the Baltic Navy. In 1819–1822, on the sloop "Discovery" participated in the circumnavigation expedition trying to find a passage from the Pacific to the Atlantic Ocean, made a description of the north-American coast—explored and described in the Bering Sea a segment of the north-western coast of Bristol Bay. In 1823, A. was promoted to Captain II rank. Until 1826, A. stayed with Kronstadt Navy, commanded the "Gangut" ship on which he participated in Navarin Battle with the Turko-Egyptian Navy. Promoted to Captain I rank, then to the rank of a Rear-admiral. In 1834, appointed Chief of staff of the Black Sea Fleet and ports. In 1837, appointed Commander of Sevastopol Port. Promoted to Vice-admiral. In 1838–1848, being in the rank of the Chairman of committees for the construction of dry docks, served as an aide to Admiral M.P.

Lazarev for Sevastopol Port development. In 1852, promoted to an Admiral. Decorated with numerous Russian and foreign orders.

Aya Cape – a cape on the southern coast of Crimea, Ukraine.

Ayancik – a village on the coast of the Black Sea, Turkey.

Ayazma, stow – (possibly, from Greek *"ayazma"*—*"consecrated, blessed"*, occupies the space between Balaklava Bight and Aya Cape, Crimea, Ukraine. The location stretching almost 3 km, is different from any other place in the SCC, and therefore treated as a special place—stow. It is characterized by the inaccessible, and often impassable precipice of the Main Crimean Ridge. Coastal blocks half-immersed in the water and rocks of varying sizes are scattered everywhere. The coast itself is encumbered with blocks and rocks, channels of stone creeping towards the sea are encountered. In places, the sea has reworked the rock and soil materials to form miniature bights and even small bights with cosy beaches: Golden, Silver, Water Nymph, Drided Fig. The largest limestone table-shaped precipice above the sea is also here. Local natural history scientists regard A.S. as one of the "seven geological miracles of the Crimea". There are famous grottoes here, some of which are submerged at the depth of 10 m. A.S. is part of the landscape reserve "Aya Cape" of republican significance, while the stow proper is declared a landscape natural monument.

Ayu-Dag – (Turk. *"Bear-Mountain"*, Gr. *"Aya"*—*"Holy"*) dome-shaped mountain and cape on the Southern Coast of Crimea, 25 km north-east of Gurzuf. The height is not too prominent, 572 m, area—5.4 km^2, length—2.5 km. Covered with thick forest and shrubs. Juts out far into the sea, structured by igneous rocks (laccolith). On the summit of A.-D. remains of ancient pagan sanctuaries and ruins of Christian churches have been found. The well-known international children's center "Artek" is at the foot of the mountain. A.-D. is frequently referred to as one of the natural mineralogical museums of the SCC. The number of minerals discovered here totals 18 (the unknown in the Crimean mineral vesuvianite was found here recently). This is the only habitat in the Crimea of a special species of forest cabbage and the only habitat of a rare fern—Chellanthes acrosticha in the Ukraine. Against the background of A.-D., rock islets Adalar are visible quite well. The former estate of the Raevskiis was built there in 1887. It is part of the republican reserve "Ayu-Dag".

Ayu-Dag (http://foto.crimean.info/data/media/1/111204-004.jpg)

"AYU-DAG" – landscape reserve, established in 1974, ARC, Ukraine. Occupies an area of 527 ha. Mountain massif of the SCC, structured by magmatic rocks. Maritime oak wood with species of Greek juniper, small-fruited strawberry tree, Turkish tereblinth, and other plants. 523 species of plants grow here, 21 of which and 16 species of animals are entered in the Ukraine's Red Data Book.

Azarov Ilya Ilyich (1902–1979) – Soviet Vice-admiral, political commissar of the Navy. From 1924—service in the Navy. In 1928—graduated from an artillery school, 1937—graduated of the Military-Political Academy. During the pre-WWII year, served in the Baltic and Pacific Navy, in 1941—officer of the Main Political Department (GPU), 1941–1943—Member of the Black Sea Military Council, participated in the defense of Odessa, Sevastopol, Novorossiisk, took part in planning the Kerch-Feodosiya Amphibian operation of 1941–1942. In 1943—Deputy Chief of GPU. In 1944 promoted to the rank of a Vice-admiral. In 1944–1947—Member of the Black Sea Military Council, in 1947–1953—Deputy Cammander-in-Chief of the 4th Fleet for political affairs. From 1953—at the central executive office of the USSR Ministry of Defense. Decorated with orders and medals. Author of a number of books on Navy activities during the Great Patriotic War (1941–1945).

Azov Cossack Force – Russian irregular force deployed in the mid-nineteenth century on the north-western coast of the Sea of Azov, around the Obitochnaya and Berda rivers (currently, the area of Zaporozhye Region, Ukraine). The force was established in 1828 and was staffed by the descendants of Zaporozhian Cossacks who, following the abolishment in May of 1775 of Zaporizhian Cich (Host), moved to the Danube mouth which at that time belonged to Turkey. At the beginning of the Russian-Turkish war of 1828–1829, these descendants, led by the koshevoi ataman

Osip Gladkii, regained their Russian allegiance and took an active part in battles, siding with the Russian army and assisting it to seize the Isakcha Fortress. The war over, a Separate Zaporizhian Force was formed from Zaporizhian Cossacks renamed in 1831 into the Azov Cossack Force and deployed on the north-western coast of the Sea of Azov, in Yekaterinoslav Province. At the end of the 1830s, the force was nearly 6,000 strong (with families) and included 12 cavalry hundreds and a flotilla (up to 30 small vessels), with a mission to guard the eastern coast of the Black Sea. The force was answerable to the governor-general of Novorossiisk and was managed by the delegated ataman and the force chancery, these both were first headquartered at the Cossack village Petrovskaya, then in Mariupol. From 1860, the Russian Government began resettling Cossacks in Kuban area, which led to serious turmoil. Therefore, on October 11 (23), 1864, the force was abolished, all officers were reckoned as nobility of Yekaterrinoslav Province, Cossacks as peasants-owners with an allotment of 9 dessiatines of land (1 dessiatine = 2.7 acres) per male head. In 1866, the abolition of the force was completed.

Azov – town and port, center of Azov District, 25 km from Rostov-on-Don, Rostov Region, Russia on the left bank of the Don River, 12.5 km from where it flows into Taganrog Bay of the Sea of Azov. Population 83,000 people (2010). One of the Russian cities whose emblem features two sturgeons (Taman, Tsaritsyn). The emblem was approved on July 3, 1967 and was changed several times afterwards. The first heraldic symbol inscribed on the banners of the Azov regiment between 1717 and 1722. In May of 1729, the emblem was presented by Earl Minich as a coat of arms of the Azov Fortress. One of the ancient cities of the Black-Sea area which in the tenth to eleventh centuries was part of the Tmutarakan Principality of Kievan Russia.

The first fortified ancient settlements in the Tanais (Don) River delta came into being more than 2,000 years ago. The official emergent year of the city is t 1067, when the Polovets Settement of Azak was founded. The settlement was named after the Azuv (Azak) Khan. Azov fortress served as protection for the Don mouth and ensured access to the sea, therefore it owners changed one another several times. During the tenth to twelfth centuries, the place of the current A. used to be occupied by a Slavic settlement that was part of the ancient Russian Tmutarakan Principality. From the thirteenth century, Azak is a town of the Golden Horde that gained a foothold in the circum-Azov area, the "great trading route" passed through it to China. In 1395, A. was destroyed by the forces of Timur (Tamerlane). Right at this moment, there takes shape on the territory of A. a colony of Genoese and Venetians who turned it into an affluent place of transfer and trade between the West and the East. This colony under the name of Tana existed until 1471, when the town was seized by the Turks who made it a solid military fortress. In 1637, A. was regained by the Don Cossacks, but in 1642 they were forced to relinquish the fortress after destroying the fortifications (these were later rebuilt by the Turks). During the famous 1642 Azov sitting, a garrison of 5,000 Cossacks for 93 days was withstanding the siege by a 250,000 Turkish force. The Azov campaigns of Peter I (1695–1696) terminated by the seizure of the Turkish fortress. This done, A. was built anew and became the base of the Russian fleet on the Sea of Azov. From 1708, A.—a town, center of Azov Province on the left bank of the Don River. After the unsuccessful Prut campaign of 1711, the town was returned to Turkey. In 1736, A. was regained by Russia. Under the 1739 Belgrade Peace Treaty, A. (together with Taganrog) was made part of the "barrier" (neutral) lands (fortifications and buildings destroyed). During the Russian-Turkish war Russia's legal rights of Azov ownership were recorded in the Kuchuk-Kainarji Peace Treaty of 1774. In 1775–1782, A.—center of Azov Province. Having become the trading quarter of Rostov District of Yekaterinoslav Province in 1810, A. missed the opportunity forever to have an emblem of its own. The status of a town was reinstated in 1926.

Of interest are the remains of the Turkish Fortress, including Alekseev Gates (1801–1805) and the foundation of the Trinity Gate, a powder magazine (1799), at present, it houses the exhibits of the regional natural history museum, among them the diorama "Azov Campaign of Peter I in 1696").

"Azov" – first ship of the Russian fleet decorated with St.George's banner flag: St. George's banner flag and St.George's ribbon worn attached to the flag pole were established in the Russian fleet by decree of April 10 (22), 1837. The first St. George's flag was awarded to the 12th Naval crew of the second Black Sea division for the act of bravery performed by the battle ship "A." in the Battle of Navarino October 8 (20), 1827 (modern-day Pylos Bay on the west coast of the Peloponnese Peninsula in the Ionian Sea). Under the terms of the 1827 London Convention, Turkey was to grant autonomy to Greece that was part of Turkey. However, this was not done, therefore the Russian, British and French squadrons were sent to the archipelago to assist the Greeks in their national-liberation struggle. On October 5, the united squadrons approached Navarino Bay, where the enemy fleet was attacked by the squadron and was defeated in battle that lasted 4 h. The 74-cannon flagship "A." commanded by Captain I rank M.P. Lazarev distinguished itself in action. It destroyed 5 Turkish ships, among them the flag-bearing fregate. Lieutenant P.S. Nakhimov, midshipman V.A. Kornilov and midshipman V.I. Istomin were among those who distinguished themselves on board "A.".

Azov (Sea of Azov) – (Lat.—"*Palus Maeotis*", Ancient. Gr.—"*Maiōtis limēn*"—"*Meotian Lake*", Ancient Rus.—"*Surozhskoe*"), "Sea of Maiotians" (Mayetians)—tribe that lived on its shore. With Romans—"Palus Meoti" ("*Meotians' Bog*"), with Turk peoples—"Azak Deniz" ("*azak*"—"*low-lying boggy terrain*"). Also—"Temerinda" ("Mother of Sea"), since Meotians believed it was a shallow water body—Mother of the large Black Sea. The list of names of the Sea of Azov contains over 400 options, most of which have to do with the name of "Azov" (Tanais). The world's shallowest sea that may be regarded as a large circum-Black Sea brackish lagoon of the Don River, in fact, this is a Black Sea Bay connected with the sea by Kerch Strait. The boundary between the seas passes along the line Takil Cape—Panagia Cape. Its area is 39,000 km^2, catchment area—630,000 km^2 (Russian territory—82 %, Ukrainian territory—18 %), mean depth 8 m, maximum depth—13.5 m. Volume of water—256 km^3. Maximum length of the sea from Arabat Arrow to the Don River Delta is 360 km, maximum width from north to south—

185 km. The contours of the Sea of Azov shores and its underwater relief, river flood plains and major deltas of the Kuban and Don rivers took shape in the conditions of submerged Azov-Kuban depression and development of the Black Sea transgression.

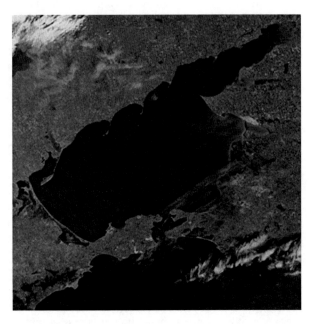

Sea of Azov (http://www.eosnap.com/lakes/sediments-in-the-sea-of-azov-russia-and-ukraine/)

The sea outline is relatively simple. The southern shore is hilly, it stretches from Kerch Strait to Arabat Arrow. Separating on the west the Sea of Azov from Sivash Lake, which is linked with the sea by Genicheskii Strait. To the east of this strait, there is Biryuchii Island, the only largest one in the Sea of Azov; in 1929, this island turned to a peninsula having connected with the shore by Fedotov sand spit. At the same time, there emerged a large, almost isolated from the sea Outlyuk brackish lagoon. The north-western shore is a steppe plain terminating near the sea in a low-lying slope dissected by rivulets and creeks. There are three large capes here: Obitochnaya, Berdyansk and Belosarai Spits. On the north-east, there is a 140 km long, largest sea bay—Taganrog Bay jutting inland, its apex being the Don delta. There are islets in the bay: Peschanyi, Cherepashka and Dolgii. On the south-east, there stretches to the distance of 100 km Kuban delta with extensive marshy and reedy banks and numerous bayous. Right after the westernmost Akhtanizov brackish lagoon of Kuban delta the bank rises. The long and relatively wide Taman Bay juts into the eastern coast. The western coast features Kerch Bay over which Mitridat Mt. towers along with Kamyshburun Bay with picturesque shores. Kerch Strait connects the Sea of Azov with the Black Sea.

Flat sea coasts give way to an even, monotonous, flat seafloor. Bottom sediments feature all kinds of silts, sand, coquina accumulations, combinations thereof. The depth gradually increase as one proceeds away from the coast. Mud volcanoes are known to exist in Temryuk Bay.

Nearly the entire river runoff into the sea (over 90 %) is provided by the Don and Kuban Rivers: 28 and 12 km^3, respectively. The rest of the runoff volume falls on smaller steppe rivers of the circum-Azov area: Molochnaya, Lozovatka, Obitochnaya, Berda, Kalmius, Yelanchik, Mius, Kagalnik, Yeya, Chelbas and Beisug. The large volume of river runoff accounts for the low salinity level of the Sea of Azov. Sometimes, saline water of Sivash Lake gets into the sea.

The water balance of the Sea of Azov, besides the river water, each year increases by around 14.2 km^3 of precipitation, evaporation amounting to around 34.3 km^3. The mean long-term data indicate that approximately 50 km^3 of sea water flows out of the Sea of Azov each year by surface flow, whereas up to 34 km^3 of Black Sea water comes into the sea via the subsurface flow. Sivash Lake receives around 1.4 km^3 of the Sea of Azov water, which get evaporated in the lake, while maximum 0.3 km^3 water leaks from Sivash Lake into the sea.

The climate of the Sea of Azov jutting far into the land is distinctly continental. It is characterized by cold winter, dry and hot summer. During the autumn-winter season, the weather is determined by the impact of the spur of the Siberian anticyclone with prevalence of easterly and north-easterly winds blowing at a speed of 4–7 m/s. A more intensive impact of this spur causes strong winds (up to 15 m/s) and is accompanied by cold air incursions. Mean air temperature of January is −1 to −5 °C, during north-easterly storms it drops to −25 to −27 °C. In spring and summer, warm, clear weather with slight winds prevails. In July, mean-monthly air temperature across the sea equals 23–25 °C, maximum exceeding 30 °C. The amount of atmospheric precipitation on the eastern coast of the sea equals 500 mm/year, on the western coast—around 300 mm/year.

Small size and shallow depths of the sea are conducive to rapid development of wind-induced wave. The waves are short, steep, at the center of the sea, they reach the height of 1–2 m sometimes up to 3 m.

Seasonal fluctuations of the sea level are mainly determined by the regime of river runoff. Seasonal variability of the sea level is characterized by level increase during the spring-summer months and decrease in autumn and winter, the range of fluctuations averaging 20 cm. The winds blowing over the sea cause significant surge-induced variations of the sea level. The highest rises of the level were noted in Taganrog up to 6 m. At other locations, 2–4 m surges are possible (Genichesk, Yeisk, Mariupol), in Kerch Strait—around 1 m.

Currents in the sea are largely wind-induced. Under the impact of westerly and south-westerly winds, there forms a counterclock-wise circulation of waters in the sea. Cyclonic circulation is formed in the case of easterly and north-easterly winds that are stronger in the eastern part of the sea. When similar winds, but stronger, are blowing in the southern part of the sea, the currents are of an anticyclonic nature. When there are slight or no winds, insignificant currents of variable directions are noted. Because weak and moderate winds are prevalent over the sea, the currents with speeds of up to 10 cm/s have maximum occurrence. When the winds are strong (15–20 cm/s), the velocities of the currents may reach 60–70 cm/s. In Kerch Strait, with northerly winds, an outflow from the Sea of Azov is observed; with the winds of southerly component, there occurs inflow of the Black Sea water into the Sea of Azov.

Ice emerges on the Sea of Azov each year. Ice coverage (area occupied by ice) depends in large measure on the kind of winter (severe, moderate, mild). In moderate winters, by early December, ice is formed in Taganrog Bay. During December, fast ice is established along the northern coast, and a little later—near the other coasts. The width of the fast ice strip—from 1.5 km in the south to 6–7 km in the north. Floating ice is produced in the central part of the sea only at the end of January—early in February, when its thickness is 30–40 cm, in Taganrog Bay—60–80 cm. In moderate winters, the sea is cleared of ice during March, at first in the southern areas and river mouths, then in the north, and finally in Taganrog Bay. The length of the ice season averages 4–5 months. When winters are unusually warm or severe, the time of ice formation and melting may be shifted by 1–2 months and even more.

In winter, almost in the entire water area the temperature of water is negative or close to zero, rising to 1–3 °C near Kerch Strait only. In summer, throughout the sea the surface water temperature is uniform: 24–25 °C. Maximum values in July-August at open sea—up to 28 °C, whereas near the shores the temperature is likely to be higher than 30 °C. The sea being so shallow is conducive to rapid spread of wind-induced and convective mixing of water down to the seabed, which tends to make vertical distribution of temperature more uniform: in most cases, temperature drop is never more than 1 °C. However, in summer, during calms, there forms a strong thermocline, which limits exchange with near-bottom water layer.

Spatial distribution of salinity in the conditions of natural inflow of river waters is rather uniform, the horizontal gradients being observed in Taganrog Bay only. At the exit from the bay, the prevalent salinity is within 6–8‰. In water area of open sea salinity is within 10–11‰. Vertically, almost in all areas gradients were only observed occasionally, largely in response to the incoming Black Sea waters. Seasonal variations do not exceed 1‰. However, they increase due to the impact of runoff distribution during the year. Taganrog Bay is filled with fresh and brackish sea waters, the boundary between which is tentatively determined by salinity of 2‰.

During the 1960–1970s, there increased fresh water withdrawals for economic needs in the basin of the Sea of Azov, which preconditioned decrease of river runoff into the sea and, accordingly, increased input of Black Sea waters. This coincided with the period of reduced water availability in the catchment area of the sea, and because all factors were at play, there began increase of salinity at the end of the 1960s. In 1976, mean salinity in the sea reached its maximum—13.7‰. In Taganrog Bay, it rose to 7–12‰. In the circum-Kerch area, especially in low-water years, salinity values were rising to 15–18‰.

The growing spread of Black Sea waters in the near-to-bottom layer of the sea led to the growth of vertical gradients of salinity and density, aggravated the conditions of near-to-bottom waters mixing and ventilation. The likelihood of oxygen lack (hypoxia) and production of unusual dead conditions to organism increased. Yet, during the 1980s, the Don River runoff increased, by the end of the 1980s, salinity declined again and at present there does not occur any rise in the salinity level of the Sea of Azov.

In the conditions of natural water regime prior to the early 1950s, the Sea of Azov exhibited a rather high bioproductivity. A large quantity of nutrients was transported to the sea with river runoff. This provided for abundant growth of phytoplankton, zooplankton and benthos. Favorable natural factors determined conditions for the development of fish fauna represented by 80 species. It was not for nothing that the ancient Greeks used to call the Sea of Azov Meotide, which means "foster-mother". In the 1930s, the total fish yield in the Sea of Azov reached 300,000 tons, more than half of this amount being valuable fish species (sturgeons, pike-perch, bream, herring, etc.). Regulation of the Don River (construction of Tsimlyansk Water Reservoir) in 1952, reduction of the runoff volume by 13–15 km^3/year, coupled with other effects of human activity in the sea basin produced grave negative changes in the marine ecosystem. Reduction of the annual runoff of the Don River, considerable decrease of the flooding water volume resulted in the shrinkage of spawning grounds areas, disturbed the conditions for the reproduction of fresh-water fish. The quantity and composition of nutrients getting into the sea has changed dramatically, and so has their distribution in the course of the year. The bulk of suspended matter settles in Tsymlyansk Water Reservoir, the quantity of such matter, brought into the sea in spring and early in the summer, has declined significantly; the input of mineral forms of phosphorus and nitrogen has dropped; instead, there increased the number of organic forms that are more difficult to assimilate by organisms. Biogenic substances reaching the sea are mainly con-sumed in Taganrog Bay and only a minor amount is taken to the open sea.

The river and sea waters are getting increasingly polluted with diverse harmful chemical agents: pesticides, phenols, in some areas of the sea—with oil products. The said ecological changes have led to an abrupt decline of bioproductivity of the sea. The food base of the fishes has declined several times as much, total yields of valuable fish species have dropped. At present, these are made up essentially of Black Sea and Sea of Azov kilkas and anchovys.

The water management situation in the sea basin is rather strained. At present, the river water input into the sea equals 28 km^3/year. Given such volume of runoff, it is possible to keep its salinity within 11.5–12‰. Further increase of water consumption in the water body basin is inadmissible as this will set of an irrevers-ible growth of salinity to the level of the Black Sea and will result in the aggravation of conditions for the life of most valuable sea organisms.

The main ports: Mariupol, Rostov-on-Don, Taganrog, Eisk, Berdyansk. After the collapse of the USSR, the Sea of Azov has virtually become a transboundary water body; at present, status of the sea has not been determined or recorded anywhere on an interstate level.

Azov campaigns of 1695–1696 – campaigns of Russian army and fleet led by Peter I during the Russian-Turkish war of 1686–1700 with a view to protecting Russia's southern lands against the attack of the Turkish and Tatar troops and occupying the

Turkish fortress Azov that closed Russia's access to the Sea of Azov and the Black Sea. Early in April of 1695, the Russian forces (around 31,000 warriors) that consisted of Streltsy (soldiers), regiments of a new "fighting formation" and manorial noblemen's cavalry set out from Moscow to Azov. To divert enemy attention from the fortress, troops headed by B.P. Sheremetev were sent to the lower reaches of the Dnieper River. On July 5 (15), Russian troops concentrated around Azov which was defended by a garrison of 7,000 soldiers. The enemy repelled two assaults causing heavy casualties to the attackers. Therefore, Peter I lifted the siege and on November 22 (December 2) Russian troops returned to Valuiki and Voronezh. As peparations for a new campaign under the guidance of Peter I were being made, the Azov fleet was established. A.Ya. Lefort was appointed Commander-in-Chief of the Azov Fleet. A.S. Shein was the Commander of the Azov Army. Peter I was in charge of overall leadership of the second campaign. On April 23–26 (May 3–6), 1696 the army set out on the campaign from the districts of Voronezh, Tambov, and Valuiki overland and by ships down the Voronezh and Don rivers. Sheremetev's cavalry again headed for the Dnieper lower reaches, but stopped at the Kolomak River. On May 27 (June 6), the main forces of the Russian fleet set out on the Sea of Azov in the Azov area and by June 12 (22) isolated the city, while the Russian army lay a siege from the land. The Turkish fleet tried to rescue Azov but failed. On June 14(24) the Turkish fleet emerged opposite the Don River mouth (6 corvettes, 17 galleys with a landing party of around 4,000 men), but having seen the Russian galleys, the fleet left for the sea. On July 17 (27), after heavy artillery fire, assault of the fortress began simulta-neously from land and sea. On July 19 (29), the garrison surrendered. As a result of successful termination of the second Azov campaign, Russia attained access to the Sea of Azov and the Black Sea, made provisions for the security of the country's southern frontiers. Because under the Constatinople Peace Treaty of 1700, for-tresses in the circum-Dnestr area were to be demolished, Russia's international standing was enhanced, Turkish neutrality on the eve of the Northern War was secured. The seizure of Azov was the first major victory of the Russian army and fleet in the struggle for access to the sea.

Azov Fleet – first regular formation of the Russian Navy instituted by Peter I in order to fight Turkey for access to the Sea of Azov and the Black Sea. In 1694, there began the construction of large ships and assembly of galleys and fire ships, using parts made in Bryansk, Preobrazhenskoe Village (near Moscow) and at other locations. By the spring of 1696, three 36-cannon ships, 23 galleys, 1,300 sail-row boats and 4 fire ships were built. The fleet, under the command of F.Ya. Lefort left Voronezh and on May 27 (June 6) entered the Sea of Azov. On June 12 (22), the Russian ships blocked the Azov Fortress in the mouth of the Don River, while the ground troops did the same from land. On July 19 (29), the fortress garrison surrendered. At the insistence of Peter I, the Boyar Duma decreed: "Let there be sea vessels". This date is regarded the official birthday of regular Russian fleet. The admiralty was transferred from Voronezh to Tavrov on the coast of the Sea of Azov, a sea port comes into being in Taganrog. During the period from 1696 to 1711,

215 ships of diverse classes were built for the Azov fleet. In the spring of 1699, Peter I for the first time in the history of Russian Navy held sea maneuvers in the vicinity of Taganrog. In August, the largest 46-cannon ship "Krepost" ('Fortress') sailed in the Black Sea and visited Constantinople with a diplomatic mission. After the Prut Treaty of 1711 and return of Azov and Taganrog to Turkey, the Azov fleet ceased to exist, its ships were disassembled or sold to Turkey.

Azov Military Flotilla – (1) A formation of Russian fleet established at the beginning of the Russian-Turkish War of 1768–1774. Under the command of the Vice-Admiral D.N. Senyavin, AMF performed successful military operations against the Turkish fleet on the Sea of Azov and the Black Sea, cooperated with ground troops, when seizing Kerch and Yenikale, repelled enemy attempts to land amphibious parties in the Crimea. In 1783, AMF was disbanded, yet its ships were included in the Black Sea fleet that was established in May of the same year.

(2) Russian flotilla with a base facility at Yeisk was configured to fight against German invaders and the White Guard. When the enemy seized the coast at the end of June of 1918, the ships were disarmed, and their personnel joined the Red Army units. In March of 1920, after Denikin's army was defeated and the Red Army reached the Azov Sea coast, the flotilla was reestablished by the staff of the South-Eastern Front (the base at Mariupol—currently, Zhdanov; from September—Taganrog; from November—Mariupol again). The flotilla included ships that were in the ports of the Sea of Azov. Flat-bottomed fishing boats and barges were reequipped as battle-boats and floating batteries, tug boats were reequipped as escort ships, fighter boats were delivered by railway. The armament, supplies and personnel came from the Baltic Fleet, Don-Azov, Volga-Caspian and other flotillas that terminated combat activity. From May 25 to September of 1920, the Don River division (former Don Flotilla of the Caucasus front) was subordinated to AMF. In May of 1920, AMF became part of the Marine Forces of the Black Sea and the Sea of Azov. There were about 70 ships and vessels (9 gun-boats, 4 floating bases, 3 mine layers, 6 escort vessels, 22 chaser cutters, 25 auxiliary vessels), 18 aircraft, for amphibian operations, a marine expeditionary division was detailed (up to 4,600 men). AMF provided fire support to troops, set up a mine-artillery position in Taganrog, placed mine barriers in Kerch Strait, dropped tactical landing parties with a view to eliminating, in association with the 9th Army, the Wrangel landing party Ulagaya; on July 9, 1920, AMF destroyed a White Guard landing party near Krivaya Kosa ('Crooked Spit'), on September 15 in combat near Obitochnaya Kosa, AMF destroyed a group of enemy ships that were transporting troops and armament. After the defeat of Wrangel in April of 1291, the ships and personnel were handed over to the Black Sea Fleet.

(3) During the Great Patriotic War (WWII), on July 22, 1941, AMF was re-established for combat actions against German Nazi invaders (main base at Mariupol; from October 9, 1941—Primorsko-Akhtarsk, at present—Primorsko-Akhtarsk; from August 3 to August 24, 1942—Novorossiisk). AMF comprised the ships of the Danube Military Flotilla, that, the Danube battles over, moved eastward across the Black Sea. AMF included squadrons of escort vessels, mining

boats, airborne tactical formation, coastal defense squadrons and marine units. The separate Kuban detachment (from May 3 to August 30, 1942) as well as the separate Don Detachment (from October 5, 1941 to July 28, 1942) based in Rostov-on-Don. The flotilla fought against German-Nazi invaders in liaison with the troops of the South and North-Caucasus Fronts, supported defense actions of the 9th and 51st Armies, took part in the Kerch-Feodosiya landing operation of 1941–1942, evacuated troops of the Crimean Front, assisted ferrying troops of the 56th Army across the Don River. For a long time, marines contained the enemy assault on Taman Peninsula. On September 5, 1942, the flotilla forces were included in Novorossiisk Defense Area (NDA) for marine operations. On February 3, 1943, AMF was reconstituted again (main base Yeisk; from September of 1943—Primorsko-Akhtarsk; from April of 1944—Temryuk). The flotilla ships took part in action on sea lines of communications, dropped tactical landing parties in Taganrog, Mariupol, Osipenko (Berdyansk). In the course of the Kerch-Eltingen Amphibian Operation, AMF dropped units of the 56th Army north of Kerch, and in January of 1944—two tactical landing parties on the coast of Kerch Peninsula. On April 20, 1944 the flotilla was disbanded, its ships handed over to the newly-established Danube Military Flotilla.

Azov Research Institute for Fisheries (AZNIIRKH) – established in Rostov-on-Don, Russia in 1958 on the basis of the Azov-Black Seas Fisheries Research Station (since 1928). From 1992, the Institute has been performing integrated studies on the Black Sea, too. The scope of its activity includes studies of aquatic systems, trends and regularities of their change in the diverse conditions of human activity, the level of anthropogenic impact and methods of improving the water quality of fishery water bodies, development of new, improvement and adoption of promising methods of monitoring the condition of water, bottom sediments and hydrobionts, regulation of the use of pesticides in fish-farming, agriculture and forestry; integrated research into marine bioresources and drafting recommendations on the development thereof, conservation and rational use of these resources; development of forecasts of likely reserves and permissible yield of major commercial fish species in the basins of the Sea of Azov and the Black Sea; resolving the problems of standardization and use of water resources, protection of young commercial fish from getting into water-intake structures; study of the mariculture problem, of natural and commercial fish reproduction (including sturgeons) and crustaceans. The institute takes part in a number of joint programs aimed at conservation of natural wealth of the Black Sea and the Sea of Azov, in the international Black Sea program (GEF-BSEP, JCZM), in the Russian-Ukrainian-Dutch Agreement on the "Sea of Azov" Project, etc.

Azov sea port – is situated on the left bank of the Don River, 14 km from its fall into the Sea of Azov. The port is connected with the sea by the Azov-Don Sea canal, 3.6–4.0 m deep (subject to hydraulic and hydrometeoconditions), the fairway being 70 m wide. The operating mode of the port is year-round (in winter, it provides icebreaker assistance to the ships). The design capacity of the port oriented to the handling of domestic Russian mineral-and-building materials-construction and bulk cargoes was in excess of 3.0 million tons. The transshipment pattern is dominated by metals, including metal scrap, timber, chemical fertilizer, coal, grain, minerals and building materials. The port specializes in minerals and building materials, packaged and piece cargoes, timber.

Azov sitting – defense of Azov by the Don Cossacks from the Turkish-Tatar troops in 1637–1642. In the spring of 1637, the Don Cossacks in association with Zaporizhians, taking advantage of the Turkish army being diverted by internal fighting in the Crimea, besieged the Azov Fortress (a garrison of 4,000 men armed with 200 cannons) and after the siege that lasted around 2 months seized it. Early in June of 1641 the Turkish-Tatar troops (over 20,000 men) backed by a strong fleet (30–40 ships and vessels) blocked the fortress. The garrison turned down the proposal to surrender made by the Turkish Sultan. Under the leadership of the Cossack chieftains O. Petrov and N. Vasiliev, the besieged rebuffed 24 assaults, making surprise sorties from the fortress, captured the enemy's big banner which later was delivered to Moscow. After the 3.5 month-long siege, the Turkish command due to heavy losses removed their troops. The Cossacks forwarded a request to the Russian Government to spread its authority on Azov. However, Russia was not yet ready for a war with Turkey; for this reason, following a direction of the Russian Government, the Cossacks left the fortress in the summer of 1642, having destroyed its fortifications.

Azov-Black Sea fishery research expedition (1922–1928) – expedition headed by N.M. Knipovich to study fishing in the Sea of Azov and the Black Sea. In association with the Don-Kuban Fishery Research Station, the Kerch Ichthyological Laboratory and some other organizations, carried out operations towards scientific provision of fisheries in the Sea of Azov and the Black Sea (within the USSR boundaries). In 1925 N.M. Knipovich's paper "On Fisheries in the Sea of Azov—Black Sea Basin" was published (based on the findings of the fishery research expedition).

Azov-Black Sea Research Institute for Fisheries and Oceanography (AzCherNIRO) – was established (based on Kerch Ichthyological Laboratory (since 1922), renamed in Kerch Fishery Research Station in 1927) in 1933 in Kerch with divisions in Rostov-on-Don, Odessa and Batumi. In the southern seas, including the Atlantic, Indian and Southern oceans AzCherNIRO has been a research base for handling complex fisheries problems in the field of oceanography and hydrobiology, ichthyology and mammalogy (study of dolphins), commercial fishing technology and economy on an integrated basis. Research effort in the Azov-Black Sea basin, ship-based and aerial fishing reconnaissance during the 1930s were the foundation conducive to fishery growth in this region in the subsequent years. In 1988 the institute was renamed in the Southern Scientific Research Institute of Marine Fisheries and Oceanography (YugNIRO).

Azov-Sivash hunting reserve – encompasses an area of 28.25 thousand ha of the circum-Azov coast. It includes Obitochnaya Spit, Biryuchii Island Spit in the Sea of Azov, Kuyuk-Tuk and Churuk Islands in Sivash. Part of Sivash Lake water area is under protection, too.

B

Baba, Cape – is the westernmost point of Anatolian part of Turkey, making it the westernmost point of whole Asia. There is a lighthouse at Cape Baba. It was called Cape Lectum in classical times.

Babugan-Yaila – the highest massif in the main Ridge of the Crimean Mountains (Mt. Roman-Kosh, 1,545 m). Quasi-plateau surface is structured by limestones. The southern slopes are covered with pine-oak woods, northern—by beech forests. At the foot of B.Y. Gurzuf and Alushta villages are located.

Bacescu Mihai (1908–1999) – outstanding Romanian hydrobiologist. Academician Professor Mihai Bacescu was born on March 28, 1908 in Brosteni, Suceava, northern Romania. His academic career began in 1928 when undertook 4 years of study at the Faculty of Biology in Iasi. During the next 6 years (1932–1938) he stayed at Agigea on the Black Sea where he dedicated his studies and research to marine biology at the Marine Biological Station, which was founded in 1926 by his advisor, Professor Ioan Borcea. This period was his first exposure to the sea and had a profound effect on his career. His doctoral dissertation, which was dedicated to the study of Romanian Mysidacea, epitomized his research at Agigea. He returned to Iasi and had a short university career during which he quickly became chair of the Department of Animal Morphology.

Soon afterward, he was nominated for a scholarship in marine biology in France (Roscoff, Banyuls-sur-mer, Paris) and Monaco by his supervisor, Professor Paul Bujor and the renowned Romanian biologist Emil Racovita, who in 1939 founded the discipline of biospeleology. In France he became highly respected by his French colleagues and carcinologists, who recognized his aptitude and talent in the study of Crustacea. In 1940 Dr. Bacescu came to Bucharest to accept an appointment as a department head, a position equivalent to professor, at the Grigore Antipa Museum of Natural History, which was founded in 1834 and was, and still remains, the premiere institution of its kind in Romania. During the next 48 years, he served the museum as a scholar, researcher, and adroit administrator.

In 1964 he was appointed director of the Grigore Antipa Museum of Natural History, a position he held with distinction until his retirement 24 years later. As

S.R. Grinevetsky et al., *The Black Sea Encyclopedia*, 75
DOI 10.1007/978-3-642-55227-4_3, © Springer-Verlag Berlin Heidelberg 2015

director, his innate talent, enthusiasm, imagination and educational and research expertise, allowed him to transform the museum into an internationally known and respected scientific and educational institution. Because of his vast knowledge of crustaceans (especially in the study of crustacean taxonomy and systematics), a renowned Romanian school of carcinologists developed during his tenure at the museum. Under his adept tutelage, his students undertook the study of a variety of different crustacean groups (Copepoda, Ostracoda, Mysidacea, Isopoda, Tanaidacea, Cumacea, Euphausiacea and Decapoda). Many of them became well-known specialists. After his retirement in 1988, he has served the museum as emeritus director, researcher and mentor until his death.

Prof. Bacescu also made a major and lasting contribution to the development of Romanian oceanography at Constantza on the Black Sea where he helped train and educate many important specialists at the Romanian Marine Research Institute. His pioneering oceanographic work on the ecology and faunistics of the Black Sea established the foundation for much of Romania's many contributions to the biological oceanography of the region.

He was elected to the Romanian Academy in 1964 and became a full member in 1992. Prof. Bacescu has served on the boards of many prestigious foreign organizations and scientific review panels. He also has received many international presentations and awards in recognition of his achievements in research and academia. During a long and productive scientific career spanning over 60 years, Prof. Bacescu published over 300 articles and monographs, most of which dealt with the taxonomy and systematics of marine malacostracan crustaceans.

Bafra – small Turkish town situated 20 km from the Black Sea and on Kizilirmak River delta, 51 km from Samsun. As a settlement, Bafra dates back to the fifth millennium BC. Population—85,000. Ruins of medieval mosque, bath-house, fortress. Water-melons and tobacco are grown.

Bafra Cape – sited to the north-west of Samsun City, Turkish Black Sea coast. The cape definitely juts out into the sea, low-lying, overgrown with wood that comes to the shore closely. Beaches stretch south-west and south-east of it. A light-house is installed 1 mile south-west of the cape.

Bafra Lake – a drainage lake sited on the left arm of the Kizilirmak River on the Turkish Black Sea coast.

Baida – sail fishing boat designed for fishing with a seine, dragnet and other devices. Was in common use on the Black Sea and Azov Sea. Length 6–7 m, width 1.7–2 m, board height 0.7–0.9 m, draught around 0.5 m, tonnage under 3 tons, crew 3–5 persons.

Baidarskaya dolina (valley) – basin in the south-western part of the Crimean Peninsula on the Chernaya River. Length around 16 km, width from 6 to 8 km. The slopes of surrounding uplands (under 500 m in height) are covered with arboreal and shrub vegetation, in the valley—fields, gardens, vineyards. The highway Yalta—Sevastopol runs across B.D. via the Baidarskie Vorota pass.

Baidarskie vorota (gates) – pass over the main ridge of the Crimean Mountains from Baidarskaya valley to the Black Sea coast. Height 739 m. Built in 1848 and are famous for their extremely dramatic position. The gates are on the very edge of a precipitous rock, almost a seaward escarp. Immediately beyond the gates, there is an imposing view of the Southern Coast of the Crimea (SCC) and the Black Sea.

Baidarskie vorota: View on the Russian Church and coast of the Black Sea (Photo by Dmytro Solovyov)

Bakalskaya Spit – sited on the north-western coast of the Black Sea, covers Karkinitsky Bay on the south-west. Embraces with a semi-ring the shallow-water Bakalskaya Bight. The spit's extreme point is Peschany Cape. Natural monument of local significance that includes the spit, Bakalskoe Lake and offshore marine complex. B.S. is of interest as a natural standard of forming a salt lake of marine origin.

Bakalskoe Lake – sited on the wide foundation of Bakalskaya Spit. Shallow-water brackish lagoon separated from the Black Sea by a sand spit. Bottom slopes are small, the bottom is covered with silt exhibiting health-giving properties.

Bakhchisaray – village (Turk. *Bahçesaray—"Garden Palace"*), situated between Simferopol and Sevastopol, Ukraine. Emerged to succeed the classical colonies Badathion and Palakion. Sited in the valley of the Churuk-Suv River ("Putrid water")—tributary of the Kacha River. Built early in the sixteenth century by the

Khan Mengli-Girei as residence of the Gireis. Mentioned for the first time as B. in 1502. Former capital city of the Crimean Khanate (early sixteenth–late seventeenth century). Thanks to the efforts of the Gireis, by seventeenth to eighteenth century B. turned into a flourishing city, a recognized center of merchants and artisans in the mountainous Crimea. There used to function a water conduit facility that supplied water to 150 water wells as well as to fountains, bath-houses, to the houses of the well-to-do city dwellers. Waste waters were used for irrigation. Bazaars were so numerous they were grouped on the basis of the prevalent goods—bread, vegetables, pickles, etc. Until now there exists the Khan's Palace elegized by Russian famous poet A.S. Pushkin. The palace was built early in the sixteenth century by the Khan Adil-Sahib-Girei, in 1736 it was burned down, built anew in the 1750s, partly remodeled by the Prince G.A. Potemkin for the arrival the Russian Empress Catherine II in 1787, was renovated in 1837 and 1857. The palace included several palatial buildings, a harem, the hall of the council and court, Falcon tower, a mosque and gardens. In B. Bakhchisaray Peace of 1681 between Russia and Turkey was concluded. During the Russian-Turkish war of 1735–1739, B. was seized by the Russian troops under the command of B.K. Minikh and was seriously demolished. In 1783, annexed to Russia, included in Novorossiisk Province, from 1902 was made part of the Province of Tavria.

Population at the end of the eighteenth century—around 6,000 people, in 1897—12,900 people (mainly, Tatars as well as Greeks, Armenians, Russians and others). Morocco leather, rifles, leather articles, soap and other products were fabricated in the city. The population grew vegetables, cultivated tobacco. The city has 35 mosques, 3 Russian Orthodox churches, 1 monastery, a synagogue, 2 Muslim religious schools (madrasahs). In B. environs, there are ruins of the medieval cave fortress city Chufut-Kale (was established in the fifth to sixth centuries, early in the sixteenth century experienced a decline, by mid-nineteenth century—was in a state of neglect), as well as the cave cities Eski-Kermen (end of the fifth to thirteenth century) and Tepe-Kermen. There is the Assumption monastery (fourteenth century) near B.

At present, B. is the center of Bakhchisaray Region of the ARC, Ukraine. Population 33,800 people. Railway station on the line Lozovaya-Sevastopol.

Bakhchisaray (http://ua.vlasenko.net/bakhchysarai/p1010088.jpg)

Bakhchisaray Peace of 1681 – Treaty of Bakhchisaray between Russia, Turkey and Crimean Khanate establishing a 20-year truce. Concluded on 13 January 1681 at Bakhchisaray. Terminated the Russian-Turkish war of 1676–1681. For the B.P. to be approved by the Sultan, a Russian Mission was sent to Istanbul in 1631 with a commission to gain the frontier along the rivers Tyasmin, Ros and Ingul, obtain Sultan's recognition of Zaporozhye as a Russian possession and push through prohibition for both the parties to settle the area between the rivers Dnieper and Bug. In the ratification instrument handed over to the Russian side in March of 1682 the Turkish terms were listed: 20-year truce; frontier to run along the Dnieper, except Kiev and Cities of Vasilkov, Tripolye, Staiki and others that remained in possession of Russia; the Russian government was not allowed to build cities (fortresses) on either side of the Dnieper; the Russian nationals were free to procure firewood and salt on the Turkish side of the Dnieper; the dwellers on either side of the Dnieper were free to settle where they wished; the prisoners of war were exchanged; Russian pilgrims were allowed to travel to Jerusalem. Russia accepted the Turkish terms. The crucial achievement of Russian diplomacy was recognition by Turkey of Kiev and its environs that belonged to Russia.

The Crimea and Turkey recognized the reunification of the Left-bank Ukraine with Russia. Kiev and the adjoining small locations in the Right-bank Ukraine were included in the Russian state, too. Under the Treaty, the Crimean Tatars were free to adhere to a nomadic lifestyle and engage in hunting in the southern steppes on either side of the Dnieper, where the Cossacks of Zaporozhye were allowed to fish in the Dnieper and its tributaries as far as the Black Sea. The Sultan and Khan were not to win over the Cossacks to their side. The Russian state recognized Yuri Khmelnitsky as the Hetman of the Right-bank Ukraine and was expected to pay annual "fisk" to the Khan. The states that signed the treaty undertook not to populate the areas between the Bug and Dniester or build fortifications there. The Sultan of Turkey and Crimean Khan pledged not to assist the enemies of the Russian state.

Balaklava – satellite city of Sevastopol, Ukraine, sited 12 km south-east of it. Population 30,000 people (2003). The city is 2.5 thousand years old. Before the 1941–1945 war, B. was connected with Sevastopol by a tramway line that had been laid in 1926. In 1957, B. became part of Sevastopol. During Sevastopol defense, B. was referred to as Minor Sevastopol. A small city sited in a cozy harbor, whose shores were populated by man from times immemorial. Scholars believe that during the eighth to seventh centuries B.C. the place was settled by Tavr people, subsequently—by Greeks. They believe it was the legendary Lamos—Port of Listrigons, who, according to ancient Greek mythology, were ogres whom Odyssey and his companions encountered during their wanderings. It will be noted that as time went on, in the nineteenth century Greek fishermen came to be referred to as listrigons. Ancient Greeks called the harbor "Symbolon-limen"—"Harbor of omens, symbols". Some researchers tend to relate such an unusual name with the fact that Tavres who used to live here (the Greeks regarded them as blood-thirsty robbers) allegedly used to make fires on the shore and by doing so managed to lure the Greek ships sailing by. When the ships approached, the orges attacked and robbed them, sacrificing the seafarers to their Goddess Virgo, whose temple presumably was on Phiolent Cape. In the first to third centuries A.D. there was a camp of Roman legionnaires here, which is proved by the Temple to Jupiter Dolikhan, protector of Romans, discovered by archeologists during the 1996 excavations, as well as by inscriptions to the Roman emperors carved on the stone, other finds of the Roman period.

In mid-sixteenth century, B. was settled by people from Genoa. They built a fortress and named it Cembalo (Tsembalo, Tsembaldo, sometimes Yamboli, Yamboly)—this way, the Genoese pronounced the Greek word "syumbolon". The fortress served the western outpost of Genoese possessions on the southern shore of the peninsula. In 1475, Cembalo City was captured by the Turks. They named the fortress their way—"Balyk-Yuva"—(Fish's Nest) which subsequently was transformed into B.

At the time of Sevastopol siege during the Crimean War, B. was the base of British troops. Balaklava harbor was used as the base of British fleet that provided all necessary supplies for the army. The British built spacious wharfs here, laid a railway line for steam locomotives linking B. with military positions, laid a subsea cable between St. George's Monastery and Varna, established a wireless communication with London and Paris. On the night of November 2 (14), 1854, the British fleet sustained heavy losses near B. resulting from a catastrophic hurricane and storm. 34 ships were smashed against rocks and sunk, among them the legendary motor steamer "Prince", which broke up completely within ten minutes and only six of her 150 crew were saved. On October 13, 1854, the Battle of Balklava was fought in the valley, not far from B.

The Crimean war over, B. for year remained an out-of-the-way settlement, whose dwellers were mainly preoccupied with fishing, grape-growing. Early in the twentieth century, B. became a rather well-known health-resort. In 1904–1905, the Russian writer Alexander Kuprin lived in B., he was fond of going to the high seas with fishermen for fishing. He wrote a whole cycle of excellent feature articles on B. and its people—"Listrigons". In 1907, Lesya Ukrainka came to B. health-resort to rest and undergo medical treatment. While in B., she completed the drama "In Puscha", worked at the poem "Ruthin and Priscille". A. Griboedov, Adam Mizkievich, V. Zhukovsky, I. Bunin and other writers visited the place, too.

During the years of the Great Patriotic War (WWII), the southern flank of the Soviet-German front ended abruptly near the ruins of the medieval Cembalo Fortress. After the war, Balaklava Harbor was used as a secret submarine base. The world's first launches of ballistic missiles from a submerged submarine were made from B. area in 1958.

At present, the harbor gradually is being turned into a yacht club. B. places of interest include the Genoese Fortress Cembalo, Twelve Apostles Cathedral (built in 1357), Naval Museum Complex "Balaklava". The winery "Golden Balka" is in B.

Balaklava: Panoramic view of the town (http://ru.wikipedia.org/wiki/Балаклава)

Balaklava engagement – On June 23, 1773, there took place a naval engagement between two Russian ships "Koron" and "Taganrog" under the command of Captain 2nd rank Dutchman I. Kingsbergen and the Turkish squadron of four ships. The Turks were trying to land an assault party in the Crimea near Balaklava but were resolutely attacked by Russian patrol ships at the Crimean coasts. During the hard-fought battle, the Turkish ships were badly damaged and were forced to retreat. B.E.—the first victory of the Russian Navy on the Back Sea. In his report, Kingsbergen left a remarkable note of Russian sailors: "With boys as good as these I would drive the devil himself from hell".

Balaklava harbor – (Turkish.—"*fish's nest*" or "*nursery pond*") is sited to the south of Sevastopol. The ancient Greek name Syumbolon-Limne—harbor of symbols as well as bight (harbor) of listrigons figures in Homer's "Odyssey". A winding, cozy, picturesque bight, concealed between precipitous rocky mountains and unseen from the Black Sea with which it is linked by a narrow, winding passage. The bight is about 1,500 m long, its width ranging from 200 to 400 m. Before the twentieth century, the bight was famous for its abundance of gray mullet and mackerel. The bight is the suite of Balaklava town.

Balaklava harbor: View on the entrance to the harbor (Photo by Dmytro Solovyov)

Balaklava heights (mountains) – Under this name reference is made in some documents to the mountains sited to the north-east of Balaklava harbor which name was the origin of B.H.

Balaklava storm – extraordinary storm on the Black Sea that occurred on November 14, 1854 in the vicinity of Balaklava. The aftermath was catastrophic: 34 British and French Naval ships were sunk, 1,500 people died, economic damage amounted to Francs 60 million. Among the sunk ships was the famous Brittish Royal Navy HMS "Prince", subsequently named "Black Prince". According to the detailed description of B.S. made by I.A. Ivashintsev and the French hydrographic engineer Keller, the storm had all characteristics of a tropical hurricane. In France, the fleet wreck gave rise to organizing the first regular weather forecast service.

Balaklava wind power station (WPS) – the first in USSR wind electric generating plant set up on Karansky Heights produced electricity in 1931. Theoretical foundations of making use of the wind energy were laid down by the Russian scientist

N.E. Zhukovsky in the second half of the nineteenth century. In 1930, the Central Aerohydrodynamic Institute (TsAGI) designed the Balaklava WPS of 10 kW capacity, at that time Europe's most powerful. A generator with vanes of 30 m in diameter was mounted in a special cage on metal supports as high as 8-storeyed structure. The construction weighed around 9 tons. During the Great Patriotic War (WWII) this unique WPS was destroyed.

"Balaklava" – naval museum complex sited in Balaklava Harbor which was a secret underground submarine base during the Soviet period. The Great Patriotic War (WWII) over, on I.V. Stalin's order the world's only underground harbor with submarine repair works was built here, top-secret facility No. 825. The construction began in the western cliff of Mt. Tavros in 1957. The first stage of the plant (submarine repair base) was commissioned in 1961. The second stage (fuel depots for the storage of petrol, oil and lubricants of 9.5 thousand tons capacity, and the arsenal) was completed in 1963. The plant was assigned the first category of stability in respect of nuclear threat. A combined canal for the entry of submarines and the entire infrastructure of the underground works made it possible to repair and service submarines independently. At the end of 1994, the Russian Black Sea Navy Command makes a decision to discard the submarine base. The last Russian submarine was removed from the harbor in February of 1995, whereupon the underground submarine harbor was fully transferred under the jurisdiction of Ukraine. By order of the Ukraine President (2002), B. became an integral part of the State Naval Museum opened in Sevastopol.

Balaklava naval museum: The underground harbor for submarines. http://ukrtrip.com/uploads/posts/1184696502_dsc05256.jpg

Balchik – a town, the Black Sea port and sea health-resort, sited 50 km north of Varna, Bulgaria. Founded by Greeks in the third century B.C. Initially called Kruna or Krounoi ("City of springs") because of the local karst springs. In the second century B.C. was renamed Dionysopolis in honor of Dionysus—god of wine and merriment, whose marble statue was thrown out onto the shore during a violent storm. One of the most ancient cities on the Black Sea coast. In classical times, was known as a major center of commerce. Minted its own coins. Got its name from the name of the Bulgarian boyar (nobleman) Balike who owned the area in question in the fourteenth century. Upon casting off the Turkish yoke in 1878, the town was part of Bulgaria. In 1913, following the Second Balkanian War, B. became the property of Romania and remained in its possession until 1940. B. has several small hotels, with rest homes around B. Not far from B., in a location sited on the terraces of the Botanical Gardens of Bulgarian Academy of Sciences, there is the former palace of the Romanian Queen Maria "Quiet nest" (1931–1938). Here over 3,000 plant species grow, more than 600 of these are rare from all over the world. There is also a scenic alpinarium with rare alpine species of plants. Since the port is in the center of the country's grain production, it specializes in grain transshipment for export. Population—12,000 (2009).

Balchik (Photo by Dmytro Solovyov)

Balchik Tuzla – lake-lagoon, 4 km from Balchik town, Bulgaria. Was produced as a result of the impact of old landslides. Famous for its balneal muds. A balneal center is here, too. The center offers the treatment of arthrosis and of locomotor system diseases. Numerous hotels. A mineral spring with water temperature of 31 °C.

Balik Lake – lake (34 km^2) located 1.5 km from the Black Sea coast in Turkey. Situated in the north-western part of the coast as far Injeburun Cape in the direction of Bafra Cape. Several small rivers fall into the lake in the south-west, the Kumjugaz River, falling into the Black Sea, flows out of the lake. Fishery is important in the lake.

Balkan Peninsula – often referred to as the Balkans is a geopolitical and cultural region of the Southeastern Europe. The region takes its name from the Balkan Mountains in Bulgaria and Serbia. The term "Balkan" itself comes from Turkish *"Balkan"*, meaning "chain of wooded mountains". The Balkan Peninsula is surrounded by water on three sides: the Adriatic Sea to the west, the Ionian, Aegean and Marmara seas to the south, and the Black Sea to the east. Its northern boundary is often given as the Danube, Sava and Kupa rivers. The Balkan Peninsula has a combined area of about 490,000 km^2. The following countries lie entirely or partially in the Balkans: Albania, Bosnia and Herzegovina, Bulgaria, Greece, Kosovo, Macedonia, Montenegro, Croatia, Serbia, Italy, Romania, Slovenia, and Turkey.

The first attested time the name "Balkan" was used in the West for the mountain range in Bulgaria was in a letter sent in 1490 to Pope Innocent VIII by Buonaccorsi Callimaco, an Italian humanist, writer and diplomat. English traveler John Morritt introduced this term into the English literature at the end of the eighteenth century, and other authors started applying the name to the wider area between the Adriatic and the Black Sea. Most of the Balkan nation-states emerged during the nineteenth and early twentieth centuries as they gained independence either from the Ottoman Empire or the Austro-Hungarian Empire. Greece in 1829, Serbia, Bulgaria and Montenegro in 1878, Romania in 1878, Albania in 1912, Croatia and Slovenia in 1918. In the twentieth century the Balkans were an arena for the First and Second Balkan Wars (1912–1913), as well as for WWI (1914–1918) and WWII (1939–1945).

The Balkans have a population of 48–71 million depending on whether the Turkish and Italian parts are counted within the peninsula. The region's principal religions are Christianity (Eastern Orthodox and Roman Catholic) and Islam. The Balkans today is a very diverse ethno-linguistic region, being home to multiple Slavic, Romance, and Turkic languages, as well as Greek, Albanian and others. Most of the states in the Balkans are predominantly urbanized. The largest cities (with population more than 1 mln) are: Istanbul (12.9 mln), Athens (3.1 mln), Bucharest (1.7 mln), Sofia (1.2 mln), Belgrade (1.2 mln).

Balkan Peninsula. http://bnr.bg/sites/en/Lifestyle/BulgariaAndEurope/Pages/2805sasedski.aspx

Bar – levee-type silt shoal (spit), made up of clastic or shell deposits, of 100–400 m in length. Is formed in the coastal zone under the impact of sea wave and streams as well as in river mouth due to siltation.

Barabulka – surmullet (*Mullus barbatus*) is a species of goatfish found in the Mediterranean Sea, North-East Atlantic Ocean from Scandinavia to Senegal, and the Black Sea. Length—10 to 20 cm. This is favored delicacies in the Mediterranean and the Black Sea countries, and in antiquity were "one of the most famous and valued fish".

Barabulka (red mullet) fried in a pan (Photo by Dmytro Solovyov)

Baranov Nickolay Mikhailovich (1836–1901) – Russian Navy officer. Studied at the Naval Cadet Corps. In 1856, promoted to a midshipman. In 1865, the Navy Ministry passed into service with the navy a rifle invented by Lieutenant B. of the Baltic Fleet. In 1860, a boarding pistol made to the specimen of B. was passed into service with the Navy. In 1871, B. was promoted to a Lieutenant–commander and appointed director of the Naval Museum. Published articles on naval tactics, whose provisions on the role of cruisers were obviously ahead of his time and were recognized much later. In 1877, at his own request, B. was sent to the Black Sea Navy as a commander of the ROPiT's refurbished steamer "Vesta". In the very first battle with the Turkish armor-plated ship "Fethi-Bulend" "Vesta" turned it to flight. After the battle, B. was promoted to a Captain 2nd rank and was designated Emperor's Aide-de-camp. I.K. Aivazovsky painted his portrait. Another piece "Battle of the Vesta" steamer with the Turkish armor-plated ship "Fethi-Bulend" was painted by the painter L. Bliniov (currently, an exhibit of the Sevastopol Black Sea Naval Museum). At the end of August of 1877, "Vesta" together with the steamer "Vladimir" transported the wounded and sick soldiers and officers from Gagry to Novorossiisk. In 1877, apparently specially for B., the rearmament of the steamer "Rossia" commenced. The heir—Crown Prince of the Russian Empire (the would-be tzar Alexander III) sacrificed the crew of his yacht for the command of the new cruiser. In December, already on the refurbished cruiser that was sailing near the Turkish coasts around Penderakli (now, Eregli), B. captured the three-mast Turkish steamer "Mersina" and brought it to Sevastopol. B.'s trophy was rather valuable: besides the captives, money, silver ore, government papers and documents, he brought an excellent, recently overhauled ship. "Mersina" was given the name of "Penderaklia" and was made part of the Russian Fleet. B. was promoted to a Captain 1st rank. In 1878, B. is appointed the commander of the oceanic liner "Rossia" purchased in Germany and converted to a cruiser. In 1881, was conferred the grade of a colonel and was appointed Governor of Kovno town. After the death of Alexander II, B.—the Mayor of St. Petersburg. Subsequently, B. governs Archangelsk Province, in 1882—Nizhegorod Province. B. died in 1901 being a member of the Government Senate. In 1906–1909, the destroyer "Captain Commander Baranov" was built in Nikolaev, its tonnage 800 tons. The destroyer was part of the Black Sea Navy until June 18, 1918 when it was sunk by its crew in Tzemes Bay (Novorossiisk) in pursuance of V.I. Lenin's order: demolish Russian ships lest they become the trophy of Kaiser Germany.

Barkas – barkaz (Dutch. "barkas"), small sailing or transport vessel widely used on the Black and Azov seas. Yawl furniture, with jib on a short horizontal bowsprit. Length 8–12 m, width 2.3–3 m, board height 1–1.3 m, draft around 0.75 m.

"Barkhatnyi sezon" – the name (in Russian) of an autumn season, most benign for treatment because the temperature regime of the air and sea makes it possible to exercise climatotherapy to the maximum extent possible, including sun baths and sea bathing. The expression "B.S." is believed to have been coined at the end of the nineteenth century on the SCC, when creative intelligentsia (artists, men of letters) spent their time at the SCC health resorts; these people, judging by their clothes, were called "velvety" public.

Bartin – the Black Sea province of Turkey in the north-western part of the country. Area 2.120 km^2, population 188,000 people (2011). Center—Bartin. The town of Bartın has very old wooden houses which are no longer found in other places. In Bartın Province is the ancient port town of Amasra (Amastris). This town is on two small fortified islands and contains many interesting old buildings and restaurants.

"Basic provisions on environment conservation, restoration and improvement in the territorial complex Black Sea—Bulgarian coast" – this program was adopted by the State Council of Bulgaria in December 1976. The basic tasks of this program included the following:

(1) Elimination of available pollution and environment protection from pollution and damage. In this context it was planned to take the integrated effective actions for resolute prevention of new pollution related to marine transport, industry, construction, agriculture and others. It was envisaged to construct treatment facilities in large cities (Burgas, Varna) and by 1985—in other seaside settlements. For improvement and enrichment of the marine and coastal environment the new industrial facilities were to be constructed beyond the seaside resort area. Special actions should be taken to protect the existing natural objects (national parks, nature reserves, etc.). It was contemplated to create new protected objects. Surveys and industrial development of power and mineral resources of the shelf and coast should include actions on environment protection.

(2) Conservation, reproduction and enrichment of biological resources. In this context the comprehensive study of changes in the marine environment was recommended. It was necessary to go on with the study of dynamics of plankton, benthos, fish populations, their reproduction, possible acclimatization of new fish species and invertebrates. The rational management of mussels and algae, control of carnivorous rapan hazardous for bivalves was recommended. The program envisaged a drastic decrease of harvesting of the valuable Pontian tree varieties, preservation of coastal forests, protection of dune vegetation controlling sand drift and forest plantings on idle lands.

(3) Rational management and conservation of natural resources. Sandy beaches covering one-third of the Black Sea coast of Bulgaria are of the highest quality. The climate, mineral spring waters and curative mud in saline lakes provide wide opportunities for health improvement. For improvement of the atmosphere in recreational zones the camps, holiday centers and other resort facilities are removed from the sandy belt; the transit transport is transferred beyond the recreation and tourism complexes.

Basistyi Nickolay Efremovich (1898–1971) – Soviet Admiral. In 1914, entered the Sevastopol Sailor Boy School. Sent to the Mining School on board the "Rion" training ship. In 1916, appointed to the "Zharkii" Destroyer as a non-commissioned officer 2nd class. As a participant of the Bosporus mine-laying operation, awarded a St. George's medal and soon promoted to a officer 1st class. In 1918, joined the

Volga Military Flotilla in Nizhny Novgorod. As a signaler of the battle-boat "Red Banner", participated in landing an amphibious party in Fort "Petrovskii" on the Caspian Sea. In 1936, entered the General Staff Military Academy. In 1937, in Spain, was adviser of a destroyers semi-flotilla commander, then adviser of the Mediterranean Fleet Commander. Upon return—Captain 2nd rank, designated chief of the department of operations of the Black Sea Staff, yet insisted on and gained appointment to a ship. In 1939–1941—commander of the Cruiser "Chervona Ukraina", from October of 1941 as a Captain 1st rank commanded a squadron of light forces made up of newest ships. Commanded the movements of ships during Odessa defense, destruction of enemy strongholds on the Black Sea coasts. Commanded the deployment of a landing party in Feodosiya during the Kerch-Feodosiya operation of 1941–1942. In 1942, promoted to a Rear-admiral. In Feodosiya, secured the support of troops, and when the army was defeated in the Crimea, controlled troops evacuation. From the end of 1942, took part in planning the Novorossiisk amphibious party landing operation. In March of 1943, appointed chief of staff of the Black Sea Squadron, 6 weeks later—Squadron Commander, and in autumn of 1944—the Chief of Staff of the Black Sea Navy. In November of 1946, took over command of the Black Sea Navy. In 1949, promoted to Admiral. In August of 1951, appointed the First Deputy Minister of the USSR Navy. In 1956, appointed Deputy Commander–in-Chief of the USSR Navy for Military and Scientific Work, in June of 1958—Inspector-Adviser of the Group of General Inspectors of the USSR Ministry of Defense. In September of 1960, resigned. Awarded numerous orders and medals. The author of memoirs "The Sea and the Coast" (1970).

Bastion – usually, pentagonal fortifications in the form of a ledge at the corners of a rectangular fencing designedo for the bombardment of the terrain in front of a fortress and along the fortress walls and the deep pits arranged around the fortress. Bastions were crucial during the period of Sevastopol defense.

Batilman – health-resort settlement in the south-western part of SCC, in Laspi Bight. Known for the summer cottages (dachas) of V.G. Korolenko, I.Ya. Bilibin, V.I. Vernadsky and other Russian writers, scholars and artists.

Batilman: View from the sea (Photo by Dmytro Solovyov)

Battle of Alma – named after the Alma River in the Crimea, north of Sevastopol, near whose mouth not far from Burlyuk Village (at present, Vilino, named after the Leiutenant-General of Aviation I.P. Vilin) the first combat between Russian forces commanded by Prince A.S. Menshikov and the army of British, French and Turkish troops commanded by the French Marshal St. Arnaud (Armand-Jacques Leroy de Saint-Arnaud, 20 August 1801–29 September 1854) and British General Lord F. Raglan (Field Marshal FitzRoy James Henry Somerset, 1st Baron Raglan, 30 September 1788–29 June 1855) took place September 20, 1854. The enemy had landed earlier near Kyzyl-Yar Lake to the south of Evpatoria during the Crimean War. In the battle of Alma the superiority of the allies manifested itself not only in numbers, but in the level of armament. The Russians lost in that battle over 5,000 men, the allies—4.3 thousand. The allies had no cavalry which did not allow them to organize active pursuit of Menshikov's army. He retreated towards Bakhchisaray, having left unprotected the road to Sevastopol. That victory made it possible for the allies to get intrenched in the Crimea and open the way to Sevastopol. B.A. demonstrated the effectiveness and fire power of new firearms, where the formerly used system of breakdown into closed columns was becoming suicidal. In the course of the battle, Russian troops for the first time spontaneously used a new battle formation—firing chain. Where there used to be the middle of the Russian positions, there stands a monument reading: "Banner to the memory of the warriors who died in the Battle of Alma". Another monument is below—to commemorate those who specially distinguished themselves in the battle of Vladimir Regiment. British gravestones are there, too.

Battle of Balaklava – On October 13, 1854, during the Crimean war of 1853–1856 in the vicinity of Balaklava, a battle between the Russian force commanded by General P.P. Liprandi and the garrison of British-Turkish troops was fought near Balaklava. The Russian command pursued the main object of attacking this main base of British troops in the Crimea as diverting the allies from Sevastopol. When attacking B., the Russians managed to seize some forts that were defended by the Turkish units. But then the attackers were stopped by the attacking British cavalry. In a bid to capitalize on its success the cavalry brigade (the brigade included great-grandfather of Winston Churchill and other representatives of British aristocratic families) headed by Lord Cardigan went on with the attack and far into positions of the Russian troops. Here, the cavalry came under canon fire of a Russian battery and then was attacked on the flank by a force of lancers commanded by Colonel Eropkin. Having lost most of his brigade, Cardigan retreated. The Russian command was unable to develop this tactical success for lack of the forces made available at Balaklava. Extra units of the allies moved urgently to assist the British. The Russians were reluctant to engage the enemy and start a new battle, so they retreated to their original positions, having destroyed the seized forts. Balaklava Battle made the allies postpone the planned assault of on Sevastopol. At the same time, it helped the Russians to better realize their weak points and buttress Balakalava, which was the Seagate of supplies for the allies at Sevastopol.

Battle of Black River – On August 4, 1855 on the banks of B.R. (10 km from Sevastopol) the Russian Army commanded by General M.D. Gorchakov mounted a battle against three French and one Sardinian divisions. The main battle was for Fedyukhin's heights and broke out on the right flank. Having fought for 6 h, the Russians lost 8,000 men in futile attacks and retreated to the starting position. The losses of the French and Sardinians were around 2,000 men. The battle of B.R. as well as other field battles of the Crimean War revealed clearly not only technical backwardness of the Russian Army, but also lack of talent of the higher military command. After the battle of B.R. the allies were able to allocate large forces to storm Sevastopol.

Battle of Sinop – the battle between the Russian and Ottoman fleet that took place on 18(30) November 1853 in the Sinop Bay during the Crimean War of 1853–1856. The Ottoman squadron (7 frigates, 3 corvettes, 2 steam frigates, 2 brigs and 2 transports, in total 510 guns) under the command of Vice-Admiral Osman Pasha and British Advisor Adolphus Slade (Mushavar Pasha) that came from Istanbul to Sinop anchored in the harbor supported by ground batteries (38 guns) and made preparations for landing near Sukhum-Kale (presently Sukhumi) and Poti. The Russian squadron (6 linear battleships and 2 frigates with 720 guns) under the command of Vice-Admiral Pavel S. Nakhimov blockaded the Ottoman fleet from the sea. Nakhimov decided to attack and destroy it in the Sinop Bay. The battle lasted for 4.5 h. In the battle the Ottomans lost 15 out of 16 vessels and over 3,000 men were killed and wounded. About 200 men were taken prisoners, including Osman Pasha and commanders of 3 vessels. The Russian casualties were 27 killed and 235 wounded. Many ships were damaged. The defeat of the Ottoman fleet weakened considerably the naval forces of Turkey and frustrated its plans of landing on the Caucasian coast. The participants of this battle were awarded the medal "In Memory of Eastern War of 1853–1856" on the Georgian ribbon.

Battle of Trapezund (1916) – the offensive of the Primorsky detachment of the Russian Caucasian Army supported by the Batumi detachment of the Black Sea Fleet against the Third Turkish Army that occurred on 12 January (5 February)–5 (18) April during World War I for seizure of the city and port of Trapezund (Trabzon). Finishing the Erzrum military operation of 1915–1916 the command of the Caucasian Army planned to storm Trapezund both from land and from sea. On 23 January (5 February) the ships of the Batumi detachment (1 battleship, 2 destroyers, 2 torpedo boats, 2 gunboats) under command of Captain 1st Rank M.M. Rimsky-Korsakov approached the mouth of the Arkhave River and suppressed the Turkish batteries with artillery fire. Under protection of the ship guns the Primorski detachment (about 15,000 men, 28 guns; commander Lieutenant-General V.N. Lyakhov) started an offensive from Arkhave to Trapezund and on 25 January (7 February) and came out to the Turkish positions at Vice. After regrouping the forces on 2(15) February the detachment resumed the attack. In February-March with the support of the amphibious landing troops the Russians occupied the cities of Atina, Mapavri, Rize, Of and Khamurgyan and by 1 (14) April they came to the Turkish fortifications on the Kara Dere River. On 25–26 March (7–8 April) 2 Kuban brigades (about 18,000 men and 12 guns) transferred here from Novorossiysk landed in Rize and Khamurgyan. They were carried by 22 transports under escort of 2 battleships, 4 cruisers, 2 aircraft transports and 19 destroyers and landed under protection of the Batumi detachment and aviation. On 31 March (13 April) the Batumi detachment was strengthened by 1 battleship and 1 destroyer from Sevastopol. On 2(15) April the Primorski detachment together with the Kuban brigades attacked and took the Turkish positions on the Kara Dere River and on 5(18) April Trapezund surrendered without fighting to the Russian troops. On the next days the Russian troops advanced along the coast as far as Buyuk Liman, thus, consolidating the achieved success. As a result, the Third Turkish Army lost its shortest link with Constantinople and from now onward the Russian command could base here the Black Sea Fleet and organize the additional base for supply of the Caucasian Army.

Battle of Yuzhnyi Bug – battle of May 20, 1788 during the Russian-Turkish war of 1787–1791 near the mouth of the Yuzhnyi Bug River between the Russian double-sloop commanded by Captain III rank R. Saken and the Turkish squadron of 30 vessels. During one of the reconnaissance trips to Ochakov, Captain Saken's double-sloop was intercepted by the Turkish ships which cut its way of retreat to the Yuzhnyi Bug mouth. Attacked by the superior Turkish forces, the double sloop began a losing combat. Unwilling to yield himself prisoner, Saken ordered that the sailors leave the vessel, while he himself stayed on and blasted the sloop together with the janissaries who had climbed aboard. In honor of this feat, a memorial plaque was installed in the church of the Naval Corps.

Battles of Anapa – (1) Campaign of the corps under the command of General Yu. B. Bibikov in winter-spring of 1790 against the Turkish fortress Anapa. The campaign was very poorly prepared. The mightiest Turkish stronghold on the Eastern shore of the Black Sea built to the designs of French engineers was protected by a 15,000 garrison. The Russian troops that approached Anapa did not have even scaling ladders. By that time, Bibikov's detachment had no horses or foodstuffs left. Instead of selling supplies to the Russians, the highlanders launched combat action against them. Despite the definitely hopeless situation, Bibikov ordered to start storming Anapa, which ended in failure. Bibikov was tried and dismissed from the army for this unsuccessful operation. Rank-and-file participants of the expedition were awarded a special soldier's silver medal with an inscription reading "For Loyalty".

(2) Seizure of Anapa on June 22, 1791 by the Kuban Corps commanded by General I.V. Gudovich. By that time, the fortress garrison had been reinforced substantially. The number of its defenders reach 25,000 people. Gudovich had good artillery which overwhelmed that of the fortress. On the night of June 21 to June 22, Russian troops, covered by artillery fire, approached the fortress berm secretly and, half an hour before break of dawn, rushed forward and seized the fortress. On Gudovich's order the mighty fortifications of Anapa were completely destroyed. A special medal "For Anapa Campaign" was made for the participants of the battle.

(3) Blockade and seizure of Anapa by the Russian troops commanded by Rear-admiral A.S. Menshikov on May 3–June 12, 1828. The fortress was under the protection of the Turkish garrison of 6,000 to 12,000 soldiers. On May 2, 1828, the Black Sea squadron under the command of Vice-admiral A.S. Graig delivered to Anapa Menshikov's landing party of 5,000 men. The next day, the town was approached overland by the detachment providing for safe landing of the amphibious party. On May 7, the ships of the Black Sea fleet began shelling the fortress. Sorties of the Turkish garrison proved to be futile. The general assault of the fortress was scheduled for June 10, but the fortress was never stormed: having no outside support, the Turks agreed to start negotiations and on June 12 surrendered. A.S. Graig was promoted to Admiral. After the Russian-Turkish war ended, Anapa became part of the Russian Empire.

Batumi, Batum – capital city of the Republic of Adjaria, Georgia, an important maritime industrial area, terminal point of the railway line Baku-Batumi, an important port on the Black Sea, a health resort. Situated in the south-eastern part of the Caucasian Black Sea coast, on the southern coast of the deep Batumi Bight, on Kakhaberi Plain adjoined on the east by Shavshet and Meskhet Ranges of the Minor Caucasus. To the south of B., the Chorokh River falls into the Black Sea; the river flows mainly in the Turkish territory, and the last 26 km—in the territory of Adjaria (Georgia). Here is the termination of the over 500 km long Caucasian Black Sea coast, one of the most popular holiday-making sites.

It is believed that the name of B. derives from the Svan word *"bat"—"stone"*. The settlement that was eventually succeeded by the city, apparently had a very long history: the Roman author Pliny (first century B.C.) mentions a port with a

Greek name *"Bathys limen"*—*"Deep Bight"*, which is an interpretation of the local name. Known to exist since the second century. In the eleventh century, the fortress Tamaristsikhe was built where B. is sited now. At the end of the thirteenth-early in the fourteenth century, B. is a possession the Mingrelian Prince Bediani; at the end of the fourteenth century—of the Prince Gurieli. From the seventeenth century—part of the Osman Empire. During the Middle Ages, the town was called Batomi, from 1878—Batum, from 1939—Batumi.

Population early in the nineteenth century—around 2,000 people. In the 1850s, the agency "Russian Company of Steamship Promotion and Trade" (ROPiT) is established in B.; B. became the terminal point of the Crimea-Caucasus and Anatolian Shipping Lines. During the Russian-Turkish war of 1877–1878, fierce battles were fought around B. In March of 1878, B. was occupied by the Russian troops under the resolution of the Berlin Congress of 1878, and B. became part of the Russian Empire. In 1878–1886, B. enjoyed the status of a free port (with the right of duty-free export and import of goods). In 1878–1887, the Mikhailovskaya Fortress was built to include the Turkish fortifications of the seventeenth–nineteenth centuries. In 1878–1883 and from 1903, B.—the center of Batumi Region (in 1883–1903—center of a district in Kutaisi Province). In 1883, the railway line Baku—Tiflis (Tbilisi)—Batum was built, B. port (third largest port of Russia on the Black Sea after Odessa and Novorossiisk) was re-equipped. In 1897–1907, the kerosene pipeline Baku-Batum. was built. The Batum Port was mainly used for promoting trade in oil and oil products as well as wheat, corn, manganese ore and other commodities. Late in the nineteenth—early in the twentieth century, B—a significant industrial center: factories and plants—tobacco, oil refineries, distilleries; railway and ship-building enterprises.

In 1905, an armed resurrection flared up in B., crushed by troops. There were major strikes in 1912–1914. During WWI, B.—one of the rear bases of the Caucasian Front (the Garrison of Mikhailovskaya Fortress—over 3,000 people) and the Black Sea naval base; trade was virtually at a standstill, industry declined.

B. population—170,000 people (2011). At present, crucial to the city economy are machine-building (machines for tea-growing and other food-processing industries), oil refinery, shipyard and zinc-plating plants, tea-packing factories, citrus-processing factory. Over the years of Soviet power, Batumi Port became one of the best on the Black Sea. It is used for importing timber, wheat, sugar, chemicals, equipment. Exported items include: oil products, tea, preserves, fruits, etc.

B. is distinguished by a warm, humid subtropical climate. Mean annual temperature is 14.5 °C, mean temperature of August +23.1 °C, of January +6.6 °C; the amount of precipitation is 2,130 mm/year, with maximum occurring in October (272 mm).

Vegetation is extremely rich and diverse, represented mainly by tropical and subtropical species, large plantations of citrus culture, groves of magnolias, palm-trees, bamboo-trees, eucalyptuses, laurel trees and bananas. The maritime park-boulevard with magnolia and palm alleys is succeeded by a wide pebble beach. Nine km from B. (near the railway station "Green Cape") there is one of the largest botanical gardens—Batumi Botanical Garden. The garden area is 111 hectares. The garden has over 5,000 species and forms of plants, typical of various regions of the world: New Zealand, Australia, Himalayas, South and North America, Mediterranean area, etc.

In B. there is a Sea Aquarium and Dolphinarium as well as a number of museums (Museum of Adjaria, Revolution Museum, I.V. Stalin's House-museum).

Batumi: View on the port (http://www.seafarersjournal.com/wp-content/uploads/2012/04/batumi. jpg)

Batumi Bay – the eastern boundary of the bay is thought to be the mouth of the Korolis-Tskali River falling into the Black Sea south-west of Green Cape. On the south-east and south, the B. is hemmed by high mountains distanced from the coastline by 1.5–2.5 km. The western coast of the B. is low-lying. Most of B.B. is flat. There is a trough along the western coast with depths ranging from 20 to 110 m; shoals are near the eastern coast. Batumi Port is at the apex of B.B., the city of Batumi lying on its shores.

Batumi botanical garden – one of the largest botanical gardens of the former USSR. Sited near to the railway station Green Cape, 9 km from Batumi (Georgia). Occupies and area of 111 ha, featuring hilly topography, the soils are krasnozems and others. Established in 1912 by the Russian botanist and geographer A.N. Krasnov. Under his plan, the gardens were to be set up on a landscape-geographical principle. The primary object of the gardens, according to Krasnov, was the acclimatization of economically valuable subtropical plants and introducing these into culture in the southern areas of Russia. After the Soviet power was established in Georgia, B.B.G. development was intensified. Under the decree of the USSR Sovnarkom dated July 30, 1925, B.B.G. was recognized as the main scientific institution of the USSR to promote subtropical crops on the Caucasian

Black Sea coast: tea, citrus culture and others. B.B.G. laid a strong emphasis on introducing in culture valuable acricultural, forest, essential-oil, ornamental and other subtropical plants. B.B.G. comprises the following floristic departments: moist tropics of Transcaucasia, New Zealand, Australia, Himalayas, East Asia, North America, South America, Mexico and Mediterranean. The collection of plants includes over 5,000 species, varieties and forms, out of this around 2,000 species of scrub plants. The herbarium contains over 40,000 sheets.

Batumi raid – (1) Combat operations of the Rioni (from May of 1877, Kobuleti) task force of Russian troops during the Russian-Turkish war of 1877–1878. Object—defeating the Turkish Corps of Darwish Pasha, seizing Batum (Batumi) and prevention of the landing of the enemy's amphibious party in the rear of the Russian troops. The Rioni body of troops (24,000 soldiers, 96 canons) began an attack on April 12 (24), 1877. Moving in the tough conditions of the mountainous-woody terrain and cross-country, the task force managed to overcome stubborn resistance of the enemy and on April 14 (26) seized the heights of Mukhaestate, Khutsubanskoe. and on May 19 (31) occupied the heights of Sameba. The second attempt to seize Batum was undertaken in January of 1878. The Kobuleti force again advanced as far as Tsikhisdziri and again retreated. Its activity tied large forces of the enemy, thereby contributing to the success of the main forces of the Caucasian Army.

(2) Attacks of the Russian mine launches against the Turkish ships at the Batumi Port during the Russian-Turkish war of 1877–1878. On the night to December 16 (28), 1877, the ship "Velikiy Knyaz Konstantin" (commander—Captain 2nd rank S.O. Makarov approached Batum covertly, launched 4 boats, of which "Chesma" and "Sinop" each had as armament one Robert Whitehead self-propelled mine (torpedo). After midnight, the mine launches penetrated in the port and attacked the Turkish armor-clad battleship "Mahmudiye" (the Turkish flagship was for many years the largest warship in the world), yet the attack proved abortive as one torpedo, having passed along the board of the battleship, jumped out onto the shore, and the other one hit against the battleship's anchor chain and exploded on the ground. On January 14(26), 1878, "Velikiy Knyaz Konstantin" repeated a raid on Batum. The launches "Chesma" and "Sinop" by hitting the Turkish armed steamer "Intibah" with two torpedoes simultaneously from a distance of about 80 m destroyed the ship. This was the first in the world recorded successful launch of torpedoes from a torpedo boat.

Batumi Region – instituted in 1878 on the territory acceded to the Russian Empire after the Russian-Turkish war of 1877–1878, by decision of the Berlin Congress of 1878—center—Batum. Population around 110,000–120,000 people. Initially, B.R. was divided into districts (Batumski, Artvinski, Adzharski). B.R. was managed by the civilian-and military administration. In 1883, B.R. was made part of Kutaisi Province, in 1903 it was again made an autonomous administrative-and-territorial unit as part of Batumski and Artvinski Districts. In 1904, area 6.1 thousand verst; population 152.9 thousand people (Georgians, Turks, Russians, Armenians, Greeks and others); 1 city, around 300 settlements. The primary occupation of the population—agriculture (80.6 % of the dwellers is engaged in agriculture). The main crops—corn, barley, wheat, rice, subtropical crops. Grape-growing, tobacco cultivation (around 20,000–25,000 puds of tobacco per annum), tea-growing. Industrial production is mainly centered in Batumsky District (this includes, leather industry, flour milling, copper smelting, metal-treatment and wood-working, oil refining, and other industries). At the end of the nineteenth century, there were 29 factories and plants in Batumi Region (2.4 thousand workers). There existed steamer traffic through Batumi Port with all ports of the Black and Azov seas. In 1883, the railway Batum-Tiflis-Baku was commissioned. There were branches of State Tiflis Commercial Bank, South-Russia Industrial Bank, Russian Bank of Foreign Trade, loan associations, 18 secular educational establishments, 19 muslim schools at the mosques.

Batumi sea trading port – sited on the south-eastern coast of the Black Sea, Georgia. Established in 1878 as a free port, which was conducive to its rapid growth. The building of a railway line (1883), wharfs for tankers and dry-cargo sea vessels (1892) and of the Baku-Batumi pipeline assisted rapid development of the port. Early in the twentieth century, the port ranked third in Russia by the volume of transshipment after Odessa and St. Petersburg. In 1923, it was accepted as a first-class port. Today one of the main links of the Transport Corridor Europe-Caucasus-Asia (TRACECA), from Batumi Port the goods are transported by the Black Sea to: the Russian ports Novorossiisk, Sochi, Taganrog; to the Ukrainian ports Iliychevsk, Odessa, Izmail; to the ports of Bulgaria—Varna; Romania—Constanţa, Turkey—Istanbul, and others.

There operate 12 deep-water wharfs with the depths up to 12 m, three of which are intended for the transshipment of oil and oil products. The port handles tankers of up to 140,000 tons. It has a container terminal of up to 50,000 TEU/year capacity; a ferry railway complex of 700,000 tons carrying capacity in compliance with the European standards; a passenger terminal. Port capacity equals 12 mln tons of oil and oil products and 3mln tons of general cargoes per year. Besides oil and oil products, non-ferrous metal ores, prominence in export is also given to tea, wine and fruits. Port imports are dominated by foodstuffs (grain, sugar, etc.) and manufactured articles. The port is serviced by a number of regular traffic lines, linking it with other ports of the Black, Azov and Mediterranean Seas; tramp vessels also call at the port. The port is the main base for the Georgian Shipping Company.

Bay – a rather wide part of the water surface (sea, lake, river, etc.) projecting into the land. In the Black Sea there are found such large B. as Burgas and Varna in Bulgaria; Mamaia in Romania; Odessa, Tendrovsky, Egorlytsky, Dzharylgachsky, Karkinitsky, Kalamitsky and Feodosia in Ukraine; Samsun and Sinop in Turkey; Taman, Taganrog, Temryuk and Arabatsky in Russia (in the Sea of Azov and the Black Sea).

Bay of Cossacks (Kazachya) – sited on the territory of Sevastopol. Adjoins Kamyshovaya Bay on the west, is part of Dvoinaya Bay, being its western bay.

Sited east of Khersones Cape. General Zoological Reserve of republican significance. Established in 1998, occupies an area of 23.3 ha. Valuable maritme nature
reserve. Rest location for numerous migratory and wintering birds. A number of
species of the animal world entered in the Red Data Book of the Ukraine has been
preserved here (swallowtail, Crimean gecko, large whip snake and others).

Beach barrier – a narrow sandy water-permeable spit, a natural bar of sand or
pebble separating a bay, lagoon or river mouth from the open sea or lake. It is
created as a result of the combined action of marine (lake) currents and tides,
aggradation processes in river mouths and tidal activity in shallow areas.

Beaching, coast-protection structures – structures intended for the protection of
the banks of water bodies from the destructive impact of waves, currents, ice
pressure and other natural factors. By nature of interaction with a water stream,
the structures are subdivided into active, using the energy of a stream for hydraulic
fill and conservation of shore deposits (on the rivers—transverse dikes, protection
dams, guide vanes; on the seas and lakes—silt-retaining groins, breakwaters), and
passive, countering a stream with the strength and stability of own design only
(on the seas—sea walls, riprap of large blocks and shaped solid masses; on rivers—
stone riprap, mats, gabions, concrete and ferroconcrete slabs).

Beisug Liman – named so after the Beisug River flowing into it in spring. The
liman is sited on the eastern coast of the Azov Sea, is separated from the sea by
Yasen Spit and two channels. Khanskoe Lake with its famous hydrogen sulfide mud
is on the northern shore of the lagoon.

Belbek Valley – formed by the Belbek River, flowing into the Black Sea. Sited
beyond the northern boundary of Sevastopol. Hilly steppe, covered with vineyards,
ends in a high cliff (shore precipice) structured by clay shales at the foot of which
sand beaches stretch both ways as far as the horizon. The precipice is not stable:
impacted by waves and ground water, there often occur landslides and cave-ins.
Fertile lands on the river bank have long been used for horticulture and are heavily
populated. The word "Belbek" is translated as "tough, strong back, spine". After the
rains or snowmelt in the upper reaches, B. turns to a shooting flow, rushing toward
the sea and bringing uprooted trees with it. Suyren Fortress is in B.V.

Belgorod-Dniestrovskiy – a town close to the coast of the Black Sea, Ukraine.
Population—50,300 (2011). It is a regional city and port situated on the right bank
of the Dniester Liman (on the Dniester estuary leading to the Black Sea) in the
Odessa Province of southwestern Ukraine, in the historical region of Bessarabia.
The city serves as an administrative center of the Belgorod-Dniestrovsky Raion.
Previous names: Ophiusa (colony of Phoenicia), Tyras (colony of Ancient Greece),
Album Castrum ("White Castle", Latin name), Cetatea Albă ("White Citadel",
Romanian name), Asperon (colony of Byzantine Empire), Maurokastron (Koine
Greek), Turla (Turkic), Montecastro/Asprokastron (colony of Genoa). The city is
known by translations of "white city".

The town became part of the Principality of Moldavia in 1359 and the fortress is enlarged and rebuilt in 1407 under Alexander the Kind and in 1440 under Stephen the Great. From 1503 to 1918 and 1940 to 1941, the city was known as Akkerman, Turkish for "white fortress". From 1918 to 1944 (with a short brief in 1940–1941), the city was known by its Romanian name of Cetatea Albă, literally "white citadel". Since 1944 the city is known as Bilhorod-Dnistrovskiy (Ukrainian), while on the Soviet geography maps often translated into its Russian equivalent of Belgorod-Dnestrovskiy, literally "white city on the Dniester River".

Meat-and-dairy plant, fish-processing factory, grape wine plant, folding-carton plant, furniture-making factory, garment factory, building materials works. There are health resorts Zatoka, Sergeevka and Lebedevka not far from B.D. on the seashore. Local climate is dry, steppe, with numerous sunny days. Good sand beaches, several kilometers in length and as wide as 750 m (Zatokinsky Beach).

Belgorod-Dniestrovskiy sea trading port – established in 1971 in Belgorod-Dniestrovskiy City (Ukraine) that stands on the coast of Dniestr Lagoon and is connected by a channel with the Black Sea. The main purpose of the port is to reduce the intensity of operations of the nearest major ports Odessa and Iliychevsk that have to deal with small-capacity fleet. The port also includes the port station Bugaz in the very estuary of the Dniestre River. The port specializes in handling the cargoes of foreign-trade and coastal ships (grain carriers, ships carrying timber, general cargoes, livestock, minerals and building materials). Intended essentially for handling general cargoes and cargoes in bulk, but is also capable of container transshipment. The port handles up to 2.5 mln tons of cargoes.

Belgrade Peace (1739) – between Russia and the Osman Empire. Concluded on 18 (29) September 1739 near Belgrade. In tandem with the separate peace treaty signed in Belgrade on (12) September 1739 by Austria and the Osman Empire, B.P. terminated the war of Russia and Austria with the Osman Empire of 1735–1739. Austria insisted on Russia's signing the peace treaty sacrificing the advantages it had gained during the war. Under the treaty, fortifications of the Azov fortress returned to Russia by the Osman Empire were to be destroyed, whereas the city itself with environs were to be declared "a barrier" between the two empires. Russia was entitled to build a fortress in the vicinity of Cherkas Island on the Don River, and the Osman Empire—in the mouth of the Kuban River (clause 3). The Greater and Minor Kabarda were declared neutral and free (clause 6). The Taganrog fortress earlier destroyed was not to be restored. Russia was not allowed to keep her fleet on the Sea of Azov or on the Black Sea (clause 3). Merchants of each of the countries were entitled to trade freely on the territory of the other country on the terms established for third-country merchants. Trade with the Osman Empire could only be conducted on board of the Turkish ships (clause 9). Russian pilgrims were guarantee free travel to Jerusalem and exemption from payments to the Sultan (clause 11). B.P. did not satisfy Russia, negotiations on specific matters of the treaty continued until 1747. The treaty was canceled by the 1774 Peace Treaty of Kuchuk-Kainarji.

Belli Grigory Grigorievich, Heinrich Heinrikhovich (?–1826) – Russian Rear-Admiral. In 1783, admitted from the British service to the Russian military service as a midshipman and sent to the Don flotilla. In 1784–1787, commanded a boat on the Azov Sea as a Lieutenant. In 1788—commander of a schooner, cruised with the fleet and took part in the Battle of Fidonisi. In 1789, promoted to a Lieutenant-commander and conducted the campaign on the Black Sea. In 1790, participated in cruising and took part in the battles at Kerch Strait and Khadjibey, and in 1791—at Kaliakra. In 1792–1798, a frigate commander on the Black Sea. In 1798, set out for the Mediterranean Sea with F.F. Ushakov's squadron, took part in liberating the islands of Tserigo (Kythira), Zante, in Corfu blockade. In 1799, promoted to a Captain 2nd rank. Led an amphibious force from the ships, seized the city of Foggia and occupied Naples. In 1800–1801, while in Naples, was attached to the coastal service. In 1802, returned to Sevastpol. In 1803, promoted to a Captain 1st rank. The next year (1804), while commanding a ship, sailed from Sevastopol to Corfu and the next year cruised in the Mediterranean Sea. In 1806, passed, with a group of ships, from Corfu to the Adriatic Sea and seized Bocca-di-Quattro Province, the Curzalo Fortress and Lis Island. In 1807, sailed in the Adriatic Sea, took part in Patras Island blockade, was promoted to a Captain-commodore. In 1808–1812, (being British, because Russia was at war with Britain) upon imperial command, stayed in Moscow, St.Petersburg and Saratov. In 1812, commanded a ship sailing on the Black Sea, in 1813—at Sevastopol Port. In 1814–1816, commanded naval barracks 59 in Sevastopol. In 1816, promoted to a Rear-admiral. In 1817–1826, commanded the 3rd naval brigade in Sevastopol. Awarded Russian and Italian orders.

Bellinsgauzen Faddey Faddeyevich (Fabian Gottlieb Thaddeus von Bellingshausen) (1778–1852) – Russian Naval personality, famous seafarer, Admiral (1843). Of noble extraction. Graduated from the Naval Cadet Corps (1797). Sailed on the ships of the Revel Squadron. In 1803, participated in the 1st Russian Circumnavigation on board the "Nadezhda" ship under the command of I.F. Kruzenstern. Upon return, served in the Baltic Navy; commanded a corvette (1809). From 1810, in the Black Sea Navy; as a frigate commander, sailed at the Caucasian coasts, did a thorough job of specifying the maps and determining the coordinates of major locations on the seashore, made a number of astronomical observations. In 1816, as a Captain 2nd rank supplemented the "Sailing directions" of 1808 with the results of the description of the Caucasian Black Sea coast, which made it possible to publish in 1817 the General Map of the Black Sea. In it, the position of the coasts was made more accurate on the basis of astronomical data, and in the offshore part of the sea soils of the seabed were designated, directions of currents were indicated (by arrows) and the depths were written down. In 1819–1821, was in charge of the round-the-world expedition that lasted 751 day on the sloops "Vostok" (under the command of B.) and "Mirny" (commanded by Lieutenant M.P. Lazarev). The object of the expedition was to obtain "knowledge about our globe to the maximum extent possible" and "discover likely proximity of the

Antarctic Pole". The expedition sailed round the Antarctic Continent, reaching the 70th degree of the south latitude, discovered 29 new islands and a coral reef.

In 1826, B. headed a group of ships that cruised in the Mediterranean Sea near the coasts of Turkey. From 1827—commander of a guards crew with which he arrived in the active army at the start of the 1828–1829 Russian-Turkish War; flying his own flag on the ship "Parmen", participated in the siege and seizure of the Turkish fortress Varna. In 1830–1831, cruised near the coasts of Kurlandia (Latvia). From 1839—commander of Kronstadt Port and military governor of Kronstadt. Under his guidance, new port facilities were built and the old facilities and forts were remodeled, the steamer works were built, a hospital was opened, much was done to provide urban amenities. From 1843—B. is also a member of the Admiralty Council. A sea in the Pacific Ocean, cape in South Sakhalin, an island in the Tuamotu Archipelago, an island in the Aral Sea, a depression in the Pacific Ocean, and other locations are named after B. There is a monument to B. in Kronstadt, and a memorial bust to him in Nikolaev.

Bellinsgauzen F.F. (http://www.navy.su/persons/02/images/bellinsgauzen_ff_00.jpg)

Belokamensk – see Inkerman.

Beluga (*Huso huso*) – one of the most valuable fishes of the sturgeons family (*Acipenseridae*) in the Caspian, Azov and Black seas. The snout is short, pointed. The mouth is big, of half-moon shape, occupies nearly the entire lower surface of the snout. Tendrils with leaf-shaped appendages along the rear rim are flattened on the sides. In the past, they used to catch specimens as long as 9 m and weighing over 1500 kg. Anadromous fish. The males reach sexual maturity at age 12–14 years, females—at age 16–18 years. Propagate also in the lower reaches of the Danube

River, as far as the Iron Gates, from February-March (at water temperature of 4–11 °C) to the end of May. Beluga is extremely prolific. Spawns from 500,000 to 4 million eggs. The fish is predatory: it feeds largely on fishes, crustaceans, mollusks, larvae of some insects. The meat and roe of B. are distinguished by excellent gustatory properties. Commercially available is fresh, frozen, smoked and dried or tinned B. Particularly valued is black caviar. Two actual catches of large B. are known to have occurred on the Balck Sea: in the 1930s on the Romanian coast (around 1,000 kg) and in 1956 between Anapa and Novorossiisk (650 kg). Today beluga is very rare in the Black Sea and Sea of Azov.

Belyavskiy Petr Evmenovich (1829–1896) – Russian hydrographer, finished Black Sea navigation company (school), took part in combat actions at the time of Sevastopol defense during the Crimean War of 1853–1856. Then, until 1860, sailed on ships on the Black and Azov seas, made navigational appraisal of the existing ports (1857–1858) and was in charge of midshipmen's practice (1859–1860). In 1861–1865, was busy making a sea description of Odessa Port. Here, he was the first of Russian hydrographers to undertake seabed drilling in the Quarantine Harbor to the depth of 24 feet (7.3 m) for the purpose of soil study. In 1865–1866, was in charge of detailed sea description in the delta arms of the Don River for subsequent dredging operations. In 1868, promoted to a Lieutenant-commander, and in 1871–1875 was the chief of the Special Survey of the North Coast in the Hydrographic expedition on the Black and Azov seas. In 1876, for successful service was promoted to a Captain 2nd rank and was attached to Russia's Ministry of Railways. Awarded Russian orders. Dismissed for retirement in 1885.

Belyi (Beloi, Bilyi) Sidor Ignatovich (Gnatovich) (1735–1788) – koshevoi chieftain of the Troop of Loyal Cossacks, Lieutenant-colonel. Was among the prominent and esteemed elders of Zaporizhye Sich. In 1774, as a troop captain, set out for St. Petersburg as the head of a delegation of Cossacks with a petition to protect the rights and possessions of the dwellers of Saporozhye, and at the time of Sich devastation by General Tekeli, was in the capital city. In 1787, B. was among the Cossack sergeant-majors who delivered an address to Catherine II when she was traveling in Novorossia; in the address, the Cossacks expressed willingness to form a new force and take part in the imminent war with Turkey. In January of 1788, after the Troop of Loyal Cossacks was formed, B. was elected koshevoi chieftain and was approved in that position by Prince Potemkin, the Commander-in-Chief of the Ekaterinoslav Army. This done, B. began organizing the troop headquarters (kosh) near the Bug River mouth. On February 27, 1788, A.V. Suvorov handed over to him Zaporozhye Banners, pernaches (pernach—kind of club), the chieftain's club. During the battle with the Turkish flotilla near Ochakov on June 11, 1788, B. was wounded lethally; he was buried with military honors in Kinburn Alexander's church.

Berdyansk – town, established in 1827, Saporizhye Region, Ukraine. Sited along the coast of the eastern part of Berdyansk Bay, northern coast of the Sea of Azov. In 1939–1958, was known as Osipenko. One of the major sea ports on the Azov Sea.

Railway station. Population—117,000 people (2011). Main industrial enterprises: "Azovkabel", Yuzhgidromash", "Azovselmash" and others; fabrication of glass fiber, hydrocarbon oils, Yuzhnyi Plant of Hydraulic machines, glass fiber works, agro-industrial combine "Azovskii", furniture and knitting factories; food-processing, light industry. Teacher-training college. The local lore and arts museums. Balneal and climate health resort established in 1902. On Berdyansk Spit of over 20 km in length, there are health resorts, children's and sports and fitness facilities, recreation facilities, boarding houses. There are interesting sculptures in B. There is the world's only monument to a small fish—Azov goby that saved the dwellers of Berdyansk from famine in the 1930s.

Berdyansk Bay – juts into the northern shore of the Azov Sea between Obitochnaya and Berdyansk Spits. The distanced between the spit terminations is 43 km. The bay shore between the spit bases is flat, then largely it is precipitous, in places transversed by balkas and ravines. There are numerous settlements on the shore of B.B. The largest cities are Primorsk and Berdyansk. The bay is open to southerly and south-westerly winds. Occasionally, these gain a strong force and produce short wave. The shores are hemmed by a shoal with depth up to 5 m, the shoal width at the bay apex reaches 7.5 km. In the western part of the bay, the bay seabed is uneven.

Beregovoe – a village sited between Foros and Katsiveli in a small picturesque natural amphitheater with excellent sand beaches at southern coast of Crimea, Ukraine. A rock with a legendary name of Iphigenia (the daughter of Agamemnon and Clytemnestra)—one of the witnesses of Upper Jurassic volcanism in the Crimea, became the symbol of B. The name of the rock is related to the ancient Greek myth about Iphigenia, the daughter of the Hellenic Tzar Agamemnon who took part in the Trojan War. The Iphigenia rock has been declared a local natural monument.

Berens Mikhail Andreevich (1879–1943) – Russian Rear Admiral (1919). In 1898, graduated from the Naval Cadet Corps. In 1900, as the Boxer Uprising in China was being quelled, B. took part in the seizure of Taku forts. In 1904, graduated from the Provisional Navigator Officers Class. During the Russian-Japanese war (1904–1905), participated in the defense of Port Artur. In 1904–1906, commanded the destroyer "Boikii", managed to escape on the destroyer from Port Artur, went to China, where he was interned. In 1904, was present golden weapon "For Courage". From 1906, served on the Baltic Sea. Was the first mate of the cruiser commander (1909–1911); commanded destroyers (1911–1916), took part in the Battle at Irben (Irbe) Strait (the Baltic Sea). In 1916–1917, commanded the battleship "Petropavlovsk". On November 16, 1917, appointed Chief of the Naval General Staff, then Chief of Staff of the Baltic Sea Mine Defense. In 1918, dismissed for retirement. Fled from Petrograd via Finland to the Far East. In White Movement: Acting Commander–in-Chief of the Navy in Vladivostok (1919–1920). Headed the departure of a squadron of ships with the Naval School in 1920 from Vladivostok to China. In August of 1920 arrived in Sevastopol at the disposal of

General P.N. Vrangel. During Civil War in Russia on the side of the Whites, commanded squadrons of ships and the Kerch base on the Azov and Black seas. One of the organizers of Navy evacuation from Sevastopol to Bizerte, Tunisia. During Navy evacuation from Sevastopol in November of 1920, commandeer of 2nd party of the squadron, junior flagman of Rear-admiral Kedrov's squadron. From 1920, commander of the Russian squadron in Bizerte. In 1924, in connection with the disbanding of Bizerte Squadron, moved to France. Awarded numerous Russian orders. Died in Tunisia.

Berezan Island – island in the north-west of the Black Sea, lies in the offshore shoal at the entrance to the Dniestr-Bugskii Lagoon, sited 12.8 km west of Ochakov and 1 km south of the entrance to Berezan Lagoon. Area aound 0.5 km^2. Maximum length from the north to the south 850 m, width—from 85 to 200 m, height up to 20 m. The island shores are precipitous. Classical authorsd used to call B. Boristhene. It is established as a result of excavations that at the end of the seventh century B.C. there was a small trading town on B.—a most ancient Greek settlement in the circum-Black Sea area, moved in the sixth century B.C. to the continent—ancient Olvia. On March 6, 1906, active participants of the Sevastopol uprising of Black Sea sailors of November, 1905 were executed by firing squad on the island: Lieutenant P.P. Schmidt, the sailors A.I. Gladkov, N.G. Antonenko, S.P. Chastik. Nowadays an obelisk stands at the point of their execution.

Berezan Liman – small shallow-water body in the north-west Black Sea area sited 46 km east of Odessa Bay. Its length 20–25 km, mean width 2–3 km, mean depth 3.3 m (maximum around 15 m). Area of water surface 60 km^2, water volume 0.2 km^3. Water exchange with the Black Sea—decisive factor of the liman water balance, the process being characterized by significant variability during a day: over 4 % of liman water volume may be involved in water exchange with the Black Sea within a single day. There is a bar at the entrance to the liman, preventing penetration of deep sea water to the liman. During positive surges, only the waters of the sea surface layer go in the water body. B.L. is inhabited by both the fishes living in the neighboring Dniestr-Bugskiy Liman, and the Black Sea immigrant species. Fish productivity of the liman is low.

Berg Moritz Anton August von (Moritz Borisovich) (1776–1860) – Russian hydrographer, Admiral, Chief Commander of the Black Sea Navy and Ports, Governor-General of Nikolaev and Sevastopol, Inspector of the Corps of Black Sea Navy Navigators, member of the Amiralty-Council. A descender of Baltic Germans. In 1791, entered the Sea Corps as a cadet, in 1795—a midshipman. From 1796 to 1805, sailed on various ships in the Baltic and German (North) seas, made treks to London and Lisbon. In 1806–1809, on the ship "Neva" of Russian-American Company sailed from Kronstadt to the coasts of Russian America. In 1809, returned to St. Petersburg overland. From 1810 to 1817, sailed on various ships over the Baltic Sea. In 1817 transferred to the Black Sea as the supervisor of Black Sea and Azov light-houses (Nikolaev). In 1818–1819, stayed in Sevastopol. In 1820, was making a description of the Black Sea. In 1820- 1824—chief of the expedition for the description of the Black Sea, Light-Houses Supervisor and Director of Sevastopol Naval School; promoted to a Captain 2nd rank. In 1825— Captain of Sevastopol port. In 1827, appointed Inspector of the Corps of Black Sea Navigators, with promotion to colonels (Nikolaev); from 1829, Major-General. In 1831 appointed chairman of the Committee for setting up dry-docks in Sevastopol. From 1832, managing director of the Hydrographic Department of the Black Sea Staff (Nikolaev). In 1838 promoted to Lieuteneant-General. In 1849, appointed member of the Admiralty-Council. In 1851 reattested to Vice-admiral and appointed Acting Chief Commander of the Black Sea Naval Staff and ports, Governor-General of Nikolaev and Sevastopol (until 1855). In 1852, promoted to an Admiral. At the time of B. in Nikolaev, for the first time the construction of the 135-canon screw-driven ships "Tsesarevich" and "Sinop" commenced. Also, B. was at the helm of the Russian Navy during the toughest years of the Russian sea history of the nineteenth century: the Crimean war and defense of Sevastopol, making provisions for the defense and supplies of maritime cities, evacuation and accommodation of the wounded, through passages of troops, etc. Awarded numerous orders.

Berlin Congress (1878) – international congress convened for the revision of the 1878 Treaty of San Stefano that terminated the Russian-Turkish war of 1877–1878 гг.; the object of B.C.was to conclude a new treaty called upon replace the Paris Peace of 1856 that virtually had become inoperative. B.C. was held in Berlin from 1 (13) June to 1(13) July 1878. The terms of the Treaty of San Stefano which stipulated considerable territorial acquisitions by Russia in Transcaucasia and enhancement of its positions on the Balkans produced active resentment of Great Britain and Austria-Hungary. On the pretext of protecting Turkey the British Government sent a naval squadron to the Sea of Marmara and conducted partial mobilization of the Navy. The Government of Austria-Hungary, in its turn, announced mobilization in Dalmatia and demanded the convening of a European conference. The participants of B.C. were: representatives of Russia, Austria-Hungary, Great Britain, Germany, Italy, Turkey and France. Some meetings (dealing with matters, relating to respective countries) were attended by representatives of Germany and Romania. Present in Berlin were also the Prime-Minister of Serbia, representatives of the Senate of Montenegro, the Irish envoy, leaders of the Armenian-Grigorian Church.

   The outcome of B.C. was the signing on July 1(13) 1878 of the Berlin Treaty, comprising 64 articles that modified some terms of the Treaty of San Stefano to the detriment of the interests of Slavic peoples of the Balkan Peninsula and Russia. Under Article 1, Bulgaria constituted itself as a self-governing principality subordinated to the Turkish Sultan whom it was to pay tribute; Bulgaria was allowed to have a Christian government and national militia. Article 2 fixed the frontiers of the new state. Art. 3–12 stipulated (1) election of a head of the principality who could not be a representative of the dynasties ruling in the great European powers and (2) the mechanism of introducing a new constitution, the procedures of mutual relations between the Bulgarian government and Turkey. Provisional government of Bulgaria remained the responsibility of the Russian Commissioner (provided he was assisted by the Turkish Commissioner and Consuls appointed by the other countries- participants of B.C.); the term of stay of Russian troops in Bulgaria was limited to 9 months (Art. 22). The areas to the south of Balkan Peninsula were to be instituted as East Rumelia Province under a direct political and military power of the Sultan on the principles of "administrative autonomy", the governor-general of the province, of Christians, was to be appointed subject to the consent of the powers (Art. 13). Turkey was not allowed to keep her troops on the territory of Bulgaria. The governor of East Rumelia was entitled to call up Turkish troops in case the country's security was threatened. Russia agreed that Austria-Hungary would be entitled to occupation and administrative management of Bosnia and Herzegovina as well as to maintaining a garrison at Novi-Bazaar sanjak (between Serbia and Herzegovina) that remain in possession of Turkey (Art. 25), Serbia (Art. 34) and Romania (Art. 43). Reaffirmed were provisions of the Treaty of San Stefano, whereby Romania gained North Dobrudzha in exchange for Bessarabia it was entitled to under the 1856 Paris Peace; Bessarabia was to be returned to Russia (Art. 45–46). Reaffirmed and broadened were the powers of the Danube Commission in which a Romanian representative was included. Freedom of navigation on the Danube was guaranteed (Art. 52), from the Iron Gates (navigation of naval ships was prohibited).

In Transcaucasia, Russia gained Ardagan, Kars and Batum (Art. 58), a new Russian-Turkish border was determined; Batum was declared a free trading port (porto-franco). Alashkert Valley and Bayazed were again returned to Turkey (art. 60). Turkey undertook to go ahead, without delay, with reforms "called for by the local needs" in the area with Armenian population, and make provisions for the security of such population. Special articles of the treaty proclaimed the principle of religious freedom and equality of civil and political rights irrespective of creed in all parts of the Ottoman Empire, as well s in Bulgaria, Montenegro, Serbia, Romania. Reaffirmed were all provisions of the 1856 Paris Peace and of the 1871 London Convention on the Black Sea straits that were not affected by the new agreement (Art. 63).

Having played a certain role in temporary stabilization of the situation on the Balkans, B.C. not only did not resolve all most acute Balkan contradictions and contradictions between the European powers, Russia and Turkey in respect of the Balkan issue, but produced new knots of mutual rivalry. B.C. failed to elaborate a procedure to control the observance of its resolutions, therefore part of the treaty articles was either ignored or modified. In 1885, Bulgaria and East Rumelia were united in a single state proclaiming full independence of Turkey; in 1886, Russia abolished the status of Batum as porto-franco. As the Balkan wars of 1912–1913 began, the effect of the Berlin Treaty actually ceased.

Bersenev Ivan Mikhailovich (circa 1745–1789) – Russian hydrographer, Captain 2nd rank, graduated from the Naval Cadet Corps in the rank of midshipman in 1765. Upon graduation, until 1770 sailed on various ships in the Baltic and White seas. In 1770, sailed in Admiral G.A. Spiridov's quadroon from the Baltic to the

Mediterranean Sea. Took part in the Battle of Chesma (1770) during the Russin-Turkish war of (1768–1774). The war over, he sailed on various ships on the Black Sea. Worked at the Admiralty-collegium, mapped areas of the Greek Archipelago (1776–1778). In 1783, B. made his first description of Akhtiar Bay, whereupon the construction of a new port and a city of Sevastopol began there. In 1785–1786, B. explored and described the Crimean coast from the Belbek River to Kinburn Spit and from Sevastopol to the Sea of Azov.

Berthier-Delagarde Alexander Lvovich (1842–1920) – Russian military engineer, General. He was born in Sevastopol in 1842. During the 1870s–1890s, under his guidance, they were building trading ports in Odessa, Kherson, Sevastopol, Yalta and Feodosiya, the first water pipelines in the cities of the Crimea, an admiralty in Sevastopol. In the course of construction work, B.-D. was making archeological observations. Known as an outstanding archeologist specializing in the study of the classical period, researcher of classical cities of the Northern circum-Black Sea area, coin collector, vice-president of the Imperial Odessa Society of History and Antiquities. B.-D. donated over Rbls. 1,000 at the end of the 1890s to finance repairs of the Genoese Fortress in Sudak.

Besikduzu – a town on the coast of the Black Sea, Turkey.

Besleti – maritime climate-balneal health-resort, sited in the valley of the Besletka River, 4 km east of Sukhumi, Abkhazia. The climate is humid subtropical, with very mild, snowless winter and very warm summer. Besides climate, the health-resort resources of B. include ground waters obtained from the drilled water wells.

Bessarabia – historic area, sited between the Black Sea and the Danube, Dniestr, Prut and Rakitna rivers. Until the early nineteenth century, B. referred only to the southern part of the Prut and Dniestr interfluves—Budzhak. During the 1st millennium B.C., it was populated by Frakians, in the 1st millennium A.D., it was invaded by Gots, Hunns, then Avars and other peoples. In the tenth to early twelfth centuries B. becomes part of the ancient Russian state, in the twelfth century—part of Galitsk Principality, in the thirteenth to mid fourteenth centuries—part of Galitsk-Volyn Principality. At the end of the fourteenth to early fifteenth centuries the area of B. adjoining the lower reaches of the Danube was disputed by Moldavian and Valashian hospodars. The Valashian Principality managed to control temporarily the area between the Prut and Danube. Until 1484, B. was part of the Moldavian Principality. As a result of Osman conquest, Belgorod Fortress and Kila with environs became Turkish administrative units (raias), in 1538, a new Turkish raia was formed in the Bessarabian territory with a center at Bendery (Tiginya) alienated from Moldavia. In the southern part of B. Tatars and Nogai people settled: these obeyed the Crimean Khanate. In 1591 and 1621, Turkish raias were set up around Izmail and Reni. After the unsuccessful Prut campaign of Peter I, Khotin raia was established The Sultan government intended to unite the North and South B. under his authority. The Russian-Turkish wars of the eighteenth to early nineteenth centuries were conducive to liberating B. from Porta. In conformity with the Ainaly-Kavak Russian-Turkish Convention of 1779, part of south B. was annexed to Moldavia, where Khotarnichi tsynut (County) was established. By early nineteenth century, B. was subdivided into three areas: Moldavian, which was part of the Moldavian Principality, Turkish, governed by Pashas, and Nogaian ruled by Budzhak Khans. After the Crimea was annexed to Russia (1783), part of the Crimean Khanate between the Dniestre and Bug also began to be governed by Budzhak Khans, and until 1791 this territory was called B. During the Russian-Turkish wars of the second half of the eighteenth to early nineteenth century, B. was ruled by the Divan of Moldavia that was headed by Russian administration subordinated to the Commander-in-Chief of the Russian Army. Under the 1812 Bucharest Peace, B. was annexed to Russia. From 1818—region, and from 1873—province of the Russian Empire (center—Kishinev). Under the 1829 Adrianopol Peace, the Danube delta was also annexed to B. In 1856, the southern part that adjoined the Danube and lower reaches of the Prut River was made part of the Moldavian Principality, which was united with Valash Principality as part of the state of Romania. Under the 1878 Berlin Treaty, South B. without the delta was returned to Russia. During WWI, the area of North B. was involved in military operations.

Bessarabian Province – transformed in 1873 from Bessarabian Region (center—Kishinev). Was subdivided into counties: Akkerman, Beltsy, Bendery, Izmail, Kishinev, Orgeevka, Soroka, Khotin. In 1897, the area was 39,000 verst; population—1.9 mln people (Moldavans, Ukraininas, Russians, Jews, Bulgarians, Germans, Turks). Population in 1910—2.4 mln people; highlanders—0.4 mln people; 12 cities, 5 trading quarters. B.P.—agriculture-oriented (early in the twentieth century, nearly 85 % of the population was engaged in agriculture). Basic crops—wheat, corn, barley, oats, potatoes, leguminous). Each year up to 100 mln puds of grain used to be exported from B.P. Stock farming played a significant role. Tobacco growing, vegeculture, fruit cropping, vine-growing and wine-making are practiced on a large scale (in 1908, around 8 mln vedros of wine was produced, 1 vedro = 10–12 l). Cottage industries include wool-cloth making, carpet-weaving, lint- and hemp-spinning, ceramics fabrication and others. Early in the twentieth century there was around 800 factories and plants (mainly small and medium) in B.P. distilleries, sugar-beet processing, tobacco factories, dairy plants, wool-cloth factories, etc. Over 6,000 flour mills, out of this, 58 large ones. Across the territory of B.P. there ran lines of the South-West Railway System, their total length being 800 verst (1911). Steamer traffic on the Danube, Prut, Dniester rivers. Major commodities of commerce—grain, products of stock farming and wine-making.

Bessarabian Region – established in 1818 on the territory of Bessarabia that became a Russian possession under the 1812 Bucharest Peace (center—Kishinev). Originally was divided into these counties: Bendery, Grechany, Kodru, Orgeen (or Kishinev), Soroka, Khotarnichan, Khotin, Tamarov (or Izmail), Yassy (or Faleshty). Under the Regulation on B.R. administration (1828), the region was divided into counties: Akkerman, Bendery, Kishinev, Leov (subsequently, Kagul), Srgeev, Soroka, Khotin, Yassy (subsequently, Beltsy) as well as Izmail Mayorate (subsequently, county). Under Adrianopol Peace of 1829, the Danube delta was included in B.R. After the Crimean War of 1853–1856, under the 1856 Paris Peace, Izmail County was alienated from B.R. (was made part of the Moldavian Principality, under the 1878 Berlin Treaty—again in the Russian Empire) and the Danube delta was alienated, too. In 1873, B.R.was transformed into Bessarabian Province.

Betta – maritime climatic health-resort and settlement 35 km to the south-east of Gelendzhik, Krasnodar Territory, Russia. Sited in a picturesque valley of a mountain river of the same name. On the south-east, is protected from winds by the offspurs of the Greater Caucasus. B. area exhibits lush vegetation (including epiobiotic pine); in the environs, there are vineyards. There is a pebble beach 300 m long.

Betta (Photo by Andrey Kostianoy)

Big fontan (fountain), Cape – sited to the north-east of Ilyichevsk Port. Marks the southern border of Odessa Bay. The shore around the cape is precipitous. The Odessa Lighthouse is installed on the cape.

Big Utrish – a small village on the coast of the Northeastern Black Sea, 15 km southward of Anapa, Russia. Recreation area, beach. Lighthouse and dolphinarium.

Biosphere reserve – (1) The representative landscape unit identified in accordance with the UNESCO Program "Man and Biosphere" for the purposes of its preservation and investigation (and/or monitoring). It may incorporate ecosystems not affected by economic activities or slightly altered, often surrounded by the lands in use. Such reserves are found in more than 60 countries. (2) Strictly protected large natural site practically not subject to man-transformed local impacts, a surrounding landscape with the century-wise processes which nature enables to reveal the spontaneous changes in the biosphere, including global anthropogenic; (3) Territory on which permanent monitoring of the anthropogenic changes in the natural environment on the basis of instrumental control of bioindicators.

Biryuchii Island – island-spit sited in the Sea of Azov. Of sedimentary and alluvial genesis. There are numerous saltlakes, wide reaches, solonchak-steppe vegetation in the low-lying part of the spit overlooking the lagoon. Birds nestle in the numerous niches on the steep rocky slopes of the island. B.I. is part of the Azov-Sivash husbandry reserve.

Black and Azov Sea Ports Association (BASPA) – international non-governmental organization established in March 1999. The Association is established under the aegis of the international Organization of the Black Sea Economic Cooperation (BSEC). The primary goals of the association—active participation of the Black Sea region ports in formulating the transportation policy of the European Union, the world community and BSEC countries in the Black Sea area as well as promotion of transport routes in the region; development of cooperation between the regional ports and international agencies; active participation of the ports in promoting priority international traffic corridors, above all between Europe and Asia. The Association places great emphasis on matters of collective security in the Black Sea, in particular, on environmental problems.

Black caviar – the roe of the sturgeons. B.C. is produced from the raw roe of great sturgeon, kaluga (*Huso dauricus*), sturgeon, stellate sturgeon, bastard sturgeon and sterlet. This is the highly nutritive, valuable and tasty food product. It contains great quantities of full-value proteins, fats and vitamins. By its calorific value this product is inferior to meat, milk and other foodstuffs. 100 g of caviar gives 280 calories to the organism. The usual caviar portion is considered one ounce (28.35 g). The common stages in production of all kinds of granular caviar, except unscreened roe, include splitting of fish with roe, removal and sorting of roe-in-sac and roe separation, making of roe parcels. The caviar quality depends on the size of roe, their color (light-gray, dark-gray and black) and fat content—the larger and lighter are the roe the higher their quality. The color and size of eggs depend on the age of female fish and its feeding. The following kinds of caviar are distinguished: great sturgeon—its color varies from light to dark gray and it is packed in a jar with a blue cover; stellate sturgeon—it is most often black and it is packed in jars with a red cover; sturgeon—blackish, golden-brown, amber-yellow, it is packed with a yellow cover. Depending on the methods of processing the caviar may granular, slightly salty in tins and strongly salted barrel, pasteurized, pressed, and unscreened. Depending on the production technology the caviar may be low-salty, fresh, low-salty pasteurized or pressed. For its better preservation the freshly prepared caviar is immediately canned. On the world market the prices of the stellate sturgeon caviar vary up to 1,500 USD/kg, the great sturgeon caviar may cost up to 10,000 USD/kg. The principal caviar producers are Russia, Iran and Azerbaijan. The achuev caviar produced from the A.S. sturgeons was valued high in Russia.

Black mullet (*Mugil cephalus*) – fish of the Mullet (*Mugilidae*) family. Its length is typically 75–80 cm, weight—2 to 3 kg, sometimes to 6 kg. This species occurs mostly in the tropical and moderate zones of the Atlantic, Pacific and Indian oceans, in the Mediterranean, Black seas and Sea of Azov. This is the pelagic fish living in stocks. In the Black Sea three localized stocks are distinguished—Crimean, Caucasian and Balkan. Its lifespan is 7–8 years. It reaches maturity at the age of 2–3 years. It grows quickly: 2-year species may be 17–39 cm in length and weigh 120–950 g, while 3-year species 34–48 cm and 700–1,800 g, respectively. The fish propagates in the coastal and open waters from June to late August. It migrates for spawning, fattening and hibernation. In April (more seldom in March) or in early May the species from the Balkan stock migrates to the Bulgarian shelf and usually to mid-June its stocks occur along the whole western coast of the Black Sea. Some fish settles in the coastal lakes for the whole summer and half of autumn, others stay in bays and freshwater sea areas. In autumn when the water cools down B.M. gradually move southward and till mid or late December it stays in the southern part of the Black Sea near Bosporus and in the Sea of Marmara. Annually the 1-year species and small stocks of 2- and 3-year fish hibernate in some the Black Sea coastal lakes and in large harbors protected from waves. B.M. has the greatest commercial significance of the whole Mullet family.

"Black Prince" – under this name, the 3,000 tons British sail-screw, iron hull steamer "Prince" went down in history. During the Crimean war, in 1854, the ship carried cargo (regulation uniform, tents and medication) for the British expeditionary corps in the Crimea. During a violent storm the ship together with two dozens other ships and craft was thrown against the sea cliffs, was shipwrecked and sank on the outer road of Balaklava harbor not far from Sevastopol, Crimea. The story of its shipwreck by early twentieth century acquired a fantastic coloring. It was widely believed that there was a huge amount of gold on board the hip—pay for the entire British Army. The steamer got the name of "Black Prince". The search for the ship began immediately after the war ended. In 1875, the search was conducted by the French, then by Italians. In 1923, a special purpose expedition for submarine works EPRON commenced a search for the steamer. At the end of 1924, some remains of the iron ship were located and part of these was lifted. No gold was found, so the work aimed at lifting the ship was terminated. In summer of 1927, the search was commenced by the Japanese diving firm "Sinkai Kogiossio Limited". The firm allegedly worked on board the located vessel, yet it too failed to find much gold. In all 7 gold coins were found, of which 4 were handed over to judge, and 3 the Japanese left to themselves. Presumably, there was no gold on board the "Prince" et al., since it had been offloaded earlier at Constantinople. An indication of this is the fact that Britain had never ventured to search "P". At the same time, France spend half a million, Italy—200,000, Japan—nearly one-fourth of a million rubles in gold. "B.P." was the subject of stories written by A.I. Kuprin, S.N. Sergeev-Tsenskii, M. Zoshchenko, E.V. Tarle and many other authors.

Black Sea – The Black Sea (together with the Sea of Azov) projecting far into the continent constitutes a most insular part of the World Ocean and is part of the basin of the Atlantic Ocean. B.S. washes the coasts of Russia, Ukraine, Romania, Bulgaria, Turkey, Georgia and Abkhazia. The sea is linked by Bosporus Strait with the Sea of Marmara, and by Kerch Strait—with the Sea of Azov. The Black Sea area is 423,000 km^2, the volume of water is 555,000 km^3, mean depth—1,315 m, maximum depth—2,258 m, the length of the coastline is 4,340 km.

The Black Sea countries

The coastline, except north and north-west, is slightly dissected. The eastern and southern coasts are steep and mountainous, western and north-western—low and flat, in places precipitous. The only large peninsula is the Crimean one. On the east, the offspurs of the Caucasus mountains come close to the sea, the Pontian Mountains stretch along the southern coasts. Around Bosporus, the coasts are low, but precipitous. In the south-west, the Balkan Mountains end abruptly right at the seashore, farther to the north there is Dobrudzha Upland, gradually giving way to lowland expanses of the wide Danube delta. The north-western and partly northern coasts are cut by lagoons: Dniester, Dnieper-Bug, separated from the sea by spits. The largest bays are in the north-west: Odessa, Karkinit, Kalamit. Besides these, worthy of mention are Samsun and Sinop Bays on the southern coast, Burgas Bay—on the western coast. Zmeinyi and Berezan islets are in the north-western part of the sea, Kefken is eastward of Bosporus.

Most of the river runoff (up to 80 %) comes to the north-western part of the sea, where major rivers bring their waters: Danube (200 km^3/year), Dnieper (50 km^3/year), Dniester (10 km^3/year), South Bug (5 km^3/year). The Inguri, Rioni, Chorokh rivers and a lot of rivulets fall into the sea on the Black Sea coast of the Caucasus. Elsewhere on the seacoast, river runoff is insignificant.

Three major structures are distinguished in the seabed relief: the shelf, continental slope and deep-water basin. The shelf accounts for up to 25 % of the seabed total area and, on the average, is confined by the depths of 100–120 m. The shelf's maximum width (over 200 km) is in the northwestern part of the sea, all of which is within the shelf zone. Nearly throughout the mountainous eastern and southern coasts of the sea, the shelf is rather narrow, just a few kilometers, while in the southwestern part of the sea it is wider (dozens of kilometers). The continental slope occupying up to 40 % of the seabed area goes down roughly to the depths of 2,000 m. The slope is quite steep and is dissected by underwater valleys and canyons. The bottom of the basin (35 %) is a flat aggraded plain, whose depths gradually increase centerward.

The Black Sea bottom and land topography (http://upload.wikimedia.org/wikipedia/commons/thumb/8/89/Black_Sea_relief_location_map.svg/2000px-Black_Sea_relief_location_map.svg.png?uselang=ru)

The Black Sea surrounded by dryland has a distinctly continental climate, which is manifest in significant variations of air temperature. Coastal relief has a great impact on climatic peculiarities of parts of the sea. For example, the northwestern part of the sea open to the effect of the air masses from the north is characterized by steppe climate (cold winter, hot, dry summer), whereas the southeastern part shielded by high mountains the climate is that of moist subtropics (abundant rainfall, hot summer and mild winter). In winter, the sea is subject to the impact of the offshoot of the Siberian anticyclone, causing the influx of cold continental air. This is accompanied by north-easterly winds, not infrequently reaching the force of a storm, by abrupt drops of air temperature, rainfall. Particularly strong north-easterly winds are typical of Novorossiisk area ("bora"). Here, masses of cold air accumulate beyond the high coastal mountains and, having overcome the mountain tops, plunge with great force downward, to the sea. The wind speed during bora reaches 30–40 m/s, bora recurrence is 20 times a year and more, essentially from November to March. When the Siberian maximum offshoot weakens in winter, the Black Sea becomes dominated by Mediterranean cyclones. These cause unstable weather with warm southwesterly winds and temperature fluctuations.

In summer, the effect of the Azores maximum spreads aver the sea, dry and hot weather prevails, thermal conditions become uniform for the entire water area. This season is dominated by slight northwesterly winds, the stormy northeasterly winds being generated seldom in the northeastern part of the sea. The lowest mean monthly air temperature in winter is noted in the northwestern part of the sea (under −5 °C), while in the east and south it increases to 6–9 °C. Minimum temperatures in the northern part of the sea reach –25 °C. . .–30 °C, in the south –5. . .–10 °C. In summer, the mean air temperature over the sea equals 23–25 °C, the maximum values being 35–37 °C.

Precipitation on the coasts is extremely non-uniform. In the southeastern of the sea, where the Caucasian ranges block the humid Mediterranean winds, precipitation is maximum (in Batumi—up to 2,500 mm/year, in Poti—1,600 mm/year). On the plain northwestern coast precipitation does not exceed 300 mm/year, near the southern and western coasts and on the SCC—600–700 mm/year.

In the water balance of the sea (km^3/year), the input component is made up of: river runoff—338, precipitation—237, water influx via Bosporus—175, via Kerch Strait—50; the output component: evaporation for the sea surface—396, runoff to the Sea of Marmara—370, to the Sea of Azov—34. The total value of water input and output averages around 800 km^3/year.

Seasonal variations of the sea level are mainly produced by the differences of river runoff input during the year. For this reason, during the warm season of the year the sea level is higher, during the cold season—lower. The value of variations is most significant where the impact of river water is maximum—30 to 40 cm. The difference between the levels of the Black Sea and the Sea of Marmara, and the type of winds around Bosporus determine seasonal variations of the water exchange via the strait. The upper Bosporus current from the Black Sea (the thickness of its layer at the entrance to the strait is around 40 m) reaches maximum in summer, its minimum being observed in autumn. The intensity of the lower Bosporus current (directed to the Black Sea) is the highest in autumn and in spring, its minimum is registered early in summer.

Wind-effect fluctuations of the sea level in the Black Sea are maximum: these have to do with the impact of steady winds. Such fluctuations are most frequent during the autumn-winter period in the western and north-western parts of the sea, where they are likely to exceed 1 m. Near the Crimean and Caucasian coasts, the positive and negative surges hardy ever exceed 30–40 cm. Their duration usually lasts 3–5 days, yet sometimes it may be longer. Seasonal density-related (steric) variations of the sea level are produced by variations during the year of temperature and salinity in water column.

Subject to the nature of winds, considerable wave develops during the autumn-winter season in the north-western, north-eastern and central parts of the sea. Prevalent waves are 0.5–1 m high, but in open water areas maximum height of the wave during very strong storms may be as high as 10 m. The most still are southern areas of the sea, where strong wave is rare, and waves higher than 3 m are hardly ever observed.

Heavy storm in Sevastopol on 11 November 2007 (Photo by Dmytro Solovyov)

Ice in the Black Sea is only formed in the narrow strip of the northwestern part of the sea (0.5–1.5 % of the sea total area). In very severe winters, fast ice along the western coasts stretches as far as Constanta, and ice floes may drift as far as Bosporus. True, over the last 150 years only five such cases have been recorded. The commencement of ice formation as a rule occurs in mid-December, and becomes quire common in February. Depending on the severity of winter, the length of the ice period varies significantly: from 130 days in severe winters to 40 days when winters are mild. Maximum ice thickness averages 15 cm, in severe winters—up to 50 cm.

Ice cover in Odessa port on 1 February 2012 (http://regonetolserfot.ru/)

Water circulation in the upper layer of the sea is directed counterclockwise, with two cyclonic gyres in the western and eastern parts of the sea and with the longcoast Rim Current. There are three areas distinguished in the Black Sea circulation system: coastal, the Rim Current area and open water areas. The nature of currents in the coastal part of the sea is characterized by considerable variability conditioned by mesoscale and smallscale eddies and meanders. The 40–80 km wide Rim Current area is right above the continental slope. The currents inside it are rather steady, the speeds on the surface equal 40–50 cm/s, sometimes exceeding 100 and even 150 cm/s (in the midstream of the flow). In the upper 100-m layer of the Rim Current, the current velocity change little, depending on the depth, and as the depth increases, they tend to slow down. In open areas of the sea, the currents are weak. The mean velocities do not exceed 5–15 cm/s on the surface and 5 cm/s at 500–1,000 m depth. The general circulation of the Black Sea is one of single direction. It is hard to judge on the nature of circulation in the water column deeper 1,000 m. Analysis of numerous satellite images of the Black Sea and hydrological data indicates the presence of diverse vortex structures, their spatial scale ranging from 10 to 100 km: coastal anticyclonic eddies, anticyclones and open sea cyclones, dipole vortex structures. The depth of vortical motions penetration reaches 300–400 m. The velocity of their movement is around 5–20 cm/s, velocity of orbital rotation—20 to 40 cm/s (maximum—45 cm/s). The lifetime of vortical structures—from several days to 6 months. The merging of anticyclones is conducive to their lengthy existence.

Satellite image of mesoscale vortical structures along the coast of the Black Sea (Image courtesy by Dmytro Solovyov)

The temperature of water on the surface of the sea in winter rises from −0.5...0 °C in the coastal areas in the north-west to 7–8 °C in the central areas and 9–10 °C in the south-eastern part of the sea. In summer, the surface water layer is as warm as 23–26 °C. During the warming period, there is formed an abrupt layer of thermocline which limits the spread of heat depthwise. Salinity in the surface layer throughout the year is minimal in the northwestern part of the sea, where the bulk of river runoff comes. In the river mouth areas, saline increases from 0–2 ‰ to 5–10 ‰, whereas on most of the water area of the open sea it equals 17.5–18.3 ‰.

During the autumn-winter season, thanks to sea waters growing cooling vertical circulation develops. By the end of the winter it embraces the layer 30 to 50 m thick in central areas and 100–150 m thick in coastal areas. Water in the northwestern part of the sea are cooled most; from there, spread on the intermediate depths all over the sea and may reach the areas most distant from the focal points of cooling. As a result of winter convection during subsequent summer warming, a cold intermediate layer is formed in the sea. It is located at 60–100 m depth all year round and is distinguished by temperature on its upper and lower boundaries—8 °C, and in the core—6.5–7.5 °C. Convective mixing in the Black Sea cannot spread deeper than 100–150 m due to greater salinity (and, hence, density) in deeper layers. In the upper layer of mixing, salinity increases slowly, then on the 100–150 m depth it increases abruptly from 18.5 to 21 ‰. This is a permanent halocline. Starting with the horizons of 150–200 m, salinity and temperature rise slowly bottom-wise due to the arrival in the deep layers of more saline and warm waters of the Sea of Marmara. At the exit from Bosporus Strait, the waters have salinity of 28–34 ‰ and temperature of 13–15 °C, but change their characteristics quickly, mixing with the Black Sea waters. In the near-bottom layer, there occurs a slight increase of temperature due to geothermal influx of heat from the sea bottom. Deep waters occupying the water column from 1,000 m to the bottom (over 40 % of the sea volume) are distinguished by great stability of temperature (8.5–9.2 °C) and salinity (22.0–22.4 ‰).

Unlike the other seas, in the Black Sea only the upper, thoroughly mixed layer of about 50 m in thickness is saturated with oxygen (7–8 ml/l). As one goes deeper, oxygen content begins to decline quickly, and at the depth of 100–150 m oxygen content is equal to zero. At the same depth there emerges hydrogen sulfide, its quantity growing with depth to 8–10 mg/l at 1,500 m depth, and farther to the bottom the content grows stable. In the centers of major gyres, where the upwelling of waters is observed, the upper boundary of the hydrogen sulfide zone is closer to the surface (70–100 m) than in the coastal areas (100–150 m). The boundary between the oxygen and hydrogen sulfide zones is occupied by the intermediate layer of their coexistence, which constitutes the lower boundary of marine life. The spread of oxygen to the deep layers is blocked by high vertical gradients of density in the zone of contact of the water masses of the Black Sea and the Sea of Marmara, which restrict convective mixing by the upper layer.

However, water exchange throughout the water column of the Black Sea does occur, albeit slowly. If there were no continuous influx of deep, more saline waters into the sea, then it would have grown fresh a long time ago due to the impact of river runoff. The deep saline waters are replenished with the lower Bosporus current all the time, they gradually go up and get mixed with the upper layers that are driven out of the sea by the upper Bosporus current. Such circulation maintains a relatively permanent salinity in the water column.

Different authors stress the following main factors impacting the vertical exchange in the sea: water upwelling in the center of gyres and water downwelling round their periphery; turbulent mixing and diffusion inside the water column; winter convection in the upper layer; wind-effected phenomena and upwelling in the coastal areas. Simultaneously, estimated time of total vertical water exchange in the sea is rather approximate: from dozens to hundreds of years. This matter needs special investigations.

Most authors tend to regard reduction of sulfates during decomposition of organic residues as the main mechanism of hydrogen sulfide formation in the Black Sea. This kind of process is possible in any water body, but hydrogen sulfide formed in it gets oxidized quickly. It does not disappear in the Black Sea due to slow water exchange and inability of its rapid oxidation in the deep layers. During upwelling, as the water reaches the upper oxygen-saturated layer, hydrogen sulfide is oxidized into sulfates. Thus, there exists in the sea the established equilibrium cycle of sulfur compounds, which is determined by the rate of water exchange and other hydrodynamic factors. The data of observations indicate the natural year-to-year fluctuations of the upper boundary of the hydrogen sulfide zone taking place differently in various areas of the sea and in different years.

The diverse flora and fauna of the Black Sea is almost entirely focused in the upper layer which is 150–200 m thick and accounts for 10–15 % of the water volume. The deep water mass devoid of oxygen is almost lifeless and is inhabited by anaerobic bacteria only. Ichthyofauna of the Black Sea used to be shaped up by representatives of varying origin and comprises around 150 fish species. Of the Black Sea food fishes, at present of significance are only anchovy, horse mackerel and sprat as well as dogfish. There dwell three species of dolphins in the Black Sea whose fishing is prohibited. The total dolphin population in the sea is around 400,000 individuals, probably much fewer.

Over the last few decades, the environmental situation in the Black Sea has deteriorated substantially. Due to intensive human activity in the sea basin and on the coast, the sea is increasingly polluted. The input of organic substances in the sea grows; concentrations of hazardous pollutants like oil products, phenols, pesticides and others in sea water have increased. The structure of biological communities has changed: the species composition of plants and animals is on the downgrade, reserves of the fishing grounds are dwindling. These adverse changes are, above all, manifest in the coastal areas of the sea where anthropogenic load is maximum. Since the end of the 1980s, fishing in the Black Sea has declined. The dolphin population is decreasing. Mediterranean seal that used to dwell in the Black Sea is now represented by a few individuals. This species faces complete extinction. All aforesaid factors are closely related to human activity.

The decline of fishing was assisted in the late 1980s by introduction of jellyfish Mnemiopsis and jellyfish Aurelia Aurita. In the sea, these predators propagated as they never had before. Intoduction of the said jellyfish has reduced significantly biomass of zooplankton calorigenic to fishes. This has impaired notably food supply and reproduction of major food fishes: anchovy and sprat.

Adverse environmental changes are manifest most in the northwestern part of the sea. The large amount of biogenic and other organic matter arriving here with continental runoff causes mass development of plankton algae. In the area of Danube runoff impact biomass of phytoplankton has gown 10 to 20-fold. Due to toxic impact of some algae during mass "blooming" destruction of fauna may be observed. Besides, during intensive growth of plankton a large number of mortified organisms settles down on the seabed, whose decomposition required dissolved oxygen. Given pronounced water stratification in summer impeding the arrival of oxygen from the surface layer, there develops deficit of oxygen (hypoxia) in the bottom layers of the sea. This is likely to result in the death of organisms (suffo- cation) which have been recurring in the northwestern part of the sea virtually each year. The unfavorable environmental situation has led to the mortification of the once extensive field of phyllophora—algae used for the preparation of agar-agar. The areas of mussel grounds have declined considerably, and the state of biological communities has deteriorated in general.

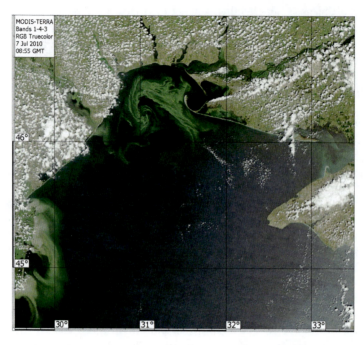

Anomalous algal bloom in the northwestern part of the Black Sea in July 2010 (Image courtesy by Dmytro Solovyov)

The Black Sea is economically important to the countries that surround it. The Black Sea is one of the powerful through-passages. The Black Sea is used for the carriage of a huge quantity of freights of the Black Sea countries, the river warer-ways linked with the sea are used, too. The Volgo-Don Canal has linked the Black Sea with the Volga River and the Caspian Sea. These water-ways are used for freight and passenger vessels traffic. Grain, coal, ore, oil and solt constitute crucial commodities of domestic traffic. The largest sea ports are: Izmail, Odessa, Ilyichevsk, Nikolaev, Kherson, Kerch, Sevastopol, Feodosiya in Ukraine; Novorossiisk, Tuapse in Russia; Poti, Batumi in Georgia; Burgas and Varna in Bulgaria; Constanţa in Romania; Istanbul, Sinop, Trabzon in Turkey. Large freight traffic of the Balkan countries is bound for the Black Sea over the Danube. The Black Sea is a major fishing area for fish, algae and mollusks. The shelf and continental slope hold good prospects for gas and oil recovery.

Its favorable conditions are conducive to the promotion of health resorts and tourism. The major climatic health resorts are: the Southern Coast of Crimea with the center in Yalta—Ukraine; the Caucasian coast—Anapa, Gelendzhik, Sochi—Russia; Gagra, Sukhumi—Abkhazia; Batumi—Georgia; Bulgarian coast—Golden Sands and Sunny Beach; Romanian coast—Mamaya.

Black Sea anchovy (khamsa) (*Engraulis encrasicholus ponticus*) – fish of the anchovy family (*Engraulidae*), has an elongate (up to 17 cm) rounded body covered

with an easily removed scale. The mouth is unusually large. The head is significantly flattened on the sides. Is common in the Black Sea. Forms two schools:
western and eastern. Lives 4–5 years. Reaches sexual maturity at I year of age. This
is a heat-loving fish, most active during warm months of the year, when it can be
encountered across the entire water area of the Black Sea in the warm water layer
(to the depth of 30 m), feeds, grows and propagates intensively. By the end of June,
spreads all over the sea, mainly in its northwestern part. At the end of September—
end of November, the western school leaves for the Crimean and Anatolian coasts
to winter. Propagates with the aid of the floating fish eggs that is spawned during
summer months. Has one of the best ratings in the Black Sea fishing. Is eaten when
salted.

Black Sea—Azov Research and Fishing Station – was established in 1919 as a
public service in Khersones.

Black Sea—Azov herring (*Alosa kessleri pontica*) – fish of the herrings family
(*Clupeidae*). Length up to 40 cm. Common in the Black and Azov Seas, leaves for
the rivers to propagate. Winters near the Crimean and Anatolian coasts. Early in
spring, herring moves northward along the Bulgarian coast. Above all, it is the
schools of mature individuals that migrate. Before mid-May, herring usually leave
the Bulgarian waters gradually for the Danube to propagate from May to mid-July
at water temperature 16–19 °C. In autumn, it moves southward, closer to the
wintering sites. Herring lives approximately 6–7 years. Reaches sexual maturity
at age 2–4 years. Herring is of great commercial value in the Black and especially
Sea of Azov. Is an object of intensive fishing at sea in April–May, in the Danube—
in June. Herring meat has excellent gustatory properties.

Black Sea—Azov sturgeon (*Acipenser giildenstadti colchicus*) – fish of the
sturgeons family (*Acipenseridae*). Common in the basins of the Black and Azov
seas. Reaches 250 cm in length; mass up to 80 kg. The snout is short, wide, rounded
in front. Males reach sexual maturity at age 8–12 years, females—at age 13–15
years. S. propagates in the Danube from April to May. This sturgeon is relatively
prolific: one female spawn from 72,000 to 840,000 fish eggs. Sturgeon is a predator,
feeding mostly on fishes, crustaceans as well as as larvae of insects and worms.
Valuable commercial species in the basins of the Black and Azov seas. Used as a
food item fresh, smoked or canned.

Black Sea—Azov White Fleet – established in 1919, played an important role in
the Crimean epic of 1919–1920 during Civil War in Russia. Took part in operations
near Nikolaev, Kherson, Ochakov.

Black Sea benthos – benthos includes organisms attached, freely lying on the
seabed, digging in, hoving over the seabed and swimming above the very seabed.
Benthic algae comprise green, blue-green, brown, red and two kinds of flower
algae—Zostera and Ruppia. In all, there are 304 benthic species, and the most
widespread of these are Phyllophora and Cystoseira. The former accounts roughly
for 95 %, the latter—for 4 % of the entire mass of seabed algae. Phyllophora is

concentrated in the northwestern part of the sea, Cystoseira is omnipresent, yet it is particularly numerous near the southern coasts of the Crimea. Its thicket is a favorite habitat of the tiny fishes of over 30 fish species. Benthic animals include diverse species of invertebrate: sandhoppers, polychaetes, crawfishes, ascidians, rhizopods, hydroids, actinias and mollusks. Of gastropods—rapa whelk, Nassa, Rissoa, Cekum, calyptraea chinensis and others. Of lamellate-branchiate—mussel, deep-sea scallop, venus, thapes and many other species of mollusks. Benthos also includes fishes living at the seabed: skates, plaice, surmullet, scorpionfish, lady-beetle, greater weever. The area of the Black Sea benthos-inhabited bottom equals 95,300 km^2, or roughly 23 % of all seabed area. There are no living organisms, except anaerobic bacteria, at the depth beyond 200 m. Two regularities of benthos distribution as the depth of the Black Sea increases manifest themselves: the lower boundary of benthos habitation in the Bosporus area going deeper due to permanent arrival of fresh water from the Sea of Marmara and elevation of this boundary in the northwestern part resulting from the rise of deep waters having a low oxygen content. On the northwestern shelf of the sea, the lower boundary of benthos habitation averages 130 m, near the Caucasus coast—at 140 m, and in the Bosporus area at 200 m.

Black Sea bibliography – is a compendium of information on literature published in 1974–1994 (volume of 364 pages) compiled in the framework of the Black Sea Environmental Programme (BSEP) under the aegis of the Cooperative Marine Science Program for the Black Sea, CoMSBLACK) funded by Intergovernmental Oceanographic Commission (IOC) of UNESCO. The Bibliography includes works on Black Sea hydrology, geology, chemistry and biology, carried out not only by Black Sea countries, but also by USA, Germany, Sweden, Italy, Great Britain and other countries. The Bibliography was published by UNDP in USA in 1995 as the first volume in the series "Black Sea Environment" edited by V.O. Mamaev (Russia), D.G. Aubrey (USA), V.N. Eremeev (Ukraine).

Black Sea Branch of Lomonosov Moscow State University – established by decision of the MSU Academic Board of March 29, 1999 with the concurrence of the supreme legislative and executive authorities of the Russian Federation and Ukraine in Sevastopol City, supported by Moscow Government and Black Sea Navy. The concept of establishing the MSU Black Sea Branch is based on the understanding of the need for further improvement of Russian-Ukrainian relations, reestablishing the uniform cultural-and-information field in the entire post-Soviet space. The primary tasks and areas of the branch activity are: training subject-matter experts of a range of professional-education and qualification levels (natural sciences, computer math, economics and management, and other fields), educa-tional activity including teaching, methodology, culture-education and upbringing activity; carrying out fundamental, applied and exploratory research. The branch's academic and administrative building is housed in the renovated historic structure of the former Lazarev Barracks.

Black Sea Branch of Lomonosov Moscow State University (http://www.hist.msu.ru/Photo/sev1.jpg)

"Black Sea—Caspian Sea" Canal – project that since the end of the nineteenth century has been toyed with on and off as one of the options to stabilize the level of the Caspian Sea. In a most general form, the project is a canal taking off the Black Sea water in an area north of Novorossiisk, run along the eastern coast of the Sea of Azov, on along Manych Depression toward Kizlar Bay of the Caspian Sea. The existing difference between the levels of the Black and Caspian Seas—around 28 m and congenial topography dividing their territories makes it possible to supply Black Sea water to the Caspian Sea by gravity flow. However, implementation of the project may have unwanted consequences to the Caspian Sea fauna.

Black Sea—Caucasus monsoon – Black Sea monsoon—in winter, prevalent northeasterly wind, in summer—winds of westerly directions in the segment of the Caucasus Black Sea coast. Seasonal change of winds is particularly pronounced on the coast between Tuapse and Batumi. In the passes and in gorges during the winter B.S.C.M. becomes similar to bora or foehn.

Black Sea coasts – are distinguished by great diversity and at the same time are only slightly dissected. There is only one large peninsula here—Crimean, and several small bays round its periphery (Karkanitsky, Kalamitsky, Feodosiisky), and on the western coast—Burgas Bay. Generally speaking, the Black Sea is mostly characterized by abrasion shores. This notwithstanding, the eastern and southern parts of the sea, featuring recent mountainous structures of alpine folding, are dominated by high mountainous abrasion shores, in the western part, where hard blocks of the margin of the ancient Russian platform and Baikal folding play the main role, there are level accumulative and abrasion-accumulative shores.

Conducive to the development of abrasion processes in the eastern part of the sea is also the extremely narrow shelf.

The north-western sea coast from the Danube Delta to Sevastopol harbor is not high. Here, the sea is approached by East-European Plain whose heights do not exceed 10 m in the south and increase to 40–50 m in the north. The plain is dissected by balkas, in places is terminated at the sea by precipices and invariable sand strips of dryland—bay-bars, separating large salt lakes and lagoons from the sea. Lagoons are particularly numerous near Odessa. Some of the lagoons are isolated from the sea perfectly, others occasionally communicate with it. Lagoons that were formed in the mouth of plentiful rivers (the Dnieper, Yuzhny Bug, Dniestr) have a permanent access to the sea. Nearly all largest lagoons (Dnieprovsko-Bugsky, Dniestrovsky, Khadjibeisky, Kuyalnitsky, Berezansky) are shallow-water; several hundred years ago, they used to be bays of 20–30 m in depth.

To the east of Sevastopol harbor, the coasts get distinctly higher. Folded structures of the Crimean Mountains stretch from Fiolent Cape to Feodosoya, along the entire SCC, first as three, then two parallel ridges, terminating as almost upright cliffs overlooking the sea. In places, from Cape Sarych to Yalta, the mountains retreat from the coast a little, their slopes become more gentle. Farther to the east, the greater range of the Crimean Mountains retreats from the coastline and gradually decreases in height, yet mountain slopes at the very shore are precipitous here, too. The coasts of Kerch Peninsula are steep almost its entire length.

The north-eastern coast of the Black Sea from Anapa to Sukhumi is predominantly high. Here, the folded spurs of the Greater Caucasus Mountain range come close, in places right up to the sea, forming upright cliffs. Terraces are outlined distinctly, in places. The mountains reach the maximum height near Sochi (up to 3,000 m), then the height is gradually reduced (to 1,000 m), and around the Kodori River the mountains get distanced from the coastline considerably. There is the large accumulative Kolkhida Lowland between the Kodori River mouth and Kobuleti town at the sea. To the south of the Rioni River mouth near to the seashore is the large Paleostomi Lake that used to be a sea bay. South of Kobuleti, the coast become mountainous again, and in Batumi area the height of some ranges exceeds 1,500 m. In the southern part of the coast, mountain spurs, often terraced, are terminated as steep ledges overlooking the sea. The southern part of the coast is dominated by accumulative ledges (Pitsunda, Sukhumsky, Adlersky, Burun-Tabiisky); these are, as a rule, formed by river deposits and are the result of complex evolution. At present, the Caucasian Black Sea coasts are subject to washout over the greater part of their length.

The southern coasts of the Black Sea are steep and upright, formed by high northern, often terraced slopes of the Pontic Mountains, strung along the coastline. Westward, the mountains gradually decrease in height, and near Bosporus Strait maximum height is 300 m. Nearly all the way, the shores are abrasion and abrasion-washout (in the east), with upright rock cliffs. It is only in small bights that sand-pebble "pocket" beaches are encountered. The main segments of accumulation are confined to the mouth of major rivers: Kizilirmak (length 1,350 km), Sakarya (824 km) and Yesilirmak (418 km). Due to flashflood debris cones of rivers in their mouths there are formed rather significant deltas that almost reach the margin of the narrow shelf. Under the impact of a strong north-western wave, the alluvial material is partly diverted eastward, which is manifest in the morphology of the deltas, above all in the formation of flank bars this side (the deltas of the Kizilirmak and Yesilirmak Rivers). West of Bosporus Strait, the shore is relatively low. Here, from Kaliakra Cape, folded structure of the Balkan Mountains (jutting out into the sea in this segment of the Black Sea coast), upright, adjoin the sea.

The western and north-western coast of the Black Sea is more low-lying, not infrequently, slightly hilly plain of varying genesis (alluvial, sea and alluvial-sea) come up to the shore. The delta of the Danube, the largest river of Western Europe is here; the delta has a complex structure. The main runoff of the Danube is currently channeled via the northern Chilia arm. Therefore, in the southern part of the delta accretion of the shore has been somewhat decelerated, in some places the shore is even washed out. Intensive use of the Danube water for irrigation by five countries through whose territories the river runs, augments the washout trend in the southern part of the delta.

Geomorphologically, the Black Sea shores throughout their length are grouped with leveled up, complex shores. These are typically characterized by alternation of accumulative and abrasion segments. Lagoon-type and abrasion-landslide shores are most widespread. The processes of abrasion are characteristic of the seashore in general and are complex. Loess and clayey shores are subject to caving in of large masses of rock, followed by shore erosion by the sea. The wall of a cliff (abrasion precipice, formed by the impact of the surf) in such locations is almost invariably upright and reaches a considerable height. Where the cave-in foundations exhibit outcrops of Pontic limestones and Meothic clays (abrasion terrace), intensive land-slide phenomena develop. Maximum number of landslides that damaged the city itself is within the segment between Dniestr Lagoon and Odessa. Landslides used to be common throughout the Caucasian Coast until coast-protection works were built.

Black Sea climate – the western and northwestern air transfer is an essential factor in formation of the climate over B.S. In winter when the anticyclone gets established over Eastern Europe the cold air masses move to the sea from the northeast. In addition, due to the active cyclonic activity over the Mediterranean Sea the warmer air drifts from the southwest to B.S. The sea itself sometimes becomes a zone of regeneration of the cyclones that came here from other regions or a zone of origin of new cyclones. By the geographical location and effect of the surrounding land we may distinguish four climatic areas in B.S.: northwestern, eastern, southeastern and southwestern. The northwestern area of the sea is the coldest. The mean annual air temperature here is about 10 °C. The winter lasts for 3 months during which snowfalls may occur and strong northeastern and north-western winds are blowing. At intrusion of cold air from the continent the temper-ature may even drop below zero (to −15 °C or −20 °C). The transition from winter to summer is rather quick despite of the fact that in early spring the weather is rather cold and windy. Beginning from April the cloudiness and air humidity tend to decrease and rainfalls occur quite seldom. At the same time the breeze circulation is formed. The summer is sunny, warm, with short rainfalls that sometimes occur during early night. The periods of calm sunny days are most frequent in August and

September and in some years even in October. Only in November when the cloudiness grows, rains become longer and winds stronger the weather turns to winter that usually starts in December.

The eastern area of the sea may be divided into two parts—northern and southern. In winter the northern subarea is more open for invasion of the cold air, thus, the climate here in this season is colder than in the southern subarea that is protected by the Caucasian range, thus, the winter here is very mild. Therefore, in the Novorossiisk area the air temperature in winter may be below −10 °C, while in the south the days with negative temperatures are rare. The summer is hot and sunny, and rainfalls are rather short. Of great interest are the climatic peculiarities of the southeastern area of B.S. Here the air is saturated with moisture. Considerable cloudiness, copious rainfalls and frequent dense mists are witnessed here. Rains fall rather evenly throughout a year. Their annual sum exceeds 1,000 mm, in other places even 2,000 mm. Regardless of the practically even distribution of precipitations within a year their sum here in the cold season is also greater than in the warm season. In autumn and winter rainfalls occur every other day, so there about 150–160 rainy days in a year. The summer is hot with the relative air humidity over 80 % which is conducive to formation of the humid tropical weather here.

The southwestern area of the sea covers the southern part of the Bulgarian water area and the adjoining water area of Turkey. Unlike the southeastern area the air humidity here is not so high due to relatively small cloudiness and the low sum of annual precipitations. However, in the southwestern area the climate in winter is also mild and negative temperatures are observed only in rare cases. The summer is sunny and hot, but not humid and the amount of precipitations is less than in the southeast.

Black Sea Cossack troops – irregular force in Russia. Established in 1787, processed by the Tzar decree in 1788 under the name of "Troops of Loyal Black Sea Cossacks". Deployed (1787–1792) on the Black Sea coast between the rivers Dniester and South Boug, subsequently in Kuban area. Took part in the Russian-Turkish war of 1787–1791. In 1792, Black Sea Cossacks forwarded a petition to the name of Catherine II, Empress of Russia regarding the allocation of Taman Land Areas "with adjoining environs" for the purpose of settlement with a view to introducing eternal and hereditary possession of the lands. В 1792–1793 гг. by Imperial decree the force was given for subsequent settlement Fanagoria and land areas between the Kuban River and the Sea of Azov (from 25,000, 12,000 were Cossacks) to perform border-guard duty. In the North Caucasus, the Cossack force built the Black Sea cordon line and founded 40 Cossack villages, of which 38 were given the traditional names of Zaporizhye Sich kurens (Kuren—unit of a Cossack troop). The center was the city of Ekaterinodar (currently, Krasnodar). In mid-nineteenth century, the force comprised 12 cavalry regiments, 9 infantry (infantrymen) battalions, 2 guard squadrons, 3 batteries and 1 cavalry-artillery company. 8,000 men were in the active military service. In 1798, a decree was issued regarding the deleting of the word "Loyal" from the force name "Troops of Loyal Black Sea Cossacks". In 1802, a provision on B.S.C.T. was passed. The troops performed with distinction in Sevastopol defense of 1854–1855, in Russian-Turkish wars. In 1860, together with part of the Caucasus combat troop, constituted complement of the Kuban Cossack Troops.

Black Sea currents – the general Black Sea circulation is driven by wind stress and buoyancy fluxes. A permanent feature of the upper layer circulation is the Rim Current, encircling the entire Black Sea and forming a large-scale cyclonic gyre, which transports water round the perimeter of the Black Sea. Direct observations of the current velocity from surface buoys suggest that the maximum speed of this stream is usually 40–50 cm/s increasing sometimes up to 80–100 cm/s. The Rim Current is concentrated above the shallow pycnocline and the volume transport of the current is estimated to be 3–4 Sv. The Rim Current is the most intense in winter-spring seasons. Winter circulation shows two gyres but in spring the current jet belts the whole basin along the bottom slope. Summer circulation attenuates significantly and in the autumn season the Rim Current usually breaks on the set of eddies.

In the central part of the Black Sea there are two smaller cyclonic gyres, occupying the eastern and western parts of the sea. Water dynamics there is very low. Outside the Rim Current, closer to the coast, numerous quasi-permanent coastal eddies are formed as a result of interaction of the Rim Current with the coastline and "wind curl" mechanism. The Black Sea circulation also manifests significant seasonal and inter-annual variability, which is controlled by atmospheric and fluvial variations.

Major features of mesoscale variability such as meandering of the Rim Current, mesoscale and smallscale anticyclonic and cyclonic eddies, filaments and jets, dipoles and tripoles, coastal upwellings could be found in the Black Sea.

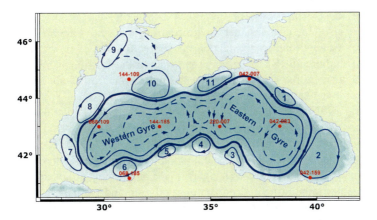

Basin-scale and mesoscale circulation in the Black Sea. *Thick line* shows the mean position of the Rim Current, *circles* illustrate places of the frequent observations of coastal anticyclonic eddies; *red points* and *numbers* mark locations of cross-over points of TOPEX/Poseidon satellite tracks and corresponding numbers of these tracks. (Source: Ginzburg A.I., Kostianoy A.G., Sheremet N. A., Lebedev S.A. Satellite altimetry applications in the Black Sea.—In: Coastal Altimetry, (Eds.) S. Vignudelli, A.G. Kostianoy, P. Cipollini, J. Benveniste, Springer-Verlag, Berlin, Heidelberg, 2011, P.367–387).

"Black Sea Day" – on the initiative of the Coordinating Center of the Black Sea Environmental Program the 31st of October was declared the B.S.D. On this very day in 1996 the ministers on environment conservation from six Black Sea countries signed the Strategic Action Plan for the Rehabilitation and Protection of the Black Sea. This Day is usually celebrated in the countries by organizing various campaigns on gathering and removal of wastes from the Black Sea beaches, actions on extension of the population knowledge on importance of keeping clean the sea and its nearby territories, etc.

Black Sea District – was established after the Caucasus war (1817–1864), was subordinated to the Kuban Chieftain assigned by order. Subsequently transformed into a province.

Black Sea dumping center – established in Odessa in 1991in view of special requirements to environmental protection while carrying out dredging operations and particularly when searching for dumping sites and disposing of dredged materials. The Center develops methods of the dredged soils: their secondary industrial utilization for consolidating the washed out areas of the shores, for construction and agricultural needs.

Black Sea Environmental Programme (BSEP) – established in September of 1993. The program is financed by the Global Environment Facility (GEF), additional resources on the basis of share participation come from the programs of EC (Program of Technical Assistance to East European Countries and Programs of Technical Assistance to CIS countries) as well a on a bilateral basis of the governments of Canada, the Netherlands, Switzerland and France. The program coordination center is in Istanbul (Turkey). The main concept of the program is to determine the existing state of the Black Sea ecological system, identify the principal causes for changes that have occurred in it and signpost the ways of improving the environmental situation. That was a stage of new quality in the process of studying the Black Sea environment because previously there was no so close and businesslike contact between the representatives of all six Black Sea countries in the area of environment.

By early 1994 there was set up in Istanbul a coordination center of the program and national coordinators were assigned (in some cases, those were ministers of nature conservation or their deputies), a working plan was agreed. The program has three main objectives: consolidate and build up the regional potential of Black Sea ecosystem management; develop and pursue the appropriate policy and legal basis for appraising, monitoring and preventing pollution and for conserving and promoting biodiversity; assist obtaining reliable investments in environment. Each of these tasks encompasses a huge sphere of activity and suggests the participation of a wide range of organizations. The BSEP steering committee includes national coordinators, representatives of the donor organizations and of non-governmental organizations. It was decided to set up a number of working groups based on the centers of activity that are national institutions, already having the basic infrastructure and personnel to coordinate work on specific tasks in the region. The government of each country agreed to accept on of such centers. The working groups themselves have at least one expert from each Black Sea country, and is free to engage more experts as required. These working groups that usually meet twice a year are oriented to practical work.

Centers of activity and their working groups: working group for fast response to emergencies is stationed in Varna (Bulgaria), for monitoring various types of pollution—in Odessa (Ukraine) and Istanbul (Turkey), for coastal zone management—in Krasnodar (Russia), for biodiversity conservation—in Batumi (Georgia), for fishing and living marine resources—in Constanta (Romania). The groups held regular working meetings at which the participants discussed materials obtained and published in various countries. This way, an interchange of information was arranged, on the strength of which experts were able to gain a general idea of the state of the Black Sea environment. Besides, there are three working groups with the headquarters at the coordination center of the program in Istanbul: data management and Geographic Information Systems (GIS); advisory board to agree on criteria of environmental quality, standards, legislation; group of experts on environmental economics.

In June of 1996, a group of experts from 14 counties drew up a resulting document—Transboundary Diagnostic Analysis (TDA). That was a comprehensive scientific evaluation of environmental problems typical of the present-day Black Sea, their causes and the steps that should be made to remedy the situation. TDA was not a political document. It is the result of almost 3 years of thorough investigations carried out by scientists who cooperated within the framework of BSEP. TDA made it possible to draw up the Strategic Action Plan for the Rehabilitation and Protection of the Black Sea. This plan signed on October 31, 1996 in Istanbul by the representatives of six Black Sea countries, namely ministers for environmental affairs, became a landmark document in which the governments of the Black Sea countries in association with broad international community undertook to adopt a pragmatic program of action based on common goals, meaning the rehabilitation and protection of the Black Sea. For the first time ever in its history the Black Sea was issued such "writ of protection". However, the Strategic Plan says that it may remain a declaration of good intents unless further action in the same vein is taken: the drawing up of Strategic Action Plans in each country and consistent implementation the reof. Including, inter alia, the involvement of broad layers of the population to handle specific tasks that may be discharged more effectively by groups of conscientious citizens and by all society. The latter, however, calls for clear understanding by all of the existing situation and willingness to rectify it. Here, were enter the domain of ecological education, ecological upbringing and ecological ethics.

Black Sea environmental series – a series of publications issued in 1993 at the request of the governments of Bulgaria, Georgia, Romania, Russia, Turkey and Ukraine. Includes studies accomplished within the framework of the Black Sea Environmental Programme. The first work published in the series was "Black Sea

Bibliography (1974–1994)", edited by V.O. Mamaev, D.G. Aubrey and V.N. Eremeev. This was followed by 5 other volumes.

Black Sea Experimental Research Station (CHENIS) – established in January 1, 1946 by transforming the Gelendzhik Coastal-Marine Expedition as an experimental oceanological station of the Institute of Oceanology of the USSR Academy of Sciences in Blue (Rybatskaya) Bay near Gelendzhik, Russia. The station received one of the first research vessels of the Institute of Oceanology "Forel" ("Trout") that had been made available by Romania. By the early 1960s, the station focused on the main research topics: first—dynamics of the coasts, study of marine borers and foulings, hydrochemistry, hydrology, seismo-acoustics and marine electronics, and subsequently—geomorphology and lithology, geochemistry of marine sediments, hydrooptics. The station became the main base of the institute for testing new samples of instruments and equipment, for making experimental and methodological studies. In 1967, under the decree of the Presidium of the USSR Academy of Sciences, CHENIS was transformed into the Southern Branch of the P.P. Shirshov Institute of Oceanology of the USSR Academy of Sciences.

Black Sea Fleet agreement – was signed on 9 June 1995 in Sochi. It stipulates creation on the basis of B.S.F. the Russian B.S. Fleet and the Ukrainian Navy with separate basing and deployment of the main Russian B.S.F. base with its headquarters in Sevastopol, transfer to the Russian Federation 81.7 % and Ukraine 18.3 % of ships and vessels from the Russian B.S.F.

Black Sea Fleet sinking – on April 23, 1918 in the face of a threat of the Crimea seizure by the German troops the RSFSR Council of People's Commissars (CPC) issued an order on relocation of the Black Sea fleet from Sevastopol to Novorossiysk. The Central Committee of the Black Sea Navy that was vested full power from early January 1918 decided to fulfill the order of RSFSR CPC. On April 29–30 Sevastopol was left by 2 battleships, 14 destroyers, 2 torpedo boats, 1 auxiliary cruiser and 10 patrol ships making the battle core of the fleet (in total about 3,500 men of the crew). On May 1–2 they joined in Novorossiysk. In Sevastopol the German troops seized old battleships, cruisers, submarines, some destroyers and other ships that were mostly inoperative. On May 11 the German Commander-in-chief in the East front laid down an ultimatum demanding from RSFSR CPC the return of ships to Sevastopol. In order to keep the Brest Peace Treaty in force RSFSR CPC had to agree and the People's Committee on Foreign Affairs sent the respective notes on May 13 and June 9. However, not willing to give the ships to the enemies RSFSR CPC decided to sink them about which a respective order was given (the directive signed by V.I. Lenin on May 28). On June 18 the battleship "Svobodnaya Rossia" ("Free Russia"), 6 destroyers and 2 torpedo boats were sunk in the Novorossiysk Bay and one more destroyer was sunk on June 19 in Tuapse. Eight patrol boats were transported via railroad to Tsaritsyn and they made the core of the future Volga Navy. From December 12, 1917 to June 4, 1918 the Black Sea fleet was under command of Admiral M.P. Sablin.

Black Sea International Shipowners Association (BINSA) – was founded in 1993 as a nongovernmental nonprofit association of shipping companies in the Black Sea, Azov, Caspian, Dnieper and Volga basins. Its purpose is to coordinate activities of the BINSA members in elaboration of the shipping policy focused on intensification and improved competitiveness of carriages, rendering assistance to the Association members in addressing the economic, financial, commercial, legal and other issues. Among the tasks of this Association there are: consolidation of the principles of free international shipping, development and strengthening of cooperation with international organizations and associations, development of merchant shipping and enhancement of its security, exchange of information, legal practices, cooperation in environmental safety and protection of the rights and interests of the BINSA members.

Black Sea jack mackerel (*Trachurus mediterraneus ponticus*) – fish of the *Carangidae* family. Reaches 26–52 cm in length. Common in the Black and Azov Seas. Reaches sexual maturity in a matter of 2 years. Lives up to the age of 14 years. Represented by two forms: big and small. The former grows fast, reaches large size, lives longer, dwells primarily in the eastern Black Sea. The small form shapes up two schools: eastern and western. Heat-loving fish, most active during the warm months (May to September), when it is common in nearly the whole of the western Black Sea, eats, grows and propagates intensively. Until late June is distributed along the western coast of the Black Sea, most fishes stay longer in the northwestern part of the sea. Early in September, jack mackerel forms schools that little by little move toward the warmest sea areas (near the southern Crimea, Caucasian and Anatolian coasts, in Bosporus and the Sea of Marmara). Jack mackerel is of primary importance to the Black Sea fishing.

Black Sea level – year-to-year variations of the Black Sea water level are determined by the correlation between the major components of the water balance: river runoff, precipitation, evaporation and water exchange through Bosporus and Kerch Strait. In the input part, the role of river runoff is most significant, besides, its year-to-year variability is rather high. However, in the Black Sea variations in the input of river waters are substantially compensated in the output part of the water exchange through Bosporus. For this reason, as a result (unlike the closed Caspian Sea), the long-term fluctuations of the Black Sea level are low, and during 1923–1995 did not exceed 15 cm, which was the positive interannual water level trend of the period. The well-defined seasonal water level variations are mainly produced by year-to-year differences of the arriving river runoff, atmospheric precipitation and evaporation from the sea surface. Under the impact of these factors, the sea level during the warm period of the year is higher, during the cold period—lower. The value of these fluctuations is most significant (30–40 cm) where the effect of continental runoff can be felt. The highest water level fluctuations in the Black Sea are those produced by wind-induced positive and negative surges associated with the impact of steady winds. These are especially frequent during the autumn-winter time in the western and northwestern parts of the sea, where these are likely to be over 1 m. In the west, strong positive surges are caused by easterly and northeasterly winds, in the north-west—by southeasterly winds. Strong negative surges in the said parts of the sea are produced by northwesterly winds. Near the Crimean and Caucasus caosts, positive and negative surges hardly ever exceed 30–40 cm. Their duration usually equals 3–5 days, sometimes it may be longer. Seasonal steric water level variations should also be mentioned: these are produced by seasonal variations of temperature and salinity in the sea column. The range of steric fluctuations of the sea level in some areas may reach 20 cm, i.e. significant value in the formation of the general Black Sea level.

Black Sea, names – emerged in hoary antiquity. The earliest of the known names is Temarun as well as the local Scythian—Akhshena, Akhshaina (ancient Iranian "akshaina"—"dark, black"). The ancient Greeks in the first century B.C. rethought this name when they began developing the sea. As a result, they arrived at a Greek name that sounded similarly: Akseinos, Aksin, Aksinos, Aksinsky Pont ("non hospitable sea"), when translated—in the form of "Pontos Melos" ("Black Sea"). Subsequently, as Greek colonization of the sea coasts spread further, the sea began to be called Euxin, Pont Euxeinos, Pont Euxinian—"hospitable sea". During the early centuries A.D., the name of the Scythian Sea is encountered, although the Scyths themselves called it "Tana" ("dark"). During the middle ages, the names Maure Talasa, Fanar-Kara-Deniz (Tukish "fanar"—"evil", "kara"—"black north", "deniz"—"sea"). One of the ancient Arabic names "Naitas" Sea (940). This name appears on Italian gloves and is encountered in numerous written sources of the thirteenth to fifteenth centuries. This name "mar Mazor" appears on the pages of the essays by Barbaro and Contarini.

Kagan Iosif used to call the Black Sea "Kustandina, Kustantinia". It is important to note that the name "Black Sea"came to be used in Constantinople and on the coast of Asia Minor concurrent with the growth of Italian commerce in the capital of the Byzanthian Empire. The emergence of the name "Black Sea"—in Greek—may be put to phonetic affinity in the sounds of the words "mare maius" (*maris maioris,* etc.), "mare Maggiore"—"Great Sea" with the Greek "mavros"—"black". In the "Travelogue" of Ignatius Smolnyanin (1389–1405), the Black Sea is also referred to as Great. In the Russian chronicles, from the "Tale of Bygone Years" the name "Pont'skoe", "Ponetskoe" Sea is encountered. Prior to the fifteenth to sixteenth centuries the name Russian Sea, Sidatskoe, Surozhskoe, Sugdeiskoe, Sugdep (after the name of the city Sugdep—at present, Sudak) the city through which the Great Silk Way was laid. As far as the name Surozhskoe is concerned, a provision must be made: either the name refers to the Sea of Azov, because the chronicle says that the Prince Mikhail Yaroslavich of Tver set out for the Orda to see Khan Uzbek in 1319, "having reached the Orda... in the mouth of the Don River, which flows into the Surozh Sea"; or this refers to the Black Sea, not the Sea of Azov, because, first, Surozh was on the coast of the Black not Azov Sea, and, second, Kerch Strait more often than not was mistaken for the Don mouth, whereas the Sea of Azov was regarded as a widening of the lower Don.

Roughly from the tenth to the twelfth century, following the Russian incursions on Tsarigrad (at present, Istanbul), "because Russians used to trample the city often" the sea was referred to as "Russian Sea". The name "Terrible Sea" was also encountered. In the thirteenth century the name "Hazar Sea" was used, too, by the name of the people who dwelled on its shores. In the fifteenth century, in the Arabic essay of Ibn Arabshah the Black Sea is unexpectedly referred to as Egyptian Sea: "the boundary of Desht land from the south—Kolzum (Khoresm, Caspian) sea, wicked and wayward, and the Egyptian Sea that came to them (Desht dwellers) from the Romanian (Byzanthian) area. The two sea almost collide, but for the Cherkess Mountains between them ...".

The origin of the sea name "Black" is related to the common practice of designating the cardinals by color, whereby black color denoted north, i.e. Black Sea—"northern sea", which correlates well with the name of the Red Sea—"southern sea". However, there exist some legends concerning the subject. According to one of these, Turkish, the name has to do with the fact that a giant's sword. As the sea tries to rid itself of the sword, it gets agitated and becomes black. There are some other hypotheses, too. One of these is associated with violent storms, when the water gets dark. Hydrologists relate the name of the Black Sea with the fact that metal objects (e.g. anchors) lowered to a certain depth turn black under the impact of hydrogen sulfide present in the depths of the sea. According to another hypothesis, after the storms there remains black silt on the shores—hence, the name.

Black Sea Naval Cooperation Task Group (BLACKSEAFOR) Agreement –
Six Black Sea countries, Turkey, Russia, Ukraine, Romania, Bulgaria and Georgia signed the Black Sea Naval Cooperation Task Group (BLACKSEAFOR) Agreement on 2 April 2001 in Istanbul. For the first time in the international practice the group of states from one region established on the basis of their naval forces the multinational group with flexible functions intended for use in emergency situations on the Black Sea solely for civil needs. Its main goals are search and rescue operations for humanitarian needs, cleaning sea mines, joint action on protecting the Black Sea environment, organizing joint exercises, good will visits and other tasks provided they are approved by all member-states. But BLACKSEAFOR can also cooperate with international organizations like the U.N. and the Organization for Security and Co-operation in Europe (OSCE) for implementation of the above tasks and taking part in peacekeeping operations if these organizations apply to the Black Sea states. The area of action of BLACKSEAFOR is the Black Sea, but this agreement envisages a possibility to go beyond this area upon consent of all member-states. The group shall comprise 4–6 vessels (one from each state) and convened whenever necessary, but at least once a year on the basis of the approved plans on implementation of concrete tasks. Any decisions concerning operational interaction of BLACKSEAFOR shall be taken on the consensus basis. The committee consisting of naval commanders of the Black Sea states (for Russia—the Commander of the Black Sea Fleet) undertakes general guidance of the affairs. The chairmanship of this committee is elected by member-states and is subject to annual rotation. The chairing state will appoint the commander of the group for a term of

1 year. The group activities are not aimed against any third state and are not targeted to creation of a military alliance against any state or a group of states.

"Black Sea Navy Day" – celebrated on May 13. On May 2(13), 1783 eleven ships of the Azov fleet entered the Akhtiar Bay (today Sevastopol) of the Black Sea. Together with the Dnieper fleet that soon joined it they made the core of the Black Sea Navy.

Black Sea Region – the notion widely used, following lapse of the Agreement on the establishment of the USSR, until now has no strict definition providing an exhaustive characterization of then region's territory. Normally, in its broad geopolitical sense it includes the countries sited on the coasts of the Black Sea: Russia, Ukraine, Romania, Bulgaria, Turkey, Georgia, and Abkhazia. A narrow interpretation is likely to confine the region to the boundaries of administrative units overlooking the Black Sea. The Black Sea is a region of historic rivalry and historic cooperation. At different times, it was a region of antagonism between the local states and a zone of their mutual trade relations. Since 1991, the current political map of the B.S.R. has taken shape; the region is closely related to the European Union, has become the area of geopolitical interests of many countries. The disintegration of the Soviet Union, emergence of globalism as a new system of economic management worldwide, fundamental structural changes of international relations being restructured radically changed the region's status in the world. The shaping up of the B.S.R. is undergoing the initial stage. Its countries are in the process of developing their statehood, transforming and restructuring the rudiments of their economies, democratizing public life. All these processes are being translated into reality at varying rates, are based on different principles, which preconditions their existing socioeconomic situation. The countries of the B.S.R that gained independence are recognized internationally, have formed each its own system of mutual ties and relations with the world at large in the spheres of policy, economy and other spheres. The sovereignization of New Independent States (NIS) has changed fundamentally the regional situation in the post-Soviet space and set off the restructuring of international relations, thereby marking the arrival of a new stage of historic development of the B.S.R. The NIS as well as Bulgaria and Romania began forging their regional policy. The geopolitical vacuum was not filled immediately, however. Oil and gas turned out to be the only unifying idea that attracted the attention of each local state to the region; besides, oil and gas proved to be the crucial economic tool in the struggle against the instability of the geopolitical situation. The B.S.R. is the zone of investment interests and at the same time—one of economic and political differences of the countries within and outside of the region. The prospects of B.S.R. growth are in many ways related to the enhancement of relations between the local states, their cooperation in joint handling of problems that are of mutual interest. Such problems as economic, environmental, demographic, political are rather numerous. To Russia and Ukraine, the B.S.R. is one of the foreign-policy priorities. This is a region of their traditional interests, extremely important to economic growth of their southern areas. The trend of rapprochement between the Mediterranean area and the Black Sea area has received a new stimulus over the last few years as a result of new emphases in the stances of Black Sea Economic Cooperation and EU.

Black Sea Region Association of Shipbuilders and Ship Repairers (BRASS) – set up in October of 1993 as a non-governmental organization. 28 organizations took part in establishing the Association: shipbuilding and ship repair plants, scientific-research institutes, commercial organizations representing the countries of the Black Sea region—Bulgaria, Russia, Romania, Turkey and Ukraine. The Association's strategic task is protection of the interests of shipbuilders and shiprepairers of the countries of the Black Sea region. The Association addresses matters of mutual information support, coordination of the pricing policy of enterprises united by the association with a view to preventing dumping.

Black Sea Regional Activity Center for Environmental Aspects of Fisheries and other Marine Living Resources (AC FOMLR) – established in 1994 on the basis of the Grigore Antipa National Institute for Marine Research and Development, Constanta, Romania. The center coordinates and facilitates the required program support and provides practical technical support for the functioning of the appropriate team of advisers of the Black Sea Commission in the field of marine ecosystem protection and rehabilitation, especially for the conservation and sustainable use of marine bioresources.

Black Sea Regional Committee (BSRC) – it was established in 1995 at the Intergovernmental Oceanographic Commission (IOC) for coordination of research in the Black Sea.

Black Sea Regional Programme in Marine Sciences and Services, IOC – the program was drafted in consonance with IOC Resolution XVIII-17 in 1996, with initial period of 4 years. It is implemented subject to coordination by the Regional Black Sea Committee. The program participants are Bulgaria, Georgia, Romania, Russia, Turkey, Ukraine. Two projects are being implemented within the framework of the Program. The first project—Global Ocean Observing System for the Black Sea—measures aimed at setting up the system of observations and forecast. The object of the project is to improve and promote regional capabilities in the field of operational oceanography, including observations, forecasting and serving the interdisciplinary users; to elaborate a plan of research for the program, a plan of scientific research for the program Black Sea GOOS. The second project— appraisal of sediments arriving in the Black Sea, mechanisms of their formation, transformation and dispersion, importance to ecology. The purpose of the project— integrated study of sediment flows, their transformation in time and space, identification of sediment deposition for appraising the ecological condition of the Black Sea ecosystems; reconstruction of recent geological history as the basis for environmental forecasting.

Black Sea road – main auto road along the Black Sea coast. Runs from Novorossiysk via the cities of Gelendzhik, Tuapse, Sochi, Gagra, Gudauta, Sukhumi, Ochamchira, Kobuleti as far as Batumi. Length 800 km (out of this, in

Russia—380 km, in Abkhazia and Georgia—420 km). Built in 1887–1905, remodeled for auto traffic during 1934 to 1950. Over the last few years, much has been done to flatten particular road segments and broaden the carriage-way. Hotels and boarding-houses for tourists are available. At Samtredia, a motor-road for Tbilisi, Baku, and Erevan branches off. Access road to Ritsa Lake and Pitsunda are available.

Black Sea salmon (*Salmo trutta labrax*) – fish of the salmons family (*Salmonidae*), reaches 90 cm in length. The body is oblong, covered with small-size tight-fitting cycloid scale. The back part is greenish, with a silvery gleam, on the sides—small black spots. Anadromous fish, dwells largely in the rivers of the Caucasus coast, enters in the Black and Azov seas for fattening.

Black Sea shad (*Alosa caspia nordmanni*) – fish of the herrings family (*Clupeidae*), has radial grooves on opercles. There is a dark spot behind the upper edge of the opercle, behind which there may be 5–8 spot more. Reaches 18 cm, rarely—22 cm in length. Common in the western part of the Black Sea. In spring, enters rivers and some brackish lakes for propagation. Has no value as a food item.

Black Sea Shipbuilding Yard – in Nikolaev City (Ukraine). In 1895, a Belgian joint-stock company commenced the building of a shipyard in Nikolaev, which was subsequently called "Society of Shipyards, Mechanical and Foundry Plants". One of the oldest shipyards. The yard was officially commissioned on October 9, 1897 as "Naval" plant (translated from French, meaning "marine"). The plant was specialized in the manufacture of military ships and vessels, ship engines, mechanisms, boilers, ship equipment, canons and ship's artillery turrets, railway cars, bridges, cranes. The yard had shops, slipways, piers and workshops with most advanced equipment at the time, and was the foremost plant for the building of steamship metal fleet on the Black Sea. The world's first submarine mine-layer "Krab" (Crab) was built here, along with the main seaborne machinery of the iron-clad battleship "Potemkin". Alongside the main orders, the shipyard built barges, ship's boilers, cranes, tram cars, steam engines, railway bridges. In 1925, the first Soviet tanker "Krasny Nikolaev" was laid down at the shipyard, in 1941, the construction of the ice-breakers "I. Stalin" and "Krasin" was completed. During the Great Patriotic War (WWII), the shipyard was evacuated and turned out products for the front. In 1949, the first postwar vessel—tanker "Kazbek" was laid down. Unique vessels, like the whaling bases "Sovetskaya Ukraina" and "Sovetskaya Rossiya", vessels for scientific expeditions and research and fishing ships of the type of "Yu.M. Shokalsky", "Akademik Knipovich" were built at the shipyard. At present, the shipyard is busy building dry-cargo vessels, large refrigerator trawlers, container carrying ships. Also, the shipyard manufactures the newest seaborne machinery and devices, pleasure craft, ship furniture. The ship is decorated with the Order of the Red Banner of Labor (1926), 2 Orders of Lenin (1949, 1977), Order of the October Revolution (1970).

"Black Sea Shipping Company" – the oldest shipping company of the Russian Empire (then of the USSR) in the Black Sea established in 1833 in Odessa. After

World War II, due to the increase in traffic volume, established separate the Black, Azov and Georgian Shipping Companies. In 1964, a tanker compartment of the Black Sea Shipping Company was organized as the Novorossiysk Shipping Company. In 1990 the Black Sea Shipping Company was the largest in Europe and second in the world. It was composed of more than 300 vessels of various types with a total tonnage of 5 mln tons. The Company did the following: organizes optimum operation of vessels, marketing, business-plan, ship's agency service, freightage; commercial operation of vessels, examination and settlement of claims, suits, marine insurance, organization and supervision of ship repair, making ship repair records. Provides legal advice and handles cases in international marine practice of defending ship owners; ensuring the safety of ships' navigation; certification of ship navigators, mechanics, electrical officers according to their positions held; drawing up plans, projects, developments, issue of reports, reviews and expertise statements in the aforesaid spheres of marine business. Since 1991, with the collapse of the USSR, the Company belongs to Ukraine. The number of vessels decreased by 2006, more than 20 times. A total collapse of the Company is observed today.

Black Sea State Biosphere Reserve – established in 1927 in the northwestern part of the Black Sea in Egorlyksky and Tendrovsky bays (USSR, Ukraine) as well as on the adjoining land and islands with a view to protecting he wintering, migratory and nestling birds and their habitats. Sited in an area between Ochakov and Perekop Isthmus, where the Black Sea coast is deeply cut in, forming numerous large and small bays. Near the shore, sand islands of varying sizes and shapes are scattered. Total area is of 63.8 ha. Prohibited in the reserve are the shooting and capture of wildlife and birds, fishing, felling trees and shrubs, hay-making, cattle-grazing, collection of medicinal plants, mush-room picking, collecting flowers, seeds, fruit, driving vehicles and other types of human activity. The strict reserve regime provides for tranquility required for the animals all year round and especially during the propagation time. The main task of the reserve is to conduct by own and invited scientists comprehensive research of the biology of animals and plants in a setting devoid of human impact, search species that may be of commercial value to people's economy, regular recording of the bird and animal population, birdbanding, studying bird migration, etc. Recorded in the reserve are over 280 species of nestling, migratory and vagrant birds. Here winter mute swan and wooper swan (population of up to 13,000 individuals), mallard duck (over 25,000 individuals), wigeon, European teal, pintail, shoveler, golden-eye, pochard and other birds. More then 100 species of birds nestle on the territory of the reserve, among them numerous insectivorous species and song birds as well as stork, heron, birds o prey—red-footed falcon, kestrel, small falcons, black kite, long-eared owl and ever rare erne. Nestling in grass are European partridge, quail, pheasant, introduced here in 1962 and great bustard that has become quite rare. Nestling on sandy islands are black-headed gull (up to 200,000 pairs) regarded he "symbol" of the reserve, terns, sand-pipers and many other birds.

There are 43 species of mammals in the reserve. These include 7 insectivorous mammals (hedge-hog, shrews), 9 species of bats, 15 species of rodents (sousliks, jerboa, mole rat, murine rodents), 8 species of predators (red fox, raccoon dog, introduced from Ussury Territory before the Great Patriotic War (1941–1945), steppe polecat, ermine, least weasel, marbled polecat , badger). Of the three species of hoofed mammals, sika deer ranks first. 20 deers were brought to the reserve in 1957 from Askania-Nova (the home of these animals is Primorsky Krai, Far-East of Russia). Also encountered are European roe deer and wild boar. The sea shoal within the reserve is the goby propagation ground, young gray mullet, plaice feeding site, plaice.

Black Sea straits – (in international law) the straits of Bosporus and Dardanelles linking the Mediterranean Sea and the Black Sea via the Sea of Marmara. Geographic position of B.S.S. determines paramount importance of their international regime to the security and economic interests of the Black Sea countries. Prior to the conclusion of the Kuchuk-Kainardzhi Peace Treaty of 1774 the regime of B.S.S. was only established by Turkey. After 1774, numerous treaties and agreements concerning the passage of foreign ships through B.S.S. were concluded (Russian-Turkish Treaties of 1799, 1805 and 1833, London Convention of 1841 and others). In 1936, the Convention on the B.S.S. regime was concluded in Montreux (Switzerland) which was signed by the USSR, Britain, Bulgaria, Greece, Turkey and other countries. This convention established free passage of merchant ships of all countries through B.S.S., procedure of passage of naval ships of the Black Sea and other countries; the convention repealed the resolutions of the Lausanne Convention of 1923 that banned fortifying B.S.S. by Turkey. The need to reconsider the 1936 Convention, whose major drawbacks had been revealed during WWII, was recognized by the Berlin (Potsdam) Convention of 1945, and was on more than one occasion stressed in the notes of the Soviet Government afterwards.

Black Sea Trade and Development Bank (BSTDB) – international non-governmental organization. Was established by 11 countries of Black Sea Economic Cooperation (BSEC): Albania, Armenia, Azerbaijan, Bulgaria, Georgia, Greece, Moldova, Romania, Russia, Turkey and Uklraine pursuant to the resolution of the 1st summit of the BSEC foreign ministers. The bank establishment agreement was signed in 1994. The priority areas of the BSTDB as a crucial tool for promoting multilateral cooperation in the region are: financing the projects that are of mutual interest to the member-countries, in particular, in the field of energy, transport and industry as well as financing the trade operations. The main office of BSTDB is in Thessaloniki, Greece. Greece, Russia and Turkey each hold 16.5 % of the shares, 13.5 % belong to Bulgaria, Romania and Ukraine, 2 %—to Azerbaijan, Albania, Armenia, Georgia and Moldova. The bank establishment agreement was ratified by all 11 countries-participants of BSTDB in 1997 and took effect the same year. As of end of 2011, BSTDB cumulative portofolio in its 11 memeber countries has reached 265 approved operations in the key sectors of infrastructure, energy, transport, manufacturing, telecommunications, financial sector and other important areas exceeding EUR 2.5 billion.

Black Sea tsunami – The Black Sea is not an area of high seismic activity, but sometimes severe earthquakes do occur here, and these may be accompanied by tsunami waves. Earthquake-induced tsunami used to be observed almost everywhere across the Black Sea areas of the Crimea and Caucasus, the wave height registered at Novorossiisk and Sevastopol was around 0.5 m. Around the Crimean Peninsula severe earthquakes that apparently produced tsunami were observed on 11October 1869, 25 July 1875, 8 January 1902, 31 May 1908 and 26 December 1919. According to the information available, the earthquake of 1941 caused the flooding of part of the land on the southern coast of the Crimea. Systematic observations of Black Sea seismicity began after the Yalta earthquakes of 26 April and 12 September 1927: these until now are regarded the most severe registered Crimean earthquakes. Both produced tsunami. The data on tsunami caused by the earthquakes near the Caucasian and Turkish coasts are even more scarce than information on the Crimean tsunami. Still, the interesting data on the two tsunami that occurred near the Caucasus coasts did leak to the press. The first tsunami is dated the first century B.C. It destroyed the ancient city of Dioscuria that used to stand where the present Sukhumi is. The second tsunami was caused by Anapa earthquake on 4 October 1905. The waves thus produced in the sea tossed up a steamer easily. In all, they distinguish three earthquakes on the Caucasus coast that caused tsunami waves: Anapa earthquake of 4 October 1905, the earthquake in the eastern part of the Black Sea of 21 October 1905, Anapa earthquake of 12 July 1966. The devastating in its aftermath Turkish earthquake of 27 December 1939 that claimed 23,000 human lives produced tsunami which in some areas reached the height of 1 m. The earthquake epicenter was in the northeastern part of Turkey. The area in which the earth tremors could be felt was of nearly elliptical shape with axes of 1,300 and 600 km. The series of strong shocks continued until 2 January 1940.

Black Sea uprising of French Navy – uprising in April of 1919 on the French Naval Ships that participated in intervention against the Soviet State. The most significant revolutionary action in foreign troops during the period of the Allied Intervention in the Russian Civil War of 1918–1920.

Black Sea whiting (*Odontogadus merlangus euxinus*) – fish of the codfishes family (*Gadidae*). In length, reaches 40 cm, more often 15–20 cm. Common in the Black Sea, enters the Sea of Azov, too. Cold-water, schooling fish. Tends to stay in the shelf zone, closer to the bottom at the depth of 80–90 m. Avoids warm water layers. Lives roughly 10–14 years. Reaches sexual maturity at age 3–4 years. Propagates almost all year round.

Black Valley – stow, 40 km north of Perekop near which in May of 1736 a heavy battle was fought between the vanguard of the Russian army and the troops of the Crimean Khan during the Russian-Turkish war of 1735–1739. When Russian army's vanguard (4,500 men) approached B.V. on May 19 (30), it was unexpectedly attacked by the Tatar cavalry (20,000 men). Having formed an infantry square, the Russian troops for 6 h repelled the enemy attacks. When a 5,000 men force of the main forces of the Russian army arrived, the Khan's troops wasted no time to retreat to Perekop.

Blavayskiy Vladimir Dmitrievich (1899–1980) – Soviet archeologist, historian and fine art expert, founder of the classical archeology sector of the Institute of the History of Material Culture and the first chief of the sector, one of the organizers of works in submarine archeology, Doctor of History, Professor. In 1917, leaves as a gold medalist the 3rd Moscow Gymnasia (upper secondary school) and enters the Social Science Department of Moscow State University (MSU) from which he graduates in 1923. Works at the State Fine Arts Museum, at the State Academy of Art Sciences (subsequently, the Academy of Art Studies), takes part in a number of archeological expeditions. It is at that time that he publishes his first works, dealing mainly with Greek vase painting. The crucial events in the life of B. of this period of his life were the study at the post-graduate course (1925–1929) and participation in Olvia excavations (1925–1926). In 1929, B. defended a Ph.D. thesis on the subject of "Black-Figure Lecythuses Prior to the fifth century B.C. from the Hellenic Cities of the North Black Sea Area". During the 1930s, B. works at State Academy of History of Material Culture at first, as a part-time staff, then as a senior scientist. It is during these years that he takes up teaching: in 1933–1936, an assistant professor of the post-graduate course at the State History Museum, in 1935–1941—an assistant professor of the Institute of post-graduate studies at the Academy of Architecture, from 1939—assistant professor of the MSU Departments of History and Philology. In the 1930s, B. conducts his first self-sustained archeological expeditions: at Kharaks (1931, 1932, 1935), at Kamysh-Burun (1933), Pantikapei (1934), Fanagoria (1936–1940). His first books are also published during that period. He works hard at the subject he regarded the principal one in his activity—"Classical Sculpture Techniques". As the Great Patriotic War (WWII) begins, B. joins the people's volunteer corps. After a grave concussion, he returns to Moscow. In 1943, B. defends a doctoral thesis on the subject "Experience of studying the classical sculpture techniques". That same year, he becomes a professor of the MSU Archeology Chair. In 1944, B. is elected the chairman of the just established sector of classical archeology of Institute of History of Material Culture and a Corresponding Member of the Academy of Architecture. During subsequent years, he heads numerous expeditions that yielded huge new factual material for the study of classical Black Sea and Mediterranean Sea areas: 1945–1949, 1952–1958—studies of Pantikapei, 1950–1951—explorations on Kerch Peninsula, 1950–1954 гг.—explorations and excavations on Taman Peninsula, 1958–1960—Soviet-Albanian expedition.

In 1957, at the initiative of B. the first post-war submarine archeological expedition was organized at the Institute of Archeology of the USSR Academy of Sciences. The expedition worked under his leadership 11 field seasons on the coasts of the Black and Azov seas. All archeologists of the expedition took a training course and were taught how to use aqualungs. In 1957, they explored the seabed of Kerch Strait. In 1959, the first-ever USSR excavations on the Black Sea of the submerged part of the capital, so-called "Asian" Bosporus—Fanagoria town were carried out. The expedition also worked at Belgorod-Dniestrovsky—in ancient Tira, Olvia, Khersones, on the Southern Coast of the Crimea, in the vicinity of Taganrog. The procedure of explorations, facilities tie-up to the shore and subsea study thereof was worked through. Concurrent with this, his several monographs are published. In 1962, B. is elected a Corresponding Member of the Hungarian Society of Classical Studies, in 1964—member of the International Council for Submarine Archeology, in 1966—Corresponding Member of the German Archeological Institute and Honorable Member of Archeological Society of Yugoslavia, in 1975—member of the Honorable International Committee of the Society of Mythraic studies. In 1958, B. was elected chairman of the Association of classical history scientists of the USSR Academy of Sciences.

B. has numerous government awards. His main works include: "Architecture of the Classical World" (1939), "The Arts of the North Circum-Black Sea Area of the Classical Epoch" (1947), "The History of Classical Painted Ceramics" (1953), "Arable Agriculture in the Classical States of the North Circum-Black Sea Area" (1953), "Essays on Military Science in the Classical States of the North Circum-Black Sea Area" (1954), "Classical Archeology of the North Circum-Black Sea Area" (1961), " Discovery of the Sunk World" (in co-authorship with G.F. Koshelenko, 1963), "Pantikapei. Essays on Archeology of the Bosporus Capital" (1964), "Classical Field Archeology" (1967), "Nature and Classical Society" (1976), "Classical Archeology and History" (1985).

Blevaka – one of three mud volcanoes around Anapa. B. is sited on Rotten Mountain, so called because the ground is ripped apart and mud surges out onto the surface. The height of one such burble, in case of large gas accumulation, may reach 32 m. In traveler's guide-books, B. is named in a more cultured manner: Plevaka or mud volcano. The volcano gushes out a white fluid, fat to the feel, together with hydrogen sulfide and balneal mud. The fluid is called "white oil" used in cosmetics industry. Unlike lagoon mud, mud-cone mud is older. It is gray and contains more useful substances than lagoon mud does.

Blooming of the sea – phenomenon of mass growth of phytoplankton changing the color of sea water. Large zones (spots, stripes) of "blooming" water are observed usually in spring in the waters rich in nutrients, on the shelf, above the submerged uplands, in locations of rising deep waters, in areas of oceanic fronts, and in high latitudes—in ice-clearings and ice-holes among the ice floes. On the Black Sea, in the northwestern part of the sea the large amount of biogenic substances coming here with the continental runoff causes mass growth of plankton algae—"B." In the area impacted by the Danube runoff, biomass of phytoplankton has grown 10–20-fold. Due to the toxic impact of some algae during mass "B", destruction of fauna may be observed. During "B.", the water becomes less transparent, grows turbid, which tells on the distribution of light in the pelagic region. Fishes and other organisms are able to avoid the "B." zones by congregating near their boundaries as "B." exerts harmful effect on them.

"Blue Line" – a system of German defense fortifications "Gotenkopf" on the Taman Peninsula during the Great Patriotic War (WWII). According to the Soviet documents, the construction of B.L. was started in January 1943 on the basis of the existing field entrenchment built by the Soviet troops in summer 1942. The main defense strip was to 6 km wide and behind it the well-fortified line extended for 40 km. The left wing of B.L. started near Verbyanaya Bar, passed over the near-Azov mouths, along the banks of the Kurka and Adagum rivers as far as the Kievsky village. Then it turned southward and crossed Varenikovsky, Moldavansky, Crimean, Nizhnebakansky and Verkhnebakansky villages. The southern wing of B.L. passed over mountains from Neberdzhaevsky village to Novorossiysk. B.L. was defended by the units of the 17th Army. The total number of the German group on Taman reached 400,000 people. Having moved from the Caucasus to Taman, the 17th Army and a part of the 1st Tank Army shortened significantly the front line and, as a result, they created denser battle lines on the peninsula. Maintaining their presence on the Taman Peninsula, the German command, on the one hand, covered Crimea and, on the other, had a base area for resuming offensive actions on the Caucasus. The German troops on Taman succeeded to draw in the considerable forces of the Red Army that were unable to take part in the spring military actions in Ukraine. On September 1943 the troops of the North-Caucasian front commanded by General I.E. Petrov started a new offensive for liberation of Novorossiysk and Taman Peninsula. In the course of these battles the Soviet troops took hold of B.L. and on October 9 forced completely the German troops from the Taman Peninsula. Liberation of Taman and Novorossiysk improved significantly the possibilities for the Black Sea fleet basing and facilitated the struggle for return of the Crimea.

"Blue Stream" natural gas pipeline – installed at the Black Sea floor at a depth over 2,000 m; the world's deepest gas pipeline. It was built for gas supply from Russia to Turkey. In March 2002 Italian firm "Saipem" constructed the first and second pipelines. This was the first venture at such depths. Its total length is more than 1,200 km and the under-the-sea stretch (from Dzhubga to Samsun) is 396 km. The annual carriage capacity of this gas transport system is 16 bill m^3. The marine stretch of the pipeline takes its origin 7 km to the northwest of Dzhubga on the Black Sea coast (Krasnodar Territory, Russia) and gets out on the Turkish coast 14 km eastward of Samsun. It was commissioned on December 20, 2002. Its official opening was on November 17, 2005. Gas is supplied along two pipelines with the diameter in the marine part being 610 mm and a wall thickness 31.8 mm.

Bluefish (*Pomatomus saltator*) – large predatory fish of the *Pomatomidae* family. It has an oblong body covered with rather small scales. It may reach 110 cm in length and 8–15 kg in weight. B. is a pelagic species living in stocks. It is very active in the warm months (from May to October). During this period B. is found mostly in the northwestern part of B.S. where it feeds intensively, grows and propagates. In autumn when the water gets cooler B. gradually migrates for hibernation to the Sea of Marmara and live there till the second half of May. Its lifespan is about 8–9 years. It reaches maturity by the age of 2 years, seldom 1 year. The spawning period lasts from the second half of June to early September. It lays eggs mostly in the northwestern part of the sea. B. feeds solely on fish (anchovy, horse mackerel, herring, mullet, mackerel) biting it through and eating only a part of it. In the coastal zone its catches may reach 800 tons, but usually 15–20 tons. For a long time B. was considered a rare species for B.S. But in 1966–1970 B. propagated so extensively that became one of the main fish species in catches. We witnessed the biological "outburst" of its population—the even equally significant and full of mystery. Similar "outburst" of large horse mackerel occurred in B.S. in the 1950s till the early 1960s. Generally, such phenomena may be attributed to a favorable combination of all factors playing a role in formation of population number.

Bogaevskiy Konstantin Fedorovich (1872–1943) – Russian Soviet painter, in whose creativity sea landscapes figured prominently, Honored Art Personality. Graduated from the Academy of Arts in St.Petersburg in 1897. Disciple of I.K. Aivazovsky. Most works of B. deal with painting the nature of native lands: East Crimean sea shore. The early paintings of B. are characterized by stylization of a classical landscape. The sea is an inalienable and mandatory element of his paintings : "Seashore", "Ancient Fortress", "Kimmeria", "Town at the Seaside", "Rocky Shore", "Tavroscythia". Many of his compositions the painter constructed against the background of a wide sea full of light—the water colors "After the Rain", "The Crimean Campagna". The creative merit of B. is artistic fixation of numerous historical places and monuments of history and architecture of the Crimea, in particular, of the ruins of ancient cities on the Crimean sea coast.

Bogolyubov Alexey Petrovich (1824—1896) – Russian painter in whose paintings significant emphasis is laid on the marine theme. After graduation from the Naval Cadet Corps, served in the Baltic Navy. During 1850–1853, was a student of St. Petersburg Academy of Arts from which he graduated as a 1st class painter. From 1853, painter of the Chief Naval Staff. During his early years as a painter was under a strong influence of I.K. Aivazovsky. In 1852, was awarded a minor gold medal for several paintings, including "The Battle of the Brig "Merkury" with two Turkish Ships", in 1853, awarded a large gold medal for 3 views of Revel (currently, Tallinn, Estonia) and the painting "View of offshore in St. Petersburg on a summer night". In 1854–1860, as a boarder of the Academy of Arts, B. worked at the studios of well-known painters in Paris and Geneva, visited Italy, Turkey, Switzerland and other countries. In 1858, was awarded the title of an academician, in 1861—of a painting professor. The author of a series of patriotic paintings on the subjects of the history of the Russian Navy that illustrated the feats of Russian seamen: "Battle of Sinop on November 18, 1853" (1856), " Battle of Frigate "Flora" at Pitsunda Cape" (1857), "Battle of Gangut" (1876)," "Breaking of Russian Galley Fleet through the Swedish One at Gange-Udd Cape (1876), "Blast of a Turkish Monitor" (1871) and others as well as numerous sea and river landscapes. As a grandson of A.N. Radishchev, B. founded in 1885 an art museum named after Radishchev, and in 1897—a drawing school in his home town Saratov.

Boklevskiy Konstantin Petrovich (1862–1928) – Russian, Soviet ship-building engineer. After graduating from the Technical Naval School in 1884, was designing and building armor-clad battleships in Nikolaev for the Black Sea Navy. In 1886–1888, he studies at the ship-building department of the Naval Academy. Took part in the building of the cruiser "Pamyat Azova" ("Memory of Azov") in St. Petersburg, organized the building of destroyers for the Black Sea Navy, in 1892–1897 supervised the building of the armor-clad destroyer "Tsarevich" ("Prince") and the Cruiser "Bayan". Worked as an aide to the Chief Engineer of St. Petersburg Port and the Chief Engineer of the plant under the Department. Took part in the construction of a series of battleships for the Pacific Navy. Made a contribution to the development of motor-boat construction. In 1898, proposed that diesel engines be used on the ships, and in 1903 made a design and built the world's first motor-vessel "Vandal". In 1902, organized Russia's first ship-building department at the St. Petersburg Polytechnical Institute and was its dean until 1923. One of the founders the "Russian Register" Company, the purpose of which was to get rid of foreign surveillance. At the initiative of B. in 1909, training courses for future engineers aerial-navigators were set up within the ship-building department of the institute, Russia's first aerodynamic laboratory was established. After the 1917 October Revolution, was the Chairman of Technological Council of the USSR Register. Chief of the Specialized Bureau for Merchant Ships Design, worked at the Naval Academy, Sovtorgflot. His main works include: "On Building Mixed-System Destroyers" (1904–1905), "Ship-Building Architecture" (1914) and others.

Bonito (*Sarda sarda*) – the fish of the mackerel family (*Scombridae*). Its body is covered by small scale. The dorsum is of the bluish color, the sides have oblique elongated dark strips. It is found in the Atlantic Ocean near the coasts of North America, Europe and Africa (mostly in the subtropical regions), in the Mediterranean. The Black Sea B. is an individiaul population. This is a warmth-loving pelagic species. For spawning and fattening it runs to B.S., in winter months—to the Sea of Marmara. It lives for 8–10 years. It grows quickly. At the age of 1 year its length is 40–45 cm; its weight is 1.2 kg. It reaches maturity at the age of 2 years, but some species at 1 year. In spring in late April at the surface water temperature being 11 °C B. migrates from the Sea of Marmara to B.S. By early June B. is usually found in the northwestern, northern and northeastern parts of B.S. where it spawns and fattens. The B. shoal stays in these parts of the sea through the whole summer. By late September B. groups into shoals and by mid- or late December it gradually moved for hibernation to the Sea of Marmara.

Bora—(Ital. *bora*, form Latin *boreas*—"*northerly wind*") – local strong, gusty cold wind streaming down onto the Black Sea coast from the adjoining not very high mountain range. More often than not, B. occurs in winter. Is produced, when arctic air encroaches into Russia's south regions and accumulates in front of the low range, when it begins flowing over the passes at a high downward speed under the impact of the pressure gradient and force of gravity. Sometimes, it exceeds 40 m/s, some of the gusts moving at a speed of over 60 m/s. B. is known best on the Dalmatian coast of the Adriatic Sea, around Novorossiisk (nord-ost), and on Baikal

Lake (sarma), on Novaya Zemlya (mountain), in the south of France (mistral), in Texas (norther), etc.

Borcea Ioan (1879–1936) – outstanding Romanian hydrobiologist who laid down the groundwork of hydrobiology in Romania, Professor of Iasi University. He examined in detail the composition and distribution of Black Sea benthos and fish fauna. In 1926, he initiated the establishment of a Marine Zoological Station at Agigea (Mamaia). The station at Agigea (near Constanta) was named after B.

Border fortifications – a system of fortifications built along the state border. In Rus their construction was started in the ninth century. In the sixteenth to seventeenth centuries they were called check lines. In the eighteenth to nineteenth centuries they consisted of border towns, fortresses and field fortifications (usually earth ramparts). These fortifications were guarded most often by the Cossacks and linear troops that located in towers or earth fortifications behind a rampart. In 1706–1708 following the order of Peter I such fortifications were built on the western border from Pskov through Smolensk to Bryansk where the fortresses were the key structures. Later the Tsaritsyn, Ukrainian, Dneprovsky, Chernomorsky, Kuban, Asov-Mozdok, Samara, Orenburg, Uisk, Irtysh, Zakamsky and other fortification lines were constructed. In the eighteenth to nineteenth centuries the total length of B.F. was about 5,000 km. In late nineteenth century it was replaced with the fortress-based system of frontier guarding that existed till World War I.

Borisov Alexander Ivanovich (?–1810) – Russian Vice-Admiral (1799). Graduated from the Naval Cadet Corps (1761–1768). Every year, participated in a Campaign on the Baltic Sea (1764–1768). In 1769–1773, sent to the Azov Military Flotilla, sailed on the Azov and Black seas. From 1775—representative of the Azov Military Flotilla in St. Petersburg. In 1776, on the frigate "Severny Orel" ("North Eagle") sailed from Kronstadt to Dardanelles, from there to Livorno, in 1777 on the frigate "Grigory" cruised from Livorno to Constantinople and back, then in 1779 returned to Kronstadt. In 1780, was attached to St. Petersburg ship's crew, in 1781–1784—to Quartermaster Expedition. In 1784, while commanding the ship "Mecheslav", sailed in the squadron of the Vice-admiral I.A. Borisov from Kronstadt to Copenhagen, in 1785–1786, commanded the same ship sailing on the Baltic Sea. Commander of the ship "Khrabry" ("Courageous") (1787–1788). From April of 1788 to 1798, worked as an adviser of the Audit expedition. General-controller (1798–1807) and member of the Admiralty-Collegium. Captain 1st rank (1789), Captain, Brigadier rank (1793), Major-General for Admiraly (1797). From May 1799, manager of the Audit expedition in St. Petersburg. In July of 1807, appointed military governor of Astrakhan Port. In November of 1808, dismissed from service. Awarded orders.

Boristhen, Borisphan, Borysthenes – see Dnieper.

Borodin Nickolay Andreevich (1861–1937) – Russian, Soviet ichthyologist. Upon graduation from the Urals Army Gymnasia (gold medalist, 1879)—entered St. Petersburg University, in 1879–1880—student of the Math Department, from

September of 1880—Dept. of Natural Sciences. From 1880—B. organizer of ichthyological work on the Ural River. In 1884—the first successful experiment in artificial insemination of sturgeon (stellate sturgeon) fish eggs. B.'s doctrine on phased evolution of fishes served the theoretical basis of elaborating the procedure of rearing the young fishes. In 1899, B. moved to St. Petersburg, where he began working as a senior expert in fish farming of the Department of Arable Farming. In 1901, B. published the book "Cossacks of the Urals and Their Fishing Practice", founded the newspaper "Uralets" ("Urals Dweller"), was editor-publisher of the "Urals Review", "Cossack Army Newsletter" (1901–1904), contributor to the "Russkie Vedomosti" ("Russian News") (from 1894) and the "Nasha Zhizn" ("Our Life"). In 1902, was elected Secretary General of the International Congress on Fishing and Fish-farming which was held in St. Petersburg. In 1900–1904, explored such fishing areas as the Azov-Don, Black-Sea—Kuban, Amy-Darya, Caspian. Exploration of the Black Sea on the principles of scientifically-based fishing was commenced by B. who headed in 1902 an expedition on the steamship "Klopachev", the launch "Druzhny" and a sail boat; the expedition explored the coast from Kerch Strait to Sochi. Studied were the conditions of fish (mainly, herring) migration from the Azov to the Black Sea and back.

Bosporan Kingdom – came into being in the fifth century B.C. on the eastern and western coasts of Bosporus Cimmerian, currently Kerch Strait, with the capital in Pantikapei. That was the largest and mightiest kingdom of all that existed on the coasts of the Black Sea in ancient times. At the time of Archeanactides and Spartakoides, B.K. was transformed into a monarchy-governed city-state that maintained trade relations with Miletian commercial centers on Meothide (at present, the Azov Sea) and in the Greek center of the empire as well as controlled grain export from the circum-Don plain areas and threatened the neighboring Scythian territories. In 107 B.C., B.K. was conquered by Mitridat VI the Evpator, who, however, was soon defeated, too. Later B.K. extended its boundaries, despite its dependency on Rome, as far as the Don River (Tanais). From the end of the first century to the early third century, B.K. was subject to invasion of the Gots; in the fourth century it was invaded by the Hunns.

Bosporus – (obviously, a Frakian name; the Greek folk-ethymological interpretation: *"bull's ford"*, *"cow's ford"*). (1). Frakian B., strait between Europe and Asia, linking Propontida (the Sea of Marmara) with the Black Sea. (2). Cimmerian B. (currently, Kerch Strait) between Khersones of Tavria (the Crimea) and Taman Peninsula, linking the Black and Azov seas.

Bosporus – strait (Turkish Karadeniz Bogazi, Greek Bosporos) between Europe and Asia, links the Black Sea and the Sea of Marmara, Turkey. The ancient Greek *"Bosporos"*—*"animal ford"*, *"bull's ford"*, *"cow's ford"*. It is possible that the animals were capable of swimming over across the strait because it was so narrow. According to another version, fording was paid for with cattle. The ancient Greek myth elates the name of the strait with Io, the daughter of the god of the rivers: the legend insists that the girl assumed the look of a cow and swam across the strait.

Subsequently, the name came to denote a "narrow strait" in general and be used with an appropriate attribute: in Herodot's writings (fifth century B.C.) the strait is called Bosporus Frakian. The existing Turkish names: Bogazichi ("Internal Strait"), Istanbul-Bogazi ("Istanbul Strait"), Karadeniz-Bogazi ("Black Sea Strait"), Marmara-Bogazi ("Strait of Marmara Sea") failed to gain popularity: in some countries, the name of "Bospor" is accepted; in Russia, France and Poland the form "Bosphor" has been in common use. Length 30 km, width from 0.7 to 3.7 km. Mean depth—65 m, maximum depth—120 m. The depth of the port on the side overlooking the Black Sea—up to 59 m, on the side overlooking the Sea of Marmara—up to 40 m. B. shores are high, steep, picturesque, with numerous bays. Many fitting natural harbors, the best of which is Golden Horn near the European coast, close to the exit to the Sea of Marmara.

Bosporus and Istanbul: View from space (NASA, 2004) (http://ru.wikipedia.org/wiki/Bosphorus)

Istanbul lies on both the shores of the strait. B. has been produced by erosion; the strait is a river valley flooded with sea water during the Quaternary Period. There are two currents in B.: the upper from the north to the south, and the lower one having a reverse direction. The upper current brings to the Sea of Marmara an average of 370 km^3/year of the Black Sea water with salinity 17–18‰, water temperature in winter 5–8 °C, in summer 23–26 °C; the cause for the current is positive fresh-water balance of the Black Sea. The lower current brings around

170 km^3 of Mediterranean Sea water per year (salinity 37–38 ‰) to the Black Sea. Water temperature in winter 10–11 °C, in summer 19–20 °C. The current is produced by the different water densities in the seas. Water exchange via the strait may change significantly due to different forcing and hydrometeorological conditions. Both the currents were for the first time studied by the Russian scholar Admiral S.O. Makarov. B. is extremely important in terms of economy, policy and military strategy not only to Turkey, but to all the Black Sea states.

Bosporus Strait: View on the Black Sea (Photo by Andrey Kostianoy)

Bosporus blockade (1914–1917) – combat actions of the Russian Black Sea Navy during WWI with a view to prohibiting passage of enemy battle ships via Bosporus Strait and deranging enemy shipments in the south and south-western areas of the Black Sea; part of the overall sea blockade of the Turkish coast. Began in the early months of the war. In 1914, the Navy planted 847 mines near Bosporus, on which the German battleship cruiser "Geben" and a Turkish mine-layer were blasted on December 13(26). In 1915-early 1916, squadrons of the Black Sea Navy operated in the southwestern part of the sea, shelling the Bosporus fortifications and coal mines at Eregli and Zonguldak. For the first time ever, the subsea mine layer "Krab" was used for laying a mine-field, on which the light German cruiser "Breslau" was blasted on July 5(18), 1915. From February of 1916, the blockade was mainly effected by means of submarines. In July of 1916, the command of the Black Sea Navy organized mass mine laying on the approaches to the strait by means of destroyers, minesweepers and the subsea mine-layer "Krab" under the cover of battleships, cruisers, destroyers. In all, 14 mine-fields (2,187 mines) were laid at Bosporus in 1916, the mine-fields were under the surveillance of the blockade reconnaissance patrol. In the course of B.B., for the first time ever, sea aircraft were used for reconnaissance and bombing Turkish facilities. Despite being full of zip actions of the Black Sea Navy, Bosporus was never blockaded completely as the depth and density of mine-laying proved insufficient. Besides, the mine-fields had no antiminesweeping devices.

Bosporus bridge – also called the First Bosporus Bridge is one of the two bridges in Istanbul, Turkey, spanning the Bosphorus Strait and connecting Europe and Asia (the other one is the Fatih Sultan Mehmet Bridge, which is called the Second Bosporus Bridge). It was opened on 30 October 1973. Its length is 1,510 m. At present, it is the 19th longest suspension bridge span in the world. The width of the cable-stayed bridge is 39 m. The bridge height over the Bosporus is 64 m. The bridge was designed by the renowned British civil engineers Sir Gilbert Roberts and William Brown. A.B. is one of the symbols of Turkey.

Bosporus bridge (Photo by Andrey Kostianoy)

Bosporus Campaign – trek of the Black Sea Naval squadron to Bosporus in May of 1915. The battleships "Tri Svyatitelya" ("Three Prelates") and "Panteleimon" shelled the shore fortifications of Bosporus. On May 4, the battleship "Rostislav" opened fire on the fortified area of Iniade (north-west of Bosporus) which was also attacked by seaplanes. Bosporus campaign culminated in a battle at the entrance to the strait between the battle cruiser "Geben", the flagship of the German-Turkish Navy on the Black Sea, and four Russian battleships. In this exchange of fire, the Russian battleship "Evstaphy" distinguished itself: its two accurate hits rendered the cruiser inoperative. That was the end of the story of single combat with Russian battleships. The trek enhanced the superiority of the Russain Navy on the Black Sea.

Bosporus Cimmerian – ancient name of the existing Kerch Strait named after the local tribe—Cimmerians. In Genoese time, used to be called Cafian Passage, derived from the Italian name of Cafa (Feodosiya) or Saint Joann Mouth—after the name of the Church of John the Forerunner in Kerch that stood on the seashore. The Turks used to call it Taman-bogaz. An unknown author has saved this legend for us: there used to live Helios, a titan-petty thief who lived and robbed people around Kerch Bay. He owned huge herds of bulls and kept enlarging these with animal he stoles from the locals. Once, seeing to flee from the pursuers on the back of a stolen bull, Helios realized that this time he would be unable to escape. Once he did, he used his thumb to "cut" a passage that separated Meothide (Greek "Mother of the Seas")—the Sea of Azov from Pont. The road to Cimmerian shepherds was thus blocked by the miraculously created Kerch Strait, named B.C. go commemorate the happily concluded Titan's theft operation. Even though they say that this is one of the 200 names known to scientists. In ancient times, there used to lie Greek (Miletian) region on the coasts of Bosporus Cimmerian, with the capital of Bosporus State Pantikapei (in Byzanty time—Bosporos) on the European and Fanagoria (Theos's colony) on the Asian coast of the Strait. B.C. was crucial to the development of trade relations with the circum-Don grain-growing areas in the Black Sea region.

Bosporus Frakian – ancient Greek name of the existing Bosporus Strait.

Botkin Sergey Petrovich (1832–1889) – Russian physician, scholar, public figure. Graduated from the Medicine Department of Moscow University (1855), underwent probation at the medical institutions of Germany, France and Austria (1855–1860). During the Crimean War of 1853–1856, worked under the guidance of N.I. Pirogov at the Bakhchisaray Sick Quarters. From 1860, therapy professor of Moscow Academy of Surgery, from 1870—Academician of Moscow Academy of Surgery and a physician-in-ordinary. The founding father of research in the development of Russian clinical medicine. Author of works on clinics and pathogenesis of cardio-vascular diseases, infectious diseases, etc. Founded Russia's first clinical and experimental laboratories.

Botkin made the first scientific substantiation of the Crimea curative factors. The dwellers and visitors of the Southern Coast of Crimea (SCC) are aware of the Botkin path in Livadia and a street in Yalta of the same name, named so to commemorate Botkin's visit to the place. Botkin visited the Crimea for the first time in 1855, when the Crimean War was in high gear. B., a recent student-turned graduate with distinction of Moscow University, he eagerly joined a group of doctors formed by N.I. Pirogov. The young physician underwent practice at the military hospitals and in typhoid barracks of Simferopol and Bakhchisaray. One of the buildings of the Crimean Medical Institute features a memorial plaque to perpetuate the stay in Simferopol of N.I. Pirogov, S.P. Botkin and the first medical nurses. In 1870, B. was awarded the title of an Academician and was the first Russian physician ever to be appointed a physician-in-ordinary of the Tzar family. His duties included escorting the Empress every summer to Livadia, the Crimea. He was one of the first to discover excellent climatic conditions of the SCC, particularly appropriate for the treatment of TBC patients. He regarded the location in Ereklik and Livadia the best. On B.'s recommendations, a sanatorium for the Empress was built at Ereklik: at present, it houses the facilities of the anti-TBC sanatorium "Mountain Health Resort". Also at B.'s initiative, a hospital was laid down on Polikur Hill, currently occupied by the I.M. Sechenov Research Institute (for climatology and climotherapy). One of the structures is until now called Botkinsky. The outstanding physician wrote: "I feel that the Crimea has a great future as a hospital station... In future, its rating is bound to be much higher than that of Montreux".

B. is the author of many priority ideas and lines of development (established the infectious nature of viral hepatite, the so-called "Botkin disease"). The founding father of field-army therapy. Chairman of St. Petersburg Society of Russian Physicians (1878–1889), member of 43 other Russian and foreign institutions and scentific societies.

Bottlenose dolphin (*Tursiops truncates*) – the largest Black Sea dolphin. Its length is approximately 2.3 m on the average (from 2 to 2.8 m), some species may reach 3.1 m. Its weight is 130–300 kg. They live for 25–30 years. The upper part of the body, head and fins are nearly black with a blue and dark-brown tint of different intensity; the belly is white. The eyes are circled by a dark strip reminding of glasses. The nose looks like a blunt beak. B.D. lives mostly in the coastal parts of oceans and seas in the Northern and Southern Hemispheres. In the Black Sea it is met more seldom than common dolphin. At exhaling they, like whales, jet out a fountain, but it is much smaller. They often jump out of water. It dives to greater depths than common dolphins. They are able to hold their breath for 10 and even 15 min. They usually move by small groups comprising several dozens of animals with a speed of 30–50 km/h. Their speed may reach 60–70 km/h. B.D. feed mostly on bottom fish, but may eat even sea fox, flatfish, ruff, not passing by the invertebrates—shrimps. The daily fish ration is 20–30 kg. Possessing the excellent acoustic sounder it can feel its "food" at a distance to 3 km. The reproduction starts from the age of 6 years. A female gives birth to one child once in 2 years; the pregnancy period—10–11 months. B.D. are easily trained and adapt easily to living in dolphinariums and oceanariums. These dolphins are most widely used by scientists over the world for various scientific experiments. They are also the gifted actors and participate in many performances.

Bottlenose dolphin (http://true-wildlife.blogspot.ru/2011/02/bottlenose-dolphin.html)

Bozkurt – a town on the coast of the Black Sea, Turkey.

Boztepe, Peninsula – sited in the central part of the Turkish Black Sea coast. The height 212 m, devoid of vegetation. Its northern shore is upright, and the eastern one—cliffy. Linked with the continent by an isthmus, on which Sinop City lies. Boztepe Cape is the southeastern termination of Boztepe Peninsula.

Brander—(from German "brand"—"fire") – (1) At the time of sail fleet, the vessel loaded with combustible and explosive substances and intended for setting the enemy ships on fire. As a rule, they usd small ships and carriers as B. When Peter I reigned, there were specially built B. in the Russian fleet. B. would be directed enemy-wise before the wind or with the stream, more often than not in foggy conditions or at night. Before B. was leaned on a different ship, the crew would set it on fire, themselves escaping in sloops. From the sixteenth century henceforth, B. was a crucial means of waging war at sea. For example, the British used B. in 1588 as they defeated the Spanish "Invincible Armada"; likewise, the Russian squadron destroyed the Turkish fleet in the Battle of Chesma of 1770.

(2) When there arrived steamships, the name of B. would be given to old vessels loaded with ballast that used to be sunk in a narrow fairway so as to block it; for example, shut down entry to a port and exit from it to the enemy ships.

Bredal Petr Petrovich (1681–1756) – Russian Vice-Admiral (1737). A Norwegian by birth. In 1703, admitted for Russian service in rowing fleet in the rank of a non-commissioned lieutenant on the recommendation of the Vice-Admiral K.I. Krujs. Participant of the Great Northern War (1700–1721). In 1705, promoted to Lieutenant. In 1710, sent to the Azov flotilla, commanded the brigantine "Lebed", while cruising from Taganrog to Kuban. In 1712, transferred to the Baltic

Sea. In 1715, commanding the frigate "Samson", seized, while cruising in ten vicinity of Vindava Port for 6 days, three Swedish privateer vessels. Was sent to Britain to take delivery of two ships bought there and brought these to Copenhagen. In 1720 was sent to Danzig, Copenhagen, Lübeck and Memel for hiring marine officers and navigators. In October of the same year was in Hamburg for purchasing ships at the shipyard of the company. In 1721–1722, sent to Spain with the news of the peace with Sweden. Then became commander of the ship "St. Alexander" that was in the vanguard of the fleet, flying the flag of Peter I. In December of 1723, was acting chief commander of the Kronstadt Military Port, on December 9, 1724, appointed director of the St. Petersburg Admiralty Office. Commander of the ship "St. Alexander", flying the flag of General-Admiral F.M. Apraksin. In 1728, took part in auditing Sestroretsk Plants, and in January of 1730, appointed Chief Commander of the Revel (Tallinn) Military Port. In January of 1732—member of the Army Marine Commission "to review and bring the fleet in due order", from April of 1733, chief commander of the renewed Archangel Port, from September of 1735—chief commander in Tavrov. In 1736, formed the Don Flotilla and brought it to Azov town area, took an active part in seizing the city, that same year, he fortified Voronezh shipyard and laid down a new one in Bryansk. Participant of the Russian-Turkish war of 1735–1739. In1737, sent to the Don expedition as the flotilla chief, participated in the Crimean Campaign and in July beat off two attacks of the Turkish fleet near Genichi and Odessa Spit. Made a map of Taganrog Harbor with measurements. In 1739, the war over, was recalled to St. Petersburg. In July of 1740, appointed member of the Admiralty-Collegium—Commander of Archangel Military Port. In 1742, commanded a squadron of ten ships during passage from the White Sea to the Baltic Sea for use in combat actions against the Swedish Fleet. Awarded the order of St. Alexander Nevsky (1741).

Brill (*Scophthalmus maeoticus maeoticus*) – (Black-Sea Turbot or Kalkan) a species of flatfish of the turbot (*Bothidae*) family. It has a rhomb-shaped body. The eyes are on the left side of the body; the coloring of the body is variable—it is usually yellow-gray or gray-brown. Its length may reach 1 m. It is widespread in the Black Sea, Bosporus and the Adriatic Sea. B. lives for 12–16 years. It reaches maturity at the age of 3 to 4 years. Its spawning period is from April through mid-June. The fish makes local migrations for fattening and growth. In March it migrates from open sea to the coastal waters and till late April stays here at depths 10–40 m. After spawning it goes to the places 60–80 m deep. In autumn it often moves to the coast for fattening. It spends the winter months at depths 70–100 m forming commercial fishing communities at depths between 70 and 80 m. In Black Sea fishery B. is the valuable fish species.

British Military Memorial Complex – monument of the period of the Crimean war. Sited on Sapun-gora (Mt. Sapun) Cathcart Hill so named in memory of the British division commander Lieutenant-General George Cathcart (1794–1854) who was killed in the Battle of Inkerman on October 24 (November 5), 1854 and was buried here. During Sevastopol siege, there was one of the British cemeteries on the hill, which, according to some data, comprised 450 common and personal graves, where the bodies of nearly 22,000 British soldiers and officers are laid to rest. There are several British generals among those who are buried here. The cemetery was established in 1882, financed by the government of Great Britain. During Sevastopol defense of 1941–1942, the cemetery was severely damaged, and after the war was in a state of utter neglect. In February of 1945, during the Yalta Conference, the cemetery was visited by the British Prime-Minister W.Churchill. In 1966, the Memorial was visited by the successor to the British throne Prince of Wales. In 1993 the first stage of the renovated cemetery was commissioned. In 2002, the "Crimean War Memorial" was opened.

Bruce Jacob Vilimovich (1670–1735) – Russian General-Field Marshal (1726), associate of Peter I. A descendant of a noble Scottish family, from 1647 his ancestors lived in Russia. Began service in "poteshny" (boy-soldier) troops (1683). Took part in the Crimean campaigns of 1687, 1689 and in Azov campaigns under Peter I (1695–1696). Was a member of Grand Embassy of 1697–1698. During Great Northern War of 1700–1721, helped Peter I in organizing the army (of artillery, mostly). Distinguished himself in seizing Narva (1704), in battles near Lesnaya Village (1708), near Poltava (1709) and in other battles. Escorted Peter I during the Prut Campaign of 1711. From 1711, General Field Armoury Master. Took an active part in restructuring field, siege and fortress artillery. From 1717, senator and president of Berg- and Manufactur-Collegiums that were in charge of Russian industry. B.—one of the best educated military figures of his time. Had encyclopaedic knowledge in the field of math, astronomy and physics. Mapped lands from Moscow to Asia Minor (1696), equipped an observatory at the School of Navigation in Moscow (1702), took part in developing Service regulations of 1716. Translated the book by the Dutch physicist and astronomer Christiaan Huygens "Cosmotheoros" and wrote the foreword to it, edited a number of works on math, astronomy and geography. Was making astronomical observations.

Brunnov Philipp Ivanovich (1797–1875) – Russian count, diplomat. Graduated from Leipzig University (1818), served at the State Collegium for Foreign Affairs. Participant of several congresses of the Holy Alliance of 1815–1833. From 1823, was in the service of Novorossiisk Governor-General M.S. Vorontsov. In 1826, took part in Akkerman Conference. In 1828–1829, chief of the diplomatic office of the Manager of Danube Principalities. Was in charge of the office of Russian representatives when the 1829 Adrianopol peace was being signed. In 1835–1839—advisor to the Minister of Foreign Affairs. In 1839–1840, envoy extraordinary and minister plenipotentiary in Stuttgart and Hessen-Darmstadt. B. was commissioned to settle differences with the British Government that arose following the conclusion by Russia of the Adrianopol and Unkar-Iskelesi Treaties. In 1840–1854—envoy to London, in 1860–1874—Ambassador to Britain. Russia's representative at the London Conference that canceled the articles of the 1856 Paris Peace Treaty on "neutralization" of the Black Sea. Resigned in 1874.

Brusilov Lev Alekseevich (1857–1909) – Russian Vice-Admiral (1908). Graduated from the Naval Junkers' Classes in Nikolaev City (1875). From 1875 to 1899, served in the Black Sea Navy. In 1876–1878, sailed on various ships in the Black and Azov seas. Participant of the Russian-Turkish war of 1877–1878. In 1878–1879, sailed on the steamship "Nizhny Novgorod" of the Volunteers' Fleet from Odessa to Sakhalin and back under a war flag. In 1879, promoted to a midshipman. Subsequently, served on the Black Sea naval ships, fulfilling the tasks of making a strategic dscription of the Black Sea coasts and ports. In 1890–1895, Flag-officer of the Commander-in-chief of the Black Sea practical squadron, Aide-de-camp and Flag-officer of Commander–in-chief of the Black Sea Navy and ports. In 1898, mine cruiser commander, sailed in the Black Sea. Awarded numerous Russian and foreign orders.

Bubyr (*Pomatoschistus minutus*) – sand goby (also known as a polewig or pollybait), marine bottom fish of Gobidae family, the head and front part of the back almost up to the first dorsal fin have no scale. Their abdominal fins are swept backwards, reach the anus. The body color is pale gray or gray-brown. B. reaches 7 cm in length. Exhibits common occurrence near the Atlantic coast of Europe, in the North, Baltic, Mediterranean seas and in parts thereof. Can be encountered near the Black Sea coasts at the depths of up to 60 m, mostly on a sandy or sand-silt seabed.

Bucharest Peace Treaty of 1812 – completed the Russian-Turkish war of 1806–1812; signed in Bucharest on 16(28) May 1812. The treaty comprised 16 open and 2 secret articles. Russia was enlarged with Bessarabia with the fortresses Khotin, Bendery, Akkerman, Kilia and Izmail (Art. 4), the Russian-Turkish border was established along the Prut River as far its confluence with the Danube, then along the Likian channel as far as the Black Sea. Moldavia and Valakhia were returned to Turkey. The boundaries from the Asian side were restored as they were prior to the war, i.e. Russia returned to Turkey all lands and fortresses seized in action. Russia remained in possession of all regions of Transcaucasia that joined Russia voluntarily as far as Arpachai, the Adjaria Mountains and the Black Sea; the only city returned to Turkey was Anapa. This way, Russia for the first time received marine bases on the Caucasian coast of the Black Sea. On the strength of a separate secret article, Russia was to be handed over a segment of the Caucasian coast "as a harbor to provide for and facilitate the delivery of military supplies and other required hardware", while the other secret article stipulated demolition of some fortifications. As a result of B.P.T., Russia provided for the autonomy of Danube Principalities by reaffirming the resolutions of the Kuchuk-Kainarji (1744) and Iasi (1791) treaties as well as the Russian-Turkish agreement on Valakhia and Moldavia of September 24, 1802. The consequence of B.P.T was that Russia retained its influence in the Danube principalities. Russia acquired the right of trade and navigation throughout the Danube stream, of military influence as far as the Prut River mouth. The treaty and some secret articles thereof were ratified by the emperor Alexander I on June 11(23), 1812 in Vilno, 1 day before the invasion of Napoleon I. The Turkish Government ratified the treaty only, but refused to ratify the secret articles (hence, the latter failed to take effect). The treaty ensured the neutrality of Turkey during the Russian Patriotic War of 1812 and enabled Russia to focus all forces on beating off the invasion of Napoleon I. The main provisions of B.P.T. were reaffirmed by the Akkerman Convention of 1826.

Bukhhta (bight) (form Germ. *"bucht"*) – small part of the ocean, sea, lake isolated from open waters by the outline of the shore or islands; small bay, protected from the wind, open for the sea, lake or storage reservoir from one side. Local conditions determine the specific hydrological regime of B. which differs from that of the waters adjoining B. As a rule, B. is a convenient place for the berthing of ships. There are several convenient B. on the Black Sea: Tsemes, Balaklava, Sinop, etc.

Buckler skate (*Raja clavata*) – fish of the *Rajidae* family. It has a flat rhomb-shaped body covered with small and large thorns. Its body length may reach 120 cm. In the adult species there are 24–32 thorns along the body midline. This species is widespread near the European coast of the Atlantic Ocean and in the Mediterranean. In the Black Sea it is found in the shelf zone. It keeps at depths from 20 to 90 m. In spring (May–June) it migrates towards a shore where it spawns at a depth of 20–50 m. It lays eggs one-by-one on aqueous plants or other substrates. It feeds mostly on shellfish and bottom fish.

Budak Liman – sometimes referred to as Shabov Liman, sited 18 km to the north-east of Burnas Cape, to the south of the entry in Dniester Liman. Separated from the Black Sea by a narrow strip of land. The north-western shore of the lagoon is elevated, precipitous and is gradually descending toward Dniester Liman. Mean salinity of the lagoon equals 26‰. It is only in the northern part of the liman (Akembet Bay), where a considerable amount of ground water arrives, salinity is perceptibly lower. Medicinal hydrogen sulfide saturated silt is deposited on the bottom of the liman.

Budishchev Ivan Matveevich (?–1828) – In 1792–1809, served on the ships of the Baltic and Black Sea Russian Navy. In 1797–1799, served as a midshipman and lieutenant. Took part in making a reconnaissance marine description of the Black Sea northern coasts and of the Kuban River lower reaches. In 1801–1802, while commanding ships, made a description of the western sea coast from Odessa to Bosporus Strait. On the strength of all materials available, made a map of the western part of the Black Sea coast in Mercator's projection. In 1809–1826— Lieutenant, Captain-lieutenant and Captain 2nd rank. Was in charge of the depot of Black Sea naval charts and continued participating in hydrographic works. During this period, drew up an "Atlas of the Charts and Plans of part of the Black Sea" (1807) and the "Azov and Black Seas Marine Guide-book".

Bug Cossack Army – irregular army, stayed on the banks of the Yuzhny Bug River and guarded Russia's south-western borders. The army was initiated by the regiment formed in Turkey in 1769 of Moldavians, Valakhs and Bulgarians to fight against Russia, but subsequently the regiment took the Russian side during the Russian-Turkish war of 1768–1774. The war over, the regiment with the families settled in Russia along the Yuzny Bug River, where it served as border guards along the Dniester River Participated in the Russian-Turkish war of 1787–1791, being part of the Ekaterinoslav Cossack Army (1787–1796). In 1800, was disbanded, but in 1803 at the request of the Cossacks was reinstated in the form of B.C.A (up to 7,000 men). Three cavalry regiments were detailed to guard the borders along the Dniester. The army was under the order-appointed chieftain of the Don Cossack Army. After the Russian-Turkish war of 1806–1812, in view of Bessarabia being annexed to Russia, B.C.A. was, in December of 1817, included in the military settlements and, together with two Ukrainian regiments, formed the Bug Lancers Division.

Bug-Dnieper-Liman Canal – sometimes referred to as Nikolaevsky sea approach canal. Begins at Berezan Island and stretches 80 km as far as Nikolaev Port. The canal comprises 13 bends: of these, 6 bends are in Dnieper Liman (liman = lagoon), the rest are in the Yuzhny Bug River. Canal width—100 m, depth—10.6 m (1967).

Bulgarian Riviera – the so-called Bulgarian Black Sea coast.

Bulancak – a town on the coast of the Black Sea, Turkey.

Bulgakov Yakov Ivanovich (1743–1809) – Russian diplomat, writer and translator. On diplomatic service from 1761. From 1764—a staff member of the embassy in Warsaw. In 1781 appointed the extraordinary envoy and minister plenipotentiary in Constantinople. Despite the Turkish urge to reconsider the Kuchuk-Kainarji Peace Treaty of 1774, B. succeeded in persuading Porta to sign the act of the Crimea, Taman and Kuban annexation to Russia. In 1783, B. made provisions for Russia to be granted the status of a most-favored nation as far as its commercial relations with Turkey were concerned. After Russia declined the ultimatum of Turkey insisting on return of the Crimea, B. was imprisoned, where he, none-the-less, managed to get hold of the planned Turkish military operations. In 1789—Ambassador to Warsaw. After the death of Catherine II, Empress of Russia, B. held no diplomatic posts.

Bulganak mud volcano field (Crimean Tatar *"bulaganak"*—*"mudy, turbid"*) – sited on Kerch Peninsula 8 km north of Kerch near Bondarenkovo Village (formerly, Bulganak). The largest in the Crimea group of 7 volcanoes, the most prominent of which are Andrusov Volcano, Vernadskogo and Obrucheva volcanoes—natural monuments of local significance.

Bulgaria, Republic of Bulgaria – state in the south-eastern part of Balkan Peninsula, in the south of Europe. Borders: on the north with Romania, on the west with Serbia and Macedonia, on the south with Greece and Turkey, on the east is washed by the Black Sea. Territory: 111,000 km^2. Population: 7.4 mln people (2011). Capital: Sofia (1.2 mln people in 2011). Other major cities: Plovdiv, Varna, Burgas. Official language: Bulgarian. Dominant religion: Christianity (orthodoxy). Monetary unit: Lev = 100 Stotinkas. State system: presidential parliamentary republic. Head of state: president, elected for 5 years. Legislative body: one-chamber parliament National Assembly (240 deputies elected for 4 years). A new administrative structure was adopted after 1999 in parallel with the decentralisation of the economic system. It includes 27 provinces and a metropolitan capital province (Sofia-Grad). All areas take their names from their respective capital cities. The provinces subdivide into 264 municipalities.

Bulgaria. Physical map (http://www.ezilon.com/maps/images/europe/Bulgaria-physical-map.gif)

The first people emerged on the territory of B. as we know it today over 500,000 years ago, and in the fourth century B.C, there were settlements of ancient Aryans. One of the tribes, Frakians, settled on this land in the fifth century B.C., created an independent state that was the homeland of the legendary leader of gladiators Spartakus. Attempts to conquer the state were made by Greek colonizers, Scythians, Persians and Macedonians; in the first century A.D., the state was seized by Romans and fell into their bondage for 400 years. From the end of the fourth century, incursions of the neighboring tribes resumed which affected the ethnic composition of the population. The Slavs who came from beyond the Danube in the seventh century assimilated the local population and from the second half of the seventh century made a union with a small group of proto-Bularians (part of the Turk-Language people who had been driven from the lower reaches of the Kuban River by Hazars). That was how the First Bulgarian Kingdom came into being in 681, which attained its highest might in the ninth to tenth centuries, being in possession of nearly the entire Balkan Peninsula. However, as early as in mid-tenth century, the Bulgarian-Slavic state entered a grave crisis and early in the eleventh century lost its independence, being subordinated to Byzantium. The resurrection against the oppressors headed by Petr and Asen was a success and managed to regain the country's independence. Thus, the Second Bugarian Kingdom came into being. The invasion of Mongolian Tatars that followed, competition with Serbia and Byzantium on the Balkans, internal strife resulted in the collapse of the Kingdom which was conquered by the Turks at the end of the fourteenth century.

This way, muslim yoke arrived, which lasted almost 500 years and was destroyed after Turkey was defeated in the Russian-Turkish war of 1877–1878. In 1908, the Third Bulgarian Kingdom came into being. But then Germany drew it into WWI on its side. Orientation to Germany continued, which eventually led to a union with Hitler. After WWII, a republic was proclaimed, and the power passed over to Communists. After the USSR collapsed early in the 1990s, B. began building a democratic society.

Natural conditions of B. are diverse: mountain relief and seashore, plains and hill country with fertile soils are convenient for arable agriculture; there are medium-height mountains covered with woods and grasslands. The mountains occupy one third of the country's territory, while two thirds is below 500 m above mean sea-level, which in the conditions of south-eastern Europe is most convenient for the arable agriculture belt. Bulgaria's latitudinal axis is the mountain system Stara Planina (Balkan Mountains) under 2,000 m in height in the west and going down to 600–800 m in the east. The mountains divide B. into two parts: northern, with the forest-free and ploughed up Danube Plain, and southern, more mountainous into which the Upper Frakian Lowland wedges. The mountain massif is dissected in two places by the deep valleys of the Iskyr and Kamchia Rivers that form gorges. There are several passes in the main mountain range of which the best known is Shipka. The middle part of Stara Planina is adjoined in the south by the Stara-Gora range. The most mountainous is the south-eastern part of the country with the Rila-Rodopi mountain massif, which includes the highest mountains and ranges of B.: Rila (the highest mountain top of B.—Musala, 2,925 m), Pirin and Rodopi. The relief of Rodopi Mountains, forest-covered on the west, is particularly diverse.

The country's climate is temperate continental, succeeded by Mediterranean climate in the extreme south. South-westerly winds cause droughts. Mean temperature of January is from −2 to 2 °C, of July up to 25 °C. Precipitation in lowlands falls in the amount of 500–600 mm/year, in the mountains 1,000–1,200 mm/year.

Chernozems account for around 3 mln ha of land, which favors the development of arable agriculture. 30 % of the territory (mountains) is covered with woods (beech, oak, hornbeam, pine, fir, spruce, filbert). The major rivers are the Danube and Maritsa. There are several national parks: Vitosha, Golden Sands, Ropotamo, Steneto and others. Encountered in the woods are red deer, fallow deer, roe, chamois, boar; in the mountains—polecat, weasel, badger, wolf, fox, squirrel, European hare, small rodents; forest-free northern area feature: polecat, hamster, mole; the species composition of birds and reptiles is quite rich. In the Black Sea near Bulgarian coast common commercial fishes are mackerel and flatfish, in the Danube River—stellate sturgeon, pike perch and carp.

The country does not have enough useful minerals, of greatest value being iron, lead-zinc and copper ores. Also available is hard and brown coal, marble, quartz sand and around 500 springs of medicinal mineral water.

Bulgarians account for 85 % of the country's population. Other nationalities include Turks (around 9 %), Greeks, Gypsies (3.4 %), Armenians, Russians and other peoples who have been integrated in the country lately. The Bulgarian Language belongs to the Indo-European family and South-Slavic group of languages. B. has very low natural growth rate of population: only 2 persons per 1,000 of inhabitants.

The culture of the Frakian period is in many ways similar to that of the Mediterranean peoples, yet it is original and has contributed to the development of classical and world civilization. The ethnonym "Bulgarians" in the ninth century used to suggest Slavic nationality, different from ancient Slavs who came from beyond the Danube. In 865, Christianity was introduced. Creation of the Slavic script by the monks from Solun (Thessaloniki) the brothers Cyrill and Methodius assisted the growth of the medieval Bulgaria. During the period of Osman yoke, national traditions were mainly kept up by monasteries that were the foci of education and script, temples of art and monuments of architecture. It was in these centers that the old Bulgarian literary school emerged, alongside the Trnovo and Kilifar Schools of Painting; these centers gave birth to writers, historians, translators, pedagogues. The unique manuscript of the list of "Interpretations of Iov" is currently kept in the Church of the Holy Sepulchre in Jerusalem. The names of Kh. Botev, Elin-Pelin, I. Vazov, G. Karaslavov, N. Vaptsarov, D. Dimov and other authors are well known in literature. Music is represented by folk ritual and worker's songs, church psalms in the ancient Bugarian Language, epic songs. Higher education progresses along the lines of adhering to the European academic tradition. The Bulgarian Academy of Sciences (BAS) is the leading scientific institution and employs most Bulgarian researchers in its numerous branches. Principal areas of research and development are energy, nanotechnology, archaeol ogy and medicine There are around 300 museums and art galleries, over 9,000 libraries in the country. Cinema industry is at an advanced stage of development.

B. is an industrial-agrarian country. Best developed is machine-building (agricultural and other machines, equipment for light and food-processing industry, computers, machine-tools, hydraulic turbines, etc.). The country is distinguished by advanced light, especially food-processing (tobacco, canned fruit) and textile industry. Ferrous and non-ferrous metallurgy, petrochemical and chemical, woodworking industry. 80 % of active population is engaged in material production. Prevalent in agriculture is plant growing: grains, leguminous crops, tobacco, gardening, grape-growing and horticulture, oil-bearing crops (ranks world's first in production and export of rose oil, Kazanlyk). B. is among the world's leading producers of canned fruit and vegetables per capita of the population. Over half of B. export is the products of industrial production (machine-tools, non-ferrous metals and ore concentrates, items of electrical-engineering, chemical, textile and other industries). 80 % of agricultural production is exported in the form of manufactured goods. B. imports are dominated by industrial raw materials, machines, equipment, ferrous metals, oil and oil products, cotton, rubber, etc.

Fishing is developed very well. Motor-road and railway transport is at an advanced stage of development in B. A ferry service between Varna and Ilyichevsk (Ukraine) has been established. The country derives good revenue from foreign tourism as a matter of tradition. Navigation with the continental countries of Europe is arranged by the Danube River. Foreign economic relations of B. with the countries of EU and worldwide are effected through the sea ports, the principal of which are Burgas and Varna. Both the ports are joint-stock companies, 100 % of their shares are owned by the state. The prospects of Bulgarian ports development (they are 24) are related to leasing the ports on concession, which will make the services market accessible to the private sector.

Bulgarian Black Sea coast – The Black Sea is around 380 km long natural eastern frontier of the country. B.B.S.C. is subdivided into two subareas: northern and southern. The northern subarea stretches from the border with Romania to Emine Cape, the southern one—from Emine Cape to the border with Turkey. Each subarea comprises several dozens of micro-areas and dozens of health-resort and tourist locations. White and golden sandy beaches occupy approximately 130 km. The region is an important center of tourism during the summer season (May–October), drawing millions of foreign and local tourists. There are plenty resources for promoting international and domestic tourism: beaches of varying size, clean coastal waters, woody hills, large plains, cozy valleys, stony slopes, picturesque river estuaries, numerous historical monuments. The characteristic feature of the coast are health-resort complexes that meet most up-to-date requirements, are located outside human settlements so as to guarantee one's full-value rest in open air with as little noise as possible. Naturally, alongside these health-resort complexes there are old health resorts, too that are still quite attractive to holiday-makers. The climate is restful, characterized by mild winter and not very hot summer. The temperature of sea water at the coast during the hottest summer months is 24–25 °C. Sun-induced water warming secures sea bathing during 5–6 months a year. One can get to B.B.S.C. by convenient auto, rail, air and water ways. Parallel to the entire sea coast, there runs a highway which is included in the all-European road network. The highways Sofia-Sliven-Burgas and Ruse-Varna link the Black Sea area with the country's internal parts as well as with foreign countries.

Bulgarian Ship Hydrodynamics Centre – sited in Varna, one of the most up-to-date hydrodynamic centers in Europe. Established in 1976. It is a national research and development Institute of the Bulgarian Academy of Sciences, performing fundamental and applied research, tests, numerical modeling and analysis in the field of the ship hydrodynamics, aerodynamics, ocean and coastal engineering, environmental protection, water transport, defence, etc. Has two towed reservoirs, each 200 m long and 16 m wide: deep-water (6.5 m) and shallow-water (1.5 m). Reservoir trolleys provide for model towing velocities of up to 6 m/s. The deep-water reservoir is furnished with a trolley capable of moving with a speed of up to

20 m/s. There is also a sea-going maneuverable reservoir sized 64×40 m for testing self-sustained models and a cavitation tunnel, with the cross-section of the working segment 0.6×0.6 m. Centre also has coastal hydraulics basins with wave generator and aerodynamic tube, as well as special open water area for ship model tests.

Bulgarian Shipping Company (Navigation Maritime Bulgare) – main shipping company of Bulgaria (Varna). Has 110 ships of 1.7 mln tons combined tonnage (1983). Established by resolutions of the Sixth People's Assembly in 1892 as "Bulgarian Steamship Company". In 1894, the first Bulgarian ships "Boris" and "Bulgaria" arrived in Varna. In 1944, when socialist revolution emerged in Bulgaria, the company only had a few small fishing vessels and large boats of 1,450 tons in total. In June 1947 the Bulgarian Steamship Company merged with the coastal navigation enterprise under the name "Navigation Maritime Bulgare", which performs foreign-trade shipments of Bulgaria, exports commercial services, takes part in the international freight market, pursues an active sea policy. At present, regular traffic lines to the countries of CIS, Mediterranean Sea, Western Europe and the Far East use up-to-date regular-line ships. Tramp ships are used to carry import and export cargoes from the ports of West Europe, America, Japan and other countries. The steamship company also has a tanker and ferry-boats. Today NAVIBULGAR employs over 2,000 qualified staff, administration and crew.

Bulgarians – nation, dominant population of Bulgaria. There is 7.25 mln (2011) B. in Bulgaria (85 % of the population. There exist considerable groups of B. outside Bulgaria, mainly resettlers of the eighteenth century, these people live, for the most part, in the south of the Ukraine and Moldavia, Greece and Turkey; some groups of B. are typical of border areas with the former Yugoslavia. Macedonians of Bulgaria tend to merge with B., being close to them by language and culture. B. as a nationality took shape mainly in the eighth century A.D. Their name stems from the Turk-language tribe of Bolgars (Bulgars) that moved to the West Black Sea area from the North Caucasus and merged with local Slavic tribes. The Bulgarian Language belongs to the southern subgroup, Indo-European language family. Most believers among B. adhere to orthodoxy. They adopted Christianity in the ninth century. Late in the fourteenth century B. fell under the power of the Turks; part of the population was converted to Islam and was under a strong influence of Turkish culture; these B.—"Turk-affected" (Pomaks), living mostly in the Rodope mountains, were not infrequently looked upon as a special people; at present, they constitute an ethnic group of B. At the end of the eighteenth century, there began the process of Bulgarian nation taking shape. The liberation struggle of B. against the Turkish yoke played a great role in this process.

Bureau of the Black and Azov Seas Weather Forecast – Ukraine's oldest Hydrometeorological Center for the Black and Azov seas. Established in 1865. In 1924, the Hydrometeorological Center was restored in Feodosiya, at which the Weather Forecast Bureau was organized. After that, the program of observations at the meteo stations and gauging posts was enlarged. The Bureau focused on servicing the marine and fish industry that operated n the Black and Azov seas and in the oceans worldwide. Hydrological sea weather forecasts, warnings of hazardous weather phenomena were prepared by the Bureau which was to relay on-line on a regular basis.

Burgas – second largest city of Bulgarian Black Sea area, a port sited far inside of Burgas Bay. The city emerged in the seventeenth century, inherited the name of the ancient Greek city-fortress Pirgos (Greek "*tower, castlen fortress*"), whose ruins are on the shore of Mandren Lake. Siten on the peninsula between Burgas Bay and surrounding three lakes: Burgas, Atanasovsko and Mandrensko. Population— 200,300 people (2011), is the fourth largest city of Bulgaria. Important seaport. Commodities exported via Burgas Port are tobacco, fresh fruit, canned fruit and vegetables, rolled metal, machines. Imported commodities include oil, coal, iron ore, ore concentrates to meet the needs of the iron-and-steel plant being set up nearby at Kremikovtzi, the Bulgaria's largest metalworking company. The port is currently being expanded through building new piers south of the old port. Major industrial, commercial and cultural center. "LUKOIL Neftochim Burgas", based in Burgas, is the largest oil refinery in Southeastern Europe and the largest industrial enterprise in Bulgaria. Owned by Russian oil company LUKOIL, the refinery has the biggest contribution among the privately owned enterprises to the country's GDP and to the state budget revenues. "LUKOIL Neftochim Burgas" is the leading producer and supplier of over 50 types of liquid fuels, petrochemicals and polymers for Bulgaria and the region and one of the leading companies in its field in Europe. Oil for the refinery comes from Russia via the oil port "Druzhba", built south of Burgas. Major base of Black Sea and oceanic fishing. Food industry, in particular, fabrication of canned fish. The Burgas Airport is the second most important in the country. The city has several museums, opera and drama theaters, several hotels, many restaurants and other tourist institutions.

Burgas-Alexandroupoli – oil pipeline planned for construction for transportation of Russian and Caspian oil from Burgas (Black Sea, Bulgaria) to Alexandroupoli (Aegean Sea, Greece) to bypass Bosporus Strait due to its being congested with ships. Pipeline length is 280 km, its carrying capacity in 35–40 mln tons of oil per year. The pipeline project was proposed in 1993–1994 by several Russian and Greek companies. In 1994, for construction of the pipeline Greece and Bulgaria signed a bilateral agreement, followed by a memorandum of cooperation, signed by Greece and Russia. In February 1998, a Greek consortium for pipeline construction named "Bapline" was established, and in May 1998, a memorandum of creation of the "Transbalkan Oil Pipeline Company" was signed. In 2000, a technical specifications and an economic evaluation of the project were prepared by the German company ILF. The agreement establishing the international project company was signed in Moscow on 18 December 2007 and the company—"Trans-Balkan Pipeline B.V." was incorporated in he Netherlands on 6 February 2008. Construction of the pipeline was scheduled to start in October 2009, and was estimated to be completed by 2011. However, in December 2011 the project was terminated by the Bulgarian government due to environmental and supply concerns.

Burgas Bay – the largest and deepest on the Bulgarian Black Sea coast: 31 km in length, 41 km in width and 25 m deep (maximum values). Its area west of the Pomorie town meridian equals 174 km^2. The bay is between Capes Emine and Maslen nos. The southern shore of the bay is elevated and precipitous, while the northern one is largely low-lying and gently-sloping. The bay shores are heavily dissected with narrow and deep bights, going in far between the jutting out capes.

Burgas Lake (Lake Vaya) – sited at the western coast of Burgas Bay, Black Sea, Bulgaria. The lake is separated from the shore by wide strips of beaches. The wide and very shallow lagoon, into which the Aitoska River and small rivers Synyrdere and Chukarska flow. Length of 9.6 km and a width of 2.5 to 5 km, area—27 km^2— the country's largest sea lake. Prior to construction of the canal linking the lake with the sea, in summer its salinity was very much like that of the sea and even exceeded it. After the control locks were built and fresh water from Mandren Lake-Storage Reservoir was supplied, in summer, the water and chemical regime of B.L. improved. Carp is common in the lake. It is one of the most productive lakes.

Burgas Province (Oblast Burgas) – administrative-territorial unit in the south-east of Bulgaria. It is the largest province by area, embracing a territory of 7,750 km^2 that is divided into 13 municipalities with a total population of 422,319 inhabitants (2009). Administrative center—Burgas City. B.R. in the north is curbed by the mountain system Stara-Planina, in the south—by the Strandja Mountains bordering with Turkey; in the east, the region is washed by the waters of the Black Sea. Major branches of industry (significant to the Republic as a whole): petrochemical, machine building (railway cars, fishing vessels, ventilators, radiators, cable fabrication), food industry (dairy, flour milling, sugar production, wine-making). Production of brown coal, copper ores, sea salt. Around 1/3 of B.R. area is arable land, the main crop cultivated is wheat; of cash crops—sunflower, sugar beet, cotton. Grape growing is advanced in the northern part. Sheep-breeding and pig-rearing are common, as well as fishing (the region accounts for 80 % of the country's sea fish yield). Burgas sand beaches are special: with dark sand, rich in iron that is heated quickly, which makes it suitable for sun-tanning from early spring to late autumn.

Bushnitsa F. – Corresponding Member of Romanian Academy of Sciences, one of the active initiator and organizer of Hydrobiology for both marine and fresh water for fish farming and fishing, was the best specialist in the field of fisheries and especially in the Romania Danube. With his direct participation a lot of research and economic measures for the integrated development of the Danube Delta was performed.

Butakov Grigoriy Ivanovich (1820–1882) – Russian Admiral (1878), naval commander, the founding father and creator of the tactics steam-powered armor-clad navy. Graduated from the Naval Cadet Corps in 1837, served in the Baltic Navy, and from Deember of 1837—in the Black Sea Navy under the command of M.P. Lazarev. In 1846–1850, while commanding the ship "Pospeshny", was attending to hydrographic work and, together with I.A. Shestakov, drew up the first systematized book on Black Sea Sailing Directions. During the Crimean war of

1853–1856, distinguished himself in Sevastopol, when defending the Malakhov Mound. At the time of Sevastopol defense of 1854–1855, commanded a group of steam-powered frigates that provided an artillery-fire support of ground troops, commanded the steamer-frigate "Vladimir" which captured in combat the Turkish steamship "Parvaz-Bakhri", made a number of successful operations on board the seam-powered frigate "Khersones". When the Vice-admiral V.A. Kornilov was killed in action, B, became Chief of staff of the Black Sea navy. In 1856, promoted to Rear-admirals and appointed the Chief Commander of the Black Sea Navy and military governor of Nikolaev and Sevastopol (until 1860). In 1860–1862, served in the Baltic Sea Navy. In 1863, published his work "New Grounds of Steamship Tactics" which gained recognition by the navies worldwide. In 1863–1867, military attaché in Britain, France and Italy. In 1867–1877, commanded a squadron of armor-clad battleships in the Baltic Sea Navy. From 1881—Chief commander of St. Petersburg Port, from 1882—member of The State Council. Developed a new school of thought based on the use of armor-clad screw-propelled ships with bolt-type guns and other technological innovations; introduced advanced methods of combat training. Awarded Russian and foreign orders. In 1863 B. was a Demidov Prize winner. His main works: "Code of Sea Signals", "Evolutionary Signals Book", "New Grounds for Steamship Tactics" (1865). A mountain of Sakhalin Island is named after B.

Bychok-Blanket (Transparent goby) (*Aphia minuta*) – marine bottom fish of *Gobiidae* family, has a glassy, semitransparent, pinkish body of up to 5 cm in length. Sometimes, it is referred to as crystal goby. This is the only representative having no attachment to the bottom (deposits fish eggs on aquatic plants only) and leading coastal-pelagic life. Feeds mainly on plankton. Common near the Atlantic coast of Europe, in the North, Baltic and Black seas. Dwells in small shoals, largely in bays.

Bychok-Kruglyak (Round goby) (*Neogobius melanostomus*) – marine bottom fish of *Gobiidae* family, the head is oval-shaped, the width slightly exceeding the height. The color of body usually from gray-brown to gray. On the first back fin, behind the fifth somactid there a big black spot, distinguishing it from other species of goby. Reaches 24 cm in length. Common in the Caspian, Azov and Black seas. Besides the sea, is encountered in nearly all coastal lakes and in the lower reaches of most Black Sea rivers—in the Dnieper and Dniester. Of all the gobies, this species is most valuable commercially. Of interest to those engaged in sport fishing.

Bychok-Kruglyash (Giant goby) (*Gobius cobitis*) – marine bottom fish of *Gobiidae* family, the head is oval-shaped, its width slightly larger than the height. The color of body is usually gray-brown or brown-yellow. Length up to 25 cm. Common near the Atlantic coast of Europe, in the Mediterranean Sea its parts thereof. Omnipresent near the Bulgarian coast, its population is greater around Maslen Nos—Michurin.

Bychok-Martovik (Knout goby or Toad goby) (*Mesogobius batrachocephalus*) – marine bottom fish of *Gobiidae* family. They call it Martovik because males usually approach the shore at the end of March. The head is large, wide and compressed from the top down. Hindhead without scale. The color of body more often than not gray-brown or brown-yellow. The body and fins (except the abdominal cupula and the anal fin) are all covered with brown spots of diverse shapes that form a marble pattern. The largest goby dwells in the Black Sea. Reaches 37 cm in length (normally, 25–28 cm). Common in the Black and Azov seas. Encountered everywhere. Hardly ever enters brackish waters. Favorite target of sporting fishermen.

Bychok-Ryzhik (Mushroom goby) (*Neogobius cephalarges*) – marine bottom fish of *Gobiidae* family, the head is large, its width larger than height. The color of body usually ranges from brown-gray to brown-yellow. Length up to 25 cm. Common in the Black and Azov seas.

Bychok-Travyanik (Grass goby) (*Gobius ophiocephalus*) – marine bottom fish of *Gobiidae* family, the body and head are slightly compressed on the sides. The abdominal cupula is underdeveloped and does not reach the anus. The body color ranges from gray-green to brown-green, sometimes is green-yellow. The back, pectoral and tail fins are covered with lengthwise, zig-zag, brown and light stripes. Reaches 25 cm in length. Common near the Atlantic coast of Europe, in the Mediterranean Sea and parts thereof, enters the Sea of Azov.

Bychok-Tsutsik (Tubenose goby) (*Proterorhinus marmoratus*) – marine bottom fish of *Gobiidae* family, the head is slighly compressed on the sides, its height greater than width. The color of body gray-brown, length up to 11 cm. Common in the Caspian, Azov and Black seas. Encountered in nearly all lakes, in estuaries and lower reaches of some rivers, and hardly ever in bays. Along the Don River. It penetrated as far as Voronezh City and even reached the Sea of Marmara, where salinity is 2 times larger that of the Black Sea.

Byzantine Empire (Fourth century—1453) – the Medieval feudal state. It has got its name from the Antique city of Byzantium where the capital was transferred and the city was named Constantinople. This was the eastern province of the Roman Empire comprising Greece, Central and Eastern Balkans, Asia Minor, Syria, Palestine and Egypt. During the reign of Diocletian (284–305) the Byzantine was ruled independently and in 395 it separated completely from Rome, first, as the Eastern Roman Empire and later on as B.E. It had the largest territory in the reign of Justinian I in the sixth century when B.E. turned into a powerful state in the Mediterranean. Seizure in 1204 of Constantinople by the participants of the Fourth Crusade led to the downfall of B.E. that was restored in 1261 under Michael VIII. Finally, B.E. lost all its lands and covered the territory a little bit larger than Constantinople. In 1453 the Ottomans conquered Constantinople that meant the end of B.E.

Byzantium – a city founded on the coast of Bosporus (presently Istanbul). It played a great political, trade and strategic role. It was founded about 660 B.C. as the Megara colony. From the end of the sixth century to 478 B.C. B. belonged to the

Persians. From the mid-fifth century B.C. it entered into the Athens Marine Union. Three times it left the union. In 378 B.C. Byzantium became the member of the Second Athens Marine Union and in 340 B.C. it successfully stood the siege of the troops led by Philipp II. After the battle at Kheronea it retained its autonomy. The climax of the city was in the fourth century B.C. In the first century B.C. Byzantium was included into the Roman Empire where it acquired great significance as a center of trade and crafts. In 330 it was renamed by Constantine I into Constantinople (the city of Constantine) and was given the status of the capital of the Byzantine Empire that was retained till its conquering by the Ottomans in 1453 who gave this city their own name—Istanbul.

Bzyb (Bzypn River) – river in Abkhazia, flows in the mountains of the West Caucasus, flows into the Black Sea. Length 110 km. Catchment area 1,410 km^2. River alimentation is mixed, snow and glacial, in the lower reaches, even rain alimentation. Ice regime is not stable. There are 10 glaciers in the basin. The volume of annual runoff (at Dzhirkhva Village 22 km from the mouth) is around 3 km^3/year. The annual march is dominated by spring (38 %) and summer (33 %) runoff. Maximum runoff is noted in May, minimum—in February. A auto road to Ritza Lake runs along the river valley.

C

Cadastre of the Black Sea coast – its preparation was started in 1945 by the Black Sea Expedition on the Eupatoria-Bakalsky sandbar and later it moved to other sections of the Soviet B.S. coast. This project was headed by V.P. Zenkovich. During this work they made 420 sea sections that helped to describe truthfully the structure of the offshore slope of B.S. for 2,240 km. The works were conducted on vessels "Forel" and "Truzhenik". The cadastre was completed in 1954 and since that time it has been kept in the archives of the P.P. Shirshov Institute of Oceanology, Russian Academy of Sciences. This cadastre was of great importance for development of the fundamental science of marine coast and construction of ports on the Black Sea. It was planned to publish the obtained results in four volumes. But only two volumes were published "Morphology and Dynamics of the Soviet coasts of the Black Sea" edited by V.P. Zenkovich.

Cargo traffic in the Azov-Black Sea Basin – the intensive cargo traffic exists in the basin. The cargo is carried both within the basin and for export to other countries. The main transported commodities are oil and petroleum products, coal, iron and manganese ores as well as agricultural products, such as grain, cotton, general cargos (products of heavy and light industries). Shuttle traffic is organized between the ports of Poti and Mariupol. From Poti the Chiatura manganese ore is delivered to the plant "Azovstal" and on the back way the Donetzk coal is delivered to meet the needs of the Transcaucasian countries. Apart from this, the coal from Donetsk is transported to the ports of the Danube, Ilyichevsk, Kherson, Nikolaev and on the back way—iron and manganese ores from Kherson and Nikolaev to Mariupol.

The regular transit of agglomerate is organized between Kamysh–Burun (Kerch) and Mariupol. On this line the ships deliver hot agglomerate with a temperature of +600 °C which is the world's unique practice. Iron ores from Krivoy Rog and manganese ores from Nikopol delivered to the ports of Nikolaev and Kherson are further transported to Danube ports Izmail and Reni from where after reloading to river vessels they are carried to the European countries.

S.R. Grinevetsky et al., *The Black Sea Encyclopedia*,
DOI 10.1007/978-3-642-55227-4_4, © Springer-Verlag Berlin Heidelberg 2015

Oil and petroleum products take a significant share in this cargo traffic. Oil is transported from Novorossyisk, Tuapse, Poti, Batumi, Odessa and others for export. Light petroleum products (produced at oil refining works in Odessa, Tuapse and Batumi) are also delivered to these ports from other regions of Russia, Ukraine, Kazakhstan and Azerbaijan. From Kherson and Odessa these light petroleum products are transported to small ports on the northwestern coast of the Black Sea, including Yalta. In the northeastern part of the basin the petroleum products are carried by tankers from Batumi, Soupsa and Tuapse to smaller ports. Fuel oil is transported by tankers from Batumi and Novorossiysk to various ports of the basin.

Agricultural products, such as grain, cotton, tobacco, tea, sugar and others are transported from port to port within the basin and worldwide. The lively cargo traffic exists on the routes connecting the ports of the B.S. basin with the European ports (Bulgaria, Romania, Turkey) and also Danube countries. Before the USSR disintegration the passenger traffic in the Azov-Black Sea basin was very intensive. Every year to 25 million passengers were transported along the Crimean-Caucasian and local lines and also along the international lines. But in the subsequent decades the passenger traffic dropped sharply. Regular cargo transportation lines connect the ports of the basin with the ports of Europe, USA, Africa, India, Ceylon, Vietnam, Indonesia, Malaysia, Yemen, Japan, Cuba, and others.

Carnelian (in the Latin meaning "*flesh-toned*") – chalcedony of various tints of red—from slightly rose to red-scarlet. This semi-precious stone is widely used in jewelry making. Well known are C. from Karadag that are met in plenty in the Koktebel Bay and nearby beaches. And quite natural that on Karadag the Western Serdolik (carnelian) Bay is found between the Baklany and Ploichaty capes, the Middle Serdolik Bay—between the Ploichaty and Tupoy capes and the Eastern Serdolik Bight—between the Tupoy cape and the Livadia Bight. In the 1950s–1960s more than 1.5 million tons of sand were excavated in the eastern part of the Koktebel beach for construction purposes and this resulted in a disastrous scouring of the beach. And the beach was completely destroyed in the early 1980s when tons of crushed stone were brought here for coastline stabilization. Today C. may be found only at good luck.

Caspian Pipeline Consortium (CPC) – is a consortium and a pipeline to transport Caspian oil from Tengiz field to the Novorossiysk-2 Marine Terminal on Russia Black Sea coast. It is also a major export route for oil from the Kashagan and Karachaganak oil fields. CPC was initially created in 1992 as a development by the Russian, Kazakhstani and Omani governments to build a dedicated pipeline from Kazakhstan to export routes in the Black Sea. Progress on the project stalled for several years until 1996 when a restructure included eight production companies in the project, among them were Chevron, Mobil, LUKoil, Royal Dutch Shell and Rosneft. BP joined the consortium in 2003. Shares were divided fifty-fifty between the three states and the eight companies. Production companies financed the construction cost of US$2.67 billion, while the Russian Federation contributed unused pipeline assets worth US$293 million. First oil was loaded onto a tanker at the Novorossiysk Marine Terminal on 13 October 2001 and the first stage of the pipeline was officially opened on 27 November 2001. Regular operations started in April 2003. On 17 December 2008, a memorandum on expanding the pipeline was signed. The CPC length is 1,510 km with five pumping stations. The marine terminal includes two single point moorings and the tank farm consists of four steel storage tanks of 100,000 m³ each. Pipeline discharge started at 350,000 barrels per day (56,000 m³/day) and then increased to 700,000 barrels per day (110,000 m³/day). In 2008, CPC transported 31.5 million tons of crude oil. An envisaged second stage will add 10 pumping stations for a total of 15. The number of tanks will increase to ten and one more mooring will be constructed in the sea. Capacity will increase to 1.3 million barrels per day (210,000 m³/day). The second stage has been estimated to cost around US$2 billion and was estimated to be completed in 2012.

Catacomb culture – it is spread widely: from the Volga River to the Lower Danube. The adepts of this culture had the specific burial rites—the dead were buried in special burial chambers called "catacombs" made under the burial mounds that had an entrance pit from which a passage branched off and led to a special cavity where the dead were buried. After burial the entrance into this chamber was most often blocked with a plate or wooden board. The explorers think that this culture was shaped beyond Crimea and only later on it became widespread on the peninsula. The adepts of C.C. most often used the available burial mounds. The main occupation of the C.C. tribes was cattle growing. However, some facts indicate that a part of the population that lived on the coast started practicing farming.

Catherine II Alekseevna (1729–1796) – the Russian Empress (1762–1796). also known as Catherine the Great. She was born in Stettin, Pomerania, Prussia as Sophie Friederike Auguste von Anhalt-Zerbst-Dornburg. This German princess in 1745 married heir to the throne future Emperor Peter III. She entered the Russian throne on June 28 (July 9), 1762 as a result of the palace coup. The domestic policy of C.II was targeted to strengthening of the feudal-serf system and its foreign policy—to strengthening of the positions of Russia on the Baltic Sea and in the Baltic Sea area, its firm establishment on the coast of the Black and Azov seas in 1783. Regardless of counteraction of France, Britain and other West European countries her foreign policy was successful and the targets were attained. In 1783 Catherine annexed Crimea. She called Crimea "the best pearl in her crown". C.II focused much attention to strengthening of the army and navy which made victorious the Russian-Turkish Wars of 1768–1774 and 1787–1791, Russian-Sweden War of 1788–1790. During C.II reign the Military Commission was set up that carried out a military reform. The army received the newest artillery and small arms. In 1796 the fleet comprised 250 ships of various classes. During the C.II reign the Russia was recognized as a great power. In 1787 C.II made a trip to Ukraine, Novorossia and Crimea. It was she who ordered to make an inscription on the gates of the new fortress of Akhtiar (Sevastopol) saying "Road to Constantinople". The escort of C.II included Prince Potyomkin, Austrian Emperor Joseph II and other high-ranking persons. C.II visited Simferopol, Sevastopol and Feodosia. Among her undoubted merits there were uniting of all Russian in a single state; protection of the southern borders of the country from Tatar raids; she was close to implementation of her third task—"to hand a lock" on the Black Sea for foreigners. For her deed for the welfare of the Russian state she was the second and the last of emperors who was dignified as "Great".

Catherine II (http://en.wikipedia.org/wiki/Catherine_the_Great)

"Catherine mile" – an original road sign put intentionally in every 10 versta (1 versta is about 1 km) along the road over which Catherine II traveled to Crimea in 1787. This was a stone tomb square in the foot and rounded on top. It was a hewn column decorated with an octagonal capital and with the straight top. Only a few of such signs are preserved in Crimea. The Old-Crimean Literary-Art Museum that was opened in 1998 demonstrates the restored "C.M.", now being the architectural monument of the republican significance.

Caucasus, Caucasian Isthmus – a territory between Black, Azov and Caspian seas extending for 720 km from 47° to 39°N, from the Kuma-Manych Depression in the north to the border of Georgia and Armenia with Turkey, and to the border of Azerbaijan and Armenia with Iran in the south. Its area is 440,000 km^2. C. is often divided into the Northern C. and Transcaucasus. The mountain system of the Greater C. is divided lengthwise into western (to Mount Elbrus), central (from

Mount Elbrus to Mount Kazbek) and eastern (to the east of Kazbek). C. has a predominantly mountain relief.

Caucasus (http://kids.britannica.com/comptons/art-54732/Caucasus)

The central position is taken by the mountain system of Greater C. the axial zone of which corresponds to the Main or Water Divide and Lateral ranges (the peaks—Mount Elbrus—5,642 m and Mount Kazbek—5,033 m). The Greater C. Range is distinguished by Alpine relief forms or cuesta with karst phenomena (Vorontsov cave, Novoafon cave, Sataplia and others). The Precaucasus begins on the northern slope of the Greater C. and extends to the Kuma-Manych Depression. It separates the Stavropol Upland (rising to 831 m) from the Kuban-Priazov and Terek-Kuma lowlands. To the south of the Greater C. there are the Kolkhida (in the west) and Kura-Araks (in the east) lowlands separating the Transcaucasian Highland that consists of the folded ridges of the Lesser C. (Mount Gyamysh, 3,724 m) and volcanic Armenian Highland (the highest peak Mount Aragats, 4,090 m), in the southeast—the folded Talysh Mountains (height to 2,492 m) and Lenkoran Lowland.

C. has rich oil and gas fields, deposits of coal, iron, manganese, copper, molybdenum, lead, zinc and other metals, springs of mineral water.

C. locates on the interface of the moderate and subtropical climatic belts. The mean temperature in January in Precaucasus varies from −2 to −5 °C, in Transcaucasus—from −6 to −3 °C on the Lenkoran Lowland. In summer the difference of the temperatures between the western and eastern parts of C. is more pronounced. The mean July temperature in the west is +23 to +24 °C and in the east +25 to +29 °C. In the mountains at the altitude of 2,000 m in January the temperature is −8 °C, in August (the warmest month) +13°C. Still higher the cold high-mountain climate is replaced with the perpetual snow belt of high ridges. The precipitations on the plains vary from 200 mm in the Kolkhida Lowland to

1,800 mm in the Kura-Araks Lowland, in the mountains 2,500 mm a year. Considerable modern glaciation (about 2,000 glaciers with a total area of 1,428 km², the largest of them are Dykh-Su, Bezengi, Karaugom, Tsanner and others). The C. rivers belong to the basins of the Caspian (Kura, Araks, Sulak, Terek and Kuma), Black (Rioni, Inguri) and Azov (Kuban) seas. Among the lakes the largest is Sevan in Armenia.

The landscapes here are quite diverse due to the existence of altitudinal belts. On the southern slope of the Greater C., on northern slopes of the Lesser C. and in the Talysh Mountains the subtropical forest landscapes are dominating. They are represented by broad-leaved and coniferous forests. The highlands of the Greater and Lesser C. and also the Armenian Highland are covered by Alpine low-grass meadows, in the most continental areas—by meadow steppes. On the highest ridges the glacial and perpetual-snow landscapes are found. Among the plain landscapes in the Precaucasus the steppes are prevailing, in the Transcaucasus—semideserts.

The mountain areas of the Greater and Lesser C. are inhabited by the forest and high-mountain fauna, including endemic species (West Caucasian and Daghestan wild goats, Caucasian black cock, Caucasian snowcock) and even genders (Prometheus' mouse) as well as animals met in Western Europe (chamois, red deer and others) and generally met species (bear, lynx, fox). The mountain fauna of the Armenian Highland is similar to that of Asia Minor (Asian gopher, Asian mountain jerboa and others). In C. numerous protected territories are available—Caucasian, Teberda, Kabardino-Balkarian and other nature reserves. C. is also one of the major resort areas. On its territory you can find the Caucasus Mineral Water resort and numerous resorts on the Black Sea coast. C. is the center of tourism and mountaineering.

Chai – (from Chinese "cha—chai" (drink), "chae—chai (in leaves), *Thea,*—genus of plants of the tea family. Evergreen shrubs, up to 3 m in height (Chinese C.) or small trees up to 10 m in height (Assam C.) with whole leaves. Four wild-growing species are known to exist (in tropical and subtropical regions of Asia) in the understory of mountain evergreen woods at the height of 1,500–2,000 m. Chinese tea (*Th. sinensis*) and Assam tea (*Th. assamica*) are cultivated in the areas of tropical, moist subtropical and southern temperate latitudes of sour or slightly sour soils only. The sum total of daily mean temperatures is not less than 4,000 °C, the minimum amount of precipitation is 500–600 mm per year. Tea is cultivated for the purpose of obtaining new shoots that are used for the preparation of a tea beverage (after the appropriate fermentation and drying). Dry leaves contain tannins that impart color to the beverage, essential oils (these determine the aroma), caffeine and other substances. The fresh leaves are rich in vitamins. They began using tea as a beverage in China as early as during the B.C. time. In Russia, tea crop came to be known in the 1890s on the Black Sea coast of the Caucasus (around Batumi). Tea crop has gained wide popularity. Chinese tea is cultivated in the southwest and in the east of Transcaucasia, on the Black Sea coast of Krasnodar Territory, in the North Caucasus. Assam tea is grown marginally in west Georgia.

Chaika – river rowing craft of Zaporizhie Cossacks in the sixteenth to seventeenth centuries tailored for sea travel. C. would normally have 12–15 pairs of oars, a detachable mast as high as 4 m, with a straight sail that would be set for leading wind. The steering oars mounted in the bow and aft ensure easy maneouvering in narrow river segments. The bottom of C. would be gouged out of a thick trunk of a willow or lime, the sides would be made higher with boards. The body would be tarred, and thick braids of reed would be tied along the sides to augment stability and unsinkability of C. when wave developed in the river. On several dozens of C. the Cossacks made daring incursions on the Turkish Black Sea ports from the Crimea to Sinop (most prominent treks dated 1589, 1604 and 1615), defeating the Turkish galleys thanks to advantage in meneouverability. C. Could be 20 m in length, 3–4 m in width and had a crew of 50–70 men. There was always a stock of weapons and food supply on board. Armament: 2–6 light cannons (swivel-guns).

Chakraz Cape – sited in the vicinity of Amasra Peninsula, on the Turkish Black Sea coast. Cape summit is an upright rock, its lower half being a steep slope with jads on top.

Chakva – urban village on the coast of the Black Sea (Adjaria, Georgia). Railway station. Tea factories, plant for citrus fruit sorting and packing, tea and citrus-growing farms. Scientific research institute for tea and subtropical crops.

"CHALLENGER" – (one who issues a challenge).

(1) British 3-mast sail-steam-powered naval corvette on which the first comprehensive round-the-world oceanographic expedition was performed in 1872–1876 marking the commencement of systems research of the ocean from board of specialized vessels, development of new technical facilities and methods of observation. Was built in 1851. Had a complete sail furniture and two steam engines. In 1872, was converted into a research vessel for hydrological, biological and meteorological investigations.

(2) In memory of "C." the name of "Glomar Challenger" was given to the up-to-date American research vessel for deep-water drilling of the ocean floor. Built in 1968. The ship's drilling equipment provides for the drilling of wells as deep as 1,000 m, the ocean depth at the point of drilling being up to 6,000 m. The ship is furnished with a dynamic-positioning system for keeping the vessel in a fixed position despite the impact of the wind, waves and currents. The use of the vessel for deep-water drilling of the ocean floor has played a significant role in increasing our knowledge of the structure of the Earth's crust and of the physical processes going on inside it. Gross tonnage 10,500 tons, length 122 m, the power of the propulsion machinery 7,360 kW, speed up to 12 knots. The vessel participated in geological surveys of the Black Sea.

Cham, Cape – juts out 12.5 km from the coast to the north-west of Bozukalle Cape and limits on the north Pershembe Bight (the most reliable place for anchorage on the Anatolian coast, Turkey). The cape is the north-eastern termination of the coast wide ledge and is formed by the mountain slope. The cape shore is precipitous and present a steep coast.

Chapega-Kulish Zakhary Alekseevich (1726–1797) – when Zaporizhye Sich was defeated in 1775, C. was Colonel. In 1878, he was among the Cossacks who had put up with dominance of Russia. C. became famous for his participation in the battles for Izmail, Bendery, was promoted to the rank of a Brigadier and was multiply decorated. C. headed the resettlement of Cossacks in the Kuban area, where he founded 40 Cossack villages, laid the foundation of the Kuban Army as well as of all settled troops.

Chatyr-Dag, Shater-Gora – quasi-plateau massif in the Main Range of the Crimean Mountains, in the vicinity of Alushta. The top height is 1,527 m. Structured by limestones. Karst forms of relief. The slopes are covered by woods of beech, hornbeam and pine, the surface of the summit exhibiting xerophytic plants.

Chauda, Cape – eastern entry cape of Feodosiya Bay, Crimea. The cape is low and flat; the shores are precipitous. C.C. is fringed by a reef passing along the shore at a distance from it in the north-western direction. There is the Chauda Light-house installed east of C.C.

Chayeli (Çayeli) – a town on the coast of the Black Sea, Turkey.

Chefaliano Mikhail Ivanovich – Captain of a Major-General rank (1799). In May of 1770, enrolled in Russian service as a volunteer in Auza port. In the Russian-Turkish war of 1768–1774, took part in landing an amphibious party on the frigate "St. Pavel" during the seizure of Keffalo fortress. In 1776–1778, cruised between Livorno and Constantinople. From 1782, continued service in the Azov flotilla, where he commanded the frigate "Trinadtsaty" ('13th'), the "St. Nicholay" frigate, the "St. Vladimir" battle ship on which he participated in the Russian-Turkish war of 1787–1791 as part of the Liman squadron near Ochakov (1787). While commanding the same ship in the squadron of Read-admiral John Paul Jones, took part in battles against the Turkish fleet in Liman, sailed between Ochakov and Kinburn. In 1789, appointed Commander-in-Chief of the rowing flotilla with which he cruised between Kherson and Ochakov. In 1790, commanded the frigate "St. Mark" in the squadron of Rear-Admiral F.F. Ushakov, sailed from Sevastopol to the Danube Mouth. In 1791, on the frigate "St. George the Victorious" took part in the battle of Kaliakria against the Turkish fleet (1791). In 1792–1798—commander of the battle ship "St. John the Forerunner" of the Black Sea navy. From 1798, commanded ship's crews in Sevastopol, in November of the same year appointed Captain over Taganrog Port. In April of 1803, dismissed from service.

"Chernoe more" ('Black Sea') – Bulgarian-Soviet team—set up in 1988 as an unofficial association of experts for the drafting of a program whose main purpose was to develop the concept of preventing the Black Sea ecological catastrophe. At the first workshop of the team "Pomorie-88" it was decided to exercise major elements of the concept in one of the disadvantaged areas of the Bulgarian shore—Burgas Bay. It was necessary to diagnose the state of Burag Bay and issue practical recommendations on minimizing the load on its environment. In 1989, seven purpose-oriented expeditions to the bay and adjoining sector of the

Black Sea were held on research vessels "Yakov Gakkel", "Akademik Boris Petrov", "Akademik Knipovich", "Professor Kolesnikov" and "Admiral Ormanov". The concluding stage of works was the workshop "Pomorie-89", at which the obtained materials were analyzed and the report "Diagnosis of the Existing State of Ecological Conditions in Burgas Bay with recommendations on nature-conservation activity" was prepared. In 1990, the monograph titled "Practical Ecology of Marine Regions of the Black Sea" was published in Ukraine on the strength of the accomplished work.

"Chernomor" – Soviet manned hyperbaric undersea station. Was developed and made by the Southern Branch of the P.P. Shirshov Institute of Oceanology (SIO), USSR Academy of Sciences in association with the DOSAAF Central Experimental Design Bureau for Specialized Equipment in 1968. "C." Was designed for 4–5 hydronauts and was intended to operate at the depths of up to 40 m, where it was possible to use conventional oxygen-nitrogen breathing mixture (i.e. air). The first experiments with "C." were made in 1968, then in 1969–1974. At that time, "C." went down for varying periods (up to 50 days) to the depths of 10–30 m in Blue Bay (near Gelendzhik) of the Black Sea. Subsequent operations with "C." were performed on the shelf of Bulgaria (until 1974) in the framework of the Bulgarian-Soviet Experiment "Shelf-Chernomor-73", then "Chernomor-74" organized by SIO USSR AS, National Oceanographic Committee of Bulgaria, and other research institutes in Bulgaria.

Chernomorets – a small village on the coast of the Black Sea, Bulgaria.

Chernomorie – name used in Bulgaria for Bulgarian circum-Black Sea area.

Chernomorskaya atherina (*Atherina mochon pontica*) – fish of *Atherinidae* family. Reaches 15 cm in length. Common in the Black and Azov seas. Pelagic schooling fish. Capable of dwelling in sea and fresh water. Age of life 3–4 years. Reaches sexual maturity at age 1 year. Propagates from April to mid-August. Each year migrates to feeding, spawning and wintering areas. At the end of autumn, C.A. forms small schools and gradually migrates to the south, and by the end of December leaves for wintering grounds—in the southern areas of the Black Sea.

Chernomorskaya briza – moist and cool summer breeze, blowing in the afternoon along the Bulgarian coast.

Chernomorskaya Cordon Line – a series of land cordons set up in the Kuban River meanders as far as Voronezh redoubt downstream as far as Bugaz almost over the distance of 300 km. The line in question was set up by Z.A. Chepega-Kulish. By 1853 C.C.L. comprised the fortress of Ekaterinodar (headquarters) and six forts.

Chernomorskaya Cossack Flotilla – established in 1787 and made up of sea Cossacks headed by Anton Golovaty. The Cossack fleet comprised 50 boats. Ther flotilla was subordinated to the chief of Limanskaya (Dnieper-Bugskaya) Flotilla Rear Admiral Charles Henry of Nassau-Siegen. The flotilla was baptized by fire in the battle of Kinburn Spit.

Chernomorskaya Oceanographic (Hydrographic) Expedition – lasted from 1923 to 1935. It was initiated by Yu.M. Shokalsky who was its first chief. Then the expedition was headed by V.A. Snezhinsky, Chief of the Hydrometeoservice of the Black Sea Naval Hydrography. In 1923, Sevastopol Biological Station of the USSR Academy of Sciences started making oceanographic sections along the line Cape Sarych in the Crimea (southern termination of the peninsula toward Inebolu Cape on the Anatolian coast of Turkey).

Chernomorskaya Coastline, Chernomorskaya Fortified Coastline – established in 1832 in the form of several forts on the sea coast from Anapa to Poti. Was made up of fortifications designed to guard the coast with a view to repressing illicit trade between the highlanders and Turks and terminating delivery of arms and ammunition to the highlanders. The Azov Cossack troops were commissioned to soldier the line. The Cossacks founded a series of forts and Cossack villages: Holy Spirit (Adler), Alexandria (Sochi), Lazarev Fort (Lazarevskoe), Golovinsky (Golovinka) and others. The official commissioning of the line took place in 1839 (1840). Part of fortress fort remains has been preserved as historic monuments. By 1853, C.S. included 25 stand-alone fortifications and forts. Of these, crucial were first-class fortifications: Anapa, Novorossiisk, Sukhum-Kale, Redut-Kale and Poti. Second-class fortifications included: Gelendzhik, Velyaminovskoe, Fort of Holy Spirit, Pitsunda, Bombary and St. Nickolas Station.

Chernomorskaya Soviet Republic – came into being in March of 1918 on the territory of Chernomorskaya Province by decision of the 3rd Congress of the Soviets of Workers', Soldiers' and Peasants' Deputies of Chernomorskaya Province (March 1–13, 1918, Tuapse) as part of the RSFSR. In May of 1918, by decision of the 3rd Extraordinary United Congress of Soviets of the Kuban (established in April of 1918) and Chernomorskaya Soviet Republics (May 28–30, Ekaterinodar) the Kuban-Chernomorskaya Soviet Republic was established as part of the RSFSR. In July of 1918, the Kuban-Chernomorskaya and Stavropol Soviet Republics were united into the North-Caucasus Soviet Republic as part of the RSFSR. Ceased to exist on January 11, 1919 in connection with the occupation of the North Caucasus by the Armed Forces of Russia's South.

Chernomorskoe – urban village, health-resort location on the coast of Akmechet Bight, Karkinitskiy Bay, center of agricultural area, the western part of the Crimean Peninsula, Ukraine. In classical time (Fourth to second centuries B.C.) this place was occupied by the ancient Greek city Kalos-Limen ("Beautiful Harbor"), whose ruins have lived to this day (curtain walls and ceramics, remains of the wall of a dwelling); in the second to third centuries A.D. the location was occupied by a settlement of an unknown name. In the eighteenth to nineteenth centuries it was called Ak-Mosque ("White Mosque") or Sheikhlar, from 1944—Chernomorskoe. Considerable reserves of natural gas have been discovered in the water area adjoining the Black Sea (Golitsynskoe Field). In late 1970s—early in the 1980s, a submerged gas pipeline was laid here for the first time. The blue fuel is supplied over the pipeline to the settlements of the Steppe Crimea. At present—republican Historic-Archaeological Reserve "Kalos-Limen".

"Chervona Ukraina" – Soviet cruiser. Formerly known as "Hetman Bogdan Khmelnitsky"—the Civil War over, was named "C.U.". Took part in the defense of Sevastopol during the war of 1941–1945, by supporting the city defenders with the fire of its batteries. Once, the cruiser was attacked by 28 enemy bombers and was heavily damaged. The struggle for cruiser unsinkability lasted 16 h, yet it did sink in the end. The batteries removed from the sunk ship and installed on the shore continued to be used effectively. In memory of the cruiser wreckage a memorial plaque is installed on Grafskaya Pier in Sevastopol, which reads: "Here, while engaging the enemy, the cruiser "Chervona Ukraina" was wrecked on November 12, 1941". After the war, the cruiser was lifted from the seabed and was commissioned as a training ship.

Chichagov Pavel Vasilievich (1767–1849) – Russian Admiral, participant of the 1812 Patriotic war. In the Navy—from 1782. Was the Aide de camp of his father—Admiral V.Ya. Chichagov. During the Russian-Swedish war of 1788–1790, commanded a battle ship. In 1791–1792,—student of a naval college in Britain. From 1802—Vice minister, Russia's minister of naval forces (1802–1811). In 1812, appointed Commander-in-Chief of the Danube Army, Chief Commander of the Black Sea Navy and Governor-General of Moldavia and Valakhia. When the Patriotic war of 1812 broke out, the southern armies were united and brought under the command of C. seeking to cut off the ways of retreat of Napoleon troops. The plan failed, however. In 1812–1813, C. commanded the 3rd Army that pursued the enemy. In February of 1813, was dismissed for retirement. An atoll in the Pacific in the Ratak group of Marshall Islands is named after C.

Chichagov P.V. (http://feb-web.ru/feb/rosarc/chg/chg-001-.htm)

Chikhachev Petr Aleksandrovich (1808–1890) – Russian geographer and geologist. Honorable member of St. Petersburg Academy of Sciences (from 1876) and of numerous other academies and societies. From 1834 to 1836, served at the Russian Embassy in Constantinople, Turkey. C.—one of the first Russian geographers who studied the countries of West Europe from the orographic, geological, paleontological and botanical standpoints. In 1834–1836, C, visited various cities of the Osman Empire and traveled to Spain, Portugal, Italy, France and other European states. In 1846–1863, undertook a number of expeditions in Asia Minor as a result of which he made detailed descriptions thereof. C. published over 50 works on physical geography of Asia Minor, the culmination of which became is monograph "Asia Minor" made up of four parts and eight volumes. In 1864, C. published the work "Bosporus and Constantinople" in Paris which was subsequently reissued in 1866 and 1877. In 1868, he published a small book titled "Pages on the East" in Paris.

Chirus – young mackerel.

Chongar Peninsula – sited in the western part of the Sea of Azov, in the north of Sivash Lake. Together with Tup-Dzhankoi Peninsula approaching the sea from the Crimea, C.P. dissects Sivash into two parts: western and eastern. Together with C.P. the bridges and filling dams connect the continent with the Crimean Peninsula. The highway and a railway line, linking Russia with the Crimea, are laid over these. C.P. often used to be the site of fierce battles when the Crimea was seized or liberated.

Chorokh, Chorukh, Coruh – river in the extreme east of Turkey and in Adjaria (Georgia). Length of the river is about 500 km (within Adjaria—20 km). Flows from North Tavra, Turkey, in the Gavur Dag Mountains, north of Erzurum. Falls into the Black Sea near Batumi. The source of alimentation in the lower reaches—rainfall. The river valley before the river falls into Ardanuch-Chaii is a deep and narrow gorge, squeezed from the north and south by mountain ranges. Downstream of this point, the Chorokh gorge widens and the river flows along Kakhaber plain over 10 km and, 4–5 km before reaching the mouth the river falls into several arms. The normal annual runoff of C. into the sea (Erge Village, 15 km from the estuary)—8.7 km^3/year. The annual march from April to July accounts for over 50 % of the annual runoff. Maximum runoff is noted in May, minimum—in August. Due to the high velocity of the flow and meandering fairway, the movement down the C. is quite risky and is facilitated by means of kayuks—narrow, sharp-nosed flat-bottom boats. The main tributary of the C. is Oltu River. Besides, the river has numerous mountain tributaries. During the rainy season, these turn to violent deep onrushes.

Chufut-Kale – fortress in the Crimea. It also used to be called "Iosafat's Valley", "Karaim Cemetery" and "Holy Oaks Grove". Translated from the Crimean Tatar Language, it means "Jude's Fortress", or Kyrk-Or, Toprak-Kala, Butmai, Gevkher-Kermen. The latter name is translated as "Treasures Fortress" ("Valuable Fortress"). Tatar ulems named it Gevkher-Kermen because in those times all gates, walls and wickets used to be decorated with precious stones. Sited on the plateau of a table mountain, its height more than 500 m above the sea level. It towers 200 m over the surrounding valleys. C.-K. is remarkable in that its prison at different times was the place of confinement of the Polish Hetnman Pototsky, Lithuanian ambassador Lez, Russian envoys Vasily Gryaznoy, Vasili Aitemirov and Prince Romodanovsky. The fate of the Russian military leader (voevoda) V,B, Sheremetev was particularly tragic: he stayed behind bars more than 20 years (from 1660 to 1681). The ransom requested for him was so large that neither the Russian Tzars, nor his relatives were able to pay it for a long time. While that man was imprisoned, four Khans succeeded one another on the Crimean throne. It was only in 1681 that the relatives finally were able to pay the ransom for the seriously-ill and blind old man, and 6 months later Sheremetev died.

Chursin Serafim Evgenievich (1906–1985) – Soviet Admiral. In the Navy from 1926. In 1931, finished The Frunze Naval Secondary School. From 1931 to 1943— on the Pacific Navy, serving on submarines. In 1944–1948—commander of submarines brigade of the Black Sea Navy. In 1948–1952—Commander-in-Chief of the Danube Military Flotilla. In 1953–1954—Commander–in–Chief of the Caspian Military Flotilla. In 1954 graduated from the Military Academy of General Staff. In 1955—Chief of Staff, 1st Deputy Commander-in-Chief of the Black Sea Navy, from 1962—Commander-in-Chief of the Black Sea Navy. Promoted to Admiral (1964). From 1968—Professor-Consultant of the Naval Academy. From 1971– retiree.

Chushka, Spit – a 15 km long sandy spit located on the northwestern tip of Taman Peninsula, eastern coast of Kerch Strait. The name is due to the numerous dolphins that like sea pigs. The spit is sandy, barring the northern part of the strait. The western shore is protected from wash-out by stone crib barriers. At the western shore of the spit, Port Kavkaz has been constructed in 1953. The port services the railway and auto ferry crossing across Kerch Strait. At the Port Kavkaz on C.S. there are plans to build a separate cargo-handing area to include the oil-loading jetties, remodel the existing and build new jetties in the existing port, build a new dry-cargo handling area for containers, general cargoes.

Chvizhepse – balneal climatic health resort location in Sochi area, Krasnodar territory, Russia. Sited on the bank of the Chvizhepse mountain river. The climate is distinguished by a great number of sunny days, clean air, low relative humidity. The primary natural curative factor is mineral water. The spring of arsenic-containing water (arsenic concentration above 5.5 mg/l) is at the height of 300 m. The Chvizhepse water ("Sochi narzan") by its composition is similar to Kislovodsk narzan, but is devoid of magnesium sulfate. The daily yield of the well equals 280 m^3. The balneary of the Sochi Research Institute of Balneal and Physiotherapy is in operation.

Cide – a village on the coast of the Black Sea, Turkey.

Cimmeria – see Crimea.

Cimmerians – the tribe that populated the area near B.S. Very little is known about this tribe and even its real existence was questioned. C. were sometimes mixed up with other peoples (e.g. by Gomer). According to Herodotus, C. made a nomadic tribe originated in the southern Russian areas. As there are no written proofs the question of their Indo-European origin remains open. The Assyrian sources contain proofs of C. staying in Armenia (Urartu) in the seventh century B.C. Approximately in 675 B.C. C. destroyed the state and in 652 B.C. conquered the capital. According to Herodotus, approximately in 600 B.C. they were defeated by tsar Aliatt. However, already in the Herodotus time (fifth century B.C.) "these were the legendary and mysterious people". C. lived in Crimea in the period from fifteenth to seventh centuries B.C. or, according to other researchers from thirteenth to twelfth to seventh centuries B.C. We are reminded of C. by such names as the Cimmerian Bosporus (Kerch Strait), Cimmerian walls, Cimmerian ford and others. One researchers link the early C. with the tribes of the catacomb culture, other—with log-house one. Judging by the Crimean materials (burials, tools) the researchers identifying C. with the late catacomb culture are closer to the truth. Obviously, C. is a collective term meaning the union of kin tribes. The Antique sources describe them as bellicose people, which fact is confirmed by archeological findings—presence of arms in burials. C. main occupations were cattle growing and farming; they lived in stone dwellings. In the seventh century B.C. the Scythian nomad tribes forced some of C. from these lands, but some of them stayed. Perhaps the remaining C. were forced by the Scythians to the mountain part of Crimea. The greater part of monuments of the Kizil-Kobinksy culture refers to this time (seventh to fourth centuries B.C.) and they were mostly found in mountain Crimea. These and many other findings (similar tools and burial rites) enable to attribute the Kizil-Kobinksy people to the latest C. forced to the mountains. According to Herodotus, C. were expelled by the Scythians from the southern Russian steppes and Crimea to Asia Minor where their traces get lost.

Citrus – evergreen tropical fruit trees or shrubs of Citrus genes (larger arboreal forms) and Fortunella (small shrubs) of the Ruta family. Characterized by elliptical relatively coarse leaves, white fragrant flowers, large sappy fruits with pulp that is divided into vectorial segments and thick rind with a high content of essential oil. Main homeland of C. is monsoon South-East Asia, C. are cultivated on a large scale in tropical and subtropical countries world-wide (in arid areas—with irrigation) between 40°N and 40°S. C. in culture are capable of enduring relatively cold winters of subtropical areas even with short frosts. Major C. are oranges, tangerines, grape-fruits, lemons. Of other C., most common are F. japonica, F. Maryarita with small yellow fruits, tangerines (C. reticulate), citrons (C. medica) with orange inedible fruits used for fragrant rind, shaddock or pomelo (C. grandis). C. are cultivated in the subtropical areas of the Caucasian Black Sea coast areas (for the most part, small-fruit Japanese tangerine-unshiu (C. unshiu), lemon, orange).

Classical cities of the Circum-Black Sea Area – came into being during Greek colonization in the sixth century B.C. on the north coast of the Black Sea. Major classical cities in the North circum-Black Sea area—Tira, Olvia; in the Crimea—Khersones, Kafa (Feodosyia), Pantikapei, around Kerch Strait and on the Caucasus Black Sea coast—Fanagoria, Hermonassa, Gorgippia, Dioskuriada, Fasis; in the Don River mouth—Tanais. At the end of the seventh century B.C., there emerged Greek trading points emporia on the coasts of the Black Sea. C.C.C. used to live their own life, tendering commercial and cultural links with their metropolitan countries. Crucial to their economies was trade with the cities of Greece and Asia Minor and with the tribes of Black Sea steppes. The cities used to import from the Aegean Sea basin wine, olive oil, metal articles, marble, ceramics, expensive textiles, etc., the major export items being grain, cattle, leather, salted fish, slaves. Cottage industries, handicrafts, agriculture and cattle-raising were just as important to their economies.

As far as the social and political structure is concerned, all C.C.C. were initially slave-owning policies. Some cities around Kerch Strait already in 480 B.C. united around Pantikapei to form the Bosporan Kingdom. C.C.C. flourished in the fourth century B.C. when they were crucial suppliers of grain and other consumer goods for many cities of Greece and Asia Minor. They maintained a close contact with local tribes, had economic and cultural impact on them, helped erode the patrimonial system, development of ownership-related differentiation and establishment of class relations.

The arts of C.C.C. have many traits characteristic of the entire art and culture of the classical world. The cities had regular planning, order was used in cult-related buildings, the type of a residential building with peristyle was common. Artistic items of all descriptions were imported in C.C.C. from Greece on a large scale: sculptures, small plastic items, gems, painted vases. At the same time, the art had features peculiar to particular locations. Portrayed in the works of art was the life of the surrounding peoples (scyths, sinds, meots, etc.), their religious ideas and rituals, ornamental wall paintings, sculptural grave-stones steles, painted ceramics (Olvian and Bosporan vases with stylized plant motives, Bosporan "water-colors"). The peculiar aspirations of C.C.C. are best manifest in the art of the Bosporan Kingdom. This was reflected in religious perceptions of C.C.C. dwellers. Along with Greek deities (Apollo, Demetra and others), they often addressed other than Greek deities, e.g. the Virgo goddess of Tavria, and others.

During the third century B.C., C.C.C. had serious economic problems due to the relaxation of economic relations with Greece and complication of the situation in the North circum-Black Sea area. The Scyths driven from the Don and circum-Dnieper steppes began exerting pressure on the cities of the Black Sea coast. In the second century B.C. the Scyth King Skilur put the weak Olvia under control. His son Palak robbed Khersones of its possessions. The dwellers of Khersones turned to the Pontian King Mitridat VI Evpator for assistance, and the city was henceforth ruled by Evpator. At the end of the second century B.C., there flared up a resurrection in Bosporan Kingdom led by Savmak. After the resurrection was crushed by the forces of Mitridat VI Evpator, Savmak became the King of Bosporan Kingdom.

Olvia, too, became one of his possessions. In his struggle against Rome (89–63 B. C.) he used the North circum-Black Sea area as the base for supplying his armies and a source of troops replacement. After Mitridat VI was killed in action, there followed a bitter struggle in Bosporus for the throne, in which the Romans and leaders of the tribes surrounding Bosporus took part. Bosporan cities were damaged and were finally ruined. During the early centuries A.D., C.C.C. once again went through a period of revival. However, Olvia, restored from ruins after being destroyed in the mid-first century B.C. failed to reach its former level of development. During the second half of the first century B.C., it was ruled by Scyth Kings, in the second century A.D. it was subjugated by the Roman Empire. Khersones alternately depended on the Bosporan Kingdom, or on the Roman Empire.

During the early ages A.D., the economic, political and cultural life of the classical cities of the north circum-Black Sea area is increasingly dominated by non-Greek, largely Sarmat, elements. In the mid-third century A.D., C.C.C. were subject to ruinous invasion of the Gots. Olvia was razed to the ground. In Bosporus, Gots and allied tribes seized power for a while and made Bosporus the base for their pirate-style incursions on the coasts of the Black and Aegean seas. Khersones, which apparently had joined the Roman Empire at the end of the third century A.D., offered a more effective than others resistance to the Gots. In the middle of the third century, there began a quick downfall of Bosporan Kingdom, the volume of trade and cottage industries declined abruptly, the economy was getting increasingly natural. At the end of the fourth century the North circum-Black Sea area became the scene of devastating aggressive campaigns of the Hunns that completed the downfall of the classical cities.

C.C.C. played an important role in the history of the peoples of East Europe as conductors of the most advanced for their time of slave-owning socioeconomic relations and culture, as mediators between these peoples and centers of the classical slave-owning world of the Mediterranean Area.

"Climatic fields, water salinity and temperature in the Black Sea" – a handbook prepared in 1987 by the Sevastopol Branch of the State Oceanographic Institute (E.N. Altman, I.F. Gertman, Z.A. Golubeva). It contains the diagrams of the average many-year distribution of water salinity and temperature in B.S. broken by months, seasons and standard levels estimated on the basis of the modern statistical methods using the materials of the regional databank created in the Branch.

Climatotherapy – a combination of treatment methods based on the rated exposure of a man's organism to the impact of the climatic and weather factors and special climatic procedures. The curative and prophylactic effect of climate on an organism may be attained thanks to some natural factors the most important of which are the altitude of an area over the sea level, remoteness from the sea, atmospheric pressure, air temperature, circulation and humidity of air, amount of precipitations, cloudiness, solar radiation intensity, including ultraviolet, and others.

Colony (from the Latin *"colonia"*) – a settlement founded by the Greeks during colonization of the Mediterranean and Black Sea coast. Greek C. was politically independent of the metropolis. The inhabitants of Roman C. were Roman citizens exercising all respective rights. In the republican period C. was founded, first, only beyond the Lacia and later—beyond Italy. They were established for consolidation of the Roman military control of the conquered territories and Romanization of the local population. C. was mostly populated with agrarian people, poor people and veterans. From the old times C. was organized for allotting land to poor people, thus, alleviating the struggle for agrarian rights that the landlords dreaded the most. C. was independent in legal terms and its administrative structure repeats the Roman one.

"Combat assault on Sapun-Gora on May 7, 1944", Diorama – commissioned on November 4, 1959. One of the major works of contemporary battle painting. Canvas size 25.5 × 5.5 m, figural plan area—83 m². Consists of two parts: canvas, the main part of the diorama, and figural plan, both constituting a whole. The diorama depicts the final stage of German fortifications being assaulted by Soviet Army on Mt. Sapun near Sevastopol on May 7, 1944, when the attackers were to cover the last few dozen meters, yet those were the hardest meters to cover. The lower floor of the diorama building houses a museum whose exhibits deal with the defense and liberation of Sevastopol. Samples of the Great Patriotic War (WWII) combat materiel are displayed on special exposition grounds.

Common Black Sea mullet (*Mullus barbatus ponticus*) – fish of the *Mullidae* family. It has an elongated body, slightly compressed on sides and covered with large scale. Its body length is approximately 20 cm. It is found everywhere near the coasts of the Black and Azov seas, in the Bosporus and the Sea of Marmara. It lives for 8–10 years. M. reaches maturity by the age of 1 year, rarely 2 years. It multiplies from early June through mid-September. In late October it starts migrating to the south and by late December it stays for hibernation in the Bosporus and the Sea of Marmara. It prefers staying near the muddy floor and with the help of two barbs M. hunts for mollusks, scuds, tiny crustaceans, small worms. In the Black Sea fishery its role is not large. This fish has many names and one of them is sultanka. It got this name because in the old times only oriental rulers—padishaks and sultans were allowed to each this fish. By tasting the Crimean people put it on the second place after flatfish.

Common dolphin (*Delphinus dephis*) – its length is 1.6 m on the average (males—from 0.9 to 2.6 m and females—from 0.8 to 2.8 m), the average weight—51 kg (40–60 kg). They differ from other Black Sea dolphins by a protruded nose reminding of a beak and by a typical fatty bulge on the forehead. Its high (to 30 cm) dorsal fin is located closer to the front end of a body, well developed, sharp and has a sickle form. The upper part of the body as well as fins, head and nose are of a dark color—from dark-gray to black-blue or black-brown and black. The sides of the animal are light-gray. The belly is white with the yellowish or light-bluish tint crossed with slanting long dark strips. C.D. is cosmopolitan. They are met in great quantities near coasts of the interior seas of the Pacific and Atlantic Oceans, including Mediterranean and the Black Sea. They usually live in great shoals. They swim fast, love to play and jump over the water surface. They do not dive deep and do not stay long under water. C.D. lifespan is 20–30 years. The propagation period of the dolphin is rather long—from June through December, but most often it propagates in the warm months—from August to October with the peak in September. Nearly every year a female dolphin after 10–11 months of pregnancy gives birth to one, rarely two calves. The length of the newborn is about 80 cm, weight—5 to 8 kg. A mother-dolphin usually feeds her child during 4 months. Her dense milk contains much fat (to 50 %) and calories. In autumn the dolphins accumulate fat that protect them from cooling in winter. C.D. usually feeds on low-value small fish, but in summer they may chase larger fish. Its food ration includes anchovy, sprat, horse mackerel and other, mostly pelagic fish.

Common kilka (*Clupeonella delicatula*) – fish of the herrings family (*Clupeidae*), elongated, significantly flattened on the sides, up to 9 cm in length and of 7–8 g mass. The dorsal part of the body is gray-green, the sides and belly—silvery-white. Common in the Sea of Azov and in the desalinized parts of the Balck Sea. One of the major commercial fishes in the Sea of Azov.

Common mackerel (*Scomber scombrus*) – is covered by small, easily detachable scale. Dorsal fins are spaced far from each other. Behind the second dorsal fin there are 4–6 finlets and behind the anal fin—4 to 5 finlets. Its back is dark-blue with cross stripes over the lateral line. Its length may reach 60 cm, weight—about 1,200 g. C.M. populates mostly in the North Atlantic Ocean, along the coast of North America (from Labrador to Cape Hatteras) and Europe, in the Mediterranean. This is a warmth-loving pelagic species. It lives for 6–8 years. It reaches maturity at one year of age. It spawns in February-May in the Sea of Marmara. After hibernation and spawning in the Sea of Marmara C.M. forms shoals and migrates northward and till July this fish spreads in the northern and northwestern parts of the B.S. In summer and autumn it keeps mainly in the northwestern part of the B.S. where it feeds extensively and grows. Usually in the second half of October the mackerel shoals start migrating to the south and in late December—early January they move for hibernation in the Sea of Marmara. They run to the B.S. only for fattening. During autumn migration in late October—early November when its population is the largest and the hydrometeorological conditions are most favorable mackerel forms considerable aggregations near Emine—Maslen Nos Cape (Bulgaria). Mackerel is a valuable commercial fish species in the B.S., its resources are varying reaching in some years more than 30,000 tons.

Common porpoise (*Phocaena phocaena*) – the smallest species of the Black Sea dolphins. Its length varies from 0.8 to 1.9 m making on the average about 1.3 m. Its weight is about 50 kg (from 30 to 90 kg). The males are smaller than females. C.P. has a blunt nose and a short, solid, not large body. It is bi-colored—the top of its body is dark-gray, nearly black, while the belly is white or light-blue. C.P. lives in northern seas as well as in the Atlantic and Pacific oceans. It is also found in the

Mediterranean and Black seas and most often in the Sea of Azov. They usually move in pairs or small groups of fewer than ten individuals. Rarely some species form brief aggregations of several hundred animals. It feeds on small pelagic and bottom fish, shrimps and other invertebrates. C.P. multiplies in July-August, seldom in September-October. A female after pregnancy for 9–11 months gives birth to one, seldom two calves, most often in the period from the second half of April to July. A calf sucks its mother during 4–6 months and after this it starts feeding independently. The dolphins which number in the B.S. several decades ago was about one million species were hunted in commercial scales first by Turkey, but later other B.S. countries joined in this. The fat of porpoise was used in manufacturing of lubrication oils, shoe polish, vitamins "Delphinol", waterproof skin and other products. Its meat was air-dried. Before the intensive fishing of porpoise the shoals numbering several hundreds and even thousands of this species were not rare in the B.S. Later the population of C.P. shrank drastically, but nowadays their number is again growing. In response to the UNESCO message fishing of dolphins in the B.S. was prohibited by the USSR in 1963 and by Bulgaria in 1966.

Common stingray or whipray (*Dasyatis pastinaca*) – bottom fish of the *Dasyatidae* family. It has a naked, flat, rhomb-shaped body ending with a long thin tail having one or two sharp bone thorns containing cells releasing poisonous secretion. It has no dorsal and tail fins. The body on top is gray or olive-brown and the belly is white. Its length is to 1–2 m and weight—4.5 to 10 kg. This species is most widespread in the Atlantic Ocean near the coasts of Europe, America and Africa and in the Mediterranean and nearby seas. This is live-bearing fish. It prefers depths of 10–80 m. In summer it comes closer to the coast. In July–August it gives birth to 4–8 youngsters. It feeds on shellfish, mollusks and small fish.

Common water snake (*Natrix tesselata*) – one of the vertebrates usually found in river deltas, freshwater and saline lagoons, coastal waters of the sea. In some lagoons, especially on the seaside of the islands in the Danube Delta, the number of snakes may be as high as two–three dozens per kilometer. The snakes feed on fish. In freshwaters these are fries of common carp, crucian carp, bream, sturgeons; in the sea—mostly bullhead. Inside snakes of the average size there were found up to 40 carp fries 2–3 cm long and small fish to 10–12 cm long. Water snakes incur significant damage to fish farms, thus, they are hunted there. In the B.S. the water snake has its own island called "Zmeiny". We can imagine how amazed were the ancient seafarers when they saw many snakes on the Achilles-Levke Island where they baked in sunrays on stones and cliffs. These were snakes that swam to the island located 37 km from the Danube Delta or maybe they made this trip on "rafts" or the cane roots that were often drifted into the sea by the Danube River. On this island the snakes found much fish, bullheads in particular, many caves and, in general, tranquility. Here they settled down and lost their ties with the Danube. The Ancient Greeks gave this island another name—Thidonisy, meaning "the island of snakes". Even today the old fishermen from Odessa and Crimea still refer to this island as Thidonisy. In all modern languages it is translated word-for-word "Island Zmeiny".

Concretions (from the Latin "*concretia*") – small nodules in the B.S. to 5–10 mm in diameter of iron and manganese hydroxides that grow most often around the shells of the *Modiola* mollusk (*Modiola phaseolina*). The largest field of the iron-manganese C. in B.S. was detected still in 1900 by Academician N.I. Andrusov in the Kalamit Bay. Later C. were found in the shelf areas near Feodosia and Yalta. Romanian geologists found C. on the Constanta shelf, while Bulgarian geologists—near the Kaliakra Cape and in the southern part of the Bulgarian shelf. The Black Sea C. contain to 26 % of iron, 5–7 % of manganese and some microelements. The knowledge of the mechanism of C. formation on the silty floor is not full, but it is thought that biological, in particular microbiological, and physico-chemical factors play their role in this process. The Black Sea C. have no commercial value, while the oceanic C. with a diameter to 15 cm that are reach not only in iron and manganese, but in copper, molybdenum, nickel and other metals are produced on a commercial scale in some regions.

Constanta – the main town and port on the B.S. in Romania. It is situated in the southeast of the country. Its population is 254,700 people (2011). C. is an administrative center of the Constanta County, Dobrudzha Region, Romania. This town was founded in the sixth century B.C. by the Greek colonists from Miletia and was called Tomi (Tomis). In 29 B.C. it was conquered by the Romans. Roman poet Publius Ovidus Naso who wrote here his well-known works, such as "Tristia" and "Epistulae ex Ponto", was on exile in these places from late 8 to 17 and died here. In 320 this city was called Constanta by the name of the sister of Roman Emperor Constantin I who was ruling that time.

C. is the center of trade routes and more than the half of the country's foreign trade traffic passes through this town. The main commodities exported via C. are oil products, grain, cement, wood and imported—machines, equipment and apatite. The major transport center and the major port in Romania. In 2003 C. became a free port. In the same year the first stage of a new container terminal was commissioned here. The port accommodates large oil storage tanks. Via pipelines it is connected with the Ploieshti oil fields. The main industries here are machine-building (ship-repair yard), food and textile. The shipyard in C. is the Romania's largest and up to date enterprise. It was founded in 1890 and since that time had been the fishery center.

The city can boast of many historical monuments of the Greek and Roman times. It has been also known as a resort since the late nineteenth century. The city has the archeological and art museums, the sea museum, the marine aquarium, and a dolphinarium. The Romanian Marine Research Institute is also found here. The Ovid's statute is installed on the Independence Square. The monument to great Roman poet Mihai Eminescu is also found in C.

Constanta County – is situated in the southeast of Romania, on the B.S. coast. This county extends over the limestone Dobrudzha Plateau passing in the north into the Tulcea Mountains (to 456 m high), part of the floodplain and vast Danube Delta. More than three-fourths of the arable lands are sown to grain crops, mostly wheat and corn; large areas are sown to barley. The major technical crops are sunflower, rape, flax, cotton, tobacco and sugar beet. The sheep and horse breeding is also developed. In the coastal areas the fishing is practiced. Building materials are produced in the quarries near the Tulcea Mountains, copper pyrites—near the Altyn-Tepe. Among the processing industries dominating are food (fish and vegetable canning factories, flour mills, etc.), machine-building (ship repair yards in Constanta, Tulcea), production of building materials (cement plants in Cernavoda, Medgidia) and production of mineral fertilizers (from 1958; plant in Navodari). Railroads connect C. with Bucharest and Bulgaria. The major sea port of the country is Constanta. The coast abounds in resort towns (Mangalia, Mamaia, Eforie, etc.).

Constantinople (from the Greek *"Konstantinupolis"*, in Old Russian *"Tsargrad"*) – in 330–1453 the capital of the Byzantine Empire. It is situated on the left coast of the Bosporus near the outlet to the Sea of Marmara. It was founded by Emperor Constantine I in 326 on the site of already existing ancient Greek colony Byzantine. Geographically its position was excellent—it made the "golden bridge between the East and West". C. was one of the major cities in the Middle Ages, a large cultural and scientific center and a super-fortress on the Bosporus coast. In 1453 it was conquered by the Turkish troops of Sultan Mehmed II and was made the capital of Turkey (till 1923). It is the historical core of Istanbul.

Constantinople (http://files.myopera.com/sanalmakina/albums/623509/Constantinople-bridge.jpg)

Constantinople Battle – a sea battle of 1043 between the fleet of the Kievan Rus' and Byzantine. The war was unleashed after the provocative attacking by the Byzantines of the Russian merchants in Constantinople (Istanbul). Seeking to force the Byzantine Emperor to abandon opposition of the Kievan Rus' and to improve further the trade relations with Byzantine Kievan Prince Yaroslav the Wise sent to Constantinople the fleet under command of his son Vladimir. The Russian fleet navigated the B.S. secretly and quite of a sudden appeared at the inlet into the Bosporus. Vladimir had a talk with Emperor Constantin XI, but the latter waived the proposals and met the Russian fleet on the flagship. The Russian boats made a line near the lighthouse at Iskrestu (Pharu) at the inlet into the Bosporus and the enemy ships also lined up. By the signal from Constantin XI the Byzantine galleys started the battle, then the battle was joined by the main forces of the Byzantine fleet that used stones, arrows and wildfire. At the peak of the battle a strong storm broke and many Russian light boats turned over and were wrecked against shore rocks or were cast ashore. The Russians suffered defeat. The remaining Russian ships commanded by Vladimir found shelter in the coastal harbors. Constantin XI sent his ships for chasing them, but quite of a sudden they were attacked by the Russian ships, surrounded and the greater part of them was destroyed. Soon Yaroslav the Wise concluded peace with Byzantine.

Constantinople "Eternal Peace" Treaty of 1720 – the "Treaty on Eternal Peace between the Russian and Ottoman Empires" that superseded the provisional Adrianople Peace of 1713. C.T. was signed on 5(16) November in Constantinople (Istanbul). It consisted of the preamble and 13 articles. It proclaimed "eternal peace" between the tsar and the sultan and confirmed the existed state border between Russia and Turkey. Once again (after 1711) Russia was entitled to have a diplomatic representative (resident) in Constantinople. The contracting parties agreed not to assist countries being hostile to any of them and to settle the border conflicts peacefully. It was confirmed that the Cossacks living on the left bank of the Dnieper were left under the Russian ruling and as the Tsar subjects they were not allowed to do harm to the people of the Porta Province, the Crimean inhabitants. Tatars and Cossacks that remained under the Turkish ruling were not allowed to take actions against the Russian subjects. The Turkish government did not protest against bringing Russian troops into Poland for protection of the government existed there and the territorial integrity of this country. Both parties were prohibited to construct fortresses between Azov and Cherkassk, in the Samara River mouth and between the Orel and Samara rivers at their inflow into the Dnieper. Both the Russian and Turkish merchants were guaranteed free trade in both states. The Russian subjects were allowed to visit the "holy places" without payment of any charges to Turkey and without a right to settle down in Turkey for living. This "Eternal Peace" settled the Russian-Turkish relations, strengthened the southern borders of Russia, and enabled to concentrate forces for successful termination of the Northern War (1700–1721).

Constantinople Peace Treaty of 1700 – the treaty between Russia and Ottoman Empire concluded in continuation of the agreement on 2 years of truce that was reached at the Karlowitz Congress of 1698–1699. It was signed on 3(13) July in Constantinople (Istanbul). The Russian delegation arrived to Istanbul on the battleship "Krepost" (Fortress), thus, stressing the fact of the Russian navy presence on the B.S. Under the conditions of this Treaty signed for 30 years onward Turkey recognized the Russia's possession of Azov with surrounding lands and the newly built fortresses (Taganrog, Pavlovsk, Mius) (Art. 4). Russia did not have to pay any more the annual tribute to the Crimean Khan (Art. 8). Turkey was returned a part of the near-Dnieper area occupied by the Russian troops with small Turkish fortresses and fortifications that had to be destroyed forthwith. The Treaty ensured the security of the borders and peoples being subjects both of the Russian state and the Ottoman Porta (Art. 5–7), exchange and setting free the captives (Art. 9); Turkey guaranteed free travel of the Russian ordinary people and monks to Jerusalem (Art. 12). The parties undertook not to build new fortifications within the border strip and not to admit any armed raids. Turkey had to return the Russian captives, to allow Russia to have a diplomatic representative in Constantinople as other states had. This Treaty ensured neutrality of Turkey at the beginning of the Northern War of 1700–1721, but it lost its force in November 1710 when Turkey declared a new war on Russia.

Constantinople Peace Treaty of 1879 – the treaty between Russia and Ottoman Empire signed on 27 January (8 February) in Constantinople (Istanbul) after the end of the Russian-Turkish War of 1877–1878. Having declared "peace and friendship" between these two countries this Treaty fixed those provisions of the San Stefano Treaty of 1878 that were annulled or altered by the treaty signed at the Berlin Congress of 1878. The Treaty also defined the war reparations to Russia (802.5 million Francs) and the payments to the nationals and organizations for the losses caused by the war (26.7 million Francs). Turkey should cover part of expenses on maintenance of the Turkish prisoners of war in Russia. Both parties obliged not to prosecute Russian and Turkish nationals who assisted the hostile armies during the war. Amnesty was granted to the Turkish nationals who took part in the liberation struggle of the Balkan peoples. All previous "treaties, conventions and obligations . . . in relation to trade, jurisdiction and status of the Russian nationals in Turkey" were restored.

Convention of Montreux – see Convention Regarding the Regime of the Turkish Straits. In some official documents it is referred to as the "Montreux Convention".

Convention on Neutrality on the Sea of 1854 – the agreement between Russia and the USA aimed, largely, against the actions of the British-French fleet during the Crimean War of 1853–1856. It was signed on 10 (22) July in Washington. This C. outlined the attitude of Russia and the USA to the rights of citizens of neutral states on the sea, acknowledged the right of the ships of all neutral countries (meaning also the US ships) to carry goods of the citizens from the countries at war (including Russia). The things and goods carried on a neutral ship even if they belonged to the citizens of warring states could not be confiscated or seized, except if they were military smuggling objects. Russia and the USA undertook to observe strictly this C. in relation to all other states that would like to subject to the proposed rules of trade and shipping. It was declared that the states acceded to this agreement would be considered its full-fledged participant. This Convention was ratified on 19 (31) October 1854 in Washington.

Convention on the Protection of the Black Sea against Pollution – was signed by the Black Sea countries on 21 April 1992 in Bucharest, Romania. This Convention consists of 30 clauses and 3 attached protocols. This Convention is applicable to the B.S. proper with the southern border set for the purposes of this C. by a line connecting Kelagra and Dalyan capes. For the purposes of this C. "the pollution of marine environment" means introduction by a man, directly or indirectly, of the substances or energy into the marine environment, including estuaries, that leads or may lead to such detrimental consequences as damage to live resources and life of the sea, hazard to the man's health, obstruction of activities on the sea, including for fishing and other lawful kinds of sea environment use, deterioration of the quality of the used marine water and recreational conditions. Bulgaria, Georgia, Romania and Russia ratified this C. in 1993, Turkey and Ukraine—in 1994. Entry into force in 1994.

Convention on the Regime of the Black Sea Straits – was signed in 1923 in Lausanne, Switzerland. It envisaged the "innocent" passage via the Straits in the peace and war time for the sea and air craft as well and demilitarization of the Bosporus and Dardanelles, i.e. dismantling of the coastal fortifications. The maximum military forces that any country could bring via the Straits into the B.S. should not exceed the forces of the largest Black Sea fleet. At the same time the states were allowed, under all circumstances, to send to the B.S. no more than three ships with the displacement not exceeding 10,000 tons each. For control of compliance with the Convention there was established the International Straits Commission with the headquarters in Constantinople and headed by a Turkish representative. Russia signed this Convention, but did not ratify it.

Convention on Wetlands of International Importance, Especially as Waterfowl Habitat – it is also known as the Ramsar Convention because it was adopted on 2 February 1971 in Ramsar, Iran. It was enforced in 1975. According to this Convention each contracting party defines the respective wetlands on its territory to be put on the List of Wetlands of International Importance. This List includes the Black Sea Biosphere Reserve. On 3 December 1982 the Protocol on Convention Alteration was signed and later on 28 May 1997—the Amendments to the original Convention.

Convention Regarding Shipping Regime on the Danube – was signed on 18 August 1948 in Belgrade by Bulgaria, Hungary, Romania, the USSR, Ukraine, Czechoslovakia and Yugoslavia. It was enforced on 11 May, 1949. This is an important international legal instrument regulating navigation over the Danube. The Convention is aimed to ensure free shipping on the Danube taking into consideration the interests and sovereign rights of the member states and, hence, at strengthening of the economic and cultural relations of these states among each other and with other countries. According to this Convention eleven states, such as Austria, Bulgaria, Hungary, Germany, Moldova, Russia, Romania, Serbia and Montenegro, Slovakia, Ukraine and Croatia being at present the contracting parties, undertake to keep their sections of the river in an appropriate condition, to carry out necessary works to ensure and improve the shipping conditions and also not to admit any obstructions and hindrances for shipping. France, Turkey, the Netherlands and Czech Republic have the status of observers at the Danube Commission.

Convention Regarding the Regime of the Turkish Straits – was signed at the International Conference (22 June–21 July 1936) in Montreux, Switzerland, on 20 July 1936 by the delegations of Bulgaria, France, Great Britain, Greece, Italy, Romania, Turkey, the USSR and Yugoslavia. It concerned the regimes of the Bosporus and Dardanelles Straits. The purpose of this Convention was defined in its preamble saying that this Convention was developed to regulate the passage and shipping in the Black Sea straits "within the framework ensuring the security of Turkey and security of the B.S. littoral states", i.e. the principle combining the full freedom of international merchant shipping with certain preference rights of the Black Sea countries in military seafaring. This provision defines that the Straits regime should take into account the geographical position of the Black Sea states and their incidental specific rights to use of the Straits as the only natural communication way of the B.S. with other water areas. In the peacetime the civil (merchant, fishing and research) vessels enjoy the full freedom of passage and navigation in the Straits, by day and by night, under any flag with any kind of cargo. Each ship depending on its displacement is charged a toll for the sanitary control and for use of the navigation fencing facilities. This Convention consisted of 5 sections (29 clauses), 4 annexes (about the rates of tolls, methods of determination of categories and tonnage of ships, etc.) and the Protocol.

Cooperative Marine Science Program for the Black Sea (CoMSBLACK) – the international program of scientific study of the B.S. Its main purpose is to validate scientifically the efficient and integrated management of the B.S. by identification of the fundamental oceanographic processes and levels contributing to the improved quality of the natural environment; assessment of the effect of anthropogenic factors and long-time climatic changes on the changing ecosystem; construction of realistic environmental models taking into account the general and regional dynamics of water exchange; development of the database of long-time observations of water exchange and changes in the composition of the bio-, geo- and chemically active substances.

CoMSBLACK was established at the meeting in Sofia (Bulgaria) in April 1991 and was implemented till 1996. The program appeared on the basis of the interim committee consisting of the scientists from the former USSR, Bulgaria, Romania, Turkey, USA and Intergovernmental Oceanographic Commission (IOC). The result of the planned meetings was organization of expeditions, including HydroBlack-91, the first quasi-synoptic study of the B.S. with participation of several vessels; CoMSBLACK-92 on study of roe and larva, evaluation of the populations of

Aurelia aurita and *Mnemiopsis leidyi* on five vessels; study in April 1993 of spring blooming caused by development of phytoplankton and others.

CoMSBLACK envisaged performance of three kinds of experiments during 5 years: general observations of the circulation in the sea for identification of the general biogeochemical frames, distribution and density of organisms, biodiversity and thermohalinic characteristics and also identification and understanding of the dynamics of the basic mixing processes; detailed investigations of the B.S. on the inter-industry and industry basis; dynamics of water exchange and currents in the coastal areas focusing on detailed field studies of the B.S. shelf.

This program played the key role in initiating the project on integration of the B.S. research data "NATO TU-Black Sea Project" implemented in the period from 1993 to late 1997 within the framework of the NATO Project "Science for Stability" designed to better understanding of the impacts playing a role in deterioration of the B.S. environment as well as various external and natural factors that may facilitate this process. This project rallied the efforts of research institutions from the littoral states and among them the Marine Hydrophysical Institute of the National Academy of Sciences of the Ukraine.

Cork oak – a group of tree species referred to the *Quercus* gender, the beech family, not shedding leaves in winter. Its height may reach 20 m, the stem diameter—1 m. A thick cork layer is formed on the stem and old branches. This is a wildly growing tree, but it is also cultivated on the Iberian and Apennine peninsulas, in the south of France, in Algeria, Morocco, and Tunisia. C.O. is also planted on the Caucasus and in Crimea. Its bark is used for production of quality cork.

Cousteau Jacques-Yves (1910–1997) – the well-known French explorer of oceans and seas, the so-called "The Silent World", French naval officer, filmmaker, innovator, scientist, a member of the Academy Francaise (1988), photographer, author and researcher who studied the sea and all forms of life in water. He developed many underwater devices, including the Aqua-Lung. In his early days he served as an officer in the French Navy. In 1936 he visited the USSR for the first time. He went to Sevastopol and Batumi on B.S. In winter 1942/1943 Cousteau and engineer Emile Gagnan invented an aqua-lung. He designed the underwater submersibles "Diving Saucer", the underwater houses of the "Precontinent" type in which C. and his colleagues spent many weeks at depths exceeding 100 m.

In 1949, Cousteau left the French Navy and in 1950 he founded the French Oceanographic Campaigns, and leased a ship called *Calypso*, which became a mobile laboratory for field research, diving and filming. In 1957 he was elected as director of the Oceanographical Museum of Monaco. In 1973, along with his two sons and Frederick Hyman, he created the Cousteau Society for the Protection of Ocean Life, Frederick Hyman being its first President; it now has more than 300,000 members. In 1978 on *Calypso* he navigated for several days in B.S. calling on Constance (Romania) and visiting the Danube Delta. He is the author of numerous books and films about the underwater world.

Cousteau J.-Y. and his "diving saucer" (http://photography.nationalgeographic.com/photography/
photo-of-the-day/cousteau-submersible-abercrombie-pod/)

Crimea – see Crimean Peninsula.

"Crimean Africa" – a natural terrain in SCC situated lower than the rock cliffs of
Baidarsky Yaily between the Aiya and Sarych capes. It got its name thanks to
abundant sunshine and vegetation, healthy and warm climate. Its present-day name
is Laspy.

Crimean Astrophysical Observatory – was organized in 1945 on the basis of the
Simeiz Branch of the Pulkovo Observatory. Rich manufacturer N.S. Maltsov initiated the astronomical observations in Crimea: in 1900 he built a small private
observatory not far from his estate, on the Koshka Mountain over the Simeiz
seaside resort. In 1908 he passed it as a gift to the Pulkovo Observatory. In 1946
the construction of a new observatory in the place of settlement Nauchny was
started. C.A.O. is the largest astronomical establishment in Crimea. It has at its
disposal the newest research facilities that enable all-round astrophysical investigations within a wide spectral range of electromagnetic radiation: from hard
gamma-quantum to meter waves of various objects of the Universe (from artificial
Earth satellites and small bodies of the Solar System to extragalactic formations).
The observatory has a reflector telescope with a mirror diameter 2.6 m, a tower solar
telescope with the main mirror sizing 1.2 m in diameter and other facilities.

Crimean Astrophysical Observatory (http://www.krimoved.ru/images/img/astrofizicheskaja-observatorija.jpg)

Crimean Autonomous Soviet Socialist Republic – the administrative-territorial unit. It was founded on 18 October 1921 as a part of the RSFSR. Crimean ASSR comprised seven districts: Simferopolsky, Sevastopolsky, Yaltinsky, Feodosyisky, Kerchensky, Dzhankoisky and Evpatorian. In 1945 it was transformed into the Crimean Region of RSFSR. In 1954 it was passed from RSFSR to the Ukrainian SSR, where as Crimean Region existed till 1991. Since 1995—Autonomous Republic of Crimea, Ukraine.

Crimean campaigns –

(1) 1556–1559. Military campaigns of the Russian and Ukrainian troops against the Crimean Khanate. They were organized to stop the Tatars raids that devastated the southern and central regions of Eastern Europe. The campaign of M.I. Rzhevsky who led the Don and Ukrainian Cossacks (1556) had the reconnaissance purpose. Rzhevsky and his troops sailed on strugs (river boats) down the Psyol River from Putivl to the Dnieper where near the Dnieper rapids they joined with the Ukrainian Cossacks sent by Prince D.I. Vishnevetsky. So, they reached the Dnieper mouth and near Ochakov seized Tatar and Ottoman captives and loot.

In spring 1558 Vishnevetsky headed the Russian-Ukrainian campaign over the Dnieper to the Perekop Isthmus. Having defeated the Tatar outpost near the Perekop fortress Vishnevetsky returned to the Dnieper and defeated the Ottomans at the fortress of Islam-Kermen. Sailing up to the Samara River mouth Vishnevetsky met the Rzhevsky's detachment on the Monastyrsky Island from where following the Tsar order he went to Moscow.

In spring 1559 voevodes D.F. Adashev and Zabolotsky with their troops came to the upper reaches of the Psyol River where they met voevode T.F. Ignatiev who came with his troops from the town of Pochep and Rzhevsky who came from Chernigov. Sailing down on the river boats over the Dnieper to Kremenchug the enlarged military unit was augmented with the Ukrainian Cossacks who joined it near the Dnieper rapids. The Russian-Ukrainian troops passed by the Ottoman fortresses in the Lower Dnieper and went into the B.S. At Ochakov the Zaporozhie Cossacks captured two Ottoman ships. In May Adashev quite unexpectedly landed in Crimea behind the Perekop Isthmus. Having met no serious resistance his unit for 3 weeks fought in Crimea having seized and plundered the towns of Kuzlev (Guzleve), Karasev (Karasu), Bakhchisaray and released captives. After return to the western coast of Crimea Adashev went to the Dnieper mouth where near Ochakov he met the Ottoman pashas and reaffirmed the peaceful intents of Ivan IV in relation to the Ottoman Sultan. Although successful, the campaigns of Vishnevetsky, Adashev and other commanders against the Crimean Khanate were stopped due to insufficiency of troops that were also involved in the Livonian War of 1558–1583 and the downfall of the Rada government and disfavor of A.F. Adashev who advocated the active foreign policy of the Russian state on its southern borders.

(2) 1686–1689. Campaigns of the Russian army commanded by Prince V.V. Golitsyn against the Crimean Khanate during the war of the "Holy League" with the Ottoman Empire. They were organized with a view to stop the Tatar and Turkish raids on the southern margins of Russia and Ukraine and to protect trade routes. If these campaigns proved victorious it was contemplated to annul the annual tributes paid to the Crimean Khan. In early May 1687 about 60,000 soldiers and noblemen cavalry and Don Cossacks were concentrated on the Merlo River. The Russian army moved to the south via Rylsk and Krasny Kut. On the Samara River it was joined by the Ukrainian Cossacks (to 50,000 people) commanded by Hetman I.S. Samoilovich. The troops made a camp in the Bolshoy Lug area, near the confluence of the Konsky Vody River, presently Konsky River, with the Dnieper. The Crimean Tatars following the order of their Khan set the steppes on fire, thus, preventing the Russian troops from crossing the Konsky Vody River. On 12 July Golitsyn received an order from Tsarevna Sophia to continue military actions and in case of their impossibility—to construct fortresses on the Samara and Orel rivers and establish there the garrisons and the ammunition stock for future campaigns. On 23 July Samoilovich on false accusations lost his hetman position. On 25 July I.S. Mazepa was elected a new hetman. On 14 August the Russian troops returned to the Merlo River where they were disbanded. The government of Tsarevna Sophia considered this campaign to be successful and awarded its participants.

In the Second C.C. (1689) about 112,000 people took part. In March the Russian troops were concentrated on the Merlo River and again under command of Golitsyn. The campaign started on 17 March. On 20 April the Cossacks (about 40,000 people) commanded by hetman Mazepa joined the

Russian army. On 13 May the Tatar unit attacked the Russian camp on the Koirka River. On 15 May the Russians collided with the Crimean Tatars not far from the village of Zelenaya Dolina. The Russian army successfully repelled the attacks of the Tatar cavalry and moved to the Kalanchik River and on 20 May reached the fortress of Perekop. Golitsyn initiated negotiations with the Khan government demanding to return all Russian captives, stop raids, give up collection of tribute, to abstain from attacking Rzech Pospolita and from assistance to the Ottoman Empire. The Khan turned down these demands. The Russians had to abandon the siege as the troops were weakened due to illnesses and lack of water. On 29 May followed by the Tatar cavalry the Russians reached the southern borders of the Russian state. On 19 June the army was disbanded. The Sophia's government welcomed Golitsyn in Moscow with festivities. Regardless of the fiasco C.C. facilitated military success of other states entering the "Holy League" and strengthening of international relations of Russia.

Crimean Government – was formed by the Tatar nationalists on 13(25) June 1918 in Simferopol after in April 1918 Crimea was occupied by the German troops in violation of the Brest Peace Treaty and liquidation of the Soviet Socialist Republic of Tavrida. It comprised among others the pro-German elements and Tatars nationalists (Chairman, Minister of Internal and Military Affairs—Lieutenant-General Matvey or Macey (in Moslem—Suleiman) Aleksandrovich Sulkevich (1856–1920) that pursued the aim to make Crimea independent through dissemination of the Moslem religion on the peninsula. On 2(15) November 1918 the German occupation troops left Crimea and the Sulkevich government passed power to C.G. formed at the congress of province chiefs, representatives of cities and regional councils. The government that incorporated the members of cadet, S.R. and RSDRP (Menshevik) parties (Chairman of the Council of Ministers and Minister of Agriculture Solomon Samoilovich Krym (see) (his original name Neiman), Minister of Justice V.D. Nabokov) that oriented to the Entente gave permission in December 1918 to location in Crimea of the Volunteer Army. It ceased to exist in late April 1919 when the Red Army liberated Crimea.

Crimean Khanate – the Tatar feudal state that existed from 1443 to 1783 in Crimea and Northern Azov area. It was formed as a result of splitting of the Golden Horde during the ruling of Devlet-Hadji Girey (1433–1466). In the 1230s the Mongol and Tatar warriors invaded Crimea and in 1239 they controlled its steppe part having destroyed the local farming and subdued the local population (Alans, Polovets, Slavs, Armenians, Greeks). From the late thirteenth century the Mongol-Tatar feudals made Crimea the permanent place of their roaming, mostly in the wintertime. At the turn of the thirteenth and fourteenth centuries a special vicegerency was formed with residency in Solkhate (Stary Krym). As a result of inter-tribal struggle in 1433 Devlet-Hadji Girey became the ruler of Crimea. In 1443 supported by the Great Principality of Lithuania he founded C.K. that was independent of the Golden Horde. He was the founder of the Girey dynasty and was in power from 1443 to 1466. The population of C.K. included the Tatars, Armenians, Greeks, Jews and others.

The Khanate was divided into principalities headed by *beys*. Khan was the supreme ruler and commander-in-chief. With the help of *beys* he could gather a numerous cavalry army. During the ruling of Menli Girey, the Ottoman troops invaded Crimea and destroyed the Genoese colonies in Prichernomorie, and from then on the Khanate entered the protection of the Ottoman Empire. In 1502 some territories of the Great Horde, the steppes from the Dniester to the Kuban were annexed to C.K. For getting booty and captives C.K. frequently raided the nearby territories, in particular Russian lands. Relying upon their troops and assistance from the rulers of Genoese colonies in Crimea and also possessing ports on the coast of the Black and Azov seas the Crimean khans controlled all trade routes to the Mediterranean and Near East countries. Very important was the trade of slaves. In the early sixteenth century the Tatars conducted first raids to the Russian lands and later organized large campaigns to Moscow (1521, 1541, 1552, 1571, etc.) that created tension on the southern borders of Russia. Till annexation of the Kazan and Astrakhan Khanates to Russia (1552–1556) the Crimean khans actively supported the anti-Moskovy policy of these khanates.

Cultural development in C.K. was connected with establishment of the Islam in Crimea in the fourteenth to fifteenth centuries. The Arab script and Moselm education were spreading in C.K. The highest religious educational establishment—medrese appeared in Solkhate. In 1500 Menli Girey ordered to construct Zindzhirly-medrese in Bakhchisaray. In the sixteenth century well-known Ottoman architect Sinan worked in Crimea. He constructed the Khan mosque (1552) in Güzleve (presently Eupatoria). In the sixteenth to eighteenth centuries the complex of the Khan palace in Bakhchisaray was constructed. The Crimean Tatar literature, first theological and later secular, appeared in Crimea. During the Livonian War of 1558–1583 the Crimean Tatars made about 20 raids to the Russian state which hampered the military actions of the Russians in the Baltic area. In 1569–1572 and in the first half of the seventeenth century the Ottoman-Crimean aggression was most intensive. Therefore, that time the struggle with C.K. was in the focus of the Russia's foreign policy. Military campaigns against C.K. were organized in 1556–1569, 1662–1669, 1687, 1689, 1735–1738; at the same time, the frontier lines were improved. From the seventeenth century the Ottoman Empire suffered a crisis and started loosing its military potency. In such situation the Ottoman sultans were unable to support the Crimean khans as before and, as a result, the *beys* power was enhanced. After annexation of Levoberezhnaya Ukraine (1654) Russia began fighting for access to the B.S. and liquidation of C.K. which led to military clashes with the Ottoman Empire. Azov Campaigns of 1695–1696 strengthened for some time the positions of Peter I in the Northern Prichernomorie, but after defeat of the Russian army in the Prut Campaign of 1711 the situation turned once again in favor of the Ottoman Empire and C.K. In 1736–1738 during the Russian-Ottoman War the Russian troops penetrated Crimea three times. The Ottomans suffered defeat in the war of 1768–1774 and the Treaty of Kuchuk-Kainardji of 1774 declared C.K. independent of Turkey. In 1783 C.K. was finally annexed to Russia. According to the Decree of 2 February 1784, the territory of the former C.K. was turned into the Tavria Region and in 1802 it was transformed into the Tavria

Province. Defeat of the Ottoman Empire in the war with Russia (1787–1791) deprived Turkey of its claims to Crimea and gave Russia control in Prichernomorie. The Bakhchisaray Palace remains the amazing cultural monument of the C.K. time.

Crimean (Tauric) Mountains – the mountains in the south of the Crimean Peninsula. Their length is about 150 km, width—to 50 km. They consist of three ridges—Southern or Main (stretching along the sea) reaching 1,545 m in height (Roman-Kosh Mountain) and two ridges reminding of cuesta with steep southern and smooth northern slopes occurring more northward: Middle (or Internal) rising to 723 m and Northern (or External) rising to 342 m. They are composed of marl, argillite shale, sandstone and limestone. The Main Ridge composed mostly of limestone consists of several table mountains (the flat top surface of which is called "yaila") with the typical karst topography. The mountain slopes are overgrown with oak, beech and pine forests, while the yaila—with the mountain xerophyte vegetation.

Crimean Mountains: Mount Ai-Petri (Photo by Dmytro Solovyov)

Crimean nature reserve – formerly the hunting grounds. The reserve regime was established here in 1923. This reserve appeared in place of the hunting grounds of the Russian Emperor family and great princes. It was protected from 1870 (the first precedent in the history of the Russian Empire). It covers the central part of Crimean Mountains. The total area of lands and coastal waters having different protection regimes takes 5 % of the peninsula territory. The area together with the branch reserve is 33,400 ha. Beech, pine and oak forests and deer, roe, moufflon and other animals living here are under protection. The branch of this reserve is "Swan's islands" located in the Karkinitsky Bay.

Crimean operation of 1944 – liberation of the Crimean Peninsula by the troops of the Fourth Ukrainian Front commanded by General F.I. Tolbukhin and Separate Maritime Army commanded by General A.I. Yeremenko with the support of the Black Sea Fleet commanded by Admiral F.S. Oktyabrsky and Azov Flotilla commanded by Rear-Admiral S.G. Gorshkov from German Army. This operation lasted for 36 days—from 8 April through 12 May 1944 and ended in victory of the Russian troops. They were opposed by the Romanian and German troops of the Seventeenth Army. A week after beginning of the offensive the Soviet troops came up to Sevastopol and on 5 May the storming of the city began. They fought most furiously for the Sapun Mountain—the key point of the German defense. On 9 May the Soviet assault units broke the German defense and rushed into the city. On 12 May the remaining German troops (21,000 people) laid down their arms on the Khersones Peninsula as the Black Sea fleet disrupted the enemy evacuation plans. The Seventeenth Army lost 140,000 people (killed, wounded, captives, drowned during evacuation). If in 1941–1942 the Germans spent 250 days for seizure of Sevastopol, then in 1944 the Soviet troops needed only 5 days for the city liberation. After recovering Crimea the Soviet Union regained control of the B.S. The casualties of the Red Army during the Crimean operation were about 85,000 people.

Crimean Peninsula, Crimea – the peninsula in the south of Ukraine. Its area is about 27,000 km^2. The territory of C.P. is shared by three administrative-territorial units of Ukraine: Autonomous Republic Crimea (the greater part), the city of Sevastopol and the Kherson Region (the northern part of the Arabat Spit with the islands).

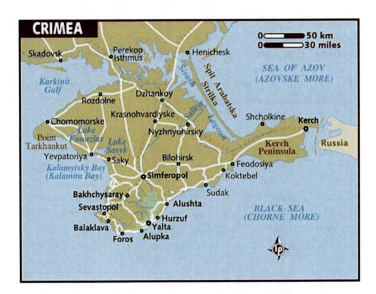

Map of Crimea (http://www.lonelyplanet.com/maps/europe/ukraine/crimea/)

So far no single interpretation of the word "Crimea" had been proposed. Some researchers think that the name "Crimea" takes its origin in the name of the city of *Qyrym* which was the capital of the Crimean province during the Golden Horde that later became applicable to the whole peninsula. Others think that the name of this peninsula originated from the Turk word *Qirim* meaning "trench" because in the ancient times the peninsula was separated from the mainland with a trench dug across the Perekop Isthmus. There are also known other Turk and Mongol names of this peninsula, such as Ada, Kyrym-Adasy that incorporate the Turk word *ada* meaning "island".

Its earliest name "Cimmeria" or the "land of Cimmerians" known in the second millennium B.C. referred to the steppe part of the peninsula stretching eastward as far as the Sea of Azov (the ancient name of the Kerch Strait is Bosporus Cimmerian). Its later name Taurica was used by historian Herodotus in the fifth century B.C. It applied to the areas inhabited by the Tauri, i.e. the southern mountain part of the peninsula (from the southern tip and nearly to Feodosia). The name Crimea came into use in the second half of the thirteenth century. It was often connected with the name of the Krym town (presently Stary Krym) that was the capital of the Crimean Khanate and this name became applicable to the whole peninsula. From the fifteenth century in all European maps C.P. is referred to as Taurica, Tauria, Taurida. Beginning from 1919 the peninsula name "Crimea, Crimean" was in wide use.

Archeological findings proved the existence of ancient farming cultures in C.P. Some researchers believe that the Cimmerians were the bearers of the Kyzylobinsky and the Tauri—of the Kemiobinsky archeological cultures (second–first millennia B.C.) taking their origin in the Bronze Age. The Tauri culture was rather archaic: at its early stages the key role was played by the stone implements. The Tauri made coins, bracelets, plates and rings of bronze. Only a few iron implements were found. Tauri fortifications were quite specific: the walls were made of dry-placed unworked stones, sometimes with tower shoulders having no internal chambers. They linked on to the rocks forming a single whole with the landscape (settlement of Uch-Bash in Inkerman, around eighth century B.C.). From the eighth to seventh centuries B.C. the Scythian nomad tribes took control of steppe C. after they forced the Tauri and part of the Cimmerians to the mountains. In the sixth to fifth centuries B.C. the Antique colonies started appearing on the C. coast, such as Kerkinitida, Mirmekia, Khersonesos. Appearance of the Greek colonies here was a logic result of the migration of the population between C. and Aegean basin countries that went on for more than one millennium. About 480 B.C. the Greek towns on the Kerch and Taman peninsulas, on the Lower Kuban and in the Eastern Azov area joined forming the Bosporus state with the center in the town of Panticapeum (modern Kerch). Among the ruling dynasties there were Archeanactides (480–438 B.C.) and Spartocus (438–109 B.C.). The population (Greeks, Sinds, Meotis, Scythians) was practicing grain growing, cattle breeding, fishery; the handcrafts were also developing. The Antique and local handicraft traditions mutually enriched each other. The Bosporus Kingdom was the main exporter of grain, cattle, fish and slaves to the cities of Greece and Asia

Minor. In its turn the Kingdom imported wines, olive oil, fabric, ceramics, weapons, jewelry and other things.

The period from the fourth to third centuries to A.D. was the peak of the economic and cultural development of the Bosporus state. In the third century B.C. the capital of the Scythian state (Neapol, near present-day Simferopol) was transferred to C. Scythian rulers were seeking control of the Bosporus state, Khersonesos and other cities of Southern C. From the third to second centuries to A.D. first the nomad Sarmatians, later the Alans started settling down in C. They assisted Khersonesos in fighting the Scythians claiming the control of C. In the second century B.C. the Bosporus faced the acute crisis aggravated by the pressing of the Scythians, Sarmatians and the Pontus state. In 107 B.C. Tsar Paerisades V, being hard-pressed by the Scythians, put himself under the ruling of Mithridates VI the Eupator which caused rebellion of the slaves and dependent population headed by Savmak. The Bosporus was included into the Pontus Kingdom.

In the first century B.C. Eastern Crimea became dependent on Rome. In the first to second centuries A.D. the Bosporus Kingdom was again at the peak of its development. The fortified settlements were constructed and the old towns were extended. After the Sarmatians started inhabiting C. some specific artistic style in jewelry (the so-called polychrome) appeared that later on was spread to Western Europe. In the late second century A.D. Bosporus Tsar Savromat II resolutely defeated the Scythians and incorporated them together with the Tauri into its state. The rulers of the Bosporus state called themselves "the tsars of the Bosporus and Tauri-Scythians". In the third to fourth centuries C. was invaded successively by the Goths and then the Huns that terminated the Antique period in the history of C. The survived population found shelter in Crimean mountains. The southwest of C. turned into a densely populated and economically developed area that maintained close relations with Kherson (Khersonesos), the only city that was not devastated. Probably, at the turn of the fifth to sixth centuries Byzantine via Kherson started development of the surrounding lands, founded outposts near the Internal Range (from the Alma River in the northeast as far as Inkerman in the southwest inhabited by descendants of the Scythians, Sarmatians, Alans and others. The fortified settlements (the so-called cave towns of Bakla, Tepe-Kermen, Kyz-Kule, Kalamita and others) gradually became the centers of lands, resident places of the local rulers and their troops. The Byzantine clergy disseminated Christinaity. In the sixth century the Goths settled down in Mountain Crimea and they gradually got mixed with the population living in the piedmonts. Goths made plundering marine raids to the coast of Asia Minor and supplied warriors to Byzantine. In the sixth century there was made an allusion to some country Dori (Doros) in Mountain Crimea. Many historians believe that its capital was first the cave town of Eski-Kermen and later, in the eighth to tenth centuries—Mangup. In the seventh century some works mentioned George, the Bishop the Chersonesos and Dorantsky. Later the Dori country had its own "Gothic bishopric". In the late seventh to early eighth centuries the Khazars conquered the southwest and east of C. In 787 Gothic Bishop John called the people of Eski-Kermen to rebel against the Khazars, but this riot was suppressed by the conquerors. In the eighth century the

Protobulgarian tribes living in the Azov area invaded Steppe Crimea. Due to a vehement struggle between the iconoclasts and the iconodules in Byzantine (eighth to early ninth centuries) the latter migrated to C. and turned it into the place of opposition to the church and secular rulers of the Byzantine Empire. During this time numerous monasteries (of Assumption and others) were built in C. In the early ninth century Byzantine established a special administrative-military district in C., the so-called Khersonesos Phema that was headed by stratig. It incorporated also the fortresses in the Khersonesos vicinity under one name—Climaty. According to emperor and writer Konstantin Bagryanorodny (tenth century), there were 9 of them. In 860 Slav Apostles Cyril and Methodius on their way to Khazaria visited C. In the late ninth century the steppe and southeastern parts of C. were invaded by the Pechenegs. In the second half of the tenth century after Price Svyatoslav Igorevich defeated the Khazar Khanate some part of Eastern C. was included into the Tmutarakan Principality. During the Russian-Byzantine war (in the 980s) after long siege the Russians captured Khersonesos and Great Kievan Prince Vladimir I Svyatoslavovich married Tsarevna Anna and about 988 he was Christianized.

In the second half of the twelfth to thirteenth centuries the frequent attacks of the nomadic Polovets weakened the C. links with the Russian principalities. In the twelfth to early thirteenth centuries C. was populated by the Greeks, Alans, Armenians, Polovets and also the descendants of the Goths, Khazars, Bulgars, Pechenegs and others. The economic and cultural relations were developing among different ethnic groups. After seizure of Constantinople by the Crusaders (1204), Khersonesos and Climaty overthrew the Byzantine ruling and started pay tribute to the Trapezund Empire. In Mountain C. the Principality of Feodoro (formerly Dori or Doros) maintained its independence. Its rulers were descendants of the noble family of Gavras-Taronits. In the thirteenth century in C. there was the area of Kyd-ker (Kyr-kor) inhabited by the Alans. According to French Sovereign G. de Ruysbroeck, in the mid-thirteenth century there were 40 castles between Khersonesos and Sudak. Crimean bishoprics were transformed into metropolitans (except Fullian) and subordinated to Constantinople patriarch (Khersonesos, Gothic, Sugdean and Bosporus).

In the thirteenth to fifteenth centuries many Italian trading posts existed in C. that were rivals in the economic and political spheres of other state formations on the peninsula territory. The Mongols invasion and raids of the Golden Horde (1223, 1238, 1248–1249, 1299, 1399) destroyed many Crimean towns and damaged farming areas around them. A part of C. became ulus (domains in possession of Mongolian khans). The Mongols and Tatars settled down in C. already in the first half of the thirteenth century. In the late thirteenth to fourteenth centuries the Tatar feudal landholding system was forming. Thus, udel (beilik) headed by the families of Yashlad (Suleshovs), Shirin, Baryn, Argyn and others appeared. In the fourteenth century the Mongol-Tatar administration was seated in the Armenian colony of Solkhat (presently Old Crimea). Soon the Tatar name Solkhat meaning Crimea was applied to the whole peninsula. In the late thirteenth to early fourteenth centuries Solkhat turned into a large trade center that attracted merchants from Near East, Central Asia, Golden Horde and Russian principalities. Local cultural traditions existed side-by-side with the Moslem art. In 1314 the mosque-medrese was built in Solkhat. In the second half of the fourteenth century due to inter-tribal wars in the Golden Horde the dependence of C. on the khans became quite nominal. In 1380 Khan Tokhtamysh concluded peace with the Genoans under which he maintained control on the southern coast of C. from Chembalo (Balaklava) to Soldaia (Sugdea, Sudak, Surozh).

In the second half of the fourteenth century the Feodoro Principality became rather powerful. Its rulers seized some part of the coast and constructed port Kalamita in the mouth of the Chernaya River (near present Sevastopol). Princes of Feodoro were relatives to the rulers of Byzantine, Trapezund and others. According to the Russian chronicles, "Gothic Prince" Stepan Vasilievich Khovra moved to Moscow in 1391 or 1402. His son Alexey, "the ruler of Feodoro and Pomorie" built in Mangup the palace (1425) and a new fortress near Kalamita (1427). In 1427 the Monastery of the Apostles (built in 686) in Partenite (near the present town of Frunzenskoye) was restored. In 1433 Alexey captured the Genoan fortress of Chembalo, but could not keep it. Feodoro maintained friendly relations with the Crimean Khanate. In 1470 the population of Feodoro was about 150,000–180,000. In 1471 Prince Isaac after visit to Caffa concluded peace with the Genoans. In 1475 the Crimean Khanate became the vassal of the Ottoman Empire. In December 1475 the Turks captured Mangup having killed almost all inhabitants and formed the judicial-administrative district (kadylyk) with the center in Mangup.

In the sixteenth to eighteenth centuries the raids of the Crimean feudal lords supported by the Ottoman Empire to the territory of Russia, Great Lithuanian Principality, later Rzech Pospolita, Moldavian Principality and others brought considerable material losses and casualties in these countries. Only in the first half of the seventeenth century over 200,000 people were captured in Russia; in 1654–1657 over 50,000 people from Ukraine. Protecting from the Tatar invasions the Russian government constructed in the sixteenth to seventeenth centuries the intercepting posts, sent troops against khans (campaign of 1559 of D.F. Adashev, unsuccessful Crimean campaigns of 1687 and 1689 of Prince V.V. Golitsyn). The Azov campaigns of 1695–1696 of Peter I weakened significantly the Crimean-Ottoman aggression, but they could not put an end to the raids to the southern areas of Ukraine and Russia. In 1736–1738 Russian troops invaded the Crimean Peninsula three times. During the Russian-Ottoman War of 1768–1774 the Russian army sized C. Under the Kuchuk-Kainardji peace of 1774 the Crimean Khanate became independent of the Ottoman Empire and was transferred under patronage of Russia.

In 1783 C. was annexed to Russia and in 1784 it became a part of the Tavria Region; in 1797–1802 C. was a part of the Novorossiysk Province; from 1802—the Tavria Province. In 1783 the Sevastopol fortress was founded. It became a large port and base of the Russian Navy on the B.S. By Decree of 22.02.1784 the Tatar nobility was granted the rights of the Russian gentry. Tatar clergy was released of taxes. In the late eighteenth century the population of C. was over 100,000. Among the inhabitants there were the Tatars, Nogais, Greeks, Armenians, Karaims, Krymchaks and others. From the late eighteenth century the Russians, Ukrainians, laters Germans, Bulgars, Hungarians, Moldavians, Lithuanians and others settled down in C. The gardening specialists from Germany were invited here to improve farming practices. In 1804 the vine-growing college was opened in Sudak. In 1812 the Nikitsky Botanical Garden was founded. In 1829 the mud-treatment centers that later became the well-known resorts were opened in Saki and later in Eupatoria. During the Crimean War of 1853–1856 C. was the main place of military actions. In the late 1870s Prince L.S. Golitsyn founded in his estate Novy Svet near Sudak the exemplary trial-production vine-growing and wine-making farm. Rich historical and cultural traditions of C. influenced positively the Russian arts and science. Being a unique region concentrating numerous archeological and architectural monuments, C. became the major center of study of the Antiquity, Christian historical monuments, cultural history of the peoples of Russia.

The traces of staying in Crimea of the ancient people of the Paleolith Time (grotto Kiik-Koba on the Zuya River 25 km eastward of Simferopol and cave camp Staroselje near Bakhchisaray) are survived. Archeologists dug out many monuments referring to the New Stone, Bronze and Early Iron ages. Unique are the exhibits dated back to these ages and later epochs that are displayed in the Crimean Historical Museum in Simferopol; the ruins of Antique Chersonesos in Sevastopol and Antique Mirmekia near Kerch; the Monastery of the Assumption built in a cave of the fourteenth century; cave towns of Eski-Kermen, Tepe-Kermen and Chufut-Kale near Bakhchisaray where the churches with fresco of the twelfth to fifteenth centuries had survived; Armenian churches of Stephan of the fourteenth century in Feodosia and Monastery Surb-Khach founded in 1340 near Stary Krym; Genoan fortresses of the fourteenth to fifteenth centuries in Feodosia, Sudak and Balaklava; the remnants of structures dated back to the Golden Horde and Crimean Khanate times—ruins of Mosque of the thirteenth century and medrese of the fourteenth century in Stary Krym, the Khan Palace of the sixteenth century, "legation doors" (1503), mosque (1740), halls, garden pavilions and the "fountain of tears" (1784) chanted by A.S. Pushkin and A. Mitskevich in their poems in Bakhchisaray, Mosques Dzhuma-Dzhami in Eupatoria (1552) and Mufti-Dzhami in Feodosia (1623); Grafsky berth and Peter & Paul Cathedral in Sevastopol (both dates back to the 1840s) with very picturesque parks; architectural monuments of the twentieth century—the former Tsar palace in Livadia and the so-called "Swallow's nest" in Miskhor and others. There are picture galleries in Feodosia, Sevastopol, Simferopol; House-Museums of Russian writers Anton Chekhov in Yalta, of M. Voloshin in Planerskoye, of A. Grin in Feodosia and Stary Krym and others. Crimea has also landscape monuments, such as Karadag (Black Mountain) near Planerskoye—the old volcanic massif with original erosional forms that is a part of the Karadag Nature Reserve; the Demerdzhi Mountain with the "valley of ghosts"; waterfalls Uchan-Su near Yalta, Dzhur-Dzhur near Alushta and others. The Crimean nature-reserve hunting grounds locate within the Main Ridge of the Crimean Mountains. In addition, the Yalta and Chernomorsky, Martjan Cape nature reserves are created. The unique Nikitsky Botanical Garden is also found on SCC.

C.P. is washed in the west and south by the Black Sea and in the east—by the Sea of Azov. In the north C.P. is connected with the mainland with the narrow Perekop

Isthmus (to 8 km wide). In the east the Kerch Peninsula is situated between B. and A. seas, while in the west the tapering part of C.P. forms the Tarkhankut Peninsula. Eastern Crimea is separated from the Taman Peninsula with the Kerch Strait. A system of shallow bays of the Sea of Azov known as Sivash is extending along the northeastern coast of C.P. It is separated from the sea with the low sandy Arabat Spit. The northern, greater part of the peninsula has a flat relief and is called Steppe Crimea, while the southern smaller part is occupied by the Crimean Mountains that run from Sevastopol as far as Feodosia and consist of three parallel ridges with the flat northern and steep southern slopes (the so-called Mountain Crimea). The southern range is the highest (Roman-Kosh Mount—1,545 m). Along the southern foot of the Southern Range a coastal strip from 2 to 8 km wide is stretching. This is the Southern Coast of Crimea (SCC). In some places the mountain outspurs reach the sea, in others—the outcrops of the parent rocks in the form of laccoliths (Ayu-Dag Mount) or of ancient volcanic massifs (Kara-Dag Mount) are found. The following mineral deposits are available here: Kerch iron ores, table salt in lakes and quarries of building stones.

The climate in Northern C.P. is moderately warm. The winter is mild (the mean temperature of January is +1 °C . . . −1 °C); the summer is hot (the mean temperature of July is +24 °C). On SCC the climate is Mediterranean characterized by a hot summer (the mean temperature of July is +24 °C) and a warm winter (the mean temperature of January is about +4 °C). Precipitations on the western slopes of the mountains are about 1,000–1,200 mm a year, in the east of the peninsula—500 to 600 mm and in the north—300 to 600 mm per year. The long, dry and sunny autumn is the best time in Crimea. The first frosts are possible in late October (on SCC—in late November). The steady snow cover is maintained only in mountains; on the plains it lays only for several days. In the warm season the breezes are dominating, in particular, on the western coast (Eupatoria). In winter the northeastern winds are prevailing. The number of sunshine hours is 2,200–2,500 a year with the minimum in January—over 70 and the maximum in July—over 340.

The rivers of C.P. are not water abundant. Many of them originate on the northern slope of the Southern Range. The Salgir River is regulated with the Simferopol reservoir. There are more than 50 salty lakes. The northern part of C.P. is mostly cultivated and only small areas with the steppe vegetation have survived. Mountains, in particular their northern slopes, are overgrown with forests: oak, beech, beech-hornbeam, in some places pine. The Crimean Nature Reserve locates on the slopes of the Southern Range. SCC has the Mediterranean vegetation including many evergreen species. The planted vegetation is also widespread here—the parks with decorative trees and shrubs, orchards, vines, tobacco plantations.

The main cities are Simferopol, Sevastopol, Kerch and others. Many seaside resort towns are found on the SCC—Yalta, Alupka, Alushta, Gurzuf and others. Availability of the favorable natural and climatic conditions as well as mineral spring waters and curative mud are conducive to development of the resort business here. More than 100 mineral water springs are known in Crimea. There are factories for bottling mineral waters. As for the balneological and mud treatment, the sulfide mud and brine from salt lakes of Adzhigol (Feodosia), Moinaksky (Eupatoria), Saksky (Saki, Eupatoria) and Chokraksky (for independent treatment) are widely applied. Rich balneo-mud resources of Crimea open bright perspectives for further resort development.

Still in the nineteenth century well-known Russian therapeutist S.P. Botkin stressed the possibilities of climatic treatment in Yalta. Construction of the resort area was started in the 1870s when local authorities, medical and public organizations and even private persons without seeking any financial support from the Tsar government initiated organization of sanatoriums in Yalta, Alupka, Eupatoria, Saki, Sevastopol, Balaklava, Sudak, Feodosia and Kerch. Celebrities, including aristocracy and high-rank businessmen, preferred building their palaces and country estates on the SCC. On 21 December 1920 V.I. Lenin signed the Decree "On Utilization of the Crimea Potential for People Treatment" that defined prospects and ways of further development of the Crimean resorts and turning Crimea into the major health-improvement area. In June 1925 the first sanatorium for peasants was opened in the former Tsar palace in Livadia and the sanatorium and pioneer camp "Artek" for children near Gurzuf. The Crimean resorts developed rather dynamically. During the Great Patriotic War (WWII) and occupation of Crimea by the Nazi the resorts were damaged badly. Thus, in Eupatoria all sanatoriums were destroyed in full or in part. In the postwar years the resorts were restored and the new sanatoriums and rest centers appeared. Already in the early 1960s about 500,000 people a year received vouchers for rest and treatment in Crimea. The sanatoriums are usually intended for the people suffering from chronic respiratory diseases, including TB, as well as the cardiovascular, locomoter and functional nervous disorders.

Crimean Primorie – the seaside climatic resort in Crimea, Ukraine, 29 km to the southwest of the railway station Feodosia and 30 km to the northeast of Sudak. It is located on the B.S. coast. The sandy-pebble beaches stretch in a strip about 3 km long and 20–30 m wide. The mountains protect this area from dominating northern and northeastern winds. This resort area has many vines, fruit orchards and decorative plants. The climate here is subtropical, of the Mediterranean type. The winter is mild, with alternating moderately frost and non-frost periods, with the unsteady snow cover. The mean temperature of January is about $-2\,°C$. The summer is very warm with prevailing hot and dry weather. The mean temperature of July is $23\,°C$. The autumn is warm, only with few clouds. The bathing season lasts from June to the first half of October. Precipitations are to 360 mm a year, the maximum—in summer. The sunshine hours are about 2,200 a year. The mild climate and warm sea are conducive for development of the climatic therapy, thalassotherapy of the respiratory (non-TB) diseases, functional disorders of the cardiovascular system and nervous system.

Crimean raids – the military campaigns of the Crimean Khanate in the sixteenth to eighteenth centuries for seizure of spoils of war and captives, receiving tribute and for political pressure on neighboring countries. They facilitated the Ottoman expansion to Eastern Europe. More than 100,000 warriors took part in the joint campaigns of the Crimean, Ottoman and other troops, and 40–60 thou warriors in the campaigns of the Crimean khans. From the early sixteenth century the territory of the Russian state suffered from regular raids the most significant of which were in 1507, 1512, 1513, 1515, 1517, 1521, 1527, 1531, 1533, 1535, 1537, 1541–1542, 1543 and 1548–1550. Only in the first half of the sixteenth century there were 48 of them; on the southern border 3–4 years of war were per each peaceful year. The Crimean units devastated the southern lands (Seversk, Tula and Ryazan lands), but the Crimean troops broke to the north of the Oka River quite seldom. After conquering of the Kazan Khanate (1522) the Russian state changed over from defense to active offensive actions in the south. In 1552, 1555, 1556, 1558 and 1559 the marches to the steppe against the "Crimean ulus" were organized. The front line of defense was moved behind the Oka. During the Lithuanian War of 1558–1583 when the Russian troops left the southern borders the raids of the Crimean khans were resumed. Large raids occurred in 1558–1560, 1562–1564, 1567, 1569 (Ottoman-Crimean march to Astrakhan) and 1570. In 1571 Crimean Khan Devlet-Girey reached Moscow and as result 36 Russian cities were destroyed. In 1572 the Ottoman-Crimean troops numbering 120,000 warriors made one more raid, but they were defeated by the Russian troops 50 km southward of Moscow. This victory and reorganization of the border guard service (1571) improved the military situation on the southern border, but some raids were still continued. Several large campaigns were made by the Crimean nobility together with the Nogai Horde (1586, 1587). The intervention of Rzech Pospolita and Sweden in the early seventeenth century broke the state defense system on the southern border of Russia. The raids became more frequent: in 1609–1611, 1614, 1615–1617 (till the Deulino Peace Treaty of 1618), in 1631 and during the Russian-Polish War of 1632–1634, in particular in 1634. The construction of the Belgorod interception line in the late 1630s–1640s strengthened considerably the southern border. In the second half of the seventeenth to eighteenth centuries the Crimean khans were more than once the allies of the Ottoman Empire making raids during the Russian-Ottoman wars or joining the Ottoman army (1677, 1678, 1711, 1736, 1769). Successes of the Russian troops in the Russian-Ottoman War of 1768–1774 put an end to C.R.

Crimean Region – on 25 June 1945 by the Order of the Presidium of the RSFSR Supreme Council the Crimean Autonomous Soviet Socialist Republic (Crimean ASSR) was transformed into K.R. within RSFSR. On 29 October 1948 the city of Sevastopol was isolated from the Crimean Region and was formalized as an independent administrative-territorial unit of the republican subordination. This situation lasted till 19 February 1954 when it was included into the Ukrainian SSR and remained till 12 February 1991. After the collapse of the USSR and the Declaration of the Independence of Ukraine, on September 4, 1991 it was adopted the Declaration of Sovereignty of the Crimea, and May 6, 1992 adopted the Constitution of the Republic of Crimea. Attempts to separate Crimea from Ukraine were halted, and in 1995 it was created the Autonomous Republic of Crimea (ARC) in Ukraine. The area of the ARC is about 27,000 km^2; population—about two million people (2008). The capital is Simferopol.

Crimean Soviet Socialist Republic – was proclaimed on 6 May 1919 after liberation of Crimea by the Red Army. The Crimean Soviet Government was established and it was headed by V.I. Lenin's junior brother—D.I. Ulianov. The republic existed for 50 days till 26 June 1919 when Crimea was seized by the Denikin troops of the White Army.

Crimean (Yalta) Summit Conference of the USSR, the USA and the United Kingdom – was held on 4–11 February 1945 in Yalta. The heads of the governments of the Soviet Union, the United Kingdom and the United States—Josef Stalin, Winston Churchill and Franklin D. Roosevelt entering the anti-Hitler coalition had met to discuss the world's postwar organization, to decide on division of Germany into occupation zones and on reparations, on participation of the USSR in the war with Japan, on postwar coordination of international relations and security and on UN establishment. The conference adopted the Declaration of Liberated Europe. The Conference was convened in the Large (White) Hall of the Livadia Palace.

Crimean (Yalta) Conference (1945) (http://sekcastillohistory20.wordpress.com/2012/02/11/this-day-in-history-11th-february-1945/)

Crimean Tatars – a people. The Crimean Tatar ethnos started its formation in the thirteenth to fifteenth centuries. The nomad groups who became vassals of the Golden Horde khans could also play a role in this process. In this way the Tatar community close to the Crimean one was shaping in Northwestern Prichernomorie. The aboriginal population of the peninsula also took part in formation of the

Crimean Tatar ethnos that was practically completed by the fifteenth century. In the mid-sixteenth century the name "Crimean Tatars" was attached to the Tatar ethnos. The language of C.T. belongs to the Turkish group of the Altai family and is divided into two main dialects entering two different groups: southwestern (Oguz) and northwestern (Kypchak). C.T. are subdivided into three sub-ethnic groups: Nogay or steppe Tatars (calling themselves "Mangyt"); "Orta Elakh", the Crimean Tatars who used to inhabit the territories between the first and second ridges of the Crimean mountains, the so-called "Tat people" and "Yaly Boylo" the people who lived on the southern coast of the peninsula. The Tatars of the piedmonts inhabited Karasubazar, Old Crimea (Solkhata), Simferopol and Bakhchisaray; the Tatars who lived on the southern coast—Yalta, Sudak and Balaklava.

Conquering of Crimea by Russia in 1771, follow-on liquidation of the khanate and final annexation of the peninsular by the Russian Empire in 1783 and abdication of Khan Shagin Girey caused exodus of the Crimean Tatars. Still in 1778 during the reign of Catherine II C.T. began to abandon their homes and move to the Ottoman Empire. In the late eighteenth century, by various estimates, the number of emigrated people varied from 800,000 to 300,000. The migration was the greatest in 1860 which was connected with the Crimean War of 1853–1856. By the early twentieth century no more than 188,000 of Crimean Tatars remained in the Tavria Province. In this period the movement for national identity of C.T. was taking shape that played later on an important role in the history of Crimea in 1917–1920. At the end of 1918, Qurultay of the Crimean Tatars was set up that declared restoration of the national statehood and that adopted the Crimea's first Constitution. By 1920 the formation of the Crimean Tatar ethnos was at its completion. After the Civil War in Russia, in October 1921, the Crimean ASSR was established as a part of RSFSR and it existed till 1945, when it was transformed into the Crimean Region of RSFSR. During the Great Patriotic War (WWII), immediately after the German troops were forced out from Crimea (the occupation of Crimea lasted from late 1941, of Kerch and Sevastopol—from May–June 1942, to April–May 1944) L.P. Beria suggested I.V. Stalin an idea on deportation of C.T. "taking into consideration their treacheries against the Soviet people". Merely for 3 days (18–21 May 1944) about 191,000 Tatars (47,000 families) were deported to Uzbekistan, Kazakhstan, Tadjikistan, and in several regions of Russia. After this the settlements, mountains, rivers, districts and district centers in Crimea bearing Tatar names (1944), villages (1945, 1948) and railway stations (1952) were given new names. In 1977–1979 C.T. started to return home. By 1990 the state concept on return of C.T. was adopted. In 1989 C.T. started in mass returning to their homeland and by 2001 their number reached 250,000. C.T. live also in Uzbekistan, Kazakhstan, Turkey, Romania, Bulgaria and Russia. Their total number estimates from 0.5 to 5 million people.

Crimean Ulus – ulus of the Golden Horde that was given this status after the Tatars conquered Crimea in 1242. The capital of C.U. was Kyrym built in the southeast of the peninsula in the valley of the Churuk-Su River.

Crimean War (1853–1856), Eastern War – it was fought between Russia, on the one side, and the alliance of Ottoman Empire, Great Britain, France and Kingdom of Sardinia, on the other. Russia has sought to weaken Turkey, to control the Black Sea Straits and to consolidate of the Russia's positions on the Balkans, in the Transcaucasus and Mediterranean. Emperor Nickolay I in a conflict with the Ottoman Empire counted on arrangements with Britain, isolation of France where in 1852 Napoleon III declared himself the Emperor, loyalty of Prussia and a benevolent position of Austria. In their turn, the ruling circles in Britain and France were seeking not to give Russia the access to the Mediterranean, avert or even eliminate its influence on the Balkans, weaken its military and political standing and also to take control of the Bosporus and Dardanelles. The Ottoman Empire counted on support of Britain and France and designed the plans on seizure of Crimea and Caucasus.

The war was caused by the dispute between the Orthodox and Catholic Churches on the right to the Holy Places in Palestine (the positions of the former were supported by Nicholas I and the positions of the latter—by Napoleon III which aggravated the Russian-French relations). In early 1853 the Russian Emperor dispatched Prince Menshikov on a special mission to Constantinople with the assignment to urge the Sultan to settle the dispute on the Holy Places in favor of the Orthodox Church and to sign a convention that would allow Russia to control the Orthodox Christians in Turkey. The tough position taken by Menshikov resulted in breakdown of negotiations. At the same time Turkey allowed the united British-French squadron to enter the Dardanelles. In response Russia broke the diplomatic relations with Turkey and in June 1853 sent its troops to Moldavia and Valakhia that were under sovereignty of the Turkish Sultan. On 27 September (9 October) the Ottoman Empire submitted an ultimatum to Russia demanding to leave these principalities and on 4 (16) October Turkey officially declared war initiating military actions on the Danube and in the Transcaucasus. On 20 October (1 November) Russia declared the war on the Ottoman Empire.

From the beginning of the war the Russian Black Sea fleet that had superior forces acted successfully on seaways of the enemy and blocked its ships in ports. On 18 (30) November the Russian squadron of ships commanded by Vice-Admiral P.S. Nakhimov destroyed the main Turkish fleet in the battle at Sinop. On the Danube theatre in November 1853 the Turkish troops made an attempt to cross the Danube near Oltenica village, but their attacks at Chetati, Zhurzhi and Kelerasha were beaten off. The Russian cannons destroyed the Ottoman Danube fleet (six gunboats and one ship).

Defeat of the Turkish troops and fleet provided Britain and France the *casus belli* for declaring war on Russia on the side of the Ottoman Empire. On 23 December 1853 (4 January 1854) the British-French fleet with the expeditionary forces onboard entered the B.S. On 9(21) February 1854 Russia broke relations with Britain and France and on 15(27) February the governments of these countries directed ultimatum to Russia demanding withdrawal of its troops from the Danube Principalities. On 28 February (12 March) they entered into military alliance with the Ottoman Empire. On 15(27) March Britain and on 16(28) March France had formally declared war on Russia. Russia declared war on these states on 11 (23) April.

With a view to forestall the enemy on the Balkans Emperor Nicholas I ordered on 11(23) March to cross the Danube near Brailov, Galatsa and Izmail. The Russian troops captured the Isakcha, Machin and Tulcha fortresses and besieged Silistria. At the same time, the fortification works on the Black Sea coast were started. On 8 (20) April the ally ships (6 battle ships, 13 frigates and 9 steamships) approached Odessa and on 10(22) April started bombing the city. However, the attempt of the allies to land was broken off by fire of the on-shore batteries (two frigates were set afire, British frigate "Tiger" was hit by Russian cannons and ran aground). In June–July the ally troops landed in Varna, while the ally fleet (34 battle ships, 55 frigates, including 4 steamships and 50 steamer frigates) blocked the Russian fleet (14 battle ships, 6 frigates and 4 steamer frigates) in Sevastopol. In view of the hostile position of Austria Chief of Staff of the Danube and Western Armies General-Field Marshall I.F. Paskevich pulled back the troops from the Danube. By September 1854 the Russian troops retreated behind the Prut River, while the Danube Principalities were occupied by the Austrians.

From autumn 1854 Crimea became the main theatre of military actions (from here is the name of this war). On 2–6(14–18) September the ally fleet (89 ships and 300 transport vessels) landed the expeditionary troops (55,000 people and 122 cannons) to the south of Eupatoria. Having met no rebuff the allies moved to the south of Sevastopol. On 8(20) September the battle at Alma took place between the ally army and the Russian army commanded by Prince Menshikov being the Commander-in-Chief of the Land and Marine Forces in Crimea. The Russians were defeated and had great casualties. The Russian troops retreated to Sevastopol and later on 12(24) December to Bakhchisaray. The ally troops approached Sevastopol from the south via Inkerman and conquered Balaklava and Kamyshovaya Bight where they organized a base for supply their troops over the sea. On 13 (25) September 1854 the siege of Sevastopol began. During a short time the soldiers, seamen and the population of the city constructed at approaches to the city the potent wooden-earth fortifications and installed their the cannons taken from the Black Sea ships. The defense was headed by Vice-Admiral V.A. Kornilov. His closest assistants were P.S. Nakhimov, Rear-Admiral V.I. Istomin, military engineer Colonel E.I. Totleben and other military commanders. In October the Russian army came up to the city and deployed on the Mekenzi heights. But all attempts to assist the Sevastopol garrison were unsuccessful. On 24 October (5 November) the Russian army suffered a new defeat in the battle of Inkerman. However, the ally troops having suffered considerable casualties had to abandon the plans of storming Sevastopol and the siege of Sevastopol began. On 14(26) January 1855 on the demand of Emperor Napoleon III the Kingdom of Sardinia joined the war on the coalition side and sent 15,000 strong corps to Crimea. On 5(17) February 1855 the Russian command attempted once more to help the Sevastopol garrison: the unit of 19,000 people commanded by S.A. Khrulev attacked Eupatoria where the 35,000 strong Ottoman corps was deployed. But the attack was broken off.

In February 1855 General Gorchakov was appointed the Commander-in-Chief of the Russian troops in Crimea.

Still in July 1854 the hostile parties of the war entered into negotiations in Vienna through the intermediary of Austria. The conditions formulated by the coalition (to prohibit Russia to have fleet on the B.S., to give up its protectorate over Moldavia and Valakhia, to abandon any claim granting it the right to interfere in Ottoman affairs on behalf of the Orthodox Christians and to annex from Russia the territories in the Danube Delta were waived by the Russian representatives and in April 1855 the negotiations were stopped. On 2(14) December 1854 Austria declared about her alliance with Britain and France, but thanks mostly to the efforts of Prince A.M. Gorchakov its declaration of the war on Russia was prevented. On 18 February (2 March) 1855 Emperor Nickolay I died. The Nickolay's son Alexander II, his successor, gradually understood that this war had no perspective.

The allies captured the cities of Kerch and Enikale (May 1855) and Kinburg Fortress (October 1855) on the B.S. In Crimea on 4(16) August 1855 the ally army defeated the Russian troops commanded by Gorchakov in the battle on the Chernaya River (to the southeast of Sevastopol). On 27 August (8 September) the French troops captured the Malakhov mound dominating over the city and on the same day the Russian garrison retreated from Sevastopol having sunk the last ships and having blasted the fortifications. With the fall of Sevastopol the military actions in Crimea were, in fact, stopped. The war revealed clearly the economic and technical backwardness of Russia. The Russian army and fleet did not have well-trained reserve forces, suffered acute shortage of armaments, ammunition, food (food supply and replenishment of the acting army in Crimea took rather long time because of insufficiency of railroads). The military expenses turned out unbearable for the budget which deficit had grown in 1853–1855 from 52.5 to 307.3 million Rubles. In such situation Russia resumed negotiations in Vienna. Because of the hostile attitude of Prussia and Sweden and in the face of menacing war from Austria, Russia, being in complete diplomatic isolation, had to admit some compromises. The final conditions of the conflict settlement were elaborated at the Paris Congress of 1856 that ended with signing on 18(30) March of the Paris Treaty.

In this war the Russian casualties were over 500,000 people, Turkish—about 400,000, French—95,000 and British—22,000 people. The military expenses of Russia made about 500 million Rubles, the expenses of the allied countries—about 600 million Rubles. Defeat of Russia in this war weakened its positions in the world and, at the same time, accelerated the reforms the most important of which was abolishing the serfdom. C.W. gave a good impetus for development of the military and naval art and technique. After the war all armies were re-equipped changing over from smooth-bore to rifle-bore weapons and the fleet—to armored steamships. Very popular became the tactics of firing lines; the positional warfare was brought into practice. In Russia the experience of C.W. was used in the naval reforms of the 1850s–1860s and the military reforms of the 1860s–1870s.

In the recent time some historians assert that C.W. was launched not by Russia, but Vatican and they call it the "Crusade". Vatican fought against the onrushing strengthening of the Eastern Christian world. That time there was a conflict between

Pope Pius IX and Napoleon III, on the one side, and the Orthodox clergy of the Jerusalem Church and Emperor Nickolay I, on the other. This conflict between the churches that had arisen in 1850 over the Holy Places in Jerusalem and Bethlehem made *casus belli* for unleashing C.W.

Curative mud (Peloids) – in this way the deposits of natural water bodies and products erupted by mud volcanoes are referred to. They have a curative effect and are used for treatment in the form of baths, wrappings and others. C.M. includes deposits of various water bodies, peat deposits in marshlands, eruption material of mud volcanoes and other natural formations consisting of water, mineral and organic substances and representing a homogeneous, thinly-dispersed plastic material that is used for treatment purposes in a heated state. C.M. is formed as a result of geological, climatic, hydrogeological, biological and other natural processes from mineral particles, organic matter (remnants of plant and animal organisms), colloidal particles of the organic and inorganic origin and water. C.M. is formed with the help of microorganisms as a result of biochemical processes in which C.M. gets enriched with the so-called biogenous components (carbon, nitrogen, sulfur, iron and other compounds) many of which (e.g. hydrogen sulfide) reveals a high therapeutic effect. By its structure C.M. is a complex physical and chemical system consisting of liquid mud, mud base and colloidal component. It comprises water and substances featuring a thinly dispersed structure, homogeneity, high heating capacity and low heat conductivity as well as high concentrations of microelements. By its origin C.M. are divided into peat, sapropel, sulfide silt and volcano. They are effective in treatment of skin, cardiovascular, nervous and other diseases. Well-known are the curative properties of the mud extracted in the Kuyalnitsky mouth near Odessa and in the salt lake (lagoon) Saki not far from Evpatoria in the Crimea. Mud for therapeutic purposes is also produced in salt lagoons of Burnas, Sjabolat and Khadzhibey in the Odessa Region and also in some lagoons in the Crimean Region and Krasnodar Territory.

D

Dagomys – a resort settlement, 10 km northwestward of Sochi, a part of Greater Sochi, Krasnodar Territory, Russia. It received its name by the Dagomys River flowing through a desolate gap. Its name is connected with *"shapsug tygje myps"* meaning "where the sun is shining".

Danube (in Ancient Greek *"Istros"*—Istr) – the second longest river after the Volga in Europe and the largest river flowing to the B.S. Its length is 2,857 km, basin area—817,000 km^2, average discharge—about 200 km^3/year (57.2 % of the annual flow of all rivers flowing into the B.S.). It takes its origin in Germany, on the eastern slopes of the Schwartzwald (Black Forest) Mountains and crosses the borders of 10 countries: Germany (7 %), Austria (10 %), Slovakia (5.9 %), Hungary (11.6 %), Croatia (4.4 %), Serbia (10.2 %), Bulgaria (5.9 %), Romania (29 %), Moldova (1.6 %) and Ukraine (3.8 %). D. has the largest international watershed basin covering partially territories of 9 additional countries: Bosnia and Herzegovina, Czech Republic, Slovenia, Montenegro, Switzerland, Italy, Poland, Macedonia and Albania. The basin length from west to east is about 1,690 km and width—820 km. The D. delta is the second largest in Europe after the Volga Delta.

S.R. Grinevetsky et al., *The Black Sea Encyclopedia*,
DOI 10.1007/978-3-642-55227-4_5, © Springer-Verlag Berlin Heidelberg 2015

Danube River map (http://upload.wikimedia.org/wikipedia/commons/f/fa/Danubemap.jpg)

It was mentioned by Ancient Greek poet Hesiod in the seventh century B.C. as Istros. This name originated from the Thracian base "*is(t)r*" meaning "flow, rush" and referred to the lower reaches of the river. The upper and middle reaches were mentioned in Ancient Roman sources as "Danuvius" (first century B.C.). This was a Celtic name and with the Celts advance eastwards this name ousted the Thracian "Istr". The form "*Danuv*" is indicative of the link of this hydronym with the Iranian "*danu*" meaning "river". The Romanian people in the villages of the Lower Danube call this river "Girlo".

In old times near modern Belgrade D. broke into two full-flowing arms, one of which flowed into the B.S. and the other (at present this is the Sava River, a Danube's tributary)—into the Adriatic Sea. The earliest and most reliable data about D. are contained in the treatises of Ancient Greek historian Herodotus (the fifth century B.C.) who wrote in the second book "History" that "the Istr River originated in the Celt lands ... and flowed crossing Europe in the middle ... Istr flowed into the Pontus Euxinia..." Its modern name this river received from the Celts who settled here in the first half of the first millennium B.C. This fast-flowing river was named *Danuvius* (quick water) from the Celt words "*danu*" meaning "quick, rushing" and "*vius*" meaning "water, river".

In the Schwartzwald Mountains in Germany near the town of Donaueschingen two mountain rivers Brigach (43 km long) and Breg (48 km long) join together at an elevation of 678 m above the sea level and flow further on as the Danube proper. On its way D. more than once sharply changes its direction. First it flows in the mountains of Germany, then crosses the Vienna depression rushing through the chains of the mountains connecting the Alps with the Carpathians, after this for over 600 km it flows over the Middle Danube lowland. Having forced its way through the range of the South Carpathian Mountains the river flows for more than 900 km over the Lower Danube lowland before emptying into the B.S. In its lower reaches D. ramifying into a dense network of arms forms an extensive delta stretching for 75 km from west to east, its width from north to south is 65 km. The apex of the delta is near the Izmail Chatal Cape 80 km from the mouth where the main D. riverbed breaks into two girlos: Kilian and Tulchinsky and 17 km more downstream the Tulchinsky girlo is divided into the Georgievsky and Sulinsky channels. Sulinsky girlo is the main navigable channel connecting D. with the sea. To ensure passage of sea ships with the draft to 7 m the Sulinsky girlo was straightened out by making ten cuts that shortened its length from 84.9 km to 63 km.

In some stretches D. makes natural borders among all Danube countries. The D. length in each littoral country varies from 1 km in Moldova to 1,075 km in Romania. The Ukrainian section of D. is 170 km long, of which 54 km is the main riverbed, 110 km the Kilian girlo and 6 km the Prorva canal connecting this girlo with the B.S. The distance (along a straight line) between the D. origin and mouth (Sulina) being 1,630 km, the meandering factor is equal to 1.7.

D. is recharged with the rain runoff from the watershed basin, melt waters of snow and glaciers and ground waters. Under natural conditions the suspended matter flow in D. is no less than 60 million tons a year. The sediment runoff of the river is one of the key factors controlling the formation of the D. vast delta. The tributaries branch off unevenly: their greater number is in the piedmont areas of the Alps and Carpathians, while on the Hungarian Lowland there is only a few of them. D. has a complicated water regime. In the upstream part the right (Alpine) tributaries affect it the most. Here the clear-cut summer maximum (due to melting of glaciers and snow in high mountains) and the winter minimum are discernible. For the rest part of D. the typical picture is spring-summer floods with the peak in May-June caused by snow melting and amplified by rain runoff. The mean water flow in the D. mouth is about 6,500 m^3/s, with the maximum being to 20,000 m^3/s and the minimum—2,000 m^3/s. Some parts of the river get frozen only during severe winters (lasting to 1.5 months). D. receives water from more than 300 tributaries and many of them are large rivers by themselves. Thirty-four tributaries are navigable. Among the largest tributaries there are: right—Inn, Drava, Sava, Morava; left—Morava, Tisa, Olt, Siret and Prut.

By physiographical conditions D. is broken into three parts: the Upper Danube (992 km long)—from the origin to the village of Genju; the Middle Danube (860 km long)—from the village of Genju to Turnu-Severin; the Lower Danube (931 km long)—from Turnu-Severin to the mouth (Sulina). The D. regime depends on the landscape zones it crosses. The Upper Danube is a typical mountain river. The Middle Danube is a plain river, except for the stretches of the Vicegrad and Iron Gates where D. forces its way through mountains. The Lower Danube is a typical plain river.

Many large and small reservoirs are constructed on D. and its tributaries. The 1950s–1960s was the period of active hydraulic construction in the D. basin.

By 1980 approximately 69 reservoirs with a capacity over 1 million m^3 each were built, thus, the D. flow was completely regulated. Among the dams there are: the largest hydropower and navigation system Djerdap-Portile-de-Fier-I (or Iron Gates) 942.9 km from the B.S. constructed in 1964–1972; Iron Gates dam II (or Djerdap-Grua-Portile-de-Fier—II) 863 km from the B.S. constructed in 1984; the Gabchikovo dam and hydropower plant constructed in 1992.

The D. river is a part of Transport Corridor No. 7 Europe—Danube—the B.S. The shipping over D. is regulated by the Danube Commission that adopted in 1948 "Convention on the Regime of Shipping over the Danube". The major ports on D. are Sulina, Braila, Galac (Romania); Izmail, Reni, Ust-Dunaisk, Kilia, Vilkovo (Ukraine); Ruse (Bulgaria); Vienna, Linz (Austria); Belgrade, Novi Sad (Serbia); Budapest (Hungary); Komarno, Bratislava (Slovakia); Regensburg (Germany); Djurdjuleshty (Moldavia). Construction in 1992 of the Main—Danube canal 171 km long and the minimum depth 2.7 m improved significantly the transit importance of D. Navigation becomes possible in the river stretch downstream Ulm to Braily and in the sea section from Sulina to Braily (170 km). The cargo is transported mainly over the Middle and Lower Danube.

There are many large cities on the D. banks, including capitals of four European countries: Austria—Vienna (population 1.7 million in 2011), Hungary—Budapest (1.7 million in 2011), Slovakia—Bratislava (0.46 million in 2012), Serbia—Belgrade (1.2 million in 2011). The capitals of three more states locate on the Danube tributaries: Zagreb (0.8 million in 2011) in Croatia on Sava River; Ljubljana (0.27 million in 2012) in Slovenia on Sava River; Sofia (1.2 million in 2011) in Bulgaria on Iskar River; Munich (1.4 million in 2011) in the Federal Land Bavaria, Germany, on Isar River. Such cities as Reni, Izmail, Kilia and Vilkovo locate in the Ukrainian section of D.

Danube River in Vienna (Photo by Andrey Kostianoy)

Great is the economic significance of D. as a transportation waterway, a source of electricity, irrigation, water supply and fishing. One of the key natural resources of D. is its fish stock. The main fish resources comprise the carps (32 species), the bullheads (13 species) and the sturgeons (6 species). At present 25–28 fish species are of commercial value in the Danube delta. The primary object of fishing is Black Sea-Azov herring. Among the migratory fish the most valuable are the sturgeons and, in particular, their most numerous species—starred sturgeon. Two sturgeon species habitate in D.—Russian and Baltic or Atlantic sturgeons. The latter is met quite seldom.

Danube Army – was formed in October 1916 by the Odessa Military District for defense of the Danube lower reaches and the southern seaside flange of the Russian-German front. It comprised also the survived Russian and Serbian troops after defeat of the Dobrudja Army. In October-December 1916 D.A. took part in military operations in the lower reaches of the Danube, Seret and Prut rivers.

Danube Biosphere Reserve – created in 1998 on the basis of the natural reserve "Danube Plavni". It belongs to the Ukrainian Academy of Sciences system. Its area is 46,400 ha with channels and internal water bodies and also a 2-km band of the B.S. water area. In February 2004 the territory of the reserve was extended by adding 1,295 ha of the Kilian region lands. In addition, 3,850 ha of lands of the Tatarbunarsky region without their withdrawal from the land users were transferred to the preserve for permanent use. Provisionally the reserve may be divided into the following zones: a nature preserve zone proper—14,904 ha, a zone with the regulated preserve regime—7,811 ha, a buffer zone (including the coastal belt of the B.S. and Zhebriyansky Bay)—19,392 ha, a zone of anthropogenic landscapes—8,146 ha. The last zone incorporates the coastal protection belt along the Ochakov girlo (442 ha) and Bystry girlo (341 ha) as well as the offshore strip of the B.S. (230 ha). Here more than 4,300 wildlife species are found. About 18.5 % of marsh plant species and 63 % of bird species registered in the Ukraine are under protection in this reserve. There are no large settlements within the reserve borders. The economic activities are confined mostly to reed cutting (part of which is exported), vegetable growing, fishing and navigation. In 1999 this reserve was put on the UNESCO List of the World Biosphere Reserves.

Danube Commission – this intergovernmental organization was set up in accordance with Clause 5 of the "Convention regarding the regime of navigation on the Danube". From 1954 it seated in Budapest, Hungary. It is called to supervise the execution of this Convention and address other matters aimed at creation of adequate technical and legal conditions for shipping over the Danube. Historically D.C. may be linked to the Paris Peace Conferences of 1856 and 1921 that outlined for the first time the international regime for ensuring free navigation on the Danube. The member states of D.C. are usually represented by ambassadors of respective countries accredited in Hungary. Twice a year D.C. holds its regular meetings. If necessary, the extraordinary meetings may be convened. Within the framework of this Commission the meetings of experts from member countries are organized that discuss technical, legal and financial issues identified in the Action

Plans approved at the meetings of D.C. At its meetings D.C. elaborates recommendations that should be incorporated into national legislations of member countries. Among the priority tasks of the Commission there are supervision of implementation of the provisions contained in the "Convention regarding the regime of navigation on the Danube"; development of a single shipping system along the whole navigable channel of the Danube with regard to specific conditions of individual stretches, basic provisions on navigation over the Danube, including basic pilot requirements; coordination of the hydrometeorological service activities on the Danube, publication of the single hydrological newsletter and hydrological forecasts (short- and long-term) for the Danube; navigation statistics for the Danube on the issues being within the terms of reference of D.C.; publication of reference books, pilots, navigation maps and atlases, etc.

D.C. cooperates closely with numerous international organizations, such as European Commission, Central Commission for Navigation on the Rhine, United Nations Economic Commission for Europe (UNECE), International Association on Protection of Navigation Over Inland Waterways, World Meteorological Organization, International Telecommunication Union, Permanent International Association of Navigation Congresses, Commission of the Oder and International Maritime Organization. By now D.C. published more than 300 publications, including periodicals, discussing various matters related to shipping over the Danube (navigation, hydraulic construction, hydrometeorology, management and ecology as well as economics and statistics).

Danube Cossack Troops – was established in 1806 in Southern Bessarabia and named Ust-Danube (Budjaksky) troops consisting of former Zaporozhie Cossacks who in 1775 resettled to Turkey and then returned to Russia. In 1807 it was liquidated; in 1828 it was restored as D.C.T. and comprised the former Budjak or Ust-Danube Cossacks who returned from Zadunaisky Secha, the Serbians, Greeks and others. They guarded the frontiers on the Danube, Prut and Black Sea coast. By the late 1850s they numbered 12,000, both men and women. In 1856 they were renamed into Novorossiysk troops. In 1868 they were disbanded and Cossacks became civilians.

Danube Delta – it is formed by numerous large and small waterways, marshes, lakes and lagoons. According to different sources, its formation started 5–11,000 years ago. Due to the sediments brought with the river D.D. is accrued at a rate no less than 45 m a year (in some places to 100 m). Many of them are connected by channels and arms of the Danube River. D.D. length from its top along the Chilia arm is 116 km and straight to the sea edge of the delta is 70–80 km. The length of the delta sea edge is about 190 km. The part of D.D. belonging to Ukraine is 830 km^2 (18 % of the total delta area) and Romania—3,370 km^2 or 82 %. The maximum depth in deltaic waterways is 39 m in the Chilia arm, 34 m in Tulcea and 26 m in Sfântu Gheorghe. D.D. takes the second place in Europe by its area—4,200 km^2 without near-delta lake-lagoons and the 22nd place in the world. The southern part of delta has the largest lagoons—Sinoe and Razim that are separated from the sea with narrow sandy strips of land. Deltaic arms form numerous islands, the biggest of which are Dranov, Sfântu Gheorghe and Letya. Before inflowing into B.S. they break down into three channels—Chilia, Sulina and Sfântu Gheorghe. D.D. is a unique combination of enormous water resources, fertile lands and warm climate which determines the significant agricultural and industrial potential of this region based on irrigated farming. D.D. is, in fact, the Ukraine's only rice-growing region. In addition, grain, forage and technical crops, fruits and vegetables are also cultivated here. In 1999 the International Coordinating Council of UNESCO's Man and the Biosphere Programme established D.D. as the single Biosphere Reserve shared by Romania and Ukraine.

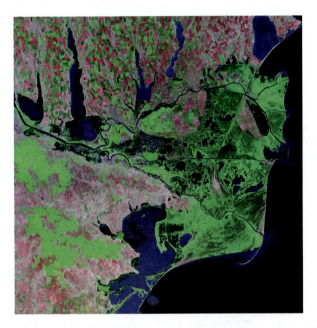

Danube Delta: View from space (http://upload.wikimedia.org/wikipedia/commons/d/d0/Danube_delta_Landsat_2000.jpeg)

"Danube Delta" (Rezervatia Biosferei Delta Dunarii) – biosphere reserve in the Romanian part of the delta. It was given the status of a nature reserve in 1938. Its area is 5,800 km^2 (1981). Apart from the Danube delta proper (3,270 km^2) the reserve incorporates also the Razim-Sinoe lake complex with an area of 1,015 km^2 (including lakes—863 km^2) located southward of the Gheorghe arm; sea coastal waters to isobath 200 m (1,030 km^2); flooded lands (421 km^2) between Tulcea and Isakchei on the right bank of Danube. Thus, the Romanian biosphere reserve stretches for 104 km along the Danube.

The greater part of the reserve is under water practically all the time. Among the aqueous vegetation the canes (or reeds, area—2,200 km^2), water lily (nymphaea) and others are prevailing here. The land is overgrown with grass vegetation (mostly meadow grasses) and forests (poplar, oak, alder and others, shrubs). The water fauna is represented by numerous freshwater and sea fish, including valuable species. The reserve has the most favorable conditions for birds—more than 300 species are found here (pelicans, swans and others). There are also the valuable species of mammals and reptiles. In the 1990s about 5,200 species of flora and fauna were found in the reserve.

Within the reserve borders there are 29 settlements where approximately 15,000 people live, including in Sulina—3,500 (2011). Their main occupations are reed harvesting, grape and fruit growing, commercial fishing, hunting (14 special zones are reserved); tourism is also developing. A special sphere in the economic activities in this biosphere reserve is navigation, first of all, along the Sulina arm (63 km long) and a part of the Tulcea arm within the reserve confines (8 km). This biosphere reserve is included into the UNESCO World Heritage Sites List.

Danube Hydrometeorological Observatory – was established in 1960 in Izmail on the basis of hydrometeorological divisions existed that time in the Soviet section of the Danube: the Danube mouth station in Vilkovo, the hydrometeorological bureau in Izmail, the meteorological station in Izmail and the basic hydrometeorological station of the Danube Naval Fleet in Izmail. The main purpose of this observatory was hydrometeorological and hydrochemical monitoring in the Ukrainian part of the Danube delta as well as study of the hydrological processes in the delta.

Danube Naval Fleet – (1) was established in 1771 during the Russian-Turkish War of 1768–1774 for support of the First Army commanded by General P.A. Rumyantsev. It was called the Liman Rowing Fleet and it fought against the strong enemy's river fleet. It supported crossing of water bodies by land troops and the actions against coastal fortresses of Turkey. In spring 1771 the building of ships on the Danube was started. The fleet also included five 24-m galliots seized from the Turks near the Tulcha fortress. In the same year the Russian troops with the support of D.N.F. captured the fortress of Zhurzha, blocked the fortresses of Tulcha and Isakcha, beat off the attempts of the Turkish troops to break through to the Danube left bank. In 1774 the Danube rowing fleet was disbanded and created anew at the beginning of the Russian-Turkish War of 1787–1791. Commanded first by Rear Admiral N.S. Mordvinov and later by Major-General I.M. Deribas it assisted the troops of General A.V. Suvorov under Kinburn, took part in storming the fortresses of Ochakov, Izmail, Kilia, Tulcha and Isakcha. In later years the Danube fleet was created for the period of military actions during the Russian-Turkish wars of 1806–1812 (it took part in defeat of the Turkish Army under Slobodzee in October-November 1811) and 1828–1829, the Crimean War of 1853–1856. During the 1877–1878 Russian-Turkish War it was officially referred to as D.N.F.; it incorporated armed steamers. The Russian mine boats on the Danube used for the first time the mine weapons against enemy ships. During World War I the Black Sea Naval unit—Special Expedition Corps (SEC) was operating on the Danube. In 1915 after Serbia defeat SEC retreated to Greece and partially to Romania where it was interned.

(2) The fleet was established in 1917 for support of the Red Guard troops with establishment of the Soviet power in Bessarabia. During the civil war its staff fought against the Austrian and German invaders.

(3) The fleet was created anew in June 1940 during liberation of Bessarabia and Northern Bukovina. It was subordinated to the Black Sea Navy command. By the beginning of the Great Patriotic War (WWII) it incorporated 5 monitors, 22 armored boats, nearly 30 guard boats, 7 mine-sweepers, 6 coastal artillery batteries, air squadron, a separate antiaircraft division, the infantry and machinegun companies. The ships of the fleet based in the ports on the Danube left bank, such as Izmail (main base), Reni, Kilia and Vilkovo. The fleet took part in defense operations on the Danube, Southern Bug, Dnieper rivers and near the Kerch Strait. On November 20, 1941 they had to leave their bases for some time. The fleet was disbanded and its ships were incorporated into the Azov Flotilla and Kerch Naval Unit. In 1944 after the Red Army came out to the Dniester the fleet was created anew for support of the offensive operations on liberation of the Danube states. That time it incorporated 50 armored boats, 18 torpedo boats, 22 mine sweepers, 16 semi-gliding boats, a separate marine regiment, an antiaircraft artillery division, and 4 artillery batteries. The fleet successfully interacted with the land forces in the offensive operations at Yassy-Kishinev, Belgrade, Budapest and Vienna, during crossing of the Dniester lagoon. The ships of the fleet fought on the Danube having passed more than 2,000 km. For its combat successes the fleet was awarded the Red Banner Order, the Nakhimov Order of first degree and the Kutuzonv Order of the second degree.

"Danube Plavni" – a nature preserve located in the southwest of the Odessa Region (Ukraine) along the Kilian arm of the Danube delta. Its area is about 15,000 ha. It was created in 1981 on the basis of the Black Sea state nature reserve for protection and study of the natural complex in the Danube Delta, wetlands of international significance. The thickets of cane, willow and sea buckthorn are growing here. Terrick, bald coot and others are nestling here. Among migratory birds there are found red-breasted goose, swans and cormorants. This is the nestling place and a place for migratory waterfowl. D.P. is crossed by migration routes of the sturgeon and Danube herring, the area of common carp spawning and fries fattening. Out of mammals the most typical here are boars, raccoon dogs, muskrats, minks and weasels.

Danube Shipping Company – see Ukrainian Danube Shipping Company.

Danube-Black Sea Canal in Romania – a navigable canal in Romania. Its length is 64 km, width 70–90 m, fairway depth 7 m. The towns of Chernovode (Romania) and Adjidja (B.S.) are linked by this canal. After its construction the Pan-European waterway from the North Sea to the B.S. (3,500 km) was formed. The idea of construction of a canal between the Danube and the B.S. had been in air since 1830, but the construction works started only in 1949, but soon they were stopped. In May 1984 there was the official opening ceremony of this canal (Chernovode—Constanta). This canal branches off from the Danube near old port Chernovode (300 km), runs over the Karasu River valley through villages of Medjidie and Basarabi. Crossing the Dobrudja plateau the canal passes the village of Strazha and ends in new port Constanta-South.

This canal is not designed for passage of large sea vessels, but only for pushed trains of six barges with the cargo-carrying capacity of 3,000 tons each with a tugboat. Two navigable locks are constructed in the canal head—in Chernovode (on the 4th km) and in Adjidja (on the 62th km). The canal is filled with fresh water and the Adjidja lock prevents the B.S. waters from intrusion into the canal. The carrying capacity of the canal is, by various sources, from 70 to 100 million tons a year. This canal is designed for transportation of the growing volumes of cargo via the Constanta port to inner regions of Romania. Four ports are situated on this canal: Chernovode and Medjidie (the 27th km on the right bank), Basarabi (the 40th km on the right bank) and Constanta. The port complex of Constanta-South is designed for receiving ships with the deadweight to 150,000–200,000 tons. It has 200 berths with the total length of 5 km and water depths from 7.0 to 22.5 m. The free economic zone is organized in the southern part of the port area. In the future the total cargo traffic via the Constanta port is contemplated to be over 230 million tons, including the Constanta-South complex—170 million tons.

Danube-Black Sea Canal in Ukraine (Bystroe Channel) – a shipway being re-constructed (opened in April 2007) in the Ukrainian part of the Danube delta of the Bystry arm (it existed in the Danube delta till 1957). It is an approach channel 6 km long, 7.65 m deep and 85 m wide on the offshore shallow side of this arm. This channel crosses the Danube Biosphere Reserve. After commissioning of this canal the economic complex in the Ukrainian part of the near-Danube area has started functioning to its full capacity. In general, the length of the deep navigable channel is 170 km and it consists of four sections: a sea approach canal 6 km long, 7.65 m deep, 85 m wide (bottom width) runs near the Bystry girlo; sea—Vilkovo section 19 km long, 60 m wide (bottom width) in the Bystry girlo and 120 m wide in the Starostambulsky and Kilian girlos with the depths of 7.26 m; Vilkovo—Izmail Chital section 94 km long, with the natural fairway width 120 m (bottom width), with the natural depths to 7.26 m; Izmail Chital—Reni section 54 km long, 120 m wide and 7.26 m deep. The last phase of this project envisages development of the shipping channel to ensure passage of ships with the draft of 7.2 m that is an alternative to the navigable Sulinsky canal on the Romanian territory.

Danube-Sasyk Canal – was built in 1980 for desalting and further supply of fresh water from Danube to the initially brackish-water reservoir Sasyk from where the water was taken for irrigation of vast lands in the Danube-Dniester interfluve. It was conceived as the first phase of a grand irrigation project with transfer of the Danube water to the Dnieper. Thus, the impressive dimensions of this canal (100 m wide or even larger than the Romanian shipping canal Constance-Chernovode) even exceeded the present-day needs. Now there is an urgent necessity to link Sasyk with the sea. The present quality of the Sasyk water is inadequate for irrigation and its use is permanently decreasing. In its turn, the decreasing water intake from the reservoir deteriorates automatically its flushing regimes, which leads to the worse water quality.

Dardanelles Strait – is a narrow strait in northwesternTurkey connecting the Aegean Sea to the Sea of Marmara. It is one of the Turkish Straits, along with the

Bosporus Strait. The strait is 61 km long, 1.2 to 6 km wide, and in average 55 m deep (maximum depth of 103 m). Water flows in both directions along the strait, from the Sea of Marmara to the Aegean via a surface current and in the opposite direction via a subsurface current. Like the Bosporus, it separates Europe from Asia. The strait is an international waterway, and together with the Bosporus, it connects the Black Sea to the Mediterranean Sea.

Dasha Sevastopolskaya (1822—1892) – Darya Lavrentievna Mikhailova (according to other sources—Darya Tkach), the famous Russia's first sister of mercy, nurse Dasha called Sevastopolskaya. She was born in Klyuchischi village near Kazan in the 1820s and grew up in Sevastopol. Her father was a sailor and died in the Sinop battle. Having become an orphan she sold her parents' house and bought a horse with a cart and white canvas for bandages. After this she went to the hottest place in the Sevastopol battle (1854–1855). She evacuated wounded soldiers, bandaged their wounds and gave them water, in other words, she worked selflessly in the military hospital. Under the guise of the man's clothing and under the name of Alexander Mikhailov she took part in the battles and made intelligence raids to the enemy's territory. Great Russian surgeon N.I. Pirogov, who himself volunteered to Sevastopol for rendering medical aid to the heroic defenders of this city, knew about her deeds very well and praised them high. D.S. was awarded the war decorations and became known in the history of the Sevastopol defense as "the first sister of mercy" and as "hero Alexander Mikhailov". She was held in high esteem and loved by the city defenders. She was awarded the Golden Medal on the Vladimir ribbon with the inscription "For hard work" and five hundred silver Rubles, i.e. she became the only representative of the lower social class who had won such award for participation in the Sevastopol defense. All her life D.S. worked in hospitals. In September 2005 in Sevastopol the foundation for the first monument to sister of mercy Dasha Sevastopolskaya was laid. The ceremony was timed to celebration of the 150th anniversary of the end of the Sevastopol defense of 1854–1855.

Dasha Sevastopolskaya (http://www.sweetstyle.ru/style/assets/images/New/star/dasha2.jpg)

Database "Black Sea" – The building up of the Black Sea oceanographic data base was commenced at the Marine Hydrophysical Institute in Sevastopol in the mid-1980s and has been under way ever since. The database brings together all available hydrological and hydrochemical data obtained in this basin by research vessels of various countries and organizations since 1890 and is on of the world's most complete oceanographic databases for the Black Sea.

De Ribas Osip (Iosif) Mikhailovich (1749–1800) – Admiral of the Russian Navy (1799). José Pascual Domingo de Ribas y Boyons was born in Naples in the family of Miguele de Ribas, the nobleman from Catalonia. He was second lieutenant of the Neapolitan guard when the Russian squadrons fighting against the Ottomans appeared on the Mediterranean. In 1772 he went on the Russian service, was promoted to Colonel. D.R. took part in negotiations that successfully ended in incorporation of Crimea into Russia. He took part in the Russian-Turkish War of 1769–1774. In 1774–1779 he served in the Land Nobility Corps in Petersburg; from 1780—Commander of the Mariupol Light Cavalry Regiment; in 1783–1784—participant of the battles for Crimea. During the Russian-Turkish War of 1787–1791 D. was the closest associate of A.V. Suvorov who praised high the military talent of D. In 1788 he fought the Turkish fleet in the Dnieper lagoon. In autumn 1789 he proposed to reinforce the Black Sea Fleet by lifting the Turkish ships drowned near Ochakov. Commanding the vanguard troops he seized the fortress of Gadzhibey. In spring 1790 he was appointed the Commander of the Black Sea Fleet. During the Tenderovsky battle he diverted the enemy's attention by maneuvers at Gadzhibey. In autumn commanding the fleet D.R. broke through the fortifications in the Danube mouth and seized the fortresses of Isakcha and Tulcha. He played an important role in storming Izmail having defeated the Turkish fleet and attacking the fortress from the river. D.R. fought in the battle at Machin. In 1791 he was among the Russian representatives at the peaceful negotiations with Turkey, took part in signing of the Iasi Peace of 1791. In the early 1790s D. as a member of the "Commission on improvement of the lands seized from the Turks" guided the construction of fortresses on the Dniester line. In 1793 he was promoted to Rear Admiral and appointed the Commander of the Black Sea rowing fleet. On the D's initiative the Russian government decided to construct "a naval harbor together with the merchant wharf" in the Gadzhibey Bay (in 1794 by the order of Catherine II D. became the chief of this project), thus, laying the foundation of the city of Odessa. Thanks to the D. efforts the population of the city was growing quickly (including, due to Greek migrants). After Emperor Paul I came on the throne (1796) D. was called back to Petersburg where he was soon appointed the member of the Admiralty Collegium. In January 1798 he was General-Kriegscommissar and from May 1799 he headed at the same time the Forestry Department. In 1799 he was promoted to Admiral. In March 1800 he was dismissed and in November of the same year called back to the service and appointed Vice-President of the Admiralty Collegium. The people of Odessa showing their respect and memory of this man called the main street in Odessa—Deribasovskaya.

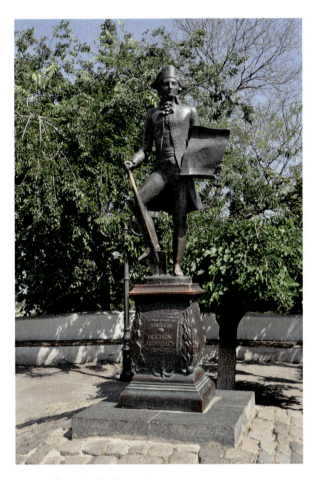

De Ribas I.M. (Photo by Evgenia Kostyanaya)

De Wollant Franz Pavlovich (François Sainte de Wollant) (1752–1818) – a statesman and military figure, Engineer-general (1810). He descended from the Brabant nobility. In 1775–1783 he took part in the War of Independence in Northern America. In 1787 he changed over from the Dutch to Russian service. During the Russian-Swedish War of 1788–1790 he participated in the Gotland marine battle (1788). In the same year as a military engineer he was transferred to the south. During the Russian-Turkish War of 1787–1791 in the rank of 1st Engineer of the Russian Army he took part in the siege and capture of Akkerman, Bendery and other fortresses, in the assault of Izmail and some other battles. In 1792–1797 D.W. headed the expedition on construction of fortresses in the south, supervised the construction of the Fanagoria, Kinburn, Perekop, Tiraspol and Oviodiopol fortresses and the port in Akhtiara (Sevastopol). He prepared the projects and controlled the construction of Ovidiopol, Tiraspol, Grigoriopol and Voznesensk. D.W. was the author of projects on buildup of Nikolaev,

Novocherkassk, Taganrog ports and some other cities in Novorossia, the author of a project and participant of construction (together with I.M. De Ribas) of the city and port of Odessa. D.W. took part in the 1794 campaign; in 1797 he became a member of the Military Collegium; in October 1797 he resigned. From July 1798 he was a member of the Water Communication Department; from 1800—a member of the expedition on road construction. In the early nineteenth century D.W. developed projects and guided the construction of the Mariinsky and Tikhvinsky water systems, some Ladoga and Onega canals, cleaning of the Dnieper rapids, completion of construction of the Oginsky canal in Belorussia and other waterworks. D.W. was one of those who initiated establishment of the Transport Engineers Corps and the Institute of Transport Engineers. From 1814 he was a member of the Cabinet of Ministers; in 1816–1818 he was an organizer of construction of the Petersburg-Moscow road.

"Death Valley" – in this way the British press referred to the Balaklava valley in the years of the Crimean War. This was the place of the famous Balaklava Battle in which the best people from the British light cavalry who descended from the most well-known families in England died.

Declaration "On state sovereignity of Crimea" – adopted on September 4, 1991 by the Session of the Supreme Council of the Crimean ASSR. On May 6, 1992 the Constitution of the Republic of Crimea was adopted and by Resolution of the Council and Presidium of Ukraine of March 1995 was abolished and the presidency in Crimea was abrogated. On October 21, 1998 the Second Session of the Supreme Rada adopted the new Constitution.

Decree of the USSR Supreme Soviet Presidium "On transfer of Crimean Region from the RSFSR to the Ukrainian SSR" dated February 19, 1954 – passed in the name of friendship of two brotherly peoples during the celebrations of the 300th anniversary of reunification of Ukraine with Russia. While making such a decision, Presidium of the USSR Supreme Soviet took into account that the Crimean Region and Ukrainian SSR had a common economy, were close territorially, had close economic and cultural relations. The decree was adopted in response to the presentation regarding the submission for approval of the Presidium of the USSR Supreme Soviet by the Presidium of the RSFSR Supreme Soviet of its own resolution dated February 5, 1954 on transfer of the Crimean Region from the RSFSR to the Ukrainian SSR as well as to the request of the Presidium of the Ukrainian SSR Supreme Soviet. Pursuant to the Presidium's resolution of February 13, 1954 requesting to transfer the Crimean Region from the Russian SFSR to the Ukrainian SSR, the meeting of the USSR Supreme Soviet passed n April 26, 1954 the Law "On transfer of the Crimean Region from the RSFSR to the Ukrainian SSR", which law thus approved the Decree of the Presidium of the USSR Supreme Soviet dated February 19, 1954 and provided for making appropriate amendments in the USSR Constitution. These amendments were only introduced on December 27, 1958 after the RSFSR Supreme Soviet passed a law that never mentioned the transfer of the Crimean Region to the Ukrainian SSR, but simply deleted it "in silence" from the list of regions making up the RSFSR. The consequences of this decision that are the responsibility of N.S. Khrushchev, the First Secretary of the CPSU Central Committee, could only be assessed by the peoples of Russia and the Autonomous Republic of Crimea after the collapse of the USSR. It is noteworthy that when the First Secretary of the Crimean Regional CPSU Committee D.S. Polyansky objected to the expediency of transferring the Crimea to Ukraine, he was removed from office immediately. Some experts on constitutional and international law doubt the legality of the original decision in the sequence of resolutions on the matter, namely of the resolution dated February 5, 1954 of the Presidium of the RSFSR Supreme Soviet because changing of the frontiers of the RSFSR territory was outside the sphere of its constitutional authority.

Defense of Odessa (1941) – lasted from 5 August to 16 October 1941 and involved the troops of the Odessa Defense Region (ODR) under command of Rear Admiral G.V. Zhukov. At first the city was defended from German Army by 34,500 people. Despite the numerical superiority of the enemy, the ODR troops comprised of the units of the Separate Primorsky Army, the Black Sea Fleet and people's volunteers for more than 2 months defended firmly the approaches to the city. Being isolated on land, the defenders of Odessa were actively supported from the sea by the Black Sea Fleet that delivered them reinforcement troops and even helped with attacks— the landing operation near Grigorievka. As there was a threat of the Germans breakthrough to Crimea the Soviet command decided to evacuate the defenders of Odessa to the Crimean Peninsula. Evacuation was conducted in good order in the period 1–16 October 1941. All defenders, machinery and armament were trans- ferred to Crimea. This was the first successful defense of the marine fort during World War II. During 73 days the defenders succeeded to incapacitate about 160,000 of soldiers and officers, about 100 tanks and 200 airplanes of the enemy. The defenders lost about 42,000 people. A special medal "For Odessa Defense" was instituted for the defenders of this city.

Denikin Anton Ivanovich (1872–1947) – Russian Major-general (1914), Lieutenant-general (1915). He finished the Nikolaev General Staff Academy (1899). Participant of the Russian-Japanese War of 1904–1905, World War I: General-quartermaster of the Eighth Army of General Brusilov. In 1914 he was appointed Commander of the Fourth Rifle Brigade ("Iron") that in 1915 was transformed into division. He took part in the battles in Galicia and Carpathian Mountains; his troops seized Lutsk; in June 1916 he captured this city once again during the so-called "Brusilov" Offensive. On September 9, 1916 D. was appointed Commander of the Eighth Army Corps on the Romanian front (September 1916— April 1917). Chief of Staff of the Supreme Commander-in-Chief (from April to May 1917). Commander of the Western Front (from May 31 to August 2, 1917). Commander of the troops of the Southwestern Front (from August to October 1917). For support of the General Kornilov's "mutiny" he was put to prison in the town of Bykhov. On November 19, 1917 he together with Kornilov and other generals escaped from prison and went to the Don. In late December 1917 Alekseev, Kornilov and D. reformed and united all military detachments entering the "Alekseev organization" into the Volunteer (White) Army. D. was appointed the deputy of General Kornilov who was the Commander of the Volunteer Army. Taking part in the First Kuban ("Ice") March he commanded the First Volunteer Division from January 30, 1918. From March 31, 1918 he was Commander of the Volunteer Army (after the Kornilov's death). From September 25, 1918 he was the Commander-in-Chief of the Volunteer Army (after the Alekseev's death). From December 26, 1918—Commander-in-Chief of the Armed Forces of Southern Russia (AFSR); from June 1, 1919—Deputy of Supreme Ruler of Russia Admiral Kolchak. By the last decree of Kolchak on January 5, 1920 D. was declared the Supreme Ruler of Russia, i.e. he became the Kolchak's successor. On March 14, 1920 during evacuation from Novorossiysk to Crimea he was the last to get onboard the torpedo boat "Captain Saken". Following the resolution of the Military Council of March 22, 1920 General Denikin turned over the command of the AFSR troops to the new Commander-in-Chief—General Vrangel. On April 4, 1920 he left Crimea for Britain on the British destroyer. In August 1922 he moved to Brussels, Belgium. Beginning from 1925 he lived in Paris, France and in November 1945 he moved to the USA where he died in the Michigan University hospital. In 2005 the D. remains were transferred to Moscow and buried at the Donskoy Monastery in Moscow.

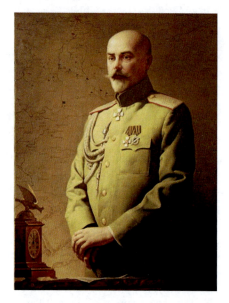

Denikin A.I. (http://alternathistory.org.ua/alternativnaya-ukraina-v-mire-denikinskoi-rossii)

"Derzky" – Bulgarian torpedo boat that earned fame because its crew, for the first time in the Bulgarian naval history, defeated the Turks during the Balkan wars of 1912–1913 having drowned the Turkish cruiser "Hamedie". D. was built in France, transported by parts to Bulgaria, assembled again and was set afloat in Varna in August 1907. On September 9, 1944 it took part in military operations and mine sweeping in the fairways in the western part of the Black Sea and in the Danube. It was in operation till 1952 and then was transferred to the Naval Museum in Varna. Its displacement was 97.5 tons, speed—22 knots. Armaments: three torpedo apparatuses and two 47-mm cannons.

Desht-i-Kipchak Khanate – an area where, according to Arab sources, the Polovets (Kipchaks) settled down and lived from 1116 to 1223. This area covered the steppes from the Dnieper to the Volga, the Precaucasus, nearly the whole Northern Prichernomorie and Crimea. In the mid-eleventh century, except for the Kerch Peninsula and Chersonesos, the whole Crimea was under control of the Polovets. The main base of the Polovets in Crimea was Sudak that was called that time Sugdea. The Polovets were ruled by the Khan that was elected by them.

"Development of a system for the Black Sea Fleet basing on the Russian Federation territory in 2005–2020" – the Federal Program envisaging construction of a new naval base for the Russian B.S. Fleet and the related infrastructure.

Devlet I Giray (1512–1577) – Crimean Khan who ruled beginning from 1551. He was an aggressive and cruel khan. He pursued the active foreign policy that implied countless military raids, first of all, on Rus'. In May 1571 he set afire Moscow. He was defeated by the Russian troops in the Molodinskaya battle in 1572. The main purpose of his military campaigns was to annex Kazan and Astrakhan, lost by the Muslim world to the Russians in the previous years. Devlet Giray's attempts were unsuccessful, but still he managed to impose annual monetary and fur tributes on some of the Russians and Ukrainians living in the south. In 1577 D.G. died of plague. He was buried near the Big Khan Mosque on the territory of the Bakhchisaray Palace that is a museum now.

Dikov Ivan Mikhailovich (1833–1914) – Russian Admiral (1905), Aide-De-Camp General (1906). He finished the Black Sea Junker School in Nikolaev (1854). In 1854–1855 as a naval cadet he showed himself well during the Sevastopol defense. In 1856 he was conferred the rank of the Midshipman. In 1859–1860 he prepared the magnetic map of the B.S. In 1860 he took part in the Caucasian war. In 1870 he commanded the ship "Prut" and in 1871 he was the assistant director of lighthouses and pilots of the Black and Azov seas. During the Russian-Turkish War of 1877–1878 he was one of the active organizers of the defense of Ochakov, Sevastopol and Kerch; commanded the special unit of the Danube Naval Fleet that supported the land troop attack of Sulin. In 1877–1878 in the rank of Captain-lieutenant D. took command of a unit of the Lower-Danube fleet. On September 27, 1877 his unit of armed ships and boats equipped with pole mines defeated the superior enemy forces at Sulin. In this battle for the first time in the world history they used the maneuvering mine fields on which the Turkish gunboat was destroyed. D. was awarded the order of St.George 4th degree, thus, having become the Russia's first naval commander who received two highest military awards (for soldiers and officers).

From 1881 he was the Commander of the study minesweeping detachment of the Black Sea Navy; in 1885–1886—of the armored frigate "Dmitry Donskoy". In

1886 D. was the Chief Mine Inspector. Then he was the Flag-captain at the Commander-in-Chief of the field army (1878–1879); a year later he became deputy chief of the Main Naval Staff. In 1888 he was promoted to Rear Admiral. In 1890 he was the Junior flagman of the B.S. Practice Squadron; in 1892–1894—the acting Senior flagman of the Black Sea Naval Division. In 1893–1896 he commanded the practice squadron of the Black Sea Navy. In 1896—the Chief commander of the Black Sea Navy and ports and the military governor of Nikolaev; in 1897–1900— Chairman of the Naval Technical Committee.

From 1898 to 1907 D. was a member of the Admiralty Council, the Chairman of the Commission on Study of Naval Operations (1904–1905). In 1905 he was promoted to Admiral and in 1906—Aide-De-Camp General. In 1906–1907 he was the permanent member of the State Defense Council; in 1907–1909—Naval Minister of the Russian Empire. At the same time D. acted as the Chief commander of the Fleet and Naval department. His authorities included direct management of the staff, combat forces and maneuver units. Supporting the idea of strong navy and radical reorganization of the naval department, he in 1905–1914 continued the naval reform. From 1909 D. was the member of the State Council. He was awarded many orders.

Dikov I.M. (http://fr.wikipedia.org/wiki/Ivan_Mikha%C3%AFlovitch_Dikov)

Dinsky Bay – is a part of the Taman Bay. It locates between the mainland and the low-lying Chushka Bar. D.B. is shallow; its shores, except of the western one, are abrupt.

Dioscuriada, Dioscuria – see Sukhumi.

Divnomorskoe – a village on the coast of the Northeastern Black Sea, 10 km southeastward of Gelendzhik, Russia. Recreation area, health resort, beach.

Djanik – the ancient province on the the the B.S. coast. At present D. is a mountain range in Turkey being a part of the Pontic Mountains. Its height is to 2,062 m.

Djiva Cape – is situated to the west of the Yesil Irmak River mouth, Eastern Anatolia, Turkey. It runs for 2.7 km to the northwest of the coast. It is composed of river sediments. The Djiva lighthouse is constructed 6.3 km to the east of the cape.

Dmitriev Prokofiy Stepanovich – Captain of the Major-General Rank (1797). He finished the Naval Cadet Corps (1761–1768). In 1765–1779 he navigated in the Baltic Sea and in 1771–1773 took part in several voyages from Arkhangelsk to Kronshtadt and back. He was directed to the Novokhoper shipbuilding yard where he commanded the construction of the frigate "Pyatyi" (Fifth) on which he cruised in the Sea of Azov and the Black Sea (1774–1778), then the frigate "Shestoy" (Sixth)—in the B.S. (1778–1779); commanded the bombardment ship "Azov" in the Taganrog port. In 1781–1782 in Kerch he commanded the Naval Crew; in 1792 supervised the building of rowing ships in Nikolaev; in 1793–1798 served in the Black Sea Admiralty Department and after its abolishment in 1798 he worked in the Office of the Commander-in-Chief of the Black Sea Navy as an advisor of the Crew Department. Captain 1st Rank (1789), Captain of the Brigadier Rank (1793).

Dmitriev Vladimir Nikolaevich (1839–1904) – organizer of climatic therapy in Russia. In 1867 he settled in Yalta and started a practice of a doctor. At the same time he studied the effect of climate and other natural factors on a man's organism. He pioneered the wine treatment. In 1878 D. published the book "Wine treatment in Yalta and on the Southern Coast of Crimea" that till now remains the classical treatise on dietology. In 1833 his other books were published, such as "Treatment with sea bathing" and "Kefir, the curative drink from the cow milk" (it was translated into French, German and Italian). For many years to come these books had been indispensable for doctors and sick people who preferred "natural methods of treatment" of various diseases. D. founded the first meteorological station on SCC having laid the foundation for study of climatic conditions. In 1890 his "Essay on climatic conditions in the Southern Coast of Crimea" was published that was awarded the Silver Medal of the Russian Geographical Society and in 1894 his work "On air treatment in Yalta" that laid the basis of the Russian aerotherapy. D. energetically supported the idea of balneology and tourism development for curative purposes. In 1891 he headed the Yalta Branch of the Crimean Mountain Club (this Club was established in Odessa in 1890) and participated in organization of a museum at this Club. He did much for organization of excursions (today this is the Yalta Traveling and Excursion Bureau). D. took much effort for implementation of many projects aimed at creating better conditions for life and rest on SCC.

Dnieper – is one of the major rivers in Eastern Europe—second by length and by the area of its basin. It is mentioned by Ancient Greek historian Herodotus in the fifth century B.C. as Borysthenes. Its total length is 2,285 km, its basin is 511,000 km^2. Its source is in the swamps of the Valday Hills. It flows over the territory of Smolensk Region, Russia (485 km—19.8 % of the watershed area), Belorussia (595 km—22.9 %), along the Belorussian-Ukrainian border (115 km) and then till its mouth it flows over Ukraine (1,090 km—57.3 %). It flows into the Dnieper lagoon of the B.S. forming a wide delta cut by several arms—the eastern shore of the Dnieper lagoon. The mean annual flow in the mouth is 52 km^3.

Dnieper watershed area (http://www.well.com/~mareev/portal/travel/4Eastern_Europe/zDnieper_ map11.jpg)

Historian Jordan (sixth century) first used the name Danaper. In the times of the Sarmatian and Goth marches in this way the lower reaches of the river were named by the Scythian tribes that inhabited the Prichernomorie steppes. The Miletian colony and merchant city of Olvia are located on the lagoon formed by D. and Southern Bug (Antique Hypanis).

D. may be divided into three parts: upper reaches from the origin to Kiev (1,333 km), middle reaches—from Kiev to Zaporozhie (621 km) and lower reaches—from Zaporozhie to the mouth (331 km). The upper reaches lie in the forest zone, the middle reaches—in the forest-steppe (as far as Kremenchug) and steppe zones, while the lower reaches—solely in the steppe zone. The headstream (from the origin to the town of Dorogobuzha) runs over the low-lying, partially waterlogged terrain overgrown with forests, and more downstream (as far as the town of Shklov)—over the hillocky terrain; the river valley here is narrow (0.5–1 km), the floodplain is sometimes lacked. Upstream of the town of Orsha there are the Kobelyaksky rapids. In a stretch Mogilev-Kiev the river valley becomes wider and its floodplain may be 14 km wide and is usually covered by inundated meadows, shrub thickets and forests. The main tributaries of D. in its upstream reaches are Drut, Berezina and Pripyat (right) and Sozh, Desna (left). The middle reaches of the D. valley are wide (6–18 km), having ancient terraces, in particular on the left bank. The right bank is steep and drops abruptly into the river. The stretch from Kanev to Kremenchug and farther downstream Dnepropetrovsk as far as Zaporozhie dam is represented by reservoirs. The main tributaries in the D. middle reaches are Sula, Psel, Vorskla (left) and Ros (right). In its lower reaches D. flows over the Prichernomorie lowland. After construction of the Kakhovka power plant the river formed a reservoir here (2,155 km^2). In this stretch D. receives Bazavluk and Ingulets tributaries (right) and the Konka River (left). Downstream Kherson D. is divided into two arms: right—Olkhovy Dnieper and left—Old Dnieper. After confluence these arms form a wide middle arm called Bakai that is separated from the Konka arm with the low-elevated island Belogrudy. After confluence of the Bakai and Konka arms the Zburievsky girlo (channel) is formed. The northern Rvach arm of D. is used for navigation.

D. has a mixed feeding, mostly from snow melting and ground waters. In the upper reaches the ice cover gets broken in early April, in the middle reaches—in mid-March and in lower reaches—in early March. The ice cover is formed from early to late December. D. with its tributaries is the main waterway for Belorussia and Ukraine. The river is navigable from the mouth to Dorogobuzha for 2,075 km. The main cargo traffic here includes wood and timber, grain, salt, coal, petroleum products, building materials. The major ports are Mogilev, Rogachev, Zhlobin, Kiev, Kanev, Cherkassy, Kremenchug, Dnepropetrovsk, Zaporozhie, Nikopol, and Kherson. Through its water systems Dnieper is connected with the rivers of the Baltic Sea basin: via the Berezina system—with Western Dvina, via the Dnieper-Neman system—with Neman, via the Dnieper-Bug—with Western Bug. The Dneprovsky, Kakhovsky, Kremenchugsky, Dneprodzerzhinsky hydropower plants are constructed on D. The population of the D. basin is 32.4 million (2001).

Dnieper Liman, Battle – a sea battle on June 17–18, 1788 in the D.L. between the Turkish fleet and the Russian rowing fleet. During the hot battles the Turkish troops despite the supporting cannon fire from Ochakov fortress were defeated and had to retreat. On their setback they got under fires of the battery on the Kinburn Bar commanded by A.V. Suvorov. Defeat of the Turkish fleet in the Dnieper Lagoon encouraged the siege of Ochakov by Russian troops. The special medal "For bravery in Ochakov Battle" was established for the participants of this battle.

Dnieper Navy Fleet – (1) Unit of the Russian Navy Fleet on the Dnieper created in 1737 for support of the Russian Army operating on the shore of the Dnieper Lagoon during the Russian-Turkish war of 1735–1739. The fleet comprised more than 650 small boats. It was liquidated in October 1739 after the end of the war.

(2) D.N.F. was formed on the Dnieper in March 1919 (with the base in Kiev) on an order from the Russian Revolutionary Council (R.R.C.) and was subordinated to the Twelfth Army of the Western Front. By spring 1920 D.N.F. consisted of approximately 40 ships and a landing unit of 650 people. The fleet was operating against the White Polish units and White Guard, took part in defeat of the gangs, supported the Red Army troops with cannon fire, landing, organized crossings and communication and laid minefields. On December 22, 1920 D.N.F. was disbanded and created anew on June 27, 1931 on the basis of the Separate Unit of Ships on the Dnieper (from October 1926) established in Kiev in October 1925. From October 1939 Pinsk became the main D.N.F. base. On June 17, 1940 D.N.F. was disbanded and instead two fleets—Pinsk and Danube were formed. During the Great Patriotic War (WWII) when the Red Army went out to the Dnieper D.N.F. was created once more on September 14, 1943 on the basis of the Navy. From October 20, 1943 it took part in combat actions against German fascists. By spring 1944 it comprised 140 ships and 2 artillery divisions. D.N.F. operating on the rivers of Dnieper, Berezina, Pripyat, Western Bug, Vistula, Oder and Spree together with the land troops took part in the defeat of enemies in Ukraine, Belorussia and Poland. After the 500-km voyage over waterways of Poland and Germany D.N.F. took part in the assault and seizure of Berlin (April-May 1945) providing fire support and crossings over rivers and canals. For combat merits and heroism of the people D.N.F. was awarded the Red Banner Order (1944) and the Ushakov Order of 1st Degree (1945); two its brigades and a division were awarded the Red Banner Order; two divisions of armored boats were conferred the rank of guards. The units and formations of D.N.F. were given the names of Pinsky, Bobruisky, Luninetsky and Berlinsky. On May 9, 1945 D.N.F. was disbanded.

Dnieper-Bug (Dneprovsky) Liman – the largest water body in the northwestern part of Prichernomorie—Kherson and Nikolaev Regions, Ukraine. It is formed by confluence of Dnieper and Southern Bug lagoons. The physiographical features enable D.B.L. division into two parts—Dneprovsky and Bugsky limans representing sea-inundated river valleys characterized by certain water, chemical and biological regimes. The northern shore of D.B.L. from the Ochakov Cape to the Dnieper mouth is largely elevated, abrupt and cut by gullies and ravines. By two pointed projections of the northern shore and the rounded projections of the opposite southern shore the lagoon is broken into three parts: western, middle and eastern. The western and eastern parts have depths no less than 5 m; the middle part has greater depths. The southern shore of D.B.L. ends with the Kinburn Bar. This shore is low-lying having in some places kuchugurs—hills.

The entrance into D.B.L. between Ochakov and Kinburn Bar is about 4 km wide and shallow. In the middle of the northern shore the Bugsky Lagoon is formed. Its area is 928 km^2, storage capacity—4.1 km^3. Its greatest length (from the Dnieper mouth to the Kinburn Strait) is 63 km; the greatest width—15 km; the average depth—4.4 m, in the central part—to 12 m. The navigable canals about 10 m deep connect the ports of Nikolaev and Kherson with the B.S. The river waters flowing into D.B.L. make 94 % of the Dnieper flow. Similar to other open lagoons in the northwestern part of Prichernomorie it has a very complicated water dynamics that is formed by flow and wind currents and also surge events. Every year the lagoon is covered with ice. The salinity level in the lagoon mouth varies from 0.05 to 16‰ (in autumn during strong southern winds). Sea waters usually occur in the bottom layer where oxygen is spent and hydrogen sulfide appears. After construction of the Kakhovka reservoir and North-Crimean canal less water from Dnieper gets into the lagoon, while the inflow of sea water increased (largely to the bottom layers). D.B.L. has great significance as a fishery water body and as a transport waterway in Ukraine. Such large industrial cities as Nikolaev, Kherson and Ochakov are situated on its shores. At present the dam stretching from Ochakov to the end of the Kinburn Bar is being constructed for creation of a reservoir for sturgeons breeding.

Dnieper-Bug Sea Port – is situated on the left shore of the Southern Bug at a distance of 26 km from Nikolaev. It was built in 1978 by the Nikolaev aluminous plant and is used for bauxite reloading and handling of general and bulk cargo. The port has 5 piers to 10 m deep and the so-called "small fleet" for auxiliary operations: piloting, berthing and others. The port is capable to receive ships with the draft to 4.5 m.

Dniester (in Ancient Greek—"Tyras") – a river in the west of Ukraine and Moldavia (some of its part makes a border between these two states). It flows into the Dniester Lagoon of the B.S. Its length is 1,360 km, the basin area 72,100 km^2; the mean annual flow is about 9.6 km^3. It originates on the northern slope of the Carpathian Mountains. In its upper reaches (as far as the town of Galich) it rushes down like a mountain river and accepts here many tributaries originating mostly on the Carpathian slopes. Downstream Galich its flow slows down, but its valley still remains narrow and deep. The most important left tributaries here are Zolotaya Lipa, Strypa, Siret and Zbruch (Sbruch). Downstream Mogilev-Podolsky the valley gets wider being narrowed only in some places by the spurs of the Volyno-Podolsky Highland coming close to the river; small rapids occur in the riverbed. Downstream the town of Rybnitsa D. flows over the Prichernomorie lowland. The valley here is 8–16 km wide. In the lower reaches the right tributaries of Reut, Byk and Botna flow into it. The ice cover is not steady, quite often it is replaced with an ice drift. During warm winters D. does not freeze at all. Ice phenomena are observed from December through March. D. is navigable to Galich. The major ports and wharfs are Mogilev-Podolsky, Soroki, Bendery and Tiraspol.

THE DNIESTER RIVER BASIN

Dniester River (http://dniester.org/wp-content/uploads/2012/03/Karta_angl.jpg)

Dniester Liman – this is the widened valley of the Dniester River, inundated Dniester mouth separated from the sea by the narrow and low sandy bar Bugaz scoured in two places with water. In the south of this bar there is a passage called the Dnestrovksy-Tsaregradsky girlo (channel). It juts into the northwestern coast of the B.S. for 41 km. The water area is 377 km^2, volume—0.57 km^3, average depth—1.5 m, axial length—43 km, width—from 4.2 to 12 km. Together with the shallow water area overgrown with aqueous vegetation having an area of 131 km^2 containing, at the mean water level, 0.19 km^3 of water the total surface water area of the lagoon becomes 508 km^2 and its volume—0.76 km^3. The river transports over 10 km^3 a year of sediments into D.L. The water balance of D.L. in conditions of free exchange with the B.S. includes the surface waters of the Dniester basin (10 km^3), sea waters coming into the lagoon via Tsaregradsky girlo (3.7 km^3), atmospheric precipitations and evaporation from the water surface (0.4 km^3). The water chemical regime is formed by interaction of the surge waves of the Dniester and the sea. The currents depend on wind directions and floods on the river. D.L. is navigable in its southern part. In winter it gets frozen. The city of Belgorod-Dnestrovsky is situated on the western shore of the lagoon, the city of Ovidiopol—on the eastern shore. On the shore of the southern channel—Bugaz port. Crayfish catches are of great commercial significance here.

Dniester Line – a series of fortresses built by the Russian government in the late nineteenth century in the south of European Russia for protection of new Russian settlements there. A decision on establishment of D.L. was taken on June 17, 1792.

Catherine II issued a special ruling on construction of D.L. Its construction became possible thanks to the Yassy Peace Treaty signed on December 29, 1791 (January 9, 1792). Overall supervision of all the works on D.L. was entrusted to A.V. Suvorov who headed the expedition on construction of southern fortresses which central office located in Kherson. Appearance of D.L. spurred the economic development of the region joined to Russia, construction of fortresses with the nearby urban infrastructure, such as Khadzhibey (Odessa), Ochakov and Voznesensk. Special attention was focused on construction of a new line along the Dniester. D.L. ran from the Dniester's tributary Mokry Yagorlyk along the Dniester to its mouths and further on along the seaside as far as Ochakov. In the surroundings of the future fortresses large contingences of troops were deployed which was a reason for settling down here of peaceful population. Specialists from abroad were actively invited by the government for construction of new fortresses and cities.

Dniester-Tsaregradsky Girlo – cuts through the southern part of the Bugaz Bar running to the Dniester Lagoon. A canal 1 mile long and 80 m wide was built through the bar that obstructed the entrance into the girlo. A lighthouse was constructed on the southern shore of D.T.G.

Dobrudzha (Dobrudscha) – a historical area. It has a hillocky terrain composed of fertile soils. It is located between the Danube River valley and delta in the north and west and the B.S. in the east. At present it comprises Northern D. which is a part of Romania, and Southern D. which belongs to Bulgaria. In the ancient times D. was inhabited by the Getic-Dacian tribes. Numerous Dacian fortifications and settlements are known. From the early second century the northeastern part of D. was absorbed into the Roman province of Lower Moesia; in the fourth century D. became a separate province of Scythia Minor. The major cities were Municipium Traiani, Tomy, Istria and Collatia. In the late sixth century D. was occupied by Slavs, in 678—by Bulgarians. The stone wall known as the Trajan rampart running from the Danube eastward to Tomy was most probably built in the tenth century by Bulgarians. It was mentioned in the mid-fourteenth century in a Latin text as "Terra Dobrodicii" meaning, possibly, "the land of Dobrodic".

Dobrynin Boris Fedorovich (1885–1951) – the well-known Russian geographer. In 1911 he graduated from the Physico-Mathematical Faculty of the Moscow University (geographical section of the natural science department) with the 1st Degree Diploma. At the Geography and Anthropology Chair he worked under the guidance of outstanding Russian geographer D.N. Anuchin and was getting ready for receiving the professor degree. In 1916 D. passed the exam for the Master degree and was approved as private-docent of the Moscow University where from this time on he taught students and carried out researches for 25 years. At the same time he lectured as an assistant at the Moscow Higher Women Courses (from 1915) and as a Professor of geography—at the Second Moscow University (later it was called V.I. Lenin Moscow State Pedagogical Institute) (1921–1930) and was also the senior researcher and division head at the Institute of Geography of the Academy of Sciences (1934–1939). From 1931 D. was Professor of geography at the Moscow University. In 1935 he was conferred the academic degree—Doctor of Geography.

In winter 1941–1942 being in evacuation in the city of Gorky D. delivered lectures at the Pedagogical Institute and from 1942 to 1950 he headed the Chair of Physical Geography in Tbilissi University (Georgia) combining this work with teaching at the Moscow University and the Moscow Region Pedagogical Institute (1944–1946). In autumn 1950 he went to live to Kiev where during one year he headed the Chair of Physical Geography in the Kiev University.

In 1908–1911 while being a student D. made several trips to Italy, France, Spain and Switzerland. As a result, his first essays about geography of these countries appeared, such as "Andalusia Cordilleras" and "Apennine Italy". In 1925 fulfilling the task of the People's Committee for Education he visited Mexico.

Much of his time he spent in field expeditions in Crimea and Caucasus. The results of his work he described in more than 20 publications and numerous articles in encyclopedias. In 1910 D. as an assistant of A.A. Kruber studied the karst phenomena in Crimea; in 1912–1915 and in 1921 he made several research trips over Crimea and Caucasus; in 1916 and in 1922–1923 he studied the relief, vegetation and landscape of Daghestan, as a result, several his works were published that became classical for their territory. In 1927 D. headed the multipurpose expedition on study of the Kerch Peninsula and in 1931–1932 and in 1935 he studied the terraces of the Black Sea coast of the Caucasus and natural landscapes of the Transcaucasus. In 1933–1934 D. studied the geomorphology of the Southern Coast of Crimea, in particular, he organized two expeditions on detailed study of the coasts of Eastern Crimea. In 1943–1944 supervised the work of a team of geographers from Georgia on physiographical zoning of Georgia. In 1946–1947 D. explored the landscapes in mountains of Adjaria.

Taking together all works of D. about the nature of Crimea we can consider this man one of the most knowledgeable experts of this peninsula. D. loved Crimea most dearly, was actively involved in activities of the Moscow Society on Crimean Studies, organized a research base of this society in Crimea and guided its activities. In the 1960s one of the P.P. Shirshov Institute of Oceanology (Moscow, Russia) research vessels was given the name of "Professor Dobrynin".

Doganyurt – a village on the coast of the Black Sea, Turkey.

Dogfish (*Squalus acanthias*) – a small thorny shark having a cylindrical body covered with small scale (also known as katran). Each dorsal fin has one well-developed thorn. The dorsum and sides are most often dark-gray, at times with small white speckles. The belly is whitish. The body length is 150–200 cm, oftener 100–125 cm; the weight is 8–12 kg. The thorny shark is found in the Atlantic and Pacific oceans, the ordinary species—in the North, Mediterranean and Black seas. They prefer living near shores. They feed on crabs, mollusks, bullheads and whiting; they are met mostly in spring and summer. The small thorny shark is the live-bearing fish. In autumn (September—November) at depths 24 to 40 m it gives birth to 16–32 "babies" being 20–30 cm in length. Its lifespan is no more than 10–12 years. It is quite numerous in B.S. The meat of D. tastes good and may be used fresh and smoked. Vitamins A and D are extracted from the liver fat. Single species of *Squalus blainvilli*, the closest kin form of the Black Sea small thorny shark, gets into B.S. via the Bosporus. The length of these sharks reaches 70 cm, the weight 4–6 kg. The local inhabitants of Crimea rendered the fat rich in vitamin A from the D. liver. At appropriate salting the D. meat can be used for balyk making that is no inferior to the sturgeon balyk. Until recently its skin was used for wood polishing. Sometimes D. is called "dog shark". Quite recently a new preparation "Katranol" was developed on the basis of the D. liver fat. It is intended for treatment of oncological diseases. D. is met most often in the Tendry area. It is not dangerous for a man. It is said that D. shoal may attack dolphins. Being an active carnivorous species D. has an amazing ability to find the fishermen nets and eat the fish entangled in them.

Katran (http://vospitatel.com.ua/zaniatia/priroda/more/images/katran%202.jpg)

Dokovaya Bight – a bight in the northern coast of the Sevastopol harbor. It is located in the mouth of the Panaiotov ravine where one of the B.S. largest dry docks was constructed for repair of large vessels. All heavy aircraft carriers of the USSR Navy and also the U.S. liner "United States" with water displacement of 53,000 tons were repaired here.

Dolgaya Spit – is situated in the eastern part of the Sea of Azov, 10 km eastward of the city of Eisk, Russia. This is the longest spit of the Eisk Peninsula. Its length is 8 km. This is the landscape monument in the Krasnodar Territory. Here one can admire the unique shell beaches, copious animal and plant world of the Sea of Azov.

Dolgorukov (Dolgorukov the Crimean) Vasily Mikhailovich (1722–1782) – Russian Prince, military chief, General-in-chief (1762). He started his military service in 1735 as a soldier, then cavalry corporal. For the bravery showed during the siege of Perekop (May 20, 1736) in the Russian-Turkish War of 1735–1739 General-Field Marshal Kh.A. Minikh, Commander-in-Chief, promoted him to ensign officer. He took part in the Russian-Swedish War of 1741–1743. In 1747–1755 he commanded the Tobolsk infantry regiment. During the Seven Years' War

(1756–1763) he distinguished himself in the battle under Kustrin (August 1758) and in the Zorndorf battle (1758). He was wounded and had to leave the army. In September 1761 he took part in storming Kolberg. In 1762 he was promoted to General-in-Chief. During the Russian-Turkish War of 1768–1774 he commanded the troops that guarded the borders with Crimea. From 1771 he commander of the army (38,000 people) sent for seizure of Crimea. Having concentrated the forces on the Mayachka River he successfully stormed the Perekop fortifications on June 14, 1771 that were defended by 50,000 Tatars and 7,000 Turks. On June 29, 1771 in the battle of Kafa he defeated nearly the 100,000 strong Turkish-Tatar army and forced them to abandon the towns of Gezlev (Evpatoria), Kerch, Enikale and Balaklava. By mid-July 1771 he ousted all Turkish troops from Crimea. D. forced Selim-Girey to flee to Constantinople. Successes of D. made possible to put loyal to Russia Saib-Girey, on the throne in the Crimean Khanate with whom D. as a representative of the Russian Empire concluded the "inseparable union". For this campaign on July 19, 1775 Prince Dolgorukov was awarded the honorable title "the Crimean". From 1780—Commander-in-Chief in Moscow, He was decorated with many orders.

Dolgy Island – is situated near the Dnieper-Bug Lagoon, Ukraine, at entrance into the Egorlytsk Bight, the largest in the group of islands comprising also Krugly, Veliky and Konsky.

Dolmen – a burial, cyclopic or megalithic monument of Tauri, the descendants of Cimmerians, from whom, in fact, the history of Crimean started. D. is a type of a chamber made of stones put upright stones supporting a large flat horizontal capstone. In Crimea D. were often surrounded by a rectangular fence made of dug-in stones. It is thought that in each chamber there were successively buried the members of one family. D. are also found on the coast of the Krasnodar Territory to the south of Gelendzhik. Unique architectural monuments of the Bronze Age.

Dolmen near Gelendzhik, Russia (Photo by Dmytro Solovyov)

Dolzhanskaya (Dolgaya) Spit – a sandy spit, the national reserve, which juts out into the Sea of Azov by more than 10 km, Eisk Region of the Krasnodar Kray, Russia. It separates the Taganrog Bay from the Sea of Azov. Very popular tourist area, especially for windsurfing.

Domozhirov Dmitry Andreevich – Russian Captain of the Major-general rank (1797). He finished the Naval Cadet Corps (1761–1768). In 1765–1769 he navigated on the ships of the Baltic Fleet. The participant of the Russian-Turkish War of 1768–1774. Showed himself well in sea battles near fortress Naoplidi-Romania and in the Chesma Bay (1770). After conclusion of the Kuchuk-Kainardji Treaty he returned to Kornshtadt. In 1778–1779 he navigated on the frigate "Legky" (Light) between Revel and Kronshtadt; in 1778–1779 prepared the description of the White Sea; in 1780–1782 made two marches on the ships of the Baltic Fleet from Arkhangelsk to Kronshtadt. After this he was commanded to Taganrog. In 1783–1786 cruised over the B.S. on the frigate "Strela" (Arrow). In September 1786 commanding the ship "Aleksandr" that sailed under D. command from Kherson to Sevastopol ran on stones, the ship was broken and cast ashore near the Tarkhankut Cape. For the loss of the ship D. was sentenced by the court to eternal exile, but he was pardoned by Catherine II. In 1788 he was appointed the Captain to Sevastopol port. In 1795–1798 in Kherson he served in the coastal fleet. In January 1802 he was dismissed. D. was awarded some orders. He was promoted in 1st Rank Captain (1789), Captain of the Brigadier Rank (1793).

Don, River – is one of the major rivers of Russia. It rises in the town of Novomoskovsk, 60 km southeast from Tula, southeast of Moscow, and flows for a distance of about 1,950 km to the Sea of Azov. Average discharge—about 950 m^3/s, watershed area—425,600 km^2. From its source, the river flows southeast to Voronezh, then southwest to its mouth. The main city on the river is Rostov-on-Don. Its main tributary is the Donets River. At its easternmost point, the Don comes near the Volga River, and the Volga-Don Canal 105 km long connecting both rivers, is a major waterway. The water level of the Don in this area is raised by the Tsimlyansk Dam, forming the Tsimlyansk Reservoir. For the next 130 km below the Tsimlyansk Dam, the sufficient water depth in the Don River is maintained by the sequence of three dam-and-ship-lock complexes: the Nikolayevsky, Konstantinovsky and Kochetovsky Ship Locks. In antiquity, the river was viewed as the border between Europe and Asia by some ancient Greek geographers. In the Book of Jubilees, it is mentioned as being part of the border, beginning with its easternmost point up to its mouth, between the allotments of sons of Noah, that of Japheth to the north and that of Shem to the south. During the times of the old Scythians, it was known in Greek as the *Tanaïs*, and has been a major trading route.

Don River (http://en.wikipedia.org/wiki/Don_River_(Russia))

Don Cossack Troops – non-regular troops and also the administrative-territorial unit in Russia. They appeared on the Don in the fifteenth century where people of different nationalities escaping from the feudal-serfdom oppression settled down. It was first mentioned in 1570 in the merit scroll of Ivan the Terrible (Ivan IV). In the late fifteenth century D.C.T. recognized the supreme power of the Russian Tsar. In the seventeenth–early eighteenth centuries regardless of numerous attempts of the Tsar to do away with the Cossack freedom D.C.T. kept their full autonomy. Till the 1730s the most important decisions were taken during the Cossack Common Assembly that elected the highest authorities—atamans. For the time of military campaigns the Common Assembly elected the march atamans who were vested the unlimited authority. Beginning from the late sixteenth century D.C.T. took part in defense of the southern borders of Russia from the Turks, Crimean Tatars and nomad tribes. Don Cossacks made raids along the Volga, Yaik (Ural), Kuma, Kuban rivers and the Caspian and Azov seas for plundering merchant caravans, Persian, Turkish and Crimean settlements. In 1637 the Cossacks seized the fortress of Azov and kept it till 1642; they took part in the Azov campaigns of 1695–1696. Beginning from 1721 D.C.T. were put under control of the Military Collegium and the Cossack self-government was liquidated.

Don Naval Fleet – (1) Detachment of the Russian fleet created in 1733 in Tavrov for support of the Russian troops with repelling the attacks of Crimean Tatars in the Don River basin and also military operations in the Sea of Azov against the Turkish fleet. It comprised 15 ferries (flat-bottomed artillery sailing ships), about 60 galleys and to 500 boats and other vessels. During the Russian-Turkish War of 1735–1739 under command of Rear Admiral (from February 1737 Vice Admiral) P.P. Bredal D.N.F. took part in seizure of Azov (June 19(30), 1736). It was disbanded in October 1739 after the Belgrade Peace Treaty with Turkey. It was created anew in 1768 under the name "Azov Naval Fleet" (A.N.F.) and comprised 132 ships. Commanded by Vice Admiral A.N. Senyavin it fought against the Turkish Fleet in the Sea of Azov and the Black Sea during the Russian-Turkish War of 1768–1774. In 1783 it was disbanded once more and its ships were incorporated into the Black Sea Fleet.

(2) The fleet was created in 1919 for assistance to the troops of the Southern Front fighting against the White Guard and invaders on the Don River (Pavlovsk base). On June 29, 1919 after seizure by the Denikin troops of the fleet base it was disbanded. On March 8, 1920 it was formed anew and called Don-Azov Naval Fleet with the base in Rostov-on-Don. It comprised 17 ships. The fleet defended the Don mouth area and the Taganrog Bay, provided artillery support to the troops, transported and landed the assault troops. On May 11 its ships were transferred to A.N.F., while the river ships formed the Don Fleet of the Caucasus Front (on May 25, it was transferred to A.N.F. as the Don River Division and in September it was subordinated to the Commander of the Don Region Troops). It was disbanded in June-July 1921 and formed anew on August 27, 1941 during the Great Patriotic War (WWII). The Separate Don Detachment of A.N.F. incorporated divisions of river gunboats, 8 armored boats and 9 semi-glider boats, 3 land batteries, an armored train and a machine-gun squadron. From October 5, 1941 to July 28, 1942 it supported the troops defending Taganrog, Rostov-on-Don, and the Don mouth. The Separate Don Unit defended river borders ensuring crossing by the troops of water bodies and covered the retreat of the troops. After fulfillment of all tasks some ships were transferred to Kuban, while the rest were destroyed.

Donetsk Region – a large industrial, cultural and scientific center of Ukraine. D.R. locates in the southeast of Ukraine. It has the sea port of Mariupol with the exit into the Sea of Azov and the Black Sea. The area of D.R. is 26,570 km^2. The main industries developed here are coal, metallurgical and machine-building. It was formed on July 2, 1932. It is situated on a flat terrain broken in the north and west by the undulating structures of the Donetsk mountain range and near-Azov highland located within the confines of the near-Black Sea steppe province. This region has a beneficial geographical location, rich mineral raw material base, developed infrastructure and high urbanization. It supports the greatest population in Ukraine (4.4 million people in 2012). The population density here is 170 people/km^2. This region includes 18 districts, 52 towns, 131 urban-type settlements, and 1121 villages (2012). The major cities are Donetsk, which is a center of the D.R. (970,000 people), Mariupol (486,000 people), Makeevka (358,000 people) and Gorlovka (279,000 people).

D.R. borders on the most densely populated and economically developed regions of Ukraine—Dnepropetrovsk, Kharkov, Zaporozhie, Lugansk, and the Rostov Region of the Russian Federation. It possesses well-developed communication lines: auto roads, railroads and aviation lines as well as the port in Mariupol opening access to the Black Sea basin. All the above facilitates business activities in D.R. About 70 % of the products manufactured in D.R. are sold on foreign markets. This region maintains trade relations with more than 130 countries. Among its main trade partners there are Russia, China, EU countries, USA and Turkey. The share of the region in the Ukrainian export is over 25 %.

There is an international airport in Donetsk. The D.R. railroad is one of the largest in Ukraine, it ensures 40 % of the general state cargo and passenger traffic. The railroad connects Donetsk with major crossing stations in Ukraine and Russia. The port in Mariupol is one of the largest in Ukraine. By 2011 the port handled over 15 million tons of cargo annually, including 50 % of ready export products. D.R. may boast of the most abundant natural resources in Ukraine (12 % of the country's whole natural wealth). D.R. completely satisfies its internal needs and supplies other regions of Ukraine with such mineral raw materials as coal, rock salt, limestone and dolomite, fireclay and porcelain clay, kaoline, mercury, asbestos, gypsum, chalk, construction and facing stone. Perspective programs are elaborated on development of the deposits of iron ore, rare-earth metals, feldspar, phosphate rock, potassium salt, bentonite clay, nifeline, graphite, semiprecious stones and others. In the north of D.R. a gas field is explored ensuring the daily yield of 70,000 m^3. Its reserve is 1,400 million m^3. At present 15 more potential gas fields are being prospected with the reserves of about 30 bill m^3.

In D.R. three world famous nonferrous metallurgical plants "Donetsk Metallurgical Plant", "Metallurgical Plant "Azovstal" and "Meriupol Metallurgical Plant named after Ilyich"; 4 metallurgical works, 2 plants producing metal tubes, tube-milling and metal rolling plants are operating. The machine-building sector of D.R. comprises more than 220 enterprises, the largest are Novokramatorsk Plant, Concern "Azovmash", Yasinovatsky Plant. In the chemical industry 15 enterprises specializing in production of mineral fertilizers, acid, soda and plastics are in operation. The coal industry includes about 100 mines and colliery groups. The capacity of coal mining enterprises is approximately 56 million tons. The electricity sector is represented by the unified complex of generating, network, repair-technical facilities. The capacity of 7 thermal power plants is about 10,000 MW and it allows not only to satisfy the region's needs, but to export power to other regions.

Rich chernozem soils in D.R. open wide possibilities for agricultural production. D.R. has 2 million ha of agricultural lands, of which 1.6 million ha are in agricultural use. These areas are composed largely of chernozem soils with the humus content to 6 %. Winter wheat, sunflower seeds and vegetables are grown here. This region also produces pork, beef, poultry meat and fish.

D.R. is one of the Ukraine's major research centers. The region has 8 institutes of the National Academy of Sciences of Ukraine, 18 higher educational establishments, 100 research and design institutes, bureaus, stations and branches.

Donuzlav – the deepest liman (lagoon) on the western coast of the Crimean Peninsula. It is 30 km long, 7 km wide; its coastline stretches for 70 km. Its depth is 4–5 m, in the shipping way 12–15 m, the maximum depth reaching 25 m. It has

10 small bays. It is separated from the sea with a narrow sand barrier 300–400 m wide and 8.2 m long. Sand is excavated from D. for construction purposes. Till 1961 D. was an inland water body; the water level in it was 90 cm lower than in the B.S. It was fed mainly with sea water that seeped through the sand bar. Hydrogen sulfide was present in the bottom silt deposits. In 1961 after construction of a canal 7 m deep the lighter water from the B.S. ran into the lake. As a result, the steady water stratification was formed in D. and the water exchange among the layers practically stopped. The modern ecosystem of D. was formed after its connection with the sea via a canal. This water body turned, in fact, into a Black Sea lagoon stretching perpendicular to the coastline for 30 km and having rugged shores. The total area of its water surface is 47.5 km^2. Water salinity is to 18 ‰.

The local landscape preserve locates here. More than 50 fish species live here. This water body has good prospects for mariculture development. The mussel and oyster farms are located here. It is planned to create artificial reefs for diving. The D. water area is used as a part of the Evpatoria port and also as a naval base of the Ukrainian Navy. The only berth in the southern part of D. handles 4.5 million tons of sand excavated here for the needs of the region. The Crimean Government is planning to construct the Crimean marine-transport-industrial complex "Donuzlav". It will comprise a port including a dry-cargo and oil areas, auxiliary and special-purpose divisions.

"Dora" – the German-made 80-cm railway cannon, the most high-power in the mankind history. It was designed and produced by the Krupp Company and named in honor of the wife of chief designer Professor Erich Muller. The first cannon was manufactured in 1941. Its full length with the tube was 32.5 m and it weighed 400 tons. In the firing position the length of the whole installation was 43 m, width 7 m and height 11.6 m. The weight of the whole system was 1,350 tons. The cannon could fire only when installed on the special dual railway tracks. "D." was firing with 7.1-ton anticoncrete and 4.8-ton high-explosive shells containing 250 kg and 700 kg of the explosive, respectively. The maximum firing range of the HE-shells was 48 km. An anticoncrete shell was capable to pierce the armor to 1 m thick, concrete—to 8 m and ground—to 32 m. "D." was making 3 shots an hour. In February 1942 "D." was sent to Crimea for support of the siege artillery. It was positioned near Bakhchisaray. In the period from June 5 to June 26 it made 53 shots, of which only 5 hit the target. Later "D." was dismounted and moved to the vicinity of Leningrad. Chief of the General Staff of Wehrmacht Colonel-General Franz Halder appraised "D." as follows: "This is a piece of art, but quite useless."

"Dora" railway cannon (http://www.vincelewis.net/dora.html)

"Druzhba" Port – locates on the B.S. coast, to the south of Burgas, one of the major industrial complexes in Bulgaria. "D." was built for handling of oil exported from the USSR. From the port the oil is pumped along a pipeline 27 km long directly to the Burgas oil refinery.

Dub – a sailing coastal cargo vessel used in the northwestern Prichernomorie and in the mouths of Don and Bug rivers in the eighteenth century. By its design this is a keel boat with rounded luffs, straight inclined stem, transom stern and a hang rudder, 2 masts. Its length is to 20 m, width—to 6–7 m, hull height—to 1.8 m, and cargo lifting capacity—to 70–100 tons. The so-called preloading dub was used for cargo transportation to large ships staying off the harbor. This was a sailing-rowing flat-bottomed vessel with a removable mast and a steering oar.

Dubivka – a sailing fishing boat used for catching fish with nets, dragnets or trotlines. Its length was 6–7.5 m, width—about 1.6 m, hull height—0.5 to 0.7 m, draft—0.3–0.4 m, cargo carrying capacity—about 2 tons, crew—3–7 people. It was rather popular in the Black and Azov seas.

Duble-Boat – a small sailing-rowing war ship designed for operation in coastal waters. Such ships were first used in the Russian fleet during the war the Turkey of 1935–1939. In the reign of Catherine II their sizes increased to 23 m. Now they had a deck and were equipped with 2 large-caliber and to 10 smaller-caliber cannons.

Dubok – a sailing-rowing fishing boat with a sprit sail that was in wide use in the Black and Azov seas. Its length was 8–10 m, width—about 2.5 m, draft—0.5–0.6 m, cargo carrying capacity—3–3.5 tons, crew—3–5 people.

Dubrovin Nikolay Fedorovich (1837–1904) – Russian war historian, General-Lieutenant (1888), Academician of the Petersburg Academy of Sciences (1890). After graduation from the Mikhailovsky Artillery Academy in 1862 he served the Senior Aid-de-camp at the artillery staff of the Separate Guard Corps; from 1864—in Petersburg District Artillery Department; from 1869—at the General Headquarters for Military History Activities; from 1882—a member of the Military Scientific Committee being at the same time a member of the Emperor's Russian Historical Society (from 1884), from 1895—a member of the Board and a treasurer of this Society; from 1895—a permanent secretary of the Petersburg Academy of Sciences; from 1896—an editor of the journal "Russian Antiquity". He was the author and editor of collections of documents on the military political history of Russia of the eighteenth–nineteenth centuries "Materials for History of the Crimean War and Sevastopol Defense" (Vols. 1–5, St-Petersburg, 1871–1874), "Reports and Sentences in the Ruling Senate in the Reign of Peter the Great" (1887–1901), "Russian History in Letters of Contemporaries (1812–1815)" (1882), "Crimea Joining to Russia" (1885–1889), "Papers of Prince G.A. Potyomkin Tavrichesky" (1893–1895). Of the works written by D. the most significant are: "History of the Crimean War" and "349 Days of Sevastopol Defense" (Vols. 1–3, written in 1872, published in 1900), "History of the Wars and Ruling of the Russians on the Caucasus" (Vols. 1–4, 1871–1878) and others. D. saw the causes of the Russia defeat in the Crimean War of 1853–1856 in backwardness of Russia and its military system.

"DUNAITRANS" – a transportation partnership (Soviet-Bulgarian) set up in 1976 by the Soviet Danube Shipping Company and the Bulgarian River Shipping Company. Among the tasks of this company there were coordination and operative management of the cargo traffic, including from foreign countries, among the Danube ports.

Duni – a seaside resort located 40 km southward of Burgas, Bulgaria, on the coast of a small bight. Here a resort complex is constructed comprising three zones: Zelenika, Marina and Pelikan. This complex combines the modern amenities with the architecture of the Bulgarian national renaissance. The sun is shining here for 280 days a year. The sand beaches and dunes are stretching for 3 km.

Dvoinaya Bight (double bight) – the westernmost bight of the Heraclean Peninsula (near Sevastopol) that, closer to the top, gets broken down into two bights: Kamyshovaya and Kazachia (Cossack). There was also the other name of this bight—Troinaya (triple) because the Kazachia Bight, in its turn, is also broken down into two bights: Solenaya and Kazachia.

Dvuyakornaya Bight (double-anchor bight) – cuts into the shore to the south of Feodosia, between the Kiik-Atlama Cape and the Ilya Cape located 7.5 km to the north-east of it. Mountain range Biyuk-Yanyshar is stretching from the isthmus connecting the Kiik-Atlama Cape with the mainland in the northwestern direction. Mountain range Tepe-Oba is running nearly parallel to the former range from the Ilya Cape.

Dyavolsky Marsh – locates in the lower reaches of the Dyavolsky River, approximately 2 km to the southwest of Primorsk, Bulgaria. Dense thickets of cane and reed divide this territory into several small lakes. The water salinity in the eastern part of the marshlands may reach 14‰.

Dynamis – the only, in the Crimean history, woman—state ruler, the granddaughter of famous Mithridates VI Eupator. In 17 B.C. after the death of her husband Asander (possibly, a distant descendant of Spartocus) she became a sole ruler of the Kingdom of Bosporus. Forced by the circumstances she had official show loyalty to Rome. But all the time she stressed her kinship to Mithridates and was doing her best to continue the traditional for him anti-Rome policy. To strengthen their positions in the Bosporus in 10 B.C. the Romans sent their nominee—Polemon of Pontus who married D. In the next 2 years they ruled together. In 8 B.C. Polemon died in an attempt to suppress the riot of a Sarmatian tribe. After this D. ruled independently till her son Aspurgus came to the throne.

Dyurso – a part of Abrau-Dyurso Village (established in 1871, population—3,500), located at the Black Sea coast in the Krasnodar region of Russia, is a part of the municipal city of Novorossiysk (located 21 km west of Novorossiysk). Abrau is on the Abrau Lake, and Durso—near the sea seven km from Abrau by the narrow road along the mountain serpentine. In Durso small strip of shingle beach, separated from both sides by rocks, behind which begins a wild beach. Seaside resort.

Dyurso (http://anapaweb.ru/more/abrau.php)

Dzhangul coast – locates to the northwest of the Olenevka village on the Tarkhankut Peninsula, Crimea. A picturesque part of the coast with a specific relief formed by landslides, avalanches, wave action with some areas overgrown with draught-resistant plants and with places of rest for migratory birds. It is the local nature reserve.

Dzhankhot (Adygean name) – a farmstead, seaside resort. It is located 20 km from Gelendzhik, Russia. This is one of the most picturesque places on the whole Krasnodar Kray coast of the B.S., Russia. D. is situated practically in the middle of a pine forest. The sea shore on which the Dzhankhotsky pine wood is growing forms a steep, in some places even abrupt cliff 40–60 m high with a narrow beach strip at its foot. The area under the cliff near D. is called "Blue abyss". Approximately two kilometers from D., near the Praskoveeevka gap on the B.S. shore there is the "Parus Rock" (Sail Rock)—the unique monument of nature. The most remarkable feature of this landmark is its proportions. While the cliff is only a little more than a meter thick, its height is about 25 m and its length about 20 m. One its edge dips into water, while the other is distanced from the shore for 10 m and is surrounded by a pebble beach.

Parus Rock (http://bravebear.ru/wp-content/uploads/2010/08/21.07.2010_parus_01.jpg)

Dzhankoy – in 1926 it acquired the status of a town; the center of the Dzhankoy Region, located in the northern part of Crimea, Ukraine. It was first mentioned in 1855 as a Tatar settlement ("Dzhankoy" meaning "nice Crimean village"). Railroad crossing point. Population—36,300 people (2012). Machine-building, cannery, butter and wine-making factories.

Dzharylgach Bight – locates to the north of the Karkinitsky Bay between the mainland and the Dzharylgach Spit. Port Skadovsk is situated on the northern coast of D.B. 8.5 miles from the Dzharylgach cape.

Dzharylgach Island – represents the eastern part of the Dzharylgach Spit, Kherson Region, Ukraine. The southern shore of the island is surrounded by a narrow shallow strip with the water depth of 5 m. The Dzharylgach lighthouse is on the eastern tip of the island. This island is a waterfowl nature reserve.

Dzharylgach Lake – a small lake on the western coast of Crimea, Ukraine.

Dzharylgach Spit – locates in the north of Karkinitsky Bay forming the Dzharylgach Bight, Kherson Region, Ukraine. The bar is low-lying. Its western part is rather narrow and has several washways; the eastern part of the bar is wide and is called Dzharylgach Island.

Dzhau-Tepe (in the language of Crimean Tatars meaning *"enemy mountain"*) – a mud volcano situated in the southwestern plain of the Kerch Peninsula at the outskirts of Vulkanovka village, 12 km southward of the Feodosia-Kerch highway. This is a large hill about 60 m high with rounded slopes; its foot is dissected by gullies. The slopes are covered by mud. D.T. was first mentioned in the early seventeenth century. The volcano came alive in 1909, 1914, 1925, 1927 and 1942.

Dzhubga – an urban-type settlement; a seaside climatic resort in the Tuapse District, Krasnodar Territory, Russia. It is located 57 km to the northwest of Tuapse on the Black Sea coast of the Caucasus in a forest (mixed coniferous-deciduous vegetation) in the valley of the Dzhubga River flowing into a bay of the same name where the Dzhubga lighthouse was constructed (on the eastern coast). It has a wide pebble-sandy beach. More southward of D. in coastal waters there are many huge underwater stones where numerous crabs, mollusks and fish live.

Dzhubga-Lazarevskoe-Sochi Natural Gas Pipeline – offshore gas pipeline in the eastern part of the Black Sea, Russia. Pipeline starts near the village Dzhubga, where it goes into the Black Sea. The pipeline laying on the bottom (depths up to 80 m) goes along the coast at a distance 5–10 km from the shore to the distribution station "Kudepsta" near Sochi. Has a landfall near the Novomikhailovskoye, Tuapse and Kudepsta villages. The pipeline is 171.6 km long. The pipeline is intended to provide a reliable gas supply to Sochi and Adler, as well as gas supply to the 2014 Winter Olympic Games facilities under construction, to give impetus to the development of gasification of Sochi and the Tuapse districts. Pipeline "D.L.S." complements the existing pipeline "Maikop-Samurskaya-Sochi" (1.5–2 billion m^3/year), which supplied the gas to the Russian Black Sea coast of the Caucasus. Construction of the pipeline "D.L.S." began in September 2009, on March 30, 2010—construction of the offshore section of the pipeline in the Black Sea near Tuapse. The gas pipeline was put into effect (land and marine areas) on June 6, 2011. The cost of construction of the gas pipeline "Dzhubga—Lazarevskoye—Sochi" was about 1 billion USD. Annual production—about 3.78 billion m^3. The design lifetime of the pipeline is 50 years.

E

East Pontiac Mountains (Kackar Mountains) – a mountain range in the north-east of Asia Minor in Turkey. The mountains extend along the B.S. coast from the valley of the Chorokh River to the Melet River. Their length is 400 km and width—to 105 m. They appeared as a result of the Alpine folding processes and are composed of crystalline, metamorphic and volcanic rocks. Their highest peak in the range is Kaçkar Dagi (3,937 m). They are intensively broken by river valleys. In some places they feature the ancient glacier forms. The northern wet slope (pre-cipitations—to 3,000 mm) is overgrown with broad-leaved (to the altitude of 400 m), mixed (to 1,250 m) and fur (to 1,900 m) forests. The southern dry slope is covered by dry Mediterranean forests and shrubs. Higher than 2,000 m the sub-Alpine and Alpine meadows are found.

Eastern Black Sea coast of Turkey – the eastern part of Turkish Prichernomorie stretching from the delta of the Kyzyl-Irmak River to the border with Georgia. In the ancient times this part was called Ponta. This area includes the provinces of Amasya, Artvin, Bayburt, Giresun, Gümüshane, Ordu, Rize, Trabzon and Tokat.

Eastern Problem – in this way the international contradictions existed in the late nineteenth to early twentieth centuries were referred to in diplomacy and historical literature. They were related to the awaited breakup of the Ottoman Empire and struggle of its peoples for overthrow of the Ottoman yoke, rivalry of the Great Powers (Austria (from 1867—Austria-Hungary), Great Britain, Prussia (from 1871—Germany), Russia, Italy, France and later on the USA) in division of the Turkish territory. As P.I. Kapnist (1830–1898), Russian poet and participant of the Crimean War, wrote: "It is already long ago that some fatal problem, like a fatal colossus scaring everybody like a night ghost, has arisen... And it was decided to call it the 'eastern problem'". The term "eastern problem" was first used at the Verona Congress (1822) of the Holy Alliance while discussing the situation established on the Balkans in connection with the national-liberation revolution in Greece (1821–1829). As a result of the wars with Turkey of 1768–1774, 1787–1791, 1806–1812 and 1828–1829 Russia extended its power over Southern Ukraine, Crimea, Bessarabia, a part of the Caucasus and ensured strong Russian presence on the B.S. coasts. Under Adrianople Peace Treaty of 1829 Russia received the Danube mouth. This sharpened the Russian-Austrian disputes, while liberation of Greece and establishment of the Danube Principalities and Serbia enhanced the Russian influence on the Balkans and this spurred the rivalry between Russia and Britain. In July 1833 Russia signed with Turkey the Hünkâr Iskelesi Treaty that obliged Turkey in the event of a war to close, on the Russia's demand, the Dardanelles for passage of all military ships, except Russian. This Treaty was a major diplomatic success of Russia. But it stirred protests from Britain and France that were accompanied by the naval strength demonstrations near the Turkish coasts. And, finally, Russia had to surrender. London Conventions of 1840 and 1841 declared the intactness of the Ottoman Empire and closed the Straits for military ships of all European countries. Russia lost the advantages gained under the Hünkâr Iskelesi Treaty. The economic and military backwardness of Russia revealed in the course of Crimean War of 1853–1856 compared to the West European states and its diplomatic isolation predetermined the decline of the Russia's influence in international affairs, including in E.P. This found its continuation in the resolutions of the 1878 Berlin Congress when Russia, regardless of the victory in the war with Turkey (1877–1878) and bewaring of a new war, but this time with the block of West European states, had to sign the Berlin Treaty (1878). Britain occupied Cyprus Island (1878) and Egypt (1882), France—Tunisia (1881), Austria-Hungary—Bosnia and Herzegovina (1878). In 1879–1882 the so-called Triple Alliance of Germany, Austria-Hungary and Italy was created that amplified the influence of these states on the Balkans. Meanwhile, establishment of the unified Romanian state and declaration of its independence (1877) became possible only due to the Russian support. As a result of Russia's victory in the Russian-Turkish War of 1877–1878, Bulgaria overthrew the Turkish yoke and the Bulgarian national state was established (1878). With the help of Russia Serbia and Montenegro won the recognition of their independence. Further struggle for re-division of the world deepened the contradictions in the Middle East. Germany started extending the sphere of its influence in this region which exacerbated its relations with Russia. In addition, activation of the Austria-Hungary policy in this region in 1908 complicated significantly the Austrian-Russian relations. Russia weakened by the defeat in the Russian-Japanese war of 1904–1905 pursued the wait-and-see policy. The widening national-liberation movements of the peoples controlled by the Turkish Sultanate made the "Eastern Problem" still more acute. The Balkan wars (1912–1913) resulted in liberation of Macedonia, Albania and Greek islands in the Aegean Sea from the Turkish ruling. During World War I the Entente countries concluded secret agreements (British-French-Russian agreements of 1915, the Sykes-Picot Treaty of 1916 and others) as a result of which Russia gained Constantinople (Istanbul) and Black Sea Straits and the Asian part of Turkey was divided among the allies.

Eastern Thrace – a historical name of the European part of Turkey.

Eastern War – see Crimean War (1853–1856).

Ebergard Andrey Avgustovich (1856–1919) – Russian Admiral (1913), Chief of Navy General Staff, Commander-in-Chief of the Black Sea Navy. Finished the Naval Training School in 1879. In 1880–1890, served on the Pacific. In 1894–1896—naval agent in Constantinople during the suppression of Boxer Uprising (or Yihetuan Movement, 1899–1901). Participant of the Russian-Japanese War. In 1908–1911 headed the Naval General Staff—supreme body for operational and strategic control of the Navy, made a considerable contribution to the re-establishment and consolidation of Russia's naval forces. In 1911—vice-Admiral, appointed commander of the naval forces (from 1914—of the Navy) of the Black Sea. In 1913 promoted to Admirals. In 1914–1916, the Black Sea Navy commanded by E. inflicted heavy losses on the enemy by supporting the maritime flank of the Caucasus front, shelling coastal fortifications, blocking Bosporus, deranging marine movements of the German-Turkish fleet. For lack of resolve in actions against the battlecruiser "Goeben" (the dwellers of Sevastopol nicknamed E. "Gebengard") E. was deposed, and replaced by Vice-Admiral A.V. Kolchak. Dismissed for retirement on December 13, 1917. In 1918, arrested by the All-Russia Extraordinary Commission (VChK), but soon released. Decorated with orders.

Ecodiving – underwater ecological tourism aimed at perception of biodiversity of the submersible world, conditions of its existence and special features enabling it to survive.

Ecosystem modeling as a management tool for the Black Sea (the NATO TU– Black Sea Project) – a regional program of international cooperation in the framework of the NATO Program "Science for Peace". The purpose of this project is to build a reliable model of the B.S. ecosystem adequately reflecting the links and interactions existing among the biological, biogeochemical and physical components of the Black Sea ecosystem for revealing the causes of its degradation and predicting the consequences of the planned actions on alleviation of the impacts and restoration of the environment equilibrium. This program is called to address the following issues: joint passing by the researchers of short-term courses, provision of all participants with the standard investigation equipment to ensure comparability of the monitoring results; elaboration of a single system of database management in the B.S. countries for handling the oceanographic data about the natural environment; application of the satellite-based information, etc. The project was implemented in 1996–1998 and coordinated by Institute of Marine Sciences (IMS) of the Middle East Technical University in Erdemli, Turkey. This program resulted in development of the first version of the database management system and approved the best models showing the dynamics of the lower trophic levels in the B.S. biological communities. A system for receiving, processing and analysis of satellite imagery was installed in IMS in Erdemli.

Efimov Mikhail Nikiforovich (1881–1919) – the first licensed Russian pilot, Ensign (1915). He finished the Railroad Technical College (1902) and after this he worked as an electrician on the railroad telegraph in Odessa. In 1908–1909 he was the champion of Russia in motorcycle racing. In summer 1909 being a member of the Odessa Aeroclub he made flights on a glider. In early 1910 he finished the Henri Farman flying school at Châlons-sur-Marne in France and was the first of the

Russian pilots who received the diploma of a pilot-aviator (No. 31). He beat the world record of Orville Wright (USA) by the flight duration with a passenger; he also worked as a trainer and a test pilot. In 1910 in Odessa he made the Russia's first flights on aircraft "Farman-4". He was one of the first in Russia to practice night flights. From late 1910 he was a trainer of the aviation school in Sevastopol; took part in air escort of the Black Sea squadron voyage, in maneuvers of the Kiev and Petersburg military detachments (1911). During World War I he went as a volunteer to the front (1915), served in reconnaissance and fighter aircraft. In early 1917 he was transferred as a flagman pilot to the hydroaviation of the Black Sea fleet. When the Russian Civil War began he served in the Red Army. E. was executed by the Denikin troops.

Efimov M.N. (http://mg1.liveinternet.ru/images/attach/c/1//51/160/51160670_Efimov.jpg)

Eforie – balneal mud and maritime climatic health resort in Romania, 14 km south of the City of Constanţa. Sited on the Black Sea coast (between the sea and Techirghiol Lake), at the height of 20–30 m. Built in 1920 by Romanian monarchs. In fact, E. comprised two independent health resorts: Eforie-Nord (northern) and Eforie-Sud (southern). E. is the second largest and one of the best-known sea health-resorts on the Romanian Black Sea coast. The climate is temperate, mild, dry; it is shaped up by the warm Black Sea and by location of the health resort in the steppe zone. Mean temperature of January is around 0 °C. The summer is very warm: the mean temperature of July is 22–23 °C. The amount of precipitation is around 400 mm per year. The number of sunshine hours is about 2,200 per year. The bathing season—from mid-June to late September. The temperature of water in June is 18 °C, in July and August 21 °C, in September 16.5 °C. The main indications for treatment are diseases of the respiratory organs, of the skin, gynecopathy. At the health-resort, they practice baths and bathing in the swimming-pools with mineral water (bittern) of Techirghiol Lake and sea water, climotherapy, sea bathing, mud therapy, etc. The patients and holiday-makers are accommodated in sanatoria and boarding-houses, hotels, rest homes. E.Sud is 5 km south of the E.-Nord health-resort. Here, they treat largely children; the main indications: rheumatism, anaemia, disbolism.

Eforie Nord – a small village on the coast of the Black Sea, Romania.

Eforie Sud – a small village on the coast of the Black Sea, Romania.

Egorlytsky Bay – is confined in the north by the Kinburn Bar and in the south by the Egorlytsky Kut Peninsula. The water depth in the bay is less than 5 m. The coasts are low and rugged. There is a group of islands at the entrance into the bay; the largest of them are the Dolgiy and Kruglyi islands. In the apex of the bay the Velikiy and Konskiy islands are situated near the shore.

Eisk – a city and a port on the Sea of Azov coast, the center of the Eisk Region, Krasnodar Territory, Russia. One of the economic centers in the Azov area. Possibly, the name of E. may be traced to the name of the Eya River. The population here is 97,000 people (2011). Railroad station. Balneological (hydrogen sulfide, sodium-chloride waters), mud and climatic resort. It was founded in 1848 by the Cossacks resettled here from the Dnieper. On March 6, 1848 on the initiative of Count M.S. Vorontsov Russian Tsar issued the decree ordering the following: "On the Sea of Azov, in the area of the Black Sea troops, near the so-called Eisk Bar to open a port and to found a city that would be called the port city of Eisk." The official opening ceremony took place on August 19 (September 1), 1848. In 1850–1860 this city was a significant exporter of agricultural products (grain, wool). But after construction of the railroad Rostov-Vladikavkaz and the sea port in Novorossiysk the E. importance as a trade and transport center got diminished. The main industries developed here include metalworking, garment, footwear, food-canning, and meat. E. is a fishery center having a fish factory and a coopering-box plant. Among other industries there are mineral and mixed forage plant, ship-repair yards, plants, and brewery works.

Eisk (http://nesiditsa.ru/wp-content/uploads/2012/09/Eysk.jpg)

Eisk Lagoon – a shallow bay in the northeast of the Sea of Azov. It is the Eya River mouth separated from the sea by the Eisk and Glafirovsky sand bars. Its length is about 24 km, width—to 13 km and depth—to 3.2 m. The city of Eisk locates on the western shore of this lagoon.

Eisk Sea Port – locates in the southeastern part of the Taganrog Bay of the Sea of Azov near the Eya River mouth. An approach canal 2.2 km long and 3.8 m deep leads the ships into the sea. The port has five cargo wharfs for reloading the building materials and general cargo. The handling capacity of the above cargo by the wharfs is 2 million tons a year. But at present the mooring facilities are in a poor condition and require complete refurbishment.

Eisk Spit (Bar) – natural sandy bar, a part of Eisk Peninsula in the Sea of Azov. This is a central resort area of Eisk town. The present length of the bar is 3 km. At the beginning of twentieth century it reached 9 km, but on 13 March 1914 a strong storm broke the bar and as a result an island was detached. Now Zelenyi (Green) Island is located at a distance of 3 km from Eisk spit.

Elchaninov Matvey Maksimovich (8.8.1756–16.8.1816) – Russian Major-General (1799). He finished the Naval Cadet Corps (1769–1775). In 1772–1777 he took part in the campaigns on the Baltic Sea; in 1778–1781—on Sea of Azov and B.S. In 1781–1784 commanding a boat and then packet-boat he cruised every year between Taganrog and Constantinople. Later on he commanded the boats "Khoper" and "St. Catherine", and frigate "Phanagoria". During the Russian-Turkish War of 1787–1791 he showed himself well in the battle with the Ottoman fleet near the Phidonisi Island (the Theydonisi marine battle of 1788). Commanding the ship "Rozhdestvo Khristovo" he took part in the battles against the Ottoman fleet in the Kerch Strait and near the fortress of Khadjibei (1790). He commended himself well in the battle near the Kaliakra Cape (1791) for which he was awarded the golden weapon (1797). In 1793–1794 he was in Belorussia where he supervised construction of the gunboats on Sozh River. In 1795–1796 commanding 50 gunboats and 5 longboats built there he led them to Nikolaev. In 1797 he was appointed the Counselor of the Moscow Admiralty Chamber. In 1798–1804 he was promoted to the Senior Counselor of the same chamber. In 1800 he conducted inspection of the Astrakhan port. In January 1805 he was dismissed.

Elenite – the resort on B.S. near the Sunny Beach, Bulgaria. Here more than 200 villas looking on the sea ascend in an amphitheatre over the sea coast. There are also several large hotels with the resort infrastructure.

Emine, Cape – cape at the Bulgarian Black Sea coast, its height is 60 m above the sea level. It is located 78 km south of Varna, 54 km north of Burgas. Strong winds blow here very often. This is the place where the Stara Planina Mountain Range reaches the Black Sea as if cutting Bulgaria into the northern and southern parts. It forms the tip of the Stara Planina. There is a light-house, whose light can be seen as far as 55 km. Not far from the light-house there are the ruins of the old Roman fortress Paleocastrum.

Cape Emine (http://zagranhome.blogspot.ru/p/blog-page_06.html)

"Empress Maria" –

(1) the 84-cannon battleship of Russian Black Sea Navy. It was built in Nikolaev in 1827. Its length was 59.7 m, width 15.6 m. Armament: 97 cannons. In August-September 1828 during the Russian-Turkish War it bombed the fortress of Varna; in February 1829 it took part in siege and taking of the fortresses Akhiollo and Sozopol. In 1833 it was a part of the squadron commanded by Russian Rear Admiral M.P. Lazarev. In 1845 it was dismantled.

(2) the 84-cannon battleship of Russian Black Sea Navy. In was built in Nikolaev in 1853. Its length was 61 m, width 17.3 m. Armament: 90 cannons. During the battle at Sinop it was the flagship of Russian Vice-Admiral P.S. Nakhimov who commanded the squadron. The cannons of this ship destroyed the Turkish frigates "Auni Allah" and "Fazli Allah". In August 1855 when the Russian troops retreated from Sevastopol this ship was sank in the roadstead. In summer 1857 during clearing of the Sevastopol Bay the ship lying at a depth of 12.8 m was inspected by U.S. divers who said that its lifting was unfeasible.

(3) the battleship of Russian Black Sea Navy. It was built in 1911 in Nikolaev on the yard "Russud" as one of four battleships: "Empress Maria", "Empress Catherine the Great", "Emperor Alexander III" and "Emperor Nikolay I" (not finished, but after February 1917 it was named "Demokratia"). Its displacement was 22,600 tons, length 168 m, width with armor 27.3 m, draft 8.3 m, speed 21 knots. Armament: 12 cannons, 4 torpedoes, crew 1,220 people. In 1915 it was commissioned and in summer 1916 the Black Sea Navy was brought under command of Vice-Admiral A. Kolchak who made "E.M." the flagship. During 1915–1916 this battleship in the first maneuver group fought in B.S. In early 1916 "E.M." took part in a major sea operation on

mass bombing of Zonguldak. On October 6, 1916 "E.M." finished urgent preparations for sea voyage. On October 7 at 6:20 the ship was shaken by a major explosion in the powder magazine in the forward turret of the main armament that was followed by 24 explosions of less magnitude. As a result, 260 seamen died immediately. Admiral A. Kolchak who arrived to the battleship immediately organized rescue of the people. Thanks to his efforts they managed to save 1,000 people. But despite the enormous efforts of the remaining crew in 48 min the ship sank at a depth 21 m. The historians are still arguing about the causes of the explosion and wreck of the ship—the suggestions vary from subversive actions of German spies to self-ignition of propellant charges. In 1916 Russian Academician A.N. Krylov initiated works on lifting "E.M.". But they were interrupted by the political events of 1917 in Russia and resumed only in 1918. On May 5 the battleship was lifted and towed to the shipyard. In 1926 the hull of the battleship was dismantled at the shipyards in Sevastopol. Later the turrets and cannons were also lifted.

Eni-Kale Fortress (New Fortress) – it was constructed by the Turks on the coast of the Kerch Strait in its narrowest place, on the steep limestone cliff where in the ancient times the Greek fortress of Parthenion was rising. Construction of this fortress took rather a long time and ended only in 1703. The construction was supervised by French engineers. In the eighteenth century when Kerch with its E. fortress was incorporated into Russia the Kerch Strait was given the name of Kerch-Enikalsky. Later on the sea canal got the same name.

Environmental impact assessment (EIA) – the activity aimed at identification and forecast of the expected impact on the natural environment, health and welfare of the people of various actions and projects, and also on subsequent interpretation and transfer of the received information; study of unfavorable consequences of the impact of the planned projects on the natural environment, health, and economic activity of the population, conditions of cultural property. EIA is a special area of the environmental expertise seeking to make a project environmentally friendly, improve its environmental characteristics, prevent, mitigate or compensate the likely economic damage, avoid costs and expenses in the course of project implementation connected with arising of some unforeseen environmental problems. There are distinguished three types of assessments: EIA of projects of some technical or economic facilities (hydroelectric dams, irrigation systems, etc.); EIA of projects of a regional scale envisaging construction of a complex of large objects or engineering systems capable to produce a combined impact on the environment over great territories; and EIA of projects connected with development of particular economic areas, creation of numerous relatively small and similar economic objects, the impact of each may be insignificant, but their combined effect shall be evaluated.

Environmental programmes in the Black Sea – Numerous environmental programmes have been performed in the Black Sea, among them: The Cooperative Marine Science Program for the Black Sea (CoMSBlack), GEF Black Sea

Environmental Programme (BSEP), European River-Ocean System Project (EROS 2000), Ecosystem Modeling as a Management Tool for the Black Sea (TU-BLACK SEA), Wave Climatology of the Turkish Coast: Measurements-Analysis Modeling (TU-WAVES), Black Sea GOOS: a Step Towards Observation and Prediction System (STOPS) (IOC Pilot Project II), Assessment of the sediment flux in Black Sea, mechanisms of formation, transformation and dispersion and Ecological significance (IOC Pilot Project II), Danube River Basin Programme, Danube Delta Project, Azov Sea Project, Black Sea Environmental Programme (Phase II), Danube River Basin (Phase II), Dnipro River Programme, Intergovernmental Oceanographic Commission (IOC)—UNESCO Black Sea Regional Programme in Marine Sciences and Services, International Atomic Energy Agency "Marine Environmental Assessment of the Black Sea Region" TC Project RER/2/003, BLASON (Black Sea Over the Neoeuxinian) French-Romanian co-operation with international participation, Southern European Seas: Assessing and Modelling Ecosystem changes (SESAME), The Ocean Data and Information Network for the Black Sea (ODINBLACKSEA), Black Sea Scene, and many others.

EPRON (Special-Purpose Underwater Works Expedition) – unit established in 1923 for lifting the sunk vessels and carrying out salvage and rescue operations in the USSR. Set up in Sevastopol at the direction of F.E. Dzerzhinskiy attached to OGPU for the search of the British steamer "Prince" ("Black Prince") that had sunk in 1854 during the Crimean war near Sevastopol. The steamer remains were discovered in Balaklava harbor, yet no gold was found there. EPRON kept lifting the sunk ships. During the first 15 years of EPRON has lifted 250 vessels. Before long, EPRON divisions were setup in Odessa, Novorossiisk, Tuapse and Kerch, subsequently in Leningrad, Murmansk, Baku, Astrakhan, Vladivostok and Khabarovsk. To facilitate control of these divisions, the Main Department of EPRON was established. The first chiefs of EPRON were L.N. Zakharov (Meier) and Rear-Admiral F.I. Krylov. The divers were trained at the Naval Diving Secondary Technical School at Balaklava. In 1931, EPRON was subordinated to the People's Commissariat of Railways and that same year was transferred to the just established Narkomvod. In 1932, EPRON was enlarged with the GOSSUDPODYEM unit, which was commissioned to handle all underwater and rescue operations for all ministries and departments. In 1939, after Narkomvod was reorganized, EPRON was left in the system of Narkomflot. When the Great Patriotic War (WWII in the USSR) broke out on June 22, 1941, EPRON became part of the Salvage and Rescue Operations Service of the USSR Navy. By the time, it had become a powerful ship-lifting and rescue unit with 28 rescue ships and tugboats, a large park of ship-lifting pontoons and other equipment, over 300 specialists, among them 600 top-rate divers. During the years of ERPRON existence, it lifted 450 vessels and ships of 437,000 tons combined tonnage. Among the vessels salvaged in 1932, freight steamer "Ilyich" (the Aegean Sea), in 1933—ice-breaker ships "Malygin" (near Spitsbergen Archipelago in the Arctic Ocean) and "Sadko" (the White Sea), in 1934—the steamers "Kuznets Lesov" (the South China Sea) and "Kharkov" (Bosporus Strait), in 1937—ice-breaker ship "Sibiryakov" (Kara Strait near Novaya

Zemlya), in 1939—steamer "Chelyuskinets" (the Baltic Sea), steamer "Petr Velikiy" (the Black Sea) and others. EPRON divers rendered considerable assistance to underwater archeologists. Awarded the Order of the Red Banner of Labor (1929). There are over 10,000 ships sunk in the Black Sea, during WWI 3,000 ships were shipwrecked, during WWII Germany and its allies lost 457 ships, the USSR Black Sea Navy lost 212 ships.

Erdemir – port on the Black Sea coast in the vicinity of Eregli town, Turkey. Built as further development of a major iron-and-steel plant. Financed by private (51 %) and state capital. The plant and industrial enterprises of the area adjoining the port will be able to use transshipment capacity of the port each year at the rate of close to 8 million tons. The port will be able to handle large-capacity ships with draft up to 20 m. The cargo turnover is currently dominated by import (roughly 65 %): iron ore (Brazil, South Africa, Canada), coal (Australia, Poland), slabs (Ukraine). The port authorities seek to attract other cargoes to the port (under consideration is the construction of a ferry terminal to enable shipments by high-capacity trucks from the Russian and Ukrainian ports and other ports of the basin).

Eregli – town, port, administrative center, sited 60 km west of Zonguldak on the slope of the upland overlooking the north-eastern part of Eregli harbor, Turkey. Population: 103,000 (2012). Established in 560 B.C. On the northern, western and southern sides, the city is surrounded by an ancient wall; on the eastern side, it ends with a precipice. There are coal mines in the city environs. Eregli harbor is the nearest anchorage to Bosporus Strait, suitable for deep draft ships. The name of the town is related to Hercules. Legend has it that it was here, in the Dzhekhen Nemagzy cave Hercules caught the three-head dog Cerberus. An iron-and-steel plant (Turkey's third largest) is also here. Eregli's special feature is cultivation of strawberry; the local variety is regarded one of the best in the world.

Eregli, Harbor – cuts into the shore between Cape Baba and Cape Cengelkaya which is 4.6 km to the south-south-east of it. The harbor shores are largely low-lying. The southern part of the eastern shore is water-logged; two rivers fall into the harbor here. The city of Eregli lies on the north-eastern shore of the harbor. The depths at the entrance to E.H. are 18–20 m and uniformly decrease as one approaches the shores.

Espiye, Bay – cuts into the coast west of Tirebolu town between the rocks of Fyryn and Cham Cape. The depth in the middle of the bay are up to 73 m. The western shore of the bay is steep and rocky, while the southern and eastern ones are sandy. Gelevara River falls into the bay.

Estuary Bar – aggradational form shaped up in the estuary of the river arm as a result of deposited silt at the point of channel flow input in the intake water body and abrupt deceleration of flow velocity. E.B. is one of the most dynamic elements of the hydrographic network of any delta which constitute the greatest obstacles to transit navigation across a river estuary. This holds good of the channel network of the Danube Delta.

"Eternal Peace Treaty" (1686) – was concluded in Moscow on May 6, 1686 between Russia and Polish-Lithuanian Commonwealth. It stressed the greater striving of both countries to formation of the anti-Ottoman coalition. This treaty confirmed the Andrusovo Treaty of 1667 under which Poland renounced Kiev, Zaporozhie Sech was passed to Russia, while Russia broke its relations with Porta and sent its troops to the Crimea. It was ratified by the Polish Sejm only in 1710.

European Danube Commission – was established subject to the 1856 Paris Peace Treaty that declared the Danube an international river. After the end of World War II in 1948 the Danube Commission was set up instead of E.D.C.

Evksinograd – a small settlement 8 km to the north of Varna, Bulgaria. The summer palace of the Bulgarian tsars that was built in 1882 by Prince Alexander I Battenberg. In the parks near this palace you can find rare plants from all over the world. The palace is now a governmental residence. Since 2007, it is also the venue of the *Operosa* annual opera festival. It is well known for its white vines and raki (Bulgarian vodka).

Evksinograd Palace (http://en.wikipedia.org/wiki/File:Euxinograd-palace-benkovski.png)

Evpatoria – a city and a sea port 65 km to the northwest of Simferopol, on the coast of the Evpatoria harbor (it incises into the coast for 1.3 km) of the Kalamitsky Bay, the Black Sea; a seaside climatic, balneo and mud resort, mostly for children in the Crimea, Ukraine. Population—124,000 (2010). It locates in the place of the former Greek colony of Kerkinitis (sixth B.C. to fourth A.D.) that was mentioned by ancient Greek historian Hecateus Miletian and "the father of history" Herodotus. The history of this city was associated with the name of Mithridates VI Eupator of Pontus. The city's modern name derives from Eupator. E. is one of the most ancient cities. The colony of Kerkinitis was later on incorporated into the Khersonesos state. Archeological excavations proved that this territory had been inhabited by the Kimmerians and Tauri before Greek colonists came here. In the second century B.C. it was conquered by the Scythians and remained under their control till third to fourth centuries A.D. when it was burnt by the Goths. In the early medieval time a fortress was again constructed here. By the late fifteenth century the Turkish fortress of Güzleve located here (its name is from the Turk "*güz*" meaning "source, spring") turned into one of the slave-trading centers in Crimea. Of medieval Güzleve only mosque Djuma-Djami (sixteenth century), the Europe's only multi-dome mosque, the Turkish baths (sixteenth century), Karaim kenas (seventeenth to eighteenth centuries), and dervish-tekie monastery (fourteenth to fifteenth centuries) had survived to the present. Struggling for the outlet to the sea the Russian troops in the eighteenth century seized this fortress twice and then renamed it into Kozlov. In 1784 after Crimea was incorporated into the Russian Empire this city was renamed into E. in honor of Mithridates VI Eupator.

E. is surrounded by water on three sides: in the south by the Kalamitsky Bay of B.S., in the east by the Crimea's largest lake of Sasyk-Sivash that comes to the city outskirts, in the west by the Moinak Lake. Numerous lagoons and salt lakes which bottoms are covered with mud are scattered around E.

The famous beaches of E. (to 100 m wide) with their fine yellow ("golden") sands stretch for many kilometers from east to west. The sea floor is smooth, gradually sloping down which is very good for bathing of children. In fact, the whole territory of the seaside resort is covered by gardens and parks. A wide asphalt-paved embankment runs along the sea parallel to the beach and ends in a big seaside park. The vicinities of E. are covered with vineyards and fruit orchards. The climate here is subtropical Mediterranean. The winter is mild, windy, without a steady snow cover. The mean air temperature in February is 1 °C; rather frosty weather is also possible. The mean air temperature in July is 23 °C. The bathing season lasts from June through September. The mean water temperature in summer is 20–22 °C. In the first half of autumn the weather is warm and sunny; in the second half—cloudy and rainy. Atmospheric precipitations—about 350 mm a year. The number of sunny hours is 2,384 a year. The climate in E. is favorable for aero-, gelio- and thalasso therapy. Already in the second half of May the sands of the beaches heat to 40–45 °C, while in July-August—to 50 °C. The main health-improving factors here are the sunshine and sea, air and sand, brine and mud of the salt lakes, as well as the mineral water of the hot springs. The sulfide mud from the Saksky Lake is used in mud treatment.

E. developed as a resort from the late nineteenth century. In 1886 the first private small mud-cure center was constructed near the Moinak Lake; later on a hotel appeared here. In 1905 the first climatic sanatorium was opened in E; in 1909–1914—a medical beach and 7 private sanatoriums. In 1915 the railroad was constructed here. Rapid construction of a resort started after the 1917 October

Revolution in Russia. The E. development was spurred further in 1936 after passing the resolution "On organization of the model children resort in Evpatoria" by the Soviet Government. By 1941 there were already 36 sanatoriums, vacation hotels and rest houses here, including 16 sanatoriums for children suffering mostly from bone-joints TB. During the Great Patriotic War and occupation of E. by the German Army the sanatoriums were destroyed, in part or in full. By 1948 they were restored and the Moinak mud-cure center was again operating. By the early 1980s more than 1.3 million people, and half of them—children, were able to have rest and treatment in E. every year. Apart from the climate and balneo therapy, mud treatment and psammotherapy, the water treatment (sea, artificial hydrogen sulfide, carbon dioxide, oxygen, radon and other baths, curative showers), remedial gymnastics, inhalation and others are practiced here.

Evpatoria (http://alta-travel.com.ua/images/1/evpatorija/001a61gt.jpg)

Evpatoria Cape – a northern entrance cape of the Kalamitsky Bay of B.S. The cape is low-lying and surrounded by reefs. Some sandy hills are rising on it. The lighthouse is constructed on this cape.

Evpatoria landing – the landing on January 5, 1942 was organized to detract the German Army and draw off a part of their forces from Sevastopol and Kerch Peninsula where the Kerch-Feodosia operation was conducted that time. In case of success of this landing the Soviet troops took hold of a very important area from where they could attack the Perekop Isthmus and cut the communications of the 11th German Army of General Erich von Manstein in Crimea. The landing troops supported by partisans and the population of this city seized the greater part of Evpatoria. The German command having duly appraised the significance of this military operation hastily redeployed here one infantry regiment and two battalions from Sevastopol. On January 8 after a hard-fought battle the landing troops were forced back, a part of them succeeded to break through the entrapment and went to the mountains to the partisans.

Evpatoria seizure – during the Crimean War on February 5, 1855 the Russian troops commanded by General S.A. Khrulev started seizure of Evpatoria. The city was occupied by the Turkish corps of Omer-Pasha that threatened the rear communications of the Russian Army in Crimea. To forestall the Turkish attack the Russian command decided to take hold of Evpatoria. Insufficiency of forces for storming this city had to be compensated by a surprise attack. But they failed to achieve this. In fear of great casualties and attack failure Khrulev ordered to stop the attack. The troops retreated to their original positions. Regardless of the failure the storming of Evpatoria paralyzed the Turkish corps that never had taken active military actions here. On the other hand, the failure of storming Evpatoria had shown that the Russian troops were unable to give adequate rebuff to the Turkish troops being least efficient ally's troops in Crimea. Quite likely that such news hastened the death of Emperor Nickolay I that occurred on February 18, 1855. Before his death he issued the last decree in which he dismissed Prince A.S. Menshikov, the commander of the Russian troops in Crimea, from this post for the failure of E.S.

Exotic species – species of plants and animals not peculiar to a given area, which were by chance or (which is less common) deliberately introduced by man and got adapted to a new habitat. E.S. are often referred to as invaders; these may be planned or not planned (accidental).

Experimental Ichthyological Station – it was founded in 1933 at the Ministry of Agriculture of Bulgaria. It continued researches conducted by the research division at the Fishery College in Varna. The station studied the biology of practically all fish species of B.S. In 1954 it was included into the Institute of Fishery and Fish Industry later renamed into the Institute of Oceanography and Fisheries. The results of researches conducted at this station were published in Proceedings of the Experimental Ichthyology Station.

Eya – a river in the Krasnodar Territory, Russia. It flows into the Eiskiy Liman of the Taganrog Bay, the Sea of Azov. Its length is 311 km. It runs in a wide valley over the Kuban-Priazov (Prikuban) Lowland. In summer it dries out. In the lower reaches it is waterlogged. Its main tributaries are Sosyka (left) and Kugo-Eya (right).

F

"Faddey Bellinsgauzen" – oceanographic research vessel, world's first, that traveled as far as the point of the South magnetic pole. Got its name in honor of the Antarctica discoverer, Russian seafarer F.F. Bellinsgauzen. "F.B." worked in the seas of the Atlantic, Indian and Pacific oceans. Called at 24 foreign ports. "F.B." made thousands of oceanographic stations. As far back as in 1968, the vessel traveled to the waters of the Antarctica, making scheduled oceanographic studies in the Drake Passage. During these operations, the Black Sea hydrographers landed on the Antarctic island Smolensk and Waterloo. In 1982–1983, for the first time in the history of the Soviet Navy "F.B." in cooperation with the expedition flagman "Admiral Vladimirskiy" made a round-the-world Antarctic cruize organized by the Black Sea Navy Hydrographic Service in honor of the 200th anniversary of the hero-city Sevastopol. The vessel left Sevastopol in December of 1982 and followed the route of the first Russian Antarctic expedition. During the 147 days of the navigation, "F.B." covered 43,500 miles, circling the globe twice. In April of 1983, the ship came back to Sevastopol.

"Faithful Black Sea Troops" (**"Troops of faithful Cossacks"**) – according to the Order of Catherine II of January 10, 1790, the Cossack troops were given the name "Faithful Cossacks" unlike the "unfaithful" Cossacks that left for Turkey. It was allowed to form the "faithful Cossacks troops" out of Cossacks from Zadunaiskiy Secha. The ataman of the camp was elected Sidor Ignatievich Belyi, the cavalry commander—Zakhariy Chepiga. Having received the rank of the Lieutenant Colonel S.I. Belyi with his troops was sent under command of A.V. Suvorov, Generalissimo of the Russian Empire. "Faithful Cossacks" showed themselves real heroes in battles under Kinburn and Ochakov. In commemoration of their victories the "Faithful Cossacks Troops" were renamed into "Faithful Black Sea troops".

Faleev Mikhail Leontyevich (circa 1740–1792) – Russian merchant, brigadier (1790) from Kremenchug. Associate and assistant of Prince G.A. Potemkin in establishing the Black Sea Navy. On Potemkin's commission, F. set up a shipyard in the mouth of Ingul River flowing into Bug Liman and founded Nikolaev town (1789), managed the shipyard on the Black Sea coast. Supervised the construction of the ship "Preobrazhenie Gospodne" (God's Transformation) in Kherson, in Nikolaev—of the frigate "Grigoriy Velikiy" (1790–1791) and other ships. During the final years of his life, served in Nikolaev as the Chief-stehr-war-commissar (1790). Was in charge of all financial matters pertaining to the Black Sea Navy construction. One of Nikolaev's central streets was named after F.

False Bosporus – a low sand coast between Terkos (Durusu) Lake and B.S., Turkey, limited in the east by the upland extending southward of the Karaburun Cape and in the west—by a chain of hills stretching along the coast. It resembles amazingly the northern entrance into the Bosporus, and the more so as the Terkos Lake has a considerable extension and confined in the south by a high bank. Therefore, this area is referred to as F.B.

Fanagoria – wine-making area on the coast of the Black Sea (today in Krasnodar Kray). Sited in the western part of Temryuk District, where around 542 B.C. natives of the ancient Greek Teos led by Phanagor, who had fled after their native city was seized by the troops of the Persian King Kir, founded a city of the same name. The city was sited on one of the islands called Fanagor. Finds of painted ceramics, amphoras and terracottas confirm the information contained in written sources indicating that the town had been founded in the third quarter of the fourth century B.C. At the time of Catherine II, a garden nursery was opened in Ekaterinodar (Krasnodar) with 25,000 vines that had been exported from the Crimea. Vine growing began to develop actively on Taman Peninsula in the 1870s. From 1875 to 1894 vineyard areas in Kuban Region increase from 218 to 2,500 dessyatines (1 dessyatina = 1.1 ha). At the end of the nineteenth century, annual crop of grapes in Kuban Region equaled 526,000 puds (1 pud = 16 kg), in Chernomor Province— 84,000 puds, and on the eve of WWI, they harvested close to 1 million puds of grapes in Kuban area. In late nineteenth century, Chernomor Province alone produced as much as 100,000 buckets (1 bucket = 10–12 l) of grape wine. In 1959, a decree was issued regarding the organizing of specialized grape-growing state farms, whereby 14 wine-making state farms were set up, and in 1983 there were 23 such state farms. Temryuk District is on the 45th parallel, which corresponds to the latitude of Bordeaux and Turin. In view of the climate features of the historic grape-growing region Fanagoria (unique wind rose shaped up by the closeness of the Black and Azov Seas; the large number of sunny days in a year: on Taman Peninsula, there are more such days than at Anapa or Sochi, a warm, protracted autumn), the classical "French" varieties of grapes are getting on quite well here: Cabernet-Sauvignon, Merlot, Chardonnay. The main zoned technical varieties of Taman Peninsula are Aligote, Rkatsiteli, Claret, Sauvignon, Saperavi as well as Champagne varieties—Traminer, Pinot-Noir, Chardonnay, Pinot-Gris, Pinot-Blanc, Sauvignon.

Fanagoria (Phanagoria), town – (Greek: Phanagoreia) classical town on Taman Peninsula that used to stand near the existing settlement Sennoi (Krasnodar Territrory, Russia). Founded in the second half of the sixth century B.C. From the fifth century B.C. was part of Bosporus State. The peak of F. development falls on the fifth–second centuries B.C. In the fourth–second centuries B.C. Fanagoria was a major economic center of the Bosporus State. In the first century B.C. was called Agrippia. In late fifth century, the town of Fanagoria had its own mint and minted silver coins. The population of F. was engaged in agriculture, fishing, handicrafts and commerce. Discovered in F. are the foundations of fortress walls, remains of residential quarters, bathing establishments, a district of pot-makers. There was a necropolis near F. The ancient town existed until the eleventh–twelfth centuries.

The area occupied by the ancient Greek F. had been settled from the time immemorial, which is indicated by archeological finds discovered in this location in the most ancient cultural layers: a fragment of flint, a silicious tip, and a fragment of a hand-axe. Many classical historians make reference to tribes that used to populate Taman Peninsula even though they tend to call it an island or even an archipelago. Thus, according to Strabon, there were five islands at the dawn of A.D. which were separated by sea gulfs, lagoons and channels of Kuban River (Greek: Gipanis) and subsequently linked with one another by silts of the Kuban Delta. At that time, the main river channel headed for the Black Sea, while several river arms—for the Sea of Azov. In the eleventh century this united island was not merged with the continent yet and was called Fanagorian Island. Arable farming, grape growing and wine-making constituted the basis of F. economy. Merchants used to import amphoras filled with Hios wine that was famous all over the Mediterranean region. In exchange, grain, meat, fish and leather were shipped from Fanagoria.

Fatsa – a town on the coast of the Black Sea, Turkey.

Fatsa, Bay – cuts into the Black Sea coast between Capes Yasun and Metropolis which is 24.6 km to the west-south-west of it. Fatsa town is on the south-western shore of the bay. Kes River falls into the bay apex. Depth reaches 165 m in the middle of the bay. There is a fisherman's village of the same name on the shore.

Fedorov Vasiliy Averyanovich (1866–?) – Russian Lieutenant-Colonel (1909), hydrographer. In 1881, was enrolled as a student of the navigator's branch of the Technological Secondary School under the Naval Department which he finished in 1885. From 1888 to 1892, was engaged in studies of Peter the Great Bay (the Sea of Japan), then sailed as a senior navigator on various vessels in the Far Eastern seas. In 1902 transferred to the Black Sea Navy. Was the custodian of the Batumi light-house and chief pilot of Batumi and Dniester-Tsaregrad course. In 1913, admitted to the Hydrography Committee as a hydrographer. A cape in Peter the Great Bay is named after F.

Fedotova Spit and Biryuchy Island – a sandy spit ending by almost an island 43 km long in total in western part of the Sea of Azov, Ukraine.

Felyuga, Feluka – (Ital. *"felucca"*, from Arabic *"fuluca"*—*"boat"*) (1). Boat on Mediterranean galleys intended to facilitate liaison with the shore and among vessels. Had 3–5 pairs of oars, a mast with a jigger. (2). Small sailing vessel for cabotage in the Mediterranean countries and for moving cargoes and fishing. Was furnished with 1–3 short masts with jiggers, sometimes with oars. Was used by Greek pirates, in which case had 6–8 cannons. (3). Sail-and-rowing or sail-motor fishing boat on the Black and Azov Seas. Had a mast with a quadrangular jigger. Length about 6 m, width about 2 m, draft—about 0.5 m, tonnage 5–6 tons.

Fenton Roger (1819—1869) – outstanding British photographer who made his name by making photographs of the Crimean War episodes that were the first photodocuments in the history of photography. Became famous thanks to his still life paintings and landscapes extremely popular during the Victorian time. In the winter of 1855 was posted to the Crimea as an official photographer of the British Government. His connections "at the top" as the founder (1853) and the first honorable secretary of the Royal Photographic Society in London helped him to secure this assignment. F. and his assistant Marcus Sparling arrived in the Crimea on board "Hecla" ship and arranged a laboratory in a van. Using the wet colloidal process adopted universally at the time, the two made around 360 photographs. Because F. and Sparling were civil servants, they confined themselves to documenting the "acceptable" sides of the campaign only. In fact, they made very few pictures of true combat operations. The failed attacks of the light cavalry brigade were, perhaps, the only defeat they managed to document photographically. Long before F. returned to Britain, his photographs from the war theater were displayed extensively in London and Paris, the most interesting photographs were printed in the "Illustrated London News" magazine.

Fenton Roger, self-portrait, February 1852 (http://www.nga.gov/exhibitions/2004/fenton/images/fenton_index_fs.shtm)

Feodosiya – (from Greek—*Theodoro*—"*God-endowed*") town and port, health-resort center, climatic and balneal health resort (Crimea, Ukraine), sited on the coast of the large Black-Sea bay in the south-east of the Crimea. Population—70,000 people (2012). In the north-east—the offspurs of the Crimean Mountains, in the north and north-east—low gently-sloping hills; solonchak vegetation is common. From the east limited by Chauda Cape, from the south-west—by Kiik-Atlama Cape. The old part of the town is sited on the slopes of Mountain Tepe-Oba, the new part—on more flat segments of the shore. Wide (up to 50 m) sand-shelly beaches, hemming the bay, give way to a very gently-sloping sandy sea bottom. The climate is of transitory nature: from subtropical Mediterranean to temperate continental. The winter is very mild; mean temperature of January–February around 0 °C; no stable snow cover is formed. In spring, dry warm weather prevails, although frosts may occur until late March. The summer is very warm, with hot dry or temperately moist weather alternating; mean temperature of July 24 °C. The autumn is warm, with little cloudiness. The bathing season lasts from July to October (temperature of sea water from 17 to 24 °C). Precipitation is around 450 mm per year. The number of sunshine hours is 2,265 per year. Dominant winds are north-westerly with a mean velocity of 4.6 m/s.

The city is famous for being a marine and balneal health resort. The main curative factors alongside climate are mineral waters and medicinal muds. There is a mineral water source Feodosiya 2 km from the town, whose sulfate-chloride sodium water (salinity 4 g/l) is bottled as a curative-table water under the name of "Feodosiiskaya" and is used for treatment through drinking. 12 km north of the town is salt Lake Adzhigol, whose sulfide silt mud is of high therapeutic value and is used for mud therapy.

Even legends fail to provide the exact data of F. establishment. Around the middle of the sixth century B.C., Greeks of Miletus, a city in Asia Minor, founded a small colony-trading station on the shore of Feodosiya Bay that grew to become an ordinary Greek city-port named Feodosiya. The city was part of Bosporus Kingdom. Along with the Greek colonies Fanagoria and Gorgippia, F. reached its prime at the time of the Bosporus King Levkon I (389–349 B.C.). From that time on, it was the westernmost border city of Bosporus Kingdom. In the middle of the first century B.C., F. gradually lost its former significance and degraded finally during the Roman time. In the fourth century A.D., the city was seized by the Hunns who destroyed it. During the fifth–sixth centuries the city gives way to a settlement called Ardabdu. In the second half of the thirteenth century, F. was seized by the Mongol Tatars, it was then that Ardabda turned to Kaffa. The fortress remains are still the evidence of the 200-year long Genoese domination: defense walls, towers, epigraphic monuments kept in the local museums. F. retained its name until 1787. The Turks renamed F. into Kuchuk-Istanbul (Istanbul Minor). In 1771, Russian troops under the command of the General-in-Chief Prince V.M. Dolgorukov seized Kaffa in action. In 1878, after the Crimea was annexed to Russia, Kaffa was included in Tavria Region and the city was returned its historic name. F. also had been called Khevleni, Khavan, Theodosiya, Constantinople Minor.

F. begins to be built-up from the end of the nineteenth century. In 1892, a railway was laid, and in 1895—the construction of port structures commenced, which made it possible to move a merchant port there from Sevastopol. The city was occupied by the forces of Nazi Germany during the Great Patriotic War of 1941–1945, sustaining significant damage to the city. The health-resort of the city adjoining the sea was built anew after. In 1954, it was transferred from Russian Federation to the administrative control of the Ukrainian Soviet Socialist Republic with the rest of Crimea. In 1982, under the Decree of the Presidium of the USSR Supreme Soviet, F. was awarded the Order of the Great Patriotic War first degree for the courage and perseverance of the city dwellers demonstrated during the war of 1941–1945.

The famous Russian painter marinist I.K. Aivazovskiy was born and lived here. He played a crucial role in the life of the city. The picture gallery with his paintings is one of the greatest cultural values of the Crimea. There are many other features in F. connected with the city eventful history. In the old part of the city there still is the medieval layout, on the Quarantine Hill—remains of the walls and towers of the Genoese fortress (fourteenth–fifteenth centuries). Besides, there are several Christian churches of the thirteenth–fourteenth centuries, the Mufti-Jamie Mosque of the sixteenth century, summer cottages and homesteads of the second half of the nineteenth century, the literary-memorial museum of Russian writer A. Grin, and museum of local lore.

Feodosiya (http://nemiga.info/ukraina/feodosiya.htm)

Feodosiya Bay – shallow-water bay in the Black Sea at the south-eastern shore of the Crimean Peninsula, between Ilya Cape and Chauda Cape distanced from it 32.5 km to the east. Length 13 km, width 31 km, depth at the entrance 20–28 m. The shores in the west are low, hemmed by sandy beaches, in the east—high and precipitous. Opposite the low Mountain Opuk, 4 km from the shore, there rise out of the water several upright rocks known as "Rock ships". These are believed to be mentioned by Homer in his "Odyssee". They are the remains of Mt. Opuk that got separated from it during the century-old onset of the sea and as a result of century-old variations of the sea level. There is Feodosiya Harbor with Feodosiya Port in the western part of the bay.

Feodosiya picture gallery – one of Crimea's oldest art museums. The gallery building is an architectural monument of the nineteenth century. Its construction is tentatively dated by the years 1845–1847. Architecturally and ornamentally, the structure is built in the spirit of Italian Renaissance villas. An exhibition hall was annexed to the main building in 1880. Construction was carried out to the design and under the supervision of famous Russian painter marinist I.K. Aivazovskiy. The official opening of the picture gallery in 1880 was timed to the artist's birthday. When Aivazovskiy was alive, the stock of paintings was replenished continually as his works were constantly exhibited in other cities of Russia and abroad. When I.K. Aivazovskiy died, the picture gallery, following his wish, becomes the property of the city. 49 pictures painted by the famous master were presented to Feodosiya as a gift.

I.K. Aivazovskiy The Black Sea Fleet in the Feodosiya Bay (1890) http://commons.wikimedia.org/wiki/File:Aivazovsky_-_Black_Sea_Fleet_in_the_Bay_of_Theodosia.jpg?uselang=ru

Feodosiya sea merchant port – was built from 1892 to 1895. Port area is 13.5 ha. The port is motorized, has a sprawling network of railway tracks. From 1992 is capable of receiving foreign vessels. The port has 2 cargo-handling areas. One—for dry cargo shipment, the other—for oil and oil products. There are two oil roads discharge jetties and one oil jetty. The northern jetty with depths up to 13.5 m is capable of handling tankers with deadweight of 80,000 tons and a draft of 12.5 m. The southern jetty—up to 80,000 tons and a draft of up to 11.3 m. Both the jetties provide for transshipment and of crude oil and light oils.

Fidonisi Island – see Zmeinyi Island.

Fidonisi Island, sea battle – On July 2(14), 1788, a sea battle at F.I. between the Russian and Turkish squadrons during the Russian-Turkish War of 1787–1791 was fought. The Russian squadron commanded by Rear-Admiral M.I. Voinovich had the object of admitting no assistance from the sea to the Turkish garrison deployed in the Ochakov fortress besieged by the Russian troops. On July 2(14), Voinovich's squadron came close to the squadron of Hasan-Pacha. The Turks had superiority in forces, fire power and in lucrative windwards position, and were the first to engage the Russians. Having shown initiative, the commander of the vanguard, captain of brigadier rank F.F. Ushakov made a bold maneuver. He ordered that the frigates should bypass the Turkish ships from the windward side and put them in a "twin fire" position, while himself on the battle ship "St. Pavel" left the squadron formation and attacked resolutely Hasan-Pacha's command ship. As a result of the battle that lasted close to 3 h, the Turkish command ship was damaged seriously. This forced Hasan-Pacha and the other ships of the Turkish squadron to leave the battle area.

Findikli – a village on the coast of the Black Sea, Turkey.

Fiolent Cape – legend has it that it is a sacred God's cape, Turk. "Filenk-Burun"— "Tiger's Cape" as there are alternating stripes of yellowish limestone and dark trachyte, the coloring of the stripes is reminiscent of tiger's skin. There are other names of the cape, too: St. George's, Chifuros. Sited in Crimea, between Sevastopol and Balaklava, approximately 12 km south-east of Khersones Cape. F.C. is structured by igneous products of the ancient volcano Fiolent. Sited on Heraclei Peninsula of the Crimea. Between George's monastery, Diana's balka and Lermontov Cape. Looks like a pyramid of stone with beaches at its foot. The ancient name of F. is Parthenium ("maiden's, virgin") is related to the ancient Greek myth on Iphigenia. The young daughter of the Hellenic King Agamemnon was taken away on a cloud by the goddess-huntress Diana to the south of Tavrida so that she could become a priestess over the sea in the Virgin's temple. The romantic myth about Iphigenia used to inspire Euripid, Eshil, German writer J.W. von Goethe, Russian poet A.S. Pushkin, Ukrainian writer Lesya Ukrainka and other writers and poets, the Austrian composer C.W.R. von Gluck, the Russian painters I.K. Aivazovskiy, V.A. Serov and others. A.S. Pushkin, I.K. Aivazovskiy, Alexander I and Nikolas II, Prince Golitsyn and many other personalities used to visit the place.

In 2005, a monument to the Holy glorified apostle Andrey the First-Called was opened here. This picturesque site on the outskirts of Sevastopol is notable because according to a legend it was here that the First-Called disciple of Christ for the first time stepped on the Crimean land and brought Evangelic Annunciation to its dwellers. As far back as in 891, the Greek seafarers founded a monastery here in the name of the holy great martyr George the Conqueror. Caught near the cape by a storm that was likely to smash their ship against the rocks, the seamen addressed their prayers to the holy George the Conqueror. Having heard their prayers, the holy God-pleaser appeared on the rock some 70 sazhen from the shore and the storm died down there and then having climbed the rock, the seamen found an icon of the Holy George on its top. They brought the sacred thing into the natural cave they had discovered on the upright shore. A church was arranged here subsequently. The most reverend of the seafarers settled near the church, having become the first brethren of the monks abode. The praying never stopped here when the Turks drove the Orthodox believers from the Crimea on a mass scale. In the eighteenth century there were left just a few brethren in the convent who were risking their lives every day, but continued their monks' exploit.

In the nineteenth century, there were household and dwelling outbuildings on the vast compound of the convent; the brethren's building, an hotel, officers' outhouses, Russian Admiral M.P. Lazarev's summer cottage, candle-making factory. The most valuable building of the convent was the Holy George the Conqueror Cathedral. Admiral F.F. Ushakov donated 3,000 Rubles by way of financing its construction. In 1929, the convent was shut down although worship services continued in one of the temples. Until 1939, an OSOAVIAKHIM sanatorium was housed here; just before the Great Patriotic War (1941–1945) broke out, the sanatorium gave way to the military-political training courses of the Black Sea Navy.

Changes in the convent's lot began early in the 1990s. In September of 1991, to commemorate celebration of the 1,100th anniversary of convent establishment, with support given by the commander-in-chief of the Black Sea Navy Admiral I.V. Kasatonov, a 7-m high cross was re-installed on the rock where the Greek seafarers allegedly had seen the Holy Martyr George. In 1994, supported by the Archbishop of Simferopol and Crimea Lazar, the then Commander-in-Chief of the Black Sea Navy Admiral E.D. Baltin transferred part of the convent to the Crimean episcopacy of the Orthodox Church, and in June of 1995, the first brethren settled in the convent. The image of an X-shaped cross with Christ's crucified disciple on it is erected on the very edge of a high and upright shore. The cross can be seen by all ships bound for Sevastopol. The monument, made by the Moscow sculptor A. Smirnov, is built with the money of the Zaporozhye enterprise JSC "Motor-Sich"—Ukraine's only and one of the world's largest manufacturer of aircraft engines.

Fiolent Cape (Photo by Dmytro Solovyov)

"Fiolent Cape" Nature Reserve – locates in the coastal zone of the Fiolent Cape in the Balaklava Region. It extends over an area is 5 ha. This is the ancient volcanic terrain with the original relief forms: numerous niches, stone chaos, abrasion arches. The wild, steep rocks about 100 m high are overgrown with juniper and other plants. It was given the status of a nature reserve in 1969 by the decision of the Sevastopol City Council and in 1984 by the joint decision of the Crimean Regional Council and the Sevastopol City Council.

Fishery Institute – was founded in 1953 in Istanbul, Turkey. It pursues a wide range of activities from hydrology, hydrochemistry, hydrobiology and commercial ichthyology to fish processing technologies and fishing facilities and economics. The Institute has two stations: on B.S. in Trabzon and on the Aegean Sea. In 1954 the Institute started publishing booklets under the common title "Institute Proceedings". In addition, the journal "Fish and Fish Industry" is published with the Institute's support.

Fishery Research Institute – was established in 1954 in Bulgaria on the basis of the Marine Biological Station and the Experimental Ichthyological Station. It carries out permanent studies of the plankton level, conducts hydrochemical measurements in the open sea near the Bulgarian coast, organizes expeditions to the western part of B.S.

Flounder *(Platichthys flesus luscus)* – the bottom fish of the *Pleuronectidae* family. It has an oval flat body covered with small scales. The eyes locate on the right side of the body, seldom on the left. The upper part of the body is gray-green, the lower—white. It is usually 30 cm long, seldom 40 cm. It lives in the Black and Azov seas in the coastal waters at depths to 70 m. It prefers brackish waters, but for feeding it migrates to the coastal lakes and river mouths. It lives for 10–12 years. The male fish reaches maturity at the age of 2 years, the females—3 years. It is spawning in cold water from January to April laying pelagic eggs. It feeds on small fish and shellfish. It is the object of commercial fishing.

Fontanka – a village on the coast of the Black Sea, Ukraine.

Footstock – (German "Fußstock" from "Fuß"—foot and "stock"—"shelf, rod"), graduated rod mounted at a water-level measuring post near the shore of the sea, lake at a river bank to watch the water level. Sometimes F. is used to denote a measuring board, 3–5 m in length, lowered from sloops and ships to measure small depths. The Kronstadt Footstock in Kronstadt near St.-Petersburg in the Gulf of Finland of the Baltic Sea is one of the oldest tide gauge station in the world. The footstock service was established here in 1707. Hydrograph M.F. Reyneke was the first who in 1840 draw a line on the base of the Blue Bridge across the Kronstadt Canal which corresponds to the mean Baltic Sea level he measured in 1825–1839. In 1871–1904 astronomer V.E. Fus established a relationship between the "zero" of the Kronstadt Footstock and level marks on land. In 1886 astronomer-geodesist F.F. Vitram installed in the bridge base a copper plate with a horizontal line denoting the "zero" of the Kronstadt Footstock. In 1898 a mareograph was installed, which measured the sea level relative to the "zero" of the Kronstadt Footstock. In 1913 Kh.F. Tonberg, Head of the Instrumental Service of Kronstadt Port, incorporated a new plate with a "zero" line, which till today serves as a reference level for all level measurements in the Former USSR. In 1946, the USSR Council of Ministers passed a decree "On introducing a uniform system of geodetic coordinates and heights on the territory of the USSR", whereby at the proposal of Yu.M. Shokalskiy the zero of the Kronstadt Footstock was regarded as the datum of heights, water levels and depths on the territory of the USSR. This system is also known as the Baltic Sea System of Heights.

The Kronstadt Footstock (http://www.panoramio.com/photo/50781901)

"For Sevastopol Defense" Medal – the first award of the Crimean War (1855). In October 1855 Alexander II arrived in Sevastopol and stayed there among the soldiers in the acting army for over a month. By its order for the Army of October 31 (November 12), 1855 the Emperor instituted a special medal for its awarding to the city defenders. He wrote: "Brave fighters of the Crimean War! Meeting them was of great pleasure for me... I am happy to see and admire you... In memory of the famous and successful defense of Sevastopol I instituted specially for the troops that defended the fortifications a silver medal on the Georgian ribbon for its wearing in a buttonhole!!" This medal was to be awarded to all (without any exceptions) defenders of Sevastopol, including women, who "... served in hospitals or were helpful during defense". The front face of the medal (28 mm in diameter) had two engraved monograms—of Nickolay I and Alexander II with the Emperor's crows

above them. The reverse face had the image of "Omniscience eye" with the
inscription "From September 13, 1854 to August 28, 1855". And over the perimeter
of the medal an inscription in large letters—"For Sevastopol Defense".

"For Sevastopol Defense" Medal (http://1exem.ru/picture/faleristika/DSCF1677a.JPG)

Foros – urban village, maritime climatic health resort 45 km south-west of Yalta
(Crimea, Ukraine). The settlement is believed to have been set up to succeed the
Genoese colony of Forya. F. is sited on the Southern Coast of Crimea near Baidar
Gates. Part of the Greater Yalta, crowning its maritime health resorts. The natural
and climatic conditions favor climatic and thalasso-therapy, climatic prophylaxis.
There are sanatorias, of which the best known is "Foros" situated in a park,
exhibiting a model of garden-and-park architecture. At a distance of 2.5 km from
Foros there are several state residences, in one of which "Zarya" the USSR
President Mikhail Gorbachev was kept during the attempted state coup in
August 1991.

Foros, Harbor – cuts into the shore at the apex of Burgas Bay, Bulgaria, west of
Foros Cape. The harbor shores are largely low-lying and sandy. The harbor is

shallow-water; the depths at the entrance to it are 7–8 m. The harbor apex is a shoal with depths under 5 m.

"Fountain of tears" – famous Bakhchisaray fountain. Originally, its carved stone slab was at the wall of the mausoleum to Dilara-Bikech—the "beautiful princess". Legend has it that she was a youthful slave of Girey-Khan, who died young. The fountain was built on Girey's order by Master Omer in 1764. Dilara-Bikech's personality has not been established finally and is shrouded by poetic legends. It was in her memory that the fountain was built.

"From the Varangians to the Greeks" Trade Route (Volkhov-Dnieper route) – the famous trade route 1,500 km long that connected in the tenth century the Baltic Sea and B.S. It went from the Varangian (Baltic) Sea along the Neva ("Nevo Lake mouth") via the Volkhov, Ilmen Lake, Lovat River and from the Baltic basin into the Black Sea basin via Usvyache, Kasple, Luchese rivers, along the upper reaches of the Western Dvina and the portage systems to the Dnieper near Gnezdovo and from here along the Dnieper via difficult Dnieper rapids and going out into B.S. near Kherson (Korsun). The southern section of this route from Kiev and further downstream Dnieper was called the Greek Route. The Dnieper mouth ended with the port of Oleshye controlled by Kiev where the boats were equipped anew and continued to navigate along the Bulgarian coast of B.S. all the way to Tsargrad (Constantinople). From the second half of the twelfth century this trade route was losing its significance because after the Crusades more convenient trade routes appeared that went from Western Europe to the Near East countries. Kiev was off these routes. In the late twelfth century after conquering of Constantinople by the Crusaders the trade relations of Rus' with Byzantine were paralyzed completely.

Frunzenskoe – urban village, maritime climatic health resort. Sited 15 km southwest of Alushta, on the Southern Coast of Crimea, Ukraine. Shielded room cold winds by the dome-shaped Mountain Ayu-Dag and the main ridge of the Crimean Mountains. The natural and climatic conditions favor climotherapy and thalassotherapy of patients with diseases of the respiratory organs, functional disorders of the nervous, cardio-vascular systems, metabolism. There are sanatorias. There are still remains of the Parthenite Basilika, part of a monastery dating back to the thirteenth–fifteenth century.

G

Gablitz Karl Ivanovich (Karl Ludwig) (1752–1821) – Russian geographer, natural scientist, nature explorer, botanic, traveler; Corresponding Member of the Petersburg Academy of Sciences (1776), Honorary Member (1796). Although he was of the German origin, from 1758 he lived in Russia. He took part in the expeditions of the Petersburg AS, in particular, in 1769 in the expedition of S.G. Gmelin over the Don basin, Volga lower reaches, Caucasus; in 1781 in the expedition over the Caspian Sea. He described some plant and animal species not known before. He was one of the pioneer explorers of Crimea, the author of the first fundamental treatise devoted to the nature of Crimea "Physical Description of the Tavria Region: Its Location and All Three Kingdoms of Nature" (1785). G. collected the original material about inanimate and live nature of the peninsula that made the basis of his unique treatise. The book includes the following sections "About all things met in the mineral kingdom", "About the plant kingdom" and "Animal kingdom". G. was the first to identify in highland of Crimea the "foremost mountains" (External range), "medium" (Internal range) and "southernmost range" (Main Crimean Range) and to refer to the "Midday" (Southern) Crimean coast as an independent geographic unit. The scientist made the first list of plants (511 varieties) growing on the peninsula. He praised high the rich nature of Crimea having stressed that "there were only a few countries that were distinguished by the same excellence as this mountain belt". In 1786 G. was presented with a garden in Sudak and later a country house in Chorgun. In 1783–1802 he held the post of Vice-Governor of Crimea. He was the founder of the Russia's first forest school in Tsarskoye Selo (1803) and Kozelsk (1805).

Gagauzs – the people with a complicated ethnic origin. By one theory these are descendants of the Polovets-Kumans who took Christianity, by the other theory these are southern Slavs who adopted the Turk language. G. settled down mostly in the south of Moldavia and Ukraine and also in Bulgaria (Southern Dobrudzha). The Gagauz language refers to the Turk group of the Altai family and is close to the Turkish language. By their cultural and everyday habits they practically do not differ from the Bulgarian people. G. are Orthodox.

S.R. Grinevetsky et al., *The Black Sea Encyclopedia*, 283
DOI 10.1007/978-3-642-55227-4_8, © Springer-Verlag Berlin Heidelberg 2015

Gagra – a town and port on the eastern coast of B.S. in Abkhazia. This is a seaside climatic and balneological resort; a railway station 80 km to the northwest of Sukhumi (capital of Abkhazia) and 60 km to the southeast of Sochi. It appeared on the terrain called Gagripsh meaning in the Abkhaz language "Holy place of Hagba (Hagaa) family". Thus, the name "Gagra" appeared. It locates in a narrow coastal strip and slopes of the Gagra Range of the Greater Caucasus. It is cut through by gaps made by the Zhoekvara, Gagripsh and other rivers. These gaps and mountain slopes are overgrown with broad-leaved (oak, beech, hornbeam and others) and coniferous forests.

In treatises of the Antique scholars G. is referred to as Triglit. This was a minute settlement on a narrow seaside strip. It had a mixed population—local Abkhaz people and visiting Greek merchants. They lived in half-dugout huts between Zhoekvara and Gagripsh rivers. A small trade harbor was made, obviously, in the mouth of the fuller-flowing Zhoekvara River. In the first century B.C. G. together with the whole Kolkhida was incorporated into the Pontus Kingdom. After being absorbed by the Roman Empire the town was given a new name—Nitika. The Romans fortified the town which was repeatedly attacked by Goths, Vorans and other invaders. In the sixth century G. became one of the places from where Christianity was spread to the western parts of Abkhazia. In the place of the ancient Abkhaz pagan temple there was built the Christian church that is still standing in the old fortress. That time the Byzantine Empire took control of the town. From the late tenth century G. became the chain connecting Georgia, Kingdom of Abkhazia and Kingdom of Tmutarakan with the Kievan Rus (Russia). It was assumed that the first Russians came here in 1017–1022. The name "Gagra" appeared for the first time on a map of 1308 made by Italian Pietro Visconti. Now this map is in the Library of Saint Mark in Venice. The Genoans and Venetians started appearing in Abkhazia in the twelfth century. In the fourteenth–fifteenth centuries founded the trading post "Santa-Sofia" where now village Alakhadzy locates, and the trading post "Pitsundo"on the Pitsunda Cape. In the sixteenth century Gagra and the rest of Abkhazia were conquered by the Ottoman Empire and the Genoan and Byzantine merchants were expelled from the Black Sea coast. The Turks turned G. into a real slave market. In the eighteenth century G. suffered a prolonged period of decline. In 1801 Eastern Georgia was incorporated into Russia. On the request of the Abkhazia ruler the Russian fleet occupied the fortress in Sukhumi. Greater Abkhazia (from the Inguri River to the Bzyb River) was officially annexed by Russia. This event was of key significance for the whole area. However, Smaller Abkhazia (from the Bzyb River to Khosta) together with the Gagra region had remained beyond the Russian influence for nearly 20 more years. Soon under the peaceful treaty Russia gained the whole eastern coast of B.S. from Anapa in the north to Poti in the south.

G. is the warmest and driest place on the Caucasian coast of B.S. The mean annual air temperature is 15 °C. The non-frost period averages here 306 days a year. The winter is mild and without snow. The mean temperature of January is 6 °C. The summer is very warm; the mean August temperature is 23 °C. The autumn is warm; the mean October temperature is 17 °C. The amount of precipitations here is about 1,400 mm a year, of which about the half falls in April–October with the minimum amount in May. The number of sunny hours is 1,830 a year. The number of days without sun is 58 a year. The bathing season here lasts from May through November (the water temperature in summer reaches 22–26 °C).

Thanks to the mild climate, picturesque coasts and rich subtropical vegetation G. soon became well-known and acquired the reputation of one of the best climatic resorts in Russia, the so-called "Russian Riviera". The celebrations on opening of this resort were organized in January 1903. However, for wide public this resort was not accessible. Its intensive development (first of all, construction of sanatoriums and other medical establishments) started only after the 1917 October revolution in Russia. On March 20, 1919 Soviet Leader V.I. Lenin signed the Decree "On State actions for medical treatment improvement" and after establishing of the Soviet power G. was turning into the "All-Russia resort". The antimalaria station was opened here and soon after periodical drainage of ponds or even their desiccation malaria was liquidated in this city. In the 1930s the resort grew and was functioning year-round. The treatment in G. was recommended to those who suffered from anemia, lung and nervous diseases as well as excessive fatigue. After the Great Patriotic War (1941–1945) the resort was developed intensively. New sanatoriums were opened. Between Old and New Gagra the new beautiful blocks of rest houses were rising. In 1971 one of the largest sanatoriums "Kavkasioni" for treatment of nervous-somatic diseases was built. It was equipped with the modern medical facilities and had cabinets of functional diagnostics, massage, remedial gymnastics, clinical and biochemical laboratories.

In the Soviet period the southern part of the resort was developed most intensively where rest houses, residential multistory buildings for the resort personnel and modern trade centers were constructed. After discovery of a thermal mineral water source near New Gagra a whole spa complex was built here. In all about 30 sanatoriums and rest houses were functioning here. For a long time climate was considered the only natural treatment factor in G. In 1962 the well drilled to a depth of 1,500–2,300 m produced thermal (43 °C) sulfide sulfate-hydrocarbonate calcium-magnesium water with the salinity of 2.2 g/l containing 27–34 mg/l of hydrogen sulfide. The well yields daily 1,100 m^3. This water is used for making baths. In addition, nearby this resort there are found considerable reserves of sapropel and peat mud that may be used in mud treatment. The resort is surrounded by very beautiful terrains: gorges of mountain rivers Zhoekvara, Gagripsh, Zhvava-Kvara, Tsikherva, Mziuri Mountain, Pitsunda Cape, Ritsa Lake and others. G. has the museum of old Abkhazian arms. Some architectural historical monuments survived in the city, such as Church of the sixth century (restored in the early twentieth century), the Genoa Fortress (dated first centuries A.D.) and the ruins of the Bzyb Fortress of the ninth–tenth centuries located not far from the city.

Gagra (http://img.liveinternet.ru/images/attach/3/16985/16985360_ScanImage37.jpg)

Gagra Range – a mountain range on the southern slope of the Greater Caucasus in Abkhazia. The range runs between the valleys of the Bzyb and Psou rivers. Its highest elevation reaches 2,736 m. The outspur of this range near G. approaches B.S. forming the picturesque view of the Gagra Range composed largely of limestone. It is cut by the Zhoekvara, Yupshara and other rivers forming the canyon-like valleys. Karst is also widespread here. The range is overgrown with broad-leaved and coniferous forests; at higher elevations one can admire the subalpine and Alpine meadows. The auto road to the Ritsa Lake runs along the valleys of the Bzyb, Yupshara and Gega rivers.

Galata Cape – the southeastern cape at an inlet into the Varna Harbor, Bulgaria. It is formed by the northeastern slopes of the Avrenska-Planina Mountains. The cape is elevated, steep, surrounded by a stony reef extending from the shore for more than 500 m. A lighthouse is constructed on the cape.

Galata Cape (http://commons.wikimedia.org/wiki/File:Cape_Galata,_Varna.JPG)

Galleass – a three-mast Turkish ship of the late sixteenth century. Its length reached 80 m. The board inclined inward had windows (ports) for 24 cannons. Up to 400 men rowed with 52 oars. About 300 fighting men could locate on its deck. The galleass (known as a "mahon" in Turkey) was developed from large merchant galleys. Converted for military use they were higher and larger than regular ("light") galleys. In the fifteenth century a type of light galleass, called the frigate, was built in southern European countries to reply the increasing challenge posed by the north African Barbary pirates in their fast galleys. The galleass was a popular type of vessel for England of Henry VIII, who had more than a dozen constructed for the English Navy during the 1530s and 1540s for his wars with France. These were four-masted vessels with a few heavy guns interspersed with the row ports on the lower deck of these vessels. In the Mediterranean, with its better weather conditions and variable winds, both galleasses and galleys continued to be use, in particular, in Venice and Turkey, long after they became obsolete elsewhere.

Galleass (http://static.diary.ru/userdir/1/1/8/3/1183608/61702071.jpg)

Galley (in Italian "galera") – a fighting row ship of the Russian fleet built by analogy with the Venetian galera. Its length—to 35–40 m, width—to 5–6.7 m, the draft—2.7 m (to 7 oarsmen per one oar, 28–32 pairs of oars). The rowing speed was to 7 knots (13 km/h). The crew could number 460 people. The oarsmen on G. were usually slaves, volunteered poor men and convicts. In the ancient times in Rus' G. was a synonym of slavery. G. had 2–3 masts with gaff-and-stay sails that could be used at leading winds. On the stern there was a deckhouse. In Russia the galley fleet including also small G. for actions in skerries—sampavei was created by Peter the Great and existed till the late eighteenth century. The Russia's first 23 G. were built in 1696 in Voronezh by Italian masters. During Peter I ruling there were built 305 G. in Russia. They took part in the Azov military campaigns.

Galera "Dvina" (http://windgammers.narod.ru/Korabli/Dvina001.jpg)

Gantiadi (former Pilenkova, today—Tsandrypsh) – an urban-type settlement, climatic resort, railway station 20 km to the northwest of Gagra, locates on the Black Sea coast of the Caucasus near the mouth of Khashupse River on the territory of the Abkhaz Republic. The piedmonts of the Gagra Range of the Greater Caucasus take origin to the northeast of G. The vegetation in the G. area consists largely of the fruit and decorative trees (plum, nut, apple, fig, palm, eucalypt and others). The climate here is subtropical, moist, and mild as this area is protected on the north by mountains. The winter is mild, snowless. The mean temperature of January is 5 °C. The summer is very warm and moist. The mean temperature of August is 22 °C. Summer heat is softened by onshore and offshore breezes. The water temperature in the sea in summer is 22–25 °C. The precipitations are about 1,550 mm a year with the maximum recorded in late autumn and winter. The mild climate is the key natural curing factor of G. that may be used in aerohelio therapy of the people with cardiovascular and nervous diseases and lung problems. Warm sea and gently sloping seashore create favorable conditions for thalassotherapy in the season from May to October. G. has many rest houses.

Garfish (*Belone belone euxini*) – the fish of the *Belonidae* family. Sometimes it is called sea pike. It has an oval needlelike body to 75 cm long covered with small scale. This is pelagic fish living in shoals. Its weight is from 50 to 300 g. It lives for 15–18 years. It propagates from late April through late August. G. is predatory fish. It feeds mostly on anchovy, silversides and other small fish. G. has usually green bones due to the presence of biliverdin, one of the bile pigments. It migrates in various seasons for spawning, fattening and hibernation. It populates in B.S. and Sea of Marmara as well as in the western part of the Sea of Azov. For spawning and fattening the fish moves to the northwestern part of the sea and by late November runs more southward. It spends winter in the Sea of Marmara. By the taste it reminds of saury. The Crimean people call it a needle, ignoring the fact that the needlefish by itself lives in the sea and has no relation to G.

Gaspra – an urban-type settlement 12 km to the southwest of Yalta and 74 km to the southeast of Sevastopol; natural climatic resort in Crimea, Ukraine. It locates on the Southern Coast of Crimea, in the piedmonts of the Main Range of the Crimean Mountains, near the foot of the Ai-Petri Mountain over the resort area of Miskhor from which G. differs by less hot summer, cooler winter and lower air humidity. The subtropical Mediterranean climate is favorable for climate therapy of chronic respiratory diseases. Sanatoriums for treatment of patients with respiratory diseases are found in G. and Koreiz directly adjoining it from the west. In 1901–1902 Russian famous writer L.N. Tolstoy had a rest and medical treatment here, thus, the sanatorium that was opened in 1922 was named "Yasnaya polyana". Here L.N. Tolstoy met writers A.P. Chekhov, A.M. Gorky, V.G. Korolenko, A.I. Kuprin and many other renowned people of that time. The sightseeing attraction of G. is the Gaspra castle built in 1831–1836. This is the most remarkable example of the Crimean Gothic-style manor. The castle is surrounded by the Kharakskiy Park being the monument of the landscape architecture of the republican significance.

Gazaria (Gatsaria) – in the thirteenth–fourteenth centuries this was the name of Crimea and nearby territories in the lower reaches of the Dniester, Dnieper rivers and Western Azov area. The Genoans used this name in their official documents and gave the name "Officium Gazarie" to the department that in 1,314 controlled their colonies in the Northern Black Sea area. This name originated from the Khazars who in the eighth–tenth centuries populated the steppes between the Volga and Dniester and also the Crimean lands. The name G. was also used in the Pope's Curia: G. of the Franciscan Vicariate of Tartar Aquilonian incorporated all lands in the Danube, Dniester and Dnieper lower reaches and a part of the Azov area to the west of Tana town (near modern Azov town on Don River), which existed in thirteenth–fifteenth centuries.

Geiden Login Petrovich (1772–1850) – Admiral. He started his service in the Dutch Fleet. In 1795 he went on the Russian service and was assigned to the Black Sea Fleet in the rank of Captain-Lieutenant under the name—Login Petrovich. In 1796 he was assigned to the Nikolaev port. In 1798–1800 G. took part in the cruise of F.F. Ushakov squadron to the Mediterranean. In 1799 commanding the brigantine "Aleksey" he navigated in the Mediterranean and took part in landing of the troops in Otranto, after this the fleet returned to Sevastopol. He commanded the frigate "Ioann Zlatoust". During the campaign of 1802 he navigated in B.S. between Constantinople and Nikolaev and made description of the coast. In 1804 he was promoted to the Captain second rank and was transferred to the Baltic Fleet. G. commanded the ship "Zachatie St. Anna" (Conception of St. Anna) in the Baltic Sea. When the Russian-Swedish war of 1808–1809 broke out he was promoted to Captain first rank and assigned to the row fleet that was prepared in Saint Petersburg and by detachments was sent to Finland. In summer 1827 with the D.N. Senyavin squadron he went to England and with a separate squadron was directed to the Mediterranean to prevent the Greek-Turkish conflict. After joining with the British and French squadrons G. took part in defeating of the Turkish-Egyptian Fleet at Navarin. For the victory in the Navarin battle he was promoted to vice-admirals. In 1828 he was appointed the commander of the second Division of the Baltic Fleet. In fact, he commanded the squadron locating in the Mediterranean. During the 1828–1829 Russian-Turkish war G. blocked the Dardanelles. In 1830 commanding the squadron of seven ships he navigated near the Peloponnese coast. Upon return to Kronshtadt he was appointed the Commander of the first fleet division. In 1833 he was given the rank of Admiral. In 1834 he was appointed the military governor of Reval (Tallinn) and in 1838—simultaneously the Commander-in-Chief of the Reval port. He founded the seamen dynasty. His two sons were the prominent marine and military figures; his grandsons and great-grandsons also served in the fleet. The name of G. was given to the atoll reef in the Marshall Island Archipelago, a bank in the Bristol Bay of the Bering Sea and a mountain on Kamchatka. He was awarded many Russian and foreign orders.

Geiden L.P. (http://www.navy.su/persons/04/images/geyden_lp_00.jpg)

Gelendzhik – a town, resort in the Krasnodar Territory, Russia. One of the most ancient settlements on the Black Sea coast. It had been known since the fourth century B.C. In the Turk language is means "white fiancé". But, possibly, it has an

Arabic origin—from "*khelendj*" meaning "poplar" that was used in the ninth century. G. is situated on the coast of the Gelendzhik Bay surrounded by high mountains. The capes protect it from open-sea storms, thus, the water in the bay warms up good and cools slowly. The water temperature in the Gelendzhik Bay is always 2–3° higher than in the open sea. Pleasant climate of G. and its convenient harbor have been attracting people from the ancient times. In the sixth century B.C. the Gelendzhik Bay was the site of a minor Greek outpost, mentioned as Torikos. Later on it was replaced by Roman, Byzantine and Genoa settlements. In the twelfth century the Byzantine colony was wiped out by the invading Mongol-Tatar troops. In the Middle Ages and up to the end of the Caucasian War the Adygey tribes of Shapsugs and Natukhays populated this area. Beginning from the late fifteenth century and for 300 years onward this area was under the Ottoman yoke. In 1831 the Russian troops under the command of General E.A. Berkhman built the G. fortress that became the outpost of the Russian troops and the naval base of the Black Sea Fleet. During the Crimean War of 1854–1856 it was blown up and abandoned. In 1857 the fortress was restored and in 1864 in its place the settlement known as Stanitsa Gelendzhikskaya appeared. In 1915 G. was given the status of a town.

In the late nineteenth century (1891–1893) Gelendzhik was developed as a spa resort. In the early twentieth century sanatoriums were constructed here for treatment of lung diseases, including tuberculosis, that had been cured in such climatic conditions even before appearance of antibiotics. In 1913 the first sanatorium for children suffering from bone tuberculosis was opened. During the Soviet power it was quickly developing as a resort. During the Great Patriotic War G. with its convenient harbor with a small port and berths became the main base for small military boats of the Black Sea fleet. This town played a very important role in setting free Novorossiysk. In 1970 G. became the resort of the country's significance. In 1996 G. was put on the list of specially protected territories of the Federal significance on the coast of B.S. and the Sea of Azov. At present more than 180 sanatoriums, holiday hotels, rest houses and tourist bases are found here. The resort zone of G. includes settlements in the coast section from Kabardinka to Arkhipo-Osipovka. The G. resorts extend in a narrow strip along the coast cut off from the rest territory by not high Markhotsky Range which spurs in some places come close to the sea and break down forming 100–200-m cliffs. Thanks to this mountain range the climate in G. differs sharply from the climate of its northwestern neighbors—Novorossiysk and Anapa. In summer the southwestern winds bring moist air, so the heat is ensured easily here. The mean annual air temperature is 13 °C. The mean temperature in January-February is 4 °C, in July-August—24 °C. Horticulture, vegetable growing and vine growing are developed here.

Gelendzhik: a view on the town and Gelendzhik Bay (http://vture.net/wp-content/uploads/0_
c32d_9a2b8077_orig.jpg)

Gelendzhik Bay – goes deep into the mainland forming a horseshoe between the Tolstyi Cape in the east and the gently sloping sandy Tonkiy Cape in the west. The shores of the bay are not elevated, but abrupt. The town of Gelendzhik locates around the bay. In the bay the bora winds are often blowing, but they are weaker here than in the Novorossiysk Bay.

Gelendzhik Coastal Marine Expedition – was organized in autumn 1947 to the Rybatskaya (Fishery) Bay (afterwards it was named Golubaya (Blue)) near Gelendzhik and to the marine part of the Ashamba River valley. This was a convenient place for creation of the experimental-methodical station. On January 1, 1949 this expedition was transformed into the Black Sea Experimental Research Station (ChENIS) of the Institute of Oceanology of USSR Academy of Sciences. At present it is the Southern Branch of the P.P.Shirshov Institute of Oceanology, Russian Academy of Sciences.

General Program of Environment Protection and Restoration in the Sea of Azov and Black Sea – it was approved by Law of Ukraine No. 2333 of 22 March 2001. This was the first national program in the Black Sea countries that was given the status of the law. It was targeted to elaboration of a state policy, strategy and plan of action aimed at prevention of the growing anthropogenic load on the natural environment of the Black and Azov seas, facilitating development of environmentally friendly activities in the Azov-Black Sea region, preservation and restoration of the biological diversity and resources of the seas, creation of favorable conditions for living, health improvement and rest of the people. The Program was planned for 2001–2010.

Genichesk – a town and port on the coast of the Genichesk Bay of the Sea of Azov, the center of the Genichesk District, Kherson Region, Ukraine. It appeared in 1781 as village Genichesk near crossing to the Arabat Spit. This name was first mentioned in Decree of Ekaterina II of February 10, 1784 devoted to construction of a blockhouse "near Enichi where there is a crossing to the Arabat Spit". The name of the future town originated from the word "*Djeniche*" that translated from the Turk means "thin" (that referred to the Arabat Spit). Initially the village was called Enichi, Djeniche, Genichi until it was finally transformed into Genichesk. In 1812 the village had two names used in parallel: Genichesk and Ust-Azovskoye. G. lands were crossed by a road along which salt excavated from Genichesk and Crimean lakes was transported to Ukraine and Southern Russia. In 1835 the postroad from G. over the Arabat Spit was built. In the same year the construction of the first port facilities was initiated. In 1837 G. received the status of a township and entered the Yuzukuisky volost, Melitopol uezd, Tavria Province. During the Crimean War of 1853–1856 G. was one of the nearest rear areas of the Russian Army. In May 1855 the enemy ships penetrated into the Sea of Azov and set afire 130 boats with food stocks staying in the strait and also grain and coal stocks on the shore. G. was subject to artillery bombardment. With time the salt production became the key business here. In the early twentieth century about 5 million poods (81,000 t) of salt were produced here giving seasonal work to over 2,000 people. G. also had iron-foundry workshops, two brick and two tile plants, a five-storey mill, fish curing plants, and a seltzer water factory. Ships from different world countries stayed off the Genichesk port. The representation offices of some foreign trade firms were opened in the town. In the early 1890s the customs and pilot branches of the Petersburg international Commercial Bank and the Azov-Don Agricultural Bank were opened in G. In the twentieth century the industry started quick development here. The town has the iron-foundry, fish canning, brick-tile plants, textile factory and food canning plants.

Genichesk Lake – Genichesk (Chongaro-Arabatsky, Prisivash) group of salt lakes. This group incorporates the Eastern Sivash and lakes on the Arabat Spit, including large ones, such as Zyablovskoe and Genicheskoe. The Eastern Sivash is separated from the Western Sivash by the Chongarsky Strait between the Chongar and Tyup-Dzhankoy peninsulas. The Genichesk group of lakes is situated in the northern part of the Arabat Spit that belongs to the Crimean Peninsula.

Genichesk Strait, Tonkiy – connects the Sea of Azov with the Sivash Bay. It separates the end of the Arabat Spit from the mainland. Its length is about 4 km, width—80–150 m, depth—to 4.6 m. The currents here depend on winds. Port Genichesk is situated on its coast.

Genoa Fortress – an outstanding fortification monument of Medieval Europe. It is built on top of a cone-shaped mountain, the ancient coral reef near Sudak, Crimea, Ukraine. The early construction period of this fortress refers to the Byzantine time, while the late, basic one, to the Genoa time (fourteenth–fifteenth centuries). Thus, it is usually called "Genoa". In 1365 Sugdea (this was the name of Sudak that time) where Venetians were ruling was captured by the Genoans. They gave a new name to the town—Solday and in 1381 started construction of a fortress that lasted for several decades onward and was completed in the early fifteenth century. In summer 1475 the fortress was captured by the Ottoman Empire. Solday was the last to surrender among other Genoa towns in Crimea. The legend said that about thousands of the fortress defenders with their Consul Christophore di Negro locked themselves in a church. But the Turks fired the building and all those inside it were burnt. This legend was confirmed by diggings conducted in 1928: many burnt human skeletons were found in the church ruins. After capturing by the Turks the Sudak fortress became an important strategic outpost in the Crimean defense. The fortress area was 29.5 ha; the total length of the surrounding walls was about 1,000 m. The height of the towers reached 15 m, the height of walls was 6 m reaching in some places 8 m and the wall thickness was 1.5–2 m. The fortress had two levels of defense: the lower one consisting of an outer solid wall reinforced with 14 towers and the upper one serving as a citadel and comprising the Consul Castle and towers connected by a wall. On the mountain top the legendary Kyz-Kule (Mainden Tower) is rising; it is also called "Dozornaya". After seizure in 1475 of the Crimean coast by the Ottoman Empire the fortress was dilapidating. Later on, following the order of Prince G.A. Potyomkin, the greater part of the fortress material was used in construction of Kirillovsky barracks. At present this is the architectural and historical nature reserve, the branch of the Ukrainian National Nature Reserve "Sofia Kievskaya". It is put on the state register of the national cultural heritage of Ukraine.

Genova Fortress in Sudak (http://uafacts.com/wp-content/uploads/2012/04/genuezskaya-krepost-1.jpg)

Geological history of the Black Sea – 50–60 million years ago a vast sea basin was extending from west to east covering Southern Europe and Middle Asia. In the west it was connected with the Atlantic Ocean and in the east—with the Pacific Ocean. It was called Tethys. Later on as a result of displacements of the Earth's crust the Tethys Sea was separated, first, from the Pacific and then from the Atlantic. In the Miocene Epoch (about 5–7 million years ago) the major mountain-formation dislocations of the Earth's crust occurred that led to formation of the Alpine, Carpathian, Balkan and Caucasus mountains. As a result, the Tethys Sea became smaller and was divided into several brakish basins (with the water salinity lower than in the sea). One of them, called by geologists Euxeinus-Caspian (Eastern Paratethys) extended from the present-day Vienna in the west to the foot of the Tien-Shan in the east and comprised the modern Black, Azov, Caspian and Aral seas. Isolated from the ocean the Euxeinus-Caspian gradually lost its salinity due to inflowing rivers (it became desalted even to a greater extent than the Caspian Sea today).

In the late Miocene—early Pliocene, i.e. 2–3 million years ago the Sarmat Sea shrank significantly, but once again its connection with the ocean was restored and its waters became saline again. Marine animals and plants appeared in it. This basin was called the Meothic Sea. During the Pliocene, 1.5–2 million years ago, the link with the ocean was broken again and the fresh Pontos Sea appeared in place of the Meothic Sea. In this period the future Black, Azov and Caspian seas were linked in the place that today locates on the territory of the Russian Plain and piedmonts of the Northern Caucasus. The saline-water fauna populated the Pontos Lake-Sea. Its representatives are still found in the Caspian and Azov seas and in the desalted areas of the Black Sea. They are referred to as the "Caspian fauna" as in the Caspian it is survived the best.

In the late Pontus as a result of the Earth's crust uplift in the Northern Caucasus area the basin was separated from the Caspian Sea. From this time onwards the Caspian, on the one side, and Black and Azov seas, on the other, developed independently although some short-time links connected them from time to time. In the Late Pliocene—Early Pleistocene, i.e. less than 1 million years ago, the Pontos Lake-Sea became smaller-size and was given the name of the Guriyskiy relict (afterwards Chaudinskiy) Lake-Sea. It contained slightly saline waters populated with relict (from the Pontos) and migratory (from the Caspian) brackish fauna. Obviously, that time the Sea of Azov did not exist as yet. The level of the Chaudinskiy Basin was lower than at present. About 400–500,000 years ago there was recorded the first case of a one-way water flow from the Caspian Sea into the Sea of Azov being a shallow estuary of the Black Sea that time, and from the latter into the Mediterranean Sea the water level in which was 10–20 m lower.

In the mid-Pleistocene period the brackish Ancient Euxeinus Basin came into existence and its was separated from the Chaudinskiy water body by a deep and rather long regression during which the sea water level dropped by 40–60 m. The Ancient Euxeinus Basin was connected via the Manych with the Caspian (early Khazar) and, obviously, had a one-way feedback with it. At the end of this epoch the communication via the Bosporus with the Mediterranean Sea was resumed. By its outlines the Ancient Euxeinus Basin reminded of the modern Black and Azov seas. In the northeast this basins was linked via the Kuma-Manych Depression with the Caspian Sea which that time was rather desalted. The fauna in the Ancient Euxeinus Basin was of the Pontus type. The next stage in the Ancient Euxeinus Basin development was the Uzunlarskiy transgression. It was transformed from the brackish water body of the Caspian type into the desalted sea basin with the water salinity close to the present one in B.S. (approximately 17‰).

During the Riss-Würm interglacial period (100–800,000 years ago) a new stage in the geological history of B.S. came. As a result of the ocean level rise the waters from the Mediterranean Sea rushed through the Bosporus and formed the so-called Karangatsky water body of Karangatsky Sea. The water salinity in it was approximately one-third higher than in modern B.S. Various species of animals and plants penetrate from the Mediterranean Sea and Atlantic Ocean into the Karangatsky Sea. Saline waters prevailed there having forced the Pontus brackish-water species into desalted estuaries, lagoons and river mouths. However, with time on the Karangatsky Sea disappeared. By the end of the Late Pleistocene (about 18–14,000 years ago) the link between the Mediterranean and Caspian seas ceased to exist and in the place of the Karangatsky Sea there was already formed the New Euxeinus Basin that was an open desalted lake-sea located at elevations of about 100 m. And once again the halophilic marine fauna and flora was disappearing, while the Pontus species that survived the heavy Karangatsky period in lagoons and river mouths came in their place in the sea. And this process lasted for about 10,000 years and after the most recent phase in the water body life came—formation of the modern B.S. This happened not at once. At first, about 7, and by some other sources, even about 5,000 years ago the link with the Mediterranean Sea and the Atlantic Ocean was restored as a result of transgression via the Bosporus and Dardanelles. The Black Sea water quickly gained in salinity and, as it believed, in 1.5–2,000 years ago the water salinity became adequate for existence of many Mediterranean species that penetrated into a new water body and today they form up to 80 % of the B.S. fauna. At the same time the old-timers, the Pontus relicts, retreated into estuaries and river mouths as it happened not once in the past.

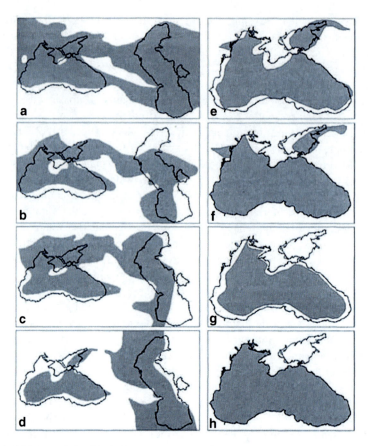

Geological history of the Black Sea: (a) Sarmat Sea, (b) Meothic Sea, (c) Pontos Lake-Sea, (d) Chaudinskiy Lake-Sea, (e) Ancient Euxeinus Basin, (f) Karangatsky Sea, (g) New Euxeinus Sea, (h) modern Black Sea (http://anapacity.com/Images/Pages/istoriya-chernogo-morya.jpg).

George Cliff – an overwater cliff forming a small island near the St. George Monastery to the east-north-east of the tip of the Fiolent Cape (southwestern Crimea) and 140 m from the coast. The name of this cliff is connected with the legend saying that in 891 the Greek ship was wrecked near Balaklava. The sailors on this ship prayed ardently asking the God for salvation. And suddenly Great Martyr George appeared before them and said that the life of sailors would be saved. The Byzantines on the ship remnants reached the cliff-island on which they found an icon. There is also another name of this island—cliff of Sacred Epiphany. Later on the white marble cross was installed on this cliff that in the 1920s disappeared. On September 14, 1991 by the 1,100 anniversary of the monastery a 7-m metal cross was installed on the George Cliff.

Georgia, Republic of – a sovereign state in the Caucasus region of Eueasia. The native Georgian name for the country is "*Sakartvelo*". The word consists of two parts. Its root "*kartveli*" is the self-name of the Georgians and the circumfix "*sa*" and "*o*" means "the area where somebody dwells". In general it may be translated as "a place where Georgians live". In Russian this country is referred to as Georgia and its peoples are called Georgians. These names were assimilated by the Russians in Oriental countries, for example, in crusader of monk Ignaty Smolnyanin to Palestine in 1389.

Its territory is 69,700 km². Population—4.47 million (2011). Its capital—Tbilisi (1.17 million people). Other large cities are Kutaisi (197,000), Batumi (126,000) and Rustavi (123,000). The official language—Georgian. The main religion—Christianity (Orthodoxy, Georgian Orthodox Church). Currency—Lari. State structure—presidential parliamentary republic. The president is elected for 5 years. The legislative body—Parliament with two houses is elected for 5 years, too. Georgia was divided into 9 regions, 1 city, and 2 autonomous Republics of Adjaria and Abkhazia, the Autonomous Area of Southern Ossetia (in 2008 Abkhazia and Southern Ossetia became independent states). These in turn were subdivided into 69 districts. The borders: in the north—with Russia, in the east—with Azerbaijan, in the south—with Azerbaijan, Armenia and Turkey, in the west—washed by B.S.

The class society appeared on the territory of modern G. in the early first millennium B.C. In the fifth century B.C. the slave-owning Kingdom of Colchis came into existence (in the west) and in the fourth–third centuries B.C.—Iberia (in the east.). The land of Colchis on B.S. had became known to the Greeks and Romans still in the ancient times thanks to the Miletian colonization, Golden Fleece and as a motherland of Medea. But Iberia that was remote from the sea became known to the Greeks and Romans only during the Third War with Mithradates and Pompey campaign to Transcaucasus (66–65 B.C.). In the late Antiquity the tribes of Laz being inheritors of Colchis and populating these areas reached such importance that Colchis was called Greek Lazika. In the sixth—early tenth centuries A.D. the states were under control of Iranian Sasanides, Byzantine and Arab Caliphate. During this time the Georgian nation was formed, in general. In the eighth—early ninth centuries the feudal kingdoms of Kakhetia, Ereti, Tao-Klardzhetsky and Abkhazia were established. In the eleventh–twelfth centuries feudal G. reached its zenith in the economic and cultural spheres. But it was short-lived because in the twelfth–fourteenth centuries the Mongol-Tatars and the Timur troops invaded the Kingdom. In the fifteenth—early nineteenth centuries the territory of the country comprised the following states and kingdoms: Kartli, Kakhetia, Imeretia, Samtskhe-Saatabago, Megrelia, Guria and Abkhazia. In the sixteenth–eighteenth centuries G. was subjugated by the feudal lords from neighboring Persian Empire and Ottoman Empire. The national-liberation movement was growing (1625—led by George Saakadze, 1659, etc.).

Under the Treaty of Georgievsk signed in 1783 Eastern Georgia received protection of Russia. In 1801 Eastern G. and in 1803–1864 Western G. were incorporated into the Russian Empire (Tiflis and Kutaisi provinces). Subsequent years witnessed rebellions against social and national oppression. The Agrarian Reform of 1864 spurred the development of capitalism. In 1918 the country was invaded by the German, Turkish and British troops. In 1921 the Soviet power was established and the Georgian Soviet Socialist Republic was formed that entered into the Transcaucasian Federation. From 1936 it entered the USSR as a union republic. After breakdown of the USSR in 1991 G. became independent. In 1993 G. joined the Commonwealth of Independent States established by the former republics of the Soviet Union. G. is a member of the regional GUAM (Georgia-Ukraine-A-zerbaijan-Moldova) organization and the Council of Europe.

The greater part of the country's territory is occupied by mountains: in the north—the Greater Caucasus Mountain ranges (to 5,201 m); in the south—Lesser Caucasus with the Colchis Lowland and Kartli Plain in between; in the east—the Alazan Plain. The climate in the western regions is subtropical and in the eastern regions is transitional from subtropical to moderate and continental in mountain areas. The mean temperature of January varies from 3 °C (Colchis) to −2 °C (Iberian Depression), of August—from 23 to 26 °C. In the mountains of Western Georgia facing the Black Sea the precipitations amount to 1,000–2,800 mm a year; in Eastern Georgia—300 to 600 mm a year. The largest rivers of the country are Kura and Rioni. G. numbers more than 25,000 small rivers with the total water reserves of 65.4 km^3. There are also 850 lakes and 43 reservoirs with the total effective capacity of 2.2 km^3. In the mountain area of the Greater Caucasus the karst phenomena are widespread; there are also many glaciers. In the Bzyb range the karst cave-abyss 1,370 m deep is discovered. Forests cover approximately two-fifth of the country's territory (in subtropics—rainy evergreen forests). Fifteen nature preserves and Tbilissi Natural Park are established in the country for nature conservation. They comprise flora—broad-leaved forests, sub-Alpine and Alpine meadows; fauna—Daghestan wild goat, chamois, deer, roedeer, bear—in Lagodekhi; in Borzhomi—coniferous-deciduous forests and among the fauna—Caucasian deer, forest wild cat, Caucasian squirrel, roedeer, bear and others.

The natural resources include the deposits of manganese, nonferrous ores, coal, oil and others. There are numerous sources of curative mineral waters. Mineral water sources (Borzhomi, Tskhaltubo, Mendzhi, Sairme) and also seaside climatic (Kobuleti and others) and mountain climatic (Abastumani, Bakuriani, Bakhmaro and others) resorts attract here many tourists, travelers and sportsmen.

Ethnic Georgians form about 84 % of Georgia's current population. The Georgians (Kartveli) consist of 17 subethnic groups—Adjarians, Gurians, Katliis, Kakhetians, Meskhi, Djavakhi, Svans and others. Other ethnic groups include Abkhazians, Armenians, Azeris, Belorussians, Bulgarians, Estonians, Germans, Greeks, Jews, Moldovans, Ossetians, Poles, Russians, Turks and Ukrainians. Georgia's Jewish community is one of the oldest Jewish communities in the world. Georgia also exhibits significant linguistic diversity: Georgian (official language), Kartvelian family, Svan, Mingrelian, and Laz languages. Georgian is a primary language of approximately 71 % of the population, with 9 % speaking Russian, 7 % Armenian, 6 %—Azeri, and 7 % other languages. The Georgian language has its own alphabet. Historically several styles of letter writing depending on the purpose have been developed.

Today 83.9 % of the population practices Eastern Orthodox Christianity, with majority of these adhering to the national Georgian Orthodox Church. Religious minorities include Muslims (9.9 %), Armenian Apostolic (3.9 %), Roman Catholic (0.8 %), other religions (0.8 %), and 0.7 % declared no religion at all in the 2002 census.

The Georgian culture, including music and poetry, had evolved over thousands of years. It's worth mentioning here the world masterpiece—the poem of Shota Rustaveli "The Knight in the Panther's Skin" (twelfth century), the well-known legends of Surkhan Orbeliani "The Wisdom of Fiction" (seventeenth century), the names of the nineteenth century poets, such as I. Chavchavadze, N. Baratashvili, A. Tsereteli and others, the antique epos (second half of the second millennium B. C.) "Amiraniani" telling about hero Amirani who brought fire to the people. The Georgian music is distinguished by its polyphony. The founder of classical music was Z.P. Paliashvili. Niko Pirosmanishvili created the paintings that survived their time. In modern G. there are also many gifted theatre and film directors and actors, many musicians.

The most ancient architectural monuments that survived to the present date back to first millennium B.C.—first century A.D., such as Acropolis in Mtskheta, the tsar residence in Armaziskhevi; Medieval Ages—Narikala, Anchiskhati, Metekhi (thirteenth century), Sioni (sixth century), Svetizkhoveli Cathedral in Mtskheta, churches in Kartli and Kutaisi. In Mtskheta the ruins of city fortifications and cemeteries survived to the present from the times of the Kartlii Kingdom (early first millennium B.C.—fifteenth century) that form the Mtskheta museum—nature preserve. In addition, the Samtavro monastery (main church of the eleventh century) is also found here. Not far from Mtskheta there is the Djvari Temple of the sixth century with the church. Of certain interest are the castle-fortress Ananuri, the Gelatsky Monastery of the twelfth century with the Main Temple (1106–1125) decorated with mosaics and the frescos of the th-thirteenth centuries preserved in other churches; the Bolnissky Sion Temple (478–493) located not far from Tbilisi; a complex of David-Garedja monasteries cut in mountains (David lavra, Dodo, Natlis-Mtsemeli monasteries and others where the frescos and portraits of historical persons survived); Alaverdi Cathedral near Telavi of the eleventh century. The decorative-handicraft art in G. is famous for its embroidery, articles made of metal, leather, wool, and ceramics.

G. is a country with transitional economics. The industrial sector (60 % of GNP) includes food (tea, wine-cognac, tobacco, mineral water), light (silk processing, cotton, wool, footwear, textile, garment), machine-building (machines), ferrous and nonferrous metallurgy, chemical and mining industries. In agriculture the plant growing is a dominating branch. The main cultivated crops here are tea, citrus, laurel, vines and nuts. Grains, potato, vegetables, fruits, industrial crops are also grown here. The share of animal husbandry (sheep-breeding, large-horned cattle, pigs) takes to 30 %. The beneficial geographical location and inclusion of the country into the International Transport Corridor on the "Silk Road" from Europe to Asia incorporate G. into the world trade and provide a possibility to use its ports for extension of the foreign economic relations. G. has several sea ports that are used both for international and national communication: Poti, Batumi and Soupsa,

which became the sea gates for Europe and other countries to the Caucasus and
Central Asia. The stable and growing cargo traffic passes through them. The
economic policy is targeted to restoration of the country's economic potential by
stabilization of domestic production and intensification of market reforms.

The resort and tourism based on favorable natural-climatic conditions and
developed infrastructure of treatment and rest are the important sector of the
G. economics. There are 103 resorts in different climatic zones in Georgia. Tourist
places include more than 2,000 mineral springs, over 12,000 historical and cultural
monuments, four of which are recognised as UNESCO World Heritage Sites
(Bagrati Cathedral and Gelati Monastery in Kutaisi, historical monuments of
Mtskheta, and Upper Svaneti).

G. has educational institutes, universities, theaters and museums, such as
S. Dzhanashia Museum of Georgia, the Arts Museum, picture gallery and others
and many libraries. Among the research establishments most well-known are the
botanical gardens and aquarium-dolphinarium in Batumi. The old part of the capital
Tbilisi preserved its narrow streets, two- and three-storey brick houses with their
peculiar outside ladders.

Map of Georgia (modified from the map shown at http://www.itlibitum.ru/MAP/EARTH/COUN
TRY/GEORGIA/index.html in accordance with separation of Abkhazia and Southern Ossetia
from Georgia in 2008)

Georgian Navy – it was created in 1998. It comprises a division of high-speed boats transferred by Turkey, Ukraine, Greece and other countries. These are patrol boats of the "Grif" type, four antimissile and two air-defense boats. In addition, the US allotted $ 27 million to the Government of Georgia for the navy construction. Using this money the shipbuilding yard in Batumi started construction of ships for the country's needs, and the Naval Academy was opened. In 2004 Greece as a grant passed the anti-missile ship from its fleet to the Georgian Navy.

Georgians (self-name "Kartveli") – a nation, the basic population of Georgia. They also live in neighboring regions of the Azerbaijan Republic. Some groups of G. settled down in the northeastern regions of Turkey—in Lazistan and in Iran—in the Isfahan Province. The Georgian language refers to the Kartveli group of the Caucasian family of languages. It has some dialects. The literary language developed on the basis of the Kartli and Kakhetia dialects. The written language appeared in the fifth century. Some G. are Christians (Georgian Church), others (Adjarians and other groups)—Sunni Moslems. The Georgian people comprised the following territorial-tribal groups: Kartlians, Kakhetians, Tushins, Khevsurs, Pshavs, Mokhevians, Mtiulians, Gudamakarians, Ingilonians, Kizikts, Djavakhs, Imeretians, Rachians, Lechkhumians, Gurians, Adjarians, Mergelians, Lazs or Chans and Svans. These groups corresponded to the regions of Georgia (Kartli, Kakhetia, Tushetia and so on). Before the October Revolution each group had their specific material culture, customs and practices that differed significantly. In the USSR as a result of transformations and national consolidation the differences among these groups smoothed out.

Georgievsky Arm (Girlo) – one of the three main arms, the southern arm, connecting the Danube River with B.S. This arm has a small delta. In the 1980s the arm length was 109.5 km. In 1981–1992 six large meanders were straighten out, as a result, the arm length became 32.5 km less and was equal to 77 km. In a point about 3.5 km from the arm inflow into the sea the long Litkov Canal branches off to the left from the arm and two channels branch off to the right. The main northern arm is 7–8 m deep. The branches of major canals Dunavets (flowing into Lake-lagoon Razelm) and Dranov (flowing into Dranov Lake) are used as a waterway for communication among sparse settlements located on its banks.

Gerasimov Aleksandr Mikhailovich (1861–1930) – Russian first rank Captain (1907), Rear Admiral (1911), Vice-Admiral (1913). In 1882 he finished the Naval Corps and in 1892 the Mikhailovsky Artillery Academy. He took part in the Russian-Japanese War of 1904–1905: senior officer of armor-clad ship "Pobeda" (Victory) in Port Arthur. After surrender of Port Arthur he was taken captive by the Japanese. After conclusion of the Portsmouth Peace Treaty (1905) he returned to Russia. He was a participant of World War I—he was the Governor of the marine fortress near Reval (Tallinn). After 1917 October Revolution in Russia he resigned. He also took part in the White Movement: from November 14, 1918 he was the Chief and Commander of ports and ships on the Caspian Sea, Petrovsk-Port; from March 21, 1919—the Assistant and Chief of the Naval Department at Commander-in-Chief General A.I. Denikin; the Commander of the river ship division (February-March 1920). In March 1920 he stayed in Constantinople. From March 29, 1920 he was the Commander of the Black Sea Fleet (at General P.N. Vrangel). In June-July 1920 G. served as the Senior naval chief at the P.N. Vrangel's envoy in the Transcaucasia, in Batumi, Georgia. Transferred into reserve. Left Crimea with the Black Sea Fleet for Biserta, Tunisia (November 1920). Director of the Naval Corps in Biserta (December 1920—March 1930). Died in Biserta.

Gerze – a town on the coast of the Black Sea, Turkey.

Gezlev – see Evpatoria.

Giresun – a province in the Black Sea Region of Turkey. It locates in the northeastern part of the country. Its area is 6,934 km^2, the population—419,000 people (2011). The provincial capital is Giresun.

Giresun – the administrative center of Giresun Province in Turkey. This is a trade port on B.S. through which agricultural products, such as hazelnuts, cherries, beans, soya and others, are exported. In the ancient times it was the capital of Kappadokiy Pontus and was called Kerasos (cherry). Its population is 101,000 (2012). It extends in an arch along the B.S. coast. It is considered the origin place of cherry. Lucius Licinius Lucullus (118—57/56 B.C.), the famous Roman military commander and politician, brought the cherry plants from here to Rome. This terrain represents a hill on a peninsula jutting into the sea. Here one can find the ruins of the Byzantine fortress built in the fourteenth century by Emperor A. Komnin.

Giresun Island – a small island (4 ha) located 1.2 km from the coast and 3.5 km eastward from Giresun town. It is the largest island on the Turkish Black Sea coast. The ruins of a roofless stone temple, fortifications, and two wine or oil presses stand on the island. The island's ancient names are Aretias, Ares, Areos Nesos and Puga. A large, black, spherical stone, located on the island and called Hamza Tasi in Turkish, is said to have magical properties. The roofless temple was in ancient times attributed to the Amazons. While the dominant plants of the island are laurels (*Laurus nobilis*) and black locusts (*Robinia pseudoacacia*), it has been reported that the island has 71 wild and introduced species of trees and herbs. It is also a wild habitat for cormorants (*Phalacrocoracidaesp sp.*) and seagulls (*Laridae sp.*). For a long time it has been preserved as a Class II historical and natural site by the Turkish Government. Thus it is not allowed to be used as a residential area. Touristic visits to the island are possible on small and medium-sized fishing and cruise boats which can be provided by tourism agencies in Giresun.

Girlo (Russian form of "gorlo" (throat)) – an arm of a river delta, a strait, a narrow water channel connecting a liman with a sea; a mouth, an underwater part of a delta; a river fairway in a river outlet to a sea (Black and Azov seas); an underwater bed, a river continuation on the surface of an underwater delta—natural shipping channel. The condition of G. controls, to a great extent, the water exchange between a liman and a sea. During storms G. may be filled with sand, and the liman–sea link can be broken.

Gizel-Dere (may be translated as *"beautiful gorge"*) – a maritime climatic resort terrain on the Black Sea coast of the Caucasus in the Tuapse District, Krasnodar Territory, Russia. It is a part of the Tuapse seaside resort zone. In the northeast this territory is confined by the outspurs of the Greater Caucasus. Here one can find very picturesque gorges and valleys overgrown with dense forests consisting largely of oak, hornbeam and chestnut. Beaches are covered with sand and small pebbles. The sea bottom is gently sloping. The climate here is characterized by a warm winter with long rainfalls and a wet summer. Summer heat is not harassing thanks to the sea breezes. There are many sunny days.

Glazenap von, Bogdan Aleksandrovich (1811–1892) – Russian Admiral. In 1824 he was promoted to naval cadet and in 1826—to midshipman. In 1824–1825 he was at sea. In 1826–1829 on the boat "Senyavin" he made a round-the-world cruise to Kamchatka and back. In 1830 he navigated in the Gulf of Finland. In 1831 he was promoted to Lieutenant. He was directed to the Army Headquarters in the Polish Kingdom. In 1832 he navigated from Kronshtadt to Danzig (Gdansk). In 1834 he was appointed the Aide-de-camp to the Chief of the Main Naval Headquarters and later transferred to the Guard Crew. In 1834–1837 on different ships he called on the Baltic ports. In 1837 he was promoted to Captain-Lieutenant. On the ship "Severnaya Zvezda" he escorted the Emperor from Odessa to Sevastopol, Yalta, Gelendzhik, Kerch and to the Abkhazia coast. For the copy of the French "Guide on Ship Navigation" translated into Russian presented to the Emperor he was granted the diamond rings from the Tsar and Great Prince Mikhail Pavlovich. In 1838 on different ships he navigated the Baltic ports and was appointed to the position of the Aid-de-camp. In 1839 he was sent to Nikolaev. He sailed with the Commander-in-Chief of the Black Sea Fleet from the ports on Taman to the Abkhazia coasts. In 1840 he was directed once again to Nikolaev. G. took part in the landing on the Abkhazia coast near the Velyaminov fort. In 1841–1845 he navigated over the Baltic Sea. In 1844 he was promoted to second rank Captain. In 1845–1846 commanding a covette he sailed from Kronshtadt to the Mediterranean in the team of Vice-Admiral F.P. Litke. In 1846 he was promoted to first rank Captain and appointed a record keeper in the Naval Committee. In 1847 he became a member of this Naval Committee. In 1848–1849 he was the Editor-in-Chief of the journal "Marine Newsletters". In 1852 he was promoted to the Rear Admiral with appointment to the Emperor's escort and approved in the position of the Director of the Naval Corps. In 1855 he was nominated the interim manager of the Hydrographic Department of the Marine Ministry. In 1860 he took the position of the Commander-in-Chief of the Nikolaev port and became the Military Governor of Nikolaev. In 1861 he was promoted to Vice-Admiral. In 1862 for his efforts in release of the Black Sea settlers attached to the Admiralty he was awarded the gold medal. In 1863 he was appointed the Commander of the troops deployed in Nikolaev. In 1869 he was promoted to Admiral. In 1871 he was appointed the member of the Admiralty Council and Aleksandrovskiy Committee for the Wounded. In 1877 he was appointed the honorary member of the Naval Academy in Nikolaev. He was awarded many Russian and foreign orders.

Global Ocean Observing System of the Black Sea (the Black Sea GOOS) – The Black Sea GOOS is a local Association formed by the Black Sea riparian countries in order to foster Operational Oceanography in the region and set up links with other regional and global organizations (with similar objectives). The Major Aims and Objectives of the Black Sea GOOS are: (1) To contribute to international planning and implementation of the GOOS and to promote it at national, regional and global level; (2) To identify regional priorities for operational oceanography; (3) To develop capacity of the regional countries and promote the level to sustain GOOS activities; (4) To provide high quality data and time series, for a better understanding of the Black Sea ecosystem; (5) To assess the economic and social benefits achieved by operational oceanography. The Co-operative Marine Science Programme for the Black Sea (CoMSBlack) (1991) was the first multinational program implemented in the Black Sea. It was recognized and supported by the Intergovernmental Oceanographic Commission (IOC) and UNESCO. The participating countries in CoMSBlack were Bulgaria, Romania, Ukraine, the Russian Federation, Turkey and the USA. On the Eighteenth Session of the IOC, the Assembly adopted a resolution (Resolution XVIII-17, UNESCO, Paris, 7–9 June 1995) which established the IOC Black Sea Regional Committee (BSRC). The resolution established the Terms of References for the BSRC and the resolution also defined the initial tasks of the BSRC for the period 1996–1997. In May 2001 in Poti (Georgia) the meeting with the participation of all Black Sea Riparian Countries, Delegations from the IOC, the EuroGOOS, the Black Sea Committee and the Black Sea Environmental Programme besides the other objectives adopted the Memorandum of Understanding of the Black Sea GOOS and elected the Black Sea Ad Hoc Steering Committee. The Black Sea GOOS Memorandum was signed by five Black Sea countries, namely Bulgaria, Georgia, the Russian Federation, Turkey and the Ukraine on July 6, 2001 in Paris. Romania followed the others and signed it on October 22, 2001 at the IOC headquarters in Paris.

"Glomar Challenger" – US research vessel for deep sea drilling. It was built in 1968. The drilling equipment installed onboard this vessel ensures drilling of wells to 1,000 m deep with the ocean depth in a drilling point reaching 6,000 m. The vessel has a system for dynamic stabilization that is capable to keep it in a fixed position when affected by strong winds, waves and currents. The principal elements of this system are: a sonar beacon installed on the seafloor near the well mouth; hydrophones placed in corners of a square under the vessel bottom and receiving signals from a sonar beacon; maneuvering devices—2 on the bow and 2 on the stern; a control system emitting signals for actuation of electric engines of maneuvering devices and engines of driving propellers. This vessel for deep sea drilling of the ocean floor played a key role in extension of our knowledge about the structure of the Earth's crust and physical processes going on inside it. The displacement of the vessel is 10,500 tons, length—122 m, capacity of a power plant—7,360 kW, speed—up to 12 knots. The vessel that was used under the US Deep Sea Drilling Program in summer 1975 made three deep sea wells in B.S.: the first—in the deepest part of the Black Sea depression, the second—50 km to the northeast of the Bosporus and the third—135 km eastward of the Akhtopol, Bulgaria. In the latter case drilling penetrated the Pliocene deposits at a depth of 3,185 m, while the thickness of the Quaternary deposits was equal to 1,075 m. This helped to conclude that the modern Black Sea depression was formed in the mid or late Pliocene. After being in operation for 15 years the *Glomar Challenger* was taken out of active duty in November 1983 and was later scrapped. Its successor, the *JOIDES Resolution*, was launched in 1985.

The Glomar Challenger (http://christofforr89.edublogs.org/files/2011/01/glomar-challenger-1h4jzoo.gif)

"Goeben" – German battle cruiser, convoy raider on B.S. It was named in honor after the German Franco-Prussian War veteran and well-known military publicist General August Karl von Goeben (1816–1880). It was set afloat in 1911, put into operation in 1912. Its draft was 22,979 tons, maximum speed—28.4 knots, cruising range—4,120 miles, 34 cannons, 4 underwater torpedo tubes, crew—1,053 people. From 1912 "G." together with light cruiser "Breslau" formed the German "Mediterranean Division". These were raiders sent by Germany to B.S. The mission of "G." pursued the aim to get maritime supremacy on B.S. Germany transferred the two ships to the Ottoman Navy on 16 August 1914. The *Goeben* was renamed *Yavuz Sultan Selim* (Terrible Sultan Selim) and the *Breslau* was renamed *Midilli*. In 1914–1917 more than once it came close to the Russian shores firing at Sevastopol, Yalta and Batumi. In May 1918 "G." called on Sevastopol when it was occupied by the Germans and witnessed the liquidation of the Black Sea Fleet of Russia. It took part in sea battles, several times was damaged by direct hits of shells and striking mines. In 1918–1926 the ship was out of use in Izmir. After reparation and reconstruction the *Yavuz* (a new name since 1930) was recommissioned in 1930, resuming her role as the flagship of the Turkish Navy and gaining better speed and fire characteristics. The four destroyers, which were needed to protect the battlecruiser, entered service between 1931 and 1932. And it was this ship that delivered in 1938 the body of General Mustafa Kemal Ataturk from Istanbul to Izmit. The *Yavuz* remained in service throughout World War II. In November 1939 she and the *Parizhskaya Kommuna* (USSR) were the only capital ships in the Black Sea region. After 1948, the ship was stationed in either İzmit or Gölcük. She was decommissioned from active service on 20 December 1950 and excluded from the Navy register on 14 November 1954. When Turkey joined NATO in 1952, the ship was assigned the hull number B70. The Turkish government offered to sell the ship to the West German government in 1963, but the offer was declined. Turkey sold the ship to M.K.E. Seyman in 1971 for scrapping. She was towed to the breakers on 7 June 1973, and the work was completed in February 1976.

SMS "Goeben" (http://www.manorhouse.clara.net/book2/goeben.jpg)

Golaya Pristan – a town, center of the Golaya Pristan District, Kherson Region, Ukraine; a wharf on the Konka River (the Dnieper arm 11 km to the southwest of Kherson). It was founded in 1709 as a Cossack village. In its name the word "*golaya*" means the lack of vegetation. The population—about 15,700 (2011). It has mechanical workshops and shipbuilding yards, a butter factory, a cane-fiber plant, peat fields. On the Golaya Pristan Lake there is a mud-bath resort Gopri.

Golden Coast – the seaside climatic resort in the Gudauta District, Republic of Abkhazia, 15 km to the northwest of the railway station "Gudauta". It is situated on the Black Sea coast of the Caucasus, on the highland passing in the north into the piedmont area of the Bzyb Range of the Greater Caucasus. The vegetation here is represented mostly by cultured plantings (palms, magnolia, and others). The climate is humid subtropical. The winter is very mild and snowless; the mean temperature of January is about 6 °C. The summer is very warm; the mean temperature in August is 23 °C. Quite frequent are sea and onshore breezes. The water temperature in summer reaches 24–25 °C. The mean annual precipitations are about 1,500 mm. The mild climate here is favorable for treatment of the cardio-vascular and nervous diseases. The resort was founded in 1948. Not far from G.C. there is the Pitsunda-Myusser nature reserve with the unique Pitsunda pine grove.

"Golden Fleece" – in Greek legends this is the pelt of the magic golden ram worshiped as a religious cult. It figures in the tale of the hero Jason and his team of Argonauts, who set out on a quest for the fleece by order of King Pelias, in order to place Jason rightfully on the throne of Iolcus in Thessaly. The story is of great antiquity and was current in the time of Homer (eighth century B.C.). It survives in various forms, among which details vary. According to the myths, Nephele received G.F. from Hermes and then passed it to her children Phrixus and Helle. On this ram Phrixus arrived to Tsar Aeetes in Colchis where he sacrificed the ram to Zeus and hang the Golden Fleece in the Ares grove where it was guarded by a dragon. According to the legend, 79 years before the Trojan War Jason, the son of Aeson who was the Tsar of Thessaly, departed for B.S. on the ship "Argo" with a crew of 56 people. Having reached Colchis he with the help of Medea, the daughter of the tsar here, had stolen the main treasure of the Colchis—G.F. Jason and Argonauts brought G.F. back to Greece (trilogy "Golden Fleece" (1821) by Austrian writer Franz Seraphicus Grillparzer).

"Golden Gates" – locate in the eastern part of Crimea, an andesite rock of the natural origin reminding by its form a stone arc. This is the result of wind and sea wave action. G.G. is a kind of the visiting card of the Kara-Dag (in Tatar *"Shaitan Kapu"* meaning "devil's gates")—the Kara-Dag Gates. A boat can sail easily under this arch.

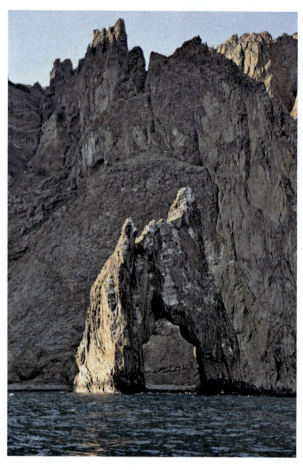

"Golden Gates" (Photo by Evgenia Kostyanaya)

Golden Horn (in Turk "*Halic*" meaning "harbor, bay") – a bay near the European shores of the southern entrance into the Bosporus Strait, Turkey. Its length is 12.2 km, width—91–222 m, depth—to 47 m. It is accessible for large sea vessels. The city of Istanbul locates on its both shores. It is mentioned by the Greek authors as "Chrisokeras"—"golden horn" ("*chrisos*" meaning "golden" and "*keras*" meaning "horn"). This name was given because of the form of the bay reminding of a horn, while its definition "golden" refer to its main richness—fish.

Golden mullet (*Mugil auratus*) – the fish of the gray mullet family (*Mugilidae*). Its body length is 55 cm, weight—1.5–2 kg. It is native to the Atlantic Ocean, the Mediterranean, Black and Azov seas. It is also well adapted in the Caspian Sea. This pelagic fish lives in schools. Every year it migrates for spawning, fattening and hibernation. It lives for 9–10 years. It reaches maturity at the age of 2–3 years. Spawning occurs from August to October both in coastal and open waters. During summer and half of autumn it feeds and grows intensively. When the water gets cool in autumn it gradually migrates southward and by the late December it reaches the southern part of B.S., the Bosporus and the Sea of Marmara for spending winter there.

Golden sands (before 1942—Uzunkum) – one of the oldest and most beautiful seaside climatic resorts on the Bulgarian coast of the Black Sea (Varna Region), 17 km to the northeast of Varna. The climate here is marine. The winter is mild, practically without snow. The mean temperature of July is 22.6 °C, of August 22 °C. The precipitations are about 500 mm a year; they fall largely in the cold period. The number of sunshine hours is 2,237 a year. Quite frequent are breezes that alleviate the summer heat. Being protected in the west by the Frangen Plateau (to 356 m high) that descends to the sea by terraces (its slopes are covered with forests and vines) and having a long beach running for 4 km (to 70 m wide) with its soft golden-yellow fine sand and smoothly sloping sea floor make this resort best suitable for climate and thalasso therapy. The water in the sea gets heated in July-August to 24 °C (the mean annual water temperature is 13.6 °C). Favorable climatic conditions, warm sea, comfortable hotels located in the vast natural park with the very picturesque whereabouts attract here many vacationers and tourists. Apart from the climatotherapy, this resort offers water treatment, remedial gymnastics and warm mineral water from local springs. There are also sport facilities, summer theatre, restaurants, cafes, and trade centers. Quite often the international scientific congresses and symposia are held here. The resort complex constructed in the 1950s is also interesting in architectural terms: the team of authors headed by architect G. Ganev was conferred in 1958 the Dimitrov Award. Three kilometers from G.S. you can visit the Aladja Monastery (it was founded in the third century) carved in a rock. It is a popular place of sightseeing. G.S. resort is connected by a picturesque panoramic highway with the resort "Druzhba" (8 km to the southwest) and Varna. One more resort Albena locates to the northeast.

Golenkin Gavriil Kozmich (Kuzmich) (before 1750–1820) – Russian Vice-Admiral (1799). In 1760 he entered the Naval Corps as a cadet, in 1763 he was promoted to naval cadet and in 1766—to midshipman. In the period from 1763 to 1769 he navigated every year in the Gulf of Finland. In 1769 in the squadron of Admiral G.A. Spiridonov he went to the Mediterranean. In 1770–1771 he was in Archipelago, took part in the battle at Napoli-di-Romania, at Chesma, in attacks of fortresses in Metelino and Budrum. In 1772 he was promoted to the Lieutenant. In 1775–1780 he served in the Kronshtadt port. In 1783 he became second rank Captain and was transferred to B.S. In 1784 he served in Kherson port. In 1785–1789 he was appointed the Commissar of the port. In 1787 he was promoted to first rank Captain and in 1789—to Brigade captain. In 1790 he took part in the battles with the Turkish fleet near the Kerch Strait and Khadzhibei. He was promoted to Captain of the major-general rank. As a commander of a ship he took part in the battle at Kaliakra. In 1792 he acted as the commander of the Sevastopol port and fleet and in 1793 he commanded the rearguard fleet on the Sevastopol sea roads and was promoted to Rear Admiral. He was a member of the Black Sea Admiralty Board. In 1794–1795 he took the post in Kherson and in 1795–1799 was the

Commander of the Kherson port. In 1799 he was promoted to Vice-Admiral and transferred from the Black Sea Fleet to the Baltic rowing fleet. In 1799–1802 he served in Kronshtadt performing the duties of the port Commander-in-Chief. In 1803 he was transferred from the rowing fleet to the ship fleet, to the red flag division. In 1805 he resigned.

Golet, Galet, Gullet (French *"gouelette"* or *"goelette"* meaning "schooner") – a sail-rowing 2-mast ship of the Russian skerry fleet in the late eighteenth—early nineteenth century. Its length was about 18–20 m; the rigging was similar to that installed on a schooner. It could have to 14 cannons aboard, was very maneuvering and was managed by a small crew. There were also transportation golets, they mostly navigated the Black and Azov seas.

Golitsyn Lev Sergeevich (1845–1915) – Russian Prince, one of the founders of winemaking in Crimea. He built the first Russian factory of quality champagne wines. In 1862 he graduated from Sorbonne (France) and in 1871—the Law Faculty of the Moscow University. In 1872–1874 he continued his education in Leipzig and Gottingen in Germany. In 1876 G. was elected the nobility head in the Murom uezd of the Vladimir Province. In 1878 he bought from Georgian Prince Kherkhadze the wild land area Paradiz (edem) covering 230 ha in Crimea at the foot of the Sokol Mountain, 7 km from Sudak. This area was called Novyi Svet. He turned his estate into an exemplary trial-production farm on winegrowing and winemaking that extended over an area more than 20 ha. He planted a nursery where he cultivated to 500 grape varieties and conducted breeding experiments. He believed that the wine quality depended, to a great extent, on the correct choice of land for growing of a certain variety of grapes (for the Crimean semidesert zone the best varieties were Aligote, Risling, Cabernet, Kokur white). He planted vines near Feodosia on 30 ha. In the early 1890s he constructed a winemaking plant in Novyi Svet having the Russia's longest cellars in the Kob-Kaya Mountain—about 3 km. In special cellar No. 4 he gathered the world's unique collection numbering more than 50,000 bottles of wine manufactured in the seventeenth—early twentieth centuries (in the early 1920s it was transferred to Massandra where it was replenished considerably and in 1990 its part was sold at the Sotheby's auction in London). In 1880—early 1890 he organized production of champagne wines applying the method of fermentation in bottles (*méthode Champenoise*), such as "Paradiz", "Novyi Svet", "Koronatsionnoye", and sparkling wines, such as "Klaret". In 1899 the greatest number of bottles with champagne (60,000) was produced in Novyi Svet. In 1903 the cellars stocked over 700,000 bottles with champagne, mostly, Golitsyn-made. These wines were supplied to different regions of the Russian Empire.

These wines won the Gold Medals at the Russian Industrial-Artistic Exhibition in Moscow (1882), at the Yalta Agricultural Exhibition (1884), at the World Exhibition in Paris (1889) and other international exhibitions. At the Industrial-Artistic Exhibition in Nizhny Novgorod (1896) the manufacturers were awarded the right to use the image of the state coat of arms on bottles. In 1900 these wines won "Grand Prix" at the World Exhibition in Paris. In 1891–1898 G. became the major winemaker of the Udelny Department (Emperor's Ministry). Under his supervision the Crimean largest wine storages were widened and constructed in Massandra, Ai-Denila, Oreanda and Livadia. In 1899 at the Ministry of Agriculture and State Property he founded six awards named after Alexander III on winegrowing and winemaking (he personally donated 100,000 rubles in their fund). G. took part in the Russian congresses on winegrowing and winemaking in Simferopol (1901), Moscow (1902), and Odessa (1903). In 1902 he was elected the Honorary Member and in 1911 the Chairman of the Winemaking Committee of the Emperor's Agricultural Society of Southern Russia. He opened his estate in Novyi Svet for excursions of scientists, entrepreneurs and students. Beginning from 1903 G. faced serious financial difficulties—his expenses on construction and operation of factories in Novyi Svet were not covered by revenues from wine sales. In 1912 Nicholas II visited G. seeking to preserve this unique farm G. donated to the Emperor a part of his estate with 113 ha of land, the wine collection (42,000 bottles of wines of the twelfth–twentieth centuries from the whole of Europe), the factory of champagne wines with cellars (presently the factory of champagne wines "Novyi Svet"), but retained the position of the chief winemaker. In Novyi Svet G. gathered an art collection that comprised numerous cutglass, glass and porcelain articles of the eighteenth century that had belonged to the Emperor's family, jewelry, Ancient Greek and ancient Ukrainian artistic ceramics, coins, paintings, and Antique sculpture.

Golitsyn L.S. (http://img1.liveinternet.ru/images/attach/c/1//62/641/62641642_Golicin_2.jpg)

Golitsyn Vasiliy Vasilievich (1643–1714) – a prince, Russian statesman and diplomat. In 1682–1689 he was the head of the *Posolsky Prikaz* (ministry of foreign affairs). The outstanding statesman who pursued the policy of consolidation of the international ties of Russia with all European countries. One of the successful events in the G. policy was conclusion of the "eternal peace" with Poland that fixed the Ukraine-Russia reunion. For attainment of these treaty he organized two campaigns against the Crimean Khanate (in 1687 and 1689), but they failed, although indirectly they prevented Khan from marching against the Russian allies. In 1689 after Peter I came on the throne G. was sent in exile for support of the Sophia government.

Golitsyn V.V. (http://www.frontier.net.ua/wp-content/gallery/other/golicin.gif)

Golovachev Viktor Filippovich (1821–1904) – Russian naval historian, acting State Counsellor (1884). He descended from a noble family. In 1838 G. finished the Naval Cadet Corps. He navigated on the ships of the Black Sea fleet. In 1839–1840 he took part in the Caucasian War. In 1848 he resigned because of illness. In 1852–1860 in the rank of the titular counselor he served in the Emperor's Public Library, in the Emperor's Ministry. From 1860 he was invited to preparation of the Russian Fleet history. Together with S.I. Elagin he collected the archive materials about the history of the fleet. In 1868–1879 he worked as a translator in the Scientific Division of the Naval Technical Committee. From 1879 he had been the assistant of the Editor-in-Chief of the "Naval Newsletters". In 1899 he was a member of the Historical Commission of the Guards Crew. G. considered the military history as a basis of the military theory on which the investigations of the military-historical events were based. He was the first in the Russian naval historiography who applied the so-called criticism-historical method (application of the military theory to analysis of the historical military facts). In his investigations G. used archive materials. The actions on the sea were considered by G. relative to the actions of land-based forces and also in the context of the course of events in the south where the Russian-Ottoman War of 1787–1791 was in progress. In his work "Chesma Battle: Its Political and Strategic Aspects and Russian Fleet in 1769" he made an attempt to disclose the social aspect of the war having stressed that it caused major material damage to the country. In his work "History of Sevastopol as the Russian Port" (Saint Petersburg, 1872) G. studied the history of the Black Sea Fleet. In his opinion Russia could ensure its development only after taking Crimea and the B.S. bases. G. thought it necessary to use the naval historiography as a means for most effective improvement of the Russian Fleet.

Golovatyi Anton Andreevich (1732–1797) – Russian military clerk and afterwards a military judge of the Black Sea Cossacks, a combat leader and experienced and educated diplomat. For some time he studied in the Kiev bursa, then in 1757 he ran away to Zaporozhie Sech, enlisted to the Cossacks (to Vesyurensky unit). In 1762 he was elected the unit ataman. In 1764 for the merits before the Zaporozhie troops he was promoted to the regimental sergeant and soon he became the military clerk. During the Russian-Turkish War of 1787–1791 the Cossacks led by G. took by assault the Berezan Island, thus, dooming the Ochakov fortress to surrender. In 1792 being already a military judge he led the Cossack delegation to the capital for handing over to Catherine II a petition asking for allotment of lands to the Black Sea Cossack troops near Taman and its whereabouts, succeeded to receive from the Empress her High consent to giving lands to the Kuban Cossacks for "eternal use". He energetically facilitated the resettlement of the Cossacks to Kuban. G. took part in construction of the 40 first kurens (units) and the military city—Ekaterinodar (presently Krasnodar). The first stage of resettlement ended on August 25, 1792. The Cossacks were headed by Sidor Belyi. The last group of the resettling Cossacks that landed on Taman in late 1793 was headed by G. In 1794 with the active support of G. the Church of Protection of Our Holy Lady was founded in Taman. G. developed the basic laws of the Kuban troops titled "Procedure of Common Use". In 1796 he commanded the fleet on the Caspian Sea during the military campaign to Persia. After death of Z.A. Chepega he was proposed to be an Ataman and on March 21, 1797 he was approved in this position. But he did not learn about his high appointment because during the Persian march on January 29, 1797 he died in Azerbaijan where he was buried and where his grave was preserved.

Golubaya (Blue) Bay – a small bay located several kilometers north-westward of the Gelendzhik Bay on the northeastern Black Sea coast, Russia. Southern Branch of P.P. Shirshov Institute of Oceanology, Russian Academy of Sciences is located here. There is a small pier for research ships and boats, and experimental works. A public beach is located outside the territory of the Institute.

Gopri – a balneological and curative-mud resort in Ukraine (Kherson Region, Golopristansky District) 18 km to the south of Kherson on the southern margins of the town of Golaya Pristan, on the left bank of the Konka River (left tributary of the Dnieper). The climate here is mildly continental. The winter is mild; the mean January temperature is about -3 °C. The summer is very warm, dry and sunny; the mean temperature in July is 23 °C. Precipitations—about 350 mm a year, falling largely in April-October. The minimum air humidity in summer is about 40 % (in July–August). There are about 2,300 sunny days a year. The main natural curative factors are availability of sulfide mud and brine from a salt lake (about 1 km long, to 750 m wide and to 1 m deep). These muds here are distinguished by a high level of calcium and magnesium carbonates and the low content of iron sulfide. Brine by its chemical composition is sodium chloride with a high content of carbonates and bicarbonates of sodium, iron, sulfur, potassium, bromine and iodine; its mineralization varies (depending on a season) from 20 to 75 g/l. The first mentions of mud curing in Gopri date back to the late nineteenth century. Apart from balneo-mud treatment (brine baths, mud applications and others) the report provides artificial mineral baths, remedial gymnastics, electrotherapy, climate therapy and others for treatment of locomotor system, peripheral nervous system, gynecological diseases. The visitors may also take outpatient treatment.

Gorchakov Aleksander Mikhailovich (1798–1883) – Russian statesman, diplomat, State Chancellor (1867), Honorary Member of the Petersburg Academy of Sciences (1856), His Highness Prince (1871). He was educated at the Tsarskoye Selo Lyceum. On leaving the lyceum G. entered the Foreign Office. In 1820–1822 he was the Secretary of Foreign Minister K.V. Nesselrode, attended the Congresses of the Holy Union. In 1822 he was the Secretary, in 1824 the First Secretary of the Embassy in London; later the Charge d'Affairs, First Secretary of the Embassy in Berlin; the Charge d'Affairs in Florence. In 1828–1833 G. was the envoy in Tuscany, from 1833—first counselor of the Embassy in Vienna. In 1841–1855 he was the Envoy Extraordinary and Plenipotentiary Minister in Stuttgart (Württemberg); at the same time (in 1850–1854) in the German Union. From 1854 he acted as and in 1855–1856 took the position of the Envoy Extraordinary in Vienna. At the Vienna Conference of Ambassadors in 1854 as a result of negotiations he prevented Austria from entering the Crimean War of 1853–1856 on the side of France. From April 1856 the Foreign Minister and from 1862 the member of the State Council. G. pursued the policy aimed at liquidation of the provisions of the 1856 Paris Treaty of Peace. In 1856 he purposefully abstained from participation in diplomatic actions against the Naples government referring to the principle of non-interference into the affairs of other countries, stressing at the same time that Russia did not waive her right of voting in international negotiations. As for the 1859 crisis in Italy (a precursor of revolution of 1859–1860), he suggested convening of a congress for peaceful settlement of this issue and when the war among Piedmont, France and Austria became inevitable he took actions preventing small German states from acceding to the Austrian policy; he insisted on the purely defensive purpose of the German Confederation.

Not without G. efforts the rapprochement between Russia and France was outlining; it began from the Stuttgart meeting in 1857 of two Emperors. In 1860 G. called for revision of the clauses of the 1856 Paris Treaty concerning the status of Christians ruled by Turkey and proposed to hold a conference to discuss this issue. Departing from the non-interference principle he condemned the policy of the Sardinian government in Italy. In 1862 the Russian-French union broke down and in its place came a union with Prussia. In 1863 G. concluded a military convention with Prussia that made it easier for the Russian Government to control the Polish rebellion of 1863–1864. He blocked the proposals of French Emperor Napoleon III on convening an international congress devoted to the problems of Central Europe. As a result of the G. policy Russia remained neutral in the wars of Prussia with Denmark (1864), Austria (1866) and France (1870–1871). Defeat of France enabled G. to successfully denounce the Black Sea clauses of the 1856 Paris Peace Treaty and to gain acknowledgement of this fact from other powers at the 1871 international conference.

G. played a key role in creation of the Triple Emperors Union (1873) He tried to use it in preparations for the war with Turkey (Reichstadt Treaty between Russia and Austria-Hungary of 1876, Russian-Austrian Convention of 1877). Opposed the exorbitant strengthening of Germany; with the 1875 circular prevented the second defeat of France. During the Russian-Turkish War of 1877–1878 he was the key figure that ensured neutrality of European powers. Successes of the Russian troops led to conclusion of the San Stefano Peace Treaty of 1878 that stirred protestations on the part of Austria-Hungary and Britain. In the face of menacing anti-Russian coalition he agreed to holding of the Berlin Congress of 1878 where he supported occupation of Bosnia and Herzegovina by Austria-Hungary. G. mostly took care about consent among the powers, about the interests of Europe insisting herewith on the exclusive right of Russia to protection of its national interests. Russia was among the first to evaluate appropriately the American and African vectors in the European policy of Russia. He resolutely discarded the interference of European powers in the civil war in the USA in 1862, supported the Federals and laid the basis for friendly relations with the USA. G. maintained person friendly relations with the outstanding foreign politicians, including Otto von Bismarck, major Turkish statesman Mehmed Fuad Pasha, which was favorable for the Russian-Turkish relations in 1856–1871. In 1882 G. resigned.

Gorchakov A.M. (http://spokusa.com.ua/images/post/390.jpg)

Gorchakov Circular – the name of diplomatic documents connected with Russian Foreign Minister A.M. Gorchakov used in literary sources. The circular of 1870 was sent on October 19(31) to the Russian diplomats in Great Britain, France, Austria-Hungary, Italy and Turkey. It informed the governments of the states that signed the Paris Peace Treaty of 1856 that Russia did not consider being bound by the clauses restricting its sovereign rights to B.S. (prohibition to have the navy on B.S., to construct fortifications). G.C. stressed that the Russian Government followed accurately all clauses of the Paris Treaty, meanwhile other powers violated them more than once. The Russian Government informed the Turkish Sultan about annulment of the supplementary convention defining the quantity and size of the military ships on B.S. After G.C. the diplomatic representatives in the same countries received the elucidating messages dwelling on some provisions of G.C. The British and Turkish Governments were notified that abrogation of restricting clauses of the Paris Treaty should not be taken as hostile actions towards Turkey. G.C. roused indignation of some European Governments, however, A.M. Gorchakov sent it at the time when France was gravely defeated in the war with Prussia, but the peace had not been concluded as yet and Prussia was interested in neutrality of Russia. Britain proposed to convene a conference for revision of the Paris Treaty. On November 20 A.M. Gorchakov informed about his readiness to take part in any meetings provided there was a preliminary consent on abolishing the clauses restricting the sovereignty of Russia. In 1871 at the London conference the convention was signed confirming the sovereign rights of Russia to B.S. (see London Conventions).

Gorchakov Mikhail Dmitrievich (1793–1861) – Russian Prince, military commander, Aide-de-camp General (1829), General of artillery (1844), member of the State Council (1856). G. entered the Russian Army in 1807 as a cadet of artillery. In the same year he was promoted to the second Lieutenant. In 1809 he served in the Separate Georgian Corps. In 1811 in the rank of the Aid-de-Camp of the Commander-in-Chief in Georgia he took part in the campaigns against Persia (1807–1813). During the Napoleonic War of 1812 he distinguished himself at Borodino battle. During the Russian-Turkish War of 1828–1829 he demonstrated bravery during crossing of the Danube (1828). From 1831 G. was General-Governor of Warsaw and the Chief of the staff of the Acting Army. In 1849 he commanded the Russian artillery in the war against the Hungarians. From the very beginning of the Crimean War (1853–1856) G. commanded the troops (3rd corps) on the Danube and Black Sea coast and later the Southern Army. On his own initiative he sent to Sevastopol some forces and supplied the Crimean Army with ammunition and food. On February 24, 1855 he was appointed the Commander-in-Chief of the Crimean Army in place of A.S. Menshikov. He showed himself a not very efficient commander and suffered a defeat in the battle on Chernaya River (on 04.08.1855). In September 1855 following his order the Sevastopol garrison retreated leaving a part of the city to the enemy. In late 1855 G. was replaced with A.N. Liders.

Gorchakov M.D. (http://www.rulex.ru/rpg/WebPict/fullpic/0009-025.jpg)

Gorele – a town on the coast of the Black Sea, Turkey.

Gorgipia – see Anapa.

Gorshkov Sergey Georgievich (1910–1988) – admiral of the USSR Navy. In 1926 he became a nondegree student of the physical-mathematical faculty of the Leningrad University. In 1927 he entered the M.V. Frunze Naval School. After finishing in 1931 of this school he served as a navigator on a destroyer of the Black Sea Fleet. In 1932 he was transferred to the Pacific Fleet. In June 1939 he was appointed the Brigade Commander of the torpedo boat squadron of the Black Sea Navy. From 1940—the commander of the cruiser brigade of the Black Sea Navy. In 1941 he finished the courses for command staff advancement at the Naval Academy. He was promoted to first Rank Captain. Commanding the detachment of assault landing ships he showed bravery during the landing operation at Grigorievka near Odessa. In 1941 he was promoted to the Rear-Admiral. He took command during the Kerch-Feodosia operation and during evacuation of the troops of the Crimean front in 1942 he commanded the defense of the Kerch Strait. After loss of bases on the Sea of Azov he organized pullout of a part of military ships and transport ships with their cargo. Using the forces of the Azov Navy, Kerch and Novorossiysk naval bases he took part in defense of the Taman Peninsula until the ships left the Sea of Azov and the troops retreated for defense of Novorossiysk. He commanded the Novorossiysk defense and was the last to leave it. In 1943–1944 he commanded the Azov Navy for the second time. In spring 1943 he commanded some landing operations. Among the major naval operations there were landings in Mariupol, Osipenko and Temryuk; support from the sea of the forces of the North-Caucasian front during liberation of the Taman Peninsula and, at last, a major operation in November 1943 on landing of the Detached Maritime Army on the Kerch Peninsula and its support on the conquered base area. In 1943 during the Kerch-Eltigen operation he commanded the preparation and landing of the marine forces and then crossing to Crimea of the troops of the 56th Army. From April 20 to December 12, 1944 he commanded the Danube Fleet that supported the Soviet troops in their attack of German Army in Eastern Europe. He commanded the actions of the fleet during forced crossing of the Dniester mouth, liberation of Bulgaria and Romania. In September 1944 he was promoted to the Vice-Admiral. In 1944 he successfully commanded the fleet during the Belgrade and Budapest operations. From 1945 he commanded the squadron of the Black Sea Navy. From November 1948 he was the Chief of the naval staff, in 1951–1955—the commanded of the Black Sea Navy. In 1955 he was appointed the First Deputy Commanded-in-Chief and in 1956— Commander-in-Chief of the Navy and USSR Deputy Naval Minister. In 1967 he was awarded the rank of the Naval Admiral of the USSR. G. advocated development of submarine fleet and guided-missile ships. He was the author of such books as "Naval Might of the State" and "In the Southern Maritime Flange. Autumn— Spring 1941" describing the army and naval operations near the coasts of the Black and Azov seas. For the achievements in the Navy development he was conferred the USSR State Award (1980) and Lenin Award (1985); he was twice Hero of the Soviet Union (1965, 1982). He was awarded many orders and medals of the Soviet

Union as well as medals of other states. In 1990 his name was given to heavy aircraft-carrying cruiser "Naval Admiral of the Soviet Union Gorshkov" (former "Baku).

Gorshkov S.G. (http://flot.com/publications/books/shelf/oceanshield/images/40.jpg)

"Gosgidrografia" – an organization established in 2000 by the Ukrainian Ministry of Transport that became the head enterprise of the Sevastopol, Odessa, Nikolaev and Kerch hydrographic regions. Its divisions carry out investigations in the Black and Azov seas, publish navigation sea maps, atlases, guidelines and reference books on navigation; disseminate operative navigation information, ensure telecommunication in coastal areas; equip the sea coast of Ukraine, nearby marine areas and inland waterways with lighthouses, navigation signs, radio engineering, satellite and other systems designed for orientation or positioning of a ship in the sea. G. has at its disposal 700 units of navigation facilities, including 35 lighthouses with the permanently living personnel, 29 automatic lighthouses, 134 luminous navigation signs, 442 buoys and stakes. In addition 11 radio beacons, several dozen other radio engineering coastal and marine navigation facilities are in operation. In the recent years the modern power-saving technologies (such as light-optical devices based on emitting diodes, solar batteries, wind-driven power installations), digital telecommunication means, control-adjustment stations of global navigational satellite systems GLONASS/GPS that may fix coordinates of a ship in the coastal zone with the accuracy to 2–10 m (Enikalsky lighthouse, Zmeinyi Islands) came into use.

Gotenland – under such name the leaders of fascist Germany planned to join Crimea directly to Reich, made it a place for rest and settlement of the Germans from Southern Tyrol.

Gothia, Crimean Gothia – a territory, kingdom. In this way I. Barbaro in the Middle Ages identified and named the eastern coast of B.S. as far as Chembalo or Balaklava. In this way the area between the Belbek River and the Ai-Todor Cape was called after 3,000 Goths came to live here who in 448 left Crimea for Italy led by the Goth Tsar Theodoric the Great. Beginning from the sixth century Goths were under the Byzantine ruling with the capital Feodoro (Dori, Doros, Mangup) being constructed not by them. In the mid-seventh century Goths were subjugated by Khazars and in 679 Khazars conquered the city of Doros. However, Goths managed to maintain their autonomy and Christianity adopted in the fourth century. In 754 Byzantine sent a bishop to Goths. In the second half of the eighth century Gothia was a vassal of Khazars and later of Byzantine again. The Mongols-Tartars invaded the Crimea and left Goths only Mangup. In 1380 G. were conquered by Genoas.

Gothian Captaincy ("Campania") – a military-state formation in Southern Crimea controlled in the twelfth–fifteenth centuries by the Republic of Genoa. From 1357 its western borders incorporated Balaklava (Cembalo) and the eastern borders—Feodosia (Caffa). On this territory the following trading posts of Genoa were established: in Foros (Fori), Alupka (Lupiko), Miskhor (Musakhori), Oreanda, Yalta, Nikita (Sikita), Gurzuf (Gorsouime), Partenit, Alushta (Lusta). This territory was governed by captains (warmen) who were in charge of defending the trading posts and sea trade routes to Italy, Middle East, Syria, Egypt and as far as China. In this period the Genoan towers were built in Balaklava. G.C. ceased to exist after capturing in 1475 of this territory by the Ottomans.

Graf Khristofor Khristoforovich – Russian first rank Captain (1801). He finished the Naval Cadet Corps (1764–1768). In 1770–1775 he took part in the campaigns on the Baltic Sea and in 1772 he served in the naval artillery. C. served in Kronshtadt (1775-1780), St.-Petersburg (1780–1783) in artillery units. In 1783 he navigated over the Baltic Sea and next year he was sent to Kazan for receiving recruits and after this was directed to Kherson. In 1785–1787 C. took part in the campaigns on B.S. He was the participant of the Russian-Turkish War of 1787–1791. Under Ochakov he served in the squadron of Rear Admiral N.S. Mordvinov (1878) and commanded bombardment battery No.1. Commanding artillery battery No.4 in Ochakov mouth he distinguished himself in the battles with the Turkish Fleet (1788). In 1789 he was sent to Kiev for depth measurements and description of the Dnieper banks needed for the naval and siege artillery. In June of the same year he was appointed the Captain in the newly established Nikolaev port. He commanded the Dnieper Fleet (1792–1793, 1794–1795), in Nikolaev—the naval coastal command (1796–1798). In 1802–1803 he commanded the ship "Mary Magdalene the Second" of the Black Sea Navy. In October 1803 he was dismissed.

"Grafskaya pristan" (Count's berth) – a monument of architecture and history, grand staircase in Sevastopol. It goes up from the wooden boat wharf. In 1787 preparing for arrival of Catherine II to Sevastopol Count M.I. Voinovich, Commander of the Black Sea Navy and ports, organized construction of a stone staircase that was called "Ekaterininskaya". But this name did not become popular and instead the name "Grafskaya" connected with the title of Voinovich—Count whose boats moored here came into use. One more name "In Honor of IIIrd International" (the official name of a wharf and area till 1941) did not come into wide use, too. In 1843 G.P. was decorated with two marble lions and four Antique figures (installed in niches of a colonnade, two of which survived to the present) of Italian sculpture Ferdinando Pelliccia (1808–1892). Such look G.P. acquired in 1846. The staircase is separated from the square by two rows of Doric columns supporting a cornice with parapet forming the grand entrance. The author of this project was British engineer John Upton who was on the service in Russia. Many historical events are connected with this wharf and numerous memorial plates remind of this. One of the plates has an inscription "In memory of the compatriots who had to leave the Motherland in November 1920"; some other reminds "Here on November 12, 1941 the cruiser "Chervona Ukraina" was drowned in a battle with the enemy". G.P. was twice destroyed and twice restored in its original look. From the wharf a wonderful view is opening on the Sevastopol, Korabelnaya and Yuzhnaya bays.

"Grafskaya pristan" (Photo by Evgenia Kostyanaya)

Gray mullet (*Mugilidae*) – the large family comprising 10 genera and about 100 species. In B.S. golden mullet, black mullet are met rather often, while red-finned mullet—more seldom. Adult G.M. is keeping closer to the shore. It often occurs in bays, lagoons and lower reaches of rivers. They feed mostly on detritus (bottom silt enriched with organic matter) and periphyton (plant and animal growth of substrate) and, to a much less extent, on benthos. The feeding G.M. moves over the sea floor at an angle of 45° and scrapes the top silt layer from it. All mullet species lay eggs in June-August at a water temperature from 16 to 25 °C. Its eggs floating on the water surface may be found very far from the coast, but the fries soon return to the coastal area. Fries may be detected if one looks attentively into the surface layer of water where they disguise excellently. Adult fish grows up in bays, lagoons, including near coasts of the Sea of Azov and in Sivash.

Great Patriotic War (1941–1945) – a term is used in Russia and former republics of the Soviet Union to describe the war period from 22 June 1941 to 9 May 1945 in the many fronts of the eastern campaign of World War II between the Soviet Union and Nazi Germany with its allies. The term *Patriotic War* refers to the Russian resistance of the French invasion of Russia under Napoleon I, which became known as the *Patriotic War of 1812*. Sometimes the war of 1812 was also referred to as *Great Patriotic War*. The phrases *Second Patriotic War* and *Great World Patriotic War* were also used during World War I in Russia. The term *Great Patriotic War* re-appeared in the Soviet newspaper *Pravda* on 23 June 1941, just a day after Nazi Germany started the war with the USSR.

Great Principality of Lithuania – a state existed in the thirteenth–sixteenth centuries on the territory of present-day Lithuania, Belorussia, part of the Ukraine and Russia. Capitals are the cities of Trakai and Vilno. In the fifteenth century it had an access to the Black Sea coast. The Karaims who lived in Crimea were on the service of the Lithuanian Princes.

Greater Caucasus – the mountain system between the Black Sea and the Caspian Sea; in its axial part, lids the Greater Caucasus or Water-Divide Mountain Range, whose northern slopes belong to Russia (Krasnodar Territory, republics: Adygeya, Karachaevo-Cherkessia, Kabardino-Balkaria, Northern Ossetia, Ingushetia, Chechnya, Daghestan), and the southern—to Georgia and Azerbaijan. The name "Caucasus" is referred to by the ancient Greek authors Eschil and Herodot, fifth century B.C. as "*Kaukasos*", in Russian, the earliest mention is quoted in"Tale of Bygone Years" (or the *Primary Chronicle* which is a history of Kievan Rus from about 850 to 1110, originally compiled in Kiev about 1113 by monk Nestor). In twelfth century G.C. was mentioned as "the Caucasus Mountains", meaning that over the 25 centuries since the first mention the name essentially has not changed. The name of "Caucasus" as generalizing the whole of the mountain system was unknown to the aborigines of the Caucasus who spoke different languages; neither the neighboring oriental peoples knew it. It is obvious that originally the name referred to Elbrus Mountain only, meaning "Snow-white Mountain", and it was not until sometime passed that it began to be applied to the entire mountain system. The attribute "Greater" opposes this mountain system to the Minor Caucasus mountains lying south of G.C., in Transcaucasia.

Greater Sochi – a city on the Russian coast of the eastern Black Sea which includes, besides Sochi City proper, health-resort settlements (from the north-west to the south-east): Magri, Makopse, Ashe, Lazarevskoe, Golovinka, Loo, Dagomys, Mamaika, Matsesta, Khosta, Kudepsta, Adler. G.S. stretches 105 km from Shepsi River to the border with Abkhazia along Psou River. Population— 500,000 people. Sochi is almost unique among larger Russian cities as having some aspects of a subtropical resort. The scenic Caucasian Mountains, pebble and sand beaches, warm sea, subtropical vegetation, numerous parks and monuments attract nuberous vacation-goers and tourists. About two million people visit Greater Sochi each summer. Sochi will host the XXII Olympic Winter Games and XI Paralympic Winter Games in February 2014, as well as the Russian Formula 1 Grand Prix from 2014 until at least 2020. It is also one of the host cities for the 2018 FIFA World Cup.

Greater Utrish, Peninsula – state reserve sited 15 km south of Anapa on the Black Sea coast, Krasnodar Territory, Russia. Features unique forests with most rare plants and amazing microclimate. As far as plants are concerned, juniper species (juniper plants) should be noted—tertiary rare species entered in the Russian Red Data Book. The landscapes of Snake Lake are beautiful. The local dolphinarium is one of Utrish's places of interest.

Greater Yalta – special administrative unit in the Crimea, Ukraine. It stretches 70 km along the Black Sea from Mt. Ayu-Dag in the east (from Gurzuf and Krasnokamenka settlements) to Foros settlement near Sarych Cape in the west. Yalta City is the administrative center of G.Y., brings together health resorts, villages, industrial settlements, agricultural associations "Massandra" and the woods of the Yalta Reserve. The baskbone of the economy: health resorts, mass holiday-making, transport, food-processing and wine-making industry as well as extraction and fabrication of building materials. Population—142,000 (2012).

Grebenskiy Cossaks – descendants of the peasants who fled from central provinces of Russia and settled down in the first half of the sixteenth century on Sunzha River, a tributary of Terek, and also descendants of the Cossacks (300 people) who in 1582 came from the Don and settled in the Grebni area (on Aktash River). Later on G.C. made their settlements along the middle and lower reaches of the Terek. G.C. took part in the Crimean campaigns of 1687 and 1689, Khiva campaign of 1717, Russian-Turkish wars of 1787–1791, 1828–1829, 1877–1878, Caucasus War, Russian-Persian War of 1826–1828, suppression of the Polish uprising of 1830–1831 and World War I.

Greek colonies on the Black Sea – were founded on the northern coast of B.S. in the sixth century B.C., in the time of appearance of the first Greek city on the Crimean Peninsula. The upper border of the epoch has not been defined as yet.

Some scientists refer the end of Antiquity to the fourth century, the time of formation of the Byzantine Empire; others—to the time of the Huns invasion and beginning of the epoch of the Great Relocation of Peoples, i.e. in the 370 s; still others believe that the Antiquity period in the Crimean history lasted till the second half of the fourth century when the greater part of the Peninsula was controlled by Byzantine. The first colonies of the Hellenes (in this way the ancient Greeks called themselves) were founded in Crimean as a result of the Great Greek Colonization— settlement of the Greek people from mainland Greece in the basins of the Mediterranean and B.S. The quickly soaring population in Greece not finding enough lands for cultivation was forced to seek new territories beyond the borders of their country. And the new settlements founded beyond Greece were called "colonies". Long before this wide-scale colonization the Greek ships visited the northern coasts of B.S. that they called Pontus Auxine that may be translated as "inhospitable sea". Maybe Hellenes were threatened by a relatively cold climate there and hostility of the local inhabitants—Tauris and Scythians. However, after the first colonies appeared here and lively trade with the local people developed the sea was renamed into Pontus Euxine meaning "hospitable sea".

The Greek settlers moved in two directions: to the north where they opened the mouths of the Istr (Danube), Tiras (Dniester), Hipanis (Southern Bug), Borisphen (Dnieper) rivers and to the east, along the Black Sea coast of Asia Minor and the Caucasian coast. On the western coast of Pont Euxin they founded such cities as Agatopolis, Appolonia, Odess, Kallatia, Tomy, Instria; on the eastern coast— Dioscuriada, Pithius, Fasis; on the southern—Heraclea, Sinopa, Amis; in the Dnieper area, in the mouths of the Dniester and Southern Bug rivers—Nikoniy, Kremniski, Tira, Olvia; in the Crimea—Kerkinitida, Khersonesos, Kalos-Limen, Feodosia; and along the Bosporus Kimmerian—Kerch Strait—Panticapaeum, Mirmskiy, Phanagoria, Nimphey, Hermonassa and many other cities.

The most important role in colonization of the Northern Prichernomorie and Crimea was played by the city of Milet being the major center of handicraft and trade. The attention of the Greek colonists was drawn to the territories along the Kerch Strait where they constructed the city of Panticapaeum that is presently Kerch. The legend says that the land for the city construction was given to the Greeks by the Scythian Tsar. Perhaps, the Scythians were interested in trade development with the Greeks, thus, they did not interfere with organization of colonies there. During the sixth century B.C. such Greek cities as Tiritaka, Kitey, Kimmerik, Mirmekiy and others appeared on the Kerch Peninsula; later on they formed the Bosporus State. Some cities were founded on the opposite bank of the Kerch Strait (Bosporus). The ancient people thought that this strait separated Europe and Asia, thus, the lands on the eastern shore were called "Asian Bosporus". The largest city of Asian Bosporus was Fanagoria founded by the Greeks from Theos and named so in honor of Fanagor, the leader of colonists. In eastern Crimea the colonists from Miletus founded Feodosia. This is the only city in Crimea that retained its Antique name. Soon the inhabitants of the Bosporus cities initiated the so-called "secondary colonization". Now they of their own accord founded numerous rural settlements on the shores of the Bosporus Strait. In the late sixth century in western Crimea the city of Kerkinitida was constructed that is now called Evpatoria. In the southwest on the Heraclean Peninsula the people of Heraclean Pontian (a city on the southern coast of B.S.) and Delos (a city on the island of the same name in the Aegean Sea) founded Chersonesos Taurian. The scientists estimated the "birth date" of Chersonesos as 422–421 B.C. Recent archeological excavations in this historical monument have revealed still more ancient findings.

Greek colonies in the northern part of the Black Sea http://commons.wikimedia.org/wiki/File: Ancient_Greek_Colonies_of_N_Black_Sea-fr.svg

"Greek Project" – the idea of Catherine II about establishment along the western borders of Russia of a Christian state uniting the peoples that had overthrown the Ottoman Empire ruling and being half-dependent on Russia. The strait issue was a part of G.P. However, the principle of addressing this issue was quite different—it assumed the establishment of a political control on the straits, the regime of the straits ensuring security of the Black Sea coasts of Russia.

Greeks – a nation native to Greece, making 95 % of its population. Originally G. were referred to as the Epirus Dorians from Illyria, but G. called themselves Hellenes. They belonged to the Indo-Europeans who approximately in 1900 B.C. came to the south of the Balkan Peninsula, while before this time they occupied the Hungarian Lowland. They migrated by several flows, and the final, the so-called Doric invasion, dated back to approximately 1200 B.C. Three main tribal groups of G. were Aeolians, Dorians and Ionians. On their invasion G. also populated the islands in the Aegean Sea near the western coast of Asia Minor. During colonization of the eighth–sixth centuries B.C. the Greek cities were founded nearly on the whole coast of B.S. and Mediterranean, first of all, in Southern Italy. With the conquests of Alexander the Great G. came to the farthest areas of Asia Minor where with time one the Hellenic states came into being. After the Romans victory over Macedonia in 148 B.C. Greece became a part of the Roman Empire. In the sixth century Greece was invaded by the Slavs, thus, the population of modern Greece includes rather a large Slav segment to which the Albanians and Aromun should be added. Beginning from the fifteenth century the struggle of G. against the Ottoman ruling facilitated consolidation of the Greek people. During the liberation war of 1821–1832 the Ottoman power was overthrown. The Greek language forms a special group of the Indo-European languages. The official language— Kapharephus is close to the ancient Greek language; the speaking language— Dimotika (the so-called new Greek language) comprises some dialects. Greek people are mostly Christians. The total G. population in the world is 14–17 million, of which 10.8 million (2011) live in Greece proper, 1.4–3 million in the USA, 0.7 million in Cyprus, 98,000 in Russia. Numerous Greek Diasporas may be also found in Turkey, Albania, Germany, Egypt, Italy, Bulgaria, Ukraine and Australia. Russian Greeks who lived in Balaklava and other cities and villages of Crimean were relocated in 1944; the Azov Greeks avoided this fate.

Green Cape – a seaside climatic resort in the Khelvachaurskiy District, Republic of Adjaria, Georgia; a railway station 9 km to the northeast of Batumi. It is situated on the Black Sea coast of the Caucasus, in the offspurs of the Meskhet Range of the Lesser Caucasus. The mountain slopes are covered with plantations of tangerines, tea and bamboo. The climate is subtropical humid. The winter is mild and snowless; the mean temperature of January is 4 °C. The summer is very warm; the mean temperature of August is 21 °C. The mean annual precipitations is about 3,000 mm. The number of sunny hours is 1,815 in a year. This resort was founded in 1904. The mild climate in combination with bathing (from May through October) is most suitable for climatic therapy. The Batumi Botanical Garden of the Georgian Academy of Sciences located near G.C. was founded in 1912 by Russian botanist A.N. Krasnov.

Green fish (*Crenilabrus ocellatus*) – fish of the Wrasse (*Labridae*) family. The length of its body is to 12 cm. It has a varying coloring, most often green or gray-green. They feed on small mollusks, worms, shellfish and algae. They have strong teeth capable to break shells of mollusks. This fish has no commercial significance. It prefers to live near cliffs with thickets of cystozeers. In the spawning period some G.F. make nests in stones. They lay to 50,000 eggs. In B.S. 8 species of this fish are found, one of them—petropsaro is put on the Red Book of Ukraine.

Greig Aleksey Samuilovich (1775–1845) – Russian military commander, Admiral (1828), member of the State Council (1833), Honorary Member of the Petersburg Academy of Sciences (1822). He studied the naval science on the British Fleet (1785–1788, 1792–1796) and on ships of the East-India Company (1789–1791). From 1796 he served on the ship "Retvizan", from 1798 he was the commander of this ship and took part in the war of Russia against France (1798–1800). In 1805 he was promoted to the Rear Admiral. During the Russian-Turkish War of 1806–1812 he participated in conquering of the Tenedos Island, in the Afon and Dardanelles battles. During the Second Archipelago Expedition in 1807 he commanded the unit of ships in the squadron of Vice-Admiral D.N. Senyavin. In 1813 he was promoted to the rank of Vice-Admiral. He took out the Russian nationality. In 1816–1833 he was the Commander-in-Chief of the Black Sea Navy and ports, military governor of Nikolaev and Sevastopol. He contributed much to strengthening of the Black Sea Fleet, both as a shipbuilder and as a ship navigator. G. organized equipment of the ships with lightning rods and cable chains; replaced a stone ballast with pig iron, brick cookrooms with iron. The cabins were provided with portholes, the tallow candles were substituted for lamps, in the lanterns glass was used instead of mica. On each deck of a ship G. installed cannons of one bore which simplified the supply of ammunition. G. paid much attention to the navigation-support facilities: preparation of new maps, provision of the ships with the best cartographic devices. He, in fact, created a new hydrographic service on B.S. In the 1820s, in addition to the existing Khersoness Lighthouse, he organized construction of two new ones— Inkerman and Tarkhankut as well as the Vorontsov Lighthouse in Odessa. In Nikolaev he deepened the fairway which made it possible to transfer here the shipbuilding yards from less convenient Kherson.

From early February 1826 G. combined the duties of a Fleet Commander with the work in Petersburg in the new Committee for Naval Education created on December 31, 1825. Having joined this Committee G., being an honest and scrupulous man, submitted a memorandum to the Emperor where he accused the top executives of the Naval Ministry of incompetence and asked to withdrawn the Black Sea Department (meaning Black Sea Fleet) from the Ministry control. Nikolay I satisfied this request and took the Black Sea Fleet under his direct control. In the Committee G. was very active and authored many projects. The Committee managed to conduct the comprehensive reorganization of the Naval Ministry and created the new fleet for the Baltic and B.S. Compared to 1803, the new fleet was more numerous and the armaments of battleships and frigates were much better; the transport ships were taken to form a separate group and for the first time steamships were included into the fleet. The Black Sea Fleet numbered 15 battleships with 120 cannons, 10 battleships with 84 cannons, 10 battleships with 60 cannons, 8 corvettes, 2 bombardment ships, 8 brigantines, 15 schooners and brigantines, 2 yachts, 20 transport ships and 5 steamers; the row ships included: 30 gunboats, 2 bombardments ships, 40 idols, 2 hospital ships and 2 yachts. In August 1827 on G. initiative the Emperor established the Naval Engineer Corps of the Black Sea Fleet. On his order of September 1, 1831 all officers and masters of the fleet should pass the annual check of their professional knowledge, including (for officers) higher mathematics and naval architecture. In addition, the officers should submit every year to the Commission their own project of a ship. Thanks to such reforming efforts and under command of G. the Black Sea Fleet became a strong and smoothly operating organism. The fleet was improved quantitatively and qualitatively. The shipyards that built modern ships, including steamers, were widened and improved. On an order of G. the Black Sea headquarters were formed for the first time. Much attention was focused on training of the staff; long sea voyages and annual maneuvers were conducted; the officers attended the lectures on the theory of shipbuilding, physics, mechanics and others; the hydrographic surveys were initiated on B.S.; the naval library was opened in Sevastopol; the first steamer "Vezuviy" (1820), the Russia's first 60-cannon frigate "Shtandart" and 14-cannon steamer "Meteor" (1825) were built. G. made various improvements in the designs of battleships, in shipbuilding and in organization of the fleet supply. On the basis of his project there were built the B.S. first 120-cannon battleship "Varshava" (it was set afloat in 1833) and the optical communication line. During the Russian-Turkish War of 1828–1829 the fleet commanded by G. successfully operated on the sea communications of the enemy during seizure of Anapa and Varna (1828). In 1833 he prepared and armed the fleet for the Bosporus expedition. As a specialist in astronomy he was elected the Honorary Member of the Petersburg Academy of Sciences. He was awarded many Russian orders.

Greig A.S. (http://img0.liveinternet.ru/images/attach/c/1//48/775/48775429_Greyg.jpg)

Grigoleti – the seaside climatic resort 12 km to the southeast of Poti and 25 km to the west of Lanchkhuti, Georgia. It locates on the Black Sea coast of the Caucasus in the western part of the Kolkhida Lowland. A wide strip of sandy beaches runs along the sea. The sands of the beach contain a high percentage of magnetic iron. The beaches are rimmed by pine forests. The climate here is humid subtropical. The winter is very mild and snowless. The mean temperature of January is 5 °C. The summer is very warm—the mean temperature of August is 23 °C. To 1,750 mm of precipitations fall here annually. The sunshine hours are 1,980 a year. The western and eastern winds are prevailing here. Mild climate makes this area most suitable for treatment of breathing organs and nervous diseases.

Grigorievskiy Liman – locates 20 km eastwards of the Odessa Harbor of B.S., Ukraine. In environmental respects this is one of the ecologically safest limans in the northwest of Prichernomorie. Mussels and other mollusks, shrimps, crabs feel comfortable here. Many bullheads and other fish populate this area. The object of commercial fishing is mussels that may be rather large (the shell length may be to 8–9 cm).

Grigorovich Ivan Konstantinovich (1853–1930) – Admiral, Naval Minister of Russia, Aid-de-Camp General of his Emperor Highness, Russian naval specialist and statesman. He served on various ships of the Baltic Fleet. In 1896–1897 being second Rank Captain G. was appointed Russian naval attaché in Britain where he represented the interests of Russia on the military-diplomatic service. In 1899 he was appointed to command the "Tsesarevich" battleship which construction was completed in France. In the same year G. now 1st Rank Captain sailed on "Tsesarevich" to Port Arthur for amplification of the Pacific squadron. In 1904 he was promoted to Rear Admiral and was appointed the commander of Port Arthur's harbor. After the Russian-Japanese War he was nominated the Chief of staff of the Black Sea Navy and ports of B.S. and from late 1906—the Commander of Libava military port on the Baltic. In 1909 by the Emperor's order he was appointed Deputy Naval Minister and soon he was promoted to the Vice Admiral.

During World War I the Naval Ministry headed by G. organized smooth functioning of industry, logistic support and training of the staff. The correctly chosen direction in the naval policy was proved by the fact that on the eve and in the course of the war 100 % of battleships, 40 % of cruisers and 30 % of destroyers were built and took part in 1941 in the Great Patriotic War. G. activities were marked by all Russian and many foreign orders. He was the member of the State Council (1914). By the resolution of the Interim Government in March 1917 he was dismissed from his office of the Naval Minister. After the 1917 October Revolution in Russia he was a member of the Naval Historical Commission being busy with generalizing the experience of the world war and military actions on the sea. He lectured at the Higher School of Water Transport. Started writing "Memoirs of the Former Naval Minister". In autumn 1924 he went abroad for medical treatment. He lived in the south of France where due to money shortage he had to sell his pictures among which marines prevailed. He died in 1930 in resort town of Menton, France, located at the border with Italy.

Grigorovich I.K. (http://www.myjulia.ru/data/cache/2011/02/28/689666_5306nothumb500.jpg)

Grinevskiy (Grin) Aleksandr Stepanovich (1880–1932) – a well-known Russian writer notable for his romantic novels who published his works under a penname "Grin". He was born in Vyatka (presently Kirov), Russia. He received a home education. G. served as a clerk in the Vyatka Town Council. When he was 16 he left for Odessa where he served a sailor on ROP&T and Voluntary Fleet. He tried many occupations—he was an actor, vagabond, worker in railway workshops, a rafter and a woodcutter. He was a deserter, political prisoner and an exile. His first story "In Italy" G. published in 1906 in "Birzhevye Vedomosti" in St.-Petersburg and signed it "Grin". In 1913 his collected works in three volumes were published. In 1919 he was called to military service, but he escaped. In 1922 he published his first book under Bolsheviks ruling. In 1930 being seriously ill G. went to live in Old Crimea. Here he started writing his new novel "Touch-Me-Not" that was not finished (later the workers of the A.S. Grin Memorial House in Feodosia continued logically the line of events and published it in 1996). His romantic novels, such as "Scarlet Sails", "She Who Runs on the Waves" (1928), and novels "Jessie and Morgiana" (1929), "The Golden Chain", "The Shining World" and "The Road to Nowhere" (1930) expressed his humanistic faith in high moral principles in a man. In 1970 the Grin Museum was opened in Feodosia and in 1971—in Old Crimea.

Grinevskiy (Grin) A.S. (http://grafskaya.com/wp-content/uploads/Grin_Feodosia_1928.jpg)

Grivas – coastal ridges in the floodplains and river deltas. Some of them may be from 0.5 to 4 m high and from 20 to 500 m wide. Practically all settlements in the Kuban delta locate on them.

"Gruzia" – a passenger ship in USSR, one of "krymchaks". It was built in 1928 in Germany on the Krupp shipyard in Kiel. Its displacement was 6,050 tons, length 116 m, width 15.5 m, draft 5.8 m, board height 7.8 m. The traveling speed was about 13 knots. The sailing range was 6,540 miles. In July 1941 (during the Great Patriotic War) "G." was equipped with anti-aircraft guns. Till October 1941 it made several cruises to Odessa. On October 14, 1941 "G." was attacked in the Odessa port by German aircraft. The ship was damaged seriously by direct hitting, but still it managed to reach Sevastopol of its own. During repair "G." was refurbished into a sanitary ship, but still it was equipped with anti-aircraft guns. On June 11, 1942 under protection of patrol ships "G." went from Novorossiysk to Sevastopol. Officially, 708 people for replacement and 526 tons of ammunition (by other sources—4,000 people and 1,300 tons of ammunition) were onboard this vessel. In the evening of June 12, "G." was attacked and the wheel was broken down. Maneuvering by engines on June 13 the vessel entered the harbor and was going to the Minnaya wharf. Just at this time there was one more attack during which a bomb breached the deck and exploded in a ship's hold with ammunition. Due to detonation the ship body was broken into two parts and the ship sank quickly. In 1949 the remains of the ship were lifted and leaving ammunition in its place they were transported to the Kazachia Bay where they were drowned once again. They returned to the problem of "G." in summer 1960 when 9,000 shells were removed from the stern part and cut for metal. The bow of the ship was left intact, was hoisted and taken farther into the sea and drowned there.

GUAM – an abbreviation of the regional union of Georgia, Ukraine, Azerbaijan and Moldavia. In late 1997 in Strasbourg (France) at the European Union Summit the four states formed the GUAM organization to address economic challenges, in particular, implementation of the TRACECA (Transport Corridor Europe-Caucasus-Asia) Project. In late April 1999 in Washington (USA) Uzbekistan signed the accession treaty because the greater part of the Silk Road passed over its territory. From this time the abbreviation turned into "GUUAM" (later on Uzbekistan left this organization). This organization was also established with a view to ensure security in this region through mutual efforts of the GUAM countries. As a result, the joint document of the GUAM countries was elaborated outlining the security model of the twenty-first century. It was offered for consideration at the OSCE Council meeting held in Vienna in 1997. The main purposes of GUAM are territorial integrity, regional cooperation, functioning of transport corridors. In May 2006 the GUAM Summit passed a resolution on changeover of the organization status from regional to international and on alteration of its name. From now on it was the GUAM Organization for Democracy and Economic Development.

Gubonin Petr Ionovich (1825–1894) – Russian outstanding entrepreneur, public figure, honorary citizen (1886). He was born in a family of a serf. In 1858 he was granted freedom and was assigned to the third guild merchant. By this time he possessed a stone quarry near Moscow. His undertaking was engaged in finishing of the embankment in Moscow, delivery of stone for construction of the Saint Isaac's Cathedral in Petersburg. In the late 1860s—first half of the 1870s he actively participated in establishment of joint stock railroad companies. In 1864 he started construction of stone bridges for the public Moscow-Kursk railroad and in 1866 he was awarded a contract on building of the Orel-Vitebsk railroad (it was completed in 1868). He participated, on the concession basis, in construction of Gryaz-Tsaritsyno (1871), Lozovaya-Sevastopol (1875) railroads in Crimea, the Perm-Ekaterinburg (1878) railroad in the Urals. G. was a member of boards of directors of some railroad companies, including Orenburg, Ural and Fastovo railroads. He founded and constructed the Petersburg (together with S.D. Bashmakov) and Moscow railroads. In cooperation with V.A. Kokorev he founded the Volga-Kama Bank (1870), "Northern Society of Insurance and Warehousing with Issuance of Warranties" (1872), the Baku Oil Company (1873). After the death of I.I. Funduklei (1804–1880), the Senator and Governor of Kiev, G. purchased his estate in Gurzuf. This was a significant event because from now on G. unrolled construction the result of which is now known as the Southern Coast of Crimea.

Gubonin P.I. (http://www.brfond.ru/book/education/ill001.jpg)

Gudauta (in the Abkhazian "Gudauta" means "valley of the Gudau River") – a seaside resort on the territory of the Abkhaz Republic. It was already mentioned in 1842 as G. village. Railway station 37 km to the northwest of Sukhumi and 43 km to the southeast of Gagra. It locates on the Black Sea coast of the Caucasus. In the north the resort is surrounded by the Bzyb Mountain Range of the Greater Caucasus. The vegetation in G. is represented mostly by cultural plantings (including the city park) where palms, magnolia, southern coniferous species, citruses and others prevail. The climate here is humid subtropical similar to that in Sukhumi, but it is distinguished by the higher air humidity and greater coolness. The winter is mild and snowless. The mean temperature in January is 5 °C. The temperature drops below 0 °C quite rarely, mostly during spontaneous intrusions of cold air masses. The spring is early with the prevailing warm and sunny weather. The summer is very warm; the mean temperature in August is 23 °C. High temperatures in summer are tolerated easily thanks to seaward and onshore breezes. The autumn is long and warm. Precipitations (1,450 mm a year) occur mostly in late autumn and winter. Mild climate, warm sea (water temperature in summer is 24–25 °C), sandy-gravel beaches, picturesque surroundings have always attracted here many visitors. In the 1890s the intensive construction of country houses was started here. But only after the 1917 October Revolution in Russia the resort capacities of G. were effectively used and construction of health-improvement establishments was initiated. The resort offers the aerohelio and thalasso therapy and a bathing season that lasts from May through November. The resort provides treatment of the cardiovascular, nervous and respiratory diseases. G. has sanatoriums, rest houses and equipped beaches. Tourist camps may be found near G. In Lykhny located 4 km northward of G. one can visit an architectural complex—the ruins of the palace, a bell tower and dome cathedral dated tenth–eleventh centuries with frescoes of the fourteenth century.

Gulripsh – an urban-type settlement, a seaside climatic resort on the Black Sea coast of the Caucasus, 12 km to the south of Sukhumi and 4 km to the northwest of the railroad station Dranda on the territory of the Abkhaz Republic. In the north and northeast the resort is surrounded by the mountain ranges of the Greater Caucasus: Bzyb, Abkhazian and Kodorsky. The vegetation in the G. region includes mostly fruit trees (citruses, vines and others) and decorative plantings. On the resort territory one can find a beautiful park. The climate here is subtropical with less air humidity than in many other resorts of the Black Sea coast of the Caucasus. There are a greater number of sunny days and no strong winds. The winter is very mild and snowless. The mean temperature of January is 6 °C. The summer is warm; the mean temperature of August is 23 °C. Sea and on-shore breezes alleviate the summer heat. Precipitations (1,400 mm a year) occur mostly in late autumn and in winter. The number of sunshine hours is 2,100 a year. The favorable climatic conditions of G. and its good prospects as a health-improvement resort have attracted many visitors here beginning from the nineteenth century. In 1902 the sanatorium for treatment of the people with TB of lungs was constructed here. The main treatment method is climatotherapy (mostly, aeroheliotherapy) that is applied for treatment of the active form of TB of lungs.

Gumista – a seaside climatic resort on the territory of the Sukhumi Region, 10 km to the northwest of Sukhumi, Republic of Abkhazia. It locates on the Black Sea coast of the Caucasus on the bank of Gumista River that flows into B.S. with several arms. The climate is humid subtropical. The winter is very mild and snowless. The mean temperature of January is 6 °C. The summer is very warm; the mean temperature of August is 24 °C. The water temperature in summer may reach 24–26 °C. Precipitations—1,550 mm a year. The number of sunshine hours is 2,120 a year. In general, the climate of G. does not differ significantly from that in Sukhumi. Favorable climatic conditions in combination with sea bathing (from May through November) may be used for development of the climatotherapy of cardiovascular, nervous and respiratory diseases. The resort was founded in 1959. The Gumista nature reserve locates to the northeast of G.

Gurnard (*Trigla lucerna*) – fish of the *Triglidae* family. It has an elongated body covered with small scale. The pectoral fins are long having three finger-like rays that the fish uses for crawling over the sea floor. The body is usually brown-red or rose-yellow. The pectoral fins are most often violet. Its length may reach 75 cm. Its usual habitat is the Atlantic coast of Europe, the northwestern and southern coasts of Africa and the Mediterranean Sea. It may live everywhere in B.S., but it is found quite rarely.

Gurzuf – an urban-type settlement, seaside climatic resort, 16 km to the northeast of Yalta. It is a part of Greater Yalta situated on the Southern Coast of Crimea,

Ukraine. It is located in a very picturesque terrain well protected from the northern and eastern winds. Historian Procopius (500–565) from the Byzantine mentioned construction of fortress in "Gorzuvita" by Emperor Justinian I populated that time by the Goths whose main occupation was farming and winemaking. Arab geographer ash-Sharif al-Idrisi (1100–1165/1166) mentioned the flourishing seaside city of "Garuru" (twelfth century). On the medieval Italian maps G. was called Grasui, Gorsanium. In the fourteenth century G. was ruled by the Genoans. In his narrative "The Journey Beyond Three Seas" A. Nikitin (fifteenth century) wrote the following: "we reached Balykas (Balaklava) and from there went to Gorzof (Gurzuf)". In the fifteenth century G. was ruled first by the Turks, then by the Greeks and after this by the Tatars. After Crimea was incorporated into Russia the G. lands were owned: in the early nineteenth century—by Duc de Richelieu (1585–1642), the founder of Odessa, then by Governor General of Novorossia M.S. Vorontsov who later on sold his estate in Gurzuf to Kiev Governor I.I. Funduklei after whose death the estate was purchased by P.I. Gubonin who built several luxurious hotels, bought country houses that were rented.

G. locates in the valley of the mountain river Avunda protected from winds by a semicircle of mountains: in the east—Ayu-Dag (Bear Mountain, 572 m high), in the north—Main Range of the Crimean Mountains (the highest peak—Roman-Kosh, 1545 m) and in the west—Nikitskaya Yaila. On the territory of this resort you can find the ancient park with coniferous and deciduous trees and shrubs. There is also a small-pebble beach running for 1 km. The climate is subtropical Mediterranean here. The winter is very mild; the mean temperature of January is 4 °C. The summer is warm, dry; the mean temperature of August is 23 °C. The autumn is also warm and long. Precipitations—550 mm a year. The number of sunshine hours is 2,200 a year. The climatic conditions here are favorable for climatotherapy and thalassotherapy and preventive treatment of respiratory diseases, functional disorders of the nervous and cardiovascular systems, metabolic disorders. The bathing season lasts from June through October. There are many sanatoriums in G. and its surroundings. In the park the house of N.N. Raevskiy, the hero of the 1812 War with Napoleon, is preserved. In this house in 1820 lived famous Russian poet A.S. Pushkin. Among the architectural masterpieces of G. is the former villa of famous Russian artist—K. A. Korovin. Not far from G. there is the Children Center "Artek".

H

Harbor oscillations on the Black Sea – the not adequately studied phenomenon observed in bays, harbors and ports. This phenomenon is observed when in some B.S. ports the vessels at berths or anchored start spontaneous motion induced by the action of some unknown force. As a result of H.O. the vessels lean on jetty walls or neighboring vessels or, on the contrary, are pressed away from the jetty walls that may even lead to breaking of mooring lines. This may happen during high waves and at calm sea alike. H.O. may last for several days. Its return is not regular and does not reveal seasonal or year-to-year variability. In some years the oscillations did not happen at all. Supposedly, such oscillations are induced by resonance of the water own oscillations in semi-closed water areas at penetration there of long-period swell waves (120–200 m) and resonance of the water oscillations in a port area with oscillations of the moored vessel. H.O. is observed most often in the ports of Tuapse (304 cases in 1964–2000, about 20 cases a year), Poti and Batumi (5.7 cases on the average a year in each port).

Hazelnut – a nut variety obtained by cross-breeding of common and Pontiac filbert. H. was bred in Turkey, where it is cultivated around the city of Giresun. H. Is commonly grown on the Black Sea coast.

Heraclea Pontica – see Eregli.

Heraclean Peninsula – such name is given to the triangle land site projecting into the sea, restricted by the line Khersones Cape—mouth of the Chernaya River—entrance into the Balaklava Bay. It is situated on the extreme southwest of the Crimean Peninsula. In the northwest and southwest this land jut is washed by the B.S. waters, the Sevastopol Bay, and in the east it is separated by the Balaklava Bay and the valleys of the Balaklava and Chernaya rivers. In the ancient times G.P. was a part of the Tauric Chersonese city-state founded by the Greeks who came here from Heraclea Pontica in Asia Minor, thus, is the name of this peninsula. In the ancient times another name—Trakhean Peninsula was in used. Perhaps it appeared because of the rough surface of a stony plain (from the Greek "$\tau\rho\alpha\chi\acute{\upsilon}\tau\eta\tau\alpha$" meaning roughness). Some researchers believe that in the ancient time H.P. was surrounded on the eastern side by a wall (rampart) and a ditch along the line mouth of Chernaya River—Kady-Koi village—Balaklava for protection of the Chersonese people from "barbarians" as the local population was called that time. In the nineteenth century the name "Iraklian" was also met.

Herodotus (Herodot) (c. 484 B.C.–c. 425 B.C.) (Greek "Herodotos" means "from Halicarnassus") – the ancient Greek historian, one of the first scholars-travelers. He took long journeys (between 455 and 444 B.C.) and studied well the littoral areas with the nearby islands in Asia Minor. He went to Egypt, Cirenu, Syrian-Phoenician coast with Cyprus, Pontus and also Hellespont, Thrace and Macedonia. Some time he lived in Athens (mid-440s B.C.—friendship with Pericles and Sophocles, interested in sophistic and natural sciences, traveling over Greek cities). In 444–443 B.C. he took part in founding the Athenian colony of Thurii in Southern Italy. During his life H. wrote the book "Histories Apodexis" or otherwise referred to as the famous "Histories" in nine books where he traced the history of relations between ancient eastern despots (Asia) and Greek slavery states (Europe) and where he included the narrative about the Greek-Persian wars. He was the first to describe the life and customs of Scythians. His works earned him the title of "the Father of History" first conferred by Cicero (106 B.C.–43 B.C.). But he could be also deservedly called the Father of Geography. His "Histories" contained many geographical descriptions that proved his wide knowledge in this science. H. described the geographical discoveries of the Greeks in Pontus Euxine, collected the vast geographical data. H. visited the Northern Prichernomorie (the Scythian lands) where he went from the city of Apolonia (Sozopol). He was the first traveler who provided more or less concrete data about B.S., although his descriptions contained naïve and even fantastic information. According to H., Pontus was the most remarkable sea. Its length was 11,100 stadia (one Greek stadia was equal to 184 m), the greatest width—3,300 stadia, while the width of the Bosporus, to which he referred as the sea mouth, was only four stadia. H. made these measurements having estimated the time for which a ship (with the known speed) passed the distance from one shore to the other. His width of the sea was close to the actual one, but the length was overestimated greatly. Behind the Pontus there was the Meotida (Maeotis) Lake (the Sea of Azov) that was considered the Mother of Pontus. According to H., the climate on the Scythian coast was severe. In winter, lasting for 8 months, the Pontus, Meotida and the whole Cimmerian Bosporus (Kerch Strait) were frozen. H. descriptions of rivers were correct, in general. The Istr (Danube) was the greatest of all known to him rivers that received the greatest number of tributaries. It was full-flowing in summer and in winter. It flowed into a marine bay through five channels. The Borisphen (Dnieper) was the second large and water-abundant river after Danube.

Herodotus (http://ademic.ru/pictures/wiki/files/72/Herodotus_Massimo_Inv124478.jpg)

Herring (*Clupeidae*) – the family of bony fish of the herring order. There are known 50 genera including 170 species. They are mostly marine fish and only a small part of them are freshwater or migratory fish. They are found mostly in the tropical, seldom in subtropical and temperate zones. They prevail in the Northern Hemisphere, but some species are found in the Arctic seas. H. is native to the Caspian, Black, Azov, Baltic, Barents and White seas. In smaller quantities they are found in other Northern and Far-Eastern seas. H. prefers living in the surface waters (pelagic) and only at times coming down to greater depths. They form large schools. They feed mostly on plankton. H. has an important commercial value.

History of oceanographic investigations of the Black Sea – in view of the favorable geographical position of B.S. this sea had been known from very old times. B.S. was mentioned in the narrations of Herodotus that survived to the present (fifth century B.C.) and in ancient Greek pilots—"periples" (fourth century B.C.). In the fifth century B.C. Herodotus described the northern shores of the sea. In the fourth century B.C. the first navigation aids—periples were prepared, while in the third century B.C.—the first map of B.S. appeared. In the ancient times Greek seafarers called B.S. Pontus Auxinus (Auxine)—the inhospitable sea, but after they got familiar with this sea and established their colonies on ots coasts they renamed the sea into Pontus Euxinus (Euxine)—a hospitable sea. In the ninth century the Russians opened the route from the Baltic Sea to the Black Sea ("from the Varangians to the Greeks") and from this time on in some sources this sea was called Russian. On some maps this name was in use till the thirteenth to fifteenth centuries. In the late twelfth century to early thirteenth century the Ottoman Empire got established on the coasts of the Pontus.

The first Russian maps and descriptions of the Black and Azov seas appeared in the sixteenth century. Later on more detailed study of the basin was initiated in the times of Peter I. Already in 1696 the hydrographic works in the sea (voyage of Russian Admiral K.I. Kryuis from Azov to Constantinople) were started. In the history of hydrographic and hydrometeorological works in B.S. we can conventionally distinguish six basic periods (stages): till the 1770s; 1770–1870; 1870–1917; 1917–1941; 1946–1991 and 1991 to the present. The period till the 1770s was characterized by the development of the general interest to the Azov-Black Sea Basin. The first hydrographic investigations of the coasts provided enough material to prepare the map of B. and A. seas that was published in 1701–1702. In the next century as a result of numerous hydrographic works conducted by Russian explorers A.N. Senyavin, P.V. Pustoshin, I.M. Bersenev, I. Billings, I.M. Budischev, F.F. Bellinsgauzen, E.P. Manganari, G.I. Butakov and others the information about B.S. was much refreshed and new atlases and maps were prepared and published, including abroad, that presented more precise delineation of the coastline, the sea depths over 500 m and bottom sediments. On the basis of these works I.M. Budischev in 1808 prepared the manuscript "Pilot or Guidebook over the Black and Azov Seas".

The investigations went on till the nineteenth century. During the expedition of E.P. Manganari (1825–1836) the detailed mapping of the coasts of the Black and Azov seas was done and in 1842 "The Atlas of the Black Sea" and in 1851 the first sea pilot were published. In the nineteenth century the meteorological observations were started: in 1801 there was opened the basin's first hydrometeorological stations (HMS) in Nikolaev, in 1808—in Kherson, in 1821—in Odessa and in 1824—in Sevastopol. The meteorological observations were interrupted in some years for various reasons and then they were resumed. The earliest reliable information about measurements of deepwater temperatures and water density dated back to 1868 when the crew of corvette "Lvitsa" commanded by V. Lapshin conducted measurements between Feodosia and Sukhumi and found that in B.S. the water density tended to grow with depth. At the end of the nineteenth century the Hydrographic Department had already several dozens of hydrometeorological stations (HMS), including those located on the Turkish coast (Sinop, Trabzon).

The new stage in the detailed and regular investigations of the basin was opened after organization in 1871 of the Hydrographic Expedition to B. and A. seas headed by Russian Captain 1st rank V.I. Zarudniy that in 1887 was transformed and from this time on this Expedition conducted only purposeful sea surveys. In this Hydrographic Expedition a special hydrological team was established headed by F.F. Vrangel that conducted serious oceanographic investigations in the northwestern part of the sea, in the Kerch Strait and near the coasts of Crimea and Caucasus. A significant event in the study of the B.S. life was organization in 1871 in Odessa of the first marine biological station that some years later was moved to Sevastopol. Its first director was Academician A.O. Kovalevskiy. At present this is the well-known Institute of Biology of the Southern Seas of the National Academy of Sciences of Ukraine, that was named after Kovalevskiy. After organization of the

Sevastopol Biological Station the studies of the B.S. fauna and flora became regular. Thus, in the twentieth century the biological investigations headed by S.A. Zernov covered a vast coastal zone. In 1909 in the northwestern part of the sea there was found a large accumulation of the red algae phyllophora well-known in publications as "Zernov's phyllophora field". In 1881–1882 outstanding Russian Admiral S.O. Makarov who was also an oceanographer, onboard of the Russian vessel "Taman" based in Istanbul conducted the detailed hydrologic surveys in the Bosporus Strait, including measurements of the water temperature, salinity, velocity and directions of currents. After interpretation of the results he found that the Bosporus had two differently-oriented currents: the upper layer current that went from B.S. into the Marmara Sea and the lower layer current that went in the opposite direction. Thus, he discovered the most amazing natural phenomenon of B.S. explaining the peculiarity of its hydrology.

In those times the first notions on the surface currents in B.S. were only shaping. It was thought that the flows of the rivers running into the sea and the wind were the main factors responsible for water transfer from east to west in the northern part of the sea and from west to east in the south of the sea. Late nineteenth to early twentieth centuries witnessed further accumulation of knowledge on hydrology of B. and A. seas. Special voyages were organized for study of the seas hydrology. Regular observations were conducted on the floating lights. New hydrometeorological stations were opened. The most important event for extension of the knowledge on general hydrology of B.S. was the first multipurpose oceanographic Black Sea expedition on gunboats "Chernomorets" (1890), "Donets" and "Zaporozhets" (1891) conducted in 1890–1891. The expedition was led by I.B. Spindler and its team also included nautical surveyor F.F. Vrangel, geologist N.I. Andrusov, chemist A.A. Lebedintsev and biologist A.A. Ostroumov. During this expedition about 200 deepwater oceanographic stations were made and the cold intermediate sea layer with a temperature lower than 8 °C was discovered. Quite unexpectedly it was found that the whole water mass deeper than 200 m contained hydrogen sulfide. Its presence was determined by the smell of deepwater samples. In 1891 the water samples were taken with a special bathometer having the gilded inside surface. The deep sea layers were devoid of any live organisms. At the entrance into the Bosporus it was found that the water coming here with the deep current from the Marmara Sea had the salinity about 34‰. This was the final proof that the deep waters of B.S. were a mix of the local and Marmara (Mediterranean) waters. Therefore, the result of this expedition was the major oceanographic discoveries.

A new stage in the Black Sea Basin investigations started after the 1917 October Revolution in Russia. In 1921 Soviet Leader V.I. Lenin signed the Decree of the People's Commissars Council of Russian Federation "On Organization of the Meteorological Service in RSFSR" that laid the basis for the hydrometeorological service of the Soviet Union. By this time in the Azov-Black Sea basin the whole network of hydrometeorological stations was in operation that conducted standard coastal observations. Investigations in the open sea were conducted by some fishing and military ships, most often as some incidental task while navigating from one port to the other. After the end of the Civil War in Russia the basin researches acquired a regular nature. In 1922 the fishery research expedition in the Azov-Black Sea basin was organized. It was headed by Russian Honorary Academician N.M. Knipovich, one of the outstanding explorers of the sea. This expedition functioned successfully till 1926. During this time it studied in detail the physical and chemical regime of sea influencing the formation of the fishery base in the basin and evaluated its fish resources. The collected data helped to verify the pattern of currents and distribution of hydrogen sulfide and oxygen in B.S. Approximately in the same time (1924–1928) on the initiative of Professor Yu.M. Shokalskiy the Black Sea oceanographic expedition was organized that became a major event in the history of B.S. exploration. During the 4 years of its work more than 1,000 deepwater stations were taken. The expedition team conducted wide-scale observations over the water temperature and salinity, the content of oxygen, phosphorus, nitrogen and other oceanographic and hydrochemical parameters. The participants of this expedition were the first in the world oceanographic practice who succeeded to take a ground sample 1.5 m high. The data on biological and hydrological conditions and also on the vertical and horizontal water exchange in the sea obtained by this expedition were very important for further development of fishery and navigation, and for better understanding of peculiarities of the water structure and physical processes in B.S.

The primary oceanographic investigations of B.S. were completed at large by the expedition headed by Yu.M. Shokalskiy and later by V.A. Snezhinskiy that conducted surveys from 1928 to 1935. The main participants of this expedition were from Sevastopol Marine Observatory and Sevastopol Biological Station. During the expedition there were made 53 cruises in which about 1,600 oceanographic stations were made, more than 2,000 biological and geological samples were taken. The materials of this expedition were used by A.D. Arkhangelskiy and N.M. Strakhov in preparation of their fundamental work on the geological structure and development history of the Black Sea Depression. The detailed investigation of the vertical hydrological structure of waters enabled a conclusion that mixing of the upper oxygen-containing and lower hydrogen-sulfide-containing water layers did really occur, although rather slowly. Chemical determinations helped to identify that hydrogen sulfide contained in deepwater layers was formed as a result of reduction of sea water sulfates by carbon of organic compounds as a result of bacterial activity. The biological surveys revealed seasonal variations of plankton and depth-wise distribution of benthos. The results of extensive expedition surveys conducted in the 1920s were used in preparation of the first generalizing work on hydrology of B.S. by N.M. Knipovich (1932). This work described a pattern of surface currents that even at present did not loose its currency (large-scale cyclonic gyres in the eastern and western parts of the sea—the so-called "Knipovich goggles").

In the late 1940s the classical works of well-known Soviet biologist V.A. Vodyanitskiy were published where, using the results of biological and hydrological observations, he validated the logical model of the vertical structure and general circulation of waters in B.S. It assumed that the water mass of the sea made something whole subject to horizontal and vertical motions from the surface to the floor. But V.A. Vodyanitskiy also admitted that the exchange between the surface and deepwater masses went slowly and evaluations of its rate were only roughly estimated.

In the Soviet Union the researches of B.S. were conducted mainly by the institutes and organizations located in Sevastopol that had at their disposal well-equipped research vessels. These were, first of all, Marine Hydrophysical Institute (R/V "Mikhail Lomonosov", R/V "Akademik Vernadskiy", R/V "Professor Kolesnikov") and Institute of Biology of the Southern Seas (IBSS) (R/V "Akademik Kovalevskiy", R/V "Professor Vodyanitskiy") of the USSR Academy of Sciences. In addition, Sevastopol was a base for research vessels of the Sevastopol Branch of the State Oceanographic Institute and Hydrographic Service of the Black Sea Fleet. Sevastopol was also the home port for research vessels of the M.V. Lomonosov Moscow State University (R/V "Akademik Petrovskiy" and R/V "Moskovskiy Universitet") which summer research cruises were combined with the practical training course of students. In the Odessa Branch of IBSS the R/V "Miklukho Maklai" had been in operation for three decades (1961–1989). It was used for detailed environmental studies in the northwestern part of the sea, including the period of environment degradation that was observed from the early 1970s.

The 1950s are characterized by more detailed study of the sea hydrology. As a result of joint efforts of the UUSR Hydrometeorological Service, hydrographic institutions, the research institutes of the USSR Academy of Sciences, Ministry of Fishery and other departments the Black Sea was covered by a dense network of synchronous multipurpose oceanographic surveys. During a year the number of oceanographic stations reached 1,000. Near Gelendzhik the Southern Branch of P.P. Shirshov Institute of Oceanology (SIO) of the USSR Academy of Sciences performed actively. In the late 1950s it organized expeditions on R/V "Akademik S. Vavilov", R/V "Akademik Shirshov" for study, among other tasks, the shelf and geological structure of the Black Sea Depression. As a result of these efforts there were prepared more accurate bathymetric, geophysical and geomorphological maps of the sea. Novorossiysk was the base of the larger research vessels of SIO, such as R/V "Vityaz" and R/V "Professor Shtokman". In the 1950s–1980s many of the mentioned expeditions and expeditions on other research vessels conducted detailed and diverse investigations of B.S. The expeditions were organized by the various institutions and organizations of the Soviet Union and later Russia, Ukraine and other Black Sea countries.

The international cooperation in this field was developing. In 1957–1959 a perceptible contribution into accumulation of on-site data was made by interdepartmental expeditions under the program of the International Geophysical Year. Such large organizations as the USSR Hydrometeorological Service, USSR Academy of Sciences and many others took part in these activities. For the first time the observations were conducted on testing grounds that enabled to trace the mesoscale variability of thermohaline structure and currents. From 1961 the USSR Hydrometeorological Service undertook systematic seasonal monitoring on several "fundamental" (the so-called "century") sections. There were six such sections in

B.S., they are Bolshoy Fontan Cape—Tarkhankut Cape; Tarkhankut Cape—Zmeinyi Island; Khersones Cape—Bosporus Strait; Sarych Cape—Inebolu Cape; Kadosh Cape—Unje; Yalta—Batumi. On each section from 9 to 20 oceanographic stations were made. The obtained monitoring results made it possible to develop understanding of the long-term and year-to-year variability of the sea regime.

In these years the Basin Hydrometeorological Service using the results of long-term observations published hydrometeorological handbooks that were very important for economy development, such as "Climate and Hydrological Atlas of the Black and Azov Seas" (1956), "Atlas of Ice of the Black and Azov Seas" (1962), "Catalog of Level Observations on the Black and Azov Seas" (1965), "Atlas of Waves and Winds on the Black Sea" (1969), "Reference Book on Hydrological Regime of the Seas and River Mouths" (1970) that in four volumes generalized the basic hydrological characteristics of the coastal zone of B. and A. seas; "Handbook on the Black Sea Climate" (1974) and others. The history of oceanographic studies in B.S. and the data accumulated by this time were included into the works of A.K. Leonov "Regional Oceanography" (1960) and in monograph of D.M. Filippov "Circulation and Structure of the Black Sea Waters" (1968).

In 1976–1978 the joint program of comprehensive oceanological studies of B.S. ("SKOICh") was implemented with participation of the basic Black Sea organizations (Marine Hydrophysical Institute (MHI), IBSS, Hydrographic Service of the Black Sea Fleet). This program included assessment of B.S. hydrology variations with regard to anthropogenic impacts. The obtained results were pooled in collections of articles. Thanks to the materials accumulated by the 1980s it became possible to change over from characteristics of the average perennial regime of B.S. to analysis of its dynamics. This was reflected in monograph of A.S. Blatov, N.P. Bultakov et al. "Variability of Hydrophysical Fields of the Black Sea" (1984) where assessments of a wide range of variability from short- to long-time were conducted for the first time. The research results were used in preparation of collective monographs of MHI under editorship of B.A. Nelepo "Integrated Research of the Black Sea" (1979) and "Comprehensive Oceanographic Studies of the Black Sea" (1980).

In Bulgaria the B.S. studies were concentrated in Varna where Institute of Fishery Resources founded in 1954 (Marine Biostation was established in 1932) and Institute of Oceanology of the Bulgarian Academy of Sciences (BAS) founded in 1973 are situated. The Bulgarian scientists focus their attention mostly on oceanography of the western B.S., including such issues as effect of the Danube runoff on the regime of this water area, oceanography of the near-Bosporus region. The international experiment "Kamchia-77, -78, -79" deserves special emphasis. It was conducted in the late 1970s under the program of the CMEA member-countries (Council of Mutual Economic Assistance) at the coastal experimental base of Institute of Oceanology BAS near Varna. Specialists from Bulgaria, Soviet Union, GDR, Poland and Romania took part in these studies. The purpose of this experiment was to study the processes of interaction and exchange in the atmosphere—hydrosphere—lithosphere system in the coastal zone of the sea. The experimental results were presented in thematic publications. A team of

Bulgarian scientists prepared also the review "Black Sea" describing principal specific features of this water body (1978, Russian version—1983).

In Romania the B.S. oceanography is studied by the Marine Research Institute located in Constanta. The shelf zone is in the focus of attention. For many years the Romanian and Ukrainian specialists have been conducting research in the Danube Delta being in the jurisdiction of both countries. In 1979–1999 the environmental monitoring of the Danube Delta was conducted.

A vast coastal zone along the southern Anatolian coast of the sea and also the water exchange in the Bosporus Strait are investigated by the Turkish oceanographers. They have at their disposal the R/V "K. Piri Reis" of Institute of Marine Sciences and Technology in Izmir and other vessels. The Turkish specialists are participants of the research cruises conducted in B.S. by Western countries and international organizations. The scientists from the USA, in particular, the Woods Hole Oceanographic Institution, draw very serious attention to the study of the B.S. nature. Still in 1969 a 7-week cruise on R/V "Atlantis II" was carried out. Defining the age of the upper sedimentary layers in the sea trough it was found that the modern sedimentation processes go on here much more intensively than on the floor of the Atlantic Ocean. In 1975 a shorter voyage of R/V "Chain" was organized during which the chronology of geological sedimentation on the basis of the ground samples was studied. In summer 1975 the special drilling vessel "Glomar Challenger" worked in B.S. It made three drillings to verify the time of the Black Sea Depression formation. In 1988 the international geological expedition on R/V "Knorr" was conducted that included five cruises. Oceanographic specialists from the USA, Turkey and European countries worked on this vessel.

From the 1980s P.P. Shirshov Institute of Oceanology (SIO) of the USSR Academy of Sciences started intensive studies of the present condition and variations in the pelagic ecosystem of B.S. under general guidance of Academician M.E. Vinogradov. Within the framework of this project the large expeditions on R/V "Vityaz" (March-April 1988) and on R/V "Dmitriy Mendeleev" (July-September 1989) were conducted. This time the attention was focused on the effect on the B.S. ecosystem of the invader—comb jellyfish *Mnemiopsis Leidyi* that propagated extensively in the sea in the late 1980s. In 1991 investigations of the B.S. ecosystem on R/V "Vityaz" were continued in the wintertime, the worse studied season of a year (February-April). The results of this program were included into the book "The Variability of the Black Sea Ecosystem" (1991). At the same time the USSR Committee on Hydrometeorology published within the framework of the Project "Seas of the USSR" the results of generalization and assessment of a broad spectrum of parameters of the B.S. regime for a long period (vol. 1—1991; vol. 2—1992; vol. 3—1996).

The present stage in the B.S. explorations is distinguished by application of the modern probes and sensors, autonomous systems, remote (satellite) methods combined with in-situ observations. The satellite data (optical, infrared, altimetric, radar and satellite tracking of drifters) are used largely by scientists in SIO RAS, MHI NASU, Russian Space Research Institute and Geophysical Center of RAS. These observations make it possible to follow the processes of large-, meso- and small-scale circulation and vortex formation, dynamics and structure of the fronts and upwelling zones, regions of higher biological productivity.

Very important for study of the condition and variability of the Black Sea ecosystems is implementation of international and regional agreements and programs supported by target-oriented financing. Thus, in April 1991 the Co-operative Marine Science Programme for the Black Sea (CoMSBlack) designed for 5 years was adopted. In 1992 the Convention on the Protection of the Black Sea Against Pollution was signed and ratified by all countries of this region. In 1993 the Global Environment Facility (GEF) sponsored the Black Sea Environment Programme (BSEP) designed for 3 years. It was targeted to rallying the forces of scientists and specialists from the Black Sea countries around the key idea of the program— identification of the present state and variations in the B.S. ecosystem, development of actions on conservation and development of its biodiversity. The International Atomic Energy Agency (IAEA) also made its contribution into environmental programs on B.S. The littoral Black Sea states acknowledged the monitoring of the radioactive sea pollution as the key issue. The target of the IAEA program is to study the presence of radionuclides in B.S. and to elucidate on the tendency in radioactive contamination. NATO is not aloof of the issues of B.S. investigation, too. Within the framework of the Program "Science for Peace" this organization assigned funds in 1993 for support of the Project "Ecosystem Modeling as a Management Tool for the Black Sea" designed for 5 years. Apart from the Black Sea countries the USA was also among its participants. One of the most important results of this activity was formation of an integrated database of hydrological, hydrochemical and hydrobiological monitoring in B.S. On the request of the UNESCO Intergovernmental Oceanographic Commission, MHI NASU prepared and published in the series "UNESCO Reports in Marine science" two books "Artificial Radioactivity of the Black Sea" and "Hydrochemistry and Dynamics of the Hydrogen-Sulphide Zone in the Black Sea".

The results of long-time oceanographic observations in B.S. were taken together in several databases that included about 100,000 hydrological stations. The main databases of oceanographic information on B.S. may be found in Russian Research Institute for Hydrometeorological Information in Obninsk, MHI NASU in Sevastopol, Russian State Oceanographic Institute and Chair of Oceanology of the Moscow State University. Numerous publications on oceanography of B.S. including several thousands of sources are systematized in reviews, monographs and specific bibliographies, in particular "Black Sea Bibliography (1974–1994)".

Due to development of the shelf resources the recent years witnessed activation of explorations in the coastal zone: conducting of local dynamic experiments in the shelf zone of the Southern Coast of Crimea (LODEX) with synchronous meso-scale CTD-surveys and installation of automatic buoy stations during a year; organization of marine testing areas and hydrographic and hydrometeorological posts in the areas of intensive navigation (Experimental Division of MHI NASU in settlement Katsiveli, port Kerch and others). For three decades the seasonal observations have been conducted on the 20-mile profile from the Danube Delta along 45°20′N at nine stations; individual meso-scale surveys are conducted with installation of automatic buoy stations near the Zmeinyi Island; a 10-mile permanent profile and coastal meso-scale surveys in the Southern Coast of Crimea (SCC) are conducted. About 200 stations are made every year. For nearly three decades a stationary oceanographic platform of MHI NASU has been operating in the sea near Katsiveli at a depth of 28 m. In the semiautomatic mode it performs the full range of meteorological observations, measurements of the sea level, waves, vertical distribution of water temperature and salinity, and current at different depths.

From 1999 to the present SIO RAS yearly carries out wide-scale, regular and integrated investigations of B.S. that enable tracking the degradation and partial restoration of its ecosystem in the presence of invaders (*Mnemiopsis Leidyi* and *Beroe Ovata*). The detailed study of the dynamics of the population of *Mnemiopsis* intruded into the Black Sea ecosystem was conducted on the basis of regular on-site observations in the eastern part of the basin. It was found that invader *Beroe* "got established" in the ecosystem and in 2000 it maintained its domination on the peak of seasonal development from late September to late November, which provided the basis for long-time invasion. Less presssure of *Mmemiopsis* (that *Beroe* eats) on the fodder zooplankton and ichthyoplankton will result in the increase of the plankton-eating fish stock. For the first time the effect of meso-scale circulation in B.S. on the composition, distribution and productivity of plankton-eating communities was demonstrated. The obtained materials provided a basis for prediction and adequate analysis of modern anthropogenic trends in the basin.

The Black Sea drifter experiments were conducted in 1999–2003 by SIO RAS and MHI NASU. The trajectories of drifters confirmed the existence of intensive meso-scale vortices both in coastal and deepwater areas of B.S. It was found that intensification of the Rim Current due to the wind effect was accompanied by a decrease of the meso-scale vortex dynamics and related horizontal (cross-shelf) water exchange, while the Rim Current relaxation results in the opposite effect. Such conclusion became possible after joint analysis of characteristics of the wind field over the Azov-Black Sea region, satellite IR-images, hydrographic observations from ships and the results of laboratory modeling of physical mechanisms of variability of macro- and meso-scale water dynamics in B.S.

By 2000 the epoch of mostly episodic observations over individual natural processes in B.S. came to an end. It was replaced by regular (satellite, ship, and autonomous) monitoring of the spatial-temporal structure of marine processes that should be interpreted with regard to their compliance with the fundamental laws. This opens wide possibilities for investigation of the problems inaccessible in the past and to development of the new outlook on traditional Black Sea oceanological problems supported by numerous reliable and adequate facts received in the course of monitoring.

Hollandia Bay – is formed in the meander of the northern coast of the Sevastopol Bay near the mouth of the Hollandia ravine that gave its name to the bay. In the eastern part of the bay there was a wharf for the boats carrying passengers within a city. In the 1940s the bay area was used as a sea aerodrome called "Sevastopol I". According to one of the versions, the name of a ravine and bay appeared by analogy with the name "Novaya Hollandia" in Petersburg. The B.S. Pilot published in 1851 by the Black Sea Hydrographic Department wrote the following: ". . . not far from Inkerman there was a garden of the Port Commander called Hollandia".

Hopa – Turkish town and port on the B.S. coast 20 km from the border with Georgia, in the valley of the Hopa River. It is famous for the trout that is specially cooked here and eaten with raca (a kind of vodka). The manganese ore is exported via the port. This port receives small vessels and "river-sea" vessels with the draft to 3 m and to 150 m long.

Hunnic Empire – about 375 A.D. the Hunns defeated the Goths, intruded into the Crimea through the Kerch Strait, marched through its territory to Chersonesos, destroyed cities and moved further to the west. In the early fifth century the ruler of H.E. was Atilla, known in Europe as the epitome of cruelty and rapacity. He united the Huns into a powerful empire that dictated its will to the peoples and cities. After defeat in Europe some of the Hunnic tribes returned to Crimea and here they settled down, including on the southern coast of Crimean as far as Chersonesos.

Hydroacoustics Division – enters into the Marine Hydrophysical Institute of NASU. It addresses the following research problem: acoustics of the ocean. H.D. conducts fundamental studies of mathematical models of interaction of acoustic fields with regard to complicated boundary conditions and spatial geometry of water environment heterogeneities; applied research and investigations of dynamics of continuous media, instrument-making, environmental safety. It is situated in Odessa.

Hydrochemical structure of the Black Sea waters – B.S. has a double-layer hydrochemical structure. Unlike other seas, in B.S. only the top well-mixed layer (0–50 m) is saturated with oxygen (7–8 ml/l). With depth the oxygen level decreases rather quickly and already at levels 100–150 m it is practically zero. And at the same depth the hydrogen sulfide appears which quantity grows quickly with the depth to 7.5–10.5 ml/l at 1,500 m depth and farther on to the bottom its level gets stabilized. In the centers of two cyclonic gyres where the upwelling process is observed the upper border of the hydrogen sulfide zone locates closer to the surface (70–100 m) than in coastal areas (100–150 m). The oxygen and hydrogen sulfide zones are separated by a thin interface layer containing both oxygen and hydrogen sulfide and representing the lower "life border" in the sea.

Oxygen spreading into deeper sea layers is prevented by the vertical density gradients restricting distribution of convective mixing by the upper layer. At the same time, in B.S. water exchange occurs involving all layers, although this is a rather slow process. Deep saline waters permanently replenished by the lower Bosporus current rise slowly and mix with the upper layers that are then entrained to the Bosporus with the upper currents. Such water transformation and circulation helps to maintain a relatively constant salinity level in the sea waters. In B.S. we can distinguish the following basic processes being responsible for the vertical exchange in the water mass: water upwelling in the centers of two cyclonic gyres and water lowering in the peripheral parts; turbulent mixing and diffusion in the water column; autumn-winter convection in the upper layer; bottom convection due to the heat released by the seabed. There are only rough estimations of the time needed for the vertical water exchange in the sea and this very important problem needs further study.

Hydrogen sulfide zone of the Black Sea – the waters of B.S. at depths over 150 m contain hydrogen sulfide, so, in these water layers there are no live organisms, except specific bacteria. Hydrogen sulfide (H_2S) is a colorless, thermally unstable (at temperatures above 400 °C it decomposes into simple elements) and noxious gas, heavier than air, smelling unpleasantly of rotten eggs. In nature it is found in oil, natural gas, volcanic gas and in hot natural water springs. By one of the versions the Black Sea got its name due to the presence of hydrogen sulfide because the anchors after long staying in water at depths more than 150 m got covered with chemical deposits of black color. And the seamen having noticed this phenomenon called this sea "Black".

The hydrogen sulfide zone of B.S. was discovered in 1890 by N.I. Andrusov. In the late nineteenth century N.D. Zelinsky assumed that hydrogen sulfide was generated in B.S. waters due to activity of specific bacteria that oxidized organic matter using oxygen from sulfates that in their turn got reduced to hydrogen sulfide. The organic matter formed as a result of photosynthesis in surface waters moves down and sediments permanently, while the sulfates are replenished mainly from the bottom inflow of saline waters via the Bosporus from the Sea of Marmara. The "biogenous" version associates the deteriorated environmental situation with the additional input of organic matter and mineral salts of anthropogenic origin transported here with domestic wastewaters and river flow. This activates all biochemical processes, including the process of bacterial sulfate reduction with formation of hydrogen sulfide in water and bottom sediments.

In the center of the western and eastern cyclonic gyres in B.S. the border of H.S.Z. is found at a depth of about 70–100 m, but closer to the coast it drops to 100–150 m. From the upper border the hydrogen sulfide concentration adds 17 µg/l with each meter. Below 500–600 m the concentration growth slows down and begins the horizon with homogeneous salinity and density in which the concentration of hydrogen sulfide is practically constant and is equal to 13 mg/l. It is thought that such bottom homogeneous layer appears as a result of action of geothermal heat rising from the sea floor that is responsible for water mixing. But it cannot be excluded that at great depths the concentration of hydrogen sulfide may grow slightly. This process is very slow and quite natural. The displacement of the upper border of H.S.Z. is caused by year-to-year and seasonal temperature fluctuations. Thus, during rather warm winters in 1999–2001 the upper limit of the hydrogen sulfide zone rose by 5–10 m. The lower the mean annual temperature on the surface and near-surface layers the greater is the oxygen stock and vice versa. That is, oxygen controls location of the upper limit of hydrogen sulfide in the sea—oxygen reacts with hydrogen sulfide and oxidizes it to sulfate. Moreover, during warm winter the surface water layers of the sea cool less than usual, hence, the process of vertical water mixing is very weak. As a result, the greater part of oxygen, and its concentration is the greatest in the surface layers, is unable to react with hydrogen sulfide. Apart from hydrogen sulfide, such biogenous elements as ammonia, phosphate and silicate also accumulate in water in H.S.Z.

Hydrographic Service – a service ensuring safe navigation on oceans, seas, lakes, water reservoirs and rivers. H.S. is established for preparation and publishing of special and comprehensive guidelines and reference books for navigation (navigation maps, pilots, tidal tables, hydrometric and bathymetric charts and atlases, and other documents); installation of navigational facilities (lighthouses, signal lights, fencing beacons on canals and shipways); warning of seafarers about changed navigational situation and regime; development of ship-borne navigational facilities and their installation on ships. In 1918 Soviet Leader V.I. Lenin signed the Decree "On the Hydrographic Service Establishment". A year later in Kerch the ichthyological station was opened on the basis of which in later years the Azov-Black Sea Fishery and Oceanography Institute was created. From 1923 to 1935 the oceanographic expedition led by Yu.M. Shokalskiy worked in B.S. on vessels "Ingul" and "Dunai". This expedition made 53 cruises over B.S. and 1,600 hydrological stations for measurement of the temperature and salinity of sea waters on all depths and at the sea bed, too. Special devices were designed for this expedition, such as the first seabed coring tube to 4 m long for taking sediments samples from the seabed. This expedition studied the underwater relief in the central part of B.S. and near the coast, bottom sediments, the composition and properties of the water mass. It also studied the oxygen and hydrogen sulfide levels in sea water. The Marine Hydrographic Service was subordinated to the Hydrographic Department of the USSR Ministry of Defense formed in 1924. In 1929 in Katsiveli near Yalta the sea hydrographic station was organized. At present this is the branch of the Marine Hydrophysical Institute of NASU in Sevastopol. Since 1947 the H.S. of the Soviet Black Sea Fleet (and now of Russian Black Sea Fleet) is located in Sevastopol, Ukraine.

Hydrographic works on the Black and Azov seas – hydrographic works on B. and A. seas were initiated by Peter I. After Azov was taken under control Admiral Kryuis, on the Peter's order, prepared the map of Don River from Voronezh to Azov and in the end he placed "Very interesting drawings of the Sea of Azov or Lake Meotic and Pontus Euxine". In 1696 the first measurements in B.S. were carried out along the Crimean coasts and near the Bosporus. Measurements were made from the ship "Krepost" (fortress) on which Ukraintsev, the Peter's envoy, went from Azov to Istanbul to the Turkish Sultan.

The first guide for navigation of B.S. is considered to be the handwritten work "Description of the Black Sea coast extending from Akhtiar to Kuban and from Akhtiar to Ovidiopol. Attachment to maps and sketches made by Captain-Commander Billings in 1797." In a small textbook the author presented a brief review of the coasts in both directions: from Sevastopol to Odessa, on the one side, and to the Kuban mouth, on the other. On 11 pages of his book "Marine Navigation Experience" published in 1804 he gave an overview of navigation over B.S. The first Russian printed pilot of B. and A. seas was prepared in 1808 by Lieutenant I.M. Budischev. It was titled "Pilot or marine guide containing description of seaways and entrances into ports, harbors on Azov and Black seas, in Bosporus and Byzantine straits together with narrations on winds and currents." Only its first part referring to A.S. was published. In the introduction to this pilot the author gave a very interesting assessment of the navigation documents: "I will not speak about the usefulness of an accurate sea map, but I must say that the service on this sea did not reach the level that could have been attained should the officials perform more effectively; trade suffered bad losses because of a great number of merchant ships wrecked every year." In 1826–1836 the hydrographic expedition headed by 1st rank Captain E.P. Manganari worked on two ships near the Russian and Turkish coasts. As a result of this expedition, in 1842 the first "Full Atlas of the Black and Azov Seas" was published, and in 1845 Manganari completed the handwritten pilot of B. and A. seas. This systemic description of both seas became a valuable contribution into the history of hydrographic science. The first pilot of B.S. that became classical was published in 1851. It was based on description of the B.S. coasts prepared in 1847 by Russian Lieutenants G.I. Butakov and I.A. Shestakov, later promoted to admirals, who commanded tenders "Pospeshnyi" and "Skoryi". The results of the previous hydrographic investigations were also used in preparation of this pilot. Description of the Caucasian coast was based on the handwritten pilot of this area made by Tarashkin, a warrant officer of the Russian Naval Pilot Corps. In 1854 A. Sukhomlinov, Second Lieutenant of the Pilot Corps, using the results of 2-year investigations of the coast on the schooner "Astralyabia", published the "Pilot of the Sea of Azov and Kerch–Enikal Strait." These two pilots provided rather ample information for the developing navigation and remained in use for many years. Only 16 years later the second revised and enlarged edition of this B.S. pilot was prepared by Lieutenant Pavlovskiy. The next, third, edition appeared in 1892. It included also the A.S. Pilot. In the afterword to this edition it was said that it was only interim, and a new edition was under preparation and would be finalized after completion of the hydrographic works that were carried out in those years. This fourth edition, revised and enlarged essentially, was published in 1903. It was the last that presented together the B.S. and A.S. pilots. This edition is kept at the Sevastopol City Museum. The collection of this museum comprises nearly the full corpora of published pilots and atlases.

Hydrological structure of the Black Sea waters – the water temperature near the surface of B.S. in winter rises from −0.5 to 0 °C in the coastal regions of the northwestern part to 7–8 °C in central regions and 9–10 °C in the southeastern part of the sea. In summer the surface waters are warmed to 23–26 °C. Only during upwelling events and surges the short-term decrease of temperature may occur (e.g. near the Southern Coast of Crimea). In the period of sea warming on the lower limit of the wind-induced water mixing the temperature "jump" layer (thermocline) is formed that restricts heat dissemination from the upper homogeneous layer. Throughout a year the surface water salinity in the northwestern part of the sea is minimal because it receives the main volume of river runoff. In the mouth areas the water salinity grows from 0–2 to 5–10‰, while in the greater part of the open sea it is equal to 17.5–18.3‰.

During a cold winter the vertical mixing is developing in the sea and at the end of winter it covers a layer from 30 to 50 m thick in the central region to 100–150 m in the coastal areas. The waters in the northwestern part of the sea cools the most from where they are carried with currents to the intermediate depths over the whole sea and may reach the farthest places from the cold source area. As a result of winter convection and with the subsequent summer warming the cold intermediate layer (CIL) is formed in the sea. It is maintained through the whole year at 60–100 m depths and has a temperature on the borders 8 °C and in the core from 6.7 to 7.5 °C. The convective mixing in B.S. cannot spread deeper than 100–150 m because of the growing salinity (and, consequently, density) in the deeper layers due to inflow there of saline waters from the Sea of Marmara. In the upper mixing layer the salinity grows slowly and then at a depth of 100–150 m it sharply increases to 17.5–18.3‰. This is the permanent layer of salinity "jump" (halocline). Beginning from 150 to 200 m depth the salinity and temperature slowly grow towards the seabed due to the input of saline warm waters from the Sea of Marmara into the deep water layers. At an outlet from the Bosporus their salinity is 28–34‰ and temperature 13–15 °C, but mixing with the Black Sea water they change their characteristics rather quickly. In the bottom layer the water temperature rise occur also due to the heat emitted by the seabed. Deep waters in the layer from 1,000 m and onward to the bottom and comprising over 40 % of the sea volume feature stable temperature (8.5–9.2 °C) and salinity (22–22.4‰).

Therefore, we can distinguish the following basic components in the vertical hydrological structures of the Black Sea waters: the upper homogeneous layer and the seasonal (summer) thermocline associated largely with the wind mixing and an annual cycle of heat input through the sea surface; the cold intermediate layer with the minimum by depth temperature that is formed in the northwest and northeast of the sea as a result of autumn-winter convection, while in other regions it is formed mostly by cold waters advected by currents; the permanent halocline—a layer with the maximum depth-related growth of salinity that occurs in the contact zone of the upper (Black Sea) and deep (Sea of Marmara) water masses; the deep layer—from 200 m to the seabed that has no seasonal variations in the hydrological character-istics and their spatial distribution is quite uniform. The processes occurring in these layers, their seasonal and year-to-year variability determine the hydrological conditions of B.S.

Hydrometeorological Center of the Black and Azov seas (HMC BAS) – was founded on June 16, 1865 in the Novorossiysk (presently Odessa) University where the Chair of physics and physical geography was opened. Professor Vasiliy Ivanovich Lapshin headed this chair. During his work as the chair head the cabinet of geography was provided with all necessary equipment and in 1865 there was opened a meteorological station that existed till 1963. The most remarkable event during the first 5 years of the chair existence was participation of Professor V.I. Lapshin in measurements on B.S. needed for preparation of a project on installation of a telegraph cable at the seabed. Professors of the chair of physics and physical geography continued their participation in expeditions on B.S. studies and in the 1880s the hydrogen sulfide contamination of the deep B.S. waters was revealed. In 1884 the Magnetic-Meteorological Observatory founded by Professor A.V. Kolossovskiy was opened at the University. HMC BAS was a successor of the Odessa Hydrometeorological Observatory and at present it is the operative production and methodical organization of the State Hydrometeorological Service of Ukraine.

Hydrometeorological observations on the Black and Azov seas – study of the hydrometeorological regime of B. and A. seas was started after incorporation of Crimea, Besarabia and Caucasus into Russia, after creation of the navy, construction of military bases, ports and lighthouses. The first marine hydrometeorological station (later on an observatory) on B.S. was founded in 1801 in Nikolaev. Then similar stations were opened in Kherson (1808) and Odessa (1821). Regular observations were conducted on stations in Simferopol (1821), Sevastopol (1824), Nikita (1826), Karadag (1831), Alma and Enissale (1833), Kerch and Takly (1863), Enikale (1865), Khersones (1865), Evpatoria (1866), Yalta (1869), Feodosia (1871), Ai-Todor (1878), Tarkhankut (1881), Ai-Petri (1895), Belogorsk (1896), and Karabi-yaila (1916). Coastal stations were organized by the Trade Ports Division, mainland stations—by other departments. The hydrometeorological observations were usually conducted on lighthouses and in ports by the Hydrographic Division of the Black Sea Fleet.

The first marine hydrometeorological station on A.S. was opened in 1870 on the Pereboinyi Island (in the Don Mouth). Later on the stations at the Taganrog and Genichesk lighthouses (1881–1882), on the Biryuchiy Peninsula (1885), on Belosaray (1893) and Enikale (1896) lighthouses started functioning. By 1900 about 45 stations have been operating on both seas. The commission on construction of commercial ports envisaged from 1893 the conduct of hydrometeorological observations on B. and A. seas. Beginning from 1908 when there was formulated a task on creation of a weather warning system for ships these works evolved most widely. The central station in Feodosia was in charge of exchange and collection of information. In 1901–1920 the pace of development of a network of marine hydrometeorological stations slowed down. During these years the following stations on B.S. were opened: in Tuapse, on the Saint-Trinity lighthouse, in Batumi, Kasperovka, Khorly, on the Dzharylgach lighthouse, in Karadag, Adler, Sochi, on the Sarych lighthouse, in port Nikolaev, in Strelets Bay and in Anapa; on A.S.: in Taganrog and Zhdanov ports, on the Peschanyi Island, in Genichesk, Margaritovka, Primorsko-Akhtarsk, Azov, in Berdyansk and Temryuk ports. In 1920 there were functioning 41 and 18 marine hydrometeorological stations and posts on B.S. and A.S., respectively.

The operation of stations and posts and regime studies were controlled by the Hydrographic Department of the Fleet or Flotilla. Stations were in the surveillance of the Marine Hydrometeorological Observatory organized in 1909 is Sevastopol on the basis of hydrometeorological station of class I (the observatory was moved here from Nikolaev where it has been operational since 1834). The stations and posts of trade ports were subordinated to the Russian Ministry of Trade and

Industry. In Feodosia, on the basis of the port hydrometeorological stations there was opened the Hydrometeorological Center that controlled the activity of civil stations and posts. The generalized data about the hydrometeorological regime of B. and A. seas were included into the pilots of these seas and placed in special publications. Thus, in 1908 there was published the wind and mist atlas of B. and A. seas, in 1915—the physiographical review of B. and A. seas. Publishing of monthly weather reviews was started. Some climatic data for the eastern part of Asia Minor were published. In 1917 the physiographical review of B.S. was published.

In 1917 the stations and services were united and subordinated to the Hydrographic Department of the Russian Black Sea Fleet and this situation lasted till 1922. After 1917 the greater part of the stations in Crimea was under control of the Naval Department. Beginning from 1921 the network of hydrological stations was transferred to the responsibility of the Party of Crimean Water Surveys. Later the civil network was transferred to the Central Department of Marine Transport (within the frame of which the Central Hydrometeorological Bureau was organized). The Hydrometeorological Center in Feodosia resumed its operation and in 1924 the weather bureau was incorporated into its structure. The programs of observations on the stations and posts became more extensive. Observations were carried out on seaways and specific surveys were carried out in port areas, in the Kerch Strait, in river deltas and certain areas of the sea. The Hydrometeorological Center of B. and A. seas (GIMEIN BAS) established a series of coastal hydrometeorological stations in Feodosia, such as Mysovoye (1925), Alushta (1926) and Chernomorskoye (1927). In 1928 the meteorological station in Pochtovy (Bazar-Dzhalga) was organized in the trial land reclamation area and in 1929—in Golubinka and Voronki.

In 1929 the Hydrometeorological Center was reorganized into the Geophysical Observatory and in 1930—into the Hydrometeorological Institute of the Black and Azov Seas that, together with civil stations, was incorporated into the Department of Hydrometeorological Service of the B. and A. seas. In the same year in Crimea the networks of different departments were taken together to from a single service. The mainland networks were transferred to the GIMECOM of Crimea (Simferopol), while marine hydrometeorological stations—to GIMEIN BAS that afterwards was reorganized into the Department of Hydrometeorological Service of the B. and A. seas and later into the Sevastopol Department of Hydrometeorological Service that united the networks of Crimea and B. and A. seas. In 1934 on the initiative of the Command of the Black Sea Navy the Department was moved from Feodosia to Sevastopol. In Feodosia there was opened the Geophysical Observatory that in 1940 was also transferred to Sevastopol. At the beginning of the Great Patriotic War in 1941 the civil and military divisions of the Hydrometeorological Service of the Black and Azov seas were united into the Department of Hydrometeorological Service (DHS) of the Black Sea Navy that was subordinated to the Head Department of DHS of the Red Army, and in localities—to the Black Sea Navy Headquarters. In summer 1944 DHS of the Black Sea Navy and the Observatory were transferred back to Sevastopol.

I

Ibr River – see Dniester River.

Ice cover in the Black Sea – is formed only in some places near the northwestern shores and only during most severe winters (e.g., 1953/1954, 1984/1985, February 2012). The Caucasian and Anatolian coasts are free of ice. In cold years the Dnieper-Bug and Dniester lagoons, lakes in the Danube Delta and on the northwestern coast get frozen. In very cold winters the Danube and, at times, even the coastal parts of the sea may be covered with ice. In historical records we can find information that in very severe winters when the sea near the Bulgarian coast got frozen the broken ice drifts even as far as the Bosporus and to Eregli. Herodotus was the first who mentioned ice cover formation: the Bosporus Cimmerian (Kerch Strait) and Meotic Sea (Sea of Azov) were often covered with a rather thick ice layer and its pieces after breaking in spring were brought to the Pontus (Black Sea). Roman poet Ovid who was exiled to Smaller Scythia (Dobrudzha) wrote that in the period from years 7 to 17 A.D. during three winters the Danube and coastal waters were frozen. Nolian (third century) remarked about frequent formation of ice cover on the Danube. A great part of B.S. was frozen in 401. Amian Marcellius wrote that nearly the whole sea was frozen and in spring the ice fields filled the Bosporus from where ice pieces got into the Sea of Marmara where they floated for nearly a month. The Byzantine sources wrote about freezing of the Bosporus in 739, 753 and 755. In 755 the ice formed in the Sea of Marmara jammed the Dardanelles. The most intensive ice formation in 762 was described by Russian Patriarch Nikiphor and chronographer Kodrin: ice fields extended in B.S. for about 185 km from the land, even near the Anatolian coast. One could simply walk over ice from Messembria (Nesebyr) to the Caucasian coast. Freezing of the Bosporus was also recorded in 928 and 934. In 1011 not only Bosporus, but a part of the Sea of Marmara was under ice. That year Syria and Egypt suffered from severe cold and ice was formed even in the Lower Nile. In 1068, according to Prince Gleb Svyatoslavovich, the northern part of B.S. got frozen. In 1232, 1621, 1669 and 1755 ice appeared near the southern coasts of B.S. and in the Bosporus. In 1813 B.S. was covered with ice from its northern coast to the southern regions of Crimea. The Bosporus was frozen in 1823,

1849 and 1862. In 1929, 1942 and 1954 ice was formed along the whole Bulgarian coast and got even into the Bosporus. In 1972 ice cover in the northwestern parts of B. and A. seas and heavy ice drift on the Danube caused formation of ice fields near the Bulgarian coast even southward of the Kaliakra Cape. But long-lasting winds blowing from the land carried ice to open sea. In other years ice and frazil ice were observed in shallow bays near the Bulgarian coast. The seaside lakes were frozen more frequent. Fresh waters freeze at 0 °C, saline waters—at lower temperatures. In oceans the water freezes at a temperature from −1.9 to −2 °C, in B.S.—at −0.9 °C if the weather is calm. At rough sea the ice crystals—brash ice are formed and in this case a water temperature may be about −1.1 or −1.2 °C.

Ice cover the northwestern part of B.S. is in good agreement with the local air temperature. A decrease (on average) of the amount of degree-days of frost in these years, the average seasonal sea ice extent decreased, with pronounced peaks in the severe winters of 1953/1954 and 1984/1985 and lows in the warm winters (for example, in 1965/1966, 1969/1970 and 1970/1971.). Given the tendency of an increase in the average of air temperature in the Black Sea in modern times, we can expect a corresponding decrease in the sea ice appearance in the north-western and north-eastern parts of the sea. This does not preclude, however, extreme events related to the sharp decrease in air temperature and strong winds, observed, for example, in February 2012, when the sea was covered with ice in Kerch, Yalta, Novorossiysk, Odessa, and Constanta. A significant reduction of time period with ice phenomena is observed in the Danube Delta in 1961–2003. During the last 50 years the probability of ice cover in the Danube Delta has decreased in three times.

Igneada – a port, a seaside resort in European Turkey located in a convenient bay on the western coast of the country. It is situated to the north of the Igneada Cape near the border with Bulgaria. The Instrandjanskiy highlands make this terrain very picturesque and protect the coast from northern winds creating here a very pleasant microclimate. The nearby hills have many caves the most well-known of which is the Dupnisa near the Sardepe village. A low-lying marshy lagoon is found near I.

Igneada Bay – locates in the northwest of Turkey near the Bulgarian state border. In the northeast it is confined by the Igneada Cape. In the top of I.B. the high Tersane Cape projects into the bay from the mid-coast. To the south and north of this cape the coast is low, marshy and fringed by a beach. The village with the same name locates near the cape. The water depth in the bay is 34–36 m.

Igneada Cape – the southern tip of the hilly peninsula surrounding the Igneada Bay on the northern side. The eastern tip of this peninsula is the Koru Cape on the high shore of which the Igneada lighthouse is constructed.

Il, Province – the administrative-territorial unit of Turkey (the country is divided into 81 ils). Turkey is subdivided in a hierarchical manner to the following administrative units: provinces (*il*), districts (*ilçe*), villages, and neighbourhoods. Each I. is headed by a governor (*vali*).

Ilyichevsk – a port and town in the Odessa Region, Ukraine. It is situated 25 km to the northeast of the Dnieper Liman on the shore of silty saline Sukhoy Liman. It was founded in 1952 as a settlement for builders of sea port Ilyichevsk that was given this name by the patronymic of Soviet Leader Vladimir Ilyich Lenin. From 1973 this was a district center, one of the most beautiful towns in the Odessa Region. Population 59,200 (2011).

Ilyichevsk Maritime Trade Port – one of the major international ports in Ukraine. It was founded in 1958 on the basis of the cargo area of the Odessa port. It is situated on the shores of Sukhoy Liman 15 km to the southwest of Odessa. The port is a part of the 9th ITC of European Union and also of the corridor "Baltic Sea—Black Sea". Train ferries connect this port with Bulgaria and Georgia. I.M.T.P. is the modern well-equipped port, one of the largest in the Black Sea region. The favorable geographical location of the port makes it an important link in the Europe–Asia trade. The loading-unloading complexes of the port incorporate 28 berths with a total length of 5.5 km. They are capable to receive any large-tonnage ships with a draft to 13 m. The total handling capacity of the port is 24 million tons of cargo a year. It specializes on reloading of heavy cargo. The specific feature of the port territory is its location on different levels in relation to the sea. The territory comprises a number of sites separated from each other by earth embankments or unused areas. The port has a specific complex for loading of railway cars going from Ilyichevst to Varna, Poti and Batumi and back on train ferries capable to transport 108 railway cars each and also a complex servicing the lines Ilyichevsk—Turkish ports, Ilyichevsk—Georgian ports on which ro-ro vessels are operating.

"Ilyichevsk-Varna" International Ferry Line – On April 23, 1975 the governments of the USSR and Bulgaria signed the Treaty on Organization of the Ferry Line between Port Ilyichevsk and Port Varna with commissioning of this line in 1978. For this purpose there were constructed four ferryboats: "Heroes of Plevna", "Heroes of Shipka"—in the USSR and "Heroes of Odessa" and "Heroes of Sevastopol"—in Bulgaria. Each of these ferryboats was capable to take onboard to 108 eight-wheel railroad cars. In addition these ferryboats could carry vehicle trailers, roll-trailers and other wheel vehicles. Special ferry complexes were constructed in Varna and Ilyichevsk. On April 11, 1978 the Treaty on Joint Operation of the Ferry Line between Port Ilyichevsk (USSR) and Port Varna (Bulgaria) was signed. The ferry line was commissioned on November 14, 1978. At present its operation is controlled by the Agreement between the Government of Ukraine and the Government of Bulgaria on joint operation of this ferry line between port Ilyichevsk (Ukraine) and port Varna (Bulgaria) signed on November 20, 1995.

Ilyin Nikolay Ivanovich (c. 1825–1892) – Russian Captain 1st rank, sea pilot, specialist on lighthouses. In 1839 he entered the Naval Cadet Corps. In 1847 in the rank of Midshipman he was appointed to serve in the Black Sea Navy where he navigated on different ships as a pilot and commander. In 1853 he took part in the battle at Sinop; in 1854–1855—in defense of Sevastopol commanding the artillery batteries. In 1856–1860 he commanded the ships of the Black Sea Fleet. In 1860 he was appointed deputy director of the B.S. and A.S. lighthouses. In 1863 he was sent to Britain to collect data about electrical lighting of beacons and about the newest achievements in lighthouse facilities. I. initiated the conversion of the lighthouse in Odessa to electrical lighting. In 1866 he was a member of the construction division of the Naval Technical Committee and was detached to the Directorate of the B.S. and A.S. lighthouses. From 1887 he was in charge of invalid hamlets near Nikolaev and Sevastopol. The "Naval Collection" published his articles about lighthouse optical facilities and about the equipment and operation of lighthouses (1860–1867). He was awarded several Russian orders.

Inal Bay – locates 8 km westward of Dzhubga. In the former USSR it was famous for its numerous rest centers belonging to large industrial enterprises and organizations. The modern resort infrastructure was created here. The wide beach is composed of pebble and boulders. The sea depth increases sharply.

Indjeburun, Indje Cape (from Turk *"Ince Burun"* meaning "thin cape") – the northernmost cape on the Anatolian coast of Turkey near Sinop. The cape is not elevated, rocky and of the reddish color. A lighthouse is constructed on it. The wharf Sarikum is located on the western shore.

Inebolu – a village on the coast of the Black Sea, Turkey.

Ingulskiy Shipyard – see Nikolaev.

Inguri, Ingur – a river in Western Georgia. In lower reaches is a border between Abkhazia and Georgia. It flows into B.S. I.R. originates as several streams in the glaciers of the Caucasus Major. In its upper course the river flows in the Svaneti Depression, more downstream it rushes through a narrow and deep gorge and then spreads over a wide valley. Near the Djvari settlement it goes out on the Colchis Lowland where the lower reaches of this river and its mouth are found. Its name most probably originates from the Ancient Georgian name of the Colchis—*Egrisi* meaning "Colchis River". The area of its basin is 4,060 km^2, the river length is 213 km. It has mixed recharge. About 8 % of the basin territory is covered by glaciers. The mean many-year flow of I. (Darcheli village, a distance from the mouth—16 km) is 4.6 km^3 a year and it varied from 7.7 km^3 in 1941 to 0.4 km^3 in 1985. The flow distribution by seasons is very uneven: about 47 % of the annual flow is in summer. The maximum flow is recorded in June-July and the minimum—in February.

Inkerman – a town (in the Turk "fortress in a cave, new fortress") in Crimea, Ukraine. Population—12,000 people (2012). It is situated in the top of the Sevastopol Bay near the inflow of Chernaya River, on the right bank of this river. In the north and east the town is surrounded by the Balaklava uplands. The limestone quarry locates in the eastern part of the town and on its basis the Inkerman plant producing building materials was constructed. In the sixth century a fortification was built here. It is thought that near I. there was the ancient Greek town Evpatoriy (Evpatoria) founded by Diophantus of Alexandria (b. between A.D. 201 and

215, d. between 285 and 299 at age 84) that later in different times was called Ktenus, Doras, Dori, Doros and Katamita. In 679 the Khazars conquered it from Goths who in the late eighth century took hold of it again. From 1204 the town had its own rulers and among them the last Byzantine Emperor Constantin XI before his coming on the throne. In the twelfth to fifteenth centuries the cave fortress Kalamita was constructed in the I. place. In 1475 it was captured by the Turks and renamed into I. In 1976 it became a town and got the name of Belokamensk (white stone) by the color of limestone. In 1991 the previous name—Inkerman was brought in use again. Administratively, Inkerman subordinates to Sevastopol, but still remains an independent town.

Inkerman Battle (Crimean War of 1853–1856) – an area in the mouth of the Chernaya River near Sevastopol where on October 24, 1854 there was a battle between Russian troops commanded by Prince A.S. Menshikov (August 26, 1787–May 1, 1869) and the British, French and Ottoman Empire Army. The ally army numbering about 63,000 people and 349 cannons, under command of British Field Marshal FitzRoy James Henry Somerset, 1st Baron Raglan (30 September 1788–29 June 1855) and French Marshal François Certain de Canrobert, usually known as François Certain-Canrobert and later simply as Maréchal Canrobert (27 June 1809–28 January 1895) was preparing for storming Sevastopol that was planned on 6 November. A.S. Menshikov, Commander-in-Chief of the Russian Army numbering 82,000 people and 282 cannons, decided to launch an attack to frustrate the designs of the allies and to urge them to raise the siege of the city (see *Sevastopol Defense of 1854–1855*). For this attack he assigned a special group commanded by General P.A. Dannenberg. The Russian troops made their major assault on the left flange by the units of Generals F.I. Soimonov and P.Ya. Pavlov with a view to disintegrate the ally army and to defeat it by parts. But this well thought plan was poorly elaborated and prepared. Broken terrain, lack of maps and also dense mist resulted in weak coordination of the attacking troops. The Russian command was practically unable to control the progress of the battle, and the Russians suffered bad losses. The outcome of the battle was decided by the attack of the French unit of 9,000 people that went to the rescue of the British troops and pushed back the Russian troops, exhausted and suffered great losses, to their initial positions on the left flange. The allies lost in this battle about 6,000 people, the Russian—more than 10,000 people. Although Menshikov failed to attain the set target this battle was very important for the future fate of Sevastopol. This battle prevented the allies from making the planned storming of the fortress and they had to undertake the winter siege.

Inkerman cave monastery (Ktenus, Kalamita) – It is said that this monastery was founded by Byzantine icon-worshippers fleeing persecution in their homeland. They built the monastery in the Monastyrskiy cliff rising on the left bank of Chernaya River. There are about 200 caves arranged in layers there. In 1926 the Inkerman monastery was closed. At present this monastery is being restored.

Inkerman cave monastery (http://crimeatraveling.ru/inkermanskij-svyato-klimentovskij-peshhernyj-muzhskoj-monastyr/)

Inkerman stone – a kind of easily worked bryozoan limestone from Inkerman. From the Antique times it had been widely used in construction. This stone had been valued for a possibility to mine it directly on the town territory or not far from it, for its strength and softness, homogeneity and permanent properties, high thermal insulation properties, long service life, decorative properties and ability to keep sharp edges after cutting. The buildings and structures made of I.S. stood intact for many centuries. Thanks to such unique construction and architectural properties of the bryozoan limestone the area between Sevastopol and the Alma-Bodrak interfluve near Simferopol may boast of many Medieval cave towns and monasteries.

Inland seas – the seas incising deeply into land and, in general, connected via one or several straits with the ocean or nearby sea. I.S. make a specific group of seas that includes the White, Baltic, Red, Mediterranean, Dead, Black, Azov, Caspian, and Aral seas. The Black and Azov seas are the water bodies isolated from the ocean to the greatest extent. I.S. are strongly influenced by surrounding land. In some I.S. their surface waters are not very saline because of the copious river flow (e.g. the Baltic, Black, Azov, and Caspian seas), others, on the contrary, feature higher salinity due to the effect of the arid climate and a weak effect of the mainland flow combined the high evaporation (e.g. Red, Mediterranean, Dead, and Aral seas).

Institute of Biology of the Southern Seas named after A.O. Kovalevskiy, National Academy of Sciences of Ukraine the oldest marine biology institution located in Sevastopol, Ukraine. In 1871 N.N. Miklukho-Maklai initiated creation of the Sevastopol Biological Station (SBS) on the basis of which this Institute was established in 1963. Appearance of such Institute enabled considerable expansion of the research and increase of the number of its divisions. This Institute incorporated the "youngest" Odessa Biological Station (created in 1954), the Karadag Biological Station (1914) and research vessels "Akademik A. Kovalevskiy", "Miklukho-Maklai", and "Professor Vodyanitskiy". A great contribution into the Institute development and shaping of its research directions was made by V.A. Vodyanitskiy who headed the Institute for 30 years (till 1968).

The principal directions of research are the following: theory of biological productivity of marine ecosystems and management of marine resources; study of the effect of pollution and other anthropogenic impacts on the sea life; protection of the sea biological resources; biological-technical tasks; issues related to development of the underwater environment by a man (acoustic and bioluminiscent interferences, biogrowths, hydrobionics). These problems are studied in 13 divisions: applied oceanology, plankton, benthos, ichthyology, theory of life forms, functioning of marine ecosystems, physiology of marine animals, algae physiology, radiation and chemical ecology, sanitary hydrobiology, biology of overgrowths, and mathematic modeling. Staff—400 people. The Karadag Branch conducts in the land and sea nature reserve subordinated to it the environmental and physiological investigations and hydrobiological works in the dolphinarium. The Odessa Branch studies the northwestern part of B.S. In 1964–1965 the first long expedition to the Central-American seas was conducted. With the active support of the Institute scientists in 1965 the Institute of Oceanology of the Academy of Sciences of Cuba was opened that has an aquarium with many representatives of the marine life of warm seas and specimens of underwater landscapes. In 1971 for the merits in development of the biological science the Institute was awarded the USSR Labor Red Banner Order. In memory of the outstanding scientists who contributed much into development of this Institute the monuments to N.N. Miklukho-Maklai and A.O. Kovalevsky are installed in front of the Institute. The Institute publishes the "Proceedings of Institute of the Southern Seas" (1985) and "Marine Ecology Journal" (since 2002).

Institute of Biology of the Southern Seas (Photo by Dmytro Solovyov)

Institute for Fish Resources – was established in 1973 on the basis of the Oceanography and Fishery Institute, Varna, Bulgaria. The Institute has at its disposal the vessel "9th September" that is used for research within the large integrated program including studies of the abiotic environment (hydrology and chemistry of water), food and raw material resources of the sea and coastal lakes. Varna Aquarium is a research unit adjoined to Institute for Fish Resources. It includes 12 scientists who conduct research related to hydrobiology, hydrochemistry, marine microbiology, ichthyology, plankton and benthos. The Varna Aquarium's library houses 30,000 volumes of specialized literature, including rare nineteenth-century books and maps. The aquarium's exhibition focuses on the Black Sea's flora and fauna which includes over 140 fish species, but also features Mediterranean and freshwater fish, exotic species from faraway areas of the world ocean, mussels and algae.

Institute for Marine Geology and Geoecology – The National Institute for Marine Geology and Geo-ecology—GeoEcoMar of Romania, Bucharest is a governmental research-development institute, co-ordinated by the Ministry of Education and Research of Romania, created by Governmental Decision in 1993 by separating and developing the Marine Geology and Sedimentology Department of the Geological Institute of Romania where the Marine Geosciences research activity started in 1968. The main activities deal with the marine, deltaic and fluvial environmental and geo-ecological studies regarding the ecosystems of the geosystem River Danube—Danube Delta—Coastal Zone—Black Sea and the environmental impact of anthropogenic structures (civil and hydro-technical works) implemented along the Danube course and in the Danube Delta, geological-geophysical survey of the Black Sea, mainly of the Romanian continental shelf, as well as of other marine areas. Bathymetric, geologic-sedimentologic, magnetometric and gravimetric mapping of the shelf area is carried out at 1:200,000 and 1:50,000 scales, land-sea interactions in the Coastal Zone and its integrated management, etc. GeoEcoMar first started its European and international co-operation with the French Institute of Marine Biogeochemistry in Montrouge and with Jacques-Yves Cousteau team. Afterwards, GeoEcoMar enlarged its scientific international participation in the EU FP4, 5 and 6 (European Centre of Excellence in FP5), GEF-UNDP Black Sea Environmental Programme, NATO projects, bilateral co-operations, etc., some-times having a central role of regional co-ordinator or national contact-point.

Institute for Marine Research (Romanian) – National Institute for Marine Research and Development "Grigore Antipa" (NIMRD) in Constanta is the leading marine research institution in Romania, as well as national coordinator and focal point with respect to international research tasks and responsibilities in the field of marine science. It appeared from merging of small research units: Marine Zoological Station "Prof. I. Borci" (Adjidja), Marine Fishery Research Station "Dr. G. Antipa" (Constanta), Marine Biological Division (Constanta) of the Biology Institute "T. Savulescu" (Bucharest). The Institute operates under co-ordination of the Ministry of Environment and Forests and its research activities are mainly oriented towards supporting adequate marine and coastal environmental management and protection. NIMRD undertakes fundamental, applied and technological development research in oceanography, marine and coastal engineering, ecology and environmental protection, and management of living resources in the Black Sea and other ocean areas. Being the technical operator of the marine monitoring network (physical, chemical and biological) and for coastal erosion survey, NIMRD hold a comprehensive volume of marine data and information which are exchanged in the framework of several international projects. The Institute has at its disposal two research vessels.

Institute for Marine Research and Oceanology – was established on 1 July 1973 in Varna at the Bulgarian Academy of Sciences. In 1985 it was reorganized into the Institute of Oceanology of the Bulgarian Academy of Sciences. The Institute was set up by the existing branch of Coastal Marine Research of the Scientific Center at the State Economic Unit (SEU) Water Transport, Departments of Hydrology and Hydrochemistry, Undewater Investigations Laboratory of the Scientific Center at the SEU Fish Industry, and several specialists from Department of Marine Meteorology and Climatology of the Institute of Hydrology and Meteorology. Its scope of activities embraced marine physics and hydrodynamics, chemistry, meteorology, climatology, geomorphology, lithodynamics, marine technique and construction design.

Institute of Marine Sciences and Management of the Istanbul University – a marine institution in Istanbul, Turkey. The Institute's mission is to collect and produce data in international standards using modern equipment and scientific methods in the fields of Marine and Maritime sciences, as well as to designate the nation's problems and needs; find scientific solutions and management strategies to preserve and defend national benefits and give graduate level education to raise well-trained specialists and researchers in the following fields of science: chemical oceanography, marine environment, marine geology and geophysics, marine management, physical oceanography and marine biology. The aim of the Institute is to become a key institute for the Black Sea and Eastern Mediterranean with internationally recognized scientists, preferred by a broad spectrum of students from different disciplines for graduate training, promoting contribution of both public and private institutions to marine research through public awareness activities, funding and international collaboration.

Institute of Marine Sciences of the Middle East Technical University (IMS-METU) – locates in Erdemli, Turkey. IMS was established in 1975 with the objectives of conducting oceanographic research and providing graduate level education in marine sciences. The Institute has four main divisions: Chemical oceanography, Marine biology and fisheries, Marine geology and geophysics, Physical oceanography. Among its main lines of research there are coastal ecology, bioresources (fish, aquaculture), microbiology, land-ocean and ocean-atmosphere interaction, water quality and pollution, specific climatic features, bottom sediments and sedimentation process, ocean circulation and paleoecology. The research accomplished by the Institute has resulted both in vastly increasing the knowledge of the seas surrounding Turkey and in establishing a data base to help in the management of the marine environment. The research is conducted on several research vessels, such as *Bilim* built in 1983 (433 register tons, crew 12 people, 14 scientists, sailing range 6,500 miles); R/V *Lamas*, a trawler ship built in 1981 that is used for biological and fishery investigations (crew four people, four scientists); R/V *Erdemli* built in 1979 used for coastal oceanographic investigations.

The Institute is located on the Mediterranean coast, near Erdemli, about 45 km west of the town of Mersin. It is a part of the Middle East Technical University, one of the leading universities in Turkey, which has its main campus in Ankara. The mission of the Institute is to provide post-graduate education in marine sciences to enable Turkish scientists to conduct research compatible with international standards; to perform systematic and integrated research utilizing both observations and modeling techniques to acquire information pertinent to the needs of Turkey and to publish the results so that they may be of use for decision makers. The programs are envisioned to provide future scientists with thorough education and training in different fields of marine science. Special emphasis is given to studying the national marine environment, in keeping with the Institute's objectives of developing and improving the marine resources of Turkey. The Institute's campus at Erdemli, houses office buildings, laboratories, computing and remote sensing facilities, a library and other services. Housing for staff and students, and harbor facility are also located on the campus.

Institute of Oceanology named after P.P. Shirshov, Russian Academy of Sciences (SIO RAS) – one of the oldest (since 1946) and the largest (staff—1,215 scientists) Russian oceanographic research center located in Moscow. SIO RAS defined its main tasks as integrated study of the World Ocean and seas of Russia proceeding from the unity of physical, chemical, biological and geological processes in them, development of the scientific basis for prediction of the Earth's climate, rational management of marine resources and ensuring safe environment for sustainable development of the mankind. SIO incorporates the Atlantic Branch in Kaliningrad, the Southern Branch in Gelendzhik on the Black Sea, Branch in Saint-Petersburg, the Northwestern Branch in Arkhangelsk, and the Caspian Branch in Astrakhan. The first director of SIO was P.P.Shirshov. Along with him actively involved in the creation and development of the new science center was well-known scientists L.A. Zenkevich, V.G. Bogorov, S.V. Bruevich, A.D. Dobrovolskiy, P.L. Bezrukov, I.D. Papanin, V.B. Shtokman and others. SIO RAS today is a team of 1,215 employees, including 105 doctors and 261 PhD scientists in different specialties. Now SIO has several full and corresponding members of the Russian Academy of Sciences.

SIO has Physical, Biological, Geological and Technical Departments and pursues several lines of research. Physical—hydrology, hydrophysics, hydrooptics, hydroacoustics; satellite remote sensing, atmosphere-ocean interaction, meteorology, climate. Biological—ecology, ecosystems, primary production, biodiversity, bioresources, commercial-value populations. Geological—mineral resources, paleooceanology, global tectonics, geophysics, geological structure and evolution of the sea floor, geophysical fields and geochemical processes. Chemical—biochemistry of organic matter, oil and gas characteristics of the World Ocean, physical and chemical condition of the sea water; chemical composition of the basic elements of ocean and sea ecosystems, processes of biochemical transformation and evolution. Marine technical facilities—technical devices for ocean studies and ocean informatics, elaboration of methods and technical facilities for long-time monitoring of the physical, chemical and biological parameters of the ocean on the basis of distributed intellectual systems of autonomous bottom, submerged and scanning stations with remote control and reading the observation results.

In the period from 1949 to 2003 the Institute's research fleet included 23 research vessels of various tonnage. The smallest of them were R/V *Geolog* and *Truzhenik* with displacement 51.4 tons each. In 1949 the Institute received the first research vessel *Vityaz* with displacement 5,700 tons. This was the vessel specially equipped for oceanographic research and many scientific discoveries that won fame to the Soviet science were made on it. At present *Vityaz* is placed in Kaliningrad and turned into the Museum of the World Ocean. Today the SIO R/V fleet based in Kaliningrad, Gelendzhik and Astrakhan incorporates three large-tonnage ships (displacement more than 6,000 tons)—*Akademik Mstislav Keldysh, Akademik Sergey Vavilov* and *Akademik Ioffe*; two medium-tonnage ships (over 1,000 tons)—*Professor Shtokman* and *Rift*; and motor boat *Ashamba* (on the Black Sea only). These ships are provided with the equipment for integrated research expeditions. The SIO fleet included also five manned deepwater vehicles—two types of

"Mir" with the submersion depth to 6 km, two types of "Pisces" capable to deep to 2 km and the vehicle "Argus" that could work at a depth to 600 m.

From 1949 the multipurpose investigations of the Black, Azov and Mediterranean seas were carried out by the vessels based in the Black Sea Experimental Research Station and later the Southern Branch (1967) of SIO, located in Blue Bay near Gelendzhik. From 1999 to the present SIO yearly carries out wide-scale, regular and integrated physical, biological, chemical and geological investigations and monitoring of B. and A. seas. The Southern Branch of SIO is used as a base for the research performed by SIO, as well as by Russian Space Research Institute RAS, Acoustic Institute RAS, Institute of Atmosphere Physics RAS, Moscow State University, and other research institutes and organizations.

P.P. Shirshov Institute of Oceanology RAS (Moscow) (http://dic.academic.ru/dic.nsf/ruwiki/131721)

Institute of Oceanology of the Bulgarian Academy of Sciences – I.O. is a leading center of marine sciences in Bulgaria. In 1985 the Institute was reorganized from Institute for Marine Research and Oceanology established in 1973 in Varna. The following research units were formed: Department of Marine Physics and the Adjacent Atmosphere, Marine Chemistry Department, Department of Hydrodynamics and Lithodynamics of the Coastal Zone (since 1989—Coastal Zone Dynamics) with Laboratory of Marine Structures, Underwater Investigations Laboratory, Tools and Equipment Laboratory. In 1978 Department of Marine Geomorphology and Quaternary Geology was set up, which in 1989 was splitted into Marine Geomorphology and Palaeogeography Department, Department of Marine Geology and Geochemistry, and Laboratory of Marine Geophysics. In 1997 these units were united into Department of Marine Geology, renamed in 1999 to Marine Geology and Archeology Department. Department of Marine Biology and Ecology was established in 1987. The I.O. research units embraces almost all fields of oceanology: marine physics, chemistry, geology and biology, coastal zone dynamics and oceanotechniques. Staff—190 persons.

I.O. has established international contacts and develops intensive international scientific collaboration with research institutes in Austria, Belgium, Cuba, France, Germany, Greece, Poland, Romania, Russia, Switzerland, Turkey, UK, Ukraine, USA, Vietnam. Since 1990, joint projects with EC, NATO, IOC—UNESCO, etc. have been initiated. The Institute took part in the Global Programme for Studying the World Ocean Impact on the Earth climate ("Razrezy"), the international programmes Global Environment and Interaction between the Danube and the North-Western Part of the Black Sea. Active international collaboration is going on with world-wide leading research centers as P.P. Shirshov Institute of Oceanology and Institute of Computing Mathematics RAS (Moscow, Russia), Woods Hole Oceanographic Institution (USA), Research Institute (Mistic, Massachusetts, USA), Oceanographic Center (Southampton, UK), IFREMER (Brest, France), Forel Institute—University of Geneva (Switzerland), Institute of Chrystalography, Mineralogy and Structural Chemistry (Vienna, Austria), Middle East Technical University (Ankara, Turkey), Institute of Marine Sciences and Technologies (Izmir, Turkey), National Center of Marine Research (Athens, Greece).

Investigations are carried out on two research vessels of the Institute *Akademik* (1,225 register tons) and *Argon* (15 tons). It also has a stationary research base in settlement Shkorpilovtsy. Among the major publications prepared by the Institute there are "Hydrology and Geology of Western Black Sea" (1979), "Geophysical Studies in the Bulgarian Zone of the Black Sea" (1980), "Interaction of the Atmosphere, Hydrosphere and Lithosphere in the Black Sea Coastal Zone ('Kamchia-77, -78, -79')" (1983), "Oil and Gas Investigations in the Black Sea Coastal Zone" (1986), "Dynamic Processes in the Coastal Regions" (1990), "Evolution of the Western Part of the Black Sea in the Neogene-Quaternary Period" (1990), "Catalog of the Oceanographic Data and Information Products in Bulgaria" (2000).

Institute of Oceanography and Fishery – this institute was established for hydrologic and hydrochemical investigations of the Black Sea coastal waters in Bulgaria in the period from 1954 to 1973. It located in Varna, Bulgaria. In 1973 it was transformed into the Institute for Fish Resources.

International Association of Maritime Universities (IAMU) – was founded in Тщмуьиук 1999. In late June 2000 the IAMU presentation assembly was conducted in Istanbul on the basis of the Maritime Faculty of the Technical University. The Association was established by seven higher maritime educational institutions from five continents. Among its founders there are the Maritime Faculty of the Istanbul Technical University (Turkey), Kobe University of Mercantile Marine (Japan), Arab Academy for Science and Technology and Maritime Transport (Egypt), Australian Maritime College (Australia), Cardiff University (Great Britain), Main Maritime Academy (USA) and World Maritime University (Sweden). Today the

Association incorporates 53 higher maritime educational institutions from different countries, including three Russian institutes and one Ukrainian—Odessa State Maritime Academy. The IAMU activities proceed along three lines: a system of maritime education and training; security management systems; global standardization of maritime education.

International Center for Black Sea Studies – it was founded in 1998 following the resolution of the Conference of the National Academies of Sciences of Member-States of Intergovernmental Organization of the Black Sea Economic Cooperation. The purpose of the Center is to enhance the role of national academies in support of the scientific and technical cooperation in the Black Sea region. The Center locates in Athens, Greece.

International Convention for the Prevention of Pollution from Ships of 1973 as modified by Protocol of 1978 (MARPOL-73/78) – the basic global international treaty covering prevention of pollution of the marine environment by ships. It comprises 20 articles, two protocols (Protocol I "Incident Reporting", Protocol II "Arbitration") and 6 Annexes. The annexes contain the regulations for the prevention of pollution by all possible hazardous substances, such as oil, noxious liquid substances in bulk, harmful substances carried in packaged form, sewage from ships, garbage from ships and air pollution. The Convention introduced the notion "special areas" in which a special regime of the marine environment protection is envisaged. These "special areas" cover the Black Sea as well as the Mediterranean, Baltic, North, Red seas, the Persian Gulf, the Antarctic area and the Northwestern European waters. In these areas due to the acknowledged technical causes related to their oceanographic and environmental conditions the special compulsory methods of the prevention of sea pollution with oil and noxious liquid substances are applicable. According to this Convention each oil tanker of 150 dwt and above and each ship of 400 dwt and above shall be subject to inspection in every 5 years. These ships shall have the International Certificate on Prevention of Pollution with Oil issued for a term no more than 5 years and a register of oil handling operations that shall record all operations with oil and tanks. The Convention imposes restrictions on discharge of oil, bilge waters and also disposal into the sea of all kinds of food and domestic wastes that are generated as a result of normal ship operation. The littoral states usually set the higher requirements to discharge of hazardous substances into the sea in their territorial waters extending these requirements to foreign ships.

International Maritime Organization (IMO) – the intergovernmental organization addressing the issues of marine shipping. It was founded in 1948 (functions from 1958). Its list of members includes over 120 states. Until 1982 it was known as the Intergovernmental Maritime Consultative Organization. The IMO's primary purpose is to develop and maintain a comprehensive regulatory framework for shipping and its remit today includes safety, environmental concerns, legal matters, technical co-operation, maritime security and the efficiency of shipping. It elaborates draft conventions, treaties and other documents and recommends them to the governments and intergovernmental organizations. IMO is governed by the Assembly of all IMO members and the Secretariat. The Assembly convenes its regular (once in 2 years) and extraordinary sessions at which it discusses the key issues, elects members to the Council, approves the budget, provides recommendations to the IMO members about conventions and regulations on safety of life at sea. The work of IMO is conducted through such committees as Marine Safety Committee (MSC), Marine Environment Protection Committee (MEPC), Legal Committee (LC). The executive body—Council comprises 32 states and guides the IMO activities between the Assembly sessions, studies and submits to the Assembly the MSC, MEPC and LC reports, elects Secretary General and maintains contacts with other organizations. There are permanent and provisional subsidiary organs. Among the permanent bodies there are the Legal Committee, the Committee on Simplification of Formalities and Technical Cooperation Committee. The secretariat staff prepares all documentation, draft reports for printing. The IMO headquarters locate in London, UK.

IMO's mission statement, as stated in Resolution A.1011(26), which sets out the Strategic Plan for the Organization for the period 2010–2015: "The mission of the International Maritime Organization (IMO) as a United Nations specialized agency is to promote safe, secure, environmentally sound, efficient and sustainable shipping through cooperation. This will be accomplished by adopting the highest practicable standards of maritime safety and security, efficiency of navigation and prevention and control of pollution from ships, as well as through consideration of the related legal matters and effective implementation of IMO's instruments with a view to their universal and uniform application."

Inviders of the Black and Azov seas – B.S. is a water body receiving many exotic species migrated from other water bodies. Some of them adapted and started propagating so successfully that provoked major environmental problems. Many natural factors favored such good adaptation in B.S. Among them the are availability of diverse habitats, both in the sea and in coastal water bodies—lagoons, limans, estuaries, river mouths; favorable feeding conditions for benthos-eating, plankton-eating and fish-eating species. The main migratory way of invaders into B.S. is via water. The organisms travel forming growths on the underwater part of ships or in ballast tanks where they get together with the water taken for increasing the draft of a ship after its unloading in a port and for restoration of its sailing capacity. Chronologically the process of invaders intrusion, accidental or intentional, into B.S. may be presented as follows. In the nineteenth century (it is difficult to give a more precise date) the *Balanus improvisus* and *Balanus eburneus* by sticking to the ship underwater part was brought into B.S. They might take their origin on the Atlantic coast of North America. Today *Balanus*, in particular their former species, become so numerous in B.S. that it is difficult to imagine that once they did not live here. We have more accurate data about invaders of the twentieth century. In 1925 the *Blackfordia virginica* was found in the coastal waters of Bulgaria. This is a jelly organism in the form of a ramified polyp to 10 cm high from which small jellyfish being a part of the zooplankton gemmate. The origin place of *Blackfordia* is the Atlantic coast of North America. At present this is a dominant species living in the brackish waters of Black and Azov seas.

Polychaete worm *Mercierella (Mercierella enigmatica)* was first detected in 1929 in the brackish waters of the Paliastomi Lake near Poti in Georgia and later on—in Gelendzhik Bay. It lives in curved limestone cylindrical tubes to 4 cm long from which the aureole of the branchial branches is projected. The tubes intertwine forming a continuous fanciful cover on stones and other underwater things and also on shells of ships. It is believed that *Mercierella* originated in the saline coastal lakes in India. In 1923 it was found in the mouth of the Seine River in France and from there it got into B.S. where it propagated copiously and intruded into the Azov and Caspian seas.

In 1933 jellyfish *Bougainvillia megas* was discovered in the Varna Lake and in the mouth of the Ropotamo River in Bulgaria. It exists mostly as a bottom form. Its appearance reminds of *Blackfordia* and it originates in the same areas. At present it is widespread in the Black and Azov seas in not very saline waters. This is a dominant species which colonies often cover like a carpet the stones, port facilities, shells of ships and water pipelines.

In 1937 a new for B.S. crab species was found in the Dnieper-Bug Liman. It was called Dutch crab (*Rhithropanopeus harrisii tridentate*) coming from the Zuiderzee Bay near the northern coast of the Netherlands. Common names include also dwarf crab, estuarine mud crab, Harris mud crab, and white-tipped mud crab. It is assumed that this crab may be brought here in 1932–1935 with the Dutch ships that called on port Nikolaev. At present the Dutch crab has a numerous population in low-saline waters of B.S. In 1948 it was found in the Azov Sea and in 1957 it was already met in the Caspian Sea. It prefers living in sandy and silt-sandy shallow waters of the sea and estuaries.

In 1946 near Novorossiysk in the mussel biocenosis there was found an unknown large gastropod—rapa (*Rapana venosa*) that came from the Sea of Japan. This is a predatory species eating oysters, mussels and other bivalves. It is thought that *Rapana* was brought to B.S. in the early 1940s by a vessel with the *Rapana* eggs stuck to the bottom. It propagated successfully, especially near the Caucasian coast and in the 1950s had eaten nearly all oysters in the Gudauta bank and then changed over to mussels and scallops populating the same areas as oysters. Later on *Rapana* started devouring the mussel communities near the Southern Coast of Crimea and then moved to the coast of Bulgaria where it also propagated extensively. It also penetrated into the southern part of the Sea of Azov and in the 1970s to the Sea of Marmara, too.

In autumn 1966 a sandy shell or *Mya (Mya arenaria)* was found. Its large up to 10 cm long whitish shells were clearly visible against the smaller local mollusks. It is thought the *Mya* larvae were transported into B.S. with ballast waters taken onboard a ship somewhere in the North Sea and, perhaps, near the coasts of North America. *Mya* became widespread in not very saline shallow areas of B.S. and A.S. forcing out the local small-size bivalves (*Lentidium mediterraneum*).

One more invader—blue crab (*Callinectes sapidus*)—one of the large-sized representatives of its order. It usually occurs near the bottom, but is also capable to rise up to the surface layers. It prefers brakish waters. This crab originated from the Atlantic coast of North America and was brought into the Sea of Marmara and Black Sea with ballast waters of ships. In B.S. the blue crab was first found in 1967 near the Bulgarian coast. In May and August 1975 two blue crabs were caught in the Kerch Strait, in 1984—in the Varna Bay. In the described cases only single individuals were revealed, but there are proofs that the blue crab has been found for many years in the Bosporus Strait.

Nudibrachia mollusks *Doridella (Doridella obscure)* are usually met near the Atlantic coast of the USA and Canada. In 1980 it was first detected in the northwestern part of B.S. In 1986 this species was found in the Varna and Burgas bays in Bulgaria, in the Kerch Strait and on the shelf of Southern Crimea.

One more newcomer is *scapharca bivalve (Scapharca inaequivalvis)*. The first young species of this mollusk that has not been met before in B.S. was found in 1968. That time its belonging to a certain species was not identified. In 1982 it was met on the Bulgarian shelf and referred to gender *Anadara granosa*. In 1984 this mollusk was found on the Romanian shelf and identified as *Scapharca inaequivalvis*, later on as *Anadara inaequivalvis*. Some more years later this mollusk was found in Zhebriyanskiy Bay in the northwestern shelf of the sea and also in the southern part of the Sea of Azov. Large-size representatives of this species may be to 5–6 cm long and weight 60–80 g.

In 1982 large (to 10–11 cm long) jelly-like organisms were found in the water body in the northwestern part of B.S. At present they are identified as comb jellyfish (*Mnemiopsis leidyi*) originated in the coastal Atlantic waters of North America. It is believed that this jellyfish got into B.S. with ballast waters. Showing a high multiplication rate and quick growth this *Mnemiopsis* propagated so successfully that by the late 1980s its biological mass in B.S. was estimated to be close to 1 billion tons. This is a typical jelly-like species (water makes 99 % of its body mass) that is known as an active carnivore eating zooplankton. With such large population and large sizes *Mnemiopsis* became a serious food rival of pelagic fish and as a consumer of their eggs and larvae—their direct enemy, too.

In 1990 in the Odessa Harbor the bottom brown algae (*Desmarestia viridis*) was found for the first time in B.S. In winter months of 1994–1995 *Desmarestia* became the mass species in the coastal zone of the harbor. Its origin place is northern seas of Europe. So far the effect of *Desmarestia* on the local fauna has not been detected.

The freshwater sunny fish (*Lepomis gibbosus*) was brought from America to Europe by aquarists. Having got into ponds and rivers it became rather numerous already in the 1930s in the Danube Delta from where it migrated to B.S. It is also met in the Dnestr and in the recent years in the Dnieper-Bug Liman.

Of the newcomers intentionally imported into B.S. only the Far-Eastern gray-millet pilengas and small mosquito fish (*Gambusia*) were successfully adapted and became mass representatives of the Black Sea fauna, although *Gambusia* adapted well only in low-saline coastal waters. The small freshwater fish *Gambusia (Gambusia holbrooki)* was resettled in the 1920s from Italy to the water bodies in the Kolkhida Lowland on the Caucasus to control malaria mosquitoes. The fish was rather euryhaline and populated many coastal areas of B.S. near river mouths.

I

Pilengas (*Mugil soiuy*) (*Liza Haematochila*) was imported intentionally as a valuable commercial fish. After special studies this kind of millet living in the low-saline waters in the Sea of Japan was recommended for adaptation in B.S. and A.S. Beginning from 1972 the batches of several hundreds of fries caught near the Amur estuary were delivered by aircraft and let out into special nursery ponds in the Shabolatskiy Liman near Odessa and in the Molochnyi Liman near Melitopol on the Sea of Azov for growing. A small portion of fries was let out directly into the Sea of Azov and in the northwestern part of B.S. It required years of persistent efforts of ichthyologists before in the early 1980s *Pilengas* started propagating in B.S. and then in the Sea of Azov and in the late 1980s it became a mass fish species in these seas and an important object of commercial fishing.

In the 1950s–1960s several thousands of grass shrimps—large shrimps from the Posiet Bay of the Sea of Japan were let out into B.S. limans (Khadjibey near Odessa, Kyziltash on the Taman Peninsula). These attempts ended in a failure.

In the period from 1965 to 1981 many thousands of fries of striped bass 5–6 cm long and weighing 20–30 g were let out in various regions of B.S. Fries were grown from the eggs delivered from the USA. There are also data that near the Utrish Cape in the Northern Caucasus single Pacific oysters were found. Nevertheless the intentional and elaborate introduction into B.S. of exotic species is inferior by its results to accidental and unintentional appearance of invaders in B.S.

In the formation of bottom biocenoses of the Sea of Azov the key role is played by B.S. invaders. After soaring propagation of *Mya areneria* in 1985–1988 its population and biomass in 1989 decreased. The minimum average annual biomass was recorded in 1990. The growth of the average annual biomass was witnessed in 1992. Taking into consideration the climatic, hydrological and hydrobiological specific features of the Sea of Azov that are close to the northwestern part of B.S. it can be assumed that in the new biocenosis *Mya areneria* will propagate more extensively in the coastal zone of the sea, in particular along sandy bars with their intensive currents and abundance of food. Accidental invasion and autoacclimatization in the Sea of Azov are of scientific and economic interest. Its further extension and adaptation may influence favorably the development of bottom biocenoses. Availability of such potent biofilter as *Mya* will facilitate water purification in the bottom layers of the sea of organic substances causing oxygen deficit and fish mortality incurring great damage to the fish population. In the water areas near settlements *Mya* will lower the water pollution.

Anadara granosa was first met in the Sea of Azov in April 1989 in the north of the Kaztepe Bay. The single individual was found on the silty-shell ground with the hydrogen sulfide smell. The conditions in A.S. are favorable for active settlement of *Anadara*. This mollusk prefers depths of 9–11 m, silty, silty-shell, silty-sandy grounds enriched rather well with organic carbon and salinity from 9 to 12‰. The acclimatized mollusks are resistant enough to variations in the oxygen regime in the bottom waters surviving in the hypoxia conditions where other mollusks die. Thus, in the recent years in summer the *Anadara* biocenosis was formed in the zone of lower oxygen concentration (less than 60 % of the saturation level). From 1989 to 1997 *Anadara* developed successfully over the whole southern part of Sea of Azov.

The average biomass of this mollusk for the sea in general has increased from 0.5 g/m^2 in 1989 to 44.2 g/m^2 in 1997, while its population from 1 to 152 species/m^2. In 1997 the density of the *Anadara* species varied from 2 to 1,205 species/m^2, while their biomass was 20.5 to 1,274.0 g/m^2.

In 1997 the jellyfish *Beroe ovata* was found in the western part of B.S. that feeds on *Mnemiopsis* imported from the same coastal waters near North America. Investigations conducted in 1999–2005 have shown that *Beroe* adapted in B.S. and found its niche in the B.S. ecosystem. Its seasonal development coincides with development of this species in the coastal waters of the North Atlantic—burst of development in late summer—early autumn and its fadeout in late autumn. Although with appearance of *Beroe* the pelagic ecosystem started restoring, but for the time of its seasonal absence under favorable conditions the *Mnemiopsis* had enough time to fain great biomass and damage essentially the pelagic ecosystem by lowering its yield. But even in such case the time of its action on the ecosystem became much shorter—it reduced to 1–2 months instead of 8–9 months before *Beroe ovata* appearance.

Iphigenia Rock – see Beregovoe.

Isachenko Boris Lavrentievich (1871–1948) – Soviet microbiologist, Academician of USSR Academy of Sciences (1946), Academician of the Ukrainian Academy of Sciences (1945). He took part in the Azov-Black Sea research-fishing expedition of 1922–1926. He discovered in the bottom silt of B.S. the mass development of *Microspira*—bacteria that actively decomposes sulfates with release of much hydrogen sulfide and this happens in the absence of proteins and fiber. *Microspira* is widespread in the bottom of B.S., it is found even at maximum depths over 2,000 m. In 1932 he supported the organization of the USSR's first Microbiological Laboratory at the Biological Station in Sevastopol. He was the author of many works on general, agricultural, technical and geological microbiology and, in particular, the microbiology of the Arctic seas and mud lakes. His name was given to an island in the Kara Sea.

Isakov (Isaakyan) Ivan (Ovanes) Stepanovich (1894–1967) – Admiral of the Soviet Union Navy, naval commander, scientist and writer. He finished the Naval Cadet School (1914–1917). In 1917–1918 he served on the ships of the Baltic Fleet. From spring 1920 he was the commander of the destroyer on the Caspian Sea. He took part in storming Enzeli. In 1920–1921 on the Baltic Sea he commanded the minesweeper, was the master's mate and later the commander of the destroyer. Due to his health condition he was transferred to B.S. In 1921–1923 he was the chief of the Southern Branch of the Supervision and Communication System of B.S., was the senior naval chief of the naval base in Batumi. In 1922–1923—Deputy Chief of the Operations Division of the Naval Headquarters on B.S. In 1924–1925 he commanded a number of destroyers, in 1925 was the Chairman of the Commission on Acceptance and Inspection of the Destroyer "Kaliakria" raised to the surface by the Special Underwater Expedition. In 1926–1928 he was deputy and later Chief of the Operations Division of the Naval Headquarters on B.S. In 1927–1928 he attended the advanced training courses for the commander staff at the Naval Academy. In 1928–1929—Deputy Chief of Staff of the Black Sea Navy on operations activities; in 1928–1931—worked in the Operations Department of the Headquarters of the Red Army; in 1932–1933—senior lecturer at the Strategy and Operations Chair of the Naval Academy. He was the author of the following works: "Landing Operation" (1934) and "Submarine Operations" (together with A.P. Alexandrov and V.A. Belli). In 1934–1935 he was the Chief of Staff of the Baltic Navy; in 1935–1936—lecturer at the Naval Academy. In 1936 he published the book "Operation of the Japanese against Qingdao in 1914" on the basis of which in 1937 he defended the dissertation for the academic degree of the candidate of naval sciences. In 1937 he was appointed the Chief of staff and in August 1937 to February 1938 he was a commander and member of the Military Council of the Baltic Fleet. In 1938 he was promoted to Flagman of the 1st rank and approved as assistant professor. In 1938 he was the member of the Chief Military Council of the Navy; in 1938–1939—the head of the Naval Academy named after K.E. Voroshilov; in 1938–1939—deputy and in 1939—first deputy of the People's Commissar of the USSR Navy. In 1940 he was promoted to the Admiral. From July to October 1941 (the beginning of the Great Patriotic War) he was the Deputy Commander-in-Chief and a member of the Military Council of the Northwestern Front. He guided the organization of Leningrad defense from the sea. Keeping the position of the Deputy Commander-in-Chief he from April 1942 was appointed the Deputy Commander-in-Chief and member of the Military Council of the North Caucasian strategic direction, North Caucasian and Transcaucasian fronts. He coordinated the action of the troops with operations of the Black Sea Fleet, the Azov and Caspian Fleet. He was one of those who commanded the battle for the Caucasus. In winter 1943 he was responsible for publication of materials on the experience of the war at sea, wrote some articles and the book "Navy in the Great Patriotic War" (1944). In 1944 he was conferred the rank of Naval Admiral. In 1944–1945 he was the member of the Governmental Commission on Preparation of Conditions for Germany Capitulation. In 1946–1947 he was Chief of Staff and Deputy Commander-in-Chief of the USSR Navy on study and application of war experience. Member of the Military Council at the Commander-in-Chief of the USSR Navy. In 1947–1954 he was the editor and in 1955–1967—member of the Editorial Board of the "Marine Atlas". From 1950 he resigned from the military service. In 1951 he was conferred the USSR State Award of the 1st degree for the scientific work "Marine Atlas". Vol. 1. "Navigation and Geographical Maps". In 1954–1956—USSR Deputy Minister of the Navy. In 1955 he was given the rank of the Admiral of the Navy of the Soviet Union and was awarded the "Marshal Star" of the Soviet Union. In May 1958 he was elected the corresponding member of the USSR Academy of Sciences. In 1958–1967 he was the General Inspector in the Inspectors' Group of the USSR Defense Ministry. From 1948 to 1967 he was the editor and consultant of the "Large Soviet Encyclopedia" on the section "Naval Science". In 1962 he published the first book "Stories about the Fleet". In 1963 he was admitted to the Union of Soviet Writers. He was the member of the Editorial Board of the journal "Oceanology" (1961–1967) and other publications. In 1965 he was conferred the Hero of the Soviet Union. He was awarded many orders and medals; he also received awards of foreign states. In 2000 a monument to I.S. Isakov was installed in Erevan, Armenia.

Istanbul – the largest city of Turkey, population—13.9 million; the only city located on two continents; extends on both shores of the Bosporus Strait near the Sea of Marmara. I. was founded about 3 millennia ago in place of the town of Ligos located on a small peninsula formed by Golden Horn, Bosporus and Sea of Marmara. The origin and the meaning of the name Ligos are unknown. In 657 B.C. a new city was constructed by the Greek colonists headed by Byzas and in his honor the city was given the name of Byzantium. After Roman Emperor Constantine made the city the new capital of the Roman Empire it was officially named Nova Roma, but more often than not it was called Constantinople or the "City of Constantine" (from the Greek "*polys*" meaning "city"). Quite often it was simply called Polis ("City"). In Rus' instead of Constantinople another name Tsargrad was in wide use in official documents, folklore and literature until seventeenth century. Later it was met only in historical sources and in poetry. Istanbul served as the capital of four empires: the Roman Empire (330–395), the Byzantine Empire (395–1204 and 1261–1453), the Latin Empire (1204–1261), and the Ottoman Empire (1453–1922) In 1453 Sultan Mehmed II captured Constantinople and proclaimed it the new capital of the Ottoman Empire with the Turkish variant of the name—Istanbul. The etymology of this name is still unclear. From the late eighteenth century the distorted name Stambul was used in European countries, including Russia. Istanbul is a city and port extending on both sides of the Bosporus in its southern part. It is the administrative center of the Istanbul Province. The climate is Mediterranean subtropical. The Bosporus Strait splits the city into the Asian (smaller) and the European part. The latter also consists of two parts divided by the Golden Horn Bay deeply incised into land (for 10 km) which banks are linked by two bridges. From 1453 to 1923 I. was the capital of Turkey, but later the capital was transferred to Ankara.

The strategic significance of the B.S. Straits facilitated development of the city and its growing role. In 1970–1973 the suspension bridge 1,560 m long was constructed over the Bosporus. This project turned out economically beneficial. Istanbul is one of the major and fastest-growing metropolitan economies in the world. It hosts the headquarters of many Turkish companies and media outlets and accounts for more than a quarter of the country's gross domestic product. Istanbul has a diverse industrial economy, producing commodities as varied as olive oil, tobacco, transport vehicles, and electronics. Its low-value-added manufacturing sector is also substantial.

11.6 million foreign visitors arrived in Istanbul in 2012, making the city the world's fifth-most-popular tourist place. Its historic center is partially listed as a UNESCO World Heritage Site. Istanbul has seventy museums (Topkapi Palace Museum, the Hagia Sophia, etc.), 17 palaces, 64 mosques, and 49 churches of historical significance.

Istanbul has eight universities, among them: Istanbul University, founded in 1453, is the oldest Turkish educational institution in the city; Istanbul Technical University, founded in 1773 as the Royal School of Naval Engineering, is the world's third-oldest university dedicated entirely to engineering sciences; Mimar Sinan Fine Arts University; Marmara University; Istanbul Medeniyet University, founded in 2010, is the newest public university. The city has about thirty prominent private institutions, for example, Robert College, founded by Christopher Robert and Cyrus Hamlin in 1863; Koç University, founded in 1992; Istanbul Commerce University; Kadir Has University; and others.

Istanbul (http://www.jetsetz.com/cheap-istanbul-flights-ist)

Istanbul Technical University – it was founded in 1773 (initially it was known as Maritime Engineering College). In late June 2000 in Istanbul the presentation assembly of the International Association of Maritime Universities was conducted on the basis of the university's maritime faculty. It possesses one of the Turkish largest libraries.

Istanbul University – one of the oldest educational institutions in Turkey. It was founded in 1453. Till the twentieth century it was the Higher Moslem School (Medrese). In 1923 after declaration of the Republic of Turkey it was reorganized in 1927. In 1933 it became the secular education institution. Its library is one of the largest in Turkey.

Istomin Vladimir Ivanovich (1809–1855) – a participant of the Sevastopol defense of 1854–1855, Russian Rear Admiral (1853). He finished the Naval Cadet Corps (1827) and was appointed to the battleship "Azov" commanded by M.P. Lazarev. In 1827 he took part in the Navarin marine battle; in 1828–1829—in operations against the Turkish Fleet in the Mediterranean, in particular, in blockade of the Dardanelles. In 1832 he was transferred to the Baltic Fleet. From 1836 I. was again sent to the Black Sea Fleet where he served on various ships. From 1850 he commanded a battleship that destroyed three Turkish ships and the central on-shore enemy battery in the battle at Sinop. For his bravery in this battle he was promoted to the Rear Admiral. From the beginning of the Sevastopol defense he commanded the 4th distance of the Malakhov Mound, he was wounded and received contusion, but continued commanding the troops that beat off the enemy attacks. For his bravery and selflessness in the defense of the Malakhov Mound he was awarded the St.-George Order of the 3rd degree. He was hit by a cannon ball and died in the Kamchatka lunette.

Istomin V.I. (http://ikrim.net/2012/0307/071470.html)

Istr, Istria River – see Danube.

Istria – the Ancient Greek colony founded in the seventh century B.C. on the Roman coast of B.S. It was located in the Danube Delta. In the Roman times it was a large port and an important economic and cultural center. In the second century A.D. many temples were constructed here in honor of Greek and Oriental Gods— Zeus, Apollo and muses Dionysus and Mitra. Here the aqueduct 30 km long, public therma, roofed market and dwelling houses were constructed. In 248 as a result of the Goths invasion I. was devastated and ruined. By the late third century I. was constructed anew, but during the new invasion in the fourth century it was destroyed again. By the fifth century the sands blocked the entrance into the harbor and formed a new lake called Sinoe. Today I. is a museum in the open air.

Istrian (Isiakov) – a harbor. It was located under the center of present-day Odessa. In the second half of the seventh century B.C.–fourth century A.D. it was the large Ancient Greek center in the Northern Prichernomorie—the Antique precursor of Odessa. In the Hellenistic times it was called Istrian Harbor, most probably recollecting that it was founded by colonists from Istria (to the south of the Danube-Istra deltas). From the late first century A.D. after settlement here of the Sarmatian tribe Isiaks the city acquired the name of Isiak Harbor. And under this name in 134 it was mentioned as a small seaside village in the "Periple of the Pontus Euxine" by Flavius Arrianus, Roman military commander and historian. Most likely, this village existed till the second half of the fourth century and disappeared after the Huns invasion.

Italian Trading Factorias (Trading Posts) – settlements and fortified places founded by colonists from Genoa, Venetia and Pisa in the thirteenth to fifteenth centuries in Prichernomorie. After the 4th Crusade and capturing by the Crusaders of Constantinople (1204) the Venetians received a free access to B.S. Having their own trade quarter in Constantinople in the first half of the thirteenth century the merchants from the Venetian Republic visited Soldaia (Sugdea, Surozh, Sudak) in Crimea, ports of the Southern Prichernomorie, in particular Amis (Samsoun, Simisso), Trabzon (Trapezund, Asia Minor). In those times the merchants from Pisa founded a small trading post (factoria) on the A.S. coast (Porto Pisano near modern Taganrog). The intensive trade colonization of Prichernomorie started in the 1260s after Genoa signed the ally treaty of Nympheum with Emperor of Nicaea Michael VIII Palaiologos (1261). This treaty provided to Genoa broad privileges, including a right of duty-free trade on the Byzantine territory and ensured it the unobstructed access to B.S. Under the treaty with the Khans of the Golden Horde Genoa annexed in the mid-1260s (on condition of recognizing the supreme power of Khan) the small port Kaffa (Feodosiya) in Crimea that soon turned into the major city and center of all holdings of Genoa in Prichernomorie.

Development of the Genoan trade colonization was facilitated by relocation here of the international trading routes from Asia to Europe (earlier they ran via Syria, Palestine and Egypt) to B.S., especially after ruining of Baghdad (1258) by the Mongols-Tatars and downfall of the last outposts of Crusaders in Near East (1291). Now the trans-Asian trade routes were crossed in Trapezund, Tana (Azak, Azov) and Kaffa. In the late thirteenth century after allotment the Byzantine Emperor of a trade quarter in Constantinople—Galata (Pera) the Genoans founded trading posts in Trapezund, Tana, Vicine (Danube mouth) and also numerous anchorage places on the eastern coast of B. and A. seas. The Venetians established trading posts in Soldaia (c. 1287–1343), Trapezund (beginning from the 1280s, established finally in 1319–1461) and Tana (from the 1320s, established finally in 1332–1475). Initially these trading posts were the centers of transit trade where elite commodities prevailed: from the East spices (pepper, ginger, cinnamon, nutmeg); pigments, frankincense, perfume and ointments; silk (crude and yarn), silk fabric, cotton; precious stones, gold, furs from Russian lands and from Caucasus, etc. Slave trade was also very significant. Among local commodities most valued were alum (the most important components of pigments for fabric exported from Pontus) and also grain and fish from Northern Prichernomorie and wines from Transcaucasus. In exchange for these products the Italian merchants brought from Western Europe wool cloth, silver, metal products (arms, bells, tower clocks), saltpeter and others. The economic and political crisis that occurred in the 1340s–1360s had changed the whole trade structure. First the trade turnover had dropped drastically and then the trading posts reoriented their trade to consumer goods, such as grain, wine, salt, fish, caviar, wood, honey, wax and leather. This was the time when the Italian merchants and, first of all, the Genoans, penetrated deep not only into the urban, but rural infrastructure of Crimea and Azov region where the Genoan nobility built their estates. In 1365 the Genoans took control of Soldaia fortress with 18 villages. In the second half of the fourteenth century they opened consulates in Chembalo (Balaklava), Sinop, Samastro (Amasra), and Sebastopolice (Sukhumi). The coast of B. and A. seas abounded in fortified and trading posts constructed by the Genoans the largest of which were Kilia, Monkastro (Belgorod-Dnestrovskiy), Illiche (in the Dnieper mouth), Djalite (Yalta), Gurzuf, Lusta (Alushta), Vosporo (Kerch), Matrega (Tmutarakan, Taman), Kopa (near Temryuk), Mapa (Anapa), Fasso (Poti), Lo Vati (Batumi), Pontiraklia (Eregli) and Varna.

The Venetians had large factorias in Tana (where their quarter neighbored on the Genoan quarter), Trapezund and Sinop (mid-fifteenth century). Descendants of the Italian factorias (not only Venetians and Genoans, but other inhabitants of Crimea as well) traded actively in the Russian lands and were known as guests-surozhans. The people from Western Europe made the smallest part of the inhabitants. The main occupation in factorias was exchange of goods that received support from local banks and agencies of the leading Italian companies. These were large centers of handicraft (fabrics, glass, ceramics, metal articles, arms, etc.) and shipbuilding (largely of sailing vessels and galleys of middle and small tonnage). The factoria was ruled by a consul (bailo, the Venetian word *bailo* derives from Latin *baiulus*, which originally meant *porter, carrier*) together with the local councils consisting of noble merchants, citizens of Genoa and Venetia and Italian people—the citizens of factorias, and special commissions (financial, forensic, military, trade, etc.). The dioceses of the Catholic Church were headed by metropolitan Vosporo, bishops of Kaffa, Chembalo, Trapezund, Semisso and others. Franciscan and Dominican monasteries were found in Kaffa, Soldaia, Chembalo, Trapezund, Semisso, Tana and others. At the same time the Armenian and Orthodox monasteries and churches with their parishes existed in factorias. But, the material welfare of factorias was constantly threatened by attacks of local rulers and trade rivals. In 1296 Kaffa was seized by the Venetian squadron, in 1308 it was destroyed by Khan Tokhtoy, but it managed to revive within a short time. In the mid-fourteenth century the strong fortifications were built around the city and its outskirts (burgi). Having successfully stood the siege of Tatars in 1344, 1346 and 1386-1387, Kaffa itself attacked the rulers of Solgat and in the fifteenth century—Mangup (Feodoro).

The Genoan and Venetian factorias were captured by the Tatars in 1343, 1395, 1410 and 1418. Less devastating were the attacks in 1442. However, Tana managed to revive many times. The Genoan factoria in Sevastopolice was destroyed in 1454 by the Ottoman Fleet and in 1455—by the Abkhaz warriors. Military conflicts of the Trapezund emperors with the Genoans in 1304, 1313–1314, 1348–1349, 1415–1418 and with the Venetians in 1375–1376 resulted more than once in evacuation of the "Latin" population of factorias. But only Ottomans managed to destroy all factorias. After seizure by Mehmed II of Constantinople (1453) the factorias were cut off from the Mediterranean markets and were gradually conquered by the Ottomans: Samastro in 1459, Sinop and Trapezund in 1461, Kaffa, Tana and all Crimean factoria in 1475.

"ITUR" Project (Italy-Turkey-Ukraine-Russia) – an underwater fiber-optic cable connecting Dzhubga settlement (Russia)—Karolina Bugaz (Odessa Region of Ukraine)—Istanbul (Turkey)—Palermo (Italy). This project was implemented in 1966. This cable has a branch "Sochi" to the government residence of the Russian President (Dzhubga-Sochi).

Ivan Island – is situated near the entrance into the Sozopol Bay, Bulgaria. The island is not large, elevated and cliffy. The small island Pyotr being simply a large rock adjoins the eastern tip of Ivan Island.

Izmail – a city, administrative center of the Izmail Region, the Odessa Oblast, Ukraine. It is situated in the southwestern part of the region, on the left bank of the Kilian arm of the Danube River, 80 km from B.S. A large port receiving all sea vessels. Population—85,000 people (2001). The city extends along the arm for 18 km. The exact time of I. foundation is unknown. According to archeological finds the settlement in the place of I. existed in the 1st millennium A.D. In the ninth to tenth centuries this place was inhabited by the Slav people. In the twelfth century the Genoans built a fortress here. In the twelfth century the lands near the Danube were a part of the Galicia-Volynia Kingdom, later they were conquered by the Mongol-Tatars and in the late fourteenth century they came under the rule of the Moldavian Kingdom. During this period the Slav settlement called Smil existed in the place of I. In the first half of the sixteenth century the Ottomans conquered this territory and they constructed the fortress here that was given the name of the Ottoman Admiral Izmail who commanded the Ottoman Navy. During the Russian-Turkish War of 1768–1774 I. was took by the Russian Corps commanded by Prince N.V. Repnin (in July 1770). From 1771 I. was the main base of the Russian Danube Fleet formation. Under the 1774 Treaty of Kuchuk-Kainarji I. was returned to Turkey. In the 1780s the European engineers were invited to supervise the fortress reconstruction. During the Russian-Turkish War of 1787–1791 the Russia troops led by A.V. Suvorov stormed this fortress and conquered it (in 1790). The governor of I. was appointed M.I. Kutuzov. Under the Treaty of Yassy of 1792 I. was returned to Turkey once again. During the Russian-Turkish War of 1806–1812 the Russian after bombardment of I. by the Danube Fleet in 1809 conquered this fortress again. In 1810 the "Tuchkov Posad" was founded near I. By the 1812

Treaty of Bucharest it was ceded to Russia with the rest of Bessarabia and became the center of the Bessarabia region (from 1873—Province). After the Crimean War of 1853–1856 by the Treaty of Paris in 1856 it was ceded to Moldavia Principality. Russia again gained control of Izmail after the Russian-Turkish War of 1877–1878 under the Treaty of San Stefano of 1878. It was ceded to Russia together with a part of Bessarabia. With the breakup of the Russian Empire in 1918 the region was occupied by the Romanian Army and in 1918–1940 it was a part of the Kingdom of Romania. In 1940 it was occupied by the Soviet Red Army and in August 1940 included in the Ukrainian SSR. During World War II (in 1941–1944) the region was occupied by the Romanian and German Army and liberated by Red Army in 1944. The Izmail Oblast was formed in 1940 and the town remained its administrative centre until the oblast was merged to the Odessa Oblast in 1954. Since August 24, 1991, Izmail has been part of independent Ukraine.

In the late nineteenth century the population of I. was 31,200 people, in 1912—34,000 (72 %—Russians, 12 %—Jews, 7 %—Moldavians as well as Greeks, Bulgarians and others); the greater part of the population worked in the port, on ships—the so-called "fish" factories. In 1903 the Russian-Danube Shipping Company was established. Railway station. Today it is the largest Ukrainian port on Danube River, a center of the food processing industry, a popular regional tourist destination. It is also a base of the Ukrainian Navy and the Ukrainian Sea Guard units operating on the river. The World Wildlife Fund's *Isles of Izmail Regional Landscape Park* is located nearby. The current estimated population is around 85,000, with ethnic Russians forming about 42.7 %, Ukrainians—38 %, Bessarabian Bulgarians—10 %, and Moldovans—4.3 %.

The ruins of the Turkish fortress survived near I. Museum of A.V. Suvorov (including the Diorama "Storm of Izmail Fortress"). The city also has the Danube Museum devoted to the life of the city and region; the Museum of the Ukrainian Danube Shipping showing the history of navigation on the Danube; the monuments built to outstanding people. Quite interesting is the architecture of the old part of the city. Today I. is the center of modern industry. This is a large beautifully built city. It has many research establishments and educations institutions.

Izmail Maritime Trade Port – one of the most well-equipped and highly-mechanized ports on the Danube, Ukraine. It is situated on the left shore of the Kilian girlo between the 85 and 94th km on the Danube. It was founded in the early nineteenth century. It is a major transport center in which the operation of the sea, river, railway and automobile transport gets intertwined. The most optimal route from the Danube European countries to the Transcaucasia, Iran and the shortest route of cargo delivery from Turkey, Greece, etc. to Russia, Baltic republics, Scandinavia go through this port. The navigation here is year-round. The port is provided with the appropriate facilities for reloading, placement and storage of the cargo delivered by sea, railroad and vehicles. The port is capable to receive ships with the maximum length of 150 m, width 30 m, draft 7 m, and deadweight of 6,000 tons. It locates on the crossing of ITC No. 7 and the Balkan branch of ITC

No. 9. IMTP has 24 piers with a total length of 2,600 m. Its cargo turnover in 2004 reached 6.6 million tons (maximum possible turnover—8.5 million tons).

Izmail Maritime Trade Port (http://www.izmport.com.ua/2011-09-12-12-51-55.html)

J

Jason – in ancient Greek mythology, he was an hero who was famous for his role as the leader of the Argonauts and their quest for the Golden Fleece in Colchis (modern Black Sea coast of Georgia). Jason appeared in various literature in the classical world of Greece and Rome, including the epic poem Argonautica and the tragedy Medea.

Jellyfish Mnemiopsis (*Mnemiopsis leidyi*) – refers to *Ctenophores* jellyfish, *Tentaculata* class, *Lobata* order, *Mnemiidae Esch* family. This invertebrate freely swimming jelly organism sizing no more than 5 cm originated from the North American coastal waters. Its body consisting of water for 96.9 % is transparent (only large specimens may be opaque), jelly-like and its form reminds of a walnut. This is an active predatory organism. Feeds on plankton and filters out practically all organic matter on the water surface. During a day it consumes food to about 40 % of its own weight. Most probably, it has got into the food chain of the biological community of B.S. with ballast waters of ships. The burst-like propagation of J.M. led to a sharp reduction of the food base and diminishment of its feeding value. Such situation affected, first of all, anchovy. J.M. was first detected in the B.S. in 1982 by underwater researchers. In the late 1980s its total biomass in B.S. reached 1 billion tons. After J.M. invasion the structure of zooplankton communities has changed significantly. The share of fish feed organisms has decreased. J.M. does not only complete for food with fish, but it also devours the fish eggs and fries. After other jellyfish *Beroe (Beroe ovata)* appeared in B.S. the population of J.M. shrank essentially.

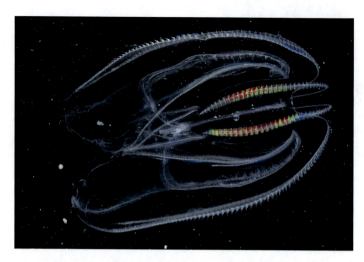

Jellyfish *Mnemiopsis leidyi* (http://images.fineartamerica.com/images-medium-large/a-sea-wal
nut-mnemiopsis-leidyi-george-grall.jpg)

Jones John Paul (1747–1792) – was a Scottish sailor and the USA first well-known
naval fighter in the American Revolution. He was born in Scotland. His naval career
was started as a shipboy at the age of 13. At 21 he was already a captain. At 26 he
became a plantation owner in Virginia, USA. In 1775 he was Lieutenant of the
newly-founded Continental Navy that later became the United States Navy. He first
hoisted the flag of a new state on merchant ship "Alfred". In 1776 commanding the
boat "Providence" he made his first cruise and captured 16 ships. The Congress
promoted J. to Captain. In 1777 he held whole England in horror. In France he
received from Louis XVI a decoration of "Ordre du Mérite Militaire" and a golden
sword. In 1781 he returned to the USA. The Congress officially acknowledges his
services to the nation. In 1783 as a representative of the Congress he was directed to
Paris for receiving prize money for the ships captured by the Americans during the
war and handed over to France. In 1787 Russian Ambassador Simolin proposed him
to enter into the service in Russia. J. went to Copenhagen where he received the
personal message from Catherine II and after this he took his final decision. From
Stockholm he went to Russia.

In 1788 J. served on B.S. First he commanded the rowing fleet operating in the
Dnieper-Bug Liman. Here he got acquainted with the Zaporozhie Cossacks who
admitted him in their ranks. In 1788 J. won the resounding victory over the Ottoman
Fleet in the Danube mouth near Ochakov. After this he went to Petersburg where he
was met as a real hero. This famous American naval officer died in 1792 in
revolutionary Paris. In 1851 his body was placed into a metal coffin filled with
cognac. The US Congress decided to transfer the remnants of J.P. Jones to America,
but this happened only in 1905. He was buried at the church of the Naval Academy
in Boston. After his death he was awarded the rank of Admiral. In the USA he is
considered the founder of the US Navy and one of its guided missile destroyers
bears the name of "Paul Jones".

Jones J.P. (http://en.wikipedia.org/wiki/File:John_Paul_Jones_by_Charles_Wilson_Peale,_c1781.jpg)

Juniper High – a slender coniferous tree with a tapering top, 8–10 m high, a kin to cypress. Its stem is often twisted and is covered with a specific band-like cracked brown-red bark. J. has the bluish-green soft needles and large oval light-violet fruits with a grayish tint. It grows in Crimea as separate trees and small groups of trees. J. is draught-resistant, indifferent to soils, grows well among rocks and even in limestone cracks. This tree releases aromatic bactericidal substances that make the air in the forest curative. This tree is under protection and put on the "Red Book of Ukraine". The nature reserve is created on the Martjan Cape.

K

Kabardinka – a resort town in the Gelendzhik resort zone. It is situated practically in the middle between Novorossiysk and Gelendzhik (21 km to the southeast of Novorossiysk and 14 km to the northwest of Gelendzhik) in the southern tip of the Tsemess Bay, Krasnodar Territory, Russia. The territory of K. had been populated since ancient times. K. appeared after conclusion of the Peace Treaty of Adrianople of 1829. The town was founded in 1836 as the Alexandria Fort that was later renamed into Kabardinskiy by the name of a military unit that made a part of the fortress garrison. K. became one of the fortresses on the Black Sea coastline. After the Caucasian War the Greeks that fled from Turkey and later on the retired soldiers came to settle here. In 1893 the Sukhumi road connected K. with Novorossiysk. From 1920 K. became a seaside resort. The population was 9,000 people. The winter is warm and summer is hot here. The mean annual temperature is +12.4 °C; precipitations—600 mm; number of sunny days—more than 200 a year. K. has the pebble and sand beach. The sea floor is sandy smoothly sloping. The bathing season lasts from June through September. There are many children health-improvement camps, several tourist bases, rest houses and small hotels in the vicinity of K. Relict pine grove. Memorial complex "Captain Zubkov Battery". Thanks to this battery during 2 years of battles for Novorossiysk during the Great Patriotic War no one German ship entered the Tsemess Bay. To the southwest of K. the Doob Cape is rising in the picturesque surroundings of which the 5-star hotel "Nadezhda" is situated. Population: 7,500 (2010).

S.R. Grinevetsky et al., *The Black Sea Encyclopedia*,
DOI 10.1007/978-3-642-55227-4_12, © Springer-Verlag Berlin Heidelberg 2015

Kabardinka (http://www.business-kuban.ru/i/pic/kabardinka.jpg)

Kacha – an urban-type settlement on the B.S. coast 5 km northward of the Kacha River mouth and 40 km from Sevastopol, Crimea, Ukraine. It was founded in 1912 in place of the farmstead Aleksandro-Mikhailovskiy due to relocation here of the Russia's first flying school (later on it became the famous Kachinsky Flying School). Population: 5,200 (2012).

Kacha River (in Turk meaning "camping") – one of the main waterways of Crimea, the third by length (69 km) and the fourth by water availability. It takes its origin on the northern slope of Babugan from confluence of the Biyuk-Uzen and Pisara rivers. In its upper reaches it forms cascades of small waterfalls. Near Shelkovichnyi village it rushes onto a wide valley. The average flow is 0.058 km^3/year. The largest tributaries (mostly in the upper reaches) are Marta, Donga, Kaspana and Stilya rivers. K. flows into B.S. It is regulated by the Zagorsky and Bakhchisaray reservoirs.

Kafes, Kaffa – see Feodosiya.

Kagalnik – a river in the Rostov Region, Russia. It flows into the Taganrog Bay of the Sea of Azov. Its length is 134 km. It runs over the Kuban-Priazov Lowland. The river is broken into several pools.

Kagul Battle – it took place on 21 July 1770 during the Russian-Ottoman War of 1768–1774 on the Kagul River (the left tributary of the Danube) near village Vulkaneshti (presently the Moldova Republic). The Russian troops under command of General P.A. Rumyantsev defeated the main forces of the Ottoman Army that was commanded by Grand Vizier Khalil-Pasha. The Turks fled, while the Tatar troops did not venture to go into action and retreated. The Turks losses were great: about 20,000 people were killed, 2,000 people were taken prisoners, 130 cannons were captured as well as the treasury of Grand Vizier. The Russian casualties were 1,500 people. As a result of this victory the Russian troops occupied Brailov, Bucharest, Izmail, Kilia, Salchi and the whole left bank of the Danube.

Kalamita – (in Greek meaning "good, beautiful cape")—a fortress built in the Middle Ages in Crimea, 6 km to the southeast of the city of Inkerman. It was constructed on a rock in the mouth of Chernaya River, in the mouth of the Severnaya Bay near Sevastopol, Ukraine. It was founded in the fifth–sixth centuries for defending the access to Chersonesos. In the south and west the fortress is protected with cliffs, while in the north and east—with a trench and a wall cut in a rock. The area surrounded by the walls was 1,500 m². In the eighth–ninth centuries a monastery was constructed here and only a complex of churches cut in the steep rock (basilica, delubrum and others) and numerous living quarters (cells) arranged terrace-like survived to the present. At the beginning of 1427 in place of the medieval fortress the rulers of the Crimean Feudal Kingdom Feodoro built a new one for defending the port opposing the Genoan colony of Cembalo (present-day Balaklava). In 1475 the Ottomans that seized Crimea destroyed the Feodoro Kingdom and its fortress was rebuilt and called Inkerman. Seeking to consolidate their position on the coast of B. and A. seas the Ottoman invaders rebuilt K. having adapted it to using firearms. They controlled this fortress till the Peace Treaty of Kuchuk-Kainarji (1774). After conclusion of this treaty K. lost its significance and decayed. K. is one of the most interesting medieval monuments in Crimea; it is a branch of the Kherson historical-archeological nature reserve "Fortress Kalamita".

Kalamitskiy (Kalamita) Bay – cuts for 13 km into the western coast of the Crimean Peninsula between Evpatoriya Cape and Lucul Cape distanced each other. The entrance width is 41 km, the depth—about 30 m. Evpatoriya Bay with port Evpatoriya projects into the coast of K.B. The northern part of the eastern coast of the bay is sandy and low, while the southern part—of the reddish color and 6–8 m wide. The Alma and Bulganak rivers flow into the bay. The coasts are low and sandy. In the coastal zone there are many salty lakes separated from the bay by sandy barriers: Sasyk (Evpatorian) Lake in the northeastern shore of the bay, Sakskoe Lake near the town of Saky, Kyzyl-Yar and Chaika lakes southward of Sakskoe Lake known for their curative effect.

Kalanchakskiy Bay (Liman) – a B.S. bay on the northern coast of the eastern part of the Karkinitskiy Bay, to the east of the Karzhinskiy Bay, Crimea, Ukraine. The Karaday Peninsula ending with Delgeltip Cape is found on the eastern side of the bay. The Kalanchakskiy islands locate southward of the bay.

Kalender Cape – locates 3.2 miles southward of the mouth of the Chorukh River, southeastern coast of the Black Sea. The cape is elevated and steep. The high Pontic Mountains confining the Chorukh River valley on the south came out to the shore near the cape. More southward of the cape there is a state border between Georgia and Turkey.

Kaliakra Cape – projects into the Black Sea for 2 km, Bulgaria. This is a narrow and rocky land on the Black Sea coast located not far from the Bulgarian town of Balchik. This is one of the most interesting natural-historic monuments on the Bulgarian coast. Although the legend connects K. with the name of 1 of 40 girls—Kaliakra, most probably the cape received its name—"Kali Akra" (Beautiful nose) from its purple coloring due to the presence of iron oxides. It was known to the Thraceans, Greeks and Romans. In the fifteenth century the Bulgarian ruler of Primorie Dobrotich built here a fortress and declared K. the capital of his lands (Karvunsky Land). During the Ottomans invasion the fortress was destroyed. The Bulgaria-Romania state border is 45 km northward of the cape. Near this cape the Russian squadron of ships commanded by Rear Admiral F.F. Ushakov defeated the Turkish Fleet under command of Kapudan-Pasha Hussein. K.C. is a nature reserve zone, so the marine vessels are not permitted to sail near K.; it is prohibited to use motor boat here; the tourists are not allowed to descend to the foot. This reserve was created to protect the Black Sea monk-seals because only several pairs of these animals survived to the present.

Kaliakra Cape (http://dic.academic.ru/pictures/wiki/files/78/Nos_Kaliakra.jpg)

Kalmius River – a river in Ukraine. Its length is 236 km, the basin area is 5,000 km². It originates on the southern slope of the Donetsk range and flows into the Sea of Azov. K. is largely of snow feeding. The ice cover is unstable and lasts from December through March. The cities of Donetsk and Mariupol (in the mouth) are situated on this river.

Kalos-Limen – see Chernomorskoe.

Kamchiya River – a river in eastern Bulgaria, flows into B.S. 34 km southward of Varna. It is formed by confluence of two rivers—Golyama-Kamchiya (Ticha) and Luda-Kamchiya originating in the eastern part of the mountain range of Stara-Planina. The length of K. (together with Golyama-Kamchiya) is 244 km; the watershed area is 5,100 km^2; the depth in the lower reaches is 4–5 m. In 40 km from its mouth K. enters into the Longoz Lowland, flooding it in spring and autumn.

Kamchiya Testing Ground – belongs to the Institute of Oceanology of the Bulgarian Academy of Sciences. It was constructed in 1977 in several kilometers from the mouth of the Kamchiya River. The central structure of the testing ground is the through trestle 250 m long reaching the depths of about 6 m. Further on along the profile the individual stationary supports are installed at depths 10, 15 and 18 m. The trestles and supports are connected via cable and radiotelephone communication lines with the on-shore laboratory. Practically all typical sites of the coastal zone were covered by investigations. During three cycles of work they managed to develop new facilities and to elaborate on the methods of measuring the absolute concentrations of suspended matter during storms from the trestle. The results of the "Kamchiya" experiments were described in three books published by the Bulgarian AS in 1980, 1982 and 1983.

Kamyshburunskiy Iron Ore Works – a large enterprise supplying metallurgical plants of the Azov region with raw material, located in Kerch, Crimea, Ukraine. Iron ore production was started in the nineteenth century. The iron ores of Kerch were used solely in cast iron production at the works in Kerch and Taganrog as well at the "Russian Providence" works. The ores are mined by the open-strip method. After extraction the ores are directed to dressing and together with coke and flux are sintered to form agglomerates that on special vessels are delivered to metallurgical plants. In 1900 on the basis of the Kerch ore deposits a plant was built that in 1927 was given the name of Pyotr Voikov who lived in Kerch. In 1939, the plant was re-named after Sergo Ordzhonikidze. Today, the company produces sinter, limestone, calcium carbide.

Kanin Vasiliy Alekseevich (1862–1927) – Russian military commander, Admiral (1916). He finished the Naval Cadet Corps (1882) and the Naval Mine Officer School (1891) and was conferred the rank of the Naval mine officer of first degree. He served on the Baltic Fleet. In 1901 he served as a miner on the flagship of the Prakticheskiy Squadron of the Black Sea Fleet. In 1902 he was appointed the senior officer at the gunboat "Chernomorets", in 1903—at the squadron armored ship "Georgiy Pobedonosets". In 1904–1907 he headed the torpedo warehouse and fire adjustment station in the port of Sevastopol. From 1907 he commanded the gunboat "Kubanets". In 1908 he was promoted to Captain 1st rank and appointed to command the battleship "Sinop". From 1911 to 1918 he served on the Baltic Fleet. In 1913 he was conferred the rank of the Rear Admiral and in 1915—the Vice Admiral. From May 1915 to September 1916 he commanded the Baltic Fleet. From September 1916 to April 1917 he was the member of the State Admiralty Council. In December 1917 he resigned. He joined the White Movement and after the German occupational troops left Crimea and Ukraine he commanded the Black Sea Fleet of the Armed Forces of Southern Russia. He emigrated from Russia and in 1920 he died in Marseilles, France.

Kappadokia – an area in Asia Minor between Halys (today—Kizilirmak) and Euphrates rivers. It was divided into Pontic K. (B.S. coast) and Inland K. Till 547 B.C. it was Lydian and later—Persian satrapy (at Cyrus II); from 302 B.C. it was independent and about 100 B.C. it was conquered by king Mithridates VI, who had wars with Rome for this territory. Rome defeated the king and K. came in dependence on Rome. From the seventeenth century it was the Roman province. K. exported alabaster, crystal glass and salt.

Karadag (in the Turk meaning "black mountain") – a mountain, the single unique isolated volcanic massif in Crimea between the Otuz valley and the large Koktebel Depression on the B.S. coast, 20 km to the southwest of Feodosia, Crimea, Ukraine. K. was formed approximately 150–160 million years ago in the Jurassic Time. The coastal part of K. was cut off as a result of strong downthrow that occurred about 50–100 million years ago. Its height reaches 723 m. K. is one of a few places on the globe where the magnetic anomaly is observed. This phenomenon was discovered in 1921 by A. Spasokukotskiy. Volcanic tuff is developed here. The seaside resort of Koktebel (formerly Planerskoe) locates at the eastern foot of this mountain. K. consists of the Coastal Range extending along the coast and the dome-shaped inland massif of the Saint Mountain. Its top is a chaotic assembly of stone towers, columns and peaks. From the sea you can admire the view of the Razboinichya Bay that is protected by the rock "Ivan Razboinik", the Pogranichnaya Bay with the magnificent andesite arch "Golden Gates" rising at its entrance and rocks Lev (lion) and Slon (elephant).

Karadag (Photo by Evgenia Kostyanaya)

Karadag Biological Research Station – was opened in 1914. Beginning from 1907 T.I. Vyazemskiy, Privat-Docent of the Moscow University, and Professor L.Z. Morokhovets contributed their own modest money to construction of this station. At different times Academician F.Yu. Levinson-Lessing, Professor I.I. Puzanov and other outstanding scientists worked here. The station belonged to the Crimean Branch of the Ukrainian Academy of Sciences. At present it is a division of Institute of Biology of the Southern Seas (IBSS) of the Ukrainian AS. The scientists carried out here regular researches of the natural landscape, fauna and flora of Crimea.

Karadag State Nature Reserve – In 1922 Academician A.P. Pavlov, the well-known geologist, addressed the Russian Scientific Congress in Moscow with a proposal to create a national park on Karadag. But the first actions on K. protection were taken only after the Great Patriotic War of 1941–1945. In 1963 K. was declared the nature monument of the republican significance and on 9 August 1979 by the resolution of the Ukrainian Council of Ministers this area was given the status of the state nature reserve and was subordinated to the Karadag Branch of IBSS of the Ukrainian AS. This was done with a view to preserve the unique mineralogical complex and to protect the rare flora and fauna of Karadag. From 1999 it became an independent organization directly subordinated to the Ukrainian AS. The nature reserve covers the whole mountain group of Karadag with an area of about 29 km^2, including 8 km^2 of the B.S. water area. The K. flora comprises 1,169 varieties of plants of which 74 are endemic, 62 are put on the Red Book of Ukraine, and 28 are in the Europe's Red List. The K. fauna comprises 3,475 species of which over 2,000 are invertebrates. Especially numerous is the insect group numbering 2,362 species, of which 61 species are put on the Red Book of Ukraine and 16 species—in the Europe's Red List. Some 4 species of the amphibians, 8 species of the reptiles, 210 species of birds and 129 species of mammals as well as 90 species of fish, dolphins and seals are found here. The extensive researches are conducted in this nature reserve with a view to study and preserve this unique landscape monument of Crimea. The territory of the reserve is closed for free visiting, only excursions led by research workers and local experienced game-keepers are allowed here.

Karaims (self-name "karailar") – the ethnic group of people speaking the Karaim Kypchak language of the Turk group of the Altai family. It has Crimean, Trakai (northern) and Galich (southern) dialects. They profess Karaimism, their sacred book is Old Testament. The Karaim written language developed on the basis of the ancient Jewish alphabet. In the thirteenth century a great number of K., mostly from the Byzantine Empire, settled in Crimea. In the late fourteenth century by the order of Great Lithuanian Prince Vitovt K. were resettled from Crimea to Troki (present Trakai, Lithuania). Their main occupation was farming, horticulture, handicrafts and trade. In 1783–1795 they were incorporated into the Russian Empire. From 1840 K. used the Jewish, Latin and Slav scripts for writing religious and secular texts. In 1837 the Decree on Tavrian Karaim Sacred Ruling was published that in 1850 was extended to K. living in western provinces. All K. communities were divided into two districts with the centers in Evpatoriya and Troki. They were headed by clergymen called "gakhams". In 1863 the rights of K. were equated with those of Orthodox which was confirmed by the circular order of the Ministry of Internal Affairs of Russia (1881). Today world population: about 2,000 from whom in Ukraine—1,200, Lithuania—240, Russia—200, Poland, Israel, USA.

Karalar Steppe – covers the northern part of the Kerch Peninsula: from the Zolotoe village on the shore of the Kazantip Bay to the Kurortnoe village on the eastern tip of the Chokrak Lake (about 11,000 ha in area). In the north its border runs over practically virgin coast of the Azov Sea, while in the south it runs into the agricultural lands of the local farms. In 1988 the nature reserve "Karalarskiy" 6,806 ha in area was created on a part of this territory. The plant communities growing here are specific of Crimea and unique for Ukraine. The rare zonal vegetation, such as fescue—feather grass—herbal steppes, is preserved here. A diverse and unique picture is created by the vegetation overgrowing the sandstone outcrops on the rocky ranges, sandy beaches on the Azov Sea coast and sand barriers of the large and practically unused Chokrak Lake, the fragments of hydrophilic steppes that use moisture from sea air. What makes this territory generally valuable is the savannah-like vegetation dominated by shrubs and ephemeral cereal grasses. Considerable areas are taken by halophytic and semidesert vegetation. In the water bodies you can find specific communities of aqueous plants, including *Ruppia spiralis.*

Karamussal (in the Turk "*kara*" meaning "black" and "*mursal*" meaning "envoy") – the Medieval Turkish cargo vessel. It had a sharp hull line, high boards and 2 masts—on the bow for square sails and on the small stern—for triangular sail. K. had a bowsprit with a jib sail. Such ships were built of sycamore wood and painted black.

Karasu – a seaside resort very popular among the people living in the economic and administrative center of Sakarya Province, Turkey. It has a beautiful beach and very picturesque whereabouts. Local population is around 24,500 which grows in the summer to 250,000 due to tourists.

Karkinitskiy Bay – a B.S. bay between the Tendrovskiy Spit and the Tarkhankut Peninsula, Crimea, Ukraine. The top of K.B. is separated from the Sivash Bay with the narrow Isthmus of Perekop. It received its name by the Greek town of Karkinita (Kerkinitis, present Evpatoria) founded on the western coast of Crimea in the fourth–fifth centuries B.C. It incises into the land for 118.5 km. K.B. is divided by the Bakalskiy Spit into two parts: the western part with the depths to 36 m with the straightened coastline and the eastern part with the depths to 10 m and the coast incised with harbors. The western and eastern parts of K.B. are connected by a narrow passage with depths from 5.8 to 13.4 m between the Dzharylgach Island and Bakalskiy Spit. In the eastern part of K.B. the so-called phyllophoric field locates at depths of 10–15 m where the commercial production of phyllophora was started after drastic reduction of the algae resources in the "major" field.

Karkinitskiy Ornithological Nature Reserve – it is located in the water area of the Karkinitsky Bay and is of the republican significance. It was established in 1978, its area is about 276 km^2. It was created for preservation of the genofund of the local flora and fauna and also natural communities specific of this water area. This is the place of rest for the migratory waterfowl in spring and the habitat for the migratory birds in winter.

Karolino-Bugaz – a village on the coast of the Black Sea, Ukraine.

Kartazzi Ivan Egorovich (1838–1903) – Russian geodetic specialist and astronomer. He started its military service as a trooper of the Volynsky Regiment. He took part in the Sevastopol defense of 1854–1855. For his excellent service in 1854 he was promoted to ensign-bearer and in 1855—to ensign. In 1858 he was sent to get education to the Nikolaev Academy of the General Staff to a geodetic department. From 1863 to 1866 he served in the observatory of the Nikolaev Academy and later in the Military-Topographic Depot. He participated in geodetic works conducted on the Northern Caucasus and was conferred the Captain rank. In 1866 he was appointed the associated astronomer of the Nikolaev Academy's Observatory. In 1868 for his achievements he was promoted to Lieutenant-Colonel and enrolled to the Military Topographers Corps as a geodetic specialist with appointment to the position of professor of higher geodesy and practical astronomy of the Nikolaev Academy of the General Staff. In 1872 he was appointed the astronomer of the Marine Astronomical Observatory in Nikolaev. From this time on he headed all major geodetic works on B.S. In 1873 together with Junior Captain K.A. Myakishev he determined the difference of longitudes among Nikolaev, Sevastopol and Kerch. In 1874 on the schooner "Novorossiysk" and ship "Prut" he determined the astronomical points on the coast of Turkey and Russia, including in Sinop, Yazon, Trapezund, Batumi, Yalta, Feodosiya, Mariupol and Taganrog. In 1875 he directly supervised the works on geodetic linking of the astronomical observatory in Nikolaev with the nearest triangulation points located near the mouth of Bug River. In 1876 under his leadership the trigonometric net was made between the astronomical observatories in Nikolaev and Odessa. In 1887 he was promoted to the actual state counselor.

Karzhinskiy Bay – is situated on the northern coast of the eastern part of the Karkinitskiy Bay, to the northeast of the Dzharylagach Bay and to the west of the Kalanchakskiy Bay.

Kasatonov Vladimir Afanasievich (1910–1989) – Admiral of the USSR Fleet. From 1927 he served in the Navy. In 1931 he graduated from the M.V. Frunze Naval College and from 1931 to December of 1932 he served as a submarine navigator and assistant commander on the Baltic Sea. In 1933–1939 he was the assistant commander, commander of a submarine, commander of a submarine division of the Pacific Fleet. In 1941 he was appointed the Head of Staff of the submarine division of the Baltic Fleet, and in this capacity he took part in the battles during the Great Patriotic War. In July 1941 he was the senior commander-operator and later the Chief of the Operations Department of the Naval Headquarters. He was in charge of study and analysis of the operational situation on the Far East and also introduction of the naval experience of other warring fleets into military training of the Pacific Navy. From December 1945 he was the Chief of Staff of the Kronshtadt Naval Defense District. In 1947–1949 he was the division head and head assistant of the Headquarters' Chief Department. From October 1949 he was the Chief of staff and first deputy Commander of the Fifth Navy and from April 1953—of the Pacific Fleet. From 1954 he commanded the 8th Navy, from December 1955—the Black Sea Navy and from February 1962—the Northern Fleet. In 1964–1971—the First Deputy of the Navy Commander-in-Chief. Fleet Admiral (1965). In the 1960s he guided development of a new generation of warships, studied the experience of fleet actions in the high-latitude regions of the World Ocean and in ice conditions of Arctic. For his great contribution into improvement of the Navy military readiness and testing of new ships he was conferred in 1966 the title of Hero of the Soviet Union. From 1974 he was the member of the General Inspectorate Group of the USSR Ministry of Defense. He wrote the books of reminiscences entitled "On the Fairways of the Fleet Service" (1994) and "Recollections about N.G. Kuznetsov". He was awarded many Soviet orders and medals and also orders of foreign states.

Kasatonov V.A. (http://viperson.ru/data/200804/kasatonovv.jpg)

Katran – see Dogfish.

Katsonis (Katsones, Kachioni) Lambros (1752–1805) – is known in Russia as Kachioni Lambro Dmitrievich, the Greek national hero who fought against the Ottoman yoke. Katsones fought as a volunteer (1770–1774) in the Russian-Turkish

War of 1768–1774. He was a commander of the volunteer Greek Fleet, organized in 1788 in Trieste, during the Russian-Turkish War of 1787–1791, a privateer. For his heroic deeds during the Russian-Turkish War he was conferred the rank of a Major and his fez was decorated with the silver embroidered hand with the words: "Under Catherine's hand". At the beginning of the war he commanded on B.S. the Russian privateer ship "Knyaz Grigory Potemkin-Tavricheskiy", later he was sent by the Tsar government to the Mediterranean for organization of privateer operations. By summer 1788 his fleet comprised 9 ships with the crews staffed with Greeks and Albanians. Here K. commanded the frigate "Minerva Severa" (Northern Minerva) (the name given in honor of Catherine II the Great, French philosophers called her so). In 1789 he won a number of victories in sea battles in the Aegean Sea. Catherine II awarded K. and conferred him the rank of Lieutenant-Colonel. Ottoman Sultan Abdul-Khamid I tried to gain K. to his side and promised him forgiveness for the shed "Ottoman blood", granting of any island in the Aegean Sea into his hereditary holding and 200,00 gold pieces. K. refused and organized new attacks of the Ottoman ships. The main battle occurred in April 1790: 30 large Ottoman ships were against 7 ships of K. In this battle K. lost 5 ships and 600 seamen. K. himself was wounded in the head. The Ottomans lost the greater part of their ships and over 3,000 seamen. Catherine II awarded K. with the Saint George Order of the 4th degree and promoted him to the rank of Colonel. The squadron of K. was restored and by summer 1791 it numbered 24 ships. After the Peace Treaty of Yassy in 1792 in which Greece was not even mentioned K. refused to disarm his ships and continued independently the struggle for liberation of his native country. As the K. ships attacked not only the Turkish, but French ships this caused the French to join with the Ottomans to try to stop K. and after this the K. squadron was defeated in 1972. The crews of the K. ships reached the Venetian lands where many were arrested. K. returned to Russia, received forgiveness and was appointed the commander of the Balaklava Greek Regiment. In 1796 Pavel I, the Emperor of Russia, signed the degree that ordered K. to go to Odessa to the Black Sea rowing fleet under command of Rear Admiral P.V. Pustoshkin. But K. went to Crimea where he purchased Panas-Chir land (in Greek "Sacred meadow") not far from Yalta and started construction of his homestead that he renamed into Livadia. He became a large industrialist, built a plant for production of grape vodka. It was said that he was poisoned by a Turkish agent. George Gordon Byron took K. as a prototype for his poem "The Corsair". Greek people esteem high their national hero and gave his name to two submarines (in 1927 and in the 1980s). The life of this remarkable person was depicted by Russian writer V. Pikul in his novel "First Listrigon of Balaklava".

Katsiveli – a resort town, scientific center located on the B.S. coast westward of Simeiz near the foot of the Koshka Mountain, Crimea, Ukraine. This is the land of A.A. Polovtsev, a husband of Count Panina, the owner of Gaspra. Few clouds, air dryness and transparency favored the organization of an astronomical center here and later of space researche. The Radio-Astronomical Laboratory of the Crimean Astrophysical Observatory locates here. It has a radio telescope with the antenna

22 m in diameter making it possible to view the sun, stars and galaxies in the centimeter and decimeter ranges. The observatory in Simeiz on the Koshka Mountain was built on the money of rich industrialist N.S. Maltsev. In 1908 he presented this scientific center to the astronomers of the observatory in Pulkovo (near St.-Petersburg) and after this it became a branch of Pulkovo. At present this town is a place of the radio-astronomical division of the Crimean Astrophysical Observatory being one of the elements of the "Interkosmos" system. This is an experimental laser station for tracking the artificial Earth's satellites. Here the radiotelescope with a parabolic antenna 22 m in diameter is installed that in October 1959 received for the first time in the mankind history the image of the Moon's back side transmitted by the Soviet automatic interplanetary station "Luna-3". In addition, the experimental division of the Marine Hydrophysical Institute (Sevastopol) with the Marine Laboratory with a storm basin, the solar engineering base of the Material Study Institute of Ukrainian AS which unique facilities are used for investigation of the sun nature are located here. Till 1917 well-known artist A.I. Kuindzhi who painted a series of sea sketches lived here. Population: 600 (2012).

Katsiveli. Radiotelescope of the Crimean Astrophysical Observatory (http://fotki.yandex.ru/users/bikogor/date/2010-08-24)

Kavarna – city-port with the millennia-long history. It is situated in the north of the Varna Bay of B.S., to the northeast of the town of Balchik, Bulgaria. The first settlement appeared here in the ancient times. It was founded by Thraceans and it was populated in succession by the Greeks, Romans, Slavs and Bulgarians. In the fifth century a Greek colony appeared in this place that later was destroyed by an earthquake. A new settlement that was constructed by the Romans was called K. A large center of rainfed farming. Several resorts are found in the vicinity of K. The historical Kaliakra Cape is located 16 km to the northeast of K.

Kavkaz (Caucasus), Port – is situated in the Temryuk Region of the Krasnodar Territory, Russia, near the settlement Kosa Chushka on the coast of the Kerch Strait. It is connected via train and passenger ferry lines with port "Krym" in Crimea, Ukraine. The line length is 3.8 km. In the cargo traffic a consideration share (over 200,000 tons annually) is taken by the chemical products (ammonia sulfate, carbamide and sulfur). The crossing over the Kerch Strait had existed since the time of the Pontus Kingdom (fourth century B.C.) and the Great Silk Road that passed in this area. In 2004 the train ferry "Kavkaz-Poti" was put into operation. On February 28, 2009 a new railway ferry line—the port "Kavkaz" (Russia)—Varna (Bulgaria) was opened, which operates by the ferry "Avangard" with 45 conventional train cars. In autumn 2010, a second ferry "Slavyanin" (capacity of 50 cars) was put in operation on this line. From Russia to Bulgaria by ferry go liquefied gas, oil products, technical oil, glass and chemical products. From Bulgaria general cargo and consumer goods come. On November 20, 2011 commercial operation of the ferry line Zonguldak (Turkey)—port "Kavkaz" has started on a ferry "ANT 2", with 80 trucks and 86 passengers. The line operates once a week. On April 2010 the presidents of Russia and Ukraine signed the Kharkiv Agreement, the articles of which provide for the construction of a bridge across the Kerch Strait.

"Kavkaz" Port (http://www.yugopolis.ru/news/incidents/2011/)

Kazachya Bay – one of the westernmost bays in Sevastopol, Crimea, Ukraine. In the west it joins the Kamyshovaya Bay forming the Dvoinaya (Troinaya) Bay. By a small peninsula going off from the gently sloping coast the bay head is divided into two parts. In the west a small shallow bay of Solenaya merges with K.B. The coasts of K.B. are elevated and gently sloping. The Ukrainian experimental maritime research center "State Oceanarium" of the Ukrainian Navy locates in this bay. On the basis of this Oceanarium the public nature reserve "Kazachya Bay" was established.

Kazantip Cape – (old meaning "bottom of depression")—a circular reef composed of bryozoan limestone located in the north of the Kerch Peninsula. It projects deeply into A.S. It separates the Arabat Bay situated westward of it and the Kazantip Bay situated to the east of it. Its elevation over the sea level is 106 m. The Antash salt lake locates near the cape. In the Soviet times permission was given to construct the Crimean Nuclear Plant on the cape, more precisely on the lake shore, but due to the efforts of the ecologists this construction project was stopped. The only things reminding of this project are the giant concrete constructions that have been left here and the nearby town of builders Schelkino that received its name in honor of Academician Kirill Ivanovich Schelkin (1911–1968), one of the "fathers" of the nuclear and hydrogen bombs and nuclear power engineering. The fans of windsurfing took fancy of this place. In the recent years the oil prospecting works had been conducted here. The cape is included into the Kazantip nature reserve.

Kazarskiy Aleksandr Ivanovich (1797–1833) – Russian naval seaman, Aide-de-camp (1829), Captain 1st rank (1831). He started his military service in 1811 as a volunteer on the Black Sea Fleet. In 1816–1819 he commanded the battle boats of the Danube Naval Fleet. From 1820 he served on the Black Sea Fleet, an officer at frigates "Evstafiy" and "Liliya", transport boat "Ingul"; in 1826–1828 he commanded the brigantine "Sopernik". During the Russian-Turkish War of 1828–1829 he showed himself well during storming of Turkish fortresses Anapa (June 1828) and Varna (September 1828). In 1829, commanding the brigantine "Mercury", he controlled successfully the non-equal battle with two Ottoman battleships "Real-Bei" and "Selimie" that had 184 cannons in total. The 4-h battle ended with retreat of the Turkish ships, while the brigantine with 22 holes in the board and 297 in masting and sails managed to reach Sevastopol. For successful actions in this battle the ship was awarded the ship's George ensign. Commanding the frigate "Pospeshnyi" he took part in cruising near the Bosporus (1829) and seizure of the Turkish fortress of Mesembria (Mesemvria) (July 1829). From 1831 he served at Russian Emperor Nickolay I. In 1839 in Sevastopol the monument to K. and the crew of "Mercury" was installed. It had the form of a trireme—a symbol of a warship in the Antiquity with the bas-relief of B. below. The pedestal had the symbolic inscription "To Kazarskiy. Example to Posterity". This was the first monument in Sevastopol.

Kazarskiy A.I. (http://upload.wikimedia.org/wikipedia/ru/e/ea/Kazarsky_a_i.jpg)

Kedrov Mikhail Aleksandrovich (1878–1945) – finished the Naval Cadet Corps (1899), Russian Rear Admiral (1920), commander of the Navy in the Gulf of Riga, assistant Minister of war and Naval minister. From 1918 he worked in London where he organized the transport for supply of the White Army. In 1920 he headed the Black Sea Fleet taking the place of gone Admiral Sablin; Chief of the Naval Department. He led the fleet from Sevastopol and other ports with the White Army units and refugees first to Gallipoli, Turkey, and then to Bizerta, Tunisia. Commander of the Russian Squadron in Bizerta (from March 1921), later he passed this post to Rear Admiral Berens. During emigration in France he was Chairman of the Naval Union. He was awarded many orders of Russia. He died in Paris.

Kefken Island – locates near the coast, 90 km eastward of the entrance into the Bosporus Strait, Turkey. The island is at a close distance to the shores of Cebeci village. In Byzantine times it was known as Daphnousia. Genoese times fortress walls and water wells are the historical heritage of the island.

Keppen Petr Ivanovich (1793–1864) – the Russian scientist in statistics, geography and ethnography. Academician (from 1843). In 1829–1834 he explored Crimea and steppes between the Dniester and Volga. The purpose of his expedition to the Southern Coast of Crimea (SCC) was "to collect for the future time the data about the survived traces of Greek colonies on SCC from the Meganomskiy (Kopsela) Cape to Aya-Burun (near Laspi) and also to collect topographic data about Tatar names of the main coastal areas, in particular those lying near the sea". The results of his archeological findings, his observations and conclusions K. took together in the book "Crimean Collection. About Ancient Relics on the Southern Coast of Crimea and Tauric Mountains" published in 1837 in Saint-Petersburg. K. is one of the founders of the Russian Geographical Society (founded in 1845). He organized regular collection of statistical data on the national composition of the Russian population on the basis of which he prepared the first "Ethnographic Map of European Russia" (1851). For this work in 1853 he was awarded the Konstantinovskiy Medal of the Russian Geographical Society. In 1848 K. wrote the work "About Population Census in Russia" (it was published in 1889). This work laid the basis for publication of "The Geographical and Statistical Dictionary of the Russian Empire". K. was buried in Bondarenkovo village to the south of Alushta where his estate located.

Keppen P.I. (http://www.rgo.ru/wp-content/uploads/2010/10/3_putevie-zametki_Keppen.jpg)

Kerch—(in ancient Rus'—Krchev) – the USSR hero city (14.09.1973) on the Crimean Peninsula, the center of the Maritime Region, Crimea, Ukraine. The port on the coast of the Kerch Strait between B. and A. seas. K. as a city starts its history in the seventh century B.C. when Greek colonists from Miletus founded a city-state on the western coast of the Cimmerian Bosporus (Kerch Strait) called Pantikapaeum that in the early fifth century B.C. till the late fourth century A.D. was the capital of the Bosporus Kingdom that lied on both sides of the strait. In the fourth century A.D. Pantikapaeum was destroyed. The city was revived under a new name—Bosporus and in the sixth century it became the Byzantine fortress. In the eighth century it got under control of the Khazars who called it Karsha or Charsha meaning "market". Its present-day name was first mentioned in the Russian chronicles of the ninth century and later it was found on the "Tmutarakan stone": "... in the year of 6576 from creation of the world (i.e. in 1068 A.D.) Prince

Gleb went over the ice of the sea from Tmutarakan to Krchev". In 1299 it was defeated by the Nogaians. In the thirteenth century it was invaded by the Mongols-Tatars, but in 1318 they left it to the Genoans who named it Cherkio. From the late fifteenth century it was in possession of Turkey under the name of Cherzeti. During this time new fortifications were constructed here. K. defended the peninsula from the Don Cossacks. During the Russian-Turkish War of 1768–1774 the fortress was sized by the Russian troops (1771). The city had 674 houses, a church, and about 50 small shops. The Turkish and Tatar population left the city. Under the 1774 Kuchuk-Kainarji Peace Treaty K. was attached to Russia and in 1775 it was incorporated into the Azov Province. There are many sources of origin of its name, but the most convincing seems the Slav name "Krk" meaning "neck or bottle neck". The customs outpost was organized in K. This city was populated with the Greeks who served under command of Count A.G. Orlov-Chesmenskiy. The civil population that numbered in 1774 about 140 people was growing. In 1777 the new fortifications were constructed in K., including the vallum and a stone dam. By the late eighteenth century this fortress lost its significance and a quarantine area was created here for the vessels entering A.S. From 1821 K. was the center of the Kerch-Enikale Region; the mooring port was opened here. In the 1820s in view of the growing trade with the Caucasus this city was rehabilitated completely. During the Crimean War of 1853–1856 K. was captured by the British landing troops and was damaged seriously by fires. K. acquired great significance as a grain-trading center. In the early twentieth century K. became a point of cargo reloading to shallow-water vessels for navigation of A.S.; the fishery center. In 1900 it was linked by a railroad with Dzhankoy (the Moscow—Sevastopol line). In the early nineteenth century archeological excavations were conducted in K. and nearby territories. In 1825 the Museum of Antiquities was founded here.

After a decline during the WWI and the Russian Civil War, the city resumed its growth in the late 1920s, with the expansion of various industries, iron ore and metallurgy. By 1939 its population had reached 104,500. During the Great Patriotoc War (1941–1945) Kerch was the site of heavy fighting between Soviet and German Armies. After several attempts, in 1942 the Germans occupied the city and the Red Army lost over 160,000 soldiers. The Adzhimushkay catacombs (mines) in the city's suburbs were the site of fighting against the occupation. Thousands of soldiers and refugees found shelter inside, and were involved in counterattacks. Many of them died underground. Later, a memorial was established on the site. The German invaders killed about 15,000 citizens and deported another 14,000 during their occupation. Kerch was liberated on 11 April 1944. Evidence of German atrocities in Kerch was presented in the Nuremberg trials, held between 20 November 1945 and 1 October 1946. After the war, the city was awarded the title of Hero City.

Today Kerch is considered as a city of metallurgists, shipbuilders and fishermen. The largest enterprises in the city are: Kerch Metallurgical Works Factory launched in 1900, Kamyshburun Iron Ore Plant, "Zaliv" ("Gulf") shipbuilding factory that produces and repairs tankers and cargo ships. Construction-materials, food processing, and light industries play a significant role in the city's economy. Kerch is also a fishing fleet base and an important processing centre for numerous fish products. Kerch has a harbour on the Kerch Strait, which makes it a key to the Sea of Azov and Black Sea, several railroad terminals and a small airport. Ferry transportation across the Kerch Strait was established in 1953, connecting Crimea and the Port "Kavkaz" in Russia. There are several ports in Kerch, including Kerch Maritime Trading Port, Kerch Maritime Fishing Port, Port "Krym"(ferry line), Kamyshburun Port.

Kerch hosts 28 schools, 9 institutes and branches of Ukrainian and Russian universities, shipbuilding and polytechnical colleges, medical school. Kerch became a popular summer resort among people of former USSR. Also, several mud-cure sources are located near the city. Despite the seaside location, the tourist importance of Kerch today is limited because of the industrial character of the city and associated pollution, the lack of beaches in town's area.

Kerch has a number of impressive architectural and historical monuments attractive for scientific tourism. The most notable of Kerch's sights are: Site of ancient settlement Pantikapaeum (fifth century B.C.–third century A.D.); Tsarskiy Kurghan (fourth century B.C.)—burial mound for one of Bosporian kings; Church of St. John the Baptist (717 A.D.); Fortress of Enikale (eighteenth century); The Great Mithridates Staircase leading on top of the Mount Mithridat (contains 428 footsteps, built in 1833–1840 under the guidance of Italian architect A. Digbi); Obelisk of Glory on the Mount Mithridat, built after World War II; Memorial of heroic guerilla warfare in Adzhimushkay catacombs; Kerch Fortress, built by the Russian military architect Totleben in the middle of nineteenth century; Sites of ancient settlements Mirmecium, Tiritaka and Nimphei. There are also some settlements gone underwater due to earthquakes; The Demetra's Crypt, a crypt with numerous frescos dated first century B.C. In 2000 the city celebrated its 2600th anniversary.

The Azov-Black Sea Research Institute for Fishery and Oceanography (AzCherNIRO, presently YugNIRO), experimental base for commercial fishery are located in Kerch. Population: 145,400 (2012).

Kerch (http://ru.ukrainecityguide.com/regions/kerch/kerch.htm)

Kerch Bridge – a railway bridge between Crimea and Taman' Peninsula that was constructed in 1944 after liberation of Crimea (USSR) from German Army. The bridge had 115 sections 27.1 m long with a main 110 m long pivot section over the shipping channel. In winter 1945 northeastern wind pushed ice from the Sea of Azov which on 18 February 1945 destroyed 32 pillars and 10 pillars more in the next days. Reconstruction of the bridge was not done, moreover the remains of pillars made shipping difficult till 1968. In 1953 a ferry line "Port Crimea—Port Kavkaz" (auto and railway transport) was open to replace the bridge. There are long-term discussions about a possibility to construct an auto bridge across the Kerch Strait with a set of projects from 230 to 900 million USD. On April 2010 the presidents of Russia Dmitry Medvedev and Ukraine Viktor Yanukovich signed the Kharkiv Agreement, the articles of which provide for the construction of a bridge across the Kerch Strait.

Kerch ferry line – in April 1944 the railway bridge over the Kerch Strait was constructed. It consisted of 115 identical spans 27.1 m long each; the 110-m span structure with the double navigation openings over the channel for passage of large ships; the jetty near the coast and the causeway. But in February 1945 as the bridge had no ice aprons the ice from the Azov Sea damaged about 30 % of the piers. The bridge was not restored, while the damaged parts were liquidated as they obstructed navigation. The ferry line was opened in 1953. It connected Crimea and the Krasnodar Territory (port "Crimea"—port "Kavkaz"). On this line four train ferries were operating, they were "Zapolyarnyi", "Severnyi", "Yuzhnyi" and "Vostochnyi". Later three car ferries were commissioned: "Kerchenskiy-1",

"Kerchenskiy-2" and "Eisk". In the late 1980s with wear-out of the train ferries first the carriage of passenger trains over the strait was stropped and later—the freight trains. In 2004 the train ferry "Annenkov" was added to the ferry fleet and the railway traffic over the strait was resumed.

Kerch Ichthyological Laboratory – was established in 1922 at the Crimean Regional Fishery Department "Glavryba" as a laboratory for applied research for support of fishing in the Azov and Black seas. In 1933 "AzCherNIRO" (Azov-Black Sea Research Institute for Fishery and Oceanography) was set up on its basis.

Kerch Iron-Ore Basin – a group of iron-ore deposits in the northern and eastern parts of the Kerch Peninsula, Crimea, Ukraine. The ore formations to 20 m thick occur amidst marine sandy-clay sediments of the Tertiary period filling the flat synclinal structures—troughs. The most well explored and studied are the deposits in the eastern part of the peninsula, in particular the Kamyshburunskiy and Eltingen-Ortelskiy, Kyz-Aulskiy, and Katerlezskiy deposits. The ores include mostly brown oolitic iron ores and are characterized by a low content of iron (28–39 %). These ores need dressing. The explored reserves are 1.6 bill tons. The ores from this basin are supplied the metallurgical works "Azovstal" in Mariupol and to the Metallurgical Works named after Voikov in Kerch.

Kerch Merchant Sea Port – its "birth date" is considered to be 10 October 1821. It is situated on the shore of the Kerch Strait connecting the Azov and Black seas. The city and port are among the most ancient cities of the world. In 2000 the city celebrated its 2600th anniversary. This port is capable to handle 2.5 million tons of freight a year. The port is provided with modern facilities, reloading machines and mechanisms that make it possible to handle general cargo (metal products, equipment and others), bulk cargo (ferroalloys, cast iron, metal scrap, steel pellets, grain), containers, and wheel vehicles. The port has 7 berths over 1,200 m in total length.

Kerch Military Operation – this was an assault on 8–20 May 1942 of the forces of the Eleventh Germany Army commanded by General Erich von Manstein against the Soviet troops commanded by General D.T. Kozlov. The 300,000 grouping of the troops of the Crimean front on the Kerch Peninsula that nearly twice exceeded the German forces was preparing to lift the siege of Sevastopol and to liberate the Crimea. But Manstein in anticipation of these actions made a preventive strike on May 8. Having the greater number of forces in the south of the Crimean Front the Germans broke the Soviet defense and the Germans landed in the rear of the Soviet troops, thus, disorganizing still more the situation. In spite of its domination on the sea the Black Sea Fleet did not provide the expected support to the defending land troops, so Manstein forces moved freely along the coast. After the Crimean Front was disintegrated the German tanks attacked the Russian positions. First the attacking units moved along the coast and then turned northward and reached the dislocation of the Soviet reserve forces and destroyed them. As a result, the basic Soviet grouping in the north of the Kerch Peninsula (47th and 51st armies) was cut off and forced to the bank of the Sivash Lake. The troops of the Crimean Front went out of control and retreated in chaos to the east. On 15 May the Germans captured Kerch. The Soviet troops numbering about 120,000 people were evacuated to the Northern Caucasus. Those who did not manage to evacuate (about 18,000 people) found shelter in the Adzhimushkay rock quarry (catacombs) where they heroically resisted the German attacks till late October 1942. The German victory "buried" the Soviet plan of Crimea liberation and, in fact, decided the fate of Sevastopol. The success of the Germans in battle at Kerch in 1942 that opened the way to new victories could be attributed to low competence of the commanding staff of the Crimean Front that enabled Manstein, without much effort, to realize this well-prepared, but risky plan.

Kerch Peninsula – the eastern part of the Crimean Peninsula. It is washed with A.S. in the north, the Kerch Strait in the east, and B.S. in the south. In the west it is connected with the Crimean Peninsula with the isthmus 17 km wide. Its area is about 3,000 km^2. The northeastern part is hilly (to 190 m high) composed mostly of limestone, clay and sandstone. The Opuk Mountain rises in the southern end of K.P. In the fifth century B.C. at the foot of this mountain there was the Greek colony of Cimmeric being a part of the Bosporus Kingdom. The ruins of ancient structures were found on the mountain top. Approximately 3 km from the Opuk Mount the rocks are rising from the sea reminding of ships under sails. There are many mud volcanoes on K.P. The southwestern part of the peninsula is flat, composed largely of Tertiary clays. The soils are chestnut, often saline. The greater part of lands is cultivated; the sagebrush-fescue steppes occur only spot-like. The large Kerch iron-ore basin is found on K.P.

Kerch Strait – runs between the Kerch (Crimean) Peninsula in the west and the Taman (Caucasian) Peninsula in the east. It connects Azov and Black seas. The strait received its name in the mid-nineteenth century by the name of Kerch town. In Antiquity this strait was called the Cimmerian Bosporus (by the Cimmerian people who inhabited Crimea) to distinguish it from the Thracian Bosporus (modern Bosporus Strait). In the Medieval Times it was called Taman-Bogazy (in the Turk meaning "Strait of Taman"). Today it makes the border between the Russian Federation and the Republic of Ukraine. Its length is about 41 km, width—from 4.5 to 15 km. The maximum depth is 18 m. The K.S. coasts in some places are low with sandy bars, in other places they are steep and cliffy. In some winters K.S. is covered with floating ice. The large city-port of Kerch locates on the western coast. The main kind of transport via K.S. is train ferry "Crimea–Caucasus". Many fish species are fished in K.S. The fishing season begins in late autumn and lasts for several months.

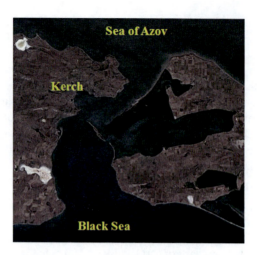

Kerch Strait (http://en.wikipedia.org/wiki/Tuzla_Island)

Kerch Strait, Sea Battle – (1) the battle in the Kerch Strait on 9–10 June 1774 between the Russian Fleet under command of Rear Admiral V.Ya. Chichagov and the Ottoman Fleet that made an attempt to break through into the Kerch Strait. Chichagov placed its fleet in the narrowest place of the strait, so the Ottomans could not take advantage of their numerical superiority. After intensive firefighting the Ottoman Fleet retreated having failed to fulfill its task. Both parties had no ship losses.

(2) the battle in the Kerch Strait on 8 July 1790 near the Takil Cape between the Russian Fleet commanded by Rear Admiral F.F. Ushakov and the Ottoman squadron commanded by Kapudan-Pasha (grand admiral) Giritli Hüseyin. This was the first battle in which Ushakov commanded the whole squadron of ships. The Ottoman ships entered into the strait for landing on Crimea. But here they met the Russian Fleet. The Ottomans taking advantage of the fair wind and artillery superiority attacked boldly the Russian ships. Ushakov, after skillful maneuvering, took the advantageous position and by the targeted short-distance shots caused the Ottoman Fleet grave damages. The Russian Fleet commander refusing from routine practices of linear battle ventured to break the ship line and concentrated the main forces in the vanguard. With darkness after the Hüseyin's ships failed to fulfill their task they left the strait. But the Ushakov's Fleet coped with its task successfully— not to let the Ottoman Fleet into A.S. and to prevent the landing of the Ottomans on the Crimean coast.

Kerch vases – the vases from the final phase of the Attic red-figure pottery production. They are generally assumed to have been produced in the second third of the fourth century B.C. The vases are thus named because a large quantity of them was found at Kerch (ancient Pantikapaion) where they were taken from Athens. K.V. were the first-class pottery produced in the Greek metropolis that time. The artists who painted them assimilated the skills of great masters and, first of all, as concerns the use of pigments. The motifs for the paintings were mostly connected with the myths about Aphrodite and Dionysus and also with the festive and everyday life of women.

Kerch-Eltigen Operation – a landing operation of the troops of the North-Caucasian Front (commanded by General I.E. Petrov) on seizure of the Kerch Peninsula that lasted from 31 October to 11 December 1943, during the Great Patriotic War. After success of the Novorossiysk-Taman operation the Soviet command put the task to seize the Kerch Peninsula and to create there a base for Crimea liberation from German Army. For this purpose two landing units went ashore. One of them landed to the northeast of Kerch, but could not take hold of the city because of the tough resistance of the Germans, but it managed to consolidate its positions on the seized area and to organize defense. The second landing unit got established to the south of Kerch, near Eltigen. After fierce battles in the early December the Germans liquidated this landing area. On 6 December the remaining landing troops attempted a breakthrough trying to reach the base to the north of Kerch. After marching for 25 km in the rear of the German positions they came to the southern outskirts of Kerch and seized the Mitridat Mountain dominating this area and organized an all-round defense. But they failed to organize interaction with the northern unit. After receiving an order about evacuation on 10 December the Eltigen unit forced its way to the coast and was transferred by sea to the Taman Peninsula. The Kerch landing area held out till the Soviet troops launched an offensive in Crimea in spring 1944. It played a very important role in seizure of the Kerch Peninsula. The Soviet Army lost in this operation more than 27,000 killed.

Kerch-Enikale Canal – was constructed by Russia in the Kerch Strait in 1874. Its width is 100 m, depth (along the axis)—9 m. The canal consists of 4 bends beginning from B.S.: Pavlovskiy, Burunskiy, Enikalskiy and Chushkinskiy. Such depth parameters did not allow the ships with deadweight 50,000 tons and more (140–200 m in length) to be loaded in full. The maximum loading of any vessel navigating this canal should not exceed 25,000 tons.

Kerch-Feodosiya Operation – the landing operation of the troops of the Transcaucasus Front commanded by General D.T. Kozlov in Kerch and Feodosia on 26 December 1941—2 January 1942. The Soviet command planned to capture the Kerch Peninsula and then liberate Crimea occupied by the German Eleventh Army of General Erich von Manstein. As the main forces of Manstein were engaged in storming Sevastopol the Soviet troops (more than 40,000 people) on 26–31 December 1941 landed near Kerch and Feodosia. The troops were supported with 43 tanks and 1,802 horses. This was the largest landing operation during the Great Patriotic War. Despite the winter storm, shortage of special landing facilities and resistance of the Germans the landing troops using the element of surprise on 29 December seized Feodosia and continued their attack in the northern direction. Commander of the German 42nd Army Corps Hans Graf von Sponeck fearing that his detachments on the Kerch Peninsula (up to 25,000 people) could be cut off gave an order to leave Kerch and retreated (for this act he was subject to court-martial and put to prison). Having seized Feodosia the Soviet command acted rather cautious and irresolutely which enabled the Germans to retreat without any problems from the Kerch Peninsula and to organize defense on the Parpachaisky isthmus. With the beginning of the K.F.O. Manstein had to stop the assault of Sevastopol (on 31 December) and to transfer some of its forces against the landing troops. On 15 January the Manstein troops made a counterattack and broke through the landing troops positions and on 18 January they seized again Feodosia. The Soviet troops retreated to the Ak-Monaisk isthmus.

Kerempe Cape – the northern tip of the western part of the Anatolian coast, Turkey. In this point the coast changes its northeastern extension for the latitudinal one. The cape is elevated. This area is remarkable for its bad weather and storms. The lighthouse is constructed near the tip of the cape.

Kerkinitida, Karkinitida, Kerkinitis – see Evpatoria.

Kessler Kar Fedorovich (1815–1881) – Russian professor, zoologist, well-known for his study of birds and ichthyofauna of the Pontus-Aral-Caspian basin. He was born in Eastern Prussia. From 1866 he headed the Zoology Chair of the Saint-Petersburg University. Such young scientists as A.O. Kovalevskiy (1866/1867 study year) and future Noble laureate I.I. Mechnikov (1868/1869 study year) worked with him as assistant professors. Later in addition to this post K. became the Dean of the Physico-Mathematical Department and in 1867—Rector of the Saint-Petersburg University. During his whole life he combined academic science and teaching with administrative and organizational activities. He established (1868) the Petersburg Society of Natural Scientists and for 11 years he was its first president (1869–1881). On his initiative the Napolitan Zoological Station in Italy became the place of permanent work for Russian zoologists. He initiated establishment of the biological station on B.S. in Sevastopol. The range of his interests was very wide. He conducted scaly studies of fauna in different regions of Russia: Crimea and Black Sea, the basins of the Volga, Neva, Ladoga and Onega rivers. He developed the first course in parasitology for the Saint-Petersburg University. K. was professor of the Kiev University, studied fish of B.S. and local fishing. His book "The Travel with Zoological Purposes over the Northern Coast of the Black Sea and Crimea in 1858" (Kiev, 1860) contains interesting facts about the nature of the Crimean Peninsula, ethnographic and everyday details.

Khadjibey Fortress – it was founded on June 10, 1793. And that time the construction of a harbor for the rowing fleet was started. The fortress and the harbor were built under direct supervision of Vice Admiral O.M. De Ribas and Colonel F.P. De Volant. Simultaneously with the fortress construction the problem of building the port and town was also raised. G.A. Potyomkin suggested that the port should be constructed in Kinburn. But as this place was unsuitable for port location the choice of Kinburn was turned down. Russian Vice Admiral N.S. Mordvinov who was appointed in 1792 the Chief Commander of the Black Sea Fleet and the Khersones Admiralty supported the choice of Ochakov for port construction. O.M. De Ribas and F.P. De Volant were ordered to inspect the Dnieper mouth, liman and coast near Ochakov. And they both came to the same opinion—Ochakov was not a good place for port construction because of poor roads, shallow waters and freezing liman. They decided that the best place for the port was Khadjibey. By the time of taking the final decision on the place of the main B.S. port Potyomkin was already dead and P.A. Zubov who became the next governor of this area supported the idea of the port construction in Khadjibey before Catherine II.

Khadjibey Resort – the balneological resort in the Odessa Region (Ukraine), 10 km to the northeast of Odessa. It is located on the southwestern coast of the Khadjibey liman. Kh.R. was one of the oldest mud curative resorts in the former USSR. Together with the Kuyalnitsky resort it had been widely known still in the early nineteenth century. The main curative factors here are mud and brine from the Khadjibey liman that was formed as a result of deposition of sediments in the mouth of the Smaller Kuyalnik River. It is situated 9.5 km to the northwest of Odessa. The liman is about 31 km long, 2.5 km wide and to 15 m deep. It is separated from the sea with the sandy-shell barrier 5 km long and 4.5 km wide. Compared to the Kuyalnitsky liman the Khadjibey liman is more water abundant and the concentration of salts in its waters is much lower. The brine of the Khadjibey liman is of the chloride-sodium-magnesium composition. The liman has sulfide muds which reserves are estimated at 13 million m^3 and that are widely applied in mud treatment.

Khadjibey Liman – locates in the northwestern part of B.S., 9.5 km to the northwest of Odessa. Its length is about 31 km, width 0.5–3.5 km, area—70 km^2 and the maximum depth—15 m. It is separated from the sea with the sandy-shell barrier 4.5 km wide. It was cut from the sea with the marine and river sediments of the Smaller Kuyalnik River. The mud of this liman is curative (Khadjibey resort).

Khaji Giray (1397–1466) – the founder of the Crimean Khanate and the Giray dynasty. The date of the Crimean Khanate establishment is considered to be 1443. By the late 1440s Melek Khaji Giray became the first in the dynasty of Crimean khans—Gireys. The people gave him the nickname "Melek" meaning "Angel". He was buried in the village of Salachik near Kyrk-Ore, the first capital of the Crimean Khanate. In 1501 the beautiful Durbe was constructed on his grave that survived to the present.

Khanskiy Lake – locates 40 km southward of Eisk on the territory of the Yasenskiy nature reserve, Krasnodar Kray, Russia. With time on it was separated from the Sea of Azov by a thin bar 100 m to 1 km wide. In the past it was a liman and was connected with the Sea of Azov via a narrow channel. The lake is saline. It is well known for its curative mud that is used in the mud therapy. This is a monument of nature in the Krasnodar Kray.

Khazaria – the area where the Khazars were roaming in the seventh–tenth centuries. This area covers Lower Volga, Don and Northern Caucasus. From the tenth century this territory ruled by the Khazars received the name of Khazar Khaganate and this name was often extended to the whole eastern part of Crimea.

Khazar Khaganate – a state headed by Khagan that existed in the mid-seventh— late tenth centuries. In 670–679 the Khazars took control of nearly the whole of Crimea, except Tauric Chersonesos. The Khazars helped Chersonesos in its struggle for independence of Byzantine. But in 711 Byzantine signed a treaty with Kh. Kh. under which Byzantine received the southern and southwestern part of Crimea with Chersonesos. In the early ninth century rich Jew Obadiah became the king of Khazaria and made Judaism the official religion. In the mid-tenth century the Khazars were forced out from Crimea by the Pechenegs who came from the East.

Khazars – the Turkic people who settled in Eastern Europe after the Hunns invasion in the fourth century and who wandered over the Western Caspian Steppe. They formed the Khazar Khaganate.

Kherson – the administrative center of the Kherson Region, a port on Dnieper River, 90 km from the Black Sea, Ukraine. It was founded in 1778 as a base of the Russian Black Sea Fleet. The name Kherson is a contraction of Chersonesos (Khersones), an ancient Greek colony in the southwestern part of Crimea. This name was only a tribute to the mode on pseudo-Greek names and had no geographical sense—in Ancient Greek *"Chersonesos"* meant "peninsula" and here no peninsula is found. This city was built by General-in-Chief I.A. Gannibal (1736–1801), the son of A.P. Gannibal, the famous moor of Peter the Great, great-grandfather of Russian famous poet A.S. Pushkin. He was also chief commander of Kherson fortress. Kh. is a sea port and an important industrial and agricultural center of the region. Its population is 338,000 (2012). Kherson has an international airport and railroad connections to Kiev and Lviv. The city hosts Kherson State Agrarian University, Kherson State University, Kherson National Technical University, and International University of Business and Law. Main historical sites are: the Church of St.Catherine, which was built in the 1780s, and has a tomb of Prince Potemkin; Jewish cemetry (Kherson has a large Jewish community, established in the mid nineteenth century); Kherson TV tower; Adzhiogol Lighthouse designed and constructed in 1911 by V.G.Shukhov.

Kherson (http://img0.liveinternet.ru/images/attach/c/7/95/222/95222360_5154319_Kherson_51. jpg)

Kherson Canal – it was constructed in the eastern part of the Dnieper Liman. It is about 40 km long and 80–100 m wide. It was built for sailing of sea vessels to the port of Kherson. The minimum water depth in the canal is 7.5 m.

Kherson Hydrobiological Station – in 1952 the so-called fleet base was established for study of the ecosystems of the Lower Dnieper and Dnieper-Bug Liman that was later renamed into Kh.H.S. The personnel of this stations worked over the whole south of Ukraine: on the Lower Bug and Lower Dniester, on the Danube and on limans in the Odessa and Nikolaev regions. In 1970s–1980s it conducted feasibility studies for construction of the Danube-Dnieper Canal and the Ochakov Delta.

Kherson Region – it is situated in the south of Ukraine, in the Prichernomorie Lowland and is washed in the south by the waters of Black Sea and Sea of Azov. Its area is 28,500 km^2. The population is 1.1 million (2013). The administrative center is Kherson. This region was established in 1944. It borders on the Nikolaev, Dnepropetrovsk and Zaporozhie regions and Crimea. In its coastal area one can find the sandy islands (Dzharylgach, Dolgiy and others), peninsulas (Chongarskiy), sand bars (Tendrovskiy, Kinburn, Arabat Spit), shallow-water bays (Sivash, Tendrovskiy, Yagorlytskiy, Dzharylgach, Korzhinskiy) as well as the Dnieper-Bug Liman. Among the significant relief forms there are valleys of the Dnieper and Ingulets rivers. The surface of the region is mostly flat, broken in some places with numerous gullies and ravines. The climate here is moderately continental. The winters are mild. The mean temperature in January varies from −3 °C in the south to −5 °C in the north. The summer is hot and dry. The mean temperature in July is from +21 °C in the north to +23 °C in the south. The precipitations are 300–400 mm a year. The largest river here is the Dnieper with its right tributary Ingulets. The greater part of the Kakhovskiy water reservoir locates in this region. The steppes here are mostly cultivated. The lands under forests and shrubs take only some 3.5 % of the whole territory. The waterlogged areas of the Dnieper valley (plavni) are overgrown with the tree and shrub vegetation. The nature reserve Askania-Nova was created in Kh.R. for preservation of the wild feather-grass and fescue steppes with their unique collection of wild animals from different world countries and the botanical garden. The greater part of the Black Sea nature reserve and the Azov-Sivash nature conservation and hunting grounds also belong to this region. Among natural deposits there are sands, limestones, various clays and peat; in the Sivash area many deposits of various salts (sodium, magnesium and others) are discovered. Kh.R. belongs to the Black Sea economic region. The most important industries here are light industry, machine-building and metalworking, power engineering, chemical, petrochemical and food.

Mild climate, mineral water sources and curative mud provide opportunities for rest and treatment here. The sulfide mud is found mostly in the salt lakes in the B.S. coastal area and nearby the Arabat Spit. Considerable mud reserves are also discovered in the Solyanoye (Gopri), Krugloe (near the settlement Novochernomorie), Liman (near the settlement Frunze) and Genichesk lakes. The most well-known resorts in this region are Gopri and Skadovsk. The coastal area of B. and A.S. has good prospects as climate therapy and recreation zones. Numerous holiday hotels and rest houses are found mostly on the B.S. coast and Arabat Spit. Among the sightseeing attractions in Kherson there are architectural, historical and cultural monuments, such as the ruins of the Ochakov and Petersburg gates, the arsenal building, the ramparts (eighteenth century), Black Sea hospital (early nineteenth century), St. Catherine (Spasskiy) and Holy Spirit cathedrals (eighteenth—early nineteenth centuries), monuments to the famous seamen—A.V. Suvorov, F.F. Ushakov, to T.G. Shevchenko, to John Howard (1726–1790; British public figure and philanthropist; buried in Kherson). Kherson has the Museum of Arts and the Museum of Regional History. The museum of the Black Sea nature reserve is located in the town of Golaya Pristan.

Kherson Sea Trade Port – it was founded in 1778 (prior to 1946 it was a berthing place). It is situated on the Dnieper bank 28 km from the river mouth, in the crossing place of river and sea routes, 95 km from the Black Sea. Before 1900 this port was the fourth largest in Russia by its export capacity. In the Soviet years the freight turnover reached 4 million tons. The design capacity of the port was 5 million tons a year. The port handled general and bulk cargo. More than 40 % of the cargo handled in the port was mineral fertilizers and chemical products and about 40 %—grain. Ferrous metal, cast iron pigs, coke, ferrous alloys, timber and peat were exported via this port.

Khersones, Khersonesos (Chersonesos) Taurica – the antique city which ruins are found in Sevastopol on the B.S. coast between Karantinnaya and Pesochnaya bays. This was an ancient Greek colony founded in 422–421 B.C. by exiled democrats from Heraclea Pontica. The city extended over a territory of 30 ha. It had about 800–900 houses and its population was 10–12,000. About 2,000 people settled outside the city walls. Kh. also possessed the nearby plain, the steppes of Taurida with the town of Kerkinitis (at present Evpatoria) and Prekrasnaya Bay (Akmechet harbor) and the southern coast as far as Feodosiya near which Kh. bordered on the Bosporus Kingdom. Kh. also owned the lands lying beyond the peninsula. The city had many rich churches and public buildings and a port harboring the merchant fleet. In the fourth century the giant wall made of formidable cut stones with corner towers and the gates was constructed around the city. The borders of environ lands were also protected with walls. In the tenth century the city was ruled by the Senate and the Council. The ruling hierarchy was as follows: the supreme executive power belonged to protevon, then went vasilevs (high priest), stratig (military chief), demiorgi (high court figure), astinom (trade and tribute inspector) and others. Any citizen may be elected by the public assembly to the Council. Apart from trade with grain, cattle, salt and fish that gave sustenance to the city the people of Kh. were also engaged in farming and vine growing. The trade was under strict control. In the fourth–first centuries B.C. and later on the city minted its own coins. Therefore, like the Bosporus Kingdom, Kh. was a significant cultural center which influence extended to the far north. Many Scythians were made citizens of Kh. and even played a prominent role in the political life of the Khersones community. The Scythians were also used by Kh. as messengers and mediators. Having originated as a Greek colony the city in the first century A.D. became a dependency of Rome and from the fourth century after breakdown of the Roman Empire it became the stronghold of the Byzantine Empire in Crimea.

For many centuries from the ancient times Kh. had flourished thanks to its location on the crossing of lively caravan roads that connected Greece, Byzantine and Rome with Scythia and strong fortifications that made the city inaccessible for the enemies who often came up to its walls. In the ninth–eleventh centuries Kh. played a very important role in development of relations between Kievan Rus' and Byzantine Empire. These relations with the powerful Ancient Rus' state became most close in the early tenth century when in 944 Prince Igor entered into a treaty with Byzantine that controlled Kh. This treaty stated that Kiev undertook to defend the "Korsun lands" from the raids of black Bulgarian tribes that roamed in the Azov area. The positions of the Kievan state became much stronger in Taurica after Great Prince Vladimir seized Kh. that was subjugate of Byzantine. Vladimir took this act in response of non-execution by the Byzantine emperors of their liabilities before Kiev. Byzantine Emperors Basil II and Constantine VIII had to conclude with Vladimir the equal treaty and to recognize the political rights of Kiev to Taurida. And the marriage of Vladimir to Basil's sister Ann after Vladimir was baptized became also a political union. Kh. was the cradle of Christianity in Rus'. During the ruling of Vladimir Svyatoslavovich Kievan Rus' became one of the most powerful states in the medieval world. Kh. that had been flourishing for many centuries was on decline in the thirteenth century when new trade routes and new trade centers appeared in Crimea, such as Kaffa (Feodosiya) and Sogdia (Sudak) that were controlled by the Venetians and Genoese. In 1299 Kh. like many other cities on the peninsula were looted by the Nogai troops of the Golden Horde. One hundred years later, in 1399, the city was burnt out and destroyed by Edigu who ruled the Golden Horde that time. In the mid-fifteenth century the life in Kh. died away completely and the last people deserted the city.

Khersones (http://image.tsn.ua/media/images/original/Oct2010/379121.jpg)

"Khersones Taurica" – the national historical and archeological nature reserve. This is a unique monument of the antique and medieval times. It was put on the list of Top 100 outstanding monuments of history and culture prepared by the World Monuments Fund. It is located on the coast of the Heraclean Peninsula between Karantinnaya and Pesochnaya bays on the territory of antique Khersonesos. During its lifetime Khersonesos-Kherson-Korsun had known the periods of flourishing and decline. More than once the city was subject to devastating earthquakes the strongest of which were in the sixth century and in the late eleventh century. The recent years showed a growing interest to Kh.T. not only as the world monument, but as a cradle of Christianity in Rus'. In the second half of the first century St. Andrey the First Called visited this place. From here he sailed up the Dnieper to a place where Kiev is standing now. Kh.T. is also connected with St. Clement, one of the first Roman popes highly esteemed by all Christians. In winter 860–861 on their mission voyage to the Khazars brothers St. Cyril and Methodius, the preachers of Christianity and creators of the ancient Slav script, stopped here.

Kh.T. is the only place on the territory of Ukraine and former USSR where their staying was fixed in scripts. The regular excavations of the city began in the second half of the nineteenth century, nearly 400 years after the city perishing. By that time the ruins were damaged irreparably. But the ruins of Kh.T. that survived to the present are the most well preserved antique monuments not only in the former USSR, but in Eastern Europe, too. And not accidentally some scientists call Kh.T. as second Troy. In 1892 K.K. Kostyushko-Valyuzhinich who was the first chief of excavations here opened the museum that was called "Storage of Local Antiquities". Today the stock of the national nature reserve numbers about 300,000 items and it is constantly replenished. The museum occupies the building of the former Khersonesos Monastery closed in 1925. It contains the greatest and most interesting collection of exhibits found during excavations here, including a marble stele with the text of the solemn oath of the Khersonesos citizens (third century B.C.), the head of a young man (fourth century B.C.), fragments of ancient Greek mosaic and many others. The Roman period (first–fourth centuries A.D.) is represented by articles of handcraft, culture and household of Medieval Khersonesos (Kherson). Many interesting things have been found on the city territory—defense fortifications in the southeastern part, the gates, parts of defense walls and towers, including the largest of them—Zenon Tower. The central street ran for about 900 m with the living quarters on both sides. The mint yard is one of the largest buildings in ancient Kh.T. that was destroyed during a fire in the third century B.C. and not restored afterwards. To the left of the main entrance there is the antique theatre (the only one in Ukraine and in Eastern Europe) that was built in the third century B.C. and stayed for about 700 years, till the fourth century when it was closed, probably, after establishment of Christianity. Much attention of the visitors is invariably drawn to the grand columns of Basilica on the seashore (early Medieval period) that is one of the most interesting temples of Khersonesos. The place where nowadays the St. Vladimir's Cathedral is rising was in ancient times the central square of the city (agora). This construction of this cathedral was started on August 23, 1861 in commemoration of Vladimir baptizing in 988 and it was finished only 27 years later. It was dedicated in 1889, in the year of the 900th anniversary of baptizing of Vladimir the Great of Kiev. This cathedral 36 m high was of the Byzantine style, bileveled. And from here Christianity came over to the Dnieper shores and to the whole of Kievan Rus'. In 1924 the cathedral was closed. Destroyed in 1942 during the siege of Sevastopol it was restored with financial support of the Kiev city administration. In 2004 the cathedral was opened for prayer services. And soon after this the beautiful monument to St. Apostle Andrey the First Called was constructed near the cathedral. It was the gift from Kiev. The national nature reserve "Kh.T." has two branches—ancient fortress of Kalamita in Inkerman and the fortress of Chembalo in Balaklava.

"Khersones Taurica", St. Vladimir's Cathedral (Photo by Andrey Kostianoy)

Khersones Cape (in the Greek *"Chersones"* meaning "cape") – in the ancient times it was known as Parthenon, Partheny and Phonary-Kap. This is the western-most cape on the Heraclean Peninsula. It is situated to the southwest of the Sevastopol Bay on the southwestern tip of the Crimean Peninsula. The cape is rocky and low-lying. The Khersones lighthouse 36 m high was constructed here in a place where in 1789–1791 the lighthouse already stood. It ensures safe approach of ships to the port of Sevastopol.

Khersones Coastal Water Nature Reserve – it covers the coast area between the Karantinnaya and Pesochnaya bays. It was given the status of a nature reserve for preservation of a part of Khersonesos Taurica that gradually submerged under water for the past millennia.

Khersones Lighthouse – see Khersones Cape.

Khmelnitskiy Bogdan (Zinoviy) Mikhailovich (1595–1657) – the Ukrainian state and military figure, hetman. He was born in the family of the chief of the Chigirinsky military and administrative area in Ukraine. He received his early education in the Kiev's fraternity school and continued his education in the Jesuit College in Lvov having remained baptized. Apart from the Little Russian language he learned Polish and Latin. Later on he learned also the Turkish and French. Upon completion of his studies Bogdan entered into service with the Cossacks. He took part in the Polish-Lithuanian War of 1620–1621 during which his father was killed in the battle of Cecora (1620) and Bogdan himself was captured and spent next 2 years in Constantinople as a prisoner. After becoming free Kh. organized sea raids of the Zaporozhie Cossacks on the Turkish cities. Well known was his raid in 1629 when the Cossack troops under command of Kh. came up to Constantinople and returned with rich loot. During the Russian-Polish War he fought against the Russian troops and was awarded with the golden saber for his bravery. After long staying in Zaporozhie, Kh. returned to his native Chigirin and was conferred the title of a sotnyk (a leader of a hundred).

An event in his private life forced Kh. to become the leader of the uprising against the Polish Government that was provoked by its policy towards the Cossacks and the Russian Orthodox population in the Rzech Pospolita. Kh. had a small estate in Subbotovo near Chigirin. Taking advantage of Bogdan absence Polish headman Chaplinsky who hated him attacked his estate, ravaged it, took with him a woman with whom Kh. lived after the death of his first wife Anna Somkovna, married her in the Catholic church and whipped one of the Bogdan's sons so hardly that he soon died. First, Kh. sought recompense in the court, but the court people only mocked him. Then he applied to the king who, as it was said, showed surprise that the Cossacks being well armed could not defend their rights by themselves. Having returned to Warsaw with nothing he decided to use arms. He secretly gathered Cossacks and spoke to them so ardently that they declared him the hetman and asked him personally, and not via envoys, to enter into negotiations with the Tatars to urge them to a union with the Cossacks. On December 11, 1647 Bogdan together with his son Timofey arrived to Zaporozhie Sech from where they headed to Crimea that was ruled that time by Islam Giray. The Khan gave them a heartily welcome, but his response in relation to the war with Poland was quite vague although Perekop murza Tugan-bey was ordered to go with Kh. without declaring a war on Poland. On April 18, 1648 Kh. returned to Sech and reported about the results of his visit to Crimea. The people of Sech welcomed him enthusiastically and elected the senior commander of the Zaporozhie Cossack troops. On April 22, 1648 Kh. started off from Zaporozhie and at some distance Tugan-bey with his Tatar troops followed him. On May 6 the Cossacks attacked the Polish cart train. The Polish troops fought bravely, but the Cossacks cut them off from water supplies and the Poles had to buy off—they left their cannons to the Cossacks. In return the Poles were vowed a free passage to Poland. But when they started retreat and on May 8 reached yar that was called Princely Bairaks, they were quite unexpectedly raided by Tugan-bey with its Tatars. They inflicted a crushing defeat on the Polish troops.

In January 1648 Kh. headed the liberation struggle of the Ukrainian people. In the battles at Zholty Vody and at Korsun the Ukrainians aided by Tatars of Tugan-bey defeated the Polish troops, but they failed to take to full advantage their victories. Kh. was awarded the rank of Hetman and made a triumphant entry into Kiev. After resuming military actions, the Ukrainian troops led by Kh. crushed the Polish troops, but being betrayed by the Crimean Khan Kh. was forced to sign unfavorable and short-lived Peace Treaty of Zborov (1649) that, in fact, only confirmed the former rights of the Malorossia Cossacks. In early June 1651 there was a battle at Berestechko that ended in the Cossack defeat. The Crimean Khan fled and when Kh. made an attempt to stop him Khan seized hetman and kept him in captive for about a month. In autumn the Treaty of Belaya Tserkov (1651) was concluded that wiped off nearly all gains of the Ukrainian people. Kh. continued his struggle and a year later he completely crushed the 20,000-strong Polish Army near

Batog. On October 1, 1653 the Zemskiy Sobor (country council) was convened in Moscow at which it was agreed that Kh. with his Cossacks would take the Moscow tsar's overlordship. On January 8, 1654 in Pereyaslavl the Cossack Rada was called that approved the decision on joining Russia.

After Malorossia accession to Russia the war with Poland was started. In spring 1654 Tsar Alexey Mikhailovich moved the troops to Lithuania and in the north Swedish King Karl X opened hostilities against Poland. It seemed that Poland was on the brink of collapse. King Jan Kasimir approached Kh., but the latter did not agree to start negotiations until Poland recognized full independence of all Ancient Rus lands. Then Jan Kasimir applied to Tsar Alexey Mikhailovich who in 1656 concluded a separate peace treaty with the Poles not even inviting Kh. to take part in negotiations. All plans of Kh. were ruined. For some time he still cherished hopes to realize them and in early 1657 he entered into an alliance treaty with Swedish King Karl X and the ruler of Transylvania, George Rakoczi. In accordance with this treaty Kh. sent the 12,000-strong Cossack troops against Poland to support the allies. The Poles informed Moscow about this and Moscow sent emissaries to Hetman. That time Kh. was already very ill, but even lying in bed he received the tsar envoys who rushed on him accusations. As a result of their insistent demands Hetman had to call back the troops sent for support of the allies. A month later Kh. ordered to call Rada in Chigirin for election of his successor. To please the old Hetman Rada elected his under-aged son Yuri his successor. Kh. died on July 27, 1657 of apoplectic shock. He was buried in the village of Subbotovo in a stone church that was built on his order. This church still exists. In 1664 Polish noble Stefan Czarniecki captured Subbotovo, burnt it and ordered the body of the hetman to be exhumed and desecrated.

Khmelnitskiy Bogdan (http://www.petrichenko.info/files/Bogdan_Khmelnitsky1.jpg)

Khobi River – a river in Georgia. Its length is 150 km. It originates on the slopes of the Megrel range and flows into B.S. In its lower reaches the riverbed is levied. Its main right tributary is Chanis-Tskhali. This river is mostly rain-fed.

Kholostyakov Georgiy Nikitich (1902–1983) – Vice Admiral in the USSR Fleet. In 1915 he worked as unskilled laborer. He took part in Civil War in Russia. In 1925 he finished the Naval Hydrographic College. From 1926 to 1938 he served on different submarines in the Far East. In 1938 he was arrested, accused of treason and condemned to 15 years in a correctional labor camp. He was sent to a labor camp on the Olga Bay shore of the Pacific. Later his case was reconsidered and he was returned his rank. In autumn 1940 he was appointed commander of the Third Brigade of submarines of the Black Sea Fleet. Later he was promoted to Captain first Rank. Kh. was appointed the Chief of submarine division of the fleet and in July 1941 he headed the Naval Base in Novorossiysk. He supported the Kerch-Feodosiya military operation in late 1941. He took part in land-based defense of Novorossiysk. After retreat of the Soviet troops he moved to Gelendzhik from where the artillery of the Novorossiysk Naval Base gunned Novorossiysk, thus, preventing the Germans from using the port. In 1942 he was promoted to Rear Admiral. In 1943 he was in charge of the landing operation in Novorossiysk. In September under Kh. command two more landing operations were conducted. The German troops left the Taman Peninsula. At night on November 1 he organized landing at Eltigen near Kerch. Regardless of the overwhelming superiority of the Germans the landing troops made strong footing on Ognennaya zemlya and defended this land for more than a month and then broke through the German positions and united with the main Soviet forces. In 1944 he acted as Commander of the Azov Flotilla replacing S.G. Gorshkov in this position. He organized two more landings—on the Tarkhankut Cape at night on January 10 and in the Kerch Bay on January 23. In December 1944 he was appointed to the Danube Flotilla and commanded its last operations. In 1950 he graduated from the General Staff Academy with a gold medal. In 1950–1951 he in the rank of Vice Admiral commanded the Caspian Military Flotilla and then received an appointment to the Pacific Ocean. In 1953–1969 Kh. was Deputy Chief of the Military Training Department of the Navy General Staff. He took other positions of importance. In 1965 he was awarded the Golden Star of the Hero of the Soviet Union. In 1969 he resigned. The memorial museum of Kh. is opened in Baranovichi, Belorussia. He was also awarded many orders and medals.

Khosta – a resort village being a part of Greater Sochi, on the Black Sea coast of the Caucasus, Russia. It is located 10 km to the southeast of Novaya Matsesta. It appeared as a resort village in the mouth of the Khosta River in the early twentieth century. It has the sandy-gravel beach. The unique yew-boxwood grove that belongs to the Caucasian nature reserve is growing here. The Khosta resort extends over the Black Sea coast of the Caucasus from the Kudepsta River in the south to the Vereschaginka River in the north for over 20 km. The summer is hot here. A great amount of precipitations falls here—1,350 mm a year. The Khosta region includes many sanatoriums of Sochi and all balneological resorts of Matsesta and

Khosta. It also comprises sources of sulfide and iodine-bromine waters in Kudepsta that are used for health improvement.

Khroni Cape – locates in the north of the Kerch Peninsula, on the coast of the Sea of Azov. This is the western tip of the northern entrance into the Kerch Bay. This is the border between the Kerch Strait and the Sea of Azov. The cape is elevated with smooth slopes. The village of Osoviny locates on its eastern slope.

Khruschev Stepan Petrovich (1791–1865) – Russian Admiral. In 1809 he graduated from the Naval Cadet Corps. In 1827 he commanded frigate "Konstantin" in the battle at Navarino. In 1828–1830 he cruised over the Mediterranean Sea and took part in blockade of the Dardanelles. In 1834 for his distinctions he was promoted to Rear Admiral. In 1838 Kh. was appointed Chief of Staff of the Black Sea Fleet and Ports. In 1842 he acted for some time as Chief Commander of the Black Sea Fleet and Ports. In 1843 he was promoted to Vice Admiral. In 1849 Kh. was appointed Commander of the Sevastopol Port and Military Governor of the city. From 1852 he was the member of the Admiralty Board. In 1855 he was conferred the rank of Admiral. He was awarded many orders.

Kiik-Atlama Cape – (in Crimean Tatar *"kiik"* means "wild" and *"atlama"* — "walk") is situated 4.1 miles eastward of the Lagerny Cape. K.A.C. is connected with the mainland with a low elongated isthmus. If viewed from the southwest and northeast it has the form of a triangle. The Ivan-Baba Island is found near the tip of this cape.

Kil – (in the Turk *"kil"* means "wool", "hair") special clay that had been used in Crimea from old times for wool degreasing. In the Middle Ages the inhabitants of Crimea widely applied kil for cleaning carpets and fabric, washing in seawater, treatment of wounds and gastric diseases. The scientific name of kil is bentonite clay (by the name of fort Benton in the USA where this clay was found and studied for the first time). Its main component is the clay mineral montmorillonite. "Earthen soap" had been extracted from old times near Sevastopol in the Sapun Mount (the Arab word *"sabun"* means soap), near the Skalistyi village in the Bodraka valley, Mylnaya (Soap) Mount near the Partizanskiy village (former Sably), on the slope near the Kurtsy village not far from Simferopol, at Prolom village near Belogorsk and in other places. The Mylnaya Mount took the top line by kil production. K. has a very interesting history. It was thanks to kil that in the early twentieth century the studies of bentonite clays were initiated in Russia and USSR. It proved to be an excellent absorber and was applied in wool degreasing, clarification of wines, juices, vegetable oil and petroleum products. The wonderful toilet soap "Marvel of Crimea" is manufactured on the basis of kil and palm oil. The kil reserves in the Kurtsovsk deposit had been depleted long ago and in its place came the Kudrinskiy deposit of bentonite clays explored in the 1980s. The Kudrinkiy kil is of a very high quality, so it can be successfully applied in medicine. K. has a curative effect on the gastrointestinal tract. At the same time K. facilitates removal of radionuclides and toxic substances from a man's organism.

Kilia – a city, port, center of the Kilian District, Odessa Region, Ukraine. It is situated 50 km downstream the city of Izmail and 47 km from B.S. The city of Kilia and Novoye Selo are mentioned in the list of the Russian cities in the Danube Delta of the late fourteenth—early fifteenth centuries. With the time Novoye Selo turned into the city of Kilia and the former city of Kilia—into a small fisherman village Kilia-Veke meaning "old Kilia" on the Romanian shore of the girlo. The name of K., in Romanian Chilia, originates from Romanian Chilie meaning "cell, secluded monastery". There is some other guess concerning the origin of this name: from the Turk *"Killi"* meaning "clayey". The population of the town is 20,800 (2011) people. It has the shipbuilding yard, essential oil and vegetable oil factories.

Kilian Arm, Kilian Girlo – the northern main arm, the direct extension of the Danube within its delta, the most water abundant (70 %) arm of all others. It is 100 km long and 0.3–0.7 km wide. The navigation along this arm is impeded due to many shallows found here. The cities of Izmail, Kilia and Vilkovo locate on its shores. The state border between Ukraine and Romania runs along K.A. The left shore of K.A. is the Odessa Region, Ukraine.

Kilyos – the resort village near the Bosporus in European Turkey. In the recent time it became more and more popular. The modern high-class hotel with a restaurant was constructed near the wonderful sandy beach. There are also several dozens of camping sites. This village is frequently visited by many inhabitants of Istanbul, in particular during weekends. There is a fourteenth century Genoese castle in the village, which was restored during the era of the Ottoman sultan Mahmud II, but it is not publicly accessible since it is located in the military zone.

Kinburn – the fortress (till 1857) located on the Kinburn bar between the Dnieper Liman and the Egorlytskiy Bay of B.S. (presently Ochakov District of the Nikolaev Region, Ukraine). It was built by the Turks in the fifteenth century. During the Russian-Ottoman Wars of the seventeenth–nineteenth centuries many battles were fought for its possession. For the first time it was seized by the Russian troops in 1735, for the second time—in 1771. In 1774 it was incorporated into Russia. During the Russian-Ottoman War of 1787–1791 the Kinburn bar with K. made a part of the Black Sea coast that was defended by the Russian troops of General-in-Chief A.V. Suvorov. At dawn of 1(12) October 1787 the Ottoman Fleet after preliminary shelling landed its troops on the bar and attempted storming of the fortress. Suvorov, supported by the cannons in the fortress and on galley "Desna", engaged a counterattack. The enemies could not resist this strike and retreated, but later restored their positions. In the second and third attacks Suvorov led the troops by himself demonstrating personal bravery. With the darkness when the Turkish Fleet sustaining great losses from bombardment by fortress cannons left its positions, while the Russians attacked the enemy landing forces and defeated them. The Turks lost 1,500 people killed, 2 ships were sunk and 2 ships seriously damaged. The Russians casualties were about 450 people killed and wounded. Suvorov was wounded twice. The victory in the battle at K. frustrated the designs of the Ottoman command and created favorable conditions for the Russians to seize the fortress of

Ochakov. The success of the Russian troops became possible due to quick and sudden actions, wise use of the reserves and a correct combination of firing and bayonet fighting.

Kinburn Bar – a sand bar, sickle-shaped, located between the Dnieper-Bug Liman and Egorlytskiy Bay of B.S. Its length is about 40 km, width—8–10 km. It is covered with feather grass steppes. Many freshwater and salty lakes are found here, so it became the nestling place of white herons, cranes and other birds. The eastern part of K.B. is waterlogged. Fishing is practiced near coasts. It is a part of the Black Sea Nature Reserve. Near K.B. there was the Peregnoinskiy host that was a kind of a port on B.S. for Zaporozhie Cossacks. In the western part of K.B. the Turks built in the fifteenth century the fortress near which a small unit headed by A.V. Suvorov defeated in 1787 the Ottoman landing regiment numbering 6,000 people. In memory of this victory the bronze monument to Generalissimo Suvorov had been "on watch" for many years on K.B.

Kinburn Strait – links the Dnieper-Bug Liman with the B.S. Its minimum width is 3.7 km, average depth—4.5–5.5 m, the maximum depth being about 20 m (in the shipping channel).

Kindgi – a seaside resort area in the Republic of Abkhazia, 20 km to the northwest of the town of Ochamchira. The climate here is humid subtropical. The winter is very mild, without snow. The mean air temperature in January is about 5 °C. The summer is warm; the mean air temperature in August is 23 °C. Atmospheric precipitation—about 1,200 mm a year. In summer the onshore and offshore breezes occur quite frequently bringing some coolness. The favorable climatic conditions here may be used for treatment of cardiovascular and nervous diseases as well as breathing organs (other than of the TB nature). A wide sandy beach and warm sea (the water temperature in summer rises to 22–24 °C) is favorable for thalassotherapy (from May through November). This resort has good prospects for development.

Kipchak Khanate – the Kipchaks or Polovets were the ancient Turkic people, descendants of the Huns and the Europeoid tribe of Dinlin who in 1116 defeated the Pechenegs and in the mid-ninth century became the sole rulers in the steppes of the Northern Prichernomorie and Crimea, except Khersonesus and Kerch Peninsula. The territory of Kipchak (Polovets) Khanate in the Arab sources is called "Desht-i-Kipchak". The Tatars-Mongols that invaded these lands in the twelfth century crushed this khanate.

Kirlangich (in the Turkic "*kirlangic*" meaning "swallow") – a high-speed sail-rowing boat in Turkey that was used by the reconnaissance services. It had 1–2 masts with fore-and-aft sails. In the eighteenth century in Russia K. were built about 22 m long, 7.6 m wide and draft of 2.4 m for sailing in B.S.

Kiten (till 1937—Urdovisa or Ordovise village meaning "Old Vise") – a seaside climatic resort to the southeast of Burgas, on the coast of the Burgas Bay, Burgas Region, Bulgaria. It is situated on the small Urdovisa Peninsula and is surrounded by scattered small bays. The climate here is marine, rather humid, with mild winter (mean temperature of January is about 3 °C) and very warm summer (mean air temperature in July is about 23 °C). This area is characterized by a great number of sunshine hours (about 2,300 a year) and frequent breezes that alleviate the summer heat. The mild climate and warm sea make this place suitable for climato- and thalassotherapy. There are two beaches here: the northern—Atliman and the southern (larger one)—Urdovisa. The resort area has a sanatorium for the people suffering from obesity, many rest houses, bungalows and camping sites. K. also has springs with mineral water.

Kiyikoy (former Midye, Medea) – a village, the Black Sea resort in European Turkey. It locates southward of Igneada on a vast plateau with a depression near the southern part of the Midye Bay. This is a village surrounded by orchards. The shores near the village are inaccessible rocks. Fishing and forestry are the main ways of living in addition to tourism in the summer. Two streams, Kazandere and Pabuçdere, surround the village in the south and the north respectively. Flowing into the Black Sea, these streams are suitable for fishing, boating and swimming. The Kasatura Bay Nature Reserve Area is 18 km south of the village along the Black Sea. The only naturally growing grove of black pine (*Pinus nigra*) in the European part Rumelia of Turkey is found at this site. On the mainland side it is surrounded by the Roman Byzantine wall which ruins survived to the present. The semi-detached Orthodox monastery is rising on a large rock to the west of the village. Population—2,077 (2010).

Kizilirmak River (in Turk meaning "Red river") – the largest river in Asia Minor that flows in the north of Turkey and brings its waters into B.S. Its Antique name is "Halys River" that, according to Strabon, originated from the Greek word meaning solonchak. The terrains crossed by the river have rich salt-bearing formations. The river length is 1,355 km; its watershed area is 75,800 km^2; its flow is 6.4 km^3/year. It takes its origin on the slopes of Kizil-Dag, crosses the Anatolian highlands and Pontic Mountains and empties into B.S. forming a wide delta. The river flows out on a plain 10 km downstream the city of Bafra. Here K. breaks into numerous arms flowing over wetlands. The K. delta is overgrown with impassable forests. It has a snow and rainfall recharge. It is most full-flowing in February-April. K. has tributaries Delice, Devrez, and Gokirmak Rivers. K. is a source of hydroelectric power and is not used for navigation. Dams on the river include the Boyabat, Altinkaya and Derbent. The river waters are withdrawn for irrigation.

Kizilirmak River Delta – the only wetlands on the Black Sea coast of Turkey and one of the five major wetlands of the country. They separate the eastern and western parts of the Black Sea coast of Turkey. The delta covers about 10,000 ha with three large lakes surrounded by a wide belt of cane thickets and marshes. Out of 420 bird species living in Turkey about 309 are found in the delta, including some quite rare ones. They include Indian tree pipit (*Anthus hodgsoni*) and little bunting (*Emberiza pusilla*). Among other species there are such world endangered species as Dalmatian pelican (*Pelecanus crispus*), pygmy cormorant (*Phalacrocorax pygmaeus*) and red-breasted goose (*Branta ruficollis*). During the spring and autumn migration the delta turns into a place of rest for thousands of migrating birds, in particular waterfowl. In winter up to 70,000 birds hibernate on the lakes here. Today the wetlands face the problems connected with drainage of soils, illegal construction of country houses and overfishing and overhunting. In order to protect these Turkey's largest wetland ecosystems the implementation of a project on ecosystem conservation in the river delta was started in July 1992. The information center on control of the environment condition was opened in Bafna, the largest settlement in the region.

Kiziltash Liman – locates in the Kuban River Delta. It is separated to B.S. with a narrow sand barrier. Its area is 164 km^2.

Klokachev Fedot Alekseevich (1732–1783) – Russian Rear Admiral. In 1745 he entered the Naval Academy. From 1747 to 1766 he served on the Baltic Sea and took part in the sea voyage to Arkhangelsk and back. He started his service as a naval cadet and was promoted to Captain first rank. In summer 1769 he commanded the ship "Severny orel" (northern eagle) and with the squadron of Admiral G.A. Spiridonov he went to the Mediterranean Sea. In spring 1770 he navigated the ship "Evropa" (Europe) from Portsmouth to the Mediterranean, took part in search of the Turkish Fleet and on June 24 he took part in the battle at Chios. In the battle at Chesma on 26 June 1770 his ship was the first to attack. In 1771 he commanded the flagship "Evropa" in the squadron of Admiral G.A. Spiridonov, took part in storming the Turkish fortresses of Pelari and Mitilena on the Lesbos Island and later returned to the Baltic. In 1776 he was appointed the commander to the ship "Iezekiil", was conferred the rank of Rear Admiral and became the commander of the Azov Fleet and Azov ports. Having taken over the fleet from A.N. Senyavin he continued efforts on development of ship building although he was already ill that time. In 1778 he undertook the general guidance on repelling the likely assault of the Turkish Fleet that entered B.S. In 1779 he was also preparing for probable appearance of the Turks. In 1780 he went on leave for 2 years due to his illness with the salary being retained. After the leave in 1782 he was directed to the Admiralty Collegium and promoted to Vice Admiral. In 1783 he took the position of the Black Sea Fleet commander. Having arrived to Kerch he assigned the officer staff to their ships, left Kerch with the basic forces and on May 2 arrived to the Akhtiar Bay and started works on its turning into the main fleet base— Sevastopol. He organized the department that comprised the deputy, crew, supply, commissariat and artillery divisions, thus, laying the foundation of the admiralty control of the southern naval forces of Russia. He was awarded many orders.

"Knipovich goggles" – two large-scale cyclonic gyres observed in the eastern and western parts of B.S. They were given the name of N.M. Knipovich. They were marked on the scheme of the Black Sea surface currents (1932) the main features of which are still recognized at present.

Knipovich Nikolay Mikhailovich (1862–1939) – Russian and Soviet hydrological specialist and zoologist, corresponding member (1927) and honorary member of the

USSR Academy of Sciences (1935), explorer of the seas in the USSR European part, the founder of the scientific fishery in the Soviet Union. In 1886 he graduated from the Physico-Mathematical Department of the Petersburg University and decided to make zoology his main occupation. In 1887 he worked at the Biological Station on Solovets Islands in the White Sea. From 1887 to 1891 he spent in prison after he was arrested for participation in the social-democratic movement. After this K. defended the thesis on the subject "Materials to Understanding of the *Ascothoracida* Group". In 1893 he was appointed the private-docent of the University and in 1894–1921 he was the senior research worker of Zoological Museum of the Academy of Sciences, Professor of zoology and public biology at the First Leningrad Medical Institute (1911–1930). From 1898 K. was the representative of Russia and Vice-President of the International Council for the Exploration of the Seas, took part in the work of various scientific committees and conferences. He was the head of many research expeditions: Murmansk expedition (1898–1901) for which the research vessel "Andrey Pervozvannyi" was specially built; several Caspian expeditions (1886, 1904, 1912–1913, 1914–1915, 1931–1932); Baltic research fishery expedition (1902). In 1902 the Russian Geographical Society awarded him the Gold Medal named after Litke for the research in the northern seas.

K. is a prominent figure among the explorers of B.S. His research activities included study of the fish fauna in the Azov-Black Sea basin. Very interesting are his research-fishery expeditions. The Azov-Black Sea expeditions conducted in 1922–1927 became an important stage in understanding the hydrological regime of the B.S. open part. These expeditions studied the seasonal changes in the temperature regime, availability and characteristics of the surface and bottom currents, salinity and transparency of sea waters, distribution of plankton and its effect on the fishery, the ice regime, depth and nature of the sea floor in the traditional fishing areas and others. These expeditions were mostly focused on biological problems, although many hydrological observations were conducted, too. The extensive results of these expeditions of the 1920s were generalized in the work "The Hydrological Explorations in the Black Sea" (1932). It, in particular, provided a scheme of surface currents which main features are still recognized at present (two large-scale cyclonic gyres in the eastern and western parts of the sea, the so-called "Knipovich goggles"). K. identified three basic zones on the basis of vertical distribution of the dissolved oxygen and hydrogen sulfide and the general hydrological structure of the water mass in B.S.: oxygen, intermediate and deep hydrogen sulfide. K. suggested that the deep layers of B.S. contained hydrogen sulfide even in the time when this sea was not connected with the Mediterranean. The results of these works were published as 10 issues of the special series entitled "Proceedings of the Azov-Black Sea Research-Fishery Expedition" (1926–1932). K. is also the author of the fundamental monograph "The Hydrology of the Seas and Brakish Waters" (1938), "The Fundamentals of the Hydrology of the European Arctic Ocean" (1960), dozens of research works on hydrology and geology of the history of the northern seas. K. was sometimes called "the best zoologist among the oceanographers and the best oceanographer among the zoologists". From the establishment of the Hydrological Institute in 1919 and till his death K. worked at this Institute as the chief of the Marine Division. For some time he was the deputy director of this Institute. He also worked in the Applied Ichthyology Division of the State Experimental Agronomy Institute. In 1929 on the K. initiative the Applied Ichthyology Division was transformed into the All-Union Fishery Institute and for some time K. was its first director. The underwater range in the Arctic Ocean was given the name of K.

Knipovich N.M. (http://upload.wikimedia.org/wikipedia/commons/c/cc/Knipovich.jpg)

"Knorr" – the US (Woods Hole Oceanographic Institution) research vessel that in May-June 1988 conducted explorations in B.S. Equipped with the modern precision measuring instruments they managed to unveil the interaction between the aerobic and anaerobic zones of the sea and to reveal essential changes in the oceanological conditions in the zone O_2–H_2S.

Research vessel "Knorr" (http://www.whoi.edu/oceanus/viewImage.do?id=64256&aid=38746)

Kobuleti – the seaside climatic resort in the Republic of Adjaria, Georgia. The center of the Kobuleti District, the railway station, 21 km to the northeast of Batumi. It is situated on a wide sandy and small-gravel beach (its length is over 10 km) being one of the best on the Caucasus. In the north and northeast the resort territory is open (here it borders on the southwestern part of the Kolkhida Lowland), while in the east and southeast it is confined by the outspurs of the Lesser Caucasus. The vegetation in K. and its surroundings is represented mostly by decorative plants, tea and bamboo plantations and pine groves. The climate here is subtropical that is characterized by abundant warmth and high humidity. The flat relief of this resort facilitates free access here of breezes that alleviate significantly the summer heat and lower the air humidity in the nighttime. The winter is very mild, without snow. The mean temperature of January is 5 °C. The summer is very warm; the mean temperature in August is 23 °C. Precipitations make 2,500 mm a year, they fall mostly in autumn and winter. The sunshine hours in a year are equal to 2,200 which is conducive to development of thalassotherapy here. The winds here are mostly weak, their velocity ranging from 1.5 to 2 m/s. Still in 1904 the private villas and mansion houses started appearing here, but there was no resort construction. After the 1917 October Revolution in Russia the resort of K. became open for wide public. From 1923 K. turned into the resort of state significance. Private country houses became sanatoriums and rest houses. For a relatively short time K. turned into a large climatic resort where about 50,000 people a year could have rest. People with cardiovascular, respiratory (other than of the TB nature), nervous and arthritis problems came here for treatment. The town has the citrus-processing factory, canning factory and a tea factory. Population—18,900 (2009). On 26 October 2010 the Parliament of Georgia adopted the law "On Kobuleti free tourist zone".

Kocaali – a town on the coast of the Black Sea, Turkey. It is a town and district of Sakarya Province in the Marmara Region of Turkey. Population—11,600 (2012).

Kocaeli, Province – the Black Sea province ("il") in Turkey. It is situated in the west of the country between B.S. and Sea of Marmara. Its area is 3,626 km², population—1.56 million people (2010). The capital is Izmit.

Kocaeli Peninsula – a peninsula in Turkey washed by the waters of B.S. to the north, Bosporus Starit to the west, and Sea of Marmara to the south. The length toward west is 90 km and the average width is about 40 km. Tectonically this is the westernmost tip of the Pontus Mountains. Plateaus and hills are to 537 m high. The southern shore is occupied by the resort zone—Anatolian Riviera.

Kodor, Kodori River – a river in the Republic of Abkhazia. Its length is 84 km, basin area—2,030 km². It is formed from confluence of the Sakeni and Gvandra rivers originating on the southwestern slopes of the Greater Caucasus. It runs in a narrow gorge and only within the last 25 km of its length it gets out onto the marine lowland. It flows into B.S. forming a delta. K. has a mixed feeding with the dominating share of rainfalls (32 %). The annual flow of the river is made up of ground waters—29 %, melt snow—21 % and melt glaciers—18 %. The mean many-year flow (at Ganakhleba village 25 km from the mouth) is equal to about 4.1 km³ a year varying from 5.7 km³ in 1941 to 2.7 km³ in 1935. During a year the greatest water flow is in spring and summer—79 %. The maximum mean many-year flow is observed in May, the minimum—in January-February.

Kokorev Vasiliy Aleksandrovich (1817–1889) – Russian prominent entrepreneur and public figure. He made a fortune, in particular during the Crimean War, so from the late 1850s he could start the trade and industrial activities. Much money he invested into construction of the Volga-Don waterway and the kerosene works in Surakhany (Baku region). He founded the Russian Shipping and Trade Company (RS&TC) (in the late nineteenth century it earned 30 % of dividends) for consolidation of the Russian Fleet positions on B.S. For development of trade relations with the Transcaucasian and Central Asian regions he established the Transcaspian Trade Partnership and later the Baku Oil Company. He took an active part in railroad construction. He contributed to organization of the Northern Society of Cargo Insurance and Stocking with Warranties and planned to set up the Company for Shipping between the Baltic Sea and America (was not realized). He took part in establishment of the Moscow Merchant Bank (1866). In 1870 together with other Moscow entrepreneurs he revived the Volga-Kama Bank. He gave money (about 2.5 million Rbls.) for construction in Moscow of the famous Kokorev town residence comprising an hotel and trade warehouses (later he passed it to the state to cover debts). He was well-known for his charitable activities. After the end of the Crimean War he organized 3 days of celebrations for the Black Sea seamen at his own expense. In 1884 on the Msta River near Vyshny Volochek he opened the Vladimir-Mariinsky refuge for Russian painters. Later the Artists House named after I.E.Repin located here. In 1861 he opened a picture gallery in a specially built house where more than 500 canvases (and half of them made by Russian painters, such as Bryullov, Aivazovskiy, Levitskiy, Borovikovskiy, Ugryumov, Matveev, Kiprenskiy and others) were demonstrated. In the late 1870s due to financial difficulties he had to sell this collection (some pictures were purchased by Alexander Aleksandrovich, the heir to the throne, merchants P.M. Tretiakov and D.P. Botkin). In 1889 K. became the honored member of the Academy of Arts.

Koktebel – a seaside resort village, Crimea, Ukraine. It is located 20 km to the southwest of Feodosiya on the coast of the Koktebel Bay. In the Turk language "kok (gek)" means "blue, light blue"; "tebe (tepe)" means "top, peak, hill" and "el"—"terrain, country", in other words we get "The country of blue peaks, the terrain of blue hills". By the twentieth century it was inhabited by the Bulgarians. And this was also the time when well-known poets, novelists, people of art started buying land sites here for construction of country houses. The founder of the Koktebel resort was renowned ophthalmologist Z.A. Yunge, Academician of the Peter Academy (1833–1898). In 1916 Russian poet and artist M.A. Voloshin settled here. M.I. Tsvetaeva and M.A. Voloshin called this place "Cimmeria"—from the name of the legendary Cimmerians who populated the B.S. coast. After expulsion in 1944 of the Crimean Tatars this place was renamed into Planerskoe because long ago the glider-pilots took fancy of this place for testing their models and holding competitions. The glider and parachute museum existed here. In 1993 this place returned its former name—Koktebel. Population—2,800 (2012). Now Koktebel is a popular seaside health resort, which has a population mainly engaged in serving guests. There is a plant for production of wines and cognacs "Koktebel". Since 2003, every year in Koktebel held the largest Ukrainian festival of contemporary ethnic and jazz music "Jazz Koktebel". Since 2006 Koktebel was chosen by the Argentine tango dancers: on May Day celebrations are held in the International Tango Camp "Crimean Vacation", and in the fall—festival "Velvet Tango", which attracts tangueros from Ukraine, Russia, Germany, Argentina, Austria, Poland, Turkey, England, Italy, France, USA, Canada and other countries.

Koktebel (Photo by Evgenia Kostyanaya)

"Koktebel" – the plant producing vintage wines and cognacs. It is located in the vicinity of the Koktebel and Otuz valleys. It is thought that vine-growing appeared here in the first centuries A.D. In the Middle Ages the Christian population practiced vine-growing in the Otuz valley and wine export gave the local people substantial income. Surozh wines had been known also in Rus'. After Crimea was incorporated into Russia the development of vine-growing and wine-making was pushed forward. In the second half of the nineteenth century the vine-growing practices were improved and acquired commercial significance, and Russian Count Vorontsov and Prince Golitsyn contributed much to this. In 1904–1906 A.E. Yunge built a winery with large basements, thus, putting wine production on a scientific basis. During the Soviet power the first sovkhoz (later it became a division of the "Sudak" sovkhoz) was organized. During the Great Patriotic War everything was destroyed. Many vines perished. In 1944 it was decided to establish a large sovkhoz named "Koktebel". At first a small winery was built, but in 1959 the construction of huge wine basements and a winery was initiated on the mountain slope not far from Planerskoe settlement. The unique works were carried out by the Moscow metro-builders. They constructed eight tunnels 50 m long, 8.4 m wide and 7.6 m high each (the lower wine storage). The upper wine storage consists of five tunnels with the same dimensions, except that they are 120–360 m long. The total length of the tunnels is 1,440 m. The wonderful dry, strong and dessert wines, cognac spirits and vintage cognacs were manufactured here following the classical technology. The vine varieties grown in these unique soil-climatic conditions enable production here of original brands of wines and cognacs. Thus, the natural conditions reigning here (the annual sum of temperatures and quantity of sunshine days) facilitate manufacturing of Madeira "Koktebel", being one of the most "sunny" wines.

Koktebel Bay, Harbor – is situated between the Malchik (Boy) Cape formed by the eastern slope of the Karadag Mountain and the Lagernyi (Camp) Cape located 4.3 km to the northeast of it. Its shape reminds of an oval half-moon. It is surrounded by not high bald mountains. The Mount of Kok-Kaya is rising in the west. The shores of the bay are elevated, but flat. The cliffs and rocks may be found on the flat shore. The resort village of Koktebel locates on the western shore.

Kolchak Alexander Vasilievich (1874–1920) – Russian Captain 1st Rank (1913), Rear Admiral, Vice-Admiral (1916), scientist in hydrology, polar explorer, commander of the Black Sea Fleet, "Supreme Ruler of the Russian State". He finished the Naval Corps (1894). In 1895 he navigated from Kronshtadt to Vladivostok on board the cruiser "Ryurik" as an assistant of the watch officer. In 1896 in Vladivostok he was transferred to the cruiser "Kreiser" of the Pacific squadron. In 1898 he was promoted to Lieutenant. Being in the Far East he, all of his own, got down to the in-depth study of oceanography and hydrology. In 1900–1902 on the schooner "Zarya" he went on a polar expedition headed by Baron E.V. Toll and there he took part in hydrographic explorations. E.V. Toll with three other members of the expedition went further north on sleighs and got lost. In 1903–1904 K. headed the rescue team and risking his life he found the materials of this expedition.

When the Russian-Japanese War began he went as a volunteer to Port Arthur (1904–1905). He succeeded in laying mines at the access to the Port Arthur Bay and also in the Amur River mouth where the Japanese cruiser "Takasago" was destroyed. In November 1904 before capitulation of the fortress he commanded the naval gun battery in the northeastern line of the Port Arthur defense. He was wounded and became a prisoner of war in Japan. In June 1905 he returned to Saint Petersburg via Japan and Canada. Upon his return he published the materials of the polar expedition. He was one of those who initiated establishment of the Naval General Staff and after its creation in 1906 he worked actively in it. In 1909 he published his basic work "Ice of the Kara and Siberian Seas".

From 1910 to 1915 he served in the Pacific and Baltic Fleet. In 1916 he was promoted to Rear Admiral and soon appointed the commander of the Black Sea Fleet with conferring him the rank of Vice-Admiral. Under his guidance the siege of the Bosporus and coal area Eregli-Zonguldak became tougher. He also organized mine laying in the enemy ports as a result the enemy ships were unable to go out into B.S.

After the February Revolution in Russia he recognized the Provisional Government and on March 12, 1917 he put the Fleet to loyalty oath to this government. He actively fought against revolutionary ideas in the Fleet. He was a proponent of the war till the final victory. When yielding to persuasions of the Baltic propagandists the sailors started disarmament of their officers he resigned from the position of the Chief of Staff of the Black Sea Fleet and went to Petrograd. He was appointed the chief of the naval mission to the USA. When he understood that the Americans gave up the idea of helping Russia in this war he decided to return to Russia via Japan where in November 1917 he learned about establishment of the Soviet power. After this he took a decision not to return to Russia. He planned to raise forces in the Far East for opposing the Bolsheviks. K. was appointed the Chief of Staff of the Russian armed forces in Manchuria, but he failed to comprehend the political situation. On his way to enlisting with the Volunteer Army in October 1918 he arrived together with British General Alfred Knox in Omsk where by the decree of the local Provisional Government he was appointed the Military and Naval Minister in the Ufa Directory Government. On 18 November he organized a military coup d'etat and the members of the Directory were arrested.

He commanded the struggle with the Soviet troops in Siberia, on the Ural and in the Far East. He advocated the idea of "the great and indivisible Russia". Thanks to his first successes he received support from the allies and was recognized "the Supreme Ruler of the Russian State" by other chiefs of the White Movement. As a result of the counter attack of the Red Army in May-June 1919 the best armies of K. had to retreat far to the east of Russia. He himself was arrested by the Czech Legion staying in Siberia and on 15 January 1920 on the Innokentievskiy station near Irkutsk they handed him over to the Socialist-Revolutionary (SR) "Political Center" that, in its turn, transferred the Admiral to the Bolshevik Military-Revolutionary Committee (MRC) in Irkutsk. At first they planned to send K. to Moscow, but upon receiving a respective order from the capital K. was executed by MRC on the ice at confluence of the Ushakovka and Angara rivers. The body of K. was disposed under the river ice. In 2004 the monument to K. was installed in Irkutsk.

Kolchak A.V. (http://www.proza.ru/pics/2008/03/14/591.jpg)

Kolchak Vasiliy Ivanovich (1837–1913) – Russian military engineer, metallurgist, the father of A.V. Kolchak. He finished the Richelieu Gymnasium in Odessa. In 1854 he went on the military service as a junker, later he was promoted to naval artillery ensign. In the Crimean War of 1853–1856 during the siege of Sevastopol he was artillery battery captain on the Malakhov mound. For his services he was decorated with the badge of the Military Order. On 26 August, 1855 he was contused, wounded in hand, taken a prisoner of war by the French and sent to the Princes Islands in the Sea of Marmara. After release from captivity he finished the 2-year courses at the Institute of the Mining Engineer Corps and was directed to the mining works on the Ural. From 1863 he was the member of the Naval Artillery Commission on acceptance of guns and shells at the Obukhov steelworks in Petersburg. In 1893 he was promoted to Major-General and after this he retired. He was the author of many scientific works on metallurgy published in "The Naval Digest" and reminiscences "The War and Captivity, 1853–1855. Recollections about Past Experience" (St.-Petersburg, 1904).

Kolkhida – the ancient name of the Caucasus, but most often it was attributed to the area along the banks of the Phasis (presently Rioni) River. This area was rich in wood, resin, flax, hemp and wax. This was also the historical name of modern Western Georgia. After the Ionic people (Miletians) founded colonies on the Black Sea coast (in the seventh century B.C.) K. was identified with the fairyland of Aia being the destination of Argonauts. That time K. was included into the system of trade relations existed in the Antiquity that incorporated the Caucasian tribes speaking presently 70 languages and many dialects. The wonderful lint fabric handwoven by the mountain people were on great demand. The capital of K.— Phasis (presently Poti) was built by Greek Themistagoras from Miletus (end of seventh—beginning of sixth century B.C.).

Kolkhida Lowland, Rioni Lowland – is situated on the B.S. coast, Georgia. It got its name from the historical and geographical area of Kolkhida or from the name of "Kolkhi", the ancient Georgian agrarian tribes that lived in the first millennium B.C. The Rioni Lowland is called by the name of the Rioni River flowing over it.

Kolkhida Monsoon – the warm eastern wind blowing from the mountains in the period from one day to a week. Sometimes it may reach the magnitude of a hurricane becoming ruinous for the vegetation.

Konstantinov (Konstantind) Kirjak Konstantinovich – Russian captain-commodore (1803). In 1783 he was conferred the rank of ensign and sent to serve in Russian Navy. Every year he took part in naval campaigns on the Baltic Sea (1783–1786). In 1784 he was promoted to midshipman. He was sent to Kiev for navigation on the galleys over the Dnieper (1786). On the galley "Bug" he escorted Catherine II on her travel over the Dnieper to the south (1787) and for this he was granted the golden watch. Commanding a galley he navigated between the Ochakov (Dnieper) Liman and Kherson (1787), took part in the battles with the Turkish Fleet in the Liman; on the large boats he cruised between Kherson and the Gadzhibei fortress where he captured the Turkish transport and 5 lansons (1789). With the rowing fleet he took part in seizure of many fortresses on the Danube, including in storming the Izmail fortress (1790). He commanded the brigantines "St. Dmitry" and "St. Peter" on B.S. (1794–1796), frigate "Advent" (1798). In 1799 he was promoted to Captain 1st rank. In November 1803 he resigned.

Kopytov Nikolay Vasilievich (1833–1901) – Russian Vice-Admiral. In 1843 he entered the naval troops of the Alexander Corps as a cadet; in 1844 he was transferred to the Naval Corps; in 1850 promoted to naval cadet. In 1851–1854 on different ships he cruised over the Baltic and North seas. In 1855 he was conferred the rank of Lieutenant. In 1857–1860 he navigated near the coasts of Korea and Japan as a commander of a corvette. In 1860 he was promoted to Captain-Lieutenant. In 1861–1862 he commanded the corvette that navigated in the Pacific. In 1863–1864, commanding the frigate "Peresvet" he sailed from Kronshtadt to the Mediterranean and back. In 1866 he was promoted to Captain 2nd rank. In 1867–1872 he cruised in the Baltic Sea. In 1871 he was promoted to Captain 2nd rank. In 1872 he was appointed the naval agent in London; in 1876— Captain of the Petersburg port. In 1882 he was promoted to Rear-Admiral and appointed a commander of a ship unit in the Pacific. In 1882–1883 K. sailed to different Pacific ports. In 1884–1890 K. commanded the squadron on the Baltic Sea. In 1888 he was promoted to Vice-Admiral and appointed the senior flagman of the Baltic Fleet. In 1890 he commanded the squadron on the Baltic Sea. In 1891 he was appointed the Commander-in-Chief of the Fleet and ports on the Black and Caspian seas and the military governor of Nikolaev. In 1891–1894 he navigated the Black Sea on different ships escorting the Emperor or General-Admiral; in 1894 on the cruiser "Pamyat Merkuria" (Memory of Mercury) he sailed with the body of demised Emperor Alexander III from Yalta to Sevastopol. He had many orders of foreign states. K. contributed much to development of shipbuilding. In autumn 1890 he suggested that the fleet needed not armor-clad ships, but cruisers.

Korakya Cape – the northeastern tip of a wide peninsula confining the Burgass Bay in the south, Bulgaria. The cape is elevated, its shores are abrupt steeply going into water. In the south the cape is cut with a deep gorge.

Koreiz – small village in Crimea, Ukraine. It is located on the Southern Coast of Crimea on the slope 14 km to the southwest of Yalta. Its population is 6,300 people (2012). This is a climatic seaside resort. K. is surrounded by vines, orchards, citrus plantations and tobacco fields.

Kornilov Vladimir Alekseevich (1806–1854) – Russian Vice-Admiral (1852), Hero of the Sevastopol Defense of 1854–1855 during the Crimean War. He finished the Naval Cadet Corps (1823). In 1824–1827 he served on a boat and later on the Baltic Fleet. From 1827 he served on the battleship "Azov", took part in the Navarinskiy battle in 1827 and in the Russian-Turkish War of 1828–1829. In 1830–1833 he served on the Baltic Fleet and from 1834—on the Black Sea Fleet where he commanded the brigantine, corvette and frigate. From 1839 he commanded the battleship "Twelve Apostles". In 1840–1846 he guided the landing operation during raids to the Caucasian coast. In 1846 he was sent to Britain where supervised construction of steamships for the Black Sea Fleet and also studied the fleet of this country and organization of its management. From 1849 he was the Chief-of-Staff of the Black Sea Fleet. K. insisted on replacement of the sailing ships for steamships and installation of new cannons on them. He wrote several manuals and instructions for the fleet and took part in preparation of the Naval Code. Commanding the unit of steamships he assisted the squadron of P.S. Nakhimov in defeating the Ottoman Fleet in the sea battle of 1853 at Sinop. In November 1853 the unit of frigates commanded by K. who stayed onboard the ship "Vladimir" took part in the battle of steamships that was the first experience for this type of vessels. During the Sevastopol defense of 1854–1855 he commanded the defense of the northern part of the city and after the Alminskiy battle, being the Chief-of-Staff of the garrison, he, in fact, was in charge of the whole defense operation. Under guidance of K. the crews of the ships sunk in the Sevastopol Bay together with the local population managed within a short time to construct a line of fortifications (redoubts, ramparts and batteries) that played an important role in giving rebuff to the numerous attacks of the British-French and Turkish troops. K. showed himself a brave and fearless man in a battle. He was deadly wounded on the Malakhov mound. The monument to K. is installed in Sevastopol displaying the moment when he deadly wounded raised slightly and gave his last order to the defenders of this city: "Fight for Sevastopol". He was buried at the Vladimir (Admiral) Cathedral in Sevastopol. He was awarded many Russian and foreign orders.

Kornilov V.A. http://www.sevmb.com/spaw2/uploads/images/2_2.jpg

Koshka Petr Markovich (1828–1882) – the participant of the Sevastopol Defense of 1854–1855, Quartermaster. He was born in the family of a serf. In May 1849 he was recruited and enrolled as a sailor of the second degree to the Thirty Unit of the Black Sea Fleet. From the very beginning of the Sevastopol defense he was appointed to the Third Bastion that defended the access to the Southern Bay and to the city center. He showed exceptional bravery and resourcefulness in the battles, took part in many raids and also acted by himself: he infiltrated in the enemy disposition, procured valuable information and brought captives and arms. The journals and newspapers wrote about his deeds; his portrait was painted. Empress Alexander Fedorovna sent him "the Cross of Blessing". In October 1855 after retreat of the defenders on the southern side of Sevastopol he was wounded and sent on a long leave. He did not have a landholding, thus, he lived in poverty and went with transports to Nikolaev, Odessa and Kherson. In 1863 re returned to the service and was assigned to the Eighth Unit of the Baltic Fleet. Several years later he returned to his native village. He was awarded two insignias of a military order, the Medal "For Sevastopol Defense". Heroic deeds of K. were depicted on the panorama painting of F.A. Rubo. On the monument to Vice-Admiral V.A. Kornilov in Sevastopol K. is shown throwing the bomb out of a trench. The bust of the legendary sailor was also installed in Sevastopol. The monument to K. was also constructed in his native village Ometintsy of the Vinnitsa Region. The origin of a saying is also connected with the name of K. (his name means "cat"). When V.A. Kornilov thanked K. for his service he answered: "Good word is pleasant even to Koshka (cat)".

Kostenko Vladimir Polievktovich (1881–1956) – Russian engineer, shipbuilder who contributed much into development of the theory of a ship and organization of shipbuilding in the USSR. In 1904 he finished the Naval Engineering School with the gold medal. After this he was appointed the assistant of a builder of the squadron armored ship "Orel" and onboard this ship he, as an engineer, navigated from the Baltic Sea to the Far East and in May 1905 he took part in the battle at Tsusima. Recollecting about the events of that time K. wrote the book "On the Ship "Orel" in Tsusima" (1955). After the end of the Russian-Japanese War (from 1906) he served the assistant of a builder of the squadron armored ship "Andrey Pervozvannyi" that was built in the shipyard of the Galernyi Island in Petersburg. In 1909–1910 K. worked under guidance of A.N. Krylov in the Design Bureau of the Naval Technical Committee. In 1912–1917 he worked in Nikolaev first the head of the Association of Shipbuilding, Mechanical and Foundry Plants and later the chief engineer of the "Naval" plant. From the first years of the Soviet power K. contributed much into organization of the shipbuilding business in the Soviet Republic. From 1919 to 1922 he was the technical chief of the shipbuilding yards in Nikolaev and in 1922–1924—the chief of the Industry Department of Ukraine. In 1926 he wrote the book "Theory of a Ship. Water Resistance to Ship Movement". That time he also invented the original form of a ship hull that reduced resistance. From 1928 he headed the Design Bureau "Sudoproject" in Leningrad and at the same time he was the chairman of the Technical Council at the Central Bureau on

Marine Shipbuilding. In 1931 he was appointed the chief engineer of "Soyuzverf" Institute. From 1932 he was in charge of design of new and rehabilitation of the old shipbuilding yards, introduced many new progressive methods into shipbuilding, such as building of ships on horizontal slipways, construction of closed and fastened slipways, wet docks, large-section assembly of ships and others. He was the Laureate of the USSR State Award.

Kotovskiy Grigoriy Ivanovich (1881–1925) – the hero of the Civil War in Russia. He was born in the family of a mechanics. In 1905 he was drafted into the army, escaped and organized in November 1905 a team of the revolting Moldavian peasants. After numerous arrests and escapes in 1907 he was sentenced to 12 years of hard labor in a colony. In early 1915 he again commanded the armed unit of peasants in Bessarabia. In 1916 he was sentenced to death that was replaced with lifelong hard labor colony. In May 1917 he was conditionally released and sent to the army on the Romanian Front; he was a member of the regiment committee of the 136 Taganrog infantry regiment. In January-March 1918 K. commanded the Tiraspol unit, from July 1919—the brigade of the 45th rifle division. Being a member of the I.E. Yakir group he took part in the heroic march from the Dniester to Zhitomir. From January 1920 K. commanded the cavalry brigade in the south, Ukraine and on the Soviet-Polish Front. From December 1920 he was the commander of the 17th cavalry division. K. was awarded 3 Red Banner Orders and the Honorary revolutionary arms. He was buried in Birzul (presently Kotovsk) in the Odessa Region, Ukraine.

Kotsebu Pavel Evstafievich (1801–1884) – Russian Infantry General (1859), Count (1874). He fought with highlanders and participated in the Persian, Turkish and Polish campaigns. From 1843 in the rank of Major-General he served in the escort of His Emperor Highness; General Quartermaster of the acting army; in 1846 he was appointed the Chief-of-Staff of the Caucasian Corps; in 1853–1855—of the III, IV and V Infantry Corpses; in 1855—Lieutenant-General, Chief-of-Staff of the Southern Army, commander of the V Infantry Corps. In 1847 he was promoted to Aide-de-Camp General. In 1856 he was the Chief-of-Staff of the First Army, Infantry General (1859), commander of the Odessa Military District. In 1862–1874 he was appointed Governor General of the Novorossiysk and Bessarabia districts. From 1863 he was the member of the State Council; in 1874–1880 he took the post of the Warsaw Governor General.

Kovalevskiy Alexander Onufrievich (1840–1901) – Russian scientist, one of the founders of the evolutionary comparative embryology and physiology of the invertebrates, one of the founders of experimental histology, Academician of the Emperor Academy of Sciences (1890). He was born in the Vitebsk Province. In 1859 he entered the Natural Science Division of the Physico-Mathematical Department of the Petersburg University. In 1860 K. went abroad to continue his education and till 1861 he studied at the Heidelberg University. Here he became a zoologist and ardent proponent of the Darwinism. In 1862 he returned to Petersburg and in 1863 passed the examination and was officially conferred the academic degree—

Candidate of Natural Sciences. K. studied development of lacelet, ascidians, combfish, holothurian and other invertebrate animals inhabiting B.S. and the Mediterranean and founded the teaching on embryo marks. Of great importance is the work of K. on combfish that provided the detailed and accurate description of all early stages of development of *Escholtzia* and others. Very interesting are also the articles of K. on development of single ascidians studied on *Phallusia mammilata* and *Cione intestinalis*. After careful studies of the history of the ascidians development only K. found out that they were really the degraded relatives of the vertebrates because in the early period of development (at the larva stage) they had the same organs as the lower vertebrates. In 1867 K. prepared the thesis for the doctoral degree having chosen the subject "Anatomy and History of *Phoronis* Development" (the lower worm about which only meager data were available). The main conclusion of his studies was that the representatives not only of some one class, but all groups of animals, the vertebrate and invertebrate alike, passed the common way of development. K. created his theory of comparative embryology proceeding from the Darwin theory about the origin of all animals on the basis of embryological materials and tested it in many experiments. K. was among the initiators of establishment of the marine research stations in Russia. He was the first director of the Sevastopol Biological Station and took this post till his death. His name was given to the Institute of Biology of the Southern Seas and the monument to him was installed in front of the institute building.

Kovalevskiy A.O. (http://img1.liveinternet.ru/images/attach/c/0//47/467/47467515_Kovalevskiy.jpg)

Kozhanov Ivan Kuzmich (1897–1938) – Russian Naval Flagman of second rank. In 1915 he finished the nonclassical college in Rostov-on-Don. In 1916 he was drafted into the Navy. In 1917 he finished the Special Naval Cadet School. He served on the cruiser "Orel" and destroyer "Bodryi" of the Siberian Fleet. From March 1918 he was the chief of the Special Detachment at the Naval Commissariat. From August 1919 he was the commander of all landing units of the Volga-Caspian

Fleet that bravely fought at Tsaritsyn and Cherny Yar. From September 1920 he headed the Moscow Expedition Division on the Sea of Azov on the Vrangel Front. From July 1921 he was the commander of the Caucasian sector of the Black and Azov seas defense, from December—a member of the Revolutionary Military Council of B. and A.S. forces. From 1922 to 1924 he was the Chief and Commissar of the Far East Navy. In 1927 he graduated from the Naval Academy. In 1927–1930 he was the Naval Attaché in Japan. In June 1931 he was appointed the commander of the Black Sea Navy (from January 1935—the Black Sea Fleet). In 1935 he became the Naval Flagman of the second rank (from 1935). He was the first who initiated the use of perspective means of warfare on the sea. He advocated the application of naval aviation, submarines and torpedo boats. He focused attention on the fleet interaction with overland troops. He was awarded many orders. In 1937 he was repressed and in 1956 he was exonerated post-mortem.

"Krab" – the world's first mine-laying submarine. It was designed by Russian engineer M.P. Naletov and built in Nikolaev. It was launched in July 1915. Its water displacement in buoyancy was 533 tons, in an underwater position—736 tons. The speed in buoyancy was 21.8 km/h, in an underwater position—13.1 km/h. The sailing range in buoyancy was 2,286 km, in an underwater position—36 km. The submergence depth was 36.6 m. Armament: 2 bow torpedo tubes, 60 marine anchored mines, one 37-mm (from 1916 to 1970-mm) cannon and 2 machineguns. The crew was 30 persons. In 1915–1916 during World War I "K." was included into the Black Sea Fleet and from the start was involved in battles. It successfully laid mines near the Bosporus and Varna. In July 1915 the Turkish gunboat "Isa Reis" and 2 days later the German cruiser "Breslau" were blown up on the mines laid by "K." near the Bosporus. In April 1919 "K." was sunk by the British troops near Sevastopol. It was lifted in October 1935 and was sent to scrap.

"Krab" (http://flot.sevastopol.info/photos/photo_submarines/sub_krab_02.jpg)

Krabbe Nikolay Karlovich (1814–1876) – Russian state and military figure, Aide-de-Camp General (1858), Admiral (1869). From 1832 he was on the military service. He finished the Naval Cadet Corps (1832). From 1832 to 1837 he served on the Baltic Fleet. In 1837 he took part in battles on the Black Sea coast of the Caucasus near the Gelendzhik fortress, Pshada, Vupan and Shapsukho rivers. In 1838 he was appointed Aide-de-Camp at the Chief of General Staff. In 1839 he participated in the landing operation of the Black Sea Fleet: he commanded the armed boat at troops landing on the Caucasian coast. In 1847 he was a member of the group that conducted explorations in the Aral Sea and in the Syrdarya River to find out whether it was possible to construct fortifications and settlements there. From 1849 he was the officer on special assignments at the Chief of General Staff. In 1853–1854 he was the Chief of Staff of the Black Sea Fleet squadron. From 1854 he was the Vice-Director and from 1856—Director of Inspection Department of the Naval Ministry. In 1860–1876 he was the Manager of the Naval Ministry. Taking this position he did a lot for provision of the Fleet with armored steamships and installation of the grooved cannons on the battle ships which was a great achievement that time. On his initiative and with his participation some legislative acts concerning the Fleet were passed among which there was the law prohibiting physical punishments. He was awarded the Russian and foreign orders. His name is given to the island and mountain in the Sea of Japan.

Krabbe N.K. (http://sochived.info/wp-content/uploads/2010/12/Krabbe.jpg)

Krasnaya Polyana – a climatic and balneological resort area in Russia (Adler District, Sochi); in 1950 it acquired the status of a village-town. It is located 52 km to the northeast of the railway station Adler and 43 km from B.S. It is a part of Greater Sochi, Krasnodar Territory. It was founded in 1869 as the Romanov settlement and was called so by the Emperor's name—it was planned to organize the Emperor's resort and quite unofficially this place was called Tsarskaya Polyana (Tsar's glade) that in 1898 was given the status of a mountain resort. The lands around this place were turned into "the reserve zone of the Tsar family". Here the hunting house for Emperor Nikolay II was constructed. In 1899 the Krasnopolyanskiy road was built that spurred further development of this mountain-climatic resort. In 1864 the end of the Caucasian War was celebrated here and in 1914 Nikolay II saluted the military parade in commemoration of 50 years passed since the end of the Caucasian War.

In 1938 this place was renamed into K.P. It is located at the foot of the southern slope of the Main Caucasian Range at an elevation of 600 m, on the right bank of the Mzymta River. The Achishkho, Aibga, Chigush and Pseashkho mountains which peaks reach 2,500–3,000 m in height and are covered with snowcaps shield this place from cold winds. The climate is humid subtropical. In winter there is much snow here and there are many sunshine days. The mean temperature in January is 1 °C, but short-time temperature drops to −20 °C and even lower are possible. The mean temperature of July varies from 20 to 22 °C. Precipitation are 1,800 mm a year. The natural health improvement factors here are not only climate, but mineral spring waters. K.P. is one of the tourist centers in Western Caucasus. The road from Adler to K.P. is one of the most beautiful on the Caucasus. On the 16th km, near the road you can see the remains of the early man encampment (Akhshtyrskiy cave), on the 33rd km—the Akhtsu gorge that is no less picturesque than the Darjal gorge. The K.P. region includes the Caucasian nature reserve. In 1973 the Museum of Caucasian Flora and Fauna was opened here. There is the Kardyvach Lake from where the Mzymta River originates. On the territory of K.P. there are the arboretum, an artificial lake with boats and a beach. From 1924 the tourist camp has been functioning here. At present this place has become popular again. In early 2000 elaboration of the master plan of the "Mountain-Marine Complex "Krasnaya Polyana" was started. The project included a settlement and a large nearly rectangular land area from the Nizhne-Imeretinskiy Lowland (this is the southern outskirts of Sochi bordering on Abkhazia) as far as the mountains located beyond K.P. The total area is over 520 km^2. The complex was divided into three parts—the maritime, piedmont and mountain. As for the mountain part, its main application has been known long ago—the mountain skiing resort with all respective attributes. Population—4,600 (2012).

Future prospects of the Krasnaya Polyana ski resort as associated with the implementation of the federal program "Development of Sochi as a mountain resort in the years 2006–2014", as well as holding the Sochi Winter Olympic Games in 2014. In addition to winter sports in the area of the village there is development of mountain tourism. In the area of Krasnaya Polyana the following Olympic competitions are scheduled: skiing, snowboarding, cross country skiing, biathlon, Nordic combined, ski jumping, bobsled, luge, skeleton, freestyle.

Krasnaya Polyana (http://kourortchernomor.ru/krasnaya-polyana-2/krasnaya-polyana.html)

Krasnodar – is a city and the administrative center of Krasnodar Territory (Kray), Russia, located on Kuban River about 150 km northeast of the Black Sea port of Novorossiysk. Population—745,000 (2010). It was founded on 12 January 1794 as Ekaterinodar, which means "Catherine's Gift" in recognition of Catherine the Great's grant of land in the Kuban region to the Black Sea (Kuban) Cossacks. After the 1917 October Revolution in Russia Ekaterinodar was renamed Krasnodar (December 1920). The origin of the city starts with a fortress built by the Cossacks in order to defend imperial borders and claim Russian ownership over Circassia, which was contested by Ottoman Empire. In the first half of the nineteenth century Ekaterinodar grew into a busy center of the Kuban Cossacks. It was granted town status in 1867. By 1888 about 45,000 people lived in the city and it became a vital trade center of southern Russia. During the Civil War in Russia the city changed hands several times between the Red Army and Volunteer Army. During the Great Patriotic War Krasnodar was occupied by the German Army between 12 August 1942 and 12 February 1943. The city sustained heavy damage in the fighting but was rebuilt and renovated after the war.

Today Krasnodar is the economic center of southern Russia. For several years, Forbes Magazine named Krasnodar the best city for business in Russia. In the industrial sector, the city has more than 130 large and medium-sized enterprises. The main industries are: agriculture and food industry (42.8 %), energy sector (13.4 %), fuel industry (10.5 %), machine construction (9.4 %), forestry and chemical industries (4 %). Tourism comprises a large part of Krasnodar's economy. There are more than 80 hotels in the city. Krasnodar is a highly developed commercial area. Krasnodar is the first in Russia in the number of malls per capita. The city has international airport and railways station.

Krasnodar (http://www.yuga.ru/media/teatralnaya_ploshhad.jpg)

Krasnodar Territory (Kray) – the subject of the Russian Federation entering into the Southern Federal District, the southernmost region of Russia. It was formed on 13 September 1937. In 1991 the Adygeya Autonomous Region withdrew from the Territory and formed the Republic of Adygea. The area of the Territory is 76,000 km^2 or 0.4 % of the Russia's territory. Its maximum length from south to north is 370 km, from east to west—375 km. In the north K.T. borders on the Rostov Region, in the east—the Stavropol Territory, in the south—Abkhazia and Karachaevo-Cherkess Republic, and in the west it has the sea border with Ukraine. The Territory is washed by the waters of two seas—B. and A. seas. The administrative center—Krasnodar (745,000 people (2010)). K.T. incorporates 44 municipal formations, including 7 cities and 37 districts. A total of 26 cities and 12 urban settlements are found in K.T. The largest cities are Sochi (343,000), Novorossiysk (242,000), Armavir (189,000), Eisk (88,000), Kropotkin (81,000), etc. The total population of K.T. is 5.33 million people (2013) making it the third among other regions of Russia. 53.5 % of the population lives in cities. More than 120 nationalities can be found here, such as the Russians 88.3 %, the Armenians 5.5 %, the Ukrainians 1.6 %, and others. The Territory has its own coat of arms and a flag. The highest legislative body is the Legislative Assembly, the highest executive body— the Administration that is formed by the governor.

The B.S. coastline is broken by the Gelendzhik and Tsemess bays, the Taman and Abrau peninsulas; the A.S. coastline—by Eisk, Beisugskiy and other limans.

The flat part of the Territory refers to the Western Pre-Caucasus; the mountain part (about one-third of the territory) to Greater Caucasus and it is represented by the high Main or Water Divided Range (to 3,360 m high) and also medium-high folded ridges and piedmonts of the so-called Black Sea Caucasus. The largest rivers are the Kuban with its tributaries Laba and Belaya, and the Mzymta River. The Tschikskoe, Shapsugskoe and Krasnodarskoe reservoirs were built to regulate the Kuban flow. In the mountains you can see the Kardyvach and Abrau lakes; on the Taman Peninsula and on the A.S. coast there are many lake-limans. The flat areas are covered with man-planted vegetation. The mountains are overgrown with broad-leaved forests, largely oak and beech, in the Tuapse-Sochi area—mostly mixed Kolkhide forests, and the dark coniferous forests, largely Caucasian fir and spruce. Still higher the sub-Alpine and Alpine meadows are extending. Vertically the forests are arranged as follows: over 800 m—beech and then coniferous forests, from 1,800 to 2,000 m—the belt of sub-Alpine and Alpine meadows. The terrain around Novorossiysk is covered with xerophilic thin forests and thickets of xerophilic shrubs. To the south of Tuapse the coast is overgrown with subtropical vegetation. The subtropical vegetation (shrubs and trees) growing in the northern part of the Black Sea coast of the Caucasus is similar to that of the Southern Coast of Crimea and Eastern Mediterranean.

In the west and north of the Territory, taking two-thirds of the K.T. area, there is the vast Kuban-Priazov or Pre-Kuban Lowland (to 120 m high), in the east—the western part of the Stavropol Upland (to 623 m high), in the southwest and in the south—the piedmonts and ranges of the Greater Caucasus: Main (to 3,261 m high), Lateral, Rock and other ranges. Coming up to the B.S. coast the mountains break down abruptly leaving only a narrow strip of the Black Sea coast. K.T. has the deposits of natural gas, oil and cement marls; in the mountains—the deposits of iron ores, table salt, gypsum, nonferrous and rare metals; in the piedmonts and on the B.S. coast—mineral springs (Goryachiy Klyuch, Matsesta and others). The climate ranges from moderate, continental to subtropical. The mean temperature of January varies from -4 °C on the plain to $+5$ °C in the south (Black Sea coast); the mean temperature of July ranges from $+22$ °C to $+24$ °C. The precipitation on the plain are 400 mm increasing from 600 to 800 mm in the piedmonts to 1,000 mm in the mountains, making 1,500 mm in the subtropical zone (Sochi) and exceeding 3,000 mm in the Krasnaya Polyana area. The northeastern winds are dominating here; in the Novorossiysk area the bora is blowing. The mountains reveal the altitudinal climatic zoning. The greater part of the rivers belongs to the A.S. basin. The Kuban River lower reaches are occupied by extensive wetlands, the so-called Kuban plavni, that stretch for about 200 km to the Kuban inflow into A.S. They are overgrown with reed thickets being of commercial significance. The largest tributaries of the Kuban are the Laba and Belaya rivers. The flat part of K.T. is composed of chernozem (black earth), the Kuban lower reaches—alluvial-meadow soils, the piedmont part—gray podzolic soils, the mountain part—moun-tain-forest gray and brown soils, and the south of the coastal area—of yellow soils. The plains are mostly cultivated. The steppe and forest-steppe vegetation is retained in the piedmont parts with the heavily rugged relief.

K.T. is an economic and administrative, agricultural and industrial region of Russia. Grain and technical crops as well as fruits and grapes are grown here. As concerns industrial production the most developed are the food industry giving over 50 % of the total industrial gross product, gas production and oil refining, production of cement, valuable timber, machine-building and light industry. The light and food industry is concentrated mainly in Krasnodar, Armavir, Kropotkin, Eisk and Tikhoretsk. The food industry includes meat processing, fat-and-oil, flour-and-cereals, canning, wine-making, tobacco, essential-oil and fish processing enterprises. The fishery enterprises locate mostly in the coastal zone of B. and A. seas (Eisk, Novorossiysk, Anapa, Tuapse, Achuev, Primorsko-Akhtarsk, Temryuk). In the textile industry the major is the worsted-cloth factory in Krasnodar. The machine-building industry is represented by machine-tool manufacturing (Krasnodar, Maikop), production of equipment for the oil industry (Krasnodar, Armavir), production of spare parts for agricultural machinery (Krasnodar, Armavir, Novorossiysk), production of transport equipment (Tikhoretsk, Novorossiysk) and others. Oilfields are developed in the Apsheron-Khadyzhenskiy, Kuban-Black Sea and Pre-Azov areas. Oil refinery plants locate in Krasnodar and Tuapse. The Territory has a network of oil and gas pipelines as well as the gas pipeline supplying gas from Kuban to the central regions of the country. The cement production is concentrated in Novorossiysk area. In the woodworking industry the plant in Maikop as well as furniture factories in Krasnodar, Armavir, Kropotkin, Labinsk and Novorossiysk should be mentioned. The large enterprise is also the tanning-extraction plant in Maikop.

Agriculture provides about the half of the gross national product of the Territory. Its main branches are grain and fruit growing, milk and meat cattle breeding and pig rearing in the Kuban-Priazov Lowland, fruit and grape growing in the piedmont areas and on the Azov-Black Sea coast. In the grain production the winter wheat and corn are the key crops. In the Kuban lower reaches large areas are given to rice cultivation for which rice irrigation systems are constructed. The main technical crops are sunflower, sugar beet, essential oil crops and tobacco. The main grape-growing region is the Black Sea—Sea of Azov coast. Tea, citrus and other subtropical crops are cultivated on the B.S. coast southward of Tuapse. Adler and Dagomys have tea factories. The famous champagne factory Abrau-Dyurso locates in the largest vine-growing region in the B. and A. sea area. K.T. is also a major region of the milk-meat cattle husbandry. Mostly large-horned cattle and pigs are grown here. Poultry breeding is also developed widely; there are several poultry farms here that are growing chicken, turkey and waterfowl. The main railroad line Kuschevskaya—Armavir—Adler crosses K.T. The mountain section from Belorechenskiy to Adler (and further on to Sukhumi) is electrified. Very important are the railroads Krasnodar—Novorossiysk and Krasnodar—Adler. A very important role in the cargo traffic is played by the marine transport (ports in Novorossiysk, Tuapse, Eisk and Sochi) and in the passenger traffic—the auto transport, in particular on the Black Sea coast.

The Black Sea coast of K.T. between Tuapse and Adler is the resort area with the center in Sochi (it incorporates the former Adler and Lazarevskoe districts of K.T.).

Resort attractions here are quite diverse: mild climate and warm sea, mountain landscapes, mineral water springs and curative mud. The resort in Eisk, resort areas of Primorsk-Akhtarsk and Temryuk are found in the Azov area. In the northern part there are such resorts as Anapa, Kabardinka and Gelendzhik. The coast section from Tuapse to Sochi, the Russia's major balneological and climatic resort area, has humid subtropical climate and rich subtropical vegetation. The bathing season lasts from June through mid-October; the mean water temperature near the coast in June exceeds 23 °C, in October 19 °C. This territory abounds in mineral water springs. Sulfide waters are used for curative purposes in Matsesta and other regions of Sochi, in the resorts Goryachiy Klyuch, Eisk, Anapa; iodine-bromine waters—in Eisk, Anapa, Gelendzhik, Tuapse, Maikop and Khadyzhenskiy Sochi (Mamaika, Lazarevskoye, Kudepsta). Curative mud which rich resources are found near Eisk, Anapa and Sochi is used in mud-cure centers and sanatoriums scattered along the coast.

Krasnodar Territory (http://www.russianworldforums.com/photos/maps/krasnodar_map.jpg)

Krasnoperekopsk – a village on the coast of the Black Sea, Ukraine.

Krasnov Andrey Nikolaevich (1862–1914) – the outstanding Russian botanist, geographer and traveler, one of the first who started development of the Russian subtropical zone. He was a participant of the Nizhegorodskiy and Poltavskiy expeditions on land assessments headed by V.V. Dokuchaev. In 1889 he defended his first thesis "History of Flora Evolution in the Southern Part of Eastern Tien-Shan". K. conducted researches in the Kharkov University where he was the first Professor of geography and created the first research geographical study room (1889–1921). In 1906 K. organized the botanical and geographical garden at the Kharkov Veterinary Institute. It became the prototype of the Botanical Garden that was organized by him some years later in Batumi. His work "Geography of Plants" (1898) deserves special mention. K. was the first in Russia who was conferred the academic degree—Doctor of Geography. He was granted this degree after public defense of his thesis (1894). He was the author of the first university textbook "Fundamentals of the Land Science" (1895–1899). In 1912 he resigned and concentrated his efforts on creation of the botanical garden in Batumi on the basis of the geographical principle. K. was keen on traveling. He explored the south of the Russian Plain, the Caucasus, Tien-Shan and Altai mountains. Twice he went on long travels to foreign countries to study tropical zones. K. conducted research on acclimatization of subtropical plants in Transcaucasus, in particular, tea. He made a great contribution to study of steppes. His doctoral thesis "Grass Steppes of the Northern Hemisphere" (1894) is a good example of the integrated physiographical description of the steppe landscape on the territory of Eurasia and North America.

The last period of his activities began in 1912; it can be called the "Batumi period". During this time he concentrated his efforts on organization of his "beloved brainchild"—the Botanical Garden in Batumi. K. was sure that the garden should become the country-wise educating center, a kind of the "live" museum". He dreamt about making a collection of the most typical and economically important tropical and subtropical plants. At the same time K. thought that this garden should be the center of experimental works on accommodation in our country of the new valuable crops that can be grown in humid subtropics. He became interested in these issues during his travels to tropics and subtropics. During 1900–1902 he conducted acclimatization experiments on the Caucasus (in the Chakva estate). K. generalized his studies of the tea culture in his book "Tea Growing Areas in Subtropics of Asia" (1897–1898). He organized publication of the journal "Russian Tropics" and was its first editor. He died while he was working in the garden, He was buried in his favorite garden in the place he selected himself about which he wrote: "Clear a pathway from my grave so that I could see Chakva surrounded by snow-capped mountains and a piece of the sea; I was the first to initiate work there; and there a piece of my heart was left. . ." The works of K. about the nature of tropics and subtropics were taken together and published in one book "In the Tropics of Asia" (1956).

Krasnov A.N. (http://sochived.info/wp-content/uploads/2010/12/Krasnov.jpg)

"Krasnyi Kavkaz", Cruiser – the first guard ship in the Black Sea Fleet that was conferred this honorary title by the order of the USSR Navy Commander of 3 April 1942. This ship named "Admiral Lazarev" was launched in Nikolaev in 1916. After its refurbishment in 1926 it got the name "Krasnyi Kavkaz". The flag was hoisted on it on 25 January 1932. Its water displacement was 9,000 tons, speed 29 knots (53.7 km/h), armament: 4 180-mm cannons, 12 100-mm cannons, 2 76-mm, 4 45-mm and 10 37-mm cannons; 4 3-tube torpedo launchers; the crew—878 people. During the Great Patriotic War it supported the landing operation in Grigorievka, evacuated troops from besieged Odessa. In 1941–1942 it took part in the Kerch-Feodosiya and in 1943 in the Novorossiysk landing operations. The crew of "Krasnyi Kavkaz" showed extraordinary courage during the Kerch-Feodosiya landing—the greatest operation in the history of this war. In the evening of 28 December 1941 the unit of ships with the landing troops onboard left Novorossiysk. The flagship was "Krasnyi Kavkaz" with landing commander Captain of 1st rank N.E. Basistyi onboard (in 1948–1951 he was the Black Sea Navy Commander). At night on 29 December the ships approached Feodosiya. At dawn there was a 6-point storm on the sea when "Krasnyi Kavkaz" started mooring at the protective jetty. Under fire of enemy cannons only 30 min were used for landing 1,853 troopers, but the cruiser was damaged by a shell. Having cleared of the jetty where it would be inevitably wrecked the ship cannoned the tank columns and firing points of the enemy. Having returned to its base the cruiser took onboard the reinforcements and again went into the stormy sea. On 4 January it moored in the port of Feodosiya. During these two voyages the ship was damaged seriously. Its stern sank deep, but the crew succeeded to take "Krasnyi Kavkaz" to Poti where it was put for repair. In August 1942 the cruiser was repaired and ready for new battles. On 23 May 1944 "Krasnyi Kavkaz" entered the Sevastopol Bay. In 1947 it was turned into a study ship. In 1950 it was used for testing the efficiency of the first Soviet anti-ship missile "Kometa".

"Krasnyi Kavkaz" (http://flot.sevastopol.info/photos/photo_cruiser/krasniy_kavkaz_02.jpg)

"Krasnyi Krym", Cruiser – a light cruiser of the Soviet Navy. When laying cruiser was given the name "Svetlana" in honor of the eponymous cruiser, heroically died 28 May 1905 in the Battle of Tsushima. He was the lead ship in a series of light cruisers of the Russian Imperial Navy. He took part in battles in the Black Sea Fleet during World War II, was awarded the title—Guard ship. On November 24, 1913 in the presence of Russian Minister of Marine light cruiser "Svetlana" was laid, but due to the unpreparedness of the shipyard and the delays in the supply of materials the actual assembly of the ship on the slipway only began on April 1, 1914. On February 5, 1925, in accordance with the order of the Marine forces of the Red Army the cruiser changed its name to "Profintern". In October 1926 actually finished the cruiser "Profintern" went to Kronstadt for dry-docking and finish outfitting works. On April 26, 1927 "Profintern" was brought to pass. Despite significant overload on the acceptance test vehicle has a top speed over 29 knots with power turbines 59,200 horsepower. According to the order of 1 July 1928 the light cruiser "Profintern" was enrolled in the Naval Forces of the Baltic Sea, and raised the naval flag of the USSR. On October 31, 1939 cruiser "Profintern" was renamed "Krasnyi Krym". The cruiser had the following principal dimensions:

overall length 158.4 m (at the waterline—154.8 m), width with armor plating 15.35 m. Main battery consisted of 15 130-mm 55-caliber guns (B-7) of 1913. Total ammunition—2,625 rounds. Torpedoes consisted of two three 533-mm torpedo tubes. Ammunition represented six torpedoes in the apparatus. On November 1943 the crew of the light cruiser "Krasnyi Krym" consisted of 48 officers, 148 petty officers and 656 rank and file sailors, in total—852 people.

On the eve of the Great Patriotic War it was held tactical reorganization of the Soviet Black Sea Fleet. As a result of the reorganization the large surface ships have been combined in the squadron based at Sevastopol and includes a battleship "Parizhskaya Komunna" (Paris Commune), detachment of light forces and the team cruisers. The cruiser "Krasnyi Krym" was added to the team cruisers. Together with the "Krasnyi Krym" in the brigade entered the light cruisers "Krasnyi Kavkaz" and "Krasnaya Ukraina", as well as 1st Division destroyers of the "Novik" and the 2nd Division destroyers like "Gnevnyi". On June 22, 1941 cruiser "Krasnyi Krym" met at the Sevastopol Marine Plant named after Ordzhonikidze, where he was under repair since May. Repair work on the cruiser were accelerated and the second half of August, the ship was commissioned. The cruiser immediately began to carry out its missions. On August 22, 1941 a detachment of ships in the cruiser "Krasnyi Krym", destroyers "Frunze" and "Dzerzhinskiy" came to the aid of the besieged Odessa. In November and December 1941 she participated the battle of Sevastopol. On December 28–30 "Krasnyi Krym" was actively involved in the Kerch-Feodosiya landing operations. On 15–25 January 1942 as part of the squadron assault the cruiser performed the landing of the troops that were part of the second and third landings near Sudak. In the second half of October 1942 the cruiser was involved in the transfer of the eighth and ninth Guards brigades from Poti to Tuapse. The transfer of these parts allow to stop the advance of the Wehrmacht in the Tuapse and stabilize the front line. During the period of the defense of the Caucasus from July to December 1942 ships of the squadron, which included "Krasnyi Krym" moved 47,848 soldiers and officers of the Soviet Army with arms and about 1 million tons of military cargo. On 3–4 February 1943 the cruiser was in a group of amphibious assault ships covering the operational area Stanichka—Yuzhnaya Ozereika near Novorossiysk. On November 5, 1944 the cruiser returned to Sevastopol. On May 31, 1949 the cruiser was transformed into a training cruiser, 7 May 1957—in the towing vessel and renamed the "OS-20", on 18 March 1958 turned into marine barracks "PKZ-144". In July 1959, the ship was removed from the lists of the ships of the Navy and commissioned to dismantle the metal. According to some reports, the ship was sunk in the late 1950s when testing new weapons.

Kremer Oscar Karlovich (1829–1904) – Russian Admiral, seafarer. In 1837 he entered the Naval Division of the Alexander Cadet Corps. In 1840 he was transferred to the Naval Corps. In 1846 he was promoted to midshipman. In 1850 on the ship "Borodino" he navigated from Arkhangelsk to Kronshtadt. In 1854–1855 he took part in the Sevastopol defense as a commandant of the Volynskiy redoubt. In 1859–1862 he made the round-the-world voyage on the clipper "Razboinik". In

1863–1864, commanding the corvette "Vityaz" he navigated to the American shores. In 1866 he was appointed the Aide-de-Camp. In 1869 he was promoted to Captain first rank. In 1870, commanding the corvette "Varyag" he sailed from Kronshtadt to Arkhangelsk, to Novaya Zemlya and to Iceland. In 1871, commanding frigate "Svetlana" he navigated to the shores of America and in 1872–1873 he made a new round-the-world voyage. In 1874 he was appointed the Captain of the port in Kronshtadt. In 1875 he was promoted to Rear-Admiral and included into an escort. In 1876 he commanded the unit of ships in the Greek waters. In 1886 he was promoted to Vice-Admiral and appointed the Chairman of the Naval Technical Committee and was conferred the rank of Aid-de-Camp General. In 1888 he was appointed the Chief of Naval General Staff. In 1890 he became commander of the acting squadron on B.S. Member of the State Council (1896). Admiral (1896). Awarded many orders. His name is given to an island in the Barents Sea.

Krieger (Vevel von Krieger) Grigoriy Alerksandrovich (1820–1881) – Russian Vice-Admiral, Director of the Hydrographic Department. In 1831 he entered the Naval Cadet Corps and finished it in 1836 in the rank of midshipman. For continuation of his education he was left in the Officer's School. In 1836–1838 he navigated on different ships over the Baltic Sea. From 1840 to 1851 he served on the Black Sea Fleet as Aid-de-Camp of Admiral M.P. Lazarev. In 1840 he was in the landing unit of Lieutenant-General Raevskiy taking part in storming Tuapse, then sailed with the Fleet commander over B.S. and the Mediterranean. In 1851–1852 he for some time headed the Black Sea hydrographic depot. During the Crimean War he was on-duty field-officer at Vice-Admiral Kornilov, then assistant of Vice-Admiral Metlin on construction of fortifications in Nikolaev and on-duty field officer of the Black Sea headquarters. From 1855 he headed the Black Sea hydrographic depot. In 1856 he was promoted to Captain 1st rank and appointed the manager of the Black Sea hydrographic division and the chief of the hydrographic depot. In 1860 he served as Vice-Director of the Hydrographic Department, then worked in the commission studying the causes of the Sea of Azov shallowing. In 1861 he was promoted to Rear-Admiral. From 1861 to 1865 he was the military and civil governor in Koven (Kaunas) and then Ekaterinoslav (Dnepropetrovsk). In 1869 he became a member of the Naval Educational Establishments Committee. From 1872 to 1874 he acted as commander of the port in Revel (Tallinn) and director of the Baltic lighthouses. In 1873 he was promoted to Vice-Admiral and in a year he was appointed Director of the Hydrographic Department.

"Krimschild" – the German award sign that was instituted on 7 July 1942 to celebrate the victorious seizure of the Crimean Peninsula by the troops under command of General Erich von Manstein. This award was conferred till October 1943. The K. should be worn on all kinds of the uniform—from the field to parade dress. For top-rank commanders the special golden shields were made and they were conferred to Manstein and commander of the Romanian Army Ion Antonescu. The shields were also awarded to all German military servicemen of the Eleventh and Third Romanian Armies. The shield was awarded to those who served in this region for at least 3 months and who took part at least in one important operation or was wounded in such. The shield was made for all branches of troops.

Kritskiy Nikolay Dmitrievich (?–after 1834) – Russian Rear-Admiral, hydrographic specialist, explorer of B. and A. seas. He is of Greek origin. In 1796 after finishing the Corps of alien coreligionists he was conferred the rank of midshipman and appointed to the Black Sea Fleet. From 1796 to 1807, commanding the gunboats he more than once participated in hydrographic works on surveying the coasts of B. and A. seas. He described the coast of A.S. from Taganrog to the Enikal Strait. He took part in the Russian-Turkish War of 1806–1812, in particular, in the battles against fortresses in Anapa, Platana and Trapezund. From 1812 to 1827 he navigated over B.S. In 1821 he prepared the inventory of the Sukhumi roadstead; in 1823 he made the layouts of the Fedonisi Island and Tendrovskiy bar; in 1824–1825 he made description of the B.S. coast from Odessa to Sevastopol and coasts of A.S. In 1826–1827, commanding the 34th Fleet crew he described the B.S. coast from Sevastopol to the Kerch Strait. During the Russian-Turkish War of 1828–1829 he took part in the siege and conquering of Anapa and Varna. On 17 August 1828, commanding the detachment of ships he got the landing troops ashore near Inodi for this operation he was awarded the golden sword. From 1828 he was the quartermaster of the Black Sea Fleet. In 1834 he left his office. He was awarded many Russian orders.

"Krivets" – bora on the B.S. coast in Romania.

Krym (Neiman) Solomon Samoilovich (1868–1936) – Russian large landholder. He finished the gymnasium in Feodosia and the Petrovsko-Razumovsky Agricultural Academy; the speaker of the zemstvo (local administration) meetings of the Feodosiya and Tavria Provinces; one of the leaders of the Tavria branch of the Kadet Party; Deputy of the Russian First and Second State Duma. From November 1918 to April 1919 K. was the Chairman of the Council of Ministers and the Minister of Agriculture in the Crimean Government.

Krymchaks – a community of the Turk-speaking Jews living in Crimea. Until 1783 in the documents of the Crimean khans they were referred to as the Jews. The name K. ("Jews-Krymchaks") first appeared in the Russian sources in 1859. The language of their religious cult, literature and community laws was Hebrew. The first mentions of Jews in Crimea were found in documents of the first century. The main occupations of K. were crafts and trade in the Antique cities. They were also on the civil and military service in the Bosporus state. The synagogue built in 909 in Caffa is one of the most ancient in Eastern Europe. From the second half of the 10th to the mid-15th centuries the Jews inhabited the cities controlled by Byzantine and the rulers of the Italian trade posts, in the first half of the fifteenth century—by viceroys of the Golden Horde, from 1443—by Crimean Khans. After resettling to the Crimean Khanate the Jews got mixed with the K. community. Different origins of K. may be traced by their family names typical of the Jews who came from the Turk-speaking countries, Asia Minor, Caucasus, Italy, Spain and others. By the early nineteenth century, after annexation of Crimea to the Russian Empire (1783) the K. population numbered only 600. In the petition submitted to Emperor Alexander I (1818) K. called themselves "the sons of Israel". They were mostly engaged in crafts, some of them in farming and grape-growing, and a few—in trade. According to the census of 1897 about 3,350 K. lived in Alushta, Yalta, Evpatoria, Kerch and other towns of Crimea.

"Krymchaks" – the name of the first passenger vessels, famous motor vessels like "Abkhazia" which in 1920–1930 were incorporated into the USSR passenger fleet. These ships were built for voyages along the Crimean-Caucasus line and over the Mediterranean. In other words, it was an attempt to revive shipping along traditional routes of ROPiT. As in 1926 the USSR had no sufficient shipbuilding capacities the Central Marine Shipbuilding Bureau (CMSB) decided to order construction of 2 out of 6 vessels in Germany. CMSB handed over the respective documentation to the Baltic Plant and to the Krupp Shipbuilding Plant "Deutsche Verft" in Kiel. During 1926–1931 the Baltic Plant built such vessels as "Abkhazia", "Adjaristan" (later renamed into "Adjaria") (1927), "Armenia" and "Ukraina" (1930), while the Germans built "Gruzia" (1928) and "Krym" (1928).

"Krym" (http://kruiznik.ru/forum/viewtopic.php?t=954&p=5925)

"Krymskaya" – the curative-table mineral water from the spring located in the resort area Saki; sold in bottles. By its composition this is the hydrocarbonate—sodium chloride water of low mineralization. It is used in treatment of the gastrointestinal and liver diseases. It may be also used as a table drink.

Kryuis Korneliy Ivanovich (1657–1727) – Russian Admiral (1721). In 1698 he was admitted by Peter I to the Russian service in the rank of Rear Admiral. He took part in the Kerch Campaign (1699) with the Peter's Fleet; in 1699–1701—in construction of the Admiralty in Voronezh, a port in Taganrog and fortifications in Azov. Commander of the squadron of the Azov Fleet (1711). Participant of the Northern War of 1700–1721. Till 1713 he commanded the Baltic Fleet and headed the defense of Saint-Petersburg during its siege by the Swedish Fleet. In 1715 he served in the Petersburg Chief Admiralty. In 1717 he was sent to Revel for organization of its defense. In December 1717 he was appointed the Vice-President of the Admiralty Collegium. In 1722 he acted as President of this Collegium. He prepared the first Russian Marine Atlas containing the maps of the Don River, Black and Azov seas (1703). He took an active part in development of the Naval Code of the Russian Fleet (1720). Buried in Amsterdam.

Kryuis K.I. (http://www.front-line.eu/260/krjuis.jpg)

Kuban – is a geographic region of Southern Russia surrounding Kuban River, on the Black and Azov seas, between the Don Steppe the Caucasus. Krasnodar Kray is often referred to as Kuban, both officially and unofficially, although the term is not exclusive to the kray and accommodates the republics of Adygea, Karachaevo-Cherkessia, and parts of Stavropol Kray and Rostov Region. The settlement of Kuban and of the adjacent Black Sea region occurred gradually for over a century, and was heavily influenced by the outcomes of the conflicts between Russia and Turkey. In the mid-eighteenth century, the area was predominantly settled by the mountainous Adyghe tribes. After the Russian-Turkish War of 1768–1774, the population of the area started to show more pro-Russian tendencies. In order to stop Turkish ambitions to use Kuban region to facilitate the return of the Crimea, Russia started to establish a set of fortifications along Kuban River in the 1770s. After the Russian annexation of the Crimea, right-bank Kuban, and Taman in 1783, Kuban River became the border of the Russian Empire. New fortresses were built on the river in the 1780s–1790s. More intensive settlement started in 1792–1794, when the Black Sea Cossack Host and Don Cossacks were re-settled to this area by the Russian government in order to strengthen the southern borders. In the end of the 18th and the beginning of the 19th centuries, the right bank of Kuban River was settled. At the same time, first settlements appear on the coast of the Black Sea and on the plain between the Kuban and Bolshaya Laba rivers. During the second half of the nineteenth century, the settlement rate intensified, and the territory was administratively organized into Kuban Oblast and Black Sea Okrug, which later became Black Sea Governorate. The location of the territory along the border had a significant effect on its administrative division, which incorporated the elements of civil and military governments.

Kuban, River – the largest river in the Northern Caucasus of Russia. It originates near Elbrus Mount from the confluence of two rivers—Ullukam and Uchkulan. K. flows into A.S. near the town of Temryuk. Its length is 907 km, basin area—about 61,000 km^2. In its upper reaches to Nevinnomyssk K. is a typical mountain river rushing through deep and narrow gorges, falling down from heights and forming rapids. After Nevinnomyssk the river goes out on a plain (Kuban-Priazovskiy Lowland) and flows over a wide floodplain; its channel abounds in islands, shallows and sandbars. In the lower reaches (from the inflow of Laba River) the river valley widens and plavni appear in the floodplain. At the 116 km from the mouth the river splits into K. proper and the Protoka arm. The recharge is mixed: glacier, snow and rainfall. The freezing regime is unsteady. In some winters the ice cover is absent. The mean annual flow near Krasnodar is 413 m^3/s. The river delta has numerous limans (350) and beginning from Krasnodar plavni are stretching. For regulation of the flow of K. and its tributaries the Tschikskiy water reservoir was constructed on Belaya River. In 1948 the Nevinnomyssk canal was built via which part of the river flow was diverted to Egorlyk River and was further used for irrigation and replenishment of Western Manych River. Not long ago K. flowed into B.S. In 1819 the Black Sea Cossacks trying to desalt the waters in the Akhtanizovskiy and Kurchnskiy limans dug two canals. Limans were desalted, but as the plain slope northwards to A.S. turned out greater than it was anticipated the Kuban water rushed there. These were first waterworks on K. This river is navigable from landing Ladozhskaya (upstream the Ust-Labinsky landing) to the mouth. The main cities located on this river are Karachaevsk, Cherkessk, Nevinnomyssk, Armavir, Kropotkin, Ust-Labinsk, Krasnodar and Temryuk.

Kuban Cossack Troops – a nonregular military detachment in Russia located along the middle and lower reaches of Kuban River with the center in the city of Ekaterinodar (now Krasnodar). It was established in 1860 on the basis of the Black Sea Cossack troops and a part of the Caucasus Cossack troops and comprised 22 cavelry regiments, 13 infantry battalions, 5 horse-artillery batteries and 1 garrison artillery unit (in December 1861 it was abolished). They were deployed in the Kuban Oblast. The troops were commanded by a chieftain nominated by the Emperor. Kuban Cossacks took part in the Caucasian War of 1817–1864. By the decree of 11.10.1864 the Azov Cossack troops were abolished and many of the Cossacks were included into K.C.T. In July 1865 the troops were awarded the George Flag "For Caucasian War" and some regiments—Georgian Flags (to the eleventh and seventeenth—"For Bravery in Turkish War, for fighting against highlanders in 1828–1829 and for conquering Western Caucasus in 1864"). In April 1866 the cavalry sotnya of K.C.T. replaced the Don Cossack troops in Zakubanie. K.C.T. participated in the Russian-Turkish War of 1877–1878; during the Akhal-Tekinskiy expedition—in storming the fortress of Geok-Tepe. In 1918 K. C.T. were disbanded.

Kuban Oblast – was formed in 1860 on the territory of the former Black Sea Cossack troops, the western part of the Caucasus Cossack troops and Trans-Kuban (the Kuban left bank). Its center was in Ekaterinodar. In the 1860s its population was 0.4 million people. Initially K.O. was divided into three districts— Ekaterinodarskiy, Eiskiy and Tamanskiy and into territories of 6 brigades, 6 regiments and 1 battalion. From 1865 in the Zakubanskiy area there were five more districts: Psekupskiy, Labinskiy, Urupskiy, Zelenchukskiy, and Elbrusskiy. The Black Sea coast of the Caucasus from Novorossiysk to the border with Abkhazia made the Black Sea District and subordinated till 1896 to the K.O. chief. In 1869 K. O. was divided into 5 districts: Batalpashingsky, Eisky, Ekaterinodarsky, Maikopsky and Temryuksky; in 1876 the Zakubansky and Kavkazsky districts were formed. In 1888 the districts were replaced with seven divisions: Batalpashinskiy (the center in Batalpashinskiy settlement), Eiskiy (town of Eisk, from 1902—settlement of Umanskiy), Ekaterinodarskiy, Kavkazskiy (settlement of Tikhoretskiy, from 1897—settlement of Kavkazskaya), Labinskiy (settlement of Armavir), Maikopskiy, Temryukskiy (from 1910—Tamanskiy; center in Temryuk, from 1897—settlement of Slavyanskiy). The region was managed by the regional board headed by a region chief who at the same time was the chieftain of the Kuban Cossack troops and chief-of-staff. In 1897 the area of K.O. was 81,200 squared verst; the population—1.9 million people (including rural—221,200 people). The population comprised about 90 % of the Russians and Ukrainians, 3.4 %—Adygs, 1.4 %—Karachaevs, 1.08 %—Germans, 1.05 %—Greeks, 0.73 %—Armenians and others. The main occupation of the people was agriculture (79 %). By the 1890s the grain growing became dominating. In 1913 by the gross grain yield K.O. became the second in Russia, by commodity grain production—the first. Among non-grain crops the most important were sunflower and tobacco; the vine-growing, horticulture and melon crop growing were developing. Out of 8.7 million desiatinas of lands in K.O. 69.2 % was in the Cossack ownership. The main industries included craftworks, such as dairies, flour-mills, breweries. In 1913 there were 7,800 plants here. Some banks had their affiliates in K.O.—Russian Foreign Trade Bank, Russian-Asian Bank, Azov-Don Bank and others. In 1875 K.O. was crossed by the Rostov-Vladikavkaz railroad; in 1887 Tikhoretskaya settlement was connected with Ekaterinodar by a railroad that in 1888 was extended as far as Novorossiysk. In the early twentieth century the railroads Sosyka-Eisk (1911), Armavir-Tuapse (1915) and others were constructed. In 1913 20 steamships and 145 non-steamships navigated over Kuban River; the "Partnership of I.I. Ditsman" was the main shipping company.

Kuban-Black Sea Soviet Republic – existed from 30 May to 6 July 1918. It was formed of the Kuban and Black Sea Soviet Republics for uniting forces to give rebuff to the foreign invaders and the White Guard. By the time of this republic formation the Soviet power got firmly established on the Northern Caucasus and Don: the Don, Stavropol and Terek Soviet Republics were created. The Congress of Soviets of the Black Sea Province that was held in March in Tuapse declared formation of the Black Sea Soviet Republic and the Congress of Soviets of the Kuban Region held in April in Ekaterinodar (now Krasnodar)—the Kuban Soviet Republic that were included into RSFSR. After this the Third Extraordinary Congress of Soviets of Kuban and Black Sea Area attended by delegates from the front (Ekaterinodar, 27–30 May 1918) decided to merge the Kuban and Black Sea Republics into a single republic of RSFSR. When the Volunteer Army of General A.I. Denikin started offensive from Mechetinskaya-Kagalnik region to Kuban (started on 23 June 1918) the First Congress of Soviets of the Northern Caucasus (5–7 July 1918) resolved to unite the Kuban-Black Sea, Terek and Stavropol Soviet Republics and to form the single North-Caucasian Soviet Republic in RSFSR.

Kuban-Priazov Lowland, Kuban Lowland – the lowland in the Western Pre-Caucasus. In the south it is confined by the piedmonts of the Greater Caucasus, Stavropol uplands, in the west—Sea of Azov and in the north—the lower reaches of the Don and Manych rivers. The southern part of occupied by the Lower Kuban valley. It is composed of pebbles and clays overlain with loess-like loams. Precipitation are 400–600 mm; the soils are black-earth. Practically the whole territory of this lowland is cultivated for growing grain and technical crops.

Kuchugury – sand dunes, ridges, hillocks running along B.S. to the north of Anapa.

Kuchuk-Kainarji Treaty of 1774 – a peace treaty signed after defeat of Turkey in the Russian-Turkish War of 1768–1774. It was concluded on 10(21) July 1774 in Bulgarian village of Kuchuk-Kainarji on the right bank of the Danube near Silistria. Ratification instruments were exchanged on 13(24) January 1775 in Constantinople. As a result of this treaty the borders of Russia were expanded as far as the Southern Bug in the southwest and the Kuban in the southeast. Under this treaty Russia received the fortresses of Kinburn (in the Dnieper mouth), Kerch and Enikale as well as Kabarda and Northern Ossetia. From now on Russia could build fortifications on the territories near the Black Sea. In exchange Russia returned to Turkey the territories captured during the war—the lands behind the Danube, Dniester, on the Northern Caucasus, in Georgia and on the islands in the Aegean Sea. The treaty proclaimed independence of the Crimean Khanate of Turkey (new Khan Shagin-Giray was of the pro-Russian standing). Both parties declared mutual freedom of trade. The Russian merchant ships could now sail freely over B.S., Danube and via the Bosporus and Dardanelles as the British, French and Turkish ships. Moldavia and Wallachia were granted autonomy within the Ottoman Empire. Russia acquired the right to protect the Orthodox Christians being subjects of Turkey, while the latter undertook to ensure freedom of the Orthodox religion over the whole Ottoman territory, to return to the Christians their lands confiscated earlier and not to prosecute them for fighting in this war on the Russian side. The Russian diplomatic mission headed by an envoy or plenipotentiary minister was opened in Constantinople. At the same time Russia got the right to appoint consuls and vice-consuls in the Turkish lands. Turkey had to pay to Russia a contribution of 4.5 million Rubles, which was much less that military expenses of Russia. The treaty gave Russia the access to B.S., ensured security of its southern borders from devastating raids of the Crimean Tatars, made possible development of the fertile steppes in the south of Russia and strengthened the influence of Russia on the Balkans. Not a single word was said in the treaty about a right of Russia to have military fleet on B.S., but at the same time the treaty did not prohibit construction of military ships. The immediate result of this treaty was annexation of Crimea to Russia (1783) and creation of the Black Sea Fleet. Striving of Turkey to revenge resulted in the Russian-Turkish War of 1787–1791 that ended with the Treaty of Yassy (1791) that was very beneficial for Russia.

Kudepsta—(former Nizhnenikolaevskoe) – the neighborhood, the separation between Khostinskiy and the Adler district of Sochi. Located on the shores of the Black Sea, at the mouth of Kudepsta River. Takes Kudepsta Valley and its tributary Zmeyka River. At the hills (range Ovsyannikova) stretched country's largest cork oak plantations. The area is rich in iodine-bromine mineral waters and curative mud, which is widely used in many spas and medical clinics. Population—17,000.

Kulakov Nikolay Mikhailovich (1908–1976) – Soviet Vice-Admiral, political figure, Hero of the Soviet Union. He served in the Navy since 1928. In 1936 he graduated from the V.I. Lenin Military-Political Academy. He served as a military commissar on the ships of the Baltic Fleet. He was the member of the Military Board of the Northern Fleet (1939) and Black Sea Fleet (1940–1943). He took part in the Sevastopol defense, military operation against Constanta, in the Kerch-Feodosiya operation of 1941–1942. Beginning from 1944 he was the Chief of the Higher Military-Political Naval Courses. In 1946–1949 he was the member of the Military Council and Deputy Commander-in-Chief of the Navy on political issues; from 1950 to 1955—the member of the Military Council of the Black Sea Fleet; in 1956–1960—Chief of the Political Division and Deputy Commander on political issues of the Kronstadt fortress (Leningrad Naval Region); in 1961–1971—Chief of the Political Division of the Leningrad Naval Base and naval educational establishments in Leningrad. Hero of the Soviet Union (7 May 1965). In 1971 he resigned. He was awarded many orders. He wrote the book of reminiscences "Trusted to the Fleet" (1985).

Kul-Oba (*"hill of ash"*) – one of the most famous Scythian mounds of the 5th-4th centuries B.C., the ancient archeological site, a part of the Panticapaeum burial mound located westward of the town of Krym (now Kerch). It was uncovered in 1830. A mount 10 m high concealed the Greek-style stone tomb made of large, accurately fit blocks. Its plan was almost square measuring 4.6 × 4.2 m. The stepped vault was 5.3 m high. A short stone corridor (dromos) led to the tomb, its ceiling was covered with stone slabs and wooden blocks. Three persons were buried here. The body of a nobleman aged 30 or 40 lay in a wooden decorated couch with his head southward. His clothing was embroidered with golden patches. His head was surmounted by a pointed hood also embroidered with golden pendants and two leaf-shaped diadems. His neck was decorated with the golden grivna rimmed by figures of Scythains on horses. His hands and legs were adorned with gold bracelets. A separate section of the couch contained weapons and a massive gold phial (a vessel for drinking) with the engraved head of Gorgon and bearded man in a pointed headgear. The Scythian weapons included a sword with a gold trim depicting the struggling animals, a bow with arrows, a whip, knives and a grind-stone. Opposite the tomb entrance there was the sarcophagus made of wood (that, perhaps, remained from other sarcophagus) in which a woman was lying. It was decorated with ivory plates with the engraved episodes from the Ancient Greek epos ("the Judgment of Paris") and scenes of Scythian hunting. The woman lay in a brocaded dress with an electrum diadem on her head and with the gold grivna on her neck. There were also found the gold bracelets and earrings, including two massive gold necklaces depicting the Athen's head being a copy of the famous statute by Phydium from Parthenon in Athens. A bronze hand mirror, silver spindle and a vessel depicting Scythian scenes were also found. At the heads of a man and a woman a slave was buried and a special niche in the wall contained horse bones, a helmet and Greek-style sheath. Several amphorae, bronze bowls and cauldrons with food remnants were placed along the walls of the tomb. It is thought that the buried man belonged to the local Scythian nobility closely tied with the rulers of the Antique cities in the Bosporus Kingdom.

Kumani Mikhail Nikolaevich (?–1889) – Russian Rear Admiral. In 1847 he started his military career as a junker on the Black Sea Fleet. In 1850–1853 in the rank of midshipman and lieutenant he sailed on the ships of the Baltic Fleet. During the Crimean War of 1853–1856 he commanded the gunboat in the rowing fleet, fought at Ochakov against the British-French ships, took part in the Sevastopol defense. In 1863 he was promoted to Captain-Lieutenant. In 1864 during the N.A. Ivashintsov expedition to the Caspian Sea he got acquainted with the means and methods of deep-water measurements. Untill 1867 he commanded ships on B.S. and later on the Baltic Sea. In 1882 he was conferred the rank of Rear Admiral. In 1885 he was appointed the administrator of Sevastopol and in 1886—the commander of the port of Sevastopol. He was awarded many Russian and foreign orders.

Kumani Nikolay Mikhailovich (1793–1869) – Russian General, hydrographic specialist, historian of the Black Sea Fleet. In 1809 he finished the Naval Cadet Corps, navigated B.S., participated in battles during the Russian-Turkish War of 1806–1812. In 1825 he was appointed the administrator of the Black Sea Depot of Maps and in 1832—the Chief of the Black Sea Hydrographic Depot. He took these positions for nearly three decades and contributed much to the hydrographic studies of B. and A. seas. In 1825 he initiated the Hydrographic Expedition and its results laid the basis for publication of the atlas of maps of B. and A. seas that for many years had been the reliable reference book for navigation of these seas. His contribution to hydrographic investigations of B. and A. seas was praized high by the command and in 1845 he was conferred the rank of Major-General. Apart from hydrographic investigations he studied the history of the Black Sea Fleet. In 1856 in the rank of Lieutenant-General he was appointed the inspector of the Black Sea Fleet. He served in this capacity till 1860 when he resigned in the rank of General. He had numerous awards. His main works: "The Hydrographic Investigations of the Black Sea", "The Black Sea Fleet Activity during Emperor Alexander I from 1801 to 1826" (1900).

Kumani Nikolay Petrovich (1730–1809) – Russian Rear Admiral. He was born on the Crete Island. In 1764 he arrived to Russia as a colonist and was sent to Astrakhan for organization of the silk and vine production. In 1769 he volunteered as a midshipman to the vessel "St. Evstafiy" in the squadron of Admiral G.A. Spiridov and went to the Mediterranean. In 1770 he participated in the Moreiskiy expedition and in the battle at Chesma; in 1771–1775 he cruised in the Archipelago and commanded smaller ships. In 1773 he was promoted to Lieutenant. In 1775, commanding the frigate "Pobeda" he sailed from the Archipelago to Kerch with the Albanian families onboard and when the ship went out from the Constantinople strait into B.S. the frigate was caught in a storm and wrecked near Balaklava. In 1776 he was sent to the Don Fleet; in 1777 he commanded the ship "Ekaterina" and navigated in A.S.; in 1778 on the same ship he was on watch near the Novopavlovskiy fortress; commanding the ship "Elan" he navigated A. and B. seas. In 1779 he commanded the ship "Ekaterina" and sailed between Taganrog and Kerch. In 1780 he delivered mail between Kerch and Constantinople; was promoted to Captain-Lieutenant. In 1781 he sailed on A.S. first commanding the ship "Ekaterina" and later galliot "Verblyud". In 1782 he was the commander on the ship "Zhurzha" in A.S. and in 1783–1784—on the same ship in the Kerch Bay. In 1785 he was transferred to the Black Sea Fleet. In 1786 he was promoted to Captain 2nd rank. In 1788 he commanded the frigate "Kinburn" on B.S.; in 1789 he was promoted to Captain 1st rank. In 1790, commanding the ship "St. John the Evangelist" he participated in the battles of the Black Sea Fleet against the Ottomans near the Kerch Strait and near Gadzhibey (was awarded the gold sword). In 1791 he commanded the ship "Preobrazhenie Gospodne" near Kaliakria. His son Mikhail Kumani was onboard this ship, too, and they side-by-side participated in the battle near Kaliakria. In 1796 he was promoted to Capitan of the Major-General rank and in 1797 he became Rear Admiral and transferred to the Baltic Fleet. He resigned due to health problems. He was awarded many orders.

Kunikov Tsezar Lvovich (1909–1943) – Hero of the Soviet Union. He started his career as a locksmith, later he finished the Moscow Machine-Building Institute. After this he took posts of the chief of the Technical Department in the Committee for Machine-Building, Committee for Heavy Machine-Building, Director of the Research Institute for Machine-Building Technology, Editor-in-Charge of the departmental newspaper "Machine-Builder". During the Great Patriotic War he volunteered to the front and was sent to the Baltic Fleet. In early 1942 K. was transferred to the Black Sea Fleet as a commander of one section of the anti-landing defense of Novorossiysk. For the battles on the Azov Sea coast and on the Taman Peninsula he was awarded the Alexander Nevskiy Order. He organized and headed the night landing on 3–4 February 1943 on Myskhako near Novorossiysk for diverting the attention of the Germans from the main landing operation near Yuzhnaya Ozereika. A small group (less than a battalion) headed by K. landed and entrenched near Stanichka. But the main landing troops could not get ashore. K. sent plainly the radio message to the headquarters: "The regiment landed without losses, move further". After this message the Germans attacked violently the landed troops. Because of a heavy storm the ships with the supporting forces could not assist K. During 2 days the landed troops headed by K. rebuffed more than 15–20 attacks. This land territory about 25 km^2 in area seized from the Germans was called "Malaya zemlya" (small land area). K. was deadly wounded and died on 12 February 1943. The inhabitants of the village of Stanichka where these landing troops fought asked to give it the name "Kunikovka".

Kunikov Ts.L. (http://upload.wikimedia.org/wikipedia/ru/5/50/Kunikov_CL.jpg)

Kurchatov Igor Vasilievich (1903–1960) – the outstanding Soviet physicist, first organizer of works in nuclear science and engineering in the USSR. He graduated from the Physico-Mathematical Department of the Crimean (Simferopol) University. He started his career in science from investigations of fluctuations of the water level in southern seas, in particular B.S. In 1924 he published his work "Application of Harmonious Analysis to Investigations of Tidal Activity in the Black Sea". After this K. started studies of seiches in the sea. In his work "Seiches in Black and Azov Seas" (1925) he demonstrated that "study of seiches in the Black Sea revealed a rather intricate picture of fluctuations. Each bay, each harbor generates their own seiches with the small period of oscillations from 10 min to 1–2 h; seiches of the whole sea are blurred by other oscillations".

Under supervision of K. the first Soviet cyclotron was constructed (1939), the spontaneous fission of the uranium nuclei was discovered (1940). In autumn 1941 K. and A.P. Aleksandrov (later Academician) working in Sevastopol managed to justify theoretically the method of ship protection from non-contact electromagnetic fuze mines and made first attempts on ship degaussing. Widely applied this mine protection technique saved dozens of ships from wreck during the Great Patriotic War. He was the founder and first director of the Institute of Nuclear Energy (from 1943). From 1960 this institute was given his name (today this is National Research Center "Kurchatov Institute"). He managed construction of the Europe's first nuclear reactor (1946), the first atomic bomb in the USSR (1949), the world's first thermonuclear bomb (1953) and nuclear power plant (1954). Academician of the USSR Academy of Sciences, thrice Hero of Socialist Labor (1949, 1951, 1954). He was awarded the USSR State Prize (1942, 1949, 1951, 1954) and Lenin Prize (1957).

In Crimea the name of K. is given to a tourist path leading from the Upper Oreanda to the top of the Ai-Nikola Mountain along which the scientist liked to walk so much.

Kurchatov I.V. (http://www.rusarm.eu/images/2009_08/kurchatov.jpg)

Kutais Province – administrative unit of Russian Empire, was formed in 1846 on a part of the territory of the former Georgian-Imeretia Province (the center—Kutais). It comprised the following districts: Kutaisskiy, Zugdidskiy, Lechkhumskiy, Ozurgetskiy, Rachinskiy, Senakskiy and Sharopanskiy. In 1866 the Sukhumskiy district was established. On the lands annexed from Turkey to Russia by the resolution of the Congress of Berlin (1878) the Karskiy and Batumskiy districts were formed, the latter in 1883 was incorporated into K.P. Some parts of K.P. maintained their traditional names: Georgia, Imeretia, Mingrelia, Svanetia and Tsebelda. In 1897 the area of K.P. was 32,000 sq. versta; the population was 1,058 million people (about 97,500 people lived in cities), including the Georgians

32.5 %, Imeretians 22.5 % and also Abkhazs 5.6 %, Turks 4.4 %, Armenians 2.7 % and Russians 1.8 %. In 1903 the Batumskiy district (Batumskiy, Artvinskiy districts and the city of Batum) and Sukhumskiy district were separated from K.P. From now on the Kutais military governor became Kutais governor. K.G. was the agrarian region in the Transcaucasus. The main occupation for the inhabitants was farming. They cultivated corn, wheat, barley, beans and lentil. The vine-growing and wimne-making, horticulture, melon crop growing and fishery were also widely practiced. K.P. had significant mineral deposits—coal, manganese, silver-lead, copper and zinc ores, oil, marble and others; the mining industry developed. In 1912 mostly small artisan and semi-artisan undertakings were operating in K.P. The Transcaucasus railroad built in 1871–1883 was very important for K.P. The trade centers were Kutais, Samtredi and Khoni. The center of export trade was port Poti via which the manganese ores, corn, nut tree and others were supplied to other countries. In K.P. there were two dioceses—Imeretinskiy and Gurian-Mengrel.

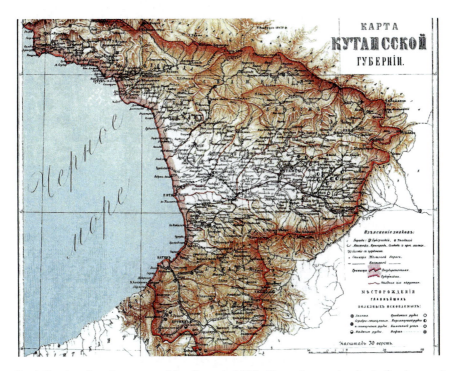

Kutais Province (http://commons.wikimedia.org/wiki/File:Kutaysskaya_guberniya.jpg?uselang=ru)

Kutuzov Mikhail Illarionovich (1745–1813) – Field Marshal of the Russian Empire and diplomat. He finished the Engineering-Artillery School. From 1776 he served in Crimea under command of A.V. Suvorov. During the Russian-Turkish War of 1787–1791 he commanded various detachments, took part in the battles of Ochakov, Akkerman, Bendery. He showed particular bravery in storming Izmail. After the end of the second Russian-Turkish War K. was sent to Constantinople as the Extraordinary and Plenipotentiary Ambassador. There regardless of the Porta wish to resume war, he managed to procure fulfillment of the Treaty of Yassy and establishment of normal relationships with Turkey. The peak of his diplomatic activities was conclusion of the Bucharest Peace Treaty with Turkey only 37 days prior to Napoleon invasion of 1812.

Kutuzov contributed much to the military history of Russia and is considered to have been one of the best Russian generals under the reign of Catherine II. He took part in three of the Russian-Turkish Wars and in the Napoleonic War, including two major battles at Austerlitz (2 December 1805) and Borodino (7 September 1812). Kutuzov is credited most with his brilliant leadership during the French invasion to Russia (1812). Under Kutuzov's command, the Russian army was defeated by the Grande Armee at the Battle of Borodino but counter-attacked once Napoleon retreated from Moscow, pushing the French out of the Russian homeland. In recognition of this, Kutuzov was awarded the title of Prince of Smolensk. A memorial was built in Moscow in 1973 to commemorate the 1812 War and Kutuzov's leadership. An order of the Soviet Union and the Russian Federation is also named after him.

Kutuzov M.I. (http://www.mutab0r.net/wp-content/uploads/2011/03/961e1b884b34817e48dcc d49f86bf1f3_1298452008_0volkovr_ptknmikutuzsmolegat.jpg)

Kuyalnitskiy Liman – a salty liman on the northwestern coast of the Odessa Bay, B.S. Its length is 28 km. A sandy barrier to 3 km wide separates it from the bay. The area of K.L. is 91 km^2; its average depth is about 3 m. The greater Kuyalnik River flows into the liman. The water temperature in summer reaches +28°C–+30 °C. The floor of the liman is covered with black mud saturated with hydrogen sulfide. The mud resort named Kuyalnik is located on this liman.

Kuznetsov Nikolay Gerasimovich (1904–1974) – Admiral of the USSR Navy. In autumn 1919 he volunteered to the North-Dvina Fleet. During the Civil War in Russia he served as a sailor. In 1921–1923 on the Baltic Sea he finished the preparatory school for naval sailors and the preparatory courses at the Naval College. In October 1926 he was appointed chief of the watch on the cruiser "Chervona Ukraina" of the Black Sea Fleet. In 1929 he was sent to the Naval Academy to the tactics faculty that he finished in 1932 with honors. In 1932–1934 he was the senior second-in-command at the cruiser "Krasnyi Kavkaz" that in 1933 became one of the best ships on B.S. In September 1934 he was the commander of the cruiser "Chervona Ukraina". In 1936–1937—the Naval Attache in Spain. Having learned the Spanish language he became the Chief naval councillor at the command of the Republican Army in Spain. From August 1937 he was the First Deputy Commander and in 1938–1939—Commander of the Pacific Fleet. In March 1939 he was appointed the Deputy Commissar of the Navy; from April 1939 to January 1947—the Naval Commissar, Deputy Minister of the USSR Navy—Naval Commander-in-Chief. In this capacity he prepared the Fleet for the Great Patriotic War. In February 1941 he ordered to increase the combatant core of the Fleet and to prepare the operations plan of the war against Germany and its allies. These efforts enabled the USSR Fleet to meet fully armed the assault of 22 June 1941. During the war he headed the Naval Commissariat. More than once he visited the battlefields. In 1945 as the Naval Commissar he took part in the Yalta and Potsdam Conferences of the heads of ally states. In 1945 he was conferred the Title of the Hero of the Soviet Union.

In 1947 he was removed from the office and later was brought to trial for passing the scientific-technical information ostensibly making the "state secret" to foreign intelligence. He was lowered to the rank of Rear Admiral. From June 1948 he was the Deputy Commander-in-Chief of the Far East Navy. From February 1950 he was the Commander of the Pacific Fleet; from 1951 he was the USSR Naval Minister and was promoted to Vice-Admiral. In 1953 he was restored in his rank—Naval Admiral and after his ministry was incorporated into the USSR Ministry of Defense he was appointed First Deputy Minister of Defense—the Naval Commander-in-Chief. Seeking to create the balanced Fleet he came into conflict with N.S. Khruschev and G.K. Zhukov. In March 1955 he was promoted to Admiral of the USSR Fleet, but after explosion on the battleship "Novorossiysk", the flagship of the Black Sea Fleet, on the Sevastopol roads K. was released of his office, and in February 1956 he was sent to resignation in the rank of Vice-Admiral. He was awarded many orders and medals. On 26 July 1988 Presidium of the USSR Supreme Council issued the order on return to N.G. Kuznetsov of his military rank of the Naval Admiral of the Soviet Union. He was the author of the books "On the Far Meridian" (1971), "The Course to Victory" (1975). His name was given to the Naval Academy of Saint-Petersburg and to the aircraft-carrying ship of the Northern Fleet.

Kuznetsov N.G. (http://savok.name/uploads/forum/images/1303845168.jpg)

Kirklareli – the Black Sea Il (province) in Turkey. It locates in the westernmost part of European Turkey on the west coast of the Black Sea. Its area is 6,550 km^2; the population is 332,800 people (2010). The center is Kirklareli. The province is bisected by the Yildiz (Istranca) mountain range. The north and northeastern parts of the province are among the least populated and developed parts of Turkey. The districts to the south and west are more populated because the land is better suited for agriculture and industrial development. The north and eastern parts of the province are dominated by forests. Forestry is an important means of living in these areas. Fishing is done along the Black Sea coast. Dupnisa Cave is a famous natural attraction and a unique geological formation within the borders of the province in the north. The 60 km long coast along the Black Sea harbors one of the most pristine and undeveloped beaches in all of Turkey. There are two Nature Reserve Areas along the coast namely Saka Gölü (Saka Lake) Nature Reserve Area to the north and Kasatura Körfezi (Kasatura Bay) Nature Reserve Area to the south. These sites are unique with their undisturbed ecosystems harboring several endangered and endemic plant and animal species.

L

Ladja, Lodja – a sail-row boat of the Eastern Slavs of the sixth–eighth centuries built for merchant voyages and military campaigns. Its length was 20 m, width—3 m. It was capable to carry 15 tons as well as 40 people and more with the food stock and outfit. It had a mast with a small square sail and one row of oars. Initially L. was hollowed out of one log, but from the tenth century it was built as a vessel. Light weight and small draft of this vessel allowed the Kievan warriors to pass easily the rapids in the Dnieper mouth and to sail over B.S. and the Caspian Sea portaging L. from the Don to Volga and back. In 907 Prince Oleg with his warriors made the voyage on L. to Constantinople (Tsargrad) and conquered it. In the twelfth century the deck L. appeared on which oarsmen located in a bilge and the warriors on the wooden deck. The identically tapered ends of L. were provided each with a steering oar that made possible to change the traveling direction without L. turning round.

Lagoon – (in the Italian *"laguna"* and in the Latin *"lacus"* meaning "lake")

(1) rather shallow natural water bodies connected with the deeper sea by a narrow strait (or straits) or separated from the sea by a sandbank or sandbar. Due to a weak connection with the sea or complete isolation L. has other than in the sea salinity (higher or lower) and specific deposits consisting mostly of river sediments. L. appears as a result of deposition of sediments on the sea floor taking the shape of an underwater bar that spreads to the shore turning into the sandbank. They occur widely over the world: on the shores of the Gulf of Mexico, B.S. (to the north of the Danube), Caspian Sea (eastern coast), on Sakhalin, Kamchatka Island, Chukotka Peninsula and in other places.—(2) bodies of water enclosed between a barrier reef and the coast of mainland or island and also bodies of water inside atolls.

Lagorio Lev Feliksovich (1827–1905) – Russian marine painter, professor of the landscape painting (1860), Honorary Member of the Academy of Arts (1900). He was born in the family of Felice Lagorio, the Naples Vice-Consul in Crimea. In 1852 he became the Russian national. He received his first lessons in art from I.K. Aivazovskiy in 1838–1840 in Feodosiya. In 1843–1850 he studied at the

S.R. Grinevetsky et al., *The Black Sea Encyclopedia*,
DOI 10.1007/978-3-642-55227-4_13, © Springer-Verlag Berlin Heidelberg 2015

St.-Petersburg Academy of Arts. In 1850 for his work "View of Luhta in the vicinity of Petersburg" he was awarded the Great Gold Medal and was granted a voyage abroad as a pensioner of the Academy of Arts. In 1851 the Academy sent him for one year to the Caucasus "for painting landscapes". In 1852 he went abroad, worked in Rome and its surroundings, visited Sorrento, Capri, and also in France (visited Normandy) and Switzerland. After his return to Russia in 1860 L. presented as a report of his voyages about 30 sketches and paintings. The paintings he created there brought him the status of professor of landscape painting bypassing the rank of Academician. L. lived and worked in Petersburg, but the summertime he spent in Crimea where in Sudak he had his art studio. In 1882 he received a government order for a series of pictures devoted to the Russian-Turkish War of 1877–1878. For making plein-air sketches he went to Asia Minor and Turkey. After these travels he painted many pictures of sea battles: "Ship "Great Prince Konstantin" with boats before attacking Turkish ships in the Sulinskiy girlo" (1877), "Defeating by ship "Great Prince Konstantin" of Turkish ships at the Bosporus. 1877" and "Sinking by boats of ship "Great Prince Konstantin" of Turkish ship "Intibach" on the Batumi roads at night of 14 January 1878" (1880). L. liked most to depict the sea and coastal life and his talent showed itself most fully in such paintings. On his pictures L. seldom showed the stormy sea. He made the first trip to the Caucasus in 1851, and during his life he often returned there. The most famous among his Caucasian paintings was the picture "Kaishaurskiy Valley" (1876). Quite often on his pictures L. showed the mountain and sea landscapes together and they attracted the greater attention of the public. In his graphical works L. depicted with great accuracy and carefulness all details of the sail vessels that he studied in his youth on the Baltic Sea. In the last years of his life L. painted landscapes. He was one of the founders and participants of the art exhibitions both in Russia and abroad.

Lagorio L.F. "Batum, 1881" (http://www.wikipaintings.org/en/lev-lagorio/)

Landlocked, inland, enclosed or semi-enclosed sea – according to the definition contained in the 1982 UN Convention on the Law of the Sea, it means a bay, basin or sea surrounded by two and more states and linked with the other sea or ocean via a narrow passage or consisting, in full or largely, of territorial seas and exclusive economic zones of two or more littoral states. Among such seas we can name Black, Azov and Baltic seas.

Landscape therapy – treatment of diseases with the beauty of nature, bubble of the sea, rustle of forest, etc. These are the key elements of balneotherapy, esthetic therapy. In this context such features as esthetic attractiveness, mystique, complexity, harmony of separate elements, detailing of landscapes acquire special importance.

Landslides – masses of mountain rocks sliding down along the slope by gravity. L. may occur on any part of a slope due to disturbance of rock equilibrium caused by the increased steepness of a slope as a result of its scouring by the sea, lake or river; loss of rock strength due to weathering or overwetting with sediments or underground waters; the effect of seismic shocks; construction and economic activities of a man conducted disregarding the geological conditions. Most often L. occurred on slopes composed of interbedding water-impermeable (clay) and water-bearing formations. The sliding surface may occur at a depth from 1 to 20 m. The landslide process may be controlled by bank strengthening and drainage structures, fixation of slopes with driven piles, etc. In the USSR the landslide research stations existed in Ukraine, on the Black Sea coast of the Caucasus.

Langeron Alexander Fedorovich (1763–1831) – Russian count, one of those who founded Odessa. He was born in Paris. In 1790 he went to the Russian military service. In December 1790 together with Duke de Richelieu he distinguished himself in storming Izmail. He took part in the battles on the Danube and Caucasus, in the Netherlands and Prussia, under Austerlitz and in the "Battle of the Nations" under Leipzig, distinguished himself in commanding the corps during seizure of Paris and in the battle near Varna. In 1815 Tsar Alexander I appointed him the Governor of Odessa and General-Governor of Novorossiysk. He realized some very important undertakings, such as establishment of the free port zone (porto-franco), publication of the first city newspaper, creation of a botanical garden, opening in 1817 of the Richelieu Lyceum, second by importance in Russia after Lyceum in Tsarskoe Selo. He died of cholera in Petersburg.

Langeron A.F. (http://img0.liveinternet.ru/images/attach/c/0//53/749/53749140_Langeron_A_F.
jpg)

Lanson – one- or two-mast sailing fishing or coaster vessel most popular on
B.S. During the Russian-Turkish War of 1787–1791 L. was called the sailing-
rowing vessels operating in the Dnieper-Bug Liman and on the Danube. They had
onboard 4–8 small guns or 1 or 2 mortars. They were used against enemy rowing
vessels, for carriage of troops and in landing operations. Its length reached 21 m,
width—6 m, draft—2.5 m.

Larger Palace of Livadiya – principal place of interest of Livadiya. The palace is
positioned on a terrace at the height of 120 m above MSL in the midst of a grand
old-time landscape park of 46 ha in area. The park was laid as early as in the 1830s
and was of a distinct two-part structure: in front of the palace is an open sun-filled
pit, punctuated by closed plantations in the rest of the territory. The interiors of
P.L. three halls only have survived. The white hall of the palace is of particular
interest: here, during February 4–12, 1945, meetings of the Crimean Conference of
the heads of great powers—participants of the anti-Hitler coalition—were held. It is
noteworthy that the structure was built to the standards of cutting-edge technology
of the time: it had a central hot-water heating system, an elevator, electric lighting.
The fire-places in the halls not only serve the purpose of decoration: they can also
be used for heating the halls, if necessary. The Livadiya park has nearly 400 different species of trees and shrubs. The park has a good volume-and-space design:
there are enough large terraces, spacious viewing grounds, providing a wide field
of view.

Larger Palace of Livadiya (Photo by Evgenia Kostyanaya)

Laspi – a seaside climatic resort in Crimea, Ukraine. It is situated on the Southern Coast of Crimea between the Sarych Cape (the southernmost point of Crimea) and the Ilyas-Kaya Mount. The climate here is subtropical, Mediterranean. The natural climatic conditions in L. where curative air is created by a rare combination of arborescent junipers and wild pistachios are very favorable for climatic therapy of respiratory organs. The village of Batiliman locates on the shore of the Laspi Bay.

Laspi Bay – locates in Crimea, Ukraine, between Capes Laspi and Aiya, and is closed by a semi-circle of steep rocks. Its depth is to 60 m.

Laspi Cape – locates 3.5 km to the northwest of the Sarych Cape. It is a rounded shore tongue formed by spurs of the mountains belonging to Baidarskaya Yaila. It is cut along the axis by a gully. L.C. makes the southwestern coast of the Laspi Bay.

"Laspi Rocks" – a nature reserve 2 km to the northwest of the Sarych Cape in Crimea, Ukraine. This is a part of the Ai-Petri Mountains known as "Sugar Heads". At the foot of the rocks the thinned thickets of the relic forests peculiar of the Southern Coast of Crimea are growing. In 1960 L.R. were given the status of a monument of nature and in 1984—state nature reserve.

Laurel (*Laurus*) – the genus of evergreen trees and shrubs of the Laurel family. They have dark-green glossy leaves and whitish-yellow flowers. Their stone fruits are of the blue-black color and contain to 25 % of fatty oils used in medicine. There are distinguished two basic varieties of L.—sweet bay (*L. nobilis*) and Canaries L. (*L. canariensis*). The most widespread is sweet bay wildly growing on the Mediterranean, in the western part of the Transcaucasus and in the south of Crimea.

It is cultivated as the essential oil, spicy (laurel leaf) and decorative plant. The L. family includes 45 genuses (about 1,100 varieties) growing mostly in the tropical and subtropical forests of South America and Asia. In the L. family there are many valuable plants, such as camphor tree, fruit tropical and subtropical tree of avocado, cinnamon tree giving cinnamon.

Lazarev Mikhail Petrovich (1788–1851) – the outstanding Russian Fleet commander and navigator, scientist, pioneer explorer of Antarctic, Admiral (1843). In 1803 he finished the Naval Marine Corps. He passed his probation in the British Fleet (1804–1808). L. took part in the Russian-Swedish War of 1808–1809 and in the Patriotic War of 1812. In 1819–1821 he commanded the vessel "Mirnyi" (Peaceful) and was F.F. Bellinsgauzen's deputy on the Russian Antarctic expedition. He was one of those who first saw the continent of Antarctica and many nearby islands. From 1826 he commanded the battleship "Azov" on which he took part in 1827 in the battle of Navarino. For the excellence in this battle the vessel "Azov" was granted the highest award—the George Flag. During the Russian-Turkish War of 1828–1829 he was in charge of the Dardanelles blockade. In 1832 he was made Chief-of-Staff of the Black Sea Fleet. In 1833–1851 he was the Commander-in-Chief of the Black Sea Fleet and B.S. ports and also military governor of Sevastopol and Nikolaev. In 1836–1840 L. organized continuous guard of the Black Sea coast of the Caucasus to put an end to attempts of the British and Turkish vessels to change the situation in this region. He did much for further development of the naval theory and practice and for improvements in construction of the Black Sea Fleet—there were built 16 battleships and more than 15 other vessels. For the first time the iron-hull ships and frigates were put afloat. During his governorship the beautiful buildings of Naval Library, Petropavlovsk Church, Dvoryanskiy (nobility) Assembly and many others were constructed in Sevastopol. Earth fortifications constructed for defending the harbor were replaced with two–three tier batteries with ramparts made of trimmed stone. Two of them—Konstantinovskiy and Mikhailovskiy survived to the present. L. proposed to use landing teams of the Black Sea Fleet for construction of on-shore fortifications. Under his surveillance the headquarters of the Black Sea Fleet developed instructions for landing of the Russian seamen. L. was remembered as a mentor to younger officers. He also tutored many talented Russian Fleet commanders, including Pavel Nakhimov, Vladimir Kornilov, Vladimir Istomin, Grigoriy Butakov and others. The geographical expeditions of L., in particular discovery together with F.F. Bellinsgauzen of Antarctica contributed much to further development of the world science. L. was the honorary member of the Russian Geographical Society. He was a founder of Novorossiysk city (12 September 1838) together with L.M. Serebryakov and N.N. Raevskiy. He was awarded many Russian and foreign orders. The name of L. was given to the coral atoll (Mataiva atoll) in the Tuamotu Archipelago in the Pacific, the island in the Aral Sea, the harbor and port in the Sea of Japan and others.

Lazarev M.P. (http://emc23.narod2.ru/istoriya/shemi_srazhenii/mp_lazarev/Lazarev.jpg?rand=267751224944269)

Lazarevskoe – a seaside resort village, a center of one of the regions of Greater Sochi, Krasnodar Territory, Russia. It is located on the Black Sea coast of the Caucasus, on the right bank of the Psezuapse River. Railway station "Lazarevskaya" is 51 km to the northwest of Sochi and 39 km to the southeast of Tuapse. The mean annual air temperature here is +13.3 °C. The beach is pebble, 40 m wide, smoothly sloping to the sea. The village received its name in honor of Russian seafarer, Commander of the Black Sea Fleet Admiral M.P. Lazarev. In 1839–1854 in place of this village there was the Lazarevskiy fort, one of the military bases on B.S. coast. In 1869 a settlement appeared here. L. has the famous shipbuilding yard where ocean yachts are constructed, the rest house, sea-water treatment center and a curative beach. Population—30,000 (2010).

Lazarevskiy Resort Area – covers the coast between the settlements of Shepsi and Mamaika. The coastline is about 110 km long. This region incorporates such resort areas as Magri, Makopse, Sovet-Kvadzhe, Ashe, Lazarevskoe, Golovinka, Loo, Uch-Dere, Dagomys and others. The climate here is humid subtropical with a very warm summer and a mild rainy winter. The mean annual rainfall here is 1,500 mm (greater than in the Central, Khosta and Adler regions); the most rainy month is December. The village of Soloniki has a source of curative-table mineral water "Lazarevskaya" that is used in treatment of gastrointestinal diseases.

Lazistan Region (called so by the name of the historical and ethnographic province of Lazistan) – a natural area in the northeast of Turkey, at the juncture of the northern marginal mountains of the Asia Minor and Armenian highlands, going out to the southeastern coast of B.S. It takes the southwestern part of the vast Mountain-Kolkhida natural area. It is situated on the northern seaside slope of the Eastern Pontic Mountains. Rainfalls occur here in winter and summer. The annual rainfall is 2,000–3,000 mm. The winter is mild, practically frostless, while the summer is hot and arid. Magnificent broad-leaved forests with dense evergreen underwood (Pontic rhododendrons, cherry laurels, holly) and lianas (ivy, silk vine) are growing here. The coastal lands are cultivated for growing tobacco, corn, grapes, orchards of hazelnut and citrus trees. The lower-elevated zone (to 400 m) is covered with broad-leaved forests (beech, hornbeam, oak, chestnut, maple); on limestones the boxwood is growing. At altitudes between 400 and 1,250 m the beech-spruce forests dominate, between 1,250 and 1,900 m—the mountain spruce forests including Caucasian and Pontic fir trees. The upper margins are made of crooked forests. Still higher (over 2,000 m) the sub-Alpine and Alpine meadows are spreading. The animal world here includes red deer, roe, chamois, boar, wolf, jackal, marten and brown bear.

Le Vasseur de Beauplan Guillaume (circa 1600–1673) – French military engineer. While being a Polish serviceman during 1630–1648, was building fortresses on the southern and south-eastern border of Rzech Pospolita. Published a work titled "Description of Ukraine" (1650). During his numerous travels, B. surveyed the Dnieper over the distance of about 1,100 km as far as the river mouth. Traced a number of rivers of the Dnieper system: the Uzh, Teterev, Irpen, Ros, Tyasmin with their numerous tributaries, being almost accurate in estimating the length of the rivers. He was rather accurate in mapping the entire South Bug, and a little less accurate in mapping its largest left tributaries—the Sinyukha and Ingul. B. mapped all considerable bends of the middle and lower Dniestre and many of its tributaries. Made a most detailed description of the lands of the Black Sea area. In November of 2005, a presentation of B.'s earliest maps of Ukraine was held in Kiev.

Lebedevka (former Burnas) – a seaside climatic and mud cure resort in Ukraine, 40 km to the southeast of Belgorod-Dnestrovskiy. It belongs to the Odessa group of resorts. It is located on a peninsula that is washed in the southeast by B.S. and the lake (liman) of Burnas. L. has sand beaches. The main natural curative features here are climate and mud of the Burnas Lake that are used in climato-thalasso and mud therapy. This resort was established in 1946. It has a sanatorium for children suffering from the respiratory diseases (non-TB type) and diseases of the locomotive system and a mud curative center.

Lebyazhie Islands – a chain of islands in the Karkinitskiy Bay locating in the northwest of the Crimean Peninsula, Ukraine. In 1949 L.I. became a branch of the Crimean Hunting Reserve and put on the Ramsar Convention on Wetlands of International Importance. In 1960 a protection area of 8,000 ha was created around the islands. L.I. is also the ornithological reserve. Many aqueous birds, including rare and endangered species, find places for rest and nesting here.

Lefort Franz Yakovlevich (1656–1699) – Russian statesman and military figure, Admiral, diplomat and close associate of Peter the Great. He was born in Geneva and was educated at the College Calvin, which is now the oldest public secondary school in Geneva. It was founded in 1559 by John Calvin. In 1675 he fought in the war of the Netherlands against France under the flag of Prince William of Orange. After this war he went to serve to Russia. In Moscow he settled down in the so-called Nemetskaya Sloboda (Germans District). In 1678 he began his military career in the Russian Army in the rank of Captain. For 2 years he served in Ukraine, took part in battles with the Turks and Tatars. In the Crimean Campaigns of 1687 and 1689 he commanded a battalion and was conferred the rank of Colonel. According to one version, in 1689 L. was one of the first who together with his troops came over to Peter I side during Peter's conflict with Tsarevna Sophia. In 1690 he was promoted to Major-General. From spring 1690 Peter I frequently visited Nemetskaya Sloboda and in September he visited the Lefort's house for the first time. Soon they developed friendly relations and L. became one of the principal organizers and participants of the military games of Peter I. In 1691 L. was promoted to the rank of Lieutenant-General and in 1693—General. In 1691 L. was put in charge of a regiment and Peter ordered construction of a sloboda for training of this regiment on the left bank of the Yauza River which would be later called Lefortovskaya Sloboda (the Lefortovo quarters, present-day Lefortovo in Moscow).

During the First Azov Campaign in 1695 L. was one of commanders of the Russian troops. He supported the idea of quick storming of Azov. In 1695 he was put in charge of the Imperial Russian Navy and in February 1696 he was appointed the Admiral, commander of the fleet that was created in Voronezh. He became the first Russian navigator in the rank of full Admiral. During the Second Azov Campaign in 1696 L. already commanded the whole Navy. In 1697–1698 he headed the Grand Embassy, a Russian diplomatic mission to Europe. As a head of this Embassy he was entitled to keep negotiations and to sign the most important documents. At the same time he sent to Peter I the information about European countries, employed specialists to the Russian Fleet and financed purchase of ship outfit. In 1698 he settled down in the palace that was built for him on the Yauza River (from 1865 it was given to the Lefortovsky Archive). In 1698–1699 this palace became a center of the Russian political and court life. Here Peter I held most important meetings and numerous court festivities.

Lefort F.Ya. (http://www.navy.su/persons/13/images/lefort_fya_00.jpg)

"Legends of Crimea" – the collections of legends and folklore stories about the far and near past of the Crimean Peninsula. The most well known are the following: three wonderful collections published in the early twentieth century by Nikandr Marks, the famous Crimean researcher of the nineteenth century; "Universal Scripts about Crimea" (1875) and "Legends of Crimea" (1883) by V.Kh. Kondraki; "Writings of Ancient Greek and Latin Authors about Scythia and Caucasus" (1893–1906) by V.V. Latyshev; "Tales, Legends and Folk Stories about Crimea" (1930) by A.K. Konchevskiy; "Tales and Legends of Crimea" (1936) by Yu. Kotsubinskiy; "Tales in Presentation of M. Kustova" (1941); "Crimean Legends" (1957); "Legends of Crimea" (1959, 1961).

"Lenin" – a passenger liner, formerly called "Simbirsk", was built before World War I in Danzig (Germany). It was 94 m long, 12 m wide, draft—5.7 m, speed to 17 knots. It was renamed in the Soviet time. The vessel navigated between Odessa and Novorossiysk. In 1941 it was retrofitted. At the beginning of the Great Patriotic War it transported food products and evacuating people. On 27 June 1941 the vessel took onboard 4,000 people (instead of 472 people and 400 tons of metal) and was ordered to go to Yalta. At 19:15 the vessels "Lenin", "Voroshilov" and"Georgia" went into the sea with the escort of the boat SKA-026. On 23:33 when "Lenin" was near the Sarych Cape a strong blow shattered the vessel. It started sinking, heeling, then the stern went up and in 7–10 min it went to the sea floor. The number of victims on the "Lenin" exceeded the number of victims on the "Titanic" (1,514 people, 15 April 1912) and "Lusitania" (1,198 people) that was attacked on 7 May 1915 by the German submarine. The vessels of "Georgia" and "Voroshilov" managed to save about 600 people from "Lenin". But the circumstances of the ship wreck have not been unveiled so far. All the documents related to this disaster were kept in secret till the mid-1990s. "Lenin" lies at a depth of 78 m approximately 4.6 km off the coast where the government estate "Zarya" is located.

Lermontovo (Lermontovka) – a village in the Tuapse district of Krasnodar Kray, Russia. Included in the Tenginka rural settlement. The village is situated on the coast of the Black Sea, 5 km east of Dzhubga on the road "Tuapse-Novorossiysk", in Tenginka Bay at the mouth of the river Shapsukho, the length of which is more than 4 km. Named in honor of Russian famous poet Mikhail Lermontov, Lieutenant Tenginka Infantry Regiment, who himself never visited this place. Tengin Regiment was garrisoned in strengthening Tenginka, built in 1838 at the mouth of the river Shapsukho. The village is recorded in the lists of villages of Krasnodar Kray on 24 September 1958. In summer, vacationers find here fine sandy beaches of Tenginka bay. The whole area around Lermontovo is filled by a huge number of hotel and entertainment complexes, water parks, medical rehabilitation centers. Population—1,122 (2010).

Lermontovskiy Resort – a seaside climatic and mud resort within the Odessa confines, Ukraine, between the city park of Langeron and Otrada, 2.5 km from the railway station "Odessa". It is a part of the Odessa resort group. Climate and mud from the Kuyalnitskiy Liman allow opportunities for rest as well as climate and thalassotherapy. The resort offers treatment with sulfide, carbon dioxide, nitrogen, radon and other baths artificially prepared on the basis of sea water; physiotherapy, remedial gymnastics and others. It has a sandy beach with all amenities, very convenient for thalassotherapy and aeroheliotherapy. The resort was opened in 1925. It also includes a rest house, a polyclinic, balneo and physiotherapy treatment center, mud spa.

Leselidze (former Ermolovsk) – a seaside climatic resort 20 km to the northwest of Gagry and 10 km to the southeast of Adler, Abkhazia. It appeared in the nineteenth century as a village of Ermolovsk named in honor of A.S. Ermolov, Minister of Agriculture, who in 1894 visited this village. Any allusions often met in the literary sources to the name of famous General Ermolov who was Commander-in-Chief during the Caucasian War are not true to fact. In 1944 the village was given the name of Soviet Commander Colonel-General K.N. Leselidze (1903–1944) who took part in the battles for the Caucasus during the Great Patriotic War. To the northeast of L. the foothills of the Greater Caucasus are stretching. The village has a wide small-gravel beach. The vegetation here consists mostly of fruit trees (grapes, nut, plum, fig, apple and others) and decorative trees (palms, laurel, eucalyptus). The climate is humid subtropical. The winter is very mild and snow-less; the mean temperature in January is 6 °C. The summer is very warm; the mean temperature of August is 23 °C. High relative humidity is registered here. The summer heat is alleviated by onshore and offshore breezes. Precipitation— 1,550 mm a year. They fall mostly in autumn and winter. The number of sunshine hours is 2,100 a year. The mild climate offers good opportunities for aerogeliotherapy. The sea water is warm—its temperature in summer reaches 22– 25 °C. The bathing season lasts from May through October. This resort was opened in 1959. Children with the cardiovascular disorders may be successfully treated here.

Leselidze Konstantin Nikolaevich (1903–1944) – Hero of the Soviet Union, the Soviet military commander, Colonel-General. In the Great Patriotic War from June 25, 1941, fighting in the battles of Minsk, Shklov, Krichev, Viaz'ma, Bryansk, Belev, Tula, in a defensive operations of Main Caucasus Range, Krasnodar-Novorossiysk, Novorossiysk (1943), Novorossiysk-Taman, Kerch-Eltigenskiy, Zhitomir-Berdichev operations. From February 1941 to August 1941 the chief of artillery of the 2nd Infantry Corps on the Western Front. From August 1941 to June 14, 1942 Deputy Commander, who is also chief of artillery of the 50th Army on the Western Front. From June 1942 to August 1942 the commander of the 3rd Infantry Corps of the Transcaucasian Front. Army troops under the command of K. Leselidze heroically defended the Caucasus, participated in the liberation of the Caucasus, in the capture of the bridgehead on the Kerch peninsula in Eltigen and in the liberation of the Right-Bank Ukraine. From August 1942 to January 1943 the commander of the 46th Army of the Transcaucasian Front. From January 1943 to

March 1943 the commander of the 47th Army of the Transcaucasian Front. From March 1943 to January 1944 the commander of the 18th Army of the North Caucasus Front, the 1st Ukrainian Front. On February 14, 1944 K. Leselidze was treated at the Central Military Hospital in Moscow, where he was evacuated from the front for the treatment of severe disease after flu complications. On February 21, 1944 K.Leselidze died in the hospital. On February 26, 1944 buried in Tbilisi Old-Veri cemetery. On February 26, 1974 reburied in Didube Pantheon of Writers and Public Figures in Tbilisi, Georgia.

Leselidze K.N. (http://en.wikipedia.org/wiki/Konstantin_Leselidze)

Levchenko Gordey Ivanovich (1897–1981) – Soviet Admiral. In 1913 he started his military service as a ship boy and a sailor in the Baltic Fleet. During the 1917 October Revolution he was among those who stormed the Winter Palace and telegraph station. From 1918 he was with the Workers-Peasant Red Fleet (RKKF). After finishing in 1922 of the Naval School he commanded different ships of the Baltic Fleet. In 1935 he was directed to the Black Sea Fleet where he commanded a division and later a brigade of destroyers. In August 1937 he was appointed the Chief of Staff and in January 1938 the Commander of the Baltic Fleet. From April 1939 he was Deputy People's Commissar of the Navy. At the beginning of the Great Patriotic War he as a representative of the Naval Command took part in defense of Odessa, Nikolaev and Sevastopol. In October–November 1941 he commanded the troops of Crimea. But after failure of the Crimean campaign of 1941 he was removed from the office and in early 1942 was reduced from Vice-Admiral to 1st rank Captain. In April 1944 he was restored in his former rank and appointed Deputy People's Commissar of the Navy once again. In 1944 he was promoted to Admiral. From March 1947 he was Deputy Commander-in-Chief of the Navy, from 1950—Deputy Naval Minister, from 1956—Deputy Naval Commander-in-Chief on Military Training. He did much for improvement of the military training in the Navy, for education and upbringing of the staff. In September 1960 he resigned.

Lighthouses of the Azov and Black Seas – many navigation facilities are installed in remote and not easily accessible places. Thus, the lighthouse of the Biryuchiy Island locates on the spit in A.S., the Chaudinskiy and Meganomskiy light-houses—in the steppe and mountain terrains. The lighthouses on the Tendrovskiy spit and Zmeinyi Island are accessible only from the sea. The construction of lighthouses where the personnel was living permanently started from the time when Russia got access to B.S. and started development of the Naval Fleet. The oldest lighthouses on B.S. are near Enikale (built in 1820), Dzharylgach (1826), Odessa (1827), on the Zmeinyi Island (1843). Many lighthouses that were destroyed during the Great Patriotic War (1941–1945) and rehabilitated in the mid-twentieth century are now equipped with the modern potent light-optical facilities that make light visible from a distance of 35–40 km. In the recent years the advanced energy-saving technologies (light-optical devices on light-emitting diodes, solar batteries, wind-powered electric installations), digital telecommunications, control-adjustment stations of global navigation satellite systems GLONASS/GPS that are capable to fix the coordinates of a ship in the coastal zone with the accuracy to 2–10 m (lighthouses at Enikale and on Zmeinyi Island) found their way to the lighthouses.

Likhachev Ivan Fedorovich (1826–1907) – Russian naval state figure, historian, Vice-Admiral (1874). He finished the Naval Cadet Corps (1843). He served on the ships of the Black and Baltic Fleets. During the Crimean War of 1853–1856 he was the flag officer at Vice Admiral V.A. Kornilov; a participant of the Sevastopol defense of 1854–1855; was in charge of crossing and communication facilities in the Sevastopol Bay. In 1857 he navigated 3 propeller corvettes from the Baltic to the Black Sea. He was the Chief of Staff at Rear-Admiral G.I. Butakov, the Naval Commander, in Nikolaev. From 1858 he was the Aid-de-Camp of Great Prince Konstantin Nikolaevich, took an active part in elaboration and implementation of the naval reform in the 1850s–1860s, he actively supported the idea of construction of armored fleet. In 1883 he resigned. L. wrote many articles on naval theory and practice. In 1888 he published his most important article "The Naval General Staff" in the journal "Russian Shipping" where he validated for the first time the need of establishment of the Naval General Staff as an operative and strategic body for fleet operation.

Likhachev I.F. (http://img1.liveinternet.ru/images/attach/c/1//60/489/60489760_lihachev_if_00.
jpg)

Lima Yuriy Stepanovich (?–1702) – first Vice-Admiral of the Russian Fleet. He
was from the Genoese family. During the reign of Fedor Alekseevich he went to
serve in the Russian Fleet. During the Chigirinskiy campaigns to Crimea of 1687
and 1689 he was the Colonel of the engineer troops. After the first Azov Campaign
of 1695 he was promoted to Vice-Admiral. In 1696 in the rank of the first Vice-
Admiral of the Russian Fleet he took part in the second Azov Campaign. At the
triumphant entry in Moscow of the troops that seized the fortress he marched at the
head of the naval regiment. Soon he returned to the land-based service and
commanded the regiment named in his honor (formerly Lefortovskiy Regiment).

Liman (from the Greek *"limen"* meaning "harbor, bay") – an extended bay with
meandering, not high, but usually steep banks. L. is most likely to occur on the
submerged banks where the sea level has risen in relation to land; this process
floods valleys of plain rivers and ravines. It is often a continuation of a river mouth
(e.g. Bug Liman). L. may be opened seaward (inlets) and may be closed, cutoff
completely from the sea with a shallow (spit, barrier bar) or may have a narrow
strait (girlo). Depending on a coast nature, terrestrial runoff, connection with the sea
the salt regime of waters and sedimentation pattern in L. vary greatly. Because of
small terrestrial runoff and arid climate the waters in L. become highly saline,
sometimes reaching the brine status and salts get deposited. If the outflow from
L. is weak or much fine earth is washed from land L. may accumulate silt deposits
possessing rich biological population and in stagnant places the water gets saturated
with hydrogen sulfide. Silts are often used for curative purposes. In humid climate
and with intensive influx the fine sands, silt stones and clays deposit in L. forming

thin seasonal layers. Quite often they are saturated with organic substances that give origin to caustobiolith deposits (coal, combustible shale, oil). Depending on isolation from the sea and evaporation intensity the water salinity in a liman may be even higher than in B.S. Thus, in particular, during dry summer the water salinity in some limans in the Odessa Region may increase to 30‰ and more, while in the Kuyalnik Liman not far from Odessa it may be even higher than 100‰. The fauna of limans includes species well adapted to considerable variations of salinity. Limans are found on many flat coasts. Most well studied are the limans on the northern coasts of the Sea of Azov and the Black Sea (Molochnyi, Miusskiy, Khadzhibeiskiy and others); their number may reach several dozens. Some of them has become closed while many hundreds years ago they were connected with the sea. The Black Sea limans are used for many purposes, including for construction of marine ports. One more important use of limans is for fattening of fries of many aqueous animals.

Liman Fleet – was established in 1787 for defending Kherson when the B.S. Fleet was divided into approximately two equal parts—Sevastopol Squadron and Liman Fleet.

Limanchik Lake – a monument of nature located 6 km to the south of Abrau Lake that is separated from the sea with a 40-m wise pebble barrier. Its length is 195 m, width 145 m, depth to 5 m. This is a freshwater and through-flow water body.

Liners of the Black Sea – large speed passenger far-going vessels. After the Great Patriotic War (1941–1945) practically the whole civil fleet of the Soviet Union was destroyed. Its restoration began, first of all, using the vessels received as a compensation for damage and, secondly, using the vessels received after division of the former merchant fleet of Germany (following the resolution of the Potsdam Conference) and, at last, using the ships that were captured on the German shipbuilding yards and that happened to be in the Russian occupation zone. As reparations the Soviet Union was handed over two high-class Romanian liners "Transilvania" and "Besarabia" that were built in 1938 in Denmark. Their displacement was 6,970 tons, speed—to 25 knots and they were capable to carry 412 passengers. But soon "Transilvania" was returned to Romania and "Besarabia" was renamed into "Ukraine" and from 1945 it started navigation along the Odessa—Crimea—Caucasus line. This vessel was in operation till 1987, during this time it had two upgrades.

In 1946 after division of the German fleet the ship "Rugen" built in 1914 went to B.S. under the name of "Ivan Susanin". It could carry 200 people. It was operated on coastal lines until 1949. After this it was used as a training ship of the Odessa Naval School and in 1960 it was discarded. One more German ship "Deutschland" under a new name "Orion" was sent to B.S. Its displacement was 557 gross register tons, speed—to 13 knots and it was capable to carry to 600 passengers. Among the captured L. there were also "Admiral Nakhimov" (former "Berlin") that perished in 1986; "Pobeda" (Victory) (former "Magdalena"); "Russia" (former "Patria") that was built in the late 1930s in Germany. The latter was the world's first liner driven by electrical diesel. Its displacement was 18,000 tons and it could carry

185 passengers. In 1945 it was confiscated by the British troops. However in 1946 it was handed over to the Soviet Union. "Russia" was operated on the Odessa—Batumi line till 1985 and after this it was sold to Pakistan for breaking. Among the ships captured in an unfinished condition and later built up in full (in 1955) there was the turboelectric ship "Marienburg" that was renamed into "Lensovet". It could take onboard 548 passengers and accommodate them in cabins of different classes and 280 more passengers to be accommodated on a deck. Its speed reached 18 knots. In 1956 "Lensovet" was transferred to B.S. where it was operated on the Odessa—Batumi line. In 1962 on the request of the Abkhaz government it was given the name of "Abkhazia" in honor of "krymchak" ship that perished heroically during the Great Patriotic War. On 22 May 1977 while navigating from Odessa to Sevastopol with the passengers onboard "Abkhazia" ran into tanker "Aktash". After this accident "Abkhazia" was in service not long and in 1980 it was transferred to Spain for breaking.

In 1950 the ship "Sobieski" that was given the name of "Gruziya" (Georgia) was purchased from Poland. This ship was built in 1939 in Britain. It could transport 778 passengers and worked on the Odessa—Batumi line from 1950 to 1975. After this it was sold to Italy for breaking.

"Gruziya" (http://img.aucland.ru/market-photos/315/9QaPOm.jpg)

In 1978 France built the motor vessel "Aivazovskiy" on an order from the USSR. It was intended for operation on the Danube shippling line Izmail—Istanbul—Yalta—Odessa. This route was a part of the tourist voyage "From Alps to Black Sea". "Aivazovskiy" should have been used in the marine part of this voyage. But the war in Yugoslavia interfered with these plans and the ship was sold. Development of the liners traffic on B.S. was most intensive in the 1960s. The first were motor vessels built on the "Northern shipyard" in Leningrad. The vessel

"Moldavia" worked on the Black Sea shipping lines (it was in operation till 1988). The vessel "Ossetia" was transferred to the Danube Shipping Company where it was in use till the early 1990s. From 1980 to 1988 the vessel "Kirgizstan" was operated by the Black Sea Shipping Company. In 1994 it was transferred to the Novorossiysk Shipping Company that operated it under the name of "Bata" till 1997. The vessel "Swanetia" navigated B.S. till 1965 and after this it was transferred to Estonia. The vessels "Uzbekistan" and "Tadjikistan" were operated by the Black Sea Shipping Company till the late 1990s. In 1960 the vessel "Estonia" was built that in 1993 was renamed into "Ekaterina II". During bora it sank near a berth in Novorossiysk. After its lifting in 2003 the vessel was towed to Turkey. The vessel "Nadezhda Krupskaya" that was built in 1965 was retrofitted in 1977 into the ambulance ship of the Black Sea Fleet and was given a new name—"Kuban". In the early 1990s it was transferred to Bulgaria.

In the 1960s–1970s the Black Sea passenger fleet was replenished with the vessels "Latvia", "Litva", "Bashkiria", "Armenia" and "Adjaria". All they were used on the regular and cruise lines. In 1964 onboard the vessel "Armenia" N.S. Khruschev visited Alexandria (Egypt) and onboard "Bashkiria"—the Scandinavian countries. In 2004 although sold to the Greek ship owner this vessel helped with evacuation of the people from one of the Thai islands after disastrous tsunami there. In 1974 the shipping company received the motor vessel "Odessa" that was capable to carry 590 passengers and could develop a speed of 19 knots. This vessel made several voyages along the Crimea—Caucasus line, but already in 1975 it was chartered by the U.S.A. After breakup of the Soviet Union "Odessa" came to Ukraine, but so far it has not been used.

Finland built turbovessels "Azerbaijan", "Belorussia", "Gruziya" and "Kazakhstan". All these ships had very comfortable cabins for passengers, numerous salons, bars, restaurants, covered swimming pools, sport and dancing areas, etc. Two decks were reserved for automobile transportation. These vessels navigated under the flag of the Soviet Union between Odessa and ports in Crimea and Caucasus. "Kazakhstan" was in use till 1991; in 1994 after disintegration of the USSR it was renamed into "Ukraine" and in 1998 it was anchored as a floating casino in Florida, USA. "Azerbaijan" changed over to the Ukrainian flag in 1991, but in 1996 it sailed under the Liberian flag and was called "Arkadia". "Gruziya" was in the Black Sea fleet till 1991 and after this it changed its owners.

In the 1970s the USSR purchased some L. from abroad, in particular, such vessels as "Fedor Shalyapin" and "Leonid Sobinov" from the British "Cunard Line". They were built in the mid-1950s and could take onboard more than 800 passengers; their speed was 19 knots. "Leonid Sobinov" was transferred to the Black Sea Shipping Company in 1973 and "Fedor Shalyapin"—in 1980. In the mid-1990s both vessels were anchored in the port of Ilyichevsk and in 1998 "Leonid Sobolev" was sent to Alang (India) for breaking and in 2005 "Fedor Shalyapin" went next. In the 1980s the Black Sea Shipping Company received a series of auto-passenger ferries built in Poland. These were "Dmitry Shostakovich" and "Lev Tolstoy" which in 1991 were handed over to Ukraine. But in 1995 "Lev Tolstoy" was registered in Liberia, while "Dmitry Shostakovich" in 2000 went under a convenient flag.

The most well-known vessels were built in GDR in the 1960s—"Taras Shev-
chenko", "Shota Rustaveli" and "Ivan Franko" that were of the same type. The
comfortable vessel "Ivan Franko" was transferred to the Black Sea Shipping
Company. It was capable to accommodate 1,464 people. In the summertime
"Ivan Franko" was operated on the Crimea—Caucasus line. It also made cruises,
but no farther than the Mediterranean. In 1991 "Ivan Franko" went over to Ukraine.
Because of arrests worldover of ships sailing under the Ukrainian flag "Ivan
Franko" stayed idling for long on the Odessa roads and in July 1997 it was sent
to Alang (India) for breaking. The vessel "Taras Shevchenko" built in 1966 was
included into the Black Sea Shipping Company, but it was operated mainly by the
foreign tourist firms earning currency for the Shipping Company. In the summertime it was used on the Crimea—Caucasus line. In 1988 it was retrofitted. However,
the USSR disintegration in 1991 interfered with its normal operation. It was
privatized by Ukraine and this became the beginning of the "Taras Shevchenko"
end. For a long time it was anchored first in Ilyichevsk and later in Odessa. The
attempts to use it were of no effect and in 2005 the ship was handed over for metal
scrap.

"Taras Shevchenko" (http://korabley.net/_nw/2/96053.jpg)

The Black Sea Shipping Company also possessed the legendary vessel "Shota
Rustaveli" that was built in 1968. Its length was 178 m, speed 18 knots. It took
onboard 650 passengers. This vessel was operated on the Mediterranean cruise lines
carrying foreign tourists. In the summertime it was used on the Crimea—Caucasus
line. After 1991 it came into the Ukrainian ownership and from 1997 it was
anchored in Ilyichevsk. In 2001 it was renamed into "Assedo" (the same name
"Odessa", but written in Latin letters back to front). In September 2003 the vessel
was sold to Alang (India) for breaking.

"Shota Rustaveli" (http://ships.at.ua/stati/SHOTA_RUSTAVELI-001.jpg)

Lipovans – a small (about 40,000 people) ethnic group of the Slavs that comprised descendants of the Russian Old Believers who fled from Russia about 200 years ago escaping from religious "hunting". They settled down in the lower reaches of the Prut River and in the Danube delta and maintained their pre-Nikon Orthodox rites. Similar to the Romanians and Ukrainians who also live in the Danube Delta, L. main occupation was fishing. Their settlements, such as Mila, Krishan, Sfyntul-George, locate on three delta arms. The descendants of the Russian Old Believers may be met not only in the Danube Delta, but also in the Romanian part of Bukovina. Here their population is not large, only some 1,800 people.

Livadiya (from Greek "*livadon*" meaning "meadow") – a seaside climatic resort in Crimea, Ukraine. This is a small village belonging to the Greater Yalta and locating 3 km from it. It is situated on the Southern Coast of Crimea, on the slope of the Mogabi Mountain. It has a fine-sand beach. L. is surrounded with vineyards. The first settlement appeared here still in the Middle Ages. After annexation of Crimea to Russia Catherine II granted these lands to the Greeks of the Balaklava battalion who guarded the southern borders. In 1834 the owner of L., the commander of the Balaklava Greek Battalion in Feodosia, sold it to Count L.S. Pototskiy, the prominent diplomat of the Russian Court. Later in 1860 his heiresses sold L. to the local department for the Tsar family. Alexander II presented it to his wife who suffered from TB. In 1861 L. became the summer residence of the Imperial family. In the 1880s the Smaller Palace was built here. In 1910–1911 in the landscape park (it was created in 1834) containing valuable species of subtropical and tropical evergreen plants the Greater Livadiya Palace in the Neo-Renaissance style was constructed. In February 1945 the Crimea Summit Meeting of the heads of ally states—USSR, USA and Great Britain was held here. As a resort L. was developed from 1925 when the world's first sanatorium for peasants was opened here. The subtropical Mediterranean climate is most favorable for climatotherapy and thalassotherapy of respiratory (non-TB), blood-circulation, and nervous functional disorders. There are several sanatoriums in L. and its surroundings. Sanatorium "Livadiya" for patients with the cardiovascular diseases locates in the Livadia Palace and surrounding old and new buildings. It is located 800 km from the sea. It has a beach and a spa center. Special foot walk paths are made in the park. Population—1,600 (2012).

London Convention (1871) – this name is adopted in the historical publications in relation to the Treaty on Amending Paris Treaty of 1856 dated 1(13) March 1871, its final version was signed in London on 3(15) March 1871 by the authorized representatives of Russia, Germany, Austria-Hungary, France, United Kingdom, Italy and Turkey. This treaty finalized the efforts of the Russian diplomacy for repudiating the restricting clauses contained in the Treaty of Paris (1856). The Franco-Prussian War of 1870–1871 that ended in the defeat of France diverted the attention of European states from the Near East problems and created favorable conditions for efforts of the Russian diplomacy seeking to revise the provisions on "neutrality" of B.S. Prussia that became stronger as a result of the war supported the claims of Russia in exchange for its neutral stand in the Franco-Prussian conflict. In October 1870 the Russian Government declared that it was not going to observe the restricting clauses. This decision caused protests from the United Kingdom and Austria-Hungary. Then Prussia (from January 1871—German Empire) initiated transfer of this issue to discussion at an international conference of all states that signed the 1856 Treaty of Paris. L.C. abrogated Clauses 11, 12 and 14 of the 1856 Treaty of Paris on "neutrality" of B.S. and also the Russian-Turkish Convention attached to the treaty restricting the quantity and tonnage of the Russian and Turkish light guard ships on B.S. (Clause 1). At the same time the invariability of other provisions of the Paris Treaty (Clause 9) was stressed. In particular the freedom of merchant shipping through the Black Sea straits (Clause 3) and prohibition of the passage of foreign military ships through them in the peacetime (Clause 2) were confirmed. On insistence of the United Kingdom and Austria-Hungary a provision was included in the last clause concerning the right of Turkey to "open straits in the peacetime for military ships of friendly and alien states" if it thinks it necessary in pursuance of the provisions of the 1856 Paris Treaty. In fact, this provision was aimed against Russia because most statements in this treaty were seeking to prevent strengthening of the Russian influence on the Balkans and Near East. Some clauses of L.C. regulated the problems of shipping over the Danube and authorities of the International Commission established in 1856 with a view to settle them.

London Conventions – two multilateral conventions on the regime of Black Sea straits concluded in London in 1840 and 1841.

First L.C. – was signed by Russia, Austria, United Kingdom, Prussia and the Ottoman Empire. The conflict between the central Ottoman Government and Muhammad Ali Pasha of Egypt in which the Ottoman Empire was immersed in 1831–1833 was revived in the late 1830s. The army sent by Ottoman Sultan against Muhammad Ali suffered defeat and the whole Ottoman Fleet sided with Sultan. Collapse of the Ottoman Empire was not in the plans of Russia. At the same time the threat of confronting the United Kingdom that was a rival of Russia in the struggle for domination in Constantinople and a wish to drive a wedge between the United Kingdom and France that openly backed Muhammad Ali forced the Russian Government to interfere into the conflict on the Sultan side and to enter into negotiations with the United Kingdom (later Austria and Prussia joined them). In addition to seeking ways to settle the Egypt crisis Russian Emperor Nikolay I and Russian Foreign Minister Count K.V. Nesselrode approached the parties with a proposal to arrange about closing the Black Sea Straits for passage of warships of all world countries (except Turkey proper). Previously the passage of foreign vessels through the Straits depended either on the Sultan's wish or was regulated by bilateral treaties of Turkey with other states. Emperor Nikolay I and Russian Foreign Minister Count K.V. Nesselrode believed that the multilateral agreement involving Great Powers will protect B.S. more reliably from intrusion of foreign fleet. The parties that signed L.C. in 1840 agreed to use all possible means of persuasion to urge Muhammad Ali to accept the peace terms they elaborated jointly. These terms envisaged sea blockage of the Egypt coast and, if necessary, joint defense of Constantinople and the straits on the Sultan's request (Clauses 2, 3). According to Clause 4, Turkey agreed to keep the Bosporus and Dardanelles closed for passage of foreign warships during the peacetime. A special protocol enclosed to L.C. allowed the Turkish Government to issue permits to passage of small foreign ships via the Straits for servicing the needs of foreign missions in Constantinople. Diplomatic and military actions of the states that signed L.C. forced the Egypt Pasha and France that backed him to make concessions: Muhammad Ali agreed to restrict his claims by keeping control over Egypt that was recognized his hereditary land.

Second L.C. – after settlement of the Egypt crisis France adopted new terms regarding the Black Sea Straits regime. On 1(13) July 1841 the representatives of Russia, Austria, France, Prussia, Ottoman Empire and United Kingdom signed a new convention that repeated the contents of Clause 4 of 1840 L.C. with the Protocol. The regime of the Black Sea Straits stated in L.C. of 1840 and 1841 existed with minor amendments till World War I, but formally—till conclusion of the Lausanne Convention on Straits (1923). The principle of closing the Straits for passage of foreign warships during the peacetime could not ensure security of the Black Sea coast of Russia. Meanwhile Russia and Turkey and later other Black Sea states were deprived of the possibility to regulate the Straits regime independently, without involvement of European states.

Longoz – wetlands in the delta of the Kamchia River, Bulgarian coast of B.S. Its area is 4,500 ha. It is covered with the dense primeval forest looking like a jungle. Many waterfowl and rare varieties of marshy and aqueous plants are found here.

Loo (from the Abazinskiy family Lou that appeared here in the thirteenth–fourteenth centuries) – a climatic seaside resort on the Black Sea coast of the Caucasus, a resort village entering into Greater Sochi, located northward of Sochi in the Lazarevskiy District of Sochi, Russia. Railway station is 18 km to the northwest of the center of Sochi. A small river flowing into B.S. has the same name. The beach is pebble; its width is to 60 m. The summer heat is alleviated by sea breezes. The sanatorium "Mountain Air" for children with the active TB forms, a curative beach and sports facilities are constructed on the high coast 3 km from the center of L. village. L. also has vacation hotels, health improvement camps, beaches, balneotherapy center, winter solarium, swimming pool filled with sea water, sport hall and summer sport grounds.

Lower Danube River Administration (LDRA) – was established in 1954 for water works and regulation of navigation over the Danube stretch from the Sulina arm mouth to Braila (the stretch of the so-called "marine Danube"). In addition, it dealt with the issues of the marine shipping in the Danube mouth. LDRA was created on a par basis by the Soviet Union and Romania. In the first 3 years of its existence the Administration addressed the tasks by joint efforts of specialists from the two countries. In mid-1957 the USSR refused to participate in the LDRA activities and from this time on the Administration was managed solely by Romania. At present LDRA conducts the bottom dredging works and ensures the channel depth of 24 f. (7.32 m) in the "marine Danube" as recommended by the Danube Commission.

Lozenets – a small village on the coast of the Black Sea, Bulgaria.

Lukull Cape (in the Turk "Ulu-Kol, Ulu-Kul" meaning "large slope") – one of the northernmost capes on the Sevastopol coast and the southern inlet cape of the Kalamit Bay. This is a cliffy cape of the reddish color well visible from the sea. The sea area near the cape is a nature reserve that enters into the Lukull coastal-aqueous complex. In 1885 a lighthouse was constructed on the cape for safe approaching the port of Sevastopol from the northern directions.

Lukull Coastal-Aqueous Complex (Reserve) – it is situated near the Lukull Cape from which it received its name. In 1972 it was given the status of a monument of nature. This is a coastal strip 6 km long and 300 m wide around the cape. It is a natural model of the sea-land interaction.

Luzanovka – a seaside climatic resort in the northeastern part of Odessa, Ukraine. It is situated on the shore of the Odessa Bay and enters into the Odessa group of resorts. The beach with its fine sand (about 2 km long and 30 m wide) is considered one of the best in Ukraine. In summer its shallow part gets heated to a temperature of 30 °C. The sea floor here is sandy, smooth, gradually sloping. The water temperature in L. is usually 1–2 °C higher than in other seaside resorts. The natural-climatic conditions in L. are most favorable for climato- and thalassotherapy.

M

Mackenzie Thomas, Foma Fomich (?–1786) – Russian Rear Admiral, born in Scotland. In 1765 he started his service as a midshipman. In 1770 on the ship "Svyatoslav" he reached the Archipelago, took part in the battles at Naples-Romagna and Chesma. In 1772 he took part in the battle at Patras in the Ionian Sea and was sent on a boat for setting the enemy ships on fire. In 1773 he was promoted to Captain-Lieutenant, in 1775 to Captain 2nd Rank for his bravery in the battle at Chesma, in 1782 to Captain of the Major-General rank, in 1783 to Rear Admiral with appointment to the Black Sea Fleet. He stayed in Sevastopol where he took part in the port construction and in 1784 he became the first Commander of the Sevastopol squadron. He died in Sevastopol.

Maeotis (from Latin "*Maeotis palus*", Ancient Greek "*Maiotis limne*" meaning presently the Sea of Azov) – the name "Maeot" could be found on the world map made by the Phoenician seafarers (approximately late seventh century B.C.). The people of Greece knew about M. notable for its considerable fish resources from the people of Miletia and other towns that established Greek colonies of Panticapaeum, Phanagoria and others on the Kerch Peninsula, Tanais in the mouth of the river with the same name.

"Magarach" Institute – was founded in 1828 on the basis of the Magarach Establishment on Experimental Wine Growing and Wine Making and a special college existed at the Nikitskiy Botanical Garden in Crimea. "Magarach" was founded according to the decree of Russian Emperor Nikolay I of 09.14.1828 and on the initiative of the Governor-General of Novorussia and Governor of Bessarabia Prince Mikhail Vorontsov. Ordinance provides for wide-ranging measures to promote horticulture, viticulture and winemaking in the Novorussia and Bessarabia. This Institute pioneered regular experiments and researches on spreading of valuable vine varieties, development of new technologies for making original wines, training of high professionals in this field. In 1963 Institute "Magarach" became the key center of scientific wine making and in 1964—of vine growing and in 1988 it was officially approved as the research center on grape growing and processing.

S.R. Grinevetsky et al., *The Black Sea Encyclopedia*,
DOI 10.1007/978-3-642-55227-4_14, © Springer-Verlag Berlin Heidelberg 2015

The experimental divisions of the Institute include two wine-making plants with the wine repository (enoteca) containing 18,600 bottles of collection wines aged from 10 to 150 years; the country's largest collection of grape varieties (more than 3,000 specimens); a refrigerator for grapes (for 1,000 tons); and a greenhouse complex (the total area is 14,000 m^2). The Institute created over 30 new highly-productive varieties among which there are Early Magarach, Bastardo Magarachskiy, Ruby Magarach, Tauria and others. They are cultivated in Ukraine, Moldavia and other vine-growing regions. These grapes were used for making of more than 20 quality wine brands, including "Muscat White", "Muscat Rose", "Sersial", "Carnelian of Tauria", "Alminskiy", "Yalta", "Swan's Nest" and others. The Institute maintains long-lasting contacts with 15 foreign countries. The original wine brands and also publications of the Institute were awarded 102 medals, most of which are golden, at USSR and international competitions. Some inventions of the Institute are patented in Spain, USA, Canada and Bulgaria. The Institute has a museum that became a kind of a model for other vine-growing and wine-making museums. Institute "Magarach" was awarded the Order of the Labor Red Banner (1979). Today this is National Institute of Grape and Wine "Magarach" (Yalta, Crimea, Ukraine)—the oldest research center run by the Ukrainian Academy of Agrarian Sciences.

Magnolia – the genus of woody plants of the *Magnoliaceae* family. The genus is named after French botanist Pierre Magnol (1638–1715). The plant has ordinary, large leaves and single flowers smelling pleasantly. There are known about 40 (by other sources—70) varieties of this plant: the deciduous varieties are found in the tropical zone of Southeast Asia and the evergreen varieties—in the Atlantic coast of North America. In subtropics of Ukraine and Russia it is cultivated as a decorative tree.

Magnolia (http://dlm6.meta.ua/pic/0/36/215/gqO48ip9zR.jpg)

Main Harbor, Main Roads – one of the old names of the Sevastopol Harbor.

Makarov Stepan Osipovich (1848–1904) – the brilliant Russian Fleet Commander, naval theorist, seafarer, oceanographer, shipbuilder, inventor and Vice-Admiral (1896). He was born in the family of a naval petty officer. He finished the Naval School in Nikolaevsk-on-Amur (1865). He navigated on the ships of the Pacific squadron commanded by Rear-Admiral A.A. Popov (1863–1864) and armored squadron of the Baltic Fleet under command of Vice Admiral G.I. Butakov (1869). He wrote a number of works that laid the foundation of the ship subdivision theory, proposed the collision mat, a device eventually adopted by almost all navy, and designed a water drainage system. In 1872 he was called back to Petersburg under command of Admiral A.A. Popov for construction of new armored fleet. From 1876 he served on the Black Sea Fleet. During the Russian-Turkish War of 1877–1878 commanding the steamship "Grand Duke Konstantin" that was converted on his initiative into a mothership for spar-torpedo boats he successfully attacked the Turkish ships having applied for the first time the Whitehead torpedoes. The Whitehead torpedo was the first self-propelled or "locomotive" torpedo ever developed. It was perfected in 1866 by Robert Whitehead from a design conceived by Giovanni Luppis of the Austro-Hungarian Navy. In 1880–1881 he was in charge of the marine section in the Akhaltekin expedition of General M.D. Skobelev.

In 1881–1882 using the Russian vessel "Taman" based in Istanbul he devoted all his free time to oceanographic studies, conducted detailed hydrological observations in the Bosporus that included measurements of water temperature, salinity, current speed and direction. He invented the "flux meter"—a device for study of current velocity. As a result of these observations he concluded that the upper current in the Strait was directed from B.S. to the Sea of Marmara, while the lower current was directed oppositely to B.S.. Thus, he discovered the most remarkable natural phenomenon of B.S. that explained specific features of its hydrological structure. In 1885 he published his books "On the Exchange of Waters Between the Black and Mediterranean Seas" and "On the Study of Permanent Sea Currents" for which he was conferred the prize of Metropolitan Makarius by the Russian Academy of Sciences for the best work in the physics and mathematics. All following investigations confirmed in full the M.'s conclusions. In 1886–1889, commanding the steam corvette "Vityaz", he circumnavigated the globe starting off from Kronstadt and sailing back over the Atlantic and Pacific oceans. During this voyage he conducted investigations in the Pacific and wrote another publication "The Vityaz and the Pacific Ocean" (vols. 1–2, 1894) which won the author the international acclaim. He was awarded the prize of the Petersburg Academy of Sciences and Golden Medal of Russian Geographic Society.

In 1890 he was appointed the Junior Flagman of the Baltic Fleet and in the same year he became the Russia's youngest Admiral and in 1891 Chief Inspector of the Naval Artillery. He invented a type of armor-piercing caps for shells and on his initiative the smokeless gunpowder came into use in the Fleet. From late 1894 he was made Commander of the Mediterranean squadron which in 1895 he led to the

Pacific due to the expected military clash with Japan. From 1896 he commanded the Baltic Fleet and in 1897 he published his most famous work "Discussion of Questions in Naval Tactics" which was translated into English, Turkish, Japanese, Spanish and Italian languages. From October 1899 he commanded the First Division of the Baltic Fleet and from December he was Chief Commander of the Kronstadt Port and the Military Governor of this city. He proposed the idea of design of an icebreaker for Arctic exploration and an expedition to the North Pole. He supervised construction of the icebreaker "Ermak" on which in 1899 and 1901 he sailed to Spitsbergen, Franz Josef Land and Novaya Zemlya. In 1901 his new book "Ermak in the Ice" was published. On the eve of the Russian-Japanese War of 1904–1905 he duly appraised a threat of a sudden attack from Japan and recommended to move ships from the inner to the outer harbor at Port-Arthur, but his warning was not taken into consideration. From February 1904 he commanded the Pacific squadron and wisely directed the ships during the Port-Arthur defense. He died on the armored ship "Petropavlovsk" that came upon a mine. Still during his lifetime S.O. Makarov was called the last outstanding Admiral of the Russian Fleet.

ВИЦЕ-АДМІРАЛЪ
С.О. МАКАРОВЪ.

Makarov S.O. (http://en.wikipedia.org/wiki/Stepan_Makarov)

Makhindjauri – a seaside climatic and balneo resort in the Republic of Adjaria, Georgia. A railway station with the same name is 6 km to the northeast of Batumi. M. is situated on the Black Sea coast of the Caucasus between the Kara-Dare and Zelenyi capes, Republic of Adjaria. The climate here is subtropical, humid. The winter is mild and without snow. The mean temperature of January is 6 °C. The summer is very warm. The mean temperature of August is 23 °C. Precipitations are abundant throughout a year, their mean annual sum is 2,700 mm. The number of sunshine hours is 1,900 a year. Tender climate provides opportunities for climate-

and thalassotherapy from May through November. M. also has mineral springs which water is used for curative baths. This resort provides treatment to the people suffering from cardiovascular, respiratory, locomotor system, peripheral nervous and gynecological diseases. This resort was founded in 1904 (resort of state significance from 1923). In May 2010, at the station Makhindjauri there was a presentation of the new project "High Speed Railway—from Tbilisi to Batumi for 3 hours".

Makopse – a seaside climatic resort 15 km to the southeast of the railway station Tuapse, Krasnodar Territory, Russia. It is located in the valley of the Makopse River in the broad-leaved forest. It has a pebble beach.

Maksimov Andrey Semenovich (1866–1951) – Russian Vice-Admiral. He finished the Naval College (1887) and Mine Officer School (1895). He took part in suppression of the Boxer Rebellion in China (1900–1901). During the Russian-Japanese War of 1904–1905 he participated in the Port-Arthur defense. He commanded the destroyer (1906–1908), the 4th (1908–1909), 7th (1909) and 5th (1909–1910) divisions of torpedo boats of the Baltic Fleet, the cruiser (1910–1913), the brigade of cruisers (1913–1914), the 2nd (1914–1915) and 1st brigades of battleships of the Baltic Fleet (1915). In summer 1915 he was appointed the chief of the mine defense on the Baltic Sea. In 1916 he was promoted to Vice-Admiral. After the 1917 February Revolution in Russia he was elected the commander of the Baltic Fleet and took this office from 4 March to 2 June 1917. From 6 September he was the Chief of Naval staff of the Supreme Commander, from 18 November—the second assistant of the Naval Minister. Went to serve on the Red Fleet. From 1918 he was the senior inspector of the Revolutionary Military Council of the Republic. From August 1920 to December 1921 he commanded the Black Sea Fleet. From 1924 he served in the Naval Inspection. He commanded the messenger vessel "Borovskiy" in the voyage to the Far East and later became the person liaison officer at the Naval Commander of the Republic. In 1927 he resigned. He was awarded many medals.

Malakhov Kurgan (mound) – the dominating height in the southeastern part of Sevastopol; one of the most important bastions at approaches to the city during the Sevastopol Defense of 1854–1855. During the Crimean War of 1853–1856 organizing the on-land city defense a defense tower was built on M.K. in autumn 1854. This tower was surrounded by fortifications of the so-called Kornilov bastion (by the name of Vice-Admiral V.A. Kornilov). In the battles for the mound Vice-Admiral V.A. Kornilov and P.S. Nakhimov who commanded the Sevastopol defense were wounded deadly. At night on 28 August (9 September) 1855 the superior French forces, but with great casualties took hold M.K. and the Russian troops had to retreat from the southern part of the city to the northern part. To commemorate the defenders of M.K. the memorial plates were installed on the defense tower, in places where Kornilov and Nakhimov were wounded, the monument over the common grave and memorial plaques on the batteries. The heroism of the Russian soldiers during M.K. storming by the enemy on 6(18) June 1855 was commemorated in the panorama "Sevastopol Defense". During the Revolution of 1905–1907 M.K. was the meeting place of workers, soldiers and sailors. In June 1905 in a gully near M.K. the representatives of ships and units of the Black Sea Fleet took a decision on starting the armed uprising. The commander of M.K. defense was Rear Admiral V.I. Istomin. Today the museum "Defense Tower of Malakhov Kurgan" is opened here.

Malakhov Kurgan (http://pulsev.com.ua/images/stories/malakhov_kurgan.jpg)

Malaya Zemlya (Minor Land) – the outpost on the coast of the Tsemess Bay (Myskhako Peninsula) near Novorossiysk. It was captured by the Soviet troops during the Yuzhnaya Ozereika operation in February 1943. On the night of 4 February 1943 the first landing party commanded by Major Ts.L. Kunikov went ashore on the western coast of the Tsemess Bay near Stanichka. After landing failure near Yuzhnaya Ozereika the major landing forces (two marine and five rifle brigades, five partisan units) were directed to the Myskhako Peninsula. Later they were supported by divisions of the 18th Army commanded by General K.N. Leselidze. The Germans having transferred here five divisions tried to liquidate this outpost. But after fierce battles the Soviet troops defended M.Z., although they failed to attain their main purpose—to seize Novorossiysk and to cut off the escape routes for the Germans via the Taman Peninsula. For 7 months the units of the 18th Army kept this outpost on the western coast of the Tsemess Bay having, in fact, paralyzed the German forces in the port of Novorossiysk. M.Z. played a very important role in takeover of Novorossiysk in September 1943. This local outpost became widely known in the 1970s because Colonel L.I. Brezhnev, the future General Secretary of the Central Committee of the Communist Party of the Soviet

Union, took part in it as political instructor and in 1978 he published the book "Malaya Zemlya".

Malaya Zemlya (http://newzz.in.ua)

Malodolinskoe (former Klein-Libental) – a seaside mud resort 25 km to the southwest of Odessa, Odessa Region, Ukraine. It refers to the Odessa group of resort areas. It is located on the northwestern coast of the Sukhoy (Malodolinskiy) liman. The main natural curative factor here is sulfide silt mud from the Kuyalnitskiy Liman. Apart from the mud the artificial mineral and gas baths are also offered for curative purposes here. The first curative establishment (mud treatment center) on the basis of the mud from the Sukhoy Liman was opened here in 1854. In 1931 the sanatorium for mud treatment of children suffering from rheumatism was constructed in M.

Malorossia – the name of Ukraine being in use in pre-revolutionary Russia (before 1917).

Maltakva – a seaside climatic resort in the Republic of Adjaria, Georgia, near the railway station Poti. It is located on the Black Sea coast of the Caucasus in the western part of the Kolchis Lowland. The climate here is subtropical, highly humid and very mild (the mean temperature of January is 6 °C) with snowless winter and very warm summer (the mean temperature of August is 23 °C). The precipitations are rather copious throughout a year; their mean annual quantity is 1,650 mm. The mild climate is used for treatment of respiratory organs, mostly in children. A wide sandy beach and a smoothly sloping seafloor make it possible to combine climatotherapy with sea bathing.

Mamaia – a seaside climatic resort in Romania, 3 km northward of the city of Constanta (administratively it is a part of this city). It locates on the sand spit between B.S. and the freshwater Siutghiol Lake. The resort was founded in 1906–1919. The climate here is mild marine. The mean temperature in January is about 0 °C, in July 22 °C. The precipitations are about 400 mm a year. The number of sunshine hours exceeds 2,200 a year. The beach extends for 10 km. The famous Murfatlar vineyards are found in the vicinity of the city. M. is not only one of the most popular resorts in Romania, but is well-known internationally. The resort offers the climatotherapy and sea bathing as well as mud treatment (mud from Techirghiol Lake in the resort with the same name). The town of Istria founded by the Greeks in the seventh century B.C. locates on the coast northward of M. This is an archeological nature reserve where one can find ruins of the fortification walls, temples, thermae and living houses dated sixth century B.C. to sixth century A.D. The monastery of the fourteenth century is located near M. At approaches to the resort there is a remarkable village that comprises houses (and there are more than 30 of them) representing different architectural styles of various parts of Romania.

Mamaia (http://www.worldalldetails.com/Pictureview/1793-Mamaia_Romania_Beach_on_a_sea son_peak.html)

Mandrensko Lake-Reservoir – is situated near the western coast of the Burgas Bay, the Black Sea coast, Bulgaria. Till 1963 it was an open liman with transparent waters. In spring the water in the liman became practically fresh, while in summer and early fall it added in salinity, in particular after long-lasting eastern winds. The highest salinity was recorded in the years with dry summer when the rivers flowing into the lake dried out. In 1963 the greater part of the lake was turned into a reservoir and the total water surface area increased nearly fourfold. At present M.L. is the largest freshwater basin on the Bulgarian coast of B.S.

Mandrensko Lake (https://en.wikipedia.org/wiki/File:Lake-mandrensko-dinev.jpg)

Mangalia – a city and balneo and climatic resort in Romania 44 km southward of the city of Constanta on the Black sea coast. M. is the southernmost seaside resort in Romania. It locates in the steppe zone abounding in mounds. The Lake of Mangalia near M. is surrounded by acacia groves. The climate here is mild marine. M. is the only place in the country with the positive temperatures in winter—the mean temperature in January is 0.5 °C. The summer is warm, the mean temperature in July is 22 °C. Precipitations are 400 mm a year. This place is distinguished by abundant sunshine (in summer to 25 sunny days a month). The number of sunshine hours is 2,500 a year. The resort offers climatotherapy, sea bathings, mineral baths and curative mud that allow opportunities for treatment of nervous, gynecological, metabolic diseases and consequences of injuries. It was founded in place of the Ancient Greek Town of Kallatis (sixth century B.C.) that was later destroyed by an earthquake and natural rise of the B.S. level. In 1925 the archeological museum "Kallatis" was created here. Its exposition includes a collection of Greek and Roman bas-reliefs, sculptures, precious articles and coins. Population—33,400 (2011).

Mangalia (http://www.hotelsmangalia.com/img/townimg.jpg)

Manganari Egor Pavlovich (1796–1859) – Russian Major-General of the Admiralty, explorer of B.S. He finished the Pilot College in Nikolaev. From 1813 he navigated on the ships of the Black Sea Fleet. His first hydrographic work was inventory of the Dnester Liman (1823). From 1825 to 1836 commanding first the brig "Nikolay" and later the yacht "Golubka" he prepared the first systemic inventory of the Dnieper Liman, B. and A. seas on the basis of astronomical posts and triangulation data. In 1836 the new "General Map of B. and A. Seas" was published based on the results of the expedition. In 1842 he prepared "The Atlas of the Black Sea" comprising 28 maps and 17 sheets with the view of coasts. In 1849 he was promoted to Major-General of the Fleet Pilots Corps (FPC) and appointed the director of the B. and A. Seas Pilot. His name was given to a cape in Sevastopol separating the Kamyshovaya and Kazachia bays. His major work was "The Atlas of the Black Sea" (1842).

Manganari Mikhail Pavlovich (1804–1887) – Russian Admiral, brother of E.P. Manganari. In 1815 he started service as a cadet on the Black Sea Fleet. In 1815–1827 he navigated over B.S. In 1821 he was promoted to midshipman and in 1828 to Lieutenant. In 1828–1829 he took part in the Russian-Turkish War. From 1830 to 1848, commanding the yacht "Golubka", schooner "Zabiyaka" and ship "Kolkhida", he made inventories and measurements near the shores of A.S., Caucasus, Crimea and Sea of Marmara which helped to create the reliable astronomic and geodetic base. In 1837 he was promoted to Captain-Lieutenant. In 1838 being the commander on the ship "Kolkhida" he navigated near the Abkhazia coasts. In 1840–1848 he conducted surveys in the Sea of Marmara mostly onboard the Turkish vessels. In 1846 he was conferred the rank of Captain 2nd Rank and in 1849—Capitan 1st Rank. By the results of surveys the Pilot of the Sea of Marmara was published in 1850 in Nikolaev. In 1854 for preparation of the Map of the Sea of Marmara Russian Geographic Society awarded him the Konstantin Medal. In 1856 he was promoted to Major-General and in 1859 renamed into Rear Admiral. In 1862 he was transferred to serve on the Black Sea Fleet. In 1863–1873 he was the commander of the Sevastopol Port and the Commander-in-Chief of the Nikolaev Port. In 1864 he was promoted to Rear Admiral. In 1869 he was directed to the eastern coast of B.S. to choose a place for port construction. In 1873–1875 he was the Commander-in-Chief of the Black Sea Fleet and Ports. In spring 1876 he was promoted to Admiral. In 1881 he was appointed once more the Commander-in-Chief of the Black Sea Fleet and Ports and Military Governor of Nikolaev. In 1882 M. was appointed a member of the Admiralty Council. His principal works are "The Pilot of the Sea of Marmara" (1850), "The Surveys of the Sea of Marmara in 1845–1848)" (1850, 1884).

Mangup, Mangup-Kale – the ancient fortified town 20 km eastward of Sevastopol. This settlement appeared in the fourth century A.D. and from the sixth century it became one of the centers of Medieval Crimea populated by the Goths. The fortress was called Dorn. In the twelfth to fifteenth centuries it was the capital of the Feodoro Principality. After the Ottoman Empire conquered Crimea (1475) M. became the administrative center of the region. The main occupation of the local population was leather tanning. In the second half of the sixteenth century the town came into decay. Some sections of the fortress walls of the sixth and fifteenth centuries, ruins of Basilica of the sixth century rebuilt in 1425, the palace of the fifteenth century, Ottoman citadel of the sixteenth century and others survived to the present. The archeological excavations have been carried out since the 1890s.

Mangup Kingdom – better known as Feodoro with the center in the town of Feodoro on the Mangup Mountain. It was established, by some sources, in the twelfth century, by other—in the thirteenth century. Allegedly M.K. incorporated the lands in the river basins of Belbek and Chernaya and also a part of the Southern Coast of Crimea. The medieval kingdom was known in the fourteenth to fifteenth centuries not only in Crimea, but far beyond it. The Mangup princes traded through the port of Avlita locating in the apex of the Sevastopol Bay with the Mediterranean countries. Among its northern partners there was Moscow. It is known that the marriage between the son of the Moscow Tsar Ivan III and the daughter of the Mangup Prince was planned, but for some reasons it was frustrated. The basic element in the coat of arms of the Mangup Prince was the double eagle that was, probably, taken from Byzantine. Prior to 1475 the Mangup princes were Orthodox and were given Christian names. In 1475 the Ottomans seized and set afire the town of Feodoro. Later it was restored and given the name of Mangup-Kale. In the Turk *"menkup"* means a loser, a person in disfavor or in trouble, while *"kale"* means fortress.

Manstein Erich von (1887–1973) – the German military commander, General Field Marshal. In 1913–1914 he studied in the Military Academy and in 1914–1918 he fought on the Eastern Front, then in Serbia and at the end of the war on the Western Front. In 1939 during the Polish campaign M. was the chief of staff of the Army Group South, later the army was transferred to France. In the Russian campaign M. distinguished himself commanding the 56th tank corps that made breakthrough from Eastern Prussia via Dvinsk (Daugavpils) to the Ilmen Lake. On 14 September 1941 he arrived in Nikolaev and took command of the 11th Army. On 30 October 1941 the famous Sevastopol defense began. M. directed 100,000-strong group against 60,000 defenders of the Sevastopol defense region. The German units rushed into the city from the north near the Mackenzie heights, but the defenders were supported by the troops that were delivered on the ships of the Black Sea Fleet and together they countered the German attack. On 26 December 1941 with the beginning of the Kerch-Feodosiya operation M. had to stop the siege of Sevastopol and transfer part of the forces against the landing troops. In January-March 1942 after some offensives the defenders of Sevastopol resumed in some areas the earlier lost positions. In May after the Soviet troops were defeated near Kerch the situation in the besieged city deteriorated significantly. M. managed to accumulate the basic forces of the 11th Army (over 200,000 people) against Sevastopol which defenders that time numbered 106,000. Still greater superiority the Germans had in tanks and aviation. On 2 June 1942 the German troops started bombardment of the city. According to M., this was the most intensive use of the German artillery during whole World War II. On 7 June the units of the 11th Army stormed Sevastopol for the third time. The battles that lasted for nearly a month without interruptions were hard-fought on both sides. On 30 June 1942 the German seized Sevastopol. This military operation had a code name "Sturgeon fishing". Near Sevastopol the Germans lost 300,000 people (150,000 people in the last storm). For the Crimean capture Colonel-General M. was promoted by Hitler to General Field Marshal. In 1944 Hitler handed M. the Knight's Cross with oak leaves. At the end of the war M. lived in his estate. After the war he was taken prisoner by the British troops. In 1949 he was sentenced by the British Military Tribunal to 18 years of imprisonment. He spent in prison 12 years and in 1962 was released due to his health condition. Later on he was invited as a consultant when the Bundeswehr was created. M. was considered the most gifted general in Wehrmacht. His war memories he published in the book "Verlorene Siege" (Lost Victories) in Germany in 1955 and translated into English in 1958.

Mapping of the Black and Azov Seas – the first information about B.S. may be found in the works of Antique geographers and authors. Homer (seventh century B.C.) mentioned that the Manych River was a strait and along it one could get from the Pontus Euxine to the Caspian Sea. Greek Hecataeus of Miletus (c. 550–c. 476 B.C.) on his map of the Earth showed many new things. He described in detail the B.S. coast likening its configuration to the "Scythian bow" which string corresponded to the southern coast, while the curved shaft—to the northern coast with the Tavria Peninsula. Such presentation of the B.S. coasts will be repeated for long by many Antique authors. Greek historian Herodotus (c. 484–425 B.C.) in his "History" named eight rivers: Istr (Danube), Tiras (Dniester), Hipanis (Southern Bug), Borisphen (Dnieper), Pantikan (most likely Ingulets), Hipacirus (possibly Kalanchak), Gerus and Tanais (Don). Only the lower reaches of these rivers at their inflow into the Pontus were known to Herodotus. Famous ancient geographer Strabon (64 B.C.–24 A.D.) traveled along the whole southern coast of B.S. as far as Armenia. In his "Geography" he referred to the Pontus as the eastern bay of the Internal (Mediterranean) sea and estimated its perimeter at 25,000 stadia (one Roman stadia was equal to 185 m). And he wondered why there was so much water in the sea: "Whether the Pontus flowed into the Internal sea or the Internal sea flowed into the Pontus?" Speculations of Strabon about the sea floor (Book 1, Chap. 3) deserve special mention. Refuting the opinion of Eratosthenes (c. 276–c. 195 B.C.) according to which the Pontus overfilled with river waters broke through a channel near the city of Byzantine and the water rushed from it into Propontidus (Sea of Marmara) and Internal Sea, Strabon believed that this occurred when the Pontus floor due to the river sediments became higher than the floor of the Internal Sea.

In the most ancient pilot "Periple of Pontus Euxinian" (second century B.C.) prepared by well-known Roman author and state figure Flavius Arrian (92–175 A.D.) on the basis of the earlier sources and his sailing from Trapezund (Trabzon) to Dioscuriada (Sevastopol) he marked many settlements. Arrian described in detail the island of Achilles (presently Zmeinyi). Claudius Ptolemy (90–168 A.D.) is considered the first who prepared the maps of the territory of future Russia. In his work "Geography" (Book 8, Chaps. 8–10) Map 8 of Europe and Map 2 of Asia covered the European and Asian Sarmatia. The hydrographic network included three rivers—Borisphene, Tanais and Ra (Volga). The map showed the Pontus (B.S.) and Meotic (A.S.) seas as well as Tauric Chersonesus and between B.S. and Caspian Sea—the Cholchis. Ptolemy influenced significantly the cartographers of the Roman Empire. Pliny the Elder (23–79 A.D.) drew much attention to B.S. and Caspian Sea with the nearby territories. The map of Peitinger that was thought to be prepared in 500 A.D. depicted the Pontus Euxinian with great distortions as well as the Meotic Lake (A.S.). It also showed Agalinus (Dniester), Selliani that flowed into the river running from the "Hipanskiy bog" (Southern Bug) and Nisanus (Dnieper).

On the Euphrates there was found a bronze Greek shield depicting a fragment of a road map with some parts of B.S.—Odessos, Bibona, Kallatis, Kherson and others. In 1021 Al-Biruni prepared a map of seven seas. It represented a circle inside which there was a smaller disk into which five bays or seas incised (one of the bays included two seas—Mediterranean and Bar Bontisa (B.S.). In 1154 Idrisi prepared the map on 17 sheets and a circular general map. It's interesting to note that on its part oriented southward where the Russian lands were depicted the Black Sea was on top. In 1318 Pietro Vesconte prepared the Atlas of sea maps among which there was the B.S. map. The configuration of B. and A. seas in it coincided with the configurations shown on the ordinary sea maps. The Atlas of Medichi prepared in 1351 included the map of B.S. that did not practically differ from other known maps of this sea. B.S. and A.S. had their usual configuration although B.S. was connected with the Caspian by a river called "Asso". The Sora (Kuban) and Don rivers flowed into A.S. From the northwest by straight lines three rivers flowed into B.S., of which the easternmost (Dnieper) had a tributary. With time the ancient periples turned into medieval sea maps. Most of them presented the coasts of B.S. and A.S. The Italian sea maps usually showed only the northern and eastern coast of B.S. In 1448 Benedictine monk Andreas Valeperger in Constanta made up the world map. The outline of B. and A. seas reminded slightly the configuration on the Ptolemy's maps or on sea maps. Among the ports of B.S. there were shown Suastopolis (Fazis) on the Caucasian coast, Album Castorum ("White camp", modern Akkerman) in the mouth of the Nester (Dniester) River.

Map of the Black Sea (1559) (http://atf.org.ua/gallery.php/gallery/6)

Map of the Black Sea (1570) (http://kraevedenie.net/2010/03/13/kart-cd/)

In Russia the creation of the Azov Fleet was accompanied by development of hydrographic works. In 1695 Colonel of the Preobrazhenskiy Regiment Yu.A. Mengden conducted topographic and geodetic surveys in the area of the Russian troops moving to A.S. In 1696 after capture of Azov Ya.V. Bryus using the results of these surveys and other cartographic materials prepared "The Map of Southern Russia" that showed B. and A. seas, the basins of the Dnieper and Don rivers. In the same year, following the order of Peter I the first marine inventory of the Taganrog Bay of A.S. from Krivaya bar to the Don River mouth was made. Using this inventory a plan was developed that became the first marine map in the Peter I time based on direct measurements. In 1699 while the ship "Krepost" (Fortress) navigated from Azov to Kerch the depth measurements were made with personal participation of Peter I. On the basis of these measurements Admiral K. Kryuis "under direct supervision of Peter I" prepared the sea map of the eastern part of A.S. Next year senior pilot Kh. Otto of the ship "Krepost" using the measurements made during navigation in 1699 from Azov to Constantinople prepared the Mercator's map at scale about 1:1,800,000. This map showed the depths along the southern coast of the Crimean Peninsula, near the entrance into the Bosporus and directly in this strait. The map was published in Amsterdam. Using the results of hydrographic works, engraver Adrian Shkhonebek from the Netherlands engraved the map "Eastern Part of the Palus Meotis Sea Presently Called the Sea of Azov". It presented the coastline of the sea, the parallel and meridional grid, the compass grid, depths, anchorage places and cities. The scale of the map was about 1:1,700,000. It was published in Moscow.

In 1701–1702 Dutch naval officer Peter Bergman prepared "The Map of the Sea of Azov and Don River from Korotoyaka to Azov and Taganrog and Dolgaya Spit". In 1702 he drew in Voronezh "The New Map of the Sea of Azov". In 1702–1704 P. Pikart published in Moscow the map "Drawing of the Black Sea from the City of Kerch to Tsar Grad" that showed the cities of Bendery, Ochakov, Taman, Trapezund and Tsargrad and included the insert depicting the Bosporus Strait with the depths marked along the fairway line. In 1703–1704 in Amsterdam "The Atlas of the Don River, Azov and Black Seas" prepared by Admiral K. Kryuis was published; it included descriptions and 17 maps. The maps were made by the cartographic materials received during naval campaigns of 1699. On its title page it was written "New Drawing Book. . ." In addition it included the interesting map of A.S. and Maotis Lake, Pontus Euxinian or Black Sea with their depths and shallows, the inflowing rivers, harbors and cities. All maps were hand colored. This Atlas is considered the first significant work in the Russian cartography.

Map of the Black Sea (1711) (http://giorgi-chachba.livejournal.com/45871.html)

Map of the Black Sea (1720) (http://raremaps.com.ua/category/maps/maps-of-the-crimea)

When any of the members of the Tsar family was going on a voyage the Academy of Sciences had to prepare the respective maps. In 1728 the Academy started publishing calendars (from 1768 they were called "menology"). Many maps included into them were original works stirring great interest. Thus, there was prepared "The Marine Map of the Black Sea" by Ya. Schmidt, "The Map of Regions along the Coasts of the Black and Caspian Seas" also by Ya. Schmidt was based on the Francendorf's map. In 1734 the Master Map of the Russian Empire prepared by I.K. Kirilov was published. On the sheet of European Russia there were depicted the Black and Azov seas. In 1739 the special navigation atlas of Louis Renard incorporated a map of B.S. made by Nicolaes Witsen.

Map of the Black Sea (1740) (http://maps-karta.blogspot.ru/2011_10_01_archive.html)

The commenced Russian-Turkish War and actions of the Russian Fleet on B.S. required reliable maps, thus, in 1755 Lieutenants I. Bersenev and L. Pustoshkin initiated surveys of its coasts. In 1785 they prepared "The Atlas of the Black Sea" comprising 11 handwritten maps. After establishment in 1768 of the Azov Fleet Admiral A.N. Senyavin undertook guidance of the surveys of the A.S. and in 1771 a new map of this sea was prepared. The works went on, and Lieutenants Elchaninov and Zaostrovskiy carried out description and measurements of the coast from Taganrog. In 1778 the Kerch Strait was described by Lieutenant Karyakin and in 1785—by Rostovtsev. However, none of these maps were published and the foreign seamen used French maps made by Jacques Nicolas Bellin in 1772 and Charles Francois Delamarche in 1785. It should be said that the Bellin's map had essential errors: the coasts of B.S. were depicted with errors to 65 miles (120.4 km), A.S.—40 miles (74 km), while the Don mouth was shown 100 miles (185 km) more eastward.

In 1784–1789 Jan van Wenzel, the Dutch traveler, using the survey results of Count I.G. Kinsbergen who was on the Russian service from 1767 to 1776, prepared the map of B.S. that was published by Van Keilen. Kinsbergen made the map of A.S. and the large map of Crimea on four sheets. In 1795 the map of the B.S. northern coast on three sheets was published following the order of Vice-Admiral Jose de Ribas. Captain I.I. Billings made the handwritten atlas on 45 sheets with the pictures of ports and coasts of B.S. In addition, there were several more maps of the northern coast and its ports. In 1797–1798 the reconnaissance marine inventory of the B.S. northern coast (from the Dnieper to Kuban) was made under guidance of Captain 1st rank I.I. Billings and with participation of I.M. Budischev, A.E. Vlito and others. These works made the basis of "The Atlas of the Black Sea" (1799) prepared under supervision of I.I. Billings. The inventory commenced by Billings was continued in 1801–1802 by Lieutenant I.M. Budischev and Midshipman N.D. Kritskiy on the western coast of B.S. from Odessa to the Bosporus Strait, while Captain-Lieutenant A.E. Vlito and P.A. Adamopulo—on the Anatolian coast from Constantinople to the Samsun Cape. Vlito and Kritskiy made inventory of the whole A.S. and the Kerch Strait. In 1804 the map appeared. In 1807 on the basis of these inventories and the Billings inventory the maps of B. and A. seas were published. In the same year "The Atlas of the Black Sea" by I.M. Budischev was published that comprised general and local maps, plans of ports, raids and bays. The maps were made in the Mercator's projection, but had different scales.

In 1804 A. Wilbrecht prepared the general map of B.S. on two sheets that was entitled "The Marine Map of the Black, Azov and Marmara Seas Based on Recent Surveys and Astronomical Observations Conducted by the Russians and French". Beginning from 1807 to 1825, in some regions of B.S. there were made reconnaissance marine inventories under guidance of I.M. Budischev, F.F. Bellinsgauzen, N.D. Kritskiy and others. The inventories covered nearly the whole coast of the sea except the eastern coast. In 1808 the first part of "The Marine Guide over the Azov and Black Seas" was published. Its content was close to the modern pilot. In 1820 following the agreement between the Russian and French governments the ship "La Chevrette" under command of captain Gautier and with participation of Captain-Lieutenant M.B. Berkh made inventory of B.S. and 2 years later the map of the sea was published. With time on the requirements of seafarers to the map accuracy had been growing. In 1825 the hydrographic expedition was organized led by Captain-Lieutenant E.P. Manganari. It made the first regular inventory of B. and A. seas. The works went on till 1836 when the Master Map of B.S. was published.

In 1832 the Hydrographic Department of the Black Sea Navy Headquarters was established that incorporated the hydrographic depot including the drawing office, printing house, library and observatory and also the instrumentation chamber. This made the basis for the future Hydrographic Department of the Black Sea Navy. In 1833 during staying of the ship squadron commanded by M.P. Lazarev in Turkey the inventory of the Bosporus and Dardanelles Straits was made by Lieutenants E.V. Putyatin and V.A. Kornilov.

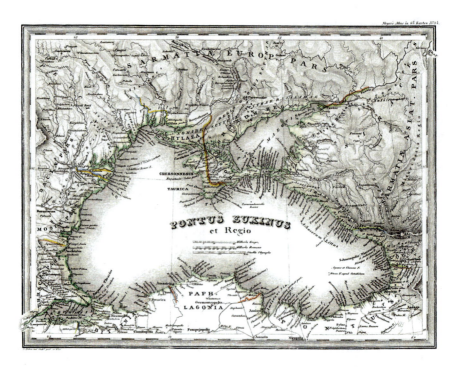

Map of the Black Sea (1840) (http://www.etomesto.ru/img_map.php?id=416)

In 1842 "The Atlas of Maps of the Black and Azov Seas" prepared on the basis of marine survey results and the maps made by E.P. Manganari was published. The Atlas had as enclosures the sketches of ports. Thanks to careful handling of the materials and the earlier prepared inventories the maps were rather truthful and remained in use till 1873. In 1847 Lieutenants G.I. Butakov and I.A. Shestakov conducted marine surveys of the coasts of the Caucasus and Crimea, the Bug River, the Dniester Liman and a part of the coast between Ochakovskiy and the Danube River Delta (1848), the Anatolian coasts of Turkey (1849) and the coast section from the Constantinople Strait to the Danube Delta. In 1851 these materials were used for publication of the Black Sea Pilot. In 1854 the Azov Sea Pilot was published; it was prepared by Second Lieutenant Sukhomlinskiy on the basis of surveys that were carried out by him in 1850–1851.

After the end of the Crimean War the merchant shipping started developing on B.S. It again required more update, detailed and accurate sea maps. In 1871 the hydrographic expedition to B.S. was organized for regular and detailed investigation of B.S. It was led by Captain 1st rank V.I. Zarudniy. As a result of surveys that were conducted till 1914 the first detailed inventory of B. and A. seas based on triangulation was carried out. By the beginning of World War I Russia could boast of the best level of knowledge of the B. and A. seas hydrography.

Marbled Crab (*Pachygrapsus marmoratus*) – a representative of the *Arthropoda* of the *Crustacea* class. It has a crust colored from blue-green to dark-brown crossed with numerous light strips reminding of mottled marble. It is most often found in the tidal zone among mussels and in algae thickets. During warm months it gets out of water and crawls among stones, in rock cracks and between breakwaters.

Marine Biological Station – the idea of such station was taking shape still in 1906, but as a research establishment it was opened only in 1932 by the State University of Sofia in Varna. It also had the marine aquarium. Until 1952 the station personnel conducted hydrological and hydrochemical studies in the coastal waters of Bulgaria, but they were quite sporadic in nature. Beginning from 1952 the station started regular qualitative and quantitative studies of phytoplankton in the Varna Bay. The results of studies were published in the "Proceedings of the Marine Biological Station".

Marine Doctrine of the Russian Federation till 2020 – it was approved by Russian President on 27 July 2001. This is the basic document defining the policy of the Russian Federation in marine activities—the national and marine policy of Russia. One of the key regional components of this policy is the policy on the Black and Azov seas. This regional policy addresses the following long-term targets: replacement of old merchant marine and mixed (river-sea) vessels, updating and further improvement of the coastal-port infrastructure; updating the legal base for functioning of the Black Sea Fleet of the Russian Federation on the Ukrainian territory, keeping the city of Sevastopol as its main base; creation of conditions, including with utilization of the region potential, for basing and use of the marine potential ensuring protection of sovereign and international rights of the Russian Federation on the Black and Azov seas, development of passenger traffic from the ports of the Krasnodar Territory to the Mediterranean countries and also ferry traffic over the Black Sea.

Marine Hydrophysical Institute (MHI) of National Academy of Sciences of Ukraine – leading marine research institute in Ukraine (Sevastopol). It was founded in 1929 on the initiative and under guidance of V.V. Shuleikin in the village of Katsiveli (Crimea). The world's first stationary marine geophysical station was constructed here to carry out systemic studies of the processes the coastal zone of the sea and in the atmosphere. In 1948 in Moscow MHI of USSR Academy of Sciences was established on the basis of the Black Sea Hydrophysical Station of USSR AS and the Marine Hydrophysical Laboratory (previously it was a department in the Institute of Theoretical Physics of USSR AS). Its director was V.V. Shuleikin. In August 1961 MHI was transferred to Ukrainian AS and in 1963 it was moved to Sevastopol. MHI comprises the following divisions: automation of oceanographic investigations; contact methods of the shelf hydrophysical investigations; ocean-atmosphere interaction; nuclear hydrophysics; turbulence; dynamics of ocean processes; oceanography; wave theory; systems analysis; sea optics; remote sensing; marine information systems and technologies. The Institute includes: Experimental Division (Katsiveli), Hydroacoustic Division (Odessa) as well as self-supporting divisions: Special Design and Technological Bureau with Trial Production Base (Sevastopol), Research-Production Center "ECOSEA-Hydrophysics"

(Sevastopol), Research and Technological Center "Shelf" (Sevastopol). The Institute studies the large-scale circulation in the World Ocean, the ocean-atmosphere interaction, turbulence of surface and deeper waters in the ocean, formation of heat stock, transfer and transformation of thermal and mechanical energy in the ocean, elaboration of the scientifically-based forecast and control of the physical condition, quality of water environment, new methods and means, including aerospace, of control of the World Ocean state, etc. At present MHI of NASU is one of the world's major oceanographic centers.

Marine Hydrophysical Institute (Photo by Dmytro Solovyov)

Marine Museum—Aquarium – was opened in Sevastopol in 1897. It was the first in Russia. In 1898 a special building was constructed for housing this museum that was designed by architect A.M. Weizan. This museum takes its origin in the Marine Biological Station in Sevastopol established in 1871 on the initiative of outstanding Russian scientists N.P. Miklukho-Maklai, I.I. Mechnikov, I.M. Sechenov and A.O. Kovalevskiy. The central basin in the aquarium is 9.2 m in diameter, 1.5 m in depth. The museum also contains 12 wall aquariums with a total volume about 7 m^3. The aquarium contains a rich collection of marine inhabitants, dozen species of Black Sea animals and plants.

Marine Observatory – is situated on the Pavlov Cape in Sevastopol and was constructed for the naval needs. Its activities may be broken into three units: Compass, Astronomy and Hydrometeorology. The Compass Unit envisages verifi-cation of compasses, their installation on ships, identification and elimination of any deviations, i.e. deviations of a magnetic arrow in a compass due to the effect of the iron and steel parts of a vessel. The Astronomy Unit includes determination of errors and rates of chronometers. The Hydrometeorology Unit conducts observations over principal characteristics of weather and sea regime.

Maritime Nature Reserve – was organized in 1927 for preservation of the wetland ecosystem and water fowl. The nature reserve covered some areas of the Northern Prichernomorie (Black Sea coastal area), Sivash and the Sea of Azov coast. It stretches for 500 km from the Kinburn Spit (Black Sea) as far as the Belosarai Spit (Sea of Azov). Till the end of 1932 the reserve was a part of the Askania-Nova Nature Reserve. It acquired the independent status from 1 January 1933. In 1937 two nature reserves—Black Sea and Azov-Sivash were established on the basis of M.N.R.

Maritime traffic regulations for the Turkish Straits and the Marmara Region – they were enforced by Turkey on 1 July 1994. The regulations outlined the rules of passage via the Straits in order to improve the safety of navigation. Special attention was focused on passage via the Straits of the vessels carrying dangerous cargo and the super-large vessels. The masters of all vessels shall inform in advance the Turkish authorities about the kind of vessel, cargo and route. Masters of the large vessels with a length of 150 m and more before fixing a voyage through the Straits must contact the Administration (Undersecretariat for Maritime Affairs, Office of the Turkish Prime Minister) and advise all necessary particulars, characteristics and the type of cargo planned to carry. The Administration taking into consideration the Straits structure, the sizes and maneuverability of a vessel and the need to ensure the safety of life, property and the environment and the shipping conditions "will advise the master about the results of consideration of these factors". Masters of vessels with dangerous cargo on board weighing 508 tons and upwards shall submit the Sailing Plan to the Traffic Control Center at least 24 h before the vessel entrance the Straits. This Sailing Plan shall include the above-mentioned information, the request for pilotage and other particulars.

Nuclear powered vessels or vessels carrying nuclear cargo or nuclear wastes before fixing of a voyage through the Straits shall receive permission for the passage of the Strait in accordance with the rules of the Undersecretariat for Maritime Affairs. Nuclear powered vessels or vessels carrying nuclear cargo or nuclear wastes before fixing of a voyage through the Straits shall receive permission from the Ministry of Environment. In both cases the masters shall comply with the respective rules. Regulations envisage compulsory invitation of pilots by the masters of Turkish vessels with a length of 150 m and more for long routes. This measure is not recommended for foreign vessels. The Administration may enforce compulsory pilotage of non-transit vessels in certain zones of the Straits and the Sea of Marmara. The vessels that without preliminary notice and without permission anchor in docks, at berths, beacons will be piloted and towed on an order from the Port Administration. The vessel that is forced to anchor due to emergency situation shall inform the Traffic Control Center thereof. The Administration will move the vessel to a safe place with pilots or tow boats. In both cases the services of pilots and tow boats are paid by the masters.

When a large vessel carrying dangerous cargo enters the Straits, the vessel with the same particulars is not permitted to enter the Straits on the opposite side until the first vessel leaves the Straits. The distance among such vessels going in one direction shall be at least 37.4 km. The distance among the vessels carrying other kinds of cargo shall be 1.5 km. The speed of vessels in the Straits shall not exceed 18.5 km/h. The Istanbul Authorities set forth that the tankers with a length over 200 m may pass through the Straits only in the daytime. The Straits may be closed for passage by the Administration for construction works, firefighting, rescue operations, pursuit of criminals, scientific and sport events.

Mariupol – a town and a port on the northeastern coast of the Sea of Azov, Ukraine. It is the tenth largest city in the Ukraine and second largest in the Donetsk Oblast— 492,000 citizens in 2010. Originally founded as a Cossack fortress, Mariupol was granted city rights in 1778. During the Great Patriotic War in 1941–1943 the city was occupied by German Army. It caused tremendous damage to the city, many people were killed. Since 1948 till 1989 the city was named Zhdanov after the Soviet politician Andrey Zhdanov. Mariupol has been a centre for the grain trade, metallurgy and heavy engineering. Today Mariupol remains a centre for industry, higher education and business. In Mariupol there are 56 industrial enterprises under various patterns of ownership. The industry of the city is diverse, among which the city's heavy industry is dominant. Mariupol is home to the major steel mills (including some globally important) and chemical plants; there is also an important seaport and a railroad junction. The largest enterprises are "Ilyich Iron and Steel Works", "Azovstal", "Azovmash", and the Mariupol Sea Trading Port. There are also shipyards, fish canneries, and the various educational institutions relating to the study of metallurgy and science, among them: Priazovskiy State Technical University, Mariupol Humanitarian University and Azovskiy Institute of Maritime Transport. In 2003, Donetsk Regional Russian Drama Theatre, the oldest theater of the region, has celebrated the 125th anniversary. A tourists interest is mainly the Sea of Azov coast. Around the city the strip of resort settlements includes: Melekino, Urzuf, Yalta, Sedovo, Bezymennoye, Sopino, Belosarai Spit, etc. The first sanatoriums are opened in a city in 1926.

Donetsk Regional Russian Drama Theatre (http://upload.wikimedia.org/wikipedia/commons/e/ed/
Theatre_night_Mariupol.jpg)

Mariupol Sea Trading Port – one of three major ports in Ukraine. It is located on
the coast of the Sea of Azov, the so-called "sea gates" of the Ukraine's largest
industrial and raw material region Donbass. The port was put into operation in
1889. By the volume of financial investments in the period 1867–1904 this
non-military port took the third place in Russia. The port territory is 67.6 ha. The
length of the mooring line is 3.2 km. The port is capable to handle over 12 million
tons of cargo a year. This is the largest and best equipped port on the Sea of Azov.
Via the Volga-Don Canal, Marian System and Belomorsko-Baltic Canal this port is
linked with all regions of Russia connected with the Volga and has access to the
Caspian, Baltic and White seas. Coal, metal, products of mechanical engineering,
varieties of ores and grains are transported from and to different cities such as
Donetsk, Kharkiv, Lugansk, and the near regions of Russia via MSTP. The regular
cargo and passenger traffic to Turkey, Greece and Israel is organized via this port.

Martyan Cape Nature Reserve – was given the status of a nature reserve in 1973.
Its total area is 240 ha, of which 120 ha are covered by the Maryan terrain, partially
by the Ai-Danil terrain and coastal zone and 120 ha are the Black Sea water area. It
is situated not far from the Nikitskiy Botanical Garden (Crimea, Ukraine). It was
established with a view to save the groves of arborescent juniper being the sample
of the Mediterranean vegetation.

Maslenyi Cape (Oil Cape) – a cape on the Bulgarian coast of the Black Sea. Allegedly, it received its name by the vessel that carried olive oil and got wrecked here. The cape is elevated and cliffy; it is surrounded by reefs.

Massandra – a village in Crimea, Ukraine. It locates on the Southern Coast of Crimea 5 km to the east of Yalta. M. has a large wine-making plant "Massandra" uniting over dozen of vine-growing farms that have plants for primary grapes processing and wine-making. M. plant was constructed and started manufacturing of the famous Crimean wines in 1894–1897. It became the center of growing of high-grade grape varieties for production of unique wines and also the creative laboratory of the classical Crimean wines. Massandra was acquired by Polish Counts Potocki in 1783. In the mid-nineteenth century, it passed to Russian Prince S.M. Vorontsov, whose father was the governor of New Russia. In 1881 Vorontsov engaged a team of French architects to design for him a palace in the Louis XIII style. He died the following year and construction work was suspended till 1889, when the palace was purchased by Russian Emperor Alexander III, who asked a Russian architect of German ancestry Maximilian Messmacher to finish the palace for his own use but he did not live to see it completed in 1900. During the Soviet years, the palace was used by Joseph Stalin as a government dacha. Population— 7,400 (2001).

Massandra Palace – the palace of Emperor Alexander III in Massandra, Crimea, Ukraine. Its construction was started in 1881 on the initiative of Prince S.M. Vorontsov as a place for picnics and hunting of the Tsar family. In 1892–1900 architect Maximilian Messmacher continued construction of this palace. Having left intact the parts of the old building he added many Baroque and modern elements. He contemplated quite different design of the park on an area of 6 ha. In 1894 Alexander III died and the palace was in oblivion till the end of the Civil War. In the Soviet time the palace was used as state dacha (country estate) and was called Stalinskiy. In 1990 it was transferred in charge of the Museum Association "Palaces and Parks of the Southern Coast of Crimea". In 1992 the Museum "Massandra Palace" was opened that is considered the best architectural monument on the Crimean coast.

Massandra Palace (http://upload.wikimedia.org/wikipedia/commons/8/89/Massandra_Front.JPG)

Matsesta – a balneological resort area 11 km from Sochi, Krasnodar Territory, Russia. It is located on the Black Sea coast of the Caucasus in the valley of the Matsesta River. The mineral water springs of M. facilitated development of Sochi as a balneological resort. The mineral waters of M. refer to thermal (from 18 to 67 °C) hydrogen sulfide chloride-sulfide sodium waters that contain iodine (to 12 mg/l), bromine (to 70 mg/l), radon (to 1 nCu/l) and others. Their salinity varies widely from 3 to 41 g/l. The waters are used largely for making curative baths, sprinkling, inhalations for treatment of blood circulation organs, central and peripheral nervous system, lomotor system, skin, gynecological and other diseases. Near the so-called "Old Matsesta" there is the "New Matsesta" located 8 km from Sochi. In 1979 the balneotherapy center was opened here. Many prominent state figures and among them I.V. Stalin who suffered from problems with joints liked to have rest in M. The country estate "Green grove" was constructed here for Stalin with the special automobile road running to the bath pavilions. It is said that it was in M. that the famous attempt against Stalin was prepared and they say that this was designed by the Japanese after their defeat on the Khasan Lake (1938). This operation was controlled by George Lyushkov, Commissar on State Security, Chief of the National Committee of Internal Affairs for the Azov—Black Sea Region, who in 1938 fled to Japan. He supervised the construction of this dacha for Stalin and knew well the location of all its objects. A group of subverters was delivered from Japan to Turkish port of Trabzon and near Batumi they crossed the state border. But the group got into an ambush and many of its members were killed. Miyagi Yotoku who entered into the group "Ramsay" headed by Richard Sorge (Soviet intelligence resident in Japan since 1933) sent to Moscow a message about the palnned attempt. Today Matsesta Resort is one of the largest medical spa complex in Russia. Founded in 1902, on the site of the medicinal sulfurous springs, the complex developed a unique, internationally recognized methods of treatment.

Matsesta River – is a river that flows into the Black Sea near Matsesta, Krasnodar Territory, Russia. The Matsesta was mentioned for the first time in history in 137 when the Roman military commander and historian Arrian (86–160 A.D.) described the coast of the Black Sea and the river Matsesta in a letter to the Roman Emperor Hadrian (76–138 A.D.). The river rises in the southern ranges of the Alek Mountains at a height of 900 m above sea level and flows through the Khostinskiy Region of Sochi into the Black Sea. The river is 17 km long, and the highest point in the river's basin is 1,003 m above sea level. The river has numerous tributaries, the largest of which are the Tsanyk and Zmeyka. The Matsesta is most famous for its sources of mineral water. The name of the river translates in the Ubykh and Adyghe languages as "fiery water", because people that immersed themselves in the water of the Matsesta found that their skin became red.

Matyushenko Afanasiy Nikolaevich (1879–1907) – one of the leaders of the revolt on the battleship "Potyomkin" (1905). From 1894 he was a worker on the railway in Kharkov, then a loader in the port of Odessa, a stoker on the merchant ship, an assistant driver on a locomotive in Vladivostok. In 1900 he was called to military service; in 1902 he finished the school of quartermasters. He was conferred the rank of petty officer and send to serve on the battleship "Potyomkin". On 14 June 1905 he was one of the first who called sailors to revolt having shot a senior officer who killed one of the leaders of the revolt—sailor G.N. Vakulenchuk. He was elected the chairman of the Ship Commission. After the revolt was defeated he in June 1905 went to live in Romania. In June 1905 the Emigrant United Sailor Committee delegated him to Geneva where he informed the leaders of the Russian revolutionary emigration about the revolt on the battleship. In 1906 he moved to the USA, in 1907—to Paris. In June 1907 he secretly went to Odessa and on 3 July 1907 he was arrested in Nikolaev and put in prison in Odessa and in 11 October 1907 he was transferred to Sevastopol. On 17 October of the same year the naval court sentenced him to death. Later the name of M. was given to one of the harbors in the Sevastopol Bay and to the former Rudolf Mountain near Sevastopol.

Mayachnyi Peninsula (lighthouse peninsula) – locates near Sevastopol, Ukraine. It is separated from the Geraclean Peninsula by a narrow isthmus formed by the apex of the Kazachia Bay in the northeast and the Black Sea coast in the southwest.

It got its name by the Chersonesus lighthouse built in 1819 on the western tip of the peninsula. During the Great Patriotic War the military airfield "Chersonesus" located here. Many researchers believe that the first Greek colony "Strabon Chersonesus" appeared here. This is proved by the ruins of the defense walls that were found in 1890 to the southwest of the apex of the Kazachia Bay.

Mechnikov Ilya Ilyich (1845–1916) – Russian biologist, pathologist, one of the founders of the evolutionary embryology, Honorary Member of the Petersburg Academy of Sciences (1902). He graduated from the Natural History Department of the Kharkov University (1864) and continued his education in Germany, then he went to Italy where he studied embryology of invertebrates. He got the Master's (1867) and Doctoral (1868) degrees and became the privat-docent of the Petersburg University. Later he was appointed the Titular Professor of the Zoology and Comparative Anatomy Faculty at the Novorossiysk University in Odessa (1870–1882). M. disagreed with the reactionary policy in education and in 1882 as an act of protest he left his office and organized a private laboratory in Odessa. In 1886, together with N.F. Gamaley he established the Russia's first bacteriological station for control of infectious diseases. In 1887 he left Russia for Paris where he was offered to work in a laboratory of the Pasteur Institute. In 1905 M. became Deputy Director of this Institute. Till the end of his life M. did not break his ties with Russia. In the Paris Laboratory many Russian scientists worked and M. himself frequently visited Russia. Until 1882 he worked mostly as zoologist. In 1911 he led the research expedition to the Kalmykia steppes to study TB dissemination. M. investigated the Black Sea sponges and coelenteretae. The investigation results led M. to development of the phagocytic immunity theory ("Immunity in Infectious Diseases" (1901), Nobel Prize in 1908 shared with Paul Ehrlich). Numerous works of M. were devoted to studies of cholera epidemiology. Philosophical reflections of M. were included into his book "Forty Years in Search of Rational Outlook" (Mechnikov 1913). M. published some biographical and reminiscence articles as well as sketches about the history of biology and medicine in Russia and abroad. M. was the Honorary Doctor of the Cambridge University (1891), Corresponding Member of the Academy of Natural Sciences in Philadelphia (1891) to mention, but a few.

Mechnikov I.I. (http://www.mechnik.spb.ru/academy/fotogallery/Left%20in%20eternity/Mechnikov.jpg)

Mediterranean Rockling (*Gaidropsarus mediterraneus*) – the fish of the cod family (*Gadidae*). Its body length reaches 30–35 cm. It is native to the Mediterranean. It is extensively met in the Black Sea. It prefers keeping near coasts, among stones and aqueous vegetation. In summer it runs to the areas with depths to 30–50 m. It propagates in November-March in coastal zones.

Mediterranization – growing number of the Mediterranean invaders (fish and invertebrates in the Black Sea). This occurs due to their adaptation to the Black Sea conditions. M. is one of characteristic changes observed presently in the biology of the Black Sea. Out of more than 500 Mediterranean species there are 109 species "striving" to live in the Black Sea. Many species run into the Black Sea only in summer and prefer spending winter and propagate in the Sea of Marmara and the Mediterranean Sea. And it was determined reliably that only about 60 species multiply in the Black Sea and they may be referred to really the Black Sea species. Among Mediterranean invaders there are such fish species as anchovy, garfish, gray mullet, greenfish, horse mackerel, goatfish, scomber, flatfish and others.

Melikhov Vasiliy Ivanovich (1794–1863) – Russian Admiral. In 1799 he entered the Naval Cadet Corps; in 1807 he was given the rank of a naval cadet. In 1808–1810 he served on different ships of the Baltic Fleet. In 1811 he was transferred to the Black Sea Fleet. In 1811–1815 he served on different ships; in 1811 his ship was wrecked near the Anapa Cape. In 1814 he was promoted to Lieutenant.

In 1817–1820 M. was the flag officer at Vice-Admiral A.S. Greig and navigated the Black Sea. In 1821 he was appointed the chief of the executive section of the secretariat of the Black Sea Department in Nikolaev. In 1822–1826 being the flag officer at Vice-Admiral A.S. Greig he cruised the Black Sea annually with the fleet. In 1823 he was conferred the rank of Captain-Lieutenant. In 1826–1830 he acted as the chief of staff at the commander-in-chief of the Black Sea Fleet. In 1828 he was promoted to Captain 2nd Rank for his courage during the Anapa siege and to Captain 1st Rank for destruction of the Turkish Fleet near Varna by a unit of rowing vessels commanded by him. In 1830 he was transferred to serve on the Baltic Fleet. In 1831–1837 he was a member of the Education Committee on the Fleet. In 1832 M. was promoted to Rear Admiral and appointed a member to the Admiralty Council. In 1836 he was appointed a member to the Naval General Auditing Council. In 1840 M. was conferred the rank of Vice-Admiral. In 1850 he prepared the Code of Marine Criminal Regulations; was included into the Committee on Development of the New Navy Regulations. In 1854 M. supervised construction of the rowing fleet comprising 62 gunboats; promoted to Admiral. In 1855 he was appointed to the State Council and sent to Nikolaev on a special secret mission. The "Marine Collection" published the article of V.I. Melikhov "Description of the Actions of the Black Sea Fleet in the War with Turkey of 1828–1829" with maps and plans.

Memorial Museum of Admiral S.O. Makarov – a unique museum that appeared as a result of joint efforts of teachers, employees and students of the Shipbuilding Institute in Nikolaev. It opened its doors for the first time in 1949, then there was long interruption and it was opened again only in 1970. Out of 256 exhibits 115 are the genuine things. They include some models of ships made in the study shops of the Institute, including the model of icebreaker "Ermak" which construction was initiated by well-known scientist and fleet commander S.O. Makarov and armor-clad ship "Petropavlovsk" on which Makarov was killed as a result of an explosion during the Russo-Japanese War of 1904–1905. The exposition also contains the articles, documents and books having relation to his life and activity.

Menshikov Alexander Sergeevich (1787–1869) – His Highness Prince, prominent Russian statesman and military commander, diplomat, Adjutant General (1817), Admiral (1833), Honorary Member of Petersburg Academy of Sciences (1831) and the Russian Academy of Sciences (1835). He was the great grandson of A.D. Menshikov. He started his service in 1805 as a junker in the Collegium of Foreign Affairs and till 1808 he was on the Russian missions in Berlin and London. In 1810 he became the aide-de-camp at the Chief of Staff of the Danube Army. From 1811 he was flugel aide-de-camp, quartermeister of the First Grenadier Division and took part in many battles of the 1812 Patriotic War. In 1823 he was transferred to the Foreign Minister. In 1824 M. retired from the army service. Living in his estate he studied the naval science. In 1826 he was called again to the diplomatic service and sent with the special mission to Persia. In 1827 he presented to the Emperor the draft of the naval department reform and was included into the Navy Formation Committee. In 1828 he was appointed the Chief of Naval

Staff (from 1831—Naval Headquarters) and was renamed into Rear Admiral and became a member of the Committee of Ministers. During the Russian-Turkish War of 1828–1829 he commanded the landing operation near Anapa and later the siege corps near Varna. In 1828 he actually managed the naval department. In 1830 he became a member of the State Council and in December 1831—General Governor of Finland. M. controlled the construction of the Saimenskiy Canal (between Saima Lake and Gulf of Finland). In 1836–1855 he was the Chief of the Naval Headquarters with the rights of the Naval Minister. M. mostly addressed the administrative and operating problems and paid small attention to technical improvements on the Fleet. In 1853 M. was sent on a special mission to Constantinople. From autumn 1853 M. was appointed the Commander-in-Chief by land and sea in Crimea. In early September 1854 M. did not dare to counteract the enemy landing in Crimea and left Eupatoria without struggle. His mistakes caused, to a great extent, the Russian Army defeat in the battle at Alma. On 24 October 1854 the Russia Army commanded by M. was defeated in the Inkerman battle. In February 1855 M. "due to bad health" was evicted from his post of the Commander-in-Chief in Crimea and was called back to Petersburg. In December 1855 he was appointed the General Governor of Kronshtadt. On 6 April 1856 he was dismissed, but retained his rank of Adjutant General and member of the State Council.

Menshikov A.S. (http://auction-ruseasons.ru/items_images/2_13.jpg)

"Merkuriy" (Mercury) – brig of the Russian Black Sea Fleet. It had 18 carronades and two small-caliber guns onboard. During the Russian-Turkish War of 1828–1829 "M." under command of Captain-Lieutenant A.I. Kazarskiy cruised together with frigate "Shtandart" and brig "Orfey", and on 14 May 1829 near the Bosporus they came upon the Turkish squadron (six battleships, two frigates, two corvettes, one brigantine, three tenders) that rushed in pursuit of the Russian ships. Two Ottoman battleships "Selimiye" (110 cannons) under the Kaputdan-Pasha flag and "Real-Bey" (75 cannons) under the Admiralty flag reached "M". Being unable to avoid the unequal battle A.I. Kazarskiy gathered his officers and they unanimously decided to take the battle and if there was a threat of brig capture to blow it up. The ship crew approved such decision, too. Brilliantly maneuvering with sails and oars A.I. Kazarskiy did not allow the enemy to make aimed fire. At the same time after accurate firing on the rigging A.I. Kazarskiy forced "Selimiye" to lay adrift and "Real-Bey" stopped chasing. The battle lasted for 4 h. "M." got 22 holes in the hull, 4 members of the crew were killed and 8 wounded. A.I. Kazarskiy himself had head contusion, but he did not leave the ship. For the exceptional bravery Emperor Nickolay I conferred "M." the highest award—the onboard George flag and pennant. The highest decree ordered to have always in the Black Sea Fleet the ship under the name of "M." or "Pamyat of Merkuriy" (Memory of Mercury) bearing the George flag as succession. In 1834 on the initiative of Admiral M.P. Lazarev the monument with the inscription "To Kazarskiy. Example to Posterity" was installed in Sevastopol on the Michmanskiy (Sailor) Boulevard. The monument was built on the money collected by seamen.

"Mikhail Kutuzov" – Soviet Cruiser, was founded on February 22, 1951 in Nikolaev shipyard. Launched on November 29, 1952, on August 9, 1954 on the cruiser raised Soviet Navy flag, enlisted in the Black Sea Fleet by Order of the Minister of Defense of the USSR on January 31, 1955. Length—210 m, width—23 m, height—52.5 m, deadweight—18,640 tons, speed—35 knots, crew—1,200; armament—four cannon towers with three 152.4 mm cannons, 12 (6×2) 100 mm cannons, two 5-funnel 533 mm torpedo tubes, 179 mines, etc. Paid a visit to Romania in 1955, Split (Yugoslavia) in 1956 and 1964, Durres (Albania) in 1956 and 1957 and to Varna (Bulgaria) in 1964, in Algeria in 1968. In June 1967, was in a war zone, carried out his mission to assist the armed forces of Egypt, and in the period from August 1 to December 31, 1968—Syria's armed forces. During the explosion of the battleship "Novorossiysk" in Sevastopol (29 October 1955), "Mikhail Kutuzov" was the closest ship to the battleship. Of the 35 people rescue team sent by the cruiser to assist the crew of "Novorossiysk", 27 sailors died. In 1987 the cruiser was transferred to the Navy Reserve. Excluded from the Navy on 3 July 1992 in Sebastopol. On 23–25 August 2001 relocated to Novorossiysk and joined the Novorossiisk Naval Base. On 28 July 2002, the Day of the Navy, the cruiser was opened as a museum ship in the passenger port of Novorossiysk.

"Mikhail Kutuzov" (Photo by Evgenia Kostyanaya)

Meromictic basin (lake) – a basin or a lake which has layers of water that do not mix due to density stratification of its waters. In contrary, in case of holomictic basin (lake) at least once each year there is a mixing of the surface and the deep waters. This mixing can be driven by strong cooling and wind, which creates waves and turbulence at the surface. The term "meromictic" was introduced by the Austrian Ingo Findenegg in 1935, apparently based on the older word "holomictic". The concepts and terminology used in describing meromictic lakes were essentially complete following some additions by G. Evelyn Hutchinson in 1937. Most of lakes are holomictic; in monomictic lakes the mixing occurs once per year; in dimictic lakes the mixing occurs twice a year (typically spring and autumn), and in polymictic lakes the mixing occurs several times a year. In meromictic basins and lakes water layers remain unmixed for years, decades, or centuries. Depending on the exact definition of "meromictic", the ratio between meromictic and holomictic basins and lakes is between 1:1,000 and 1:3,000. The Black Sea is the world's largest M.B. and the largest water body with anaerobic waters on the globe. Here seasonal vertical circulation spreads only to a depth of halocline. As a result, there is no oxygen in the stagnant waters below halocline. This spurs the process of anaerobic bacterial reduction of sulfates that generates hydrogen sulfide and, therefore, everything alive here, including bacteria, perishes. In the Black Sea such waters occur from the depths of 130–150 m.

Metlin Nikolay Fedorovich (1804–1884) – Russian Admiral. In 1819 being a cadet of the Naval Corps he was promoted to midshipman and in 1821 to warrant officer. In 1829 he was sent to serve on the Black Sea Fleet. From 1829 to 1836 he took part in the military operations on the Black Sea. In 1836 he was promoted to Captain-Lieutenant. In 1838 M. commanded the brig and navigated over the Archipelago, Marmara, Mediterranean and Black seas. As the Chief of staff in the squadron commanded by Vice-Admiral M.P. Lazarev he cruised near the Abkhazia coasts and was promoted to Captain 2nd Rank (1838). In 1839–1840, commanding the frigate "Brailov" he navigated near the Caucasian coast. M. took part in landing operations at the rivers of Subashi and Sochi (1839) and Tuapse (1840). In 1840 he was promoted to Captain 1st Rank. In 1841–1849 M. cruised over the Black Sea regularly. In 1849 he was conferred the rank of Rear Admiral and appointed to the Fourth Fleet Division. In 1850–1851 on the flag ship he navigated near the eastern coasts of the Black Sea. In 1851 he was appointed the acting supply chief of the Black Sea Fleet and the Black Sea ports. In 1854 he was appointed the acting Chief of Staff at the Commander-in-Chief of the Black Sea Fleet and Ports. In 1855 he was promoted to Vice-Admiral and was appointed the Head of the Naval Unit and Military Governor of Nikolaev. He was given control of the Danube Fleet. M. directly managed the quartermaster division and was assigned the rights of the Commander-in-Chief of the Black Sea Fleet. In 1856 he became a member of the Admiralty Council and in 1857—the temporary Naval Minister. In 1858 M. was promoted to Admiral. In 1860 he was appointed a member of the State Council and dismissed from the office of the Naval Minister.

MHI Experimental Division of National Academy of Sciences of Ukraine (NASU) – its objective: provision of the Marine Hydrophysical Institute with regular field studies and experimental base in the zone of recreation and major marine transport corridor near the Southern Coast of the Crimea for promoting the system of geoecological monitoring of the coastal and offshore waters; development and introduction on the basis of the marine experimental test site of new processes to control the state of environment, biota and bottom sediments; development of the scientific-and-technical base of operational oceanography and of locally-distributed interdisciplinary technical network of geoecological monitoring, including the stationary oceanographic platform, small research vessels, hydrographic and hydrometeorological stations; establishment and development of a subregional marine scientific-information center for collection and interchange of data within the network of a global system of the Black Sea observation. Constitutes a part of MHI NASU, sited at Katsiveli village. Crimea, Ukraine.

Michurin (till 1934—Vasiliko, till 1950 and after 1991—Tsarevo) – a seaside climatic resort in Bulgaria (Burgas Region), 70 km to the southeast of the city of Burgas. The climate here is warm with a mild (mean temperature in January is 3.2 °C) wet winter and very warm (mean temperature of July 23 °C) dry and sunny summer. Precipitation is 682 mm a year with the maximum in December and the minimum in August. This terrain is shielded from western winds by the Bosna upland (to 500 m high, the Strandja mountain system) which slopes are overgrown largely with deciduous forests (Oriental beech, oak, ash and others). Tender climate, warm sea and sandy beaches allow opportunities for climato- and thalassotherapy here. This resort provides treatment for the people having nonspecific diseases of breathing organs, functional disorders of the nervous system. In M. you can find the ruins of the Medieval fortification wall, several ancient buildings, and a church (1831). In the vicinity of M. there are other seaside resorts, such as Kiten (to the northwest) and Akhtopol (to the southeast). Population—5,900 (2009).

Midia – port and shipyard on the coast of the Black Sea, Romania. The Port of Midia is located approximately 20 km north of Constanta. It is one of the satellite ports of Constanta and was designed and built to serve the adjacent industrial and petrochemical facilities. The port covers 834 ha of which 234 ha is land and 600 ha is water. There are 14 berths (11 operational berths, 3 berths belong to the Constanta Shipyard) with a total length of 2.24 km. The port depths are 9 m which allows access to tankers having a 8.5 m maximum draught and 20,000 DWT. The Port of Midia is mainly used for the supply of crude oil for the nearby Petromidia Refinery.

"Miklukho-Maklai" – a research vessel of the Odessa Branch of the Institute of Biology of the Southern Seas. It was named in honor of the renowned Russian traveler and anthropologist N.N. Miklukho-Maklai. From 1961 to 1989 the ship navigated the Black Sea. The expeditions on this vessel discovered considerable changes in the Black Sea ecosystem; spreading of the water bloom zones, changes of their color and transparency; oxygen deficit in bottom layers in the shelf zone; mass death of bottom aqueous animals, degradation of the Zernov's phyllophora field and others.

Mikryukov Viktor Matveevich (1807–1875) – Russian Vice-Admiral. In 1822 he was given the rank of midshipman and in 1826—warrant officer. In 1822–1827 he sailed on different ships over B.S. In 1828 on the ship "Ioan Zlatoust" (John Chrysostom) he took part in the battles against Turkey. In 1830 on the same ship he transported the landing troops and cannons from Varna and Kovarna to the Black Sea ports and cruised near the Abkhazia coast. In 1831 he was promoted to Lieutenant. M. took part in seizure of Gelendzhik, commanded a transport ship and cruised near the Abkhazia coast. In 1832–1838 he navigated on different ships to various ports on the Black Sea and cruised near the Abkhazia coast. In 1839–1842 he commanded the lugger. In 1843 he was promoted to Captain-Lieutenant. In 1843–1845 M. cruised over the Black Sea on different ships. In 1845–1849 he commanded a steamer. In 1850 he was conferred Captain 2nd Rank. M. commanded the corvette "Orest" (1850–1851) and "Chesma" (1852–1853). Commanding the ship "Chesma" he took part in the battle at Sinop and in 1853 was promoted to Captain 1st Rank. In 1854–1855 he was assigned to the garrison of Sevastopol commanding the 1st and 2nd outposts. In 1857 he was transferred to the port of Astrakhan and after its liquidation in 1867 M. was transferred to the Black Sea Fleet and in 1868 was conferred the rank of Vice-Admiral.

Military-Historical Museum of the Black Sea Navy of the Russian Federation –
(also known as Museum of the Red Banner Black Sea Fleet) one of the oldest
museums which exposition shows the history of creation of the Black Sea Fleet in
Sevastopol. It was founded on 11 April 1869 on the initiative of the participants of
the Sevastopol Defense of 1854–1855 and on the money collected by subscription.
For organization of the museum in Saint Petersburg the Special Committee headed
by E.I. Totleben was established and in Sevastopol—the commission was headed
by Vice-Admiral P.I. Kislinskiy. Collection of donations into the museum's fund
was organized. General E.I. Totleben provided his house for the museum and its
five rooms accommodated the first exposition. On the 25th anniversary, in 1895 the
Museum moved to its own building (today this is an architectural monument). In
1895 architect Academician A.M. Kochetov designed the building especially for
this museum. The façade of the building decorated with an exquisite stone carving
has in the center of its fronton the figures "349" meaning the number of the days of
the heroic defense, as well as the decorative banners, arms and naval attributes cast
in bronze and made by sculpture B.V. Eduard. The cannons and mortars dating back
to the defense time are installed near the museum. After the 1917 October Revolution the museum collection was extended having received some historical materials related to the Fleet of the young Soviet Republic and its development during
prewar 5-year periods. During the Great Patriotic War (1941–1945) the most
valuable exhibits were evacuated to Tuapse and later to Ulianovsk, but the museum
itself continued its work and its collection was replenished with the relics taken
from the battlefields. In the years of the German occupation the Nazi destroyed
some part of the museum building. The museum was opened again for visitors only
in 1948. It possesses more than 2,500 exhibits that tell us about the struggle of
Russia for access to the Black Sea, creation of the Fleet and the Russian-Turkish
wars of the eighteenth century. A great part of the exposition is devoted to the
heroic defense of Sevastopol when the joint British-French Fleet comprising 89 battleships and 300 transport vessels started landing here its 62,000-strong army, while
the Russian troops in Crimea numbered that time no more than 33,000. The
museum displays the pictures of I.K. Aivazovskiy and I.P. Krasovskiy depicting
the events of this war. Among the exhibits there are busts and personal belongings
of the commanders of this defense—Admiral V.A. Kornilov, Admiral
P.S. Nakhimov and other military figures, surgeon N.I. Pirogov, the novel "Sevastopol Stories" by L.N. Tolstoy who took part in this defense (the book was
published in 1856). The revolutionary events on the Black Sea Fleet and its
participation in World War I are also widely represented. The museum demonstrates the sculpture of Lieutenant P.P. Shmidt who commanded the revolutionary
squadron, the letters he wrote to his son and also the models of the battleship
"Potyomkin" and cruiser "Ochakov". In the museum you can acquaint with the
materials about struggle of the Black Sea mariners for the Soviet power in the south
of the country, about creation of the Soviet Black Sea Fleet. The exposition tells
about battles of the Fleet during the Great Patriotic War (1941–1945). Among the
exhibits there are photos from the fronts, personal things, war awards, portraits of
commanders, models of ships and submarines delivering replacement troops,

ammunition and food to Sevastopol during the 250-day defense. Numerous materials tell us about revival of the Black Sea Fleet and restoration of the city in the postwar years. Near the Museum in its yard there are installed cannons and siege mortars, including captured ones, weapons of the period of the Great Patriotic War (1941–1945) and other specimens of military weaponry. Not far from the Museum one can see the Church of St. Michael the Archangel. During the first defense of the city it was the garrison church. On the outer walls of the church the marble plates are installed naming the regiments that defended Sevastopol and among them the Arkhangelsk, Moscow, Ural, Poltava, Lvov, Minsk and other regiments. The church houses one of the Museum divisions. The museum has its affiliate on the Malakhov Mound where it locates inside a defense tower.

Military-Historical Museum of the Black Sea Navy of the Russian Federation (Photo by Elena Kostyanaya)

Miskhor (meaning *"middle village"*) – a seaside climatic resort 12 km to the southwest of Yalta, Crimea, Ukraine. It is a part of Greater Yalta. It is thought that M. appeared in the Middle Ages. In the Russian literature this name was first mentioned by Academician P.S. Pallas. The territory of M. extending along the sea for 7 km is overgrown with subtropical evergreen trees and shrubs. M. has a well-equipped treatment beach made of small pebbles. The main curative factor here is subtropical climate of the Mediterranean type. M. is the warmest place on the Southern Coast of Crimea. The winter is very mild here. The mean temperature of February is 4 °C. The summer is very warm, arid and sunny. The mean temperature of July is about 25 °C. The autumn is warm and long. The bathing season lasts from June through October. The number of precipitation here is about 500 mm a year. There are about 2,300 sunshine hours a year. The greater part of M. is covered by a park representing a monument of the garden-park architecture. It was established in the late eighteenth century. Its area is over 20 ha. Natural and climatic conditions here are favorable for climate and thalasso-therapy of the diseases of breathing organs, functional disorders of the nervous and cardiovascular systems. One of the remarkable features of M. is a palace "Swallow's nest". M. merges with other resort townlets, such as Gaspra and Koreiz.

Mithridat Mountain – rises near the city of Kerch, Crimea, Ukraine. This mountain is named after the Pontus Tsar Mithridates VI Eupator (132–63 B.C.) who ruled the countries on both coasts of the Black Sea. Besieged on this mountain by the enemies he took poison. But as for many years dreading of being poisoned he took microdoses of different poisons he did not die. After this he ordered the leader of his own guardsmen to kill him.

Molochnaya – a river in the Zaporozhie Region. Its length is 197 km, the basin area is 3,450 km^2. It originates on the Priazov Upland and flows into the Molochnyi Liman of the Sea of Azov. It is meandering in its lower reaches. The average flow near the village of Terpenie is 1.52 m^3/s. In some years it may dry out and freeze. The M. basin has many ponds. The cities of Tokmak, Molochansk and Melitopol are situated on M.

Monk seal (*Monachus monachus—M. albiventris*) – like all other *Pinnipedia* they live in water and on seashores. The Black Sea seal may reach 3.5 m in length and 250 kg in weight. It is capable to develop speed to 25 km/h, dive to a depth of 300–350 m and stay under water for 25–30 min. The dorsal part of the body is colored gray-brown, while the belly is whitish from where it got its name. M.S. is not found any more in its traditional places of habitat. In the early nineteenth century it could be found on the Black Sea coast and in some places it was even hunted, but only for getting its skin and fat. The seal prefers staying near rocky coasts with numerous caves where the animals can propagate. The female after 10–11 months of pregnancy gives birth usually to one pup (most often in early spring) who for 3–4 months feeds on the mother's milk and then starts its independent life. The seals live for 30–40 years. They feed mainly on fish preferring larger species. At present M.S. is so rarely met in the Black Sea that there is a threat of its complete disappearance. It is put on the International Red Book. In Bulgaria the seal is called "Bulgarian water bear". The oceanariums in Crimea have not a single species of this seal. In 1950 M.S. was seen last near Zmeinyi Island. In the 1990s single animals were observed in the nature reserve "Danube plavni". Odd colonies are still living near Fogankent in Turkey where in 1994 two adult seals were noticed.

Monk seal (http://www.oceanology.ru/wp-content/uploads/2010/01/mediterranean-monk-seal-monachus-monachus-big.jpg)

Monton Alexander Ivanovich (1861–?) – Major-General of the Hydrographic Corps, explorer of the Black Sea. He finished the Naval Engineering College. In 1879 he started his service on the Black Sea. From 1884 he worked first in the Hydrographic Expedition of the Black Sea Fleet and then in the Separate Surveys Division of the Black Sea. In 1886 in the rank of Second Lieutenant he was engaged in topographic surveys on the southern coast of the Karkinit Bay. In 1890 he conducted the plane-table survey of the Belosarai Spit. In 1891 being the team head he undertook depth measurements near the southwestern coast of Crimea and in the Evpatoria Bay. In 1892–1894 he carried out the triangulation surveys in the Sea of Azov to the east of Mariupol and in the southwestern part of the Taganrog Bay from Eisk to Yasenskiy Spit. In 1895 M. took part in development of the geodetic control grid on the Caucasian coast from Gagra pass and further southward as far as Sukhumi. In 1896 he was appointed the commander of the floating lighthouse "Sandy Island" (Black Sea) and in 1906—the Director of the lighthouses and pilots of the White Sea. In 1912 he was again transferred to the Black Sea as deputy director of lighthouses and pilots of the Black and Azov seas. In 1913 he became a member of the Hydrographic Corps and in 1923 he was appointed the Chief of the Pilot Unit of Ubekochernaz.

Monument to sunk ships in Novorossiysk – In memory of the death of the Black Sea Fleet squadron in Tsemess Bay on 18 June 1918, a magnificent ensemble was created at the 12th km of the Sukhumi road near Novorossiysk. On the observation deck facing the bay a 12-m high granite sculpture of a kneeling, grieving sailor was erected. And in the sea a few hundred meters from the shore, mast-mounted obelisk with the flags of metal covered with colored enamel, signifying the Flag Code historical signal by the ships going down to the sea depths: "those who are perishing, but do not give up" as a reminder to future generations of the sacred duty to the Motherland. According to the Brest Treaty in 1918, signed by the Soviet Government, the Black Sea Fleet with all the infrastructure should be transmitted by the Central Powers (Germany, Austria-Hungary, the Ottoman Empire, the Bulgarian Empire). In connection with the occupation of the German-Austrian forces of Ukraine and Crimea after the signing of the Brest-Litovsk Treaty Soviet Government gave an indication of the relocation of the Black Sea Fleet from Sevastopol to Novorossiysk. In early May 1918 18 warships moved to Novorossiysk two battleships ("Svobodnaya Rossiya" and "Volya"), ten destroyers of the "Novik" type, and six destroyers. Because of the threat of capture of Novorossiysk by German invaders, which appeared on the Taman Peninsula, a decision on the issue of the Black Sea Fleet ships had to be taken immediately. At the request of V.I. Lenin, Chief of the Naval General Staff of the EA Berens has prepared a report on the status of the Black Sea Fleet. This document has undergone a thorough discussion of the Supreme Military Council of the Russian Republic. Concise resolution of V.I. Lenin on the need of sinking of the Black Sea Fleet in Novorossiysk in 1918: "In view of the hopelessness of the situation proved by the highest military authorities, the Navy must be destroyed immediately. Chairman of the Council of People's Commissars, V. Ulyanov (Lenin)."

Monument to sunk ships in Novorossiysk (http://shturman-tof.ru/Morskay/morskie_pamytniki/02_chf_satopl_2novorossiysk.htm)

Monument to sunk ships in Sevastopol – was installed in Sevastopol (Crimea, Ukraine) near the Primorsky boulevard on an artificial island 10 m from the shore. It represents a column with the Corinthian capital crowned with an eagle with outspread wings keeping a laurel wreath. The monument was constructed in 1905 by the 50th anniversary of the Sevastopol defense. On marble plate there was etched the phrase "In memory of the ships sank in 1854–1855 to obstruct the entrance on the roadstead". The author of the monument was sculptor A.G. Adamson from Estonia and architect V.A. Feldman. This is a kind of the Sevastopol emblem.

Monument to sunk ships in Sevastopol (Photo by Andrey Kostianoy)

Mordvinov Nikolay Semenovich (1754–1845) – Russian Admiral. He started his service in 1766 in the rank of midshipman. In the period 1766–1769 he navigated a lot over the Baltic Sea and was promoted to warrant officer (1768). In 1769–1770 he commanded the regal yacht "Schastie" (Happiness). In 1770 he was appointed Flugel Aide-de-Camp of Admiral S.I. Mordvinov, in 1771—General Aid-de-Camp of the Major rank to Admiral Charles Knowles. In 1774 he was promoted to Captain-Lieutenant. M. was sent to Britain to serve as a volunteer on the British vessels. In 1774–1777 he cruised near the American coasts. In 1781 he was promoted to Captain 1st Rank and appointed the commander to the ship "Saint George the Victorous". In 1782–1784 commanding the ship "Tsar Konstantin" in the squadron of Admiral V.Ya. Chichagov he sailed as far as Livorno and back. In 1783 he was promoted to Captain 2nd Rank. In 1785 he was appointed the senior member of the Black Sea Admiralty Board, went to Kherson and took the office of the Black Sea Commander. In 1787 he was conferred the rank of Rear Admiral. In the war with Turkey, commanding the Liman Squadron of eight ships he defended the Dnieper mouth from the enemies. In 1788 he was in charge of preparation of ships and rowing ships for the military campaign. In 1789 he was dismissed and in 1790 was recalled to service in the same rank. In 1792 he was promoted to Vice-Admiral and appointed the Chairman of the Black Sea Admiralty Board in Nikolaev and Commander of the Black Sea Fleet and Ports. In 1797 he was given the rank of Admiral and in 1798 he was called from Nikolaev to the capital. In 1799 he was dismissed. In 1801 M. was made President of the Admiralty Collegiums and had to attend the meetings of the Board established at the Tsar Court. On 8 September 1802 he was appointed the Navy Minister of the Russian Empire, but took this office only till 28 December.

Mordvinov N.S. (http://nikolaev-moscow.at.ua/_ph/51/2/261474337.jpg)

Morozova-Vodyanitskaya Nina Vasilievna (1893–1954) – Soviet biologist. She graduated in 1915 from the Physico-Mathematical Faculty of the Kharkov Higher Female Courses specializing as algologist-hydrobiologist. From 1915 to 1921 she was the assistant at the Botany Chair of the Kharkov Higher Female Courses. From 1919 to 1922 she was a teacher at a secondary school. She took part in establishment of the Biological Station in Novorossiysk and conducted researches of marine macrophytes there. From 1922 to 1931 she studied the species composition of bottom algae in the Novorossiysk Harbor (Tsemess Bay) and conducted detailed research of algae ecology. In 1923 she published her first work about the "*pediastrum*" gender systematics where she showed that the law of homologous series evolved by N.I. Vavilov was fully applicable to this gender of algae. In 1934 she started studies of phytoplankton in the Black Sea. In 1936 she published her works about photosynthesis of benthos algae in the Karkinit Bay and near Karadag. In the period 1939–1941 she was Professor of the Hydrobiology Chair at the State University of Rostov and was at the same time the senior scientist at the Biological Station in Novorossiysk. During the Great Patriotic War (1941–1945) she worked at the Institute of Evolutionary Morphology of the USSR Academy of Sciences. In 1944 she returned to the Black Sea studies and worked at the Sevastopol Biological Station of USSR AS and from January 1945—at the Novorossiysk Biological Station named after V.M. Arnoldi. She resumed her studies of phytoplankton in the Black Sea and received a full picture of distribution of the phytoplankton population and biomass in the water mass of the Black Sea in its different regions. More detailed investigations were conducted in the Sevastopol Bay. In 1954 she published her basic work "Phytoplankton of the Black Sea" for which she was awarded the Order of Lenin.

Morskoe Village – formerly Kapsikhor, located 16 km westward of Sudak on the Black Sea coast on the Sudak-Alushta route in the Kapsikhor Valley, Crimea, Ukraine. The name "*Kapsikhor*" meaning "scorched land" most likely originated from the hills surrounding the valley that even in spring are covered with scorched grass. After Crimea was incorporated into Russia the Kapsikhor Valley was passed to V.S. Popov, the head of the G. Potyomkin Chancellery (in 1791 Potyomkin died "on hands" of Popov and his niece.) The resort history of this place is long. The patients were treated here with sun and sandy baths. In 1910 the colony of country houses, the so-called "dachas", appeared in Morskoe. Writer Leonid Andreev and famous Russian bass Fedor Shalyapin lived here. In 1981 the legendary Russian music group "Kino" of Victor Tsoy was organized here. At present this is the well-known resort with many rest houses and a children camp.

Mosquito fish (*Gambusia affinis*) – fish of the *Poeciliidae* family. The length of male species—to 4 cm, of female species—to 7.5 cm. Their place of origin—North America. They have adapted in many regions, including in the south of Ukraine, on the Caucasus, in Central Asia where they are used for malaria control (it eats the mosquito larvae). M.F. was imported for malaria mosquito control first to Italy and from there in 1925 to Abkhazia (Sukhumi) and from there in the 1930s to Uzbekistan. They naturalized practically in all plain water bodies in the Aral Sea basin no matter how different they were. It propagated well and became the mass small fish both in aryks and large rivers. This live-bearing fish reaches its maturity already at the age of 3–4 months having by this time a body length of 2 cm. The reproduction period lasts from March through November when the water temperature is no lower than 14 °C. During a year one female is capable to produce to seven generations. They eat wigglers of mosquito, including malaria, adult insects, algae, cyclops, copepods and other small organisms. It has no commercial significance, but it is very effective in malaria control.

Mud salse – a small-size mud volcano. Its form depends on the density of the erupted mud. If the mud is watery the cone is not formed because mud flows down over the ground surface. The gas released by M.S. consists mostly of gaseous hydrocarbons (methane is prevailing), but also contains carbon dioxide (CO_2) and at times carbon oxide (CO) and nitrogen (N_2), but in small quantities. The salse waters and mud contains iodine (I) and bromine (Br). The gas composition in salses is not the same everywhere. M.S. are found on the territory of Crimea (Ukraine), Taman Peninsula in Krasnodar Territory (Russia), in Azerbaijan and Turkmenistan.

Mud volcano – a geological formation that erupts, permanently or periodically, mud and gases often mixed with water and oil to the ground surface. They are found mainly in the oil-bearing and volcanic areas. M.V. on the Taman Peninsula are the only live volcanoes on the Caucasus. Among them—"Tizdar", "Plevak" on the shores of the Sea of Azov; "Karabetova Hill", "Shapursky" in Taman.

Mud volcano on the Sea of Azov coast (http://www.azovskoe.com/image/dostoprim/dostoprim1. jpg)

Munnich Christofor Antonovich (Burchard Christoph Graf von Munnich) (1683–1767) – Russian military and state figure, General-Field Marshall (1732). Originally he was a German soldier-engineer. Prior to 1735 he supervised military-engineering works in Russia, took highest military and administrative positions. The best features of M. as a military commander were revealed during the Russian-Turkish War of 1735–1739 when he commanded the troops in Crimea and Bessarabia. At the same time he was the military governor of Belgorod and Kiev. In 1736 he seized the fortifications at Perekop and occupied Bakhchisaray.

However, the shortage of water, food and forage supplies forced him to leave Crimea. In 1737 he took by storm the fortress of Ochakov and in 1738 he tried, but without success, to break through the Turkish-Tatar positions on the Dniester. On 17 August 1739 he defeated the army of Crimean Khan Veli-Pasha in the battle at Stavuchany and occupied the fortress of Khotin and nearly the whole Moldavia. In his military deeds M. showed himself an energetic, resolute, efficient and brave man who did his best to realize everything he contemplated. He initiated some reforms in the Russian Army: establishment of the first cavalry guard corps, assignment of new staffs to the field and garrison units for the war and peace time, organization of 20 detachments for protection of the southern borders of Russia from the raids of Crimean Tatars having increased the number of fortresses from 31 to 82 and others. He was buried in his estate, in the village of Lunia near Tartu, Estonia. For his service he was awarded the orders and golden arms. He was the author of the book "Notes of Field Marshall Munnich" (1774).

Burchard Christoph Graf von Münnich Rußisch-Kaiserlicher General-Feld-Marschal, der Heil. Apostels Andreä, der Heil. Alexander Newsky und der weißen Adler Ordens Ritter.

Munnich   Ch.A.   (http://upload.wikimedia.org/wikipedia/commons/7/71/Stenglin_M%C3%BCnnich_engraving_after_Buchholtz_1760s.jpg)

Museum of shipbuilding and fleet history – a branch of the regional museum in Nikolaev, Ukraine. Its exposition demonstrates development of the shipbuilding industry in the south. It was opened in 1978. The museum possesses about 3,000 items and documents placed in 14 halls. They show how in the late eighteenth century the shipbuilding was taking shape. Among exhibits there are documents and models of various ships, including the model of the Black Sea first cargo-passenger paddle steamer "Odessa", the materials about the Crimean War of 1853–1856, account construction of metal ships, armored and other ships, the first submarines. The models of vessels that took part in the Revolution of 1905—battleship "Potyomkin" and torpedo boat No. 267 found their place in the museum. It shows the work of shipbuilders of Nikolaev during the Civil War and the Great Patriotic War, development of the shipbuilding from the first days to the present. The museum also displays the models of transportation and fishing vessels built in the south of the country. The exposition has its open-air part—in front of the museum one can see cannons, mines, torpedoes and other military equipment removed from ships.

Museum of the Marine Fleet of Ukraine – one of the largest museums on the Black Sea. It shows the most important stages in the fleet history, revolutionary and combat traditions of the seamen, their participation in the Great Patriotic War of 1941–1945. It is located in the building of the former British Club built in 1842 in Odessa, Ukraine. The museum was opened in 1965 in Odessa. It demonstrates over 100,000 exhibits among which there is a large collection of models of ships and vessels of the Russian and Soviet fleet, including the model of the flagship of Peter I "Ingermanland", the model of the steamer "Diana" that was commanded in 1899–1903 by P.P. Shmidt, the model of the frigate "Arkhimed"—the first Russian ship with a propeller screw. The museum displays the "Marine Code" edited by Peter I and published in 1720; medals commemorating the most eminent historical events on the sea; navigation instruments of the Peter epoch, such as sundial, compass, astrolabe and others. The museum demonstrates the engine telegraph from the icebreaker "Ermak". The history of the whale hunting is represented by the models of whaling ships and whaling processing floating bases, a large diorama "Whale hunting", the whaling gun and fishing equipment. A small hall reconstitutes the command bridge with all its facilities: the chart room with modern navigation instruments, the wireless room and the pilot room that contains compass, radar, gyropilot and other devices. Through the glass of this hall you can see the diorama—Odessa port at night with the Vorontsov Lighthouse and marine terminal with the lights of the city in the background.

Myakishev Konstantin Andreevich (1834–1898) – Russian Major-General of the Fleet Pilots Corps, hydrographic specialist, geodesist and astronomer. In 1845 he entered the First Pilot Semi-Crew (later it was renamed into Naval Technical College). After graduation from the College in 1854 he was given the rank of ensign of the Pilots Corps of the Baltic Fleet and transferred to the Officer Classes of the Naval Cadet Corps. In 1858 after finishing these Classes he was appointed the Deputy Chief of the Caspian hydrographic expedition and was in charge of its astronomical unit. Under guidance of well-known hydrographer N.A. Ivashintsov he took part in the works on preparation of maps and layouts of the Caspian Sea. From 1867 he was the astronomer assistant at the Naval Observatory in Kronstadt.

As an experienced geodesist and astronomer M. for 8 years carried out works on defining the geographical position of the Black Sea ports in Russia and Turkey. During this time he fixed the position of 38 astronomical points, established the geodetic connection of Nikolaev with the ports of Sevastopol, Evpatoria, Yalta, Feodosiya, Kerch, Mariupol, Taganrog, Perekop and Kherson. The determined coordinates laid the basis for all further survey works on the Black and Azov seas. In 1875 as a work supervisor he took part in determination of altitudes over the sea level of the most important geodetic points in Crimea and on the Caucasus. From 1880 to 1884 he headed the works on geodetic connection of forts on the Kronstadt road and on determination of geographical coordinates of the most important points in the Onega Bay in the White Sea. In 1885–1891 he headed a Special Survey of the Black Sea. In 1891 he resigned.

Myrtle (*Myrtaceae*) – the family of woody, mostly evergreen plants. It incorporates 75 genders (about 3,500 varieties). They largely grow in America, Australia, on the Polynesia Islands, on the Mediterranean. Eucalyptus, myrtle and other varieties are also widely cultivated on the Black Sea coast of the Caucasus. M. gives valuable wood, spices (clover trees). Some M. varieties are used for draining of terrains (eucalyptus), some others as decorative plants (myrtle).

Myskhako – a cape in the Black Sea and a village near Novorossiysk, Russia. Population—8,000 (2010). During the Great Patriotic War the Soviet landing troops captured in 1943 the base area near M. that was later called "Malaya Zemlya". The climate is subtropical Mediterranean. The average water temperature in summer is 20 °C, in winter +6 °C. Sunny days—230 a year. Summer is hot, the average temperature is +28 °C, rainfall is low. The swimming season begins in June and lasts until October. Winter is mild, with an average temperature of +3 °C. In winter predominantly north-easterly winds blow, the so-called "nord-ost". Wind gusts reached 120 km/h or more. Beach near Myskhako is rocky cliffs, sometimes gravel, sand. Since 1869 there is a development of viticulture. At the village locates one of the oldest wine-producing companies in Russia—JSC "Myskhako". Season grape harvest begins in late August and continues through November. Very interesting tour and wine tasting. From 2005 to the end of November M. hosts the annual holiday of new wine "Beaujolais Nouveau." Also in the village is developing tourist industry.

Mzymta – a river in the Krasnodar Territory, Russia. Its length is 82 km, basin area is 855 km^2. It originates on the southern slope of the Greater Caucasus Range, runs through the Kardyvach Lake and flows into the Black Sea near Adler. A rafting river. M. valley is very picturesque. The Krasnopolyanskiy hydropower station and the settlement Krasnaya Polyana locate on M.

N

Nagatkin Ivan Ivanovich (?–?) – a marine team leader. In 1751 he graduated from the Naval Academy. In 1754–1779 N. served on the ships and in land-based units of the Baltic Fleet. In 1770 being Captain 1st Rank he headed the first Russian expedition on reconnaissance and inventory of the Dnieper, Dnestr and Danube rivers.

Nakhimov Pavel Stepanovich (1802–1855) – Russian Fleet commander, Admiral (1855). In 1818 he finished the Naval Cadet Corps and till 1821 he served on the Baltic Fleet. In 1822–1825 he made the round-the-globe voyage on the ship commanded by M.P. Lazarev. In 1827 he took part in the battle at Navarino that ended in total defeat of the Turkish-Egyptian Fleet. During the Russian-Turkish War of 1828–1829 he commanded the corvette in the siege of the Dardanelles and after returning to Kronshtadt he was appointed the commander of the frigate "Pallada". In 1834 he was sent to the Black Sea as a commander of the battleship "Silistria" on which he cruised near the Caucasian coasts. In 1845 in the rank of Rear Admiral Nakhimov commanded the ship brigade and from 1852 in the rank of Vice-Admiral—the ship division. During the Crimean War of 1853–1856 he commanded the squadron of the Black Sea Fleet. In November 1853 the Nakhimov's squadron blocked the main forces of the Turkish Fleet in the Sinop Bay (Turkey) and annihilated the ships and the landing troops, thus, preventing seizure of the Caucasus. In 1854–1856 Nakhimov successfully headed the heroic defense of Sevastopol. He was the squadron commander and from February 1855— the commander of the Sevastopol Port and the Military Governor of Nikolaev. He was in charge of formation of marine battalions in Sevastopol, construction of batteries, formation and training of reserve forces, military actions along the main strike, took part in person in repelling assaults, focused much attention on the Fleet—army interaction, efficient use of artillery, construction of engineering structures, medical and logistic support. On 10 July 1855 he was fatally wounded on the Malakhov Kurgan (Mound) and later the memorial plaque was installed here. N. was buried in Sevastopol in the Vladimir Cathedral near other outstanding Russian Admirals, such as M.P. Lazarev, V.I. Istomin, V.A. Kornilov. The

S.R. Grinevetsky et al., *The Black Sea Encyclopedia*,
DOI 10.1007/978-3-642-55227-4_15, © Springer-Verlag Berlin Heidelberg 2015

monument was opened to Nakhimov in Sevastopol. In 1944 the Nakhimov Order (with two degrees) and the Nakhimov Medal were established. His name was given to the cruisers of the Russian and Soviet Navy, to the Black Sea liner and the naval college in Saint-Petersburg. The monument to P.S. Nakhimov was constructed on the central square in Sevastopol by the design of Academician N.V. Tomskiy. This monument reminds of the other Nakhimov monument constructed by I.N. Shreder by the design of artist A.A. Bildering that was opened in 1898 commemorating the 45th anniversary of the victory at Sinop and that was destroyed in 1928 on the pretext that it was a monument to the "tsar's admiral". In 1932 on its pedestal the bronze monument to Lenin was put that during the German occupation was taken away to Germany by the Nazi for remelting. The monument to P.S. Nakhimov as we know it now was installed in 1959.

Nakhimov P.S. (http://ote4estvo.ru/uploads/1266845967_artlib_gallery-94868-b.jpg)

Monument to P.S. Nakhimov (http://commons.wikimedia.org/wiki/File:SevaNahimov.jpg?uselang=ru)

Nassau-Siegen Karl Genrikh Nikolay Otton (Charles Henry of Nassau-Siegen, 1743–1808) – Hessen Prince, Admiral of the Russian Fleet (1790). At the age of 15 he went to serve in the French Army and took part in the Seven Year War (1756–1763). In 1786 he moved to Russia and started his service in the rank of Rear-Admiral. He made friends with G.A. Potyomkin and was presented to Russian Empress Catherine II. He took part in preparation of her visit to Crimea (1787). In 1788 he was sent to the Black Sea Fleet and participated in the Russian-Turkish War of 1787–1891. Commanding the Dnieper Row Fleet in June 1788 near the Kinburn Spit at Ochakov he won victory over the Turkish Fleet. In 1789 he was appointed the commander of the Baltic Rowing Fleet. He took part in the Russian-Swedish War (1788–1790) and won the First Battle of Svensksund against the Swedes in 1789 and again in 1790 during the Battle of Bjorkosund. But on 10 July 1790 Nassau-Siegen suffered a disastrous defeat in the Second Battle of Svensksund, in which he lost one third of his fleet and 7,400 men. After the Peace Treaty with Sweden signed in 1790, despite this defeat, the Empress marked his bravery by conferring him the rank of Admiral. He never recovered from the defeat and left Russia in 1792 for the Rhine, to fight the French Revolution. But after the Peace of Amiens in 1802, he returned to France, where he solicited without success a position in Napoleon's Army. He returned to Russia to die in his estate in Tynna in 1808.

National Center for Space Facilities Control and Testing of Ukraine – it is located near Evpatoria, Crimea. Its establishment and many achievements are connected with the Soviet period. It has the quantum-optical system "Sazhen" that first measured the distance from the Earth to the Moon, the unique radiotelescope with the antenna 70 m in diameter (it was constructed in the period 1973–1978) and operating range 10 bill km, the antenna complexes for spaceships control.

"Natural Environment of the Black Sea" – the information bulletin of the Black Sea Ecological Program.

Naval orders and medals (Russia) – decorations awarded to the seamen of the Russian Navy for special exploits. They have their own statute (regulations) specifying who and for what exploits may be awarded, description of the award, special rules of awarding an order or medal.

Medal "For Bravery at Seizure of Ochakov" (1788).

Medal "For Turkish War" (1829). It has a form of a hexagonal cross on the Georgian ribbon. It was intended for officers and sailors who participated in the Russian-Turkish War of 1828–1829.

Medal "For Sevastopol Defense" (1855) on the Georgian ribbon that was awarded to all seamen participated in defense of the city from 13 September 1854 to 28 August 1855 and also citizens of Sevastopol, including women.

Medal "In Memory of War of 1853–1856" (1856). The Fleet officers who participated in the battle at Sinop, military actions on Kamchatka and in the Port of Petropavlovsk were awarded the medal on the Georgian ribbon; those who were not directly involved in battles, but who stayed on the territory with the martial law regime—on the Andreevsky ribbon, other Naval officers—on the Vladimir ribbon.

Medal "In Memory of the Russian-Turkish War of 1877–1878" (1878). The officers and lower-rank seamen who participated in the battles on the Black Sea were awarded the medal of light bronze, while those who stayed on the territory with the martial law regime—of dark bronze.

Medal "For the 50th Anniversary of the Sevastopol Defense" (1904) on the Georgian ribbon. It was awarded to the veterans of the city defense.

"Order of Ushakov" of the 2nd degree (1944)—it was awarded to the Naval officers for successful development and realization of naval operations resulted in victory over the enemy superior in number.

"Order of Nakhimov" of the 2nd degree (1944) was awarded to the Naval officers for successful development and realization of naval operations that involved rebuff to the enemy's attacks, causing the enemy considerable damage and saving many of its own forces.

"Medal of Ushakov" (1944) on a ribbon with dark-blue, white and light-blue strips. It was awarded to the troopers, master sergeants and sergeants of the Navy for bravery in the battles with the USSR enemies on sea.

"Medal of Nakhimov" (1944) on a white-blue ribbon. It was awarded to the troopers, master sergeants and sergeants of the Navy and also to the persons who did not serve in the Navy, but who at risk of their lives facilitated the fulfillment of military tasks of ships and fleet units.

Medals for exploits during the Great Patriotic War, defense, liberation and seizure of cities and territories—"For Odessa Defense" (22.12.1942), "For Sevastopol Defense" (22.12.1942).

Navy Museum in Varna (Bulgaria) – The exposition tells about the history of the national seafaring. The basis of this collection was made by the Russian naval officers being on the Bulgarian service. It was opened in 1923 as "Marine Museum". In 1955 it was passed to subordination of the Ministry of National Defense of Bulgaria and received its present name. The exhibits of the historical division tell about the Russian-Ottoman Wars of the eighteenth to nineteenth centuries, about development of the Bulgarian Naval Fleet from 1879. The Museum shows rather comprehensively the participation of the Bulgarian Navy in World Wars I and II. The exposition shows the development of the Bulgarian Navy and shipbuilding at the present stage. It displays the smaller ships for oceanographic research, in particular, bathyscaph "Shelf". There is a collection comprising models of ships of various classes and types. Among other exhibits there is a bow figure of the sail boat of the nineteenth century, the lighthouse optics, anchors, mines and others. In the park near the Museum the famous Bulgarian torpedo boat "Derzkiy" and artillery weapons are installed.

Neapolis – Scythian Neapolis, the capital of the late Scythian state that located to the southeast of Simferopol, Crimea. It was on the peak of prosperity in the second century B.C. when Iranian Tsar Skilur ruled over the Crimean steppes. Descriptions of N. were found in the poetic hymn devoted to chieftain Diophantus. The story about the long forgotten city was the main idea of the diaries of Russian botanist Kh. Steven that had been never published. He visited N. ruins in 1827. In the late 1850s excavations were conducted here by A.S. Uvarov and in 1890 by archeologist and Orientalist Professor N.N. Veselovskiy. They uncovered a structure built on a hill which configuration reminded of an isosceles triangle about 1 km long. The fortress was protected on two sides by steep natural rocks and on the third side—by a wall 600 shags (strides) long. N. borders on the Salgir River valley and the deep gully called "Sobachia balka" (Dog's Gully). In the third century A.D. Neapolis was seized and destroyed halfway by the Goths. The fortress ruins were partially restored in the early fourth century by the order of Byzantine Emperor Justinian the Great. In the Golden Horde time impoverished N. turned into Tatar fort Kermenchik ("small fortress"). Beginning from the fifteenth century one small settlement Ak-Mechet survived in place of Neapolis.

Neberdzhay Water Resrvoir – built on the river Neberdzhay (Lipki) of 8.2 km from its source. Situated on the northern slope of Markotkhskiy Range, 10 km north-east from Novorossiysk and serves as a main source of water for the city. Surface—0.75 km^2, volume—0.0081 km^3. The lake is about 1 km long and 500 m wide. Water supply to Novorossiysk can be up to 26,700 m^3/day. The reservoir is equipped with artificial water retaining structure in the form of a dam of 250 m long. Lake level is at 182 m above sea level. First stage of the reservoir was built in 1959, the second stage—in 1963. In 1999, the project of reconstruction of the reservoir was announced, but no significant work under the project has been undertaken. In 2002, the reservoir was filled to 100 %, which was of concern to some local residents about possible flooding of neighboring villages. In autumn 2011, it was reported on plans for large-scale reconstruction of the reservoir.

Nebug – small village on the coast of the Northeastern Black Sea, 10 km north-westward of Tuapse, Russia. The village was founded in 1864 as a Cossack village Nebugskaya of the Shapsugskiy shore battalion. In that year, 50 of the Cossacks and their families from the ship landed and settled a mile from the beach. Today this is a rereation area with several resorts, hotels, dolphinarium, the first aqua park in Russia, beach. Population—3,600 (2002).

Nemirov Congress (1737) – peace negotiations between Russia and Austria, on the one part, and Turkey, on the other, during the Russian-Turkish War of 1735–1739. It was held on 16 (27) August—11 (22) November in the Ukrainian area of Nemirov (presently an urban settlement in the Vinnitsa Region). The Russian delegation claimed from Turkey the annulment of all previous Russian-Turkish treaties and their replacement with new ones; passing of the Kuban, Crimea and the lands between the Don and Danube river mouths to Russia in order to ensure peace in the Russian state; granting independence to Moldavia and Wallachia under the Russian protectorate to create a barrier protecting Russia and Turkey against mutual conflicts; recognition by Turkey of the Emperor title of the Russian tsars; free navigation of Russian merchant ships over the Black Sea; fulfillment by Turkey with good faith the provisions of the Karlowitz Congress (1698–1699) concerning immunity of Rzech Pospolita. Turkey waived these claims. Austria did not support them, too, and in its turn it laid claims to a part of Moldavia, Wallachia, Serbia and Bosnia. Russia was not going to support the enormous and ungrounded claims of Austria. Using the Austrian-Russian contradictions Turkey already prior to official closing of the Congress moved its troops against Austrian forces on the Balkans. The Congress closed without coming to terms. No less important here was the role of France that did her best to keep the Turkish Government from serious conces-sions to the allies pushing it to war continuation. In 1739 Austria after being defeated despite the success of the Russian troops signed on 21 August (1 September) 1739 the separate treaty with Turkey. In 1739 the Russian Army conquered Khotin, Yassy and liberated Moldavia. After Austria withdrew from the war Russia signed with Turkey the Treaty of Belgrade in 1739.

Nemitz Alexander Vasilievich (1879–1967) – Russian Vice-Admiral. In 1900 he finished the Naval Cadet Corps and in the rank of warrant officer was sent to the Black Sea Fleet. After suppression of the revolt on the battleship "Potyomkin" he refused to fire into seamen from the transport ship "Prut" and as a lawyer defended other participants of the revolt and succeeded to mitigate the punishment. From 1910 he lectured at the Marine Academy in Nikolaev and at the same time studied in it. In 1913–1914 he was the Chief of the Second (Black Sea) Operating Unit of the Naval General Staff and from 1914—field officer of the Naval Department. He received an appointment to the operating fleet. In 1915 he was appointed to the Black Sea Fleet to command the gunboat "Donets", in 1916—the commander of destroyers division, in 1917—the commander of the mine division. During the service on the Black Sea Fleet in 1916–1917 N. took part in joint actions of the Batumi ship unit and the Primorskiy troops of the Caucasian Front, in landing operations on the Trapezund direction and in raids on the Anatolian coast and enemy communications. From August to December 1917 he commanded the Black Sea Fleet. N. was promoted to Rear Admiral. Regardless of the revolutionary ideas spreading on the fleet he tried to maintain the fleet combat capacity and to continue combat operations. He went to sea for the last time as the commander on 19–24 October 1917 with 12 ships. During this voyage the destroyers in the Igneada Bay wrecked the Turkish torpedo boat, damaged two transport ships, forced three minesweepers to come ashore. In March 1919 he voluntary went to serve to the Red Army. He was appointed the Chief of the Navy of the Odessa Military District. During the breakthrough of a group of troops of the Southern Front from entrapment he was the Chief of Staff in this group and although he was wounded he continued his duties. From February 1920 to November 1921 N. commanded the Navy of the young Soviet Republic. He took part in the defeat of the Vrangel Army in Crimea. Before forcing the White Army from Crimea he succeeded formation of the single Navy of the Black and Azov seas subordinated directly to the commander of the Republic's Navy. These forces were needed both for defense of the bays and ports and likewise for attacks. In particular, he initiated creation of the Naval Expedition Division and the transport fleet for its transfer. In August 1920 this division was used as the counter-landing troops for total defeat of the Vrangel forces landed on Taman and in September during defense of Mariupol, the main base of the Azov Navy. N. was one of those who organized interaction of the naval forces with the troops of the Southern Front during liberation of Crimea. After capture of Crimea when there were certain difficulties with the supply of the 100,000-strong army N. led three divisions over the ice to Kuban. In 1921 feeling unwell after receiving wounds he applied for quitting the office of the commander of the Republic's Navy. In 1921–1940 he was an envoy on special missions at the Chairman of the Revolutionary Military Council and deputy inspector of the Navy Council of the Red Army. In 1936 he was conferred the rank of Flagman 1st Rank, in 1941—Vice-Admiral. During the Great Patriotic War (1941–1945) N. published the works "Military and Geographical Description of the Theatre of Military Actions in the Lower Danube" (1942) and "Waterway of the Danube" (1944).

They helped the Danube Fleet to operate more efficiently. In 1940–1947 he was Professor at the Chair of Strategy and Tactics in the Naval Academy and in the Air Forces Academy. In March 1947 he resigned. After this he went to Sevastopol where he worked first in the museum and later conducted hydgoraphic researches on the Black Sea Fleet.

Nemitz A.V. (http://upload.wikimedia.org/wikipedia/commons/9/9e/Photo_Alex_V_Nemitz.png?uselang=ru)

Nesebar – until 1934 "Mesembria". This is the city-museum and seaside climatic resort in Bulgaria (Burgas Region) located to the northeast of the resort area of Pomorie and 39 km from Burgas. It is situated on the Black Sea coast on the small rocky peninsula of Nesebar, 4 km from the resort "Sunny Beach". One of the most popular tourist attractions in Bulgaria. The climate here is Mediterranean. The winter is mild, practically without snow. The mean temperature of January is 2.4 °C. The summer is very warm and sunny; the mean temperature of July is 23 °C. Precipitation is 420 mm a year. In the north and west the resort is protected from cold winds with the outspurs of the Stara-Planina Mountains. Tender climate, warm sea and very good fine-sand beaches allow opportunities for climato- and thalassotherapy. The sanatoriums and rest houses are found here. This is one of the most ancient cities on the Black Sea coast. It was founded by the Greeks who came here about 510 B.C. from the settlement known as Megara. This Greek colony was built in place of the Thracean town of Menas (Meneapol) that existed from the early 1st millennium B.C. The remains from the Hellenistic period include the ruins of the fortification walls and towers of the Acropolis, Theatre and the Temple of

Apollo. The town fell under the Roman rule in 72 B.C. In the early first century great Roman poet Ovid on his way to exile to Tomy (presently Constanta in Romania) visited this town and wrote that Mesembria had two ports. By the end of the fourth century Mesembria became the Byzantine fortress. It was developing as a trade intermediary between Thrace on which land it was founded and the Greek world lying on the other side of Proponide. In the years of prosperity the temples of Dionysus, Zeus, Hera, Asclepius and Apollo were constructed. Bronze and silver coins were minted in the city. In the early ninth century the town was conquered by Bulgarian Khan Krum who gave the town a new name—N. In the ninth century the town was incorporated into the Bulgarian Empire. In 1366 it was seized and ravaged by the Crusaders and after this sold to Byzantine. Only after restoration of the Second Bulgarian Empire it became Bulgarian again. In 1453 the town was captured by the Turks and remained under their control till liberation of Bulgaria from the Ottoman rule. The greater part of the Antique town was destroyed by the sea. Once the territory of the peninsula was 40 ha and now it is only 24 ha. The new quarters of the town were built on the mainland. In 1956 Nesebar was given the status of a city-museum. In 1958–1959 the resort complex "Sunny Beach" was constructed northward of Nesebar. Its abundance of historic buildings prompted UNESCO to include Nesebar in its list of World Heritage Sites in 1983. Population—11,600 (2009).

Nesebar   (http://4.bp.blogspot.com/-k9nhn238WA8/TiVVDG-slrI/AAAAAAAACbE/
HsMrS5PBPYA/s640/nessebar+22.jpg)

Nesebar Peninsula – locates on the Black Sea coast, Bulgaria. It is 850 m long and 300 m wide. The peninsula is rocky, with steep shores. A narrow isthmus 400 m long strengthened with an embankment is connecting the peninsula with the mainland. A not high upland comes to the isthmus in the west. The town of Nesebar locates in the north of the peninsula.

Nestinarstvo – a folk ritual performed in the southern Prichenomorie of Bulgaria. It involves a barefoot dance on smouldering embers. Bulgarian ethnographers think that it took origin in the ritual dances of the Thraceans performed in honor of Dyonisus—the god of wine and festivity. N. is performed in the villages of Brodilovo, Kosti and Balgari. Every year on 21–22 May—the day of St. Konstantin and St. Helen the festivities are organized that are crowned with the barefoot fire-dancing.

Neuston – the collective term for the organisms that populate a thin surface layer of sea water less than 5 cm thick. They are very important for the life of seas and oceans. N. was discovered and studied in the Odessa Branch of the Institute of the Southern Seas. Here Yu.P. Zaitsev and others formulated the basic concept of the science on neuston.

New Rome – see Istanbul.

Nightingale Florence (1820–1910) – British nurse who became the first military-field nurse, an organizer and chief of a nurse team during the Crimean War, a public figure. From the age of 17 she devoted her life to caring after the sick and wounded. In 1854 on the proposal of the British Ministry of War N. and a staff of 38 women volunteer nurses were sent to Turkey where they organized the on-field hospital for the wounded and sick soldiers of the British Army. As a result, the death rate of the wounded dropped from 44 to 2.2 %. In May-June 1855 she went to Balaklava for inspection of all British hospitals. In 1860 in London N. opened the world's first School for Nurses and till 1872 she was the expert of the British Army in medical care of the sick and wounded. She wrote a number of books about nursing the sick and wounded and some of them were translated into Russian. After her death the League of International Red Cross in 1912 established the Nightingale Medal for nurses that is conferred once in 2 years: on 12 August—the birth date of N. This day is now the Medical Nurse Day. The vesica piscis-shaped medal is composed of gold and silver-gilt and bears a portrait of Florence Nightingale surrounded by the words "Ad memoriam Florence Nightingale 1820–1910". On the reverse, the name of the recipient and the date of the award are engraved, surrounded by the inscription "Pro vera misericordia et cara humanitate perennis décor universalis" (true and loving humanitarianism—a lasting general propriety). It is awarded for "Exceptional courage and devotion to the wounded, sick or disabled or to civilian victims of a conflict or disaster" or "Exemplary services or a creative and pioneering spirit in the areas of public health or nursing education". Among those who received this medal there are over 25 Soviet nurses and hygiene instructors. N. is depicted on the

10-pound British banknote. Henry Wadsworth Longfellow devoted his poem "Santa Filomena" (1857).

Nightingale   F.   (http://www.flyingcoloursmaths.co.uk/wp-content/uploads/2013/05/Florence-Nightingale.jpg)

Nikitin Afanasiy (?–1472) – the Russian explorer, merchant from the city of Tver. In 1466–1472 he traveled to India and Persia. In spring 1469 on a caravan of ships he arrived to Ormus (on the coast of the Persian Gulf) from where onboard an Arab merchant vessel he sailed over the Arabian Sea as far as India where he lived for 3 years and traveled extensively. On the way back via Afghanistan, Persia, Iraq and Turkey N. reached Trapezund, crossed the Black Sea and in 1472 arrived in Kafu (Feodosiya). In autumn 1472 on his way home he died not far from Smolensk. He described his trip in a narrative known as "The Journey Beyond Three Seas" that is a valuable literary-historical feature. In the homeland of N. in the city of Tver the bronze monument to him was erected on the bank of the Volga River.

Nikitin A. (http://tverigrad.ru/wp-content/uploads/2013/)

Nikitskiy Botanical Garden – one of the oldest botanical gardens in the world, located close to Yalta, Crimea, Ukraine. Its area is about 880 ha. In 1812 the first trees brought here from other continents were planted near the Greek settlement of Nikita. The founder and the first director of the garden was Russian botanist of Swedish descent Christian Steven. The collection of the garden numbers about 240,000 varieties, species and forms of plants. Among the most unique and valuable features of this garden there are bamboo grove planted in 1912, the 1,000-year old pistachio obtrusifolious, 700-year old olive tree, 500-year old European yew, California redwood planted in 1840, the 100-year old alley of tapered cypresses, the giant redwood-dendron 33 m high with the stem 1.7 m in diameter. The main divisions of the garden are flora, nature protection, dendrology and decorative gardening, fruit trees, subtropical and nut-bearing trees, new technical plants, mobilization and introduction of plant resources, plant biochemistry, plant physiology, plant agroecology and feeding, plant protection, cytogenetics, research and organization. The garden has established wide scientific contacts with similar establishments in foreign countries. The garden is the part of the Ukrainian Academy of Agrarian Sciences.

Nikitskiy botanical garden (http://tvil.ru/media/storage/resource/cache/20110401//100x100/1301658127_s1.jpg)

Nikolaev – (Mykolaiv) the center of the Nikolaev Region, Ukraine, on the left bank of the Yuzhnyi Bug River at its confluence with the Ingul River; a port 80 km from the Black Sea. This is also the railway station of the same name. It was founded in 1789 as the shipbuilding center in the south of Russia by the Russian Governor General of Novorossiya Prince Grogoriy Potyomkin, initially as a shipyard called simply a "New Shipyard on the Ingul River". Prince Potyomkin signed an order to construct a shipyard on 27 August 1789, which is considered to be the city's birth date. The shipyard was to undertake the repair of naval ships in the Russian-Turkish War of 1787–1792. The history of the city has always been closely connected to ship building and repair. The Russian Empire's Black Sea Navy Headquarters was in Nikolaev for more than 100 years until the Russian Navy moved it to Sevastopol. In 1862 the commercial port and customs office started functioning. Grain and iron ore were transported via N. By the late nineteenth century, Nikolaev's port ranked third in the Russian Empire, after Saint Petersburg and Odessa, in terms of trade with foreign countries. In the post-Great Patriotic War (1941–1945) period Nikolaev became one of shipbuilding centers of the USSR, with three shipyards: "Black Sea", "61 Kommunara", and "Okean". The Varvarovskiy drawbridge connects the city with the right bank of the Yuzhnyi Bug. The city has the following industries: construction materials, shipbuilding (plant named after Nosenko), machine-building (plants: road machinery—asphalt laying machines and others, flour-making equipment, repair-mechanical), food (confectionary, pasta, meat-packing, refrigerating plants), light (leather-footwear, knitting, perfume-glass factories). The Ukraine's only plant of lubricating and filtering equipment ("Dormashina") is also operating in N. The city has pedagogical and shipbuilding, medical and music institutes as well as shipbuilding, railway transport, construction and other colleges, among them: Admiral Makarov National University of Shipbuilding, Petro Mohyla Black Sea State University, Mykolaiv Pedagogical Institute and Mykolaiv State Agrarian University. There are also several art museums and theaters in the city: the Academic Ukrainian Theater of Drama and Musical Comedy, the Mykolaiv State Puppet Theater, the Mykolaiv Academic Art Russian Drama Theater, and the Mykolaiv Oblast Philharmonic Orchestra. Mykolaiv has the following museums: Mykolaiv Regional Museum of Local History, Museum of Shipbuilding and Fleet, Museum of the World War II Partisan Movement, V.V. Vereschagin Art Museum. Population—500,000 (2011).

Nikolaev (http://en.wikipedia.org/wiki/Mykolaiv)

Nikolaev base area (landing) – the base area in the Black Sea port of Nikolaev captured on 26 March 1944 by a unit of the Soviet landing troops (68 soldiers) led by Senior Lieutenant K.F. Olshanskiy. This base area was needed to support the troops of the Third Ukrainian Front (commanded by General R.Ya. Malinovskiy) that fought for conquering Nikolaev and to prevent the German troops from destroying port facilities. For 2 days the unit of landing troops fought vehemently having rebuffed 18 attacks and having killed over 700 German soldiers and officers. Nearly the whole unit of landing troops died, but they succeeded to keep their positions till the main forces reached the port. All landing participants were awarded the title of Hero of the Soviet Union.

Nikolaev (Mykolaiv) Oblast – is situated in the south of Ukraine, in the basin of the Yuzhnyi Bug lower reaches. The area of N.O. is 24,600 km^2 (4.1 % of the Ukraine territory). It was formed in 1937. Its administrative center is Nikolaev

(Mykolaiv). In the west, north and east N.O. borders on four other regions (oblasts) of Ukraine: Odessa, Dnipropetrovsk, Kirovograd and Kherson. .In the south it is washed by the waters of the Black Sea and its limans. The coastline runs for several hundreds of kilometers. The greater part of the region is covered by the Prichernomorie Lowland. In the north it has the Pridneprovskiy Highland (to 240 m high) cut by gullies, ravines and river valleys.

The climate in N.O. is moderately continental with the mild low-snow winter and hot arid summer. The mean annual temperatures in summer are +21.1 °C, in winter −2 °C. Annual precipitation varies from 330 mm in the south to 450 mm in the north. The northern part of the region is composed of ordinary black earth, while the southern part of black earth, chestnut and dark-chestnut soils. The region has 85 rivers that refer to the basins of the Yuzhnyi Bug and Dnieper. The largest of them are Yuzhnyi Bug 257 km long within the region, Ingul—179 km, Ingulets— 96 km. The following limans incise deeply into land: Dnieper-Bug (63 km), Tiligulskiy (60 km), Bugskiy (42 km) and Berezanskiy (26 km). The region abounds in ponds and reservoirs with a total water area over 13,000 ha. Rivers and ponds are used mostly for irrigation of agricultural lands and for fishery.

The population in the region is 1.25 million (2006), of which about 76 % is urban population. In the population structure over 75 % are Ukrainians and 20 % are Russians. At the same time the Moldavians, Bulgarians, Greeks and other peoples are also living here. Administratively N.O. is divided into 19 districts. There are 9 cities, 17 urban settlements and 901 villages. Of all cities five cities are of the regional significance, such as Yuzhnoukrainsk, Pervomaisk, Voznesensk, Ochakov and Mykolaiv.

The mineral deposits of the region are represented by considerable reserves of building materials: stone, granites in a wide color range featuring high decorative properties, sawn stone, raw cement, clay and tilestone, sand. Of commercial significance are also deposits of limestone, kaoline, road-building materials, etc. The total area of the forest stock reaches 70,000 ha, including 38,000 ha under forests. The prevailing tree species found here are oak, pine, white acacia, poplar and others.

Among the natural recreational resources there are sea sandy beaches running for more than 70 km, the picturesque landscapes on the banks of the Yuzhnyi Bug and numerous reservoirs, the mineral water sources yielding to 1,000 m^3 a day, the curative mud, in particular in the Tiligulskiy and Beikushanskiy limans with the reserves over 2 million m^3.

The economic potential of N.O. makes a part of the Ukraine's economic complex. It is characterized by the potent industries, developed agricultural sector, dense transport network and developed port facilities. The leading industry is N.O. is shipbuilding. The major enterprises in this industry are "Chernomorskiy Shipbuilding Plant" and State Enterprise "Shipbuilding Plant named after 61 Communars". They manufacture tankers, dry-cargo vessels, container ships, refrigerating ships, trawlers, floating hotels, boats, etc. The nonferrous metallurgy is also a key industry in the region. It is represented by the Europe's largest aluminum plant "Nikolaevskiy alumina refinery". This plant is capable to produce one million tons of alumina a year. N.O. also has some electrical-engineering and electronic plants. In the total light industry output of Ukraine the region's share is as large as 8.8 %. This industry is represented by large garment factories.

The region has developed agriculture. The area of agricultural lands is 2.01 million ha, of which 1.7 million ha are plowlands. Agriculture here specializes on plant growing and animal husbandry. The main cultivated plants are grain and technical (sunflower, sugar beet), vegetable and melon crops. Horticulture and vine-growing are also developed. A high level of plant growing in which great share is taken by forage crops is a good basis for meat and milk animal rearing.

The region possesses a dense transport system including railway, marine, river, automobile, aviation and pipeline transport. The operating length of railways is 700 km, of which 33.6 % are electrified. The automobile transport plays an important role in inter-city and inter-district carriages. Nikolaev is crossed by highways running to Odessa, Kiev and Crimea that make a part of the International Transport Corridor: Eurasian and the Black Sea. The length of general-purpose automobile roads is 4,700 km, including hard-paved—4,600 km. Nikolaev is a large transportation center with its marine and river ports: Nikolaev Sea Trade Port, Sea Port "Oktyabrsk", Sea Port Dnieper-Bug, and Nikolaev River Port. The international airport in Nikolaev is among the largest and most well equipped airports in the south of Ukraine. Pipelines transport every year about 2.1 million tons of ammonia via the territory of the region. The length of the ammonia pipeline is 444 km. The region possesses considerable export potential. The main items of export are alumina, ships, mechanical equipment, leather, plant products, etc. Imported are mostly bauxites, oil products, chemical products.

N.O. has many scientific and cultural establishments: 28 higher educational institutions and universities, eight branches and faculties of the National University "Kievo-Mogilyanskiy Academy", Kiev Slav University, Law Institute, Odessa State University, Kiev State Institute of Culture and others. Here you can find three professional theatres, over 570 cultural clubs, etc. The region has over 120 territories and natural features with the status of nature reserves, regional landscape parks "Kinburn Spit" and "Granite-Steppe Pribuzhie", a part of the Black Sea Biosphere Nature reserve, about 19 parks, etc. The Zoo in Nikolaev was founded 100 years ago and is considered one of the most beautiful in Ukraine. Recreational centers locate on the Black Sea coast and on the shores of its limans, in the resort zones of village of Koblevo, cities of Ochakov and Nikolaev.

Nikolaev Plants and Shipyards Joint Stock Company – this is a shipbuilding company that appeared as a result of construction in 1895–1898 in Nikolaev (Kherson Province) of the shipyard with repair workshops. The company could build simultaneously four large armored ships or eight medium-size cruisers. The shipyard had also special workshops producing artillery turrets, shells, locomotives, railway cars, railway bridges. It was the only large plant on the Black Sea building war ships as in the late nineteenth century the shipbuilding in the south of Russia was practically non-existent, save small plants "Vadon" in Kherson and "Bellino-Fenderich" in Odessa.

Nikolaev River Port – is situated on the Yuznyi Bug River not far from the Nikolaev Sea Trade Port. The river port takes an area of 185,000 m² of the river water area; the berth length is 800 m. The port is equipped with modern facilities that makes it possible to handle export and import cargo here. The port is capable to handle to six million tons of cargo. The port undertakes passenger carriages over local lines and can receive tourist and excursion vessels cruising over the Dnieper, Danube rivers and navigating to the Crimean ports.

Nikolaev Sea Trade Port – locates near the city of Nikolaev (Mykolaiv), Ukraine, on a large peninsula on the left bank of the Yuzhnyi Bug, 80 km from the Black Sea. This is one of the oldest ports in Ukraine. It was founded in 1862. This port is a very important transport center in the south of Ukraine. It is connected with the Black Sea via the canal going over the Dnieper-Bug Liman and the Yuzhnyi Bug River. The canal length from the port of Nikolaev to Ochakov port is 79 km, its depth is 10.6 m and the width in the bed is 100 m. Vessels 215 m long and 30 m wide can navigate along the canal. The vessels more than 187 m long may pass along the canal with a towboat. The port area is 227 ha. This port supports the year-round navigation. Main export cargoes of the port of Nikolaev are steel scrap, steel products, grain, logs, fertilizers (white and pink chloride of potassium). The main cargoes imported are: refrigerated cargoes (meat, fish, fruits and suchlike), containerised cargoes, oil and oil products. Annual freight turnover at Nikolaev is about 1.5–2 million tonnes. There are 15 piers at the port: piers Nos. 1–3 serve for transportation of passengers, piers Nos. 4–14 are used for cargo-handling, pier No. 15 is the oil-handling point, treating small tankers.

Nikolaev Shipbuilding University named after Admiral S.O. Makarov – was established in 1920 in Nikolaev, Ukraine. It was awarded the Order of the Labor Red Banner (1970). Structure—Shipbuilding Institute (6 departments), Institute of Computer, Engineering and Technological Sciences (8 departments), Mechanical Engineering Institute (10 departments), Institute of Automatic Control and Electrical Engineering (7 departments), Institute of Humanities (8 departments), Institute of Extramural and Distant Learning (8 departments), two faculties, and 6 branches of other institutes. University continues to develop the international relationships with Germany, China, France, the Netherlands, USA, Iran, Norway, etc., the general scientific workings out are conducted, leading experts, post-graduate students and students trained abroad. During existence, the University has prepared about 55,000 engineers—ship builders, machine engineers, electricians, economists, programmers.

Nikonov Andrey Ivanovich (1811–1891) – Russian Admiral. In 1828 he entered the Naval Cadet Corps. In 1831–1835 onboard brig "Paris" he navigated from Kronstadt to Malta, cruised in the Archipelago and Mediterranean and arrived to Sevastopol. In 1838–1839 he sailed along the eastern coast of the Black Sea and took part in landings near Subashi and Psezuap. In 1841 he cruised near the coast of Abkhazia and later navigated between Sevastopol and Odessa. In 1842 N. navigated over the Black and Azov seas. He was sent from Nikolaev to Saint-Petersburg with the junkers of the Black Sea Fleet. In 1846–1848 N. navigated with the squadron near the eastern coast of the Black Sea. In 1848 and 1850 he commanded the frigate "Khersones" of the Black Sea Fleet. In 1852–1854 commanding the corvette "Oriadna", he navigated over the Black and Mediterranean seas. In 1854 he was appointed commander to the ship "Varna" in the Sevastopol roads. In 1855 he was assigned to the garrison of Sevastopol. In 1863 he was transferred to Odessa as a captain of the quarantine port. In 1865 he was commander of the Sevastopol Port. In 1867 he was assigned to the Black Sea Fleet staff. In 1868 he was appointed the chairman of the naval court in the Nikolaev Port, in 1876—the governor, commander of the port and superintendant of Sevastopol. Later he was commander of the city defense controlling the on-land batteries, minefields, ships and troops. He did much for improvement of Sevastopol defense and for active defense of the ships during the Russian-Turkish War of 1877–1878. In 1891 he was promoted to Admiral.

Noctiluca (*Noctiluca miliaris*) – a representative of the unicellular mobile pyrophytic algae. They possess bioluminescence—an ability to fluoresce. This "live gleam" has the unlimited range of tints, but most often it is silver-white. The longest and most intense fluoresce of the sea is observed in summer and autumn. When N. propagates intensively the fishermen call this phenomenon of sea gleaming "ekamos". It helps them to catch fish in dark moonless nights when fish shoals while on the move leave a gleaming trace. After being filtered from water and dried out they will still gleam with cold light. Such fluorescence is created by the substance that the scientists called "luciferin" in honor of Lucifer—the Master of Hell.

Nordman Alexander Davidovich (1803–1866) – Russian zoologist. In 1821 he entered the university in the Finnish city of Abo (Turku) and graduated from it in 1827 with the academic degree of Doctor of Philosophy. In 1831 he became the member of the Moscow Society of Natural Scientists. In 1832 N. worked as professor at the Natural History Chair of the Richelieu Lyceum in Odessa. At the same time he was appointed the Director of the Odessa Botanical Garden. In 1833 with the expedition N. visited the southern area near the Black Sea in 1835 and 1842 he visited Crimea once more travelling along the Black Sea coast. In 1847 he went to Crimea, Sevastopol and Balaklava for collection of marine fish specimens. In 1848 he left Odessa and went to work in the Helsingfors (Helsinki) University in Finland. In 1859 N. was elected the corresponding member of the Russian Academy of Sciences. In 1860 he went to Crimean for the last time and again visited Odessa to take the herbarium of Ch. Steven. N. was the acknowledged expert of the Black Sea fauna.

North Crimean Canal – it was built for irrigation and water supply of the southwestern part of the Kherson Region and the steppes of Northern Crimea as well as for water supply of the Kerch industrial region, in particular the cities of Kerch and Feodosiya. Its construction was started in 1962. From 1971 this canal supplied water to Feodosiya and Kerch. This canal supports irrigation of an area of about 1.4 million ha, including over 0.8 million ha of arable lands. About 40,000 ha are included into rice rotation. In the irrigated zone 660,000 ha are supplied with water. The source of irrigation water is the Dnieper River. In 1956 the headworks were constructed for water intake from the Kakhovka Reservoir. They were designed for water discharge of 380 m³/s. The length of the main canal is about 400 km. In its upper and middle sections (from Kakhovka to Dzhankoy) 210 km long the water flows by gravity. In other canal sections the three-step water lift to a height of about 120 m is ensured. The water discharge at the canal head is 294 m³/s, in its tail part—3 m³/s. The main canal has four branches that form separate irrigation systems: Razdolnenskiy, Azovskiy, Krasnogvardeiskiy and Chernomorskiy. The total length of the canal with its branches is 690 km. More than 200 km of the main canal and main distribution canals are coated with concrete.

Northeastern Monsoon of Western Transcaucasia (Kolkhida Monsoon) – strong northeastern winds prevailing in October-May in the western areas of Transcaucasia and winds from the Black Sea dominating in June. The winter northeastern winds are connected with the cyclones passing over the sea. However, the "overflow" of cold air masses over the central parts of the Caucasian ridge is observed quite seldom. The northeastern monsoon is more prominent in December and March. It becomes stronger due to development of the local anticyclone over the Armenian uplands.

Novikov Modest Dmitrievich (before 1841–1893) – Russian Vice-Admiral. In 1841 he entered as a cadet the Naval Corps. In 1847 after cruising in the North Sea he was promoted to warrant officer and assigned to the Black Sea Fleet. In 1848–1850 N. cruised on different ships near the eastern coast of the Black Sea. In 1851–1852 onboard the brig "Phemistocles" he sailed from Odessa to Constantinople and then to Archipelago and back. In 1853 he cruised near the Caucasian coast. In 1854–1855 N. was on the staff of the Sevastopol garrison, took part in nine raids and was contused on the 6th bastion. In 1869 he was dismissed. In 1876 N. returned to the Fleet. In 1886–1888 he commanded the unit of the Black Sea Fleet. In 1891 N. was conferred the rank of Vice-Admiral and appointed the senior flagman of the Black Sea Fleet.

Novoazovsk – a village on the northeastern coast of the Sea of Azov, Ukraine.

Novomikhailovskiy – a small seaside resort village in the Tuapse District, Krasnodar Territory, Russia. It was founded in 1864. It is situated 33 km to the northwest of the railway station Tuapse, on the road Tuapse—Novorossiysk. It is a part of the Tuapse resort zone. It is located on the Black Sea coast in the picturesque valley of the mountain river of Nechepsukho. The beach here is small-pebble or sandy in some places. The seafloor is sandy, very good for bathing. The All-Russia Children Center "Orlenok" locates here (from 1960). This center first appeared as the pioneer camp "Solnechnyi". Today this is a cluster of seven camps three of which are working year-round. On an area of 300 ha you can find a school, the palace of culture and sports, indoor swimming pool, the house of aviation and aeronautics, and other facilities. There are also recreational centers here. Population—10,200 (2010).

Novorossiya – the official name of Southern Ukraine on the coast of the Black and Azov seas that was in use from the second half of the eighteenth century and to 1917. N. appeared during colonization in the eighteenth to nineteenth centuries of the steppes near the Black Sea by the Russians, Ukrainians, Serbs, Bulgarians, Greeks and Germans. N. is synonymous with the official name of Novorossiysk Province and Novorossiysk General Government (in nineteenth century centered in Odessa). In a broad sense—historical territory attached to the Russian Empire, including the Kherson, Ekaterinoslavl, Tauride, Bessarabia, Stavropol Province, as well as the Kuban region and the area of the Don Cossacks.

Novorossiysk – the city in the Krasnodar Territory, Russia; the port on the bank of the non-freezing Tsemes (Novorossiysk) Bay of the Black Sea; railway station. It is believed that the first people settled in the vicinity of Novorossiysk back in the Paleolithic era. In V B.C. on the site of modern Novorossiysk originated Bata—a commercial Greek city. In II B.C. it was destroyed by nomadic Alans. In the thirteenth century Tsemes Bay belongs to the Golden Horde. At the mouth of Tsemes River a fortress Batario was built by the Genoese. In 1453 the Ottoman Turks took Constantinople by storm, and Taman and Tsemes Bay moved into Turkish territory. In 1829 according to the Treaty of Adrianople Tsemess Bay area has moved from Turkey to Russia. On 12 September 1838 the ships of the Russian squadron entered the Tsemes Bay, and 5816 men under the command of General N.N. Raevskiy and Admiral M.P. Lazarev landed on the ruins of Turkish fortress. This day is now celebrated as the birthday of the city. In 1838 a fort was constructed in the place where Tsemes River flowed into the Black Sea. Apart from Raevskiy and Lazarev, the third founder of the city is rightly considered Admiral L.M. Serebryakov. Thanks to his unceasing efforts Novorossiysk was built and became such beautiful city. In 1839 the name "Novorossiysk" appeared meaning "city of Novorossiya"—in this way from the late eighteenth century all vast expanses of the steppes near the Black Sea that were liberated from the Turkish ruling and incorporated into Russia were referred to. After connection of Zakubanie (lands behind the Kuban River) this name was applied to this territory, too. The military settlement grew quickly and in 1850 the seaside city-port appeared here. One more great impetus for quick development of this settlement was given by geological surveys that helped to discover enormous reserves of marlstones—raw material for production of quality cement.

On 27 September 1866 Novorossiysk becomes the center of the Black Sea Region (Black Sea coast from Taman to Georgia until 1896). The city then there were 430 residents. In 1882 it was commissioned a first cement plant, in 1888 first train came to Novorossiysk, in 1893 close to Novorossiysk grain elevator it was put into operation the world's first three-phase power station. From 1896 to 1920 Novorossiysk was the center of the Black Sea Province. From the late nineteenth century oil was exported from here. The oil terminals were owned by the French businessmen who founded here JSC "Russian Standard". In 1905 during the first Russian Revolution the first Soviet of the Workers' Deputies was formed in the city and the Novorossiysk Republic was established, but it existed for only 2 weeks. In

1918 the main part of the Black Sea Fleet was sank in the Tsemes Bay and the Germans did not get a single Russian warship. In 1920 this city was the last base of the defeated the White Army and from here the first Russian emigrants leave Russia in March 1920.

"Exodus". A monument to evacuation of the White Army from Novorossiysk in March 1920 (Photo by Evgenia Kostyanaya)

During the Great Patriotic War (1941–1945) N. was one of the most important outposts in the battles for the Caucasus and its regions abounding in oil resources. In 1942–1943 the greater part of the city was occupied by the Germans. Very fierce battles were fought near N. The heroic defense of this city lasted for 225 days up to 16 September 1943. The battles in the area called "Malaya Zemlya" are included as separate exploits into the history of that time. After the war the city was built anew.

The economic development of the city is connected with opening of the marl deposits (raw material for cement production), construction of a railway line linking the city with the Northern Caucasus and the sea port for grain export. N. is also an industrial city depending on steel, food processing and furniture manufacturing. This is a major center of the cement industry (the first cement plant called "Zvezda" (now "Proletariy") was built in 1882, the other plant "Tsel" (now "Oktyabr")—in 1898). N. also has the plants producing asbestos-cement boards, spare parts for agricultural machinery, railway car and ship repair workshops, machine-building plant, light and food industries, including mill elevator. The city development was spurred significantly by construction of the Russia's largest Novorossiysk Sea Trade Port for handling oil, cement, grain, timber, container cargo. It is served by the country's largest tanker fleet. In 1964 the oil terminal of Sheskharis was commissioned. On 14 September 1973 N. was conferred the status of "Hero City".

City stretches 25 km along the Tsemes Bay like amphitheater, which is surrounded by the mountains of the North Caucasus. In the south-western part the city is bordered by mountains of Navagirskiy Ridge extending from Anapa. The highest point of the ridge close to Novorossiysk is Mount Koldun (Wizard) (447 m). On the north Markotkhskiy Mountain Range presses the city to the sea and stretches 50 km to the south-east up to Gelendzhik. The highest point of the ridge is Sakharnaya Golova (Sugarloaf) Mount (558 m). The mountains protect the city from cold air masses coming from the continent. Narrow amd small Tsemes River flows through the industrial part of the city. In the south-eastern part of the city there is Solenoe (Salt) Lake, which is popularly known as liman. Sudzhukskaya Spit separates it from the sea. At 14 km from the Novorossiysk there is the largest freshwater lake in the Krasnodar region Abrau Lake.

Novorossiysk is located in a seismic zone. From 1799 to 2013, there have been about 20 significant earthquakes. The climate in the area of Novorossiysk is close to the Mediterranean. In winter, the landscape is dominated by temperate air masses in the summer—tropical. The average air temperature ranges in Novorossiysk from 4.3 °C in January to 24.8 °C in July. The average sea temperature ranges in Novorossiysk Bay in the winter from 7 °C to 12 °C, in summer—between 20 °C and 25 °C. In the summer, especially in August, in the coastal zone thunderstorms, tornadoes and heavy rains may occur and can cause flooding (8 August 2002, 7 July 2012).

Novorossiysk is the center of the cement industry in the south of Russia (5 cement plants), created on the basis of large deposits of high quality marl. The city has the headquarters of a major cement producer "Novorostsement" and cement plants of the group of companies "Inteko"—Verkhnebakanskiy cement plant and "Atakaytsement". Developed machinery, ship repair, radio factory, construction materials, steel industry, timber, food, railway industry, grain and fuel oil terminals. Novorossiysk is one of the major wine-producing centers of Russia. Local agro produces table and sparkling wines (Abrau-Dyurso). In the mid-1970s in Novorossiysk a first in the USSR plant for "Pepsi-Cola" production was built.

Novorossiysk—Russia's largest port on the Black Sea, the turnover of all the terminals of the port in 2009 amounted to 123.6 million tonnes. Port provides maritime trade activities of Russia with Europe, Middle East, Africa, Asia, South and North America. The largest stevedoring company—"Novorossiysk Sea Trade Port". The city hosts the Novorossiysk Shipping Company—one of the largest in Russia. There are timber and passenger ports also. Novorossiysk has a major railway station (terminus) providing delivery and handling of import and export cargo via port. There is a passenger train station, which provides connections to major cities in Russia.

Novorossiysk has State Maritime University named after Admiral Ushakov and Novorossiysk Polytechnic Institute—a branch of Kuban State University of Technology. There are also 15 branches of institutions located in other cities. N. has several museums, theatres, planetarium, parks, and stadium. At the entrance into the port on the Myskhako hill you can see the monument to the defenders of the legendary "Malaya Zemlya"—a memorial complex "Defense Line" and other features. In 1994 Russian Government passed the Resolution "On formation of the Novorossiysk naval region and creation of the naval base on the Krasnodar territory". Population—251,000 (2013).

Novorossiysk city (Photo by Andrey Kostianoy)

"Novorossiysk" – Soviet battleship, flagman of the Black Sea Fleet after WWII. Formerly named "Giulio Cesare" (built in 1914), one of the largest ships of the Italian Fleet till 1948, it was passed to the Soviet Union after World War II as a reparation with other 33 battleships and auxiliary ships. It was handed over to the USSR on 6 February 1949 in the Albanian port of Valona (presently Vlora). On 26 February the Soviet crew navigated it to Sevastopol. On March 5, 1949 the battleship was given the name "N." On 28 October 1955 "N." returned to Sevastopol from a voyage and anchored near marine hospital 100 m from the coast. At 1:29 on the night of 29 October a powerful explosion shattered the ship. Many thought that there were two explosions. A large hole was made in the ship bottom from board to board 5–6 m wide with a total area of 150 m^2. Eight decks were broken through. At 4:15 in the morning "N." listed to port, stayed in this way for several moments, then turned over and only bottom was visible on the water surface. According to the veterans of this battleship, 608 people died, of which 50 to 100 were killed by the explosion. This was the gravest disaster in the postwar history of the USSR Navy. On 2 May 1957 the battleship was lifted from a depth of 18 m and transported to the Kazachia Bay for cutting. The cause of the loss of this ship is still unknown. But there are many versions: the German mine left from the wartime; actions of the Italian subversive divers from a group of Prince Valerio Borghese as an act of revenge for transfer of this ship to the Soviet Union; attack of supersmall submarines X-51 and X-52 (British); or actions of some other forces wishing collapse of the Black Sea Fleet. It's strange enough, but battleship "N." perished in the same place where in 1916 the ship "Empress Maria" first exploded and then turned over. This history repeated itself 39 years later. The disaster was investigated by the special governmental commission headed by the Deputy Chairman of the USSR Council of Ministers. According to his report "it can be asserted that the accident on 29 October 1955 occurred after explosion of a charge located on the seafloor of the Sevastopol Bay in place of the battleship anchorage." The weight of the charge in the TNT equivalent was 1,000–1,100 kg. The conclusions of this commission are contested till now. On 22 August 2013 veteran of Italian division of frogmen "Gamma" Ugo D'Esposito acknowledged that the Italian military exploded the Soviet battleship "Novorossiysk".

"Novorossiysk" battleship in Sevastopol (http://republic.com.ua/article/19741-old.html)

Novorossiysk Bay – see Tsemes Bay.

Novorossiysk bora (Nord-Ost) – strong northeast winds (from Latin "*borea*" meaning "northern wind") in the Novorossiysk area on the Black Sea. It occurs during the invasion of cold air from the North Caucasus Plateau to the Black Sea coast. In this case, the mass of cold air being transshipped through the mountains, looks like giant clouds of cotton candy. In the event of Nord-Ost, a sharp decrease of temperature (in a few hours the temperature can drop by 10–15°). Bora usually lasts for about 46 days a year reaching the maximum intensity from November to March. Half of this period the wind speed is no less than 20 m/s. The maximum speed of the northeastern wind during bora is 40 m/s, on the Markotkhskiy Ridge— to 60 m/s. An episode of bora may last for 1–3 days, at times even a week. Novorossiysk Bay becomes unnavigable. Several strong bora events are listed below:

1848—six ships of the Black Sea Fleet squadron wrecked from a strong Nord-Ost, 57 sailors died.

1993—several ships, including fishing sank in Tsemes Bay, five sailors died.

1997—maximum wind speed gusts reached 62 m/s, air temperature dropped to −20 °C, schools, public transport did not work.

December 2002—as a result of storm and severe icing (air temperature dropped to −15 °C) two ships sank in the port of Novorossiysk.

February 2012—a powerful hurricane with wind gusts up to 45 m/s left without electricity for more than 150,000 residents of Novorossiysk and Gelendzhik. Wind damaged the roof of 63 houses in the city. Governor of the Krasnodar region Alexander Tkachev, who visited the city on 8 February 2012, called Novorossiysk "ice apocalypse".

In addition to Novorossiysk, the bora phenomenon is also observed on the Mediterranean coast of France, on the Adriatic coast of Croatia (from Rijeka to Split), Novaya Zemlya in the Barents Sea, the Urals, and even in Antarctica.

Novorossiysk bora (http://pics.livejournal.com/ardelive/pic/000a527d)

Novorossiysk Cement Plants – a major center of cement industry in Russia. The plants are located in Novorossiysk and nearby villages Gayduk and Tonnelnaya, Krasnodar Kray. They act on the basis of the world's largest deposits of marl. The consortium includes plants: "Proletariy", "Oktyabr", "Pervomaiskiy" and "Pobeda Oktyabrya." Novorossiysk first cement factory "Zvezda" (now the "Proletariy") was founded by the Black Sea Society "Portlandcement" in 1882. In 1898 the second plant "Tsep'" (now the "Oktyabr") was put into operation. It was followed by the construction of a series of cement plants until 1912 in the vicinity of Novorossiysk: "Solntse", "Beton", "Titan", "Skala", "Pobeda", "Orel", "Atlas").

By 1917 in total there were 10 low-power plants, equipped with primitive shaft furnace construction by firms "Ditch" and "Schneider" in which natural marl was burnt. After the Civil War of 1918–1920, mills have been restored and expanded, renovated and modernized. In 1940, cement production increased by more than 2.5 times in comparison with 1913. During the Great Patriotic War of 1941–1945, plants were destroyed. Much of the Novorossiysk was temporarily occupied by the German Army. Almost a year lasted heroic defense of the plant "Oktyabr", which was at the frontline. After liberation of the city in September 1943 the cement plants were restored on a new technical basis. By the early 1970's 61 old cement kilns were replaced by 16 modern powerful rotating ones. In 1973 the plants have produced 4.45 million tons of high-quality cement. Today, the city has the headquarters of a major cement producer "Novorostsement" and cement plants of the group of companies "Inteko"—Verkhnebakanskiy cement plant and "Atakaytsement".

Novorossiysk cement plant "Proletariy" in (Photo by Andrey Kostianoy)

Novorossiysk Grain Terminal – was constructed for the implementation of an investment project for transshipment of grains and oilseeds in the Novorossiysk port with a capacity of 3.6 million tons per year. Construction of the terminal began in 2004 and was put into operation in May 2008. The terminal is equipped with a high-speed high-performance equipment, applicable for the first time in Russia. Modern process control system uses an innovative method of dust control by oil which is applied for the first time in Russia. Elevators have a total capacity for storing grain of 120,000 tons; a transport gallery of grain delivery to ships has a total capacity—1,600 tons/h. In 2009 N.G.T. has exported 5.083 million tons of cereals.

Novorossiysk Landing Operation 1943 – amphibious operation on 10–16 September 1943 organised by the Soviet Black Sea Fleet in Novorossiysk during Novorossiysk-Taman operation against German Army during the Great Patriotic War of 1941–1945. As one of the largest Soviet amphibious operations, Novorossiysk landing operation went down in history as one of the most well planned and prepared by the Soviet Army, and carried out jointly by the Army and Navy. It showed that with careful preparation of the landing operations their success is possible even on a heavily fortified coast.

Novorossiysk Marine Biological Station – was established in 1920 on the initiative of A.L. Bening, Director of the Volga Biological Station (in Saratov), and V.M. Arnoldi, the prominent professor, botanist, who became its first director (1920–1921). In 1923 the station subordinated to the Higher Council of the National Economy (VSNKh). That time it was visited by the Black Sea—Azov research-fishery expedition headed by N.M. Knipovich who appreciated its activity and agreed to become the Chairman of the Scientific Council of the station. In 1934 the station was included into the USSR Academy of Sciences. In 1935 it was transferred to the Rostov University and after this the Biological Institute was set up there. In 1938 the station became a part of the Zoological Institute of USSR Academy of Sciences as a laboratory. Today, this is Novorossiysk Training and Research Marine Biological Center—a branch of Kuban State University (Krasnodar). It was renamed from the Novorossiysk Marine Biological Station of the KubSU in 2000. It is one of the oldest marine research laboratories in Russia. The Center is engaged in the study of sea water, sediments, marine ecosystems in conditions of chronic anthropogenic pressure, the influence of various sources of pollution on the development of marine organisms, the study of the processes of accumulation of toxins in the biological organisms of the Black Sea basin. The Center data bank has a unique long-term hydrological, hydrochemical, and hydro-biological set of data (1966–2009). The main tasks of the Center is still organizing and conducting monitoring studies on the state of marine, natural, drinking water, soil, sediment, marine biota in the areas of the Southern Federal District.

Novorossiysk Marine State Academy named after F.F. Ushakov – Russian State maritime educational institution of higher professional education in Novorossiysk founded in 1975, is the largest institution on water transport and the only higher education institution in the South of Russia, producing marine specialists for shipping companies, ship-building and ship-repair yards, ports and transport terminals. Academy trains students in 14 specialities. It includes 5 faculties, 36 departments, Southern Regional Center for Continuing Professional Education, branches in the cities of Rostov-on-Don, Sevastopol, and Astrakhan. Academy has teaching and laboratory facilities in three buildings, more than 500 laboratories and classrooms, a reading room with 160 seats, a library with a book stock of about 300,000 copies. Academy has more than 300 teachers, among them 29 professors and doctors, 8 academicians, 198 associate professors and candidates of science. Today, the Academy trains more than 4,000 students.

Novorossiysk Sea Port – the largest foreign trade port on the Black Sea in Russia that ensures reloading of the most important Russian cargoes. It is situated on the shore of the Tsemes Bay in the northeastern part of the Black Sea. Novorossiysk wasfounded in 1838, but already on 30 June 1845 Russian Emperor ordered to construct a port in Novorossiysk on the northeastern coast of the Black Sea for receiving ships coming from Russian and abroad. Maritime trade has become a major activity in the city. Already in 1846 Novorossiysk visited 102 Russian and 17 foreign ship. During this period, the port exported rye, wheat, honey, fish, tobacco, and sugar. In 1887, the total turnover of the port amounted to 120,000 tons having only one wooden jetty operated by "Russian Standard" Company. A significant factor in the development of sea port of Novorossiysk was the construction of the railway line Novorossiysk—Tikhoretskaya in 1888.

By 1940, the N.S.P. consists of four loading and unloading areas, cement piers, coastal area, eastern waterfront. Port area occupied 106.5 ha. Number of berths was 41 with a total length of 4,686 m. The total economic turnover of the port for 1940 reached 1.554 million tons; 416 ships and 27,564 railroad cars were handled in the port. During the Great Patriotic War (1941–1945) fights in the port area have caused significant damage to the material base of the port. On 16 September 1943 the city was liberated. Of the 40 berths, only one survived, but it was badly damaged. All warehouses, cranes, handling equipmen were destroyed. In November 1943 National Defense Commission makes a decision on the first stage of the restoration of the N.S.P. The official start of the seaport after the devastation caused during the war is 1 October 1944.

Today, about 80 companies work in the N.S.P., among them: (1) The group of companies "Novorossiysk Commercial Sea Port" (NCSP), including NCSP, Oil terminal "Sheskharis", "Novoroslesexport", "Novorossiysk Ship Repair Yard", "IPP", "Novorossiysk Grain Terminal", "NCSP Fleet"; (2) "Caspian Pipeline Consortium—R" (CPC-R); (3) "Novorossiysk node transport company" (NUTEP), (4) "Stroykomplekt"; (5) Base of fleet maintenance; (6) "Novorossiysk management of State Sea Resque Service of Russia". In the port of Novorossiysk and military harbor there is also Novorossiysk Naval Base. Total turnover of the port in 2010 amounted to 117 million tons. The Port Passenger Terminal in the summer provides regular cruises of passengers on hydrofoils ships to Sochi and Yalta.

The main commercial company on cargo handling in the port of N. is "Novorossiysk Commercial Sea Port" set up in 1992. In 2006 it was put on the RF list of strategic enterprises. Non-freezing of the Tsemess Bay in winter and a system of piers and breakwaters provide an opportunity for year-round navigation in the port. By the volume of cargo turnover NCSP is one of the major port operator in Russia. This company handles 30 % of all cargo transported via the Russia ports and 20 % of the total volume of oil and oil products exported from Russia. The port has more than 80 berths with a total length of 13.5 km. Here a special terminal for large-tonnage containers, the only in the south of Russia, is located. The port is divided into 4 cargo areas—"Eastern", "Central", "Western" and oil-loading (oil harbor "Sheskharis") and includes 20 cargo berths specially constructed for handling of coal, cement, grain, crude sugar, metal, mineral deposits and general cargoes, containers as well as crude oil and oil products. NCSP has the most well developed and effectively operating railway network compared to other ports that is capable to handle over 500 railways cars a day.

Novorossiysk commercial sea port (Photo by Andrey Kostianoy)

Novorossiysk Naval Base – was first organized in July 1920. It included the naval port, 36 ships, the sea aviation unit, Novorossiysk fortified area, fleet semi-crew, ship-lifting team and different on-land organizations. The ships and personnel of the base took part in the defeat of the White Army of Vrangel in Crimea and in liberation by the Red Army of Transcaucasus. After the Civil War in Russia the base was reduced significantly. In late 1930 due to aggravation of the international situation the base was gradually extended. In 1940 it included the unit of guard ships for protection of the marine area, the coastal artillery unit and various supporting services. A division of submarines and a detachment of torpedo boats based here. From the very beginning of the Great Patriotic War (1941–1945) the ships and units of the base participated in war operations on the Black Sea. They realized and supported marine transportation, evacuated the wounded from besieged Sevastopol and from the Kerch Peninsula; later they took part in the

Kerch-Feodosiya landing operation of 1941–1942. On 17 August 1942 as a part of the Novorossiysk defense area the base defended Novorossiysk. In early September 1942 when there was a threat of seizure by the German fascists of Novorossiysk the base was evacuated to Gelendzhik. In early February 1943 the ships and personnel of the base took part in naval landing and seizure of the base area to the south of Novorossiysk (Myskhako, Stanichka), they carried various cargo for the troops of the 18th Army during 7-month heroic defense of Malaya Zemlya. In the period from 9 September to 9 October 1943 the forces of the base took part in war actions of the North-Caucasian front and the Black Sea Fleet on forcing out the Germans from Novorossiysk and Taman Peninsula. During the Crimean operation of 1944 this base supported the actions of the Black Sea Fleet and Azov Navy as well as sea carriages among the ports of the Caucasus and Crimea. After the war in 1946 the base was disbanded.

The decision on the establishment of Novorossiysk Naval Base was taken by the Government of the Russian Federation on 8 September 1994. Pursuant to Ordnance of the Navy General Headquarters of 21.02.1997, beginning from August 1997 the Department of the Novorossiysk Naval Area was transformed into Department of the Novorossiysk Naval Base. The zone of responsibility of the base is over 18,000 sq. miles of the water area extending from the borders with Georgia to the border with Ukraine on the Black and Azov seas.

Novorossiysk Sheskharis Oil Terminal – one of the largest oil terminals in Russia. Designing of the oil complex was started in 1960 on the basis the Resolution of the USSR Council of Ministers on 14 June 1956 and 24 December 1957, and of the Council of Ministers of the RSFSR of 12 September 1959. Construction of the terminal began in 1961. On 19 October 1964 the first tanker "Likhoslavl" took on board 37,000 tons of oil. Simultaneously with the construction of "Sheskharis" oil terminal, in 1962 construction of 20 reinforced concrete oil tanks in Grushovaya began. Most of these facilities were built in 1967. In 1978, it received an access to the sea via the pipeline through which ships could be loaded directly."Grushovaya" oil base became one of Russia's largest oil depots. "Sheskharis" oil terminal is able to receive tankers up to 250,000 tons. Since 2012 N.S.O.T. includes "Grushovaya" oil base and represents a single oil complex with an export capacities 40–50 million tons a year.

Novorossiysk Sheskharis oil terminal (Photo by Andrey Kostianoy)

Novorossiysk Republic – was founded in 1905 by the workers of Novorossiysk supported by the Soviet of the Workers' Deputies. It existed for only 2 weeks: 12–25 December 1905. This was the first attempt to form of a local power body. It was suppressed by the Tsar's troops.

Novorossiysk Shipping Company (Novoship) – is the largest shipping company in the South of Russia. Founded in 1964 as the tanker branch of the Black Sea Shipping Company, N.S.C. was separated from the parent company on 20 January 1967, and became the independent enterprise with the aim of effective management of bulk-oil fleet and ports on the Russian coast of the Black Sea. On 10 November 1992, it was registered as a Joint Stock Company. At that time the fleet consisted of 101 vessels totaling 4.7 million deadweight tons and 14.3 years old on average. In 1993 Novoship started development of an independent Fleet Renewal Program. In 1993–2000, N.S.C. had invested about 854 million dollars in purchase and construction of 34 modern vessels aggregating 1.37 million deadweight tons. These vessels became the basis for one of the youngest tanker fleets in the industry: nowadays Novoship owns and operates 51 modern vessels representing 4.5 million deadweight tons. The Company employs nearly 4,000 seafarers and shore-based personnel. In December 2007 Novoship became part of the "Sovcomflot" Group.

Novorossiysk shipping company (Photo by Andrey Kostianoy)

Novorossiysk Timber Port – JSC "Novoroslesexport" (NLE)—a large universal port, loading capacities which are divided into two specialized terminals: container and timber terminal. Located on the northeastern coast of the Black Sea, in the Novorossiysk Commercial Sea Port—the largest trading port in the south of Russia. JSC "NLE" processes the ships throughout the year, including container ships of class PANAMAX. Has significant areas for temporary storage, moorings, access roads, well-developed technical base and experienced staff. This allows to quickly process incoming cargo flows and carry out various types of shipping goods around the world. The main activities of the company—the transshipment of containers and timber cargo. Port area—65 ha, the water area—15.3 ha, 11 berths with a total length—1,373 m, depth at the berths—6.5–13.9 m, cold store—5,000 tons, 19 railway tracks of the total length of 7,500 m.

Novosilskiy Fedor Mikhailovich (1808–1892) – Russian Admiral. In 1818 he entered the Naval Cadet Corps. In 1827–1828 N. navigated onboard the brig "Commerstraks" over the Baltic Sea. In 1828 he was transferred to the Black Sea Fleet. In 1829 on the brig "Merkuriy" he cruised near the Bosporus and took part in a battle with two Turkish ships. On the frigate "Pospeshnyi" he cruised over the

Black Sea, on the frigate "Flora" he took part in the battles at fortresses of Agatopol and Inadoy. In 1832 onboard the ship "Empress Catherine II" he navigated between Sevastopol and Ochakov. In 1833 he sailed to the roads of Buyuk-Dere with the expedition to help sultan to suppress the rebellion of the Egyptian Pasha. Together with the landing troops he came to Feodosiya. In 1835 and 1836 he commanded the brig "Merkuriy"; in 1837 and 1838 he cruised near the eastern coast of the Black Sea, took part in seizure of Sochi and Tuapse. In 1839–1848 as a commander of the ship "Three Sanctifiers" he navigated the Black Sea every year. In 1850–1852 as a flagman he cruised over the Black Sea. In 1853 after joining the squadron of P.S. Nakhimov he participated in destruction of the enemy squadron at Sinop (1853). In 1854 N. was the chief of the second defense line in Sevastopol. For some time he acted as a commander of the Sevastopol Port and Military Governor. He was appointed the chief of troops assigned to the military post in Nikolaev.

Novosilskiy F.M. (http://samlib.ru/img/w/werjuzhskij_n_a/gubernator/stb7.jpg)

Novyi (New) Afon – a seaside village, climatic resort in the Gudauta District, Republic of Abkhazia. The railway station with the same name locates 18 km to the northwest of Sukhumi. It is situated on the Black Sea coast of the Caucasus, on the mountain slopes overgrown with rich subtropical vegetation. It appeared in place of the Thracean colony that existed from the early first millennium B.C. The Russian Orthodox Monastery of St. Panteleimon, built on the Mount Afon (meaning in Greek "calm, desolate") on the Khalkidiki Peninsula in Northern Greece, founded in 1875 its daughter community in Russia. The place for it was found on the Black Sea coast in Abkhazia over the Psyrtsakh gorge (in the Abkhazian it means "silver fir spring"). This monastery was blessed as Novoafonskiy Simono-Kananitskiy and the settlement that appeared nearby was named "Novyi Afon". During the Soviet ruling this "religious" name was substituted for Abkhazian Psyrtsakh, by the name of a river flowing near the monastery walls. In 1948 the settlement was given the Georgian name—Akhali-Afoni ("New Afon"); in 1967 the original Russian name "Novyi Afon" was officially restored. Population—1,500 (2011).

The landscapes of N.A. are very diverse. The coastal zone locates at altitudes to 80 m, while the rest part of the resort—on the slopes of the Bzyb Range of the Greater Caucasus (at altitudes to 600 m). The ridges of the Greater Caucasus protect this resort from northern winds. A large park is stretching along the coastline of the resort where a wide variety of decorative subtropical vegetation can be found, such as palms, laurel, eucalypt, magnolia, oleander and others. The ponds with weeping willows locate near the very edge of the sea. The whole territory of N.A. is crossed by cypress alleys. The climate here is humid subtropical. The winter is very soft, without snow. The mean temperature of January is 6 °C. The temperatures below 0 °C are quite rare. The spring comes early and it is usually warn and sunny. The summer is very warm. Precipitation is 1,600 mm a year, the maximum being in late autumn and winter. Despite rather high humidity the summer heat is endured easily thanks to onshore and offshore breezes. The favorable climate of this place allows opportunities for aeroheliotherapy and thalassotherapy. The bathing season lasts from May through November; the temperature of sea water in summer is 24–26 °C. In the karst valley of the Tskhvara River 8 km from N.A. there are mineral water springs. This resort provides treatment of the people with the cardiovascular and respiratory diseases. The resort was founded in 1922. In 1961 there was discovered the Novyi Afon or Anakopean Cave being strikingly beautiful. There is a legend saying that the Orthodoxy preachers—Apostles Andrey the First Called and Simon Kananit found a hideout in one of the caves feeing from their pursuers.

Novyi Afon (http://www.blacksea-tours.ru/novyj-afon)

Novyi Svet – a small village on the southern coast of Crimea, Ukraine. The village was founded at the end of nineteenth century by efforts of one of the founders of Russian winemaking, Prince Lev Golitsyn. Prince founded here in his estate, a plant for the production of sparkling wines. Since the local nature favored winemaking, and the efforts of the Prince at the plant maintained the traditional technology of champagne, the wine came out of exceptional quality. First, the village was called Paradise, but in 1912 Russian Emperor Nickolay II visited the estate of Prince Golitsyn, who renamed Paradise in Novyi Svet (New World). In the village there are two surviving palaces of Golitsyn, in one of which there is a museum, while the other is notable for its Mauritanian architecture (depicted on the coat of arms of the village). Continues to operate a factory of sparkling wines. Population -1,100 (2012).

Nudelman Alexander Emmanuilovich (1912–1996) – Russian designer of arms. He was born in Odessa. N. graduated from the Polytechnics Institute. He was the Chief Designer of the Design Bureau OKB-16. He designed the best aircraft large-bore cannons used during World War II (KS-37, NS-23, NS-45). Creation of the most potent quick-firing aircraft cannon NS-37 at the height of the war was decisive for the Russian aircraft becoming dominating in air. During the whole war the Germans could not invent such powerful cannon. Hitting of aircraft with only one shell sent from such cannon was sufficient to destroy it. During the Great Patriotic War more than 8,000 cannons NS-37 were produced. N. also supervised construction of the USSR's first radio-guided anti-tank complex "Falanga", the USSR's first self-propelled air-defense system "Strela" (Arrow) and the world's first complex of the tank's controlled armament "Kobra". In 1962 he defended the doctoral theses where he developed the principles of design and construction of the automatic cannons of the next generation. He was the Laureate of the Stalin, three State and Lenin Awards; twice the Hero of Socialist Labor. The monument to him is constructed in Odessa.

O

Obitochnaya Spit – a sandy spit 20 km long in the northern part of the Sea of Azov, Ukraine.

Obreskov Alexey Mikhailovich (1718–1787) – Russian diplomat. From 1740 he was the resident of the Russian Embassy in Turkey and from 1752—resident in Turkey. He took part in negotiations on settlement of border conflicts between the Crimean Khanate and the Zaporozhie Cossacks. Due to his efforts Turkey took a neutral stand in the Seven Years' War (1756–1763). At the beginning of the Russian-Turkish War of 1768 he was arrested together with the whole staff of the mission. He took part in conclusion of the Kuchuk-Kainarji Peace Treaty. At the end of his life he works at the Collegium of Foreign Affairs.

Obzor – a village locating to the north of the Sunny Beach on the Black Sea coast between Varna and Burgas on the slopes of the Stara-Planina Mountains, Bulgaria. It was founded by Ancient Greeks. In Antiquity it was known as Heliopolis—"the town of the Sun". In the Byzantine times the emperors had their residence here. The ruins of the Roman Church of Jupiter can be found in the center of Obzor.

Oceanological Center of NASU – was established in 1999 in Sevastopol. It united the leading marine organizations of Ukraine. It incorporates the Marine Hydrophysical Institute (MHI) with such departments as experimental department (Katseveli), acoustic department (Odessa); Research Production Center "ECOSI—Hydrophysics" (Sevastopol); Research Center for Management of Natural Resources of the Shelf "Shelf" (Sevastopol); the Institute of Biology of the Southern Seas with such departments as Odessa Branch (Odessa), Karadag Nature Reserve (Karadag), Experimental Station "Batiliman" and Research Production Enterprise "Tekhryba" (Sevastopol); the Operation Center of the International Ocean Institute; State Research Production Enterprise on Underwater Technologies "Marine Technologies" (Sevastopol).

Ochakov – a city, center of the Ochakov District, Nikolaev Region, Ukraine. It is situated on the northern coast at the entrance into the Dnieper-Bug Liman. It was founded in the fifteenth century as a fortress, first Tatar and later Turkish, named Ozu-Kale (from where is its present name "Ochakov"). In the sixth to seventh centuries B.C. the ancient Greek colony of Alektor located here. In the fourteenth century the Ochakov area was incorporated into the Crimean Khanate. In 1492 in place of the destroyed fortress of Dashev Crimean Khan Mengli-Girey ordered to construct a new fortress of Kara-Kermen or Ozu-Kale (this is a year of Ochakov foundation). After the Crimean Khanate came under control of Turkey (in the late fifteenth century) the fortress was given a new name—Achi-Kale (Ochakov) and soon it was turned into a naval base—the stronghold of the Ottoman ruling in the Northern Prichernomorie. During the Russian-Turkish War of 1735–1739, on 2 (13) July 1737 the Russian troops seized Ochakov. But under the Peace Treaty of Belgrade of 1739 O. was left under the control of Turkey. In the Russian-Turkish War of 1787–1791 the Russian Fleet defeated the Turkish Fleet at Ochakov and in June 1788 the Russian Army commanded by G.A. Potyomkin besieged the fortress and on 6(17) December stormed it. Under the Treaty of Yassy of 1791 this fortress passed to Russia and was exploded. It was restored by the Russians during the war of 1877–1878 against Turkey. During World War I the fortresses of Ochakov and Kinburn had to protect the Dnieper-Bug Liman from the sea.

Today there are food, canned fish, oil-pressing and other industries are developed here. There is the Military-Historical Museum named after A.V. Suvorov in the city. Population—14,700 (2012).

"Ochakov" – a cruiser of the Black Sea Fleet. Its construction was started in 1901 and it was set afloat in Sevastopol in 1902. Its displacement was 6,645 tons, draftline length—132.5 m, speed—22 knots, crew—570 people. Armament: 12 of 152 mm cannons, 12 of 75 mm cannons, 12 of smaller cannons and 2 underwater torpedo tubes. In October 1905 after publishing of the Tsar Manifesto promising to the people the basic civil rights the Bolshevik fraction of Russian Social-Democratic Workers Party (RSDWP) in Sevastopol demanded discharge of political prisoners and improvement of conditions of sailors and workers. The city was engulfed by spontaneous rebelions and the military organization of RSDWP decided to lead the revolt. On 8(21) November the disorders among the sailors occurred on the cruiser. On the night of 13(26) November the crew banished the officers, elected the new commanding staff and delegates to the Soviet of the Sailor, Soldier and Worker Deputies, established contacts with the mutinous sailors of the Fleet, soldiers of the garrison and workers of Sevastopol. On 14(27) November the delegation invited Lieutenant P.P. Shmidt elected the Fleet commander to take command of the ship. The cruiser turned into the headquarters of uprising on ships and on the shore. The crews of torpedo boats "Svirepyi", No. 265, 268 and 270 and mine cruiser "Griden" joined "Ochakov" that stood at the exit from the Sevastopol Bay. The red flags were also hoisted on torpedo boats "Zorkiy" and "Zavetnyi", gunboat "Uralets", mine-layer "Bug", study ship "Dnestr" and 4 port vessels that stood in the Yuzhnyi Harbor. The sailors headed by P.P. Shmidt on the torpedo boat "Svirepyi" went along the squadron ships calling their crews to rebel. They released the participants of the rebellion on the armored ship "Potyomkin" that were kept on the floating prison "Prut" and captured the armored ship itself (by that time renamed into "Panteleimon"). However, the attempts of P.P. Shmidt to persuade other ships of the squadron to rise in rebellion ended in failure. The command of the Black Sea Fleet sent ships and troops for suppression of rebellion. On 15 November, on expiration of 2-h surrender ultimatum "Ochakov" was shelled from fortress and ship cannons. The cruiser was set on fire and it was shaken by explosions. P.P. Shmidt was arrested. In February 1906 the rebels among which there were 41 people from the "Ochakov" crew were tried. Shmidt, Antonenko, Gladkov and Chastnik were sentenced to death and on 6 March they were shot down on the Berezan Island. In 1907 the cruiser was renamed into "Kagul". By early 1910 it was restored completely. In 1912–1913 it was a part of the international squadron in Constantinople. During World War I it took part in the battles. After the 1917 February Revolution in Russia it was returned its previous name "Ochakov". The crew of the cruiser fought for the Soviet power during the Civil War. In 1918 when the ship was staying in Sevastopol the White Guard captured it and renamed into "General Kornilov" and in 1920 together with the remaining Fleet it was taken to Bizerte (Tunisia). In 1924 the cruiser was inspected by the Soviet commission headed by A.N. Krylov. But the planned transfer of the ships did not occur and in 1933 "Ochakov" was cut into metal in Brest (France).

Ochakov Battle – (1) Russian-Turkish War of 1735–1739. Ochakov was the main Turkish fortress in Northern Prichernomorie. On 30 July 1737 the Russian Army commanded by General—Field Marshall B.K. Munnich besieged the fortress. Its garrison together with the families numbered 17,000 people. On 1 July the fortress was shelled and on the next day the Russian troops stormed it. The long siege was impossible as the Russian troops had not enough supplies of food and fodder. Having approached the fortress the troops stopped in front of a deep trench. The attack failed. Under the violent fire from the fortress the troops being unable to move forward started to move back. The Turks left the fortress and started killing the retreating soldiers. Thanks to the accurate and intensive fire of the Russian ginners the city by that time was enveloped in fires and the gunpowder stores started exploding. Escaping from the fire the Turks left the city and retreated seaward. Having found that the marine gates were open and poorly protected the hussars together with the Cossacks rushed through them into the fortress. This forced the Ocahkov Governor to surrender. Four thousand people were took captives. The rest died mostly in fire and explosions. The Russian casualties were about 4,000. Because of food shortage B.K. Munnich could not use to the advantage this success and retreated to Ukraine having left 9,000 garrison in the fortress.

(2) Took place on 14–30 October 1737. The city was besieged by 50,000 Crimean-Turkish Army (20,000 of Turks and 30,000 of Crimean Tatars) commanded by Ientsh-Ali-Pasha and Khan Mengli-Girey. On 17 October 6,000 of janissary stormed the fortress, but received a strong rebuff. The storming on 25 and 29 October also ended in a failure. The Ochakov garrison not only defended itself, but also organized bold raids. Having lost during storming about 10,000

people and nearly the same number due to plague epidemics Ientsh-Ali-Pasha had to raise the siege and retreat to Bendery. The casualties of the Ochakov garrison were 2,000. In 1738 the Russians had to leave the fortress because plague was spreading in the steppes that caused death of the two-thirds of the Ochakov garrison. Not waiting for the plague to kill the remaining troops the Russian command ordered to leave Ochakov.

(3) Siege and seizure of Ochakov by the Russian troops commanded by General— Field Marshall G.A. Potyomkin on 1 July—6 December 1788 during the Russian-Turkish War of 1787–1791. The fortress was defended by 15,000 strong Turkish garrison commanded by Hassan-Pasha. After defeat of the Turkish Fleet by the Russian squadron in the Dnieper Liman the fortress was blocked. Potyomkin organized a passive siege that lasted for 5 months. In spite of the blockade the fortress garrison was not going to surrender. It acted energetically making numerous raids. When the winter came, about 40–50 people died every day from frosts. The cavalry lost all horses because of fodder shortage. The soldiers who lived in dug-out huts and were afraid to get frozen in the barren steppe asked the commander to start storming as soon as possible. All this urged Potyomkin to take energetic actions. And at the beginning of winter he decided to start storming. On 6 December 1788 when the air temperature dropped to minus 23 °C the striking unit of 15,000 soldiers in six columns started assault of the Ochakov fortifications. Having got across the trench and rampart the Russian troops rushed into the city where fierce battles were waged. The other part of the troops fought its way into Ochakov from the less fortified side, over the ice of the frozen harbor. In this battle about two-thirds of the Ochakov garrison died and 4,500 were taken captives. The Russian losses in this storming were about 3,000 people. After this attack Potyomkin ordered to destroy Ochakov being the last mighty outpost of the Ottoman Empire in Northern Prichernomorie. In honor of this victory the golden cross "For the Service and Bravery" was made for the officers participating in this battle and the special silver medal "For Bravery in Ochakov Storming"—for the lower-rank servicemen.

Ochamchira – a village, center of the Ochamchira Region, Republic of Abkhazia. It is situated on the Black Sea coast in the mouth of the Galidzga River. This is a railway junction. The tobacco fermentation plant, tea factory and other enterprises were found here. Ochamchira Bay is the naval base for Russian Coast Guard patrol ships. Population—5,300 (2011).

Odess, Odessos – presently Varna, Bulgaria. This was the Miletian colony in Thrace founded in the early sixth century B.C. About 50 B.C. it was seized and ruined by the Goths. In the times of the Roman ruling Odess belonged to the province of Lower Mezia. Some Antique monuments, including thermae, survived to the present.

Odessa – a city, center of the Odessa Region, Ukraine. It is situated on the coast of the Khadjibey (Odessa) Bay and is the marine gates of Ukraine. This is a large industrial, scientific and cultural center; a well-known resort area; the Black Sea port located 30 km from the Dniester mouth. O. is the junction of automobile and railway roads. Initially the fort of Kachibey located here that was built by Lithuanian Prince Vitovt and was ceded to the Grant Dutchy of Lithuania. The first mention about O. was found in the Polish Chronicles of 1415. In 1480 the Ottomans conquered Kachibey and renamed it into Gadzhibey (Khadjibey, Adzhibey). In 1764 the fortress of Eni-Dunya was built near the village of Khadjibey. This was a small fortress with two round towers and the tower gates. The height of the walls reached 7–8 m. In 1787 after beginning of the Russian-Turkish War the significance of this fortress had grown enormously. On 14 September 1789 the fortress was conquered by the troops commanded by Major General Jose de Ribas. Russia formally gained possession of Khadjibey as a result of the Treaty of Yassy of 1791. O. was founded on an order from Russian Empress Catherine II. 1794 is considered the year of O. foundation. According to one of the stories, the name "O." originates from Ancient Greek Odessos. It was thought that it located not far from the Khadjibey Bay and this circumstance played its role in renaming of Khadjibey. The very beneficial geographical location facilitated the vehement growth of O.

The second half of the seventeenth century and the eighteenth century were marked by important victories of the Russian and Ukrainian peoples in their joint struggle for liberation of Northern Prichernomorie from the Turkish and Tatar invaders during the Crimean Campaigns of 1687 and 1689 and the Turkish Wars of 1736–1739 and 1768–1774. The final stage of this struggle was the Russian-Turkish War of 1787–1791 when the Russian troops led by outstanding commander A.V. Suvorov won several victories—at Kinburn, Fokshany, Rymnik, Ochakov, etc. The young Black Sea Fleet commanded by F.F. Ushakov was also victorious. In the battle on 14(26) September 1789 the Russian troops commanded by General I.V. Gudovich and the unit of the Ukrainian Cossacks headed by Anton Golovatyi and Zakhariy Chepiga stormed and seized the fortress of Eni-Dunya that was on the way to Izmail. In December 1790 there was one more brilliant victory—seizure of Izmail. The Russian-Turkish War of 1787–1791 ended with signing of the Treaty of Yassy under which Turkey refused from its claims to Crimea, while Russia was returned the territories between the Yuzhny Bug and Dniester rivers where Khadjibey located. The carts of migrants moved to the liberated territories where the villages came to life again and new cities appeared, such as Kherson, Nikolaev, Tiraspol. Under guidance of A.V. Suvorov there were constructed new fortifications among which, in particular, were the fortresses of Tiraspol and Ovidiopol.

In 1793 the fortress on the bank of the Khadjibey Bay was also constructed for protection of the coastal area. The garrison consisted of 2,000 people. After defeat of Bendery in 1811 the fortress lost its significance. On 27 May 1794 Catherine II issued the ruling on construction of a new port city in place of Khadjibey. In 1968 the marine passenger terminal was constructed in this historical place. A memorial plate is attached to the wall of the terminal saying "Here on 22 August (2 September) 1794 the construction of the first structures of the future port and city of Odessa was started". This date became the birth date of a new port that on 8 February 1795 was given the name "Odessa". In 1803 Tzar Alexander I appointed Duc de Richelieu the Governor of Odessa. And rather soon O. became the third largest and most significant city in the Empire.

Appearance of O. being the southern "window to Europe" was very important for the Russian state. The city grew at an overwhelming pace having no precedence that time. The export of grain through the port of Odessa gave it the leading positions in the economics of the country. From 1819 the import traffic had increased significantly because O. was given the status of a "free harbor", the city of *porto-franco* (it was abolished in 1859). The status of *porto-franco* meant the very low taxes on the imported commodities or no taxes at all. Therefore, the prices of such commodities were much lower in O. than in other cities. In the second half of the nineteenth century Novorossiya (New Russia)—in this way the lands of Northern Prichernomorie liberated from the Turkish invaders were officially referred to—was turned into one of the most prosperous provinces of the Russian Empire and O.—into a large city with the population of 100,000 (by 1850).

Defense of O. during the Crimean War of 1853–1856 brought this city the first military honors. In 1854 for successful rebuff of the attempts of the British and French troops to land in Crimea O. was awarded the Certificate of Honor. The city was built according to the plan prepared by engineer Franz Pavlovich De Wollant (François Sainte de Wollant), Colonel of the Russian Army, and signed by A.V. Suvorov in 1793. The plan was amended several times by De Wollant himself, Jose de Ribas as well as other architects and engineers, but its main concept was kept intact and realized.

During the Great Patriotic War (1941–1945), Odessa fought against German and Romanian troops from August 5 to October 16, 1941. Since 13 August 1941 Odessa was completely blocked from land. Despite the land blockade the enemy failed to break the resistance of the defenders—Soviet troops were evacuated and transferred to increase 51th Special Army, defending the Crimea. In 1941–1944 Odessa was occupied by Romanian troops and was part of Transnistria. In early 1944, due to the advance of the Red Army German troops entered Odessa, and the Romanian administration eliminated. On 10 April 1944 Odessa was liberated by Red Army. During the occupation, the population of the city of Odessa has actively resisted the invaders. During the years of occupation, tens of thousands of civilians were executed in Odessa.

In the Soviet time much was made on rehabilitation and improvement of the city outskirts that were turned into modern beautiful districts of the city. Today Odessa is the biggest sea port in Ukraine, transporting grain, sugar, coal, petroleum products, cement, metals, jute, wood, products of machine-building industry. Naval base (in the past) and the fishing fleet. A major railway center. The main industries—oil processing, machine building, metalworking, light industry, food, timber, agricultural, and chemical industries.

Odessa (Photo by Evgenia Kostyanaya)

Odessa is a cultural, educational and scientific center. There are: Odessa National University named after I.I. Mechnikov, founded on 1(13) May 1865, which is the ancestor of many of today's institutions of higher education in Odessa, based on its faculties. Among them—Odessa State Medical University, Odessa National University of Economics, National University "Odessa Law Academy", Odessa State Agrarian University. Odessa National Polytechnic University, founded in 1918, is also the ancestor of many high schools in the city: Odessa National Maritime University, Odessa State Academy of Civil Engineering and Architecture, Odessa State Academy of Refrigeration, Odessa National Academy of Telecommunications, Odessa National Academy of Food Technologies. In Odessa there is a number of research institutes and laboratories, including the oldest research institute of higher education in Ukraine—Research Institute of Physics of

Odessa National University named after I.I. Mechnikov. A large area above the sea shore is covered by the Botanical Garden of the Odessa National University named after I.I. Mechnikov.

A prominent place in the cultural life of the city and the country belongs to Odessa National Academic Theater of Opera and Ballet. Its building, built in the years 1884–1887, has a unique architectural structure. There are 10 other theaters in the city. There are many memorial places connected with historical events. The city is a popular tourist destination with many seaside climatic and mud resorts (Kuyalnitskiy, Khadjibey, Lermontovskiy and others).

Odessa is often called "Odessa-Mother", "Southern Palmira", "Black Sea Babylon", "South Capital" or "Small Paris". Population −1.35 mln (2012).

Odessa National Academic Theater of Opera and Ballet (Photo by Evgenia Kostyanaya)

Odessa Battle (Crimean War of 1853–1856) – bombardment of Odessa on 10 April 1854 by the British-French squadron of ships commanded by French Admiral Ferdinand Alphonse Hamelin carrying 1,760 cannons onboard. The city was defended on the seaside by hastily built batteries having at their disposal 103 cannons. The most tough artillery confrontation was between the 6th battery commanded by warrant officer Schegolev (4 cannons and 28 gunners) and 12 British-French ships that came up most close to the coast in this place (1,500 m). Regardless of the overwhelming superiority of the allies in fire power (350 cannons against 4), Schegolev exchanged fire during 6 h and succeeded to set on fire 2 frigates. Only after the greater part of the gunners was killed and 2 cannons were destroyed Schegolev plugged the remaining cannons, lined the remaining gunners and under the drum beats they retreated to the city. For their bravery all gunners of the 6th battery were awarded the George Crosses and warrant officer Schegolev was conferred the rank of Junior Captain. The attempt of landing of the British troops on this day was stopped by the fire of on-land cannons. On 12 April having failed to achieve serious success the British-French squadron left the Odessa region. Bombardment of Odessa urged the Russian command to keep large forces in this region that made easier the landing of ally troops on the Crimean Peninsula.

Odessa Botanical Garden – it was founded in 1819 on the initiative of General-Governor A. Langeron by well-known specialist on rose cultivation Charles Dessemet. Originally the garden was created for supply of the Odessa population with the planting material—seedlings of fruit and decorative plants. When the director of the garden was A.D. Nordman he organized the Central Gardening School that also comprised the sericulture division. The garden was situated near the French boulevard. About 500 varieties of flora of Southern Russia were systemically arranged and grown here. The exchange of plant material with botanical gardens of Petersburg, Berlin, Vienna and Paris was organized. Beginning from 1841 the garden conducted regular meteorological observations of atmospheric pressure, air temperature and precipitation. The garden was the experimental base of the Odessa State University for study of plant acclimatization and landscaping of the Prichernomorie steppes. Today the main direction of the O.B.G. activity is protection, research and enrichment of the plants; research work on introduction, breeding, and efficient use of plants; promotion, training and education on botany, conservation, crop production, ornamental horticulture, park construction and landscape architecture.

Odessa Commercial Sea Port – is situated in the southwestern part of the Odessa Bay on the artificial territory of 141 ha in area. The largest port in the Black and Azov sea basin. It locates in the northwestern part of the Black Sea on the historically established trade routes between East and West. It merges with the city of Odessa—the major industrial, cultural and resort center of Ukraine. The port configuration that was mainly shaped in 1870–1890 has survived to the present. In those years the Reidovyi, Novy, Zavodskoy and Neftyanoy jetties, Potapovskiy and Androsovskiy quays (they were called by the names of contractors that built them) were constructed. The Port development was spurred significantly by transition of the marine fleet to steam engines and construction in 1865 of the railroad that connected Odessa with the key economic centers of Ukraine, Caucasus as well as with Moscow and Petersburg.

The Port was restored from ruins twice. In 1918–1919 when the Austrian-German troops, British-French invaders and White Guard seized Odessa the port facilities were in disastrous conditions. Still greater damage was incurred to the Port by the fascists during temporary occupation of Odessa during the Great Patriotic War. While leaving Odessa in April 1944 they turned the Port into ruins. By 1950 the Port was restored. In 1967 for great exploits the Port of Odessa was awarded the Order of Lenin.

Today, the Port consists of the Karantinnyi, Novyi, Kabotazhnyi, Prakticheskiy, Zavodskoy, Rabochiy and Neftyanoy harbors. The Port numbers over 50 wharfs. The harbors are protected with piers and breakwaters. The Port is capable to receive large cruise passenger vessels and has the modern marine terminal. This is one of the principal bases for development of sea carriages and tourism. The Port has 54 protected berths with depths to 8–13 m. Six berths are capable to handle tankers to 330 m long with draft to 13 m. The passenger complex with its five berths is designed for receiving passenger vessels to 240 m long and with draft to 11 m.

The port's technical capacities allow handling of more than 21 mln tons of dry and 25 mln tons of bulk cargoes annually. Container terminals provide handling of over 900,000 TEU per year. The passenger terminal is capable to serve up to four million tourists a year. Such types of cargoes as oil and oil products in bulk, liquefied gas, tropical and vegetable oils, technical oils, containers of all types and sizes, ferrous and nonferrous metals, ore, pig-iron, raw sugar in bulk, grains in bulk, perishables in containers, various cargoes in bags, boxes, packages, big-bags and integrated cargo units, motor transport are handled in the port. Cargoes potentially dangerous environmentally is an exception. Within the port territory there are eight production-handling terminals for dry cargoes handling, passenger terminal, oil harbour and two container terminals, terminals for vegetative and technical oils processing, specialized berths for Ro-Ro type vessels and handling of grains.

On 23 March 2000 the Supreme Rada of Ukraine passed the law "On Special (Free) Economic Zones on the Territory of the Odessa Commercial Sea Port". This zone was established for 25 years within the artificially backfilled and washed-in territory of the Karantinyi jetty. The area of this zone is 32.5 ha. According to this law a special customs regime is applied in this zone. The Port of Odessa is connected with Pan-European Transport Corridor No. 9. The volumes of container handling are growing especially quickly in the Odessa Port.

Odessa was the starting point of the popular passenger Crimea-Caucasus lines to Ochakov, Nikolaev, Kherson, Rostov-on-Don, Izmail and also permanent lines abroad Odessa—Varna, Odessa—Nesebr, Odessa—Marseilles, Odessa—Venice, the Near East line with calling on the main ports of Romania, Bulgaria, Turkey, Greece, Egypt, Syria, Lebanon and Cyprus. Since 1990s regular passenger lines and cruises on the Black Sea ceased completely due to absence of Ukrainian and Russian tourist liners. In summer 2013 the first regular cruise line Odessa-Sevastopol-Novorossiysk-Sochi-Feodosiya-Yalta was put in operation. Now Odessa is visited frequently by cruises coming from the Mediterranean Sea.

Odessa Commercial Sea Port (Photo by Elena Kostyanaya)

Odessa group of resorts – includes climatic and balneo resorts, such as Arkadia, Bolshoy Fontan, Karolino-Bugaz-Zatoka, Kuyalnitskiy (Kuyalnik), Luzanovka, Lebedevka, Lermontovskiy, Malodolinskoe, Kholodnaya Balka, Khadjibey, Chernomorka and the resort area Shabo. These resorts locate along the Black Sea coast within Odessa and to southwest of it and also near Kuyalnitskiy, Khadjibey and Sukhoy (Malodolinskiy) saline limans, mostly in the coastal areas of the Prichernomorie Lowland stretching for 70 km from Luzanovka to the Karolino-Bugaz Spit. The climate typical of the seaside steppe areas is prevailing here. Warm waters near the Black Sea coast (to 24–26 °C in summer), small waves and wide sandy and pebble beaches provide excellent opportunities for thalassotherapy. The best period for bathing is June-August. The important curative factor of the Odessa resorts—sea air saturated with salts of chloride, bromine and iodine makes possible the aeroiontherapy here. Apart from the climate, one more effective curative factor found here is liman and lake mud. Liman brine is used in bath treatment. In some resorts the mineral waters produced from wells are also used in treatment. The Odessa resorts are intended for treatment of patients suffering from chronic loco-motor, nervous, gynecological, respiratory diseases. The resorts in Odessa have existed for over 170 years. A great role in resort development was played by the Odessa Balneological Society that was founded in 1867. Most popular were the famous limans—Kuyalnitskiy, Khadjibey and Sukhoy (Malodolinskiy). After the 1917 October Revolution in Russia, in June 1919, notwithstanding the very hard times for the country there were opened several sanatoriums on the Kuyalnitskiy and Khadjibey limans as well as a mud bath treatment center in Kholodnaya Balka. In 1921 the resort Arkadia was constructed and, in general, the mud and climatic resorts were quickly developing that time. During the Great Patriotic War the occupation of Odessa by the Germans that lasted from October 1941 to April 1944 incurred great damage to the resorts, but they were restored rather quickly after the war.

Odessa Hydrometeorological Institute (OHMI) – it was founded on 1 May 1932 following the resolution of the Government of Ukrainian SSR in Kharkov that was the capital of Ukraine that time. Initially it was called the Kharkov Engineering Hydrometeorological Institute. At the beginning of the Great Patriotic War the Institute was evacuated to the city of Ashkhabad (Turkmen SSR) where it continued its work till August 1944. By the decision of the USSR Council of People's Commissars of 4 July 1944 the Kharkov Engineering Hydrometeorological Institute was transferred from Ashkhabad to Odessa and renamed into the Odessa Hydrometeorological Institute (OHMI). At present OHMI is the leading institution in the system of higher hydrometeorological education of Ukraine, the basic institute of the Ministry of Ecology and Natural Resources of Ukraine. It prepares specialists for the following areas "Hydrometeorology", "Ecology", "Management" and "Computer Sciences". On the basis of OHMI there were established such study complexes as Ukrainian Center for Hydrometeorological Education and Educational-Research Center "Ecology" which, together with OHMI, incorporate the Kharkov and Kherson Hydrometeorological Colleges, Kiev Geological Survey College, State Institute for Personnel Qualification Advancement and Retraining of the Ministry of Ecology and Natural Resources of Ukraine, Odessa Branch of Institute of Biology of the Southern Seas of National Academy of Sciences of Ukraine. Within the OHMI framework the research-methodological commission on hydrometeorology of the Ukrainian Ministry of Education and Science is operating. OHMI has at its disposal one of the region's best material and technical base. The computer center, study laboratories provided with the modern electronic hydrometeorological equipment and automatic complexes for environment monitoring, personal computers were organized to support the education process and researches. Functioning are also the study bureau of meteorological and hydrological forecasts, the study bureau of reception and interpretation of satellite information, practical research-study bases on the Dniester River (village of Mayaki) (for hydrologists and hydroecologists), on the Black Sea coast near Otrada (for oceanologists and meteorologists), in the village of Chernomorka (for meteorologists, radiometeorologists and agrometeorologists).

Odessa Limans – the name of three brakish limans: Kuyalnitskiy, Khadjibey and Tiligulskiy locating near Odessa.

Odessa National Maritime University – was founded in 1930 as Odessa Institute of Water Transport Engineers. In 1945 it was renamed into the Odessa Institute of Marine Fleet Engineers and in 1994 into the Odessa State Maritime University. In

2002 by the decree of the Ukrainian president the University was given the status of national. This is the Ukraine's only educational institution that prepares all kinds of specialists for water transport. The University also has a technical library with more than 700,000 books, including the unique collection of books on sea-related professions. The students are provided an opportunity to have their practical courses in large shipping companies, sea and river merchant ports and in other organizations of marine and river transport of Ukraine. The university consists of seven faculties and 2 centers: Shipbuilding, Marine Engineering, Mechanization of ports, Sea transport and offshore facilities, Transportation technologies and systems, Economics and Management, Law, Centre of Postgraduate Education and Training, Center of preliminary training of young people. 418 teachers, including 59 professors and doctors of sciences and 210 candidates of sciences and associate professors work at 33 departments of the university. Number of students—5,340.

Odessa National Maritime Academy – it was opened in 1954 as the Higher Engineering Maritime College being one of the world's leading centers of higher maritime education. It prepares qualified naval specialists for the marine, river, fishery fleet and also for organizations of marine transport. It has a study and research complex consisting of the following faculties and centers: Sea Navigation Faculty, Maritime and Inland Waterway Navigation Faculty, Engineering Faculty, Electrical Engineering and Radio Electronics Faculty, Maritime Law Faculty, Azov Maritime Institute of O.N.M.A. and Izmail Faculty, Military Training Department, Foreign Students Dean Office Professionals are trained in 45 departments by 350 teachers, among them 74 doctors of science and professors, 175 candidates of science and assistant professors. Nowadays cooperation with foreign partners (maritime educational institutions, scientific establishments and scientific and engineering institutions, shipping and crewing companies), as well as upgrading the quality of training process and meeting the international standards take an important place in the work of Odesa National Maritime Academy.

Odessa Observatory – in 1865 Professor V.I. Lapshin organized regular meteorological observations at the Chair of Physics and Physical Earth Sciences of the Novorossiysk University in Odessa, thus, laying the basis for future O.O. The observation results were published in the newspaper "Odessa Newsletters" and were sent via telegraph to Petersburg and Paris. The essential contribution into the work of O.O. was made by A.V. Klossovskiy who in 1883 became its head. Understanding the importance of climatic data for Southern Ukraine being a rich agricultural region he, working in this observatory, not only created a network of meteorological stations in the southwest of Russia, but also organized studies of marine environment parameters. Regular hydrological observations initiated in Odessa in 1874 near the Richelieu Lighthouse were gradually extending. Observations were started by the sea level gauge of the Vorontsovskiy Lighthouse in the Karantinnaya Harbor. Among the most interesting works written by A.V. Klossovskiy before 1895 A.I. Voeikov noted "The strong winds and storms in the east of the Black Sea", "Key moments in the farming history", "Climate of Odessa", "Dynamics of meteorological elements in Kiev", "Precipitation in

Southwestern Russia, their distribution and forecast", "On organization of the physico-geographical investigations in the southwest of Russia", "On temperature fluctuations in the coastal zone of the Black and Azov Seas" and others. Oceanographic investigations in Odessa became wider after opening in 1913 of the hydrometeorological station of category I near the Karantinnyi jetty. One of the tasks of this station was study of the northwestern part of the Black Sea, servicing of various organizations, first of all, shipping company, organization and control of operation of ship meteorological stations.

Odessa Region (Oblast) – was formed on 27 February 1932 in Ukraine. In 1956 the Izmail Region was included into it. O.R. has an area of 33,310 km^2, its population is 2.4 mln people (2013), of which 1.6 mln—urban and 0.8 mln—rural population. The Region includes 26 districts, 19 cities, 33 urban-type settlements. The administrative center of the Region is Odessa.

O.R. is the far southwestern area of Ukraine. It borders on the Vinnitsa, Kirovograd, Nikolaev Regions, on Romania and Moldavia. The greater part of the territory is covered by the Prichernomorie Lowland gradually dropping into the Black Sea. It is broken by numerous rivers, ravines, gullies that impart a rolling nature to its relief. The eastern part of the coast is of the liman type and only the Dnestrovskiy Liman is connected with the sea. The Kuyalnitskiy, Khadjibey and other limans are separated from the sea with the sandy-shell bars. The western part of the coast between the Dnestrovskiy Liman and the Kilian gyrlo of the Danube abounds in lakes: Sasyk, Shagany, Alibey, Budakskoe and others. In the northeast of the Region there is the Volyno-Podolskiy Upland (rising to 270 m) that is cut by deep gullies and ravines. Among the mineral reserves there are found the deposits of construction materials—granite, gneiss, limestone and others.

The climate here is warm and dry. The mean temperature of January is from −2 °C to −5 °C, of July from +21 to +23 °C. The precipitation is 300–400 mm a year. The largest rivers are Yuzhnyi Bug, Dniester and Danube. In the south of the Region there are small rivers that dry out in summer. Many lakes (including Katlabukh, Yalpug, Kagul and others) are found in the Dniester and Danube river valleys. The central and northern parts of the Region have many ponds with a total area 4,500 ha. The Danube delta and Dniester plavni are waterlogged in some places. The Region lies in the steppe and forest-steppe zone. The soils here are largely black-earth, in the north podzolic, while in the south and southwest they turn into dark chestnut. Meadow soils are developed in the river valleys. Some parts of watersheds and valleys of rivers, gullies are overgrown with forests. The main varieties found here are oak, beech, ash and linden. In the plavni the willow, black alder, thickets of cane and reed are growing.

The indigenous population is the Ukrainians, while the Russians, Moldavians and other nationalities are also living here. The largest cities are Odessa, Izmail, Belgorod-Dnestrovskiy, Ilyichevsk and others.

O.R. is an economic and administrative region. This is mostly the agricultural region being a major producer of grain and sugar. Fishing is developed on the Black Sea and on the Danube. The Region provides nearly the half of the fish catches of

Ukraine. The main industries here include machine-building and metal working, ship repair and food processing. Of minor significance are textile, garment, and footwear industries. Developed is also the industry of building materials. Navigation is developed along the Danube and Dniester rivers. The main sea ports are Odessa, Ilyichevsk, Yuzhnyi and Izmail. In the Odessa port reloading of oil and oil products makes two-thirds of the cargo turnover. The Ilyichevsk Port has a railway terminal that services crossings to Varna and Poti. Here the Ukraine's largest container terminal is found. The major port on handling chemical cargo is Yuzhnyi. The food industry takes the greatest share in industrial production of O.R. The Region has processing industries, while the mining industries are lacked. The industry services agriculture and merchant fleet. O.R. is the main producer of grapes in Ukraine. An important role is played by the foreign trade.

Odessa-Brody Oil Pipeline – it was built in 2001 for transit of the Caspian oil to Europe. Its length is 667 km, the carrying capacity at the first stage is 9 mln tons a year and the maximum—40 mln tons a year. The oil pipeline was operated in the reverse mode delivering Russian oil to Odessa for its further export through the port of Yuzhnyi. Ukraine is considering prolongation of this oil pipeline as far as Gdansk in Poland.

"Ognennaya Liniya " (hot line) – the name of the sea route Kerch-Mariupol along which special vessels carried hot agglomerate.

"Ognennaya Zemlya" – in this way a small base area near the fishery village of Eltingen (presently Georgievskoe near Kerch, Crimea, Ukraine) was called in 1943. This was the landing place of the Soviet troops who during 40 days and nights kept this area rebuffing the German attacks.

Oil pipeline – a complex of structures for pumping crude or treated oil between two points on the land and on the sea to great distances (up to thousands kilometers) from its production place to the places of refining or transportation. O.P. consists of a pipeline proper, pumping stations, communication means and accessory facilities. In the Prichernomorie area oil is transported via pipelines Baku—Supsa (Georgia) (926 km), Baku—Novorossiysk (1,411 km), Tengiz—Novorossiysk (1,589 km), Baku (Azerbaijan)—Tbilisi (Georgia)—Ceyhan (Turkey) (1,730 km).

Oil pollution – there were many accidents on the Black Sea that resulted in oil spills. Thus, in 1986 in the port of Ilyichevsk the tanker "Uzhgorod" spilled 40 tons of oil; in 1997 in the port of Odessa the ship "Athenian Faith" (under the Maltese flag) spilled 50 tons of oil; in November 2007 due to a severe strom in the Kerch Strait a small tanker "Volgoneft-139" broke apart and spilled 1,000 tons of fuel oil (masut). So far oil pollution on the Black Sea has not reached the scale of environmental disaster. In addition oil gets into the sea with storm wastewaters and river waters. There are several natural oil seepages from the sea bottom. About 200–250 oil spills (generally released from ships) are yearly detected by satellite methods in the Black Sea. According to some estimates, in total the Black Sea receives every year about 100,000 tons of oil and oil products.

Oktyabrsk special sea port – was founded in 1965. It locates 25 km from the city of Nikolaev on the left bank of the Dnieper-Bug Liman, Ukraine. One of the most modern ports in Ukraine. It is designed for handling oil and ore carrying vessels, container vessels, ferries. The Port has 7 deepwater berths with a total length of 1.9 km. The navigation is year-round, but in winter only vessels certified as ice-breakers are permitted for shipping. The Port is capable to handle to 1 mln tons of general cargo. It can receive vessels 215 m long and with freshwater draft 10.4 m.

Oktyabrskiy (Ivanov) Filipp Sergeevich (1899–1969) – Soviet Admiral. In December 1918 he volunteered to serve on the Baltic Fleet. From 1920 to 1938 he served on various ships of the Baltic and Pacific Fleet. In 1939–1943 he commanded the Black Sea Fleet. In July 1940 he participated in formation of the Danube Fleet. In 1941 he was conferred the rank of Rear Admiral. By the beginning of the Great Patriotic War (1941–1945) due to his efforts the Black Sea Fleet was ready for the war and it was the first unit that met the enemies fully armed. During the war he was responsible for laying mines, formation of sea brigades, repair of ships and evacuation of the population. He organized the aircraft and ship attacks on the oil product warehouses in Romania. At approaching of the Germans he focused his efforts on defense of Odessa and later Sevastopol. In November 1941 he was appointed the commander of the Sevastopol defense region. Near Sevastopol he organized support of the land troops by the Fleet. O. took part in development of the Kerch-Feodosiya operation and commanded its realization. He ensured supply of Sevastopol and actions on enemy communications. After defeat of the troops of the Crimean Fleet in May 1942 he organized their transfer to the Taman Peninsula. On 1 July 1942 on the last plane he left Sevastopol that was impossible to defend any more. He commanded the operations of the Black Sea Fleet from the command posts on the Caucasus. After the unsuccessful attempt of landing near Novorossiysk in February 1943 he was dismissed from the Black Sea Fleet. In 1943–1944 he commanded the Amur (River) Fleet. In 1944–1948 he was again appointed the commander of the Black Sea Fleet. In 1944 he was promoted to Admiral. He took part in development of the military campaign of 1944 on the Black Sea and operation on liberation of Crimea. He commanded the actions of the Black Sea Fleet and Danube Fleet during the Yassy-Kishenev operation. In 1946 he was appointed a member of the Supreme Military Council of the Armed Forces keeping the post of a Fleet commander. In December 1948 he became Deputy Commander-in-Chief of the Navy. In April 1952 he was appointed the Chief of the Navy Research Center. In 1954–1957 he retired. In 1957–1960 he was the director of the Black Sea Higher Naval School named after P.S. Nakhimov in Sevastopol. In September 1960 he was included into the group of general inspectors of the USSR Ministry of Defense. He was awarded many orders and the title of the Hero of the Soviet Union.

Oktyabrskiy F.S. (http://litopys.net/img/thisday/Julay/08-07/oktyabrsky_1938.jpg)

Olginka – a village on the Black Sea in Tuapse Region, Krasnodar Kray, Russia. Olginskaya village was founded in 1864 as part of the coastal Shapsugskiy battalion (regiment).The village is located in the valleys of the river Tu and its tributary Kabak, flowing into Olginskiy Bay of the Black Sea. "Olginka", "Kuban", "Agriya", "Orbit" and "Gamma" health centers are located in Olginka. Population—2,250 (2010).

Olive – trees of the *Oleaceae* family. There are known to 60 varieties of this tree. It originates from Near East. Of economic importance is cultural O.—an evergreen subtropical fruit tree 4 to 12 m high. The flesh of its fruits contains to 56 % of oil. The fruits are used for eating (fresh and canned) and for making olive (Provence) and olive kernel oil. It is cultivated on all continents in subtropical regions. About 88 % of O. plantations are found in Spain, South France, Italy, Greece, Turkey and other Mediterranean countries. It is also grown in Crimea and on the Black Sea coast of the Caucasus, in Azerbaijan and Turkmenistan.

Olvia – (from the Greek "rich, happy") the Miletian colony near the mouth of Gipanis (Bug) and Borisphen (Dnieper) of the fifth to fourth centuries B.C. This was the prosperous trade (with wheat and slaves) and fishery center. It had connections with the Scythians. This slavery city-state was founded in Antiquity in the early sixth century B.C. The territory of O. (presently the Ochakov and Nikolaev Regions) is a nature reserve. In 60 A.D. O. was conquered by Dacian King Burebista. Excavations of the Russian and Soviet archeologists led by B.F. Farmakovskiy revealed the ruins of the fort walls and towers, double gates, agora and temple devoted to two Gods—Appollo Delphinium and Zeus. The excavations of O. are going on.

"On annexation of the Crimean Peninsula, Taman Island and the whole Kuban side into the Russian Empire" – the manifesto issued by Catherine II on 8 April 1783. For many years the main rival of Russia in the struggle for the Crimean Peninsula was Turkey that controlled Crimea. However, the last Crimean Khan abdicated and fled under protection of the Russian Army, while the Tatar nobility that was hostile towards Russia took refuge in the Ottoman Empire. This immersed Crimea into the economic and political chaos that facilitated bloodless annexation of Crimean into Russia.

"On use of the Crimea for treatment of laborers" – a decree signed by V.I. Lenin on 21 December 1920. After this decree the Crimean Peninsula was gradually turned into the health improvement resort for the whole country. The Central Department for Crimean Resorts was established in Simferopol. Its first head was D.I. Ulyanov. The full text of the decree is engraved on the monument constructed in his honor in Yalta at the entrance into the Primorskiy Park.

Opuk – a cape near Feodosiya, Crimea, Ukraine. The cape is wide and obtuse. The mountain of Opuk 185 m high is rising 1.5 km northward of the cape. The slopes of the mountain are smoothly running down to the sea and end with the O. cape. The Ship-Stone Rock rises from the sea 4 km to the southwest of the cape.

Orbeli Ruben Abgarovich (1880–1943) – the founder of the Soviet underwater archeology. In 1903 he graduated from the Law Department of the Petersburg University and was left to work at the Chair of Civil Law. In 1904 he was sent to Germany for continuation of his education and in 1906 at the Jena University he defended thesis and was awarded the academic degree of Doctor of Law. In the same year he defended thesis for the magister degree at the Petersburg University. After the 1917 October Revolution in Russia he worked at the Tambov State University (from 1918—Professor of this University). In the period from 1924 to 1934 he worked in the USSR Academy of Sciences and was elected a member of the Leningrad Soviet of the workers, peasant and red army deputies. In 1934 he started studying the history of diving for ship raising and the emergency-rescue works. His studies covered a large period from the Antiquity to the twentieth century. Knowledge of ancient classic and European languages and extensive erudition permitted O. to cope brilliantly with this task and to prepare a series of articles some of which were published in journals "EPRON" and "Ship Raising". He studied with great interest the works of Leonardo da Vinci devoted to the methods of underwater swimming. In 1935 O. became a member of the scientific-research council of the Chief Department of EPRON. In 1937 he became interested in underwater archeology. He prepared a large plan of actions on study of the submerged cities of Northern Prichernomorie, strengthening of banks for preservation of the dilapidating monuments, preparation of the hydroarcheological map of the country. With the help of the EPRON's divers he conducted investigations in Olvia and in the village of Sabatinovka on Yuzhnyi Bug where the famous

Sabatinovka boat was raised from the river bed (presently it can be found in the Central Naval Museum in Saint-Petersburg). In 1939 he conducted underwater archeological works in Feodosiya, Koktebel and Kerch. The results of his numerous works were published in the book "Investigations and Researches" (1947).

Ordu – the Black sea Il (province) in Turkey. Its area is 6,142 km^2, population— 719,200 (2010). Center—Ordu. Its adjacent provinces are Samsun to the northwest, Tokat to the southwest, Sivas to the south, and Giresun to the east. Ordu province has a strip of the Black Sea coast and the hills behind, historically an agricultural and fishing area and in recent years, tourism has seen an increase, mainly visitors from Russia and Georgia, as Ordu boasts some of the best beaches, rivers, and lush, green mountains on the Black Sea coast. Walking in the high pastures is now a popular excusrion for Turkish holidaymakers. Ordu is famous for hazelnuts. Turkey as a whole produces about 70 % of the world's hazelnuts, and over 50 % of those come from Ordu.

Ordu—(from ancient Greek "Cotyora") – a town and port on the Black Sea coast; administrative center of the Ordu Il, Turkey. It is located 150 km eastward of Samsun. It lies at the crossing point of highways on the route Trabzon-Sivas. In the eighth century B.C. the settlement of Cotyora was founded in this area as one of a string of colonies along the Black Sea coast established by the Miletians. The area came under the control of the Danishmends, then the Seljuk Turks in 1214 and 1228 and Haci Emir Ogulları Beyligi in 1346. Then it passed to the dominion of the Ottomans in 1461 with the Empire of Trabzon. The modern city was founded by the Ottomans as Bayramlı near Eskipazar as a military outpost 5 km west of Ordu. In 1869, the name was changed to Ordu. Today, it is known for manufacturing of copper dishware, oil-pressing, tea and timber-working industries. The hazelnuts and soy beans are cultivated here. Fishing of hakes, mackerel and anchovy is also developed here. Today the city is the centre of a large hazelnut processing industry, including *Saǧra*, one of the largest Turkish hazelnut processors and exporters, and *Fiskobirlik*, the largest hazelnut co-op in the world. The *Saǧra* factory shop, selling many varieties of chocolate-covered hazelnuts, is one of the town's attractions. The Boztepe aerial tramway is another popular attraction which is set to become a modern symbol for the city. The ruins of the ancient Greek city of Cotyora may be found here. Population—147,900 (2012).

Organization of the Black Sea Economic Cooperation (BSEC) – was founded on 25 June 1992 in Istanbul. Its founders are 11 states of this region, such as Republic of Azerbaijan, Republic of Albania, Republic of Armenia, Republic of Bulgaria, Greek Republic, Georgia, Republic of Moldova, Romania, Russian Federation, Turkish Republic and Ukraine. According to the BSEC Statutes the organization has a status of the "Observer" that is given to Austrian Republic, Germany, Arab Republic of Egypt, Israel, Italian Republic, Republic of Poland, Republic of Slovenia, Tunis Republic and French Republic. Some international organizations also take part in BSEC activities—UN Economic Commission for Europe (UNECE), UN Economic and Social Commission for Asia and the Pacific

(UN ESCAP), European Conference of the Ministers of Transport (ECMT), European Commission (EC), Parliamentary Assembly of the Black Sea Economic Cooperation (PABSEC), Business Council for Black Sea Economic Cooperation (BCBSEC), Black Sea Bank for Trade and Development (BSBTD) as well as representatives of nongovernmental organizations, such as Black Sea Regional Association of Shipbuilders and Ship Repairers (BRASS), Black Sea International Shipowners Association (BINSA), Black and Azov Seas Ports Association (BASPA). According to the Statutes, the BSEC member countries cooperate in the following areas: trade and economic development, banks and finance; communication; power engineering; transport, agriculture and agroindustrial complex; healthcare and pharmaceutics; environment protection; tourism; science and technology; exchange of statistical data and economic information; cooperation among customs and other border bodies; public relations; control of organized crime, illegal turnover of narcotics, arms and radioactive materials, all acts of terrorism and illegal migration or in any other related areas as decided by the BSEC Council. The main permanently functioning decision-making body is the Council of Foreign Ministers of member countries. Some concrete functions on improvement of regional contacts are performed by the Regional Energy Center (Varna), International Center for Black Sea Studies (Athens), Balkan Centre for Small and Medium-sized Enterprises (Bucharest), Black Sea Trade and Development Bank (Thessaloniki) and others. Among large-scale projects of BSEC there are: construction of the ring road around the Black Sea connected to the trans-European highways, united energy system of the BSEC countries, trans-regional fiber-optic communication line (Palermo-Istanbul-Odessa-Novorossiysk) and others.

Orlov Nikolay Alekseevich (1786–1861) – Prince, Russian military and state figure, diplomat. In 1801 he went to service in the Collegium of Foreign Affairs, took part in the war of 1812. The efforts of O. resulted in signing in 1829 of the Adrianopol Peace Treaty. During negotiations he behaved with great subtlety and skill. In 1833 he signed the Treaty of Hunkar Iskelesi on the conditions very beneficial for Russia. In 1856 he was the first Russian envoy at the Paris Congress that put an end to the Crimean War.

"Orlyonok"—Russian Children's Center "Orlyonok" (Eaglet) – Federal State Educational Center, the largest in the modern Russia, for recreation and health of children, which annually receives about 20,000 childrens aged from 11 to 16 years. Located next to the villages of Novomikhailovskoe and Plyakho, Tuapse District of Krasnodar Kray, Russia. Founded in 1959.

Territory is more than 300 ha and extends along an arc of sandy beaches in the Golubaya (Blue) Bay from an unnamed creek in the north to Cape Guavga in the south. From the West it is washed by the Black Sea, and on the East by low wooded hills. "O" from east to west is crossed by several unnamed streams and Plyakho River. The climate is mainly determined by the warm Black Sea and is close to subtropical. The average daily temperature for May to October is 20–24 °C. The most favorable period for swimming (the water temperature above 17 °C) lasts from mid-May to mid-October. The annual rainfall is 1,000–1,300 mm. The Center unites 9 children camps. More than 1,500 employees of the children's center provide an annual vacation of about 20,000 children from 89 regions of Russia, CIS and foreign countries.

Ostroumov Aleksey Aleksandrovich (1858–1923) – Russian biologist. He studied genesis of the Black Sea fauna. Having studied the fauna of the Bosporus, Sea of Marmara and low-salinity areas of the Black Sea coast (limans, river mouths), Sea of Azov and others, O. concluded that the Black Sea fauna included relic species— the representatives of the ancient geological epochs that survived in low-salinity waters of the Black Sea till the present and the Mediterranean species that got into the Black Sea after its connection with the Mediterranean. O. investigations clarified the natural history and development of fauna in the Black, Azov and Marmara seas, Bosporus and Dardanelles about which only scanty information was available. O. found out the peculiarities in the fauna distribution in the Bosporus depending on salinity of the Bosporus water currents. He also revealed that near the Bosporus the Mediterranean species were living at the Black Sea floor and explained this by the greater salinity of waters in the lower Bosporus current. He was the first to attempt unveiling the regularities in benthos distribution in the Black Sea and to identify 6 belts (zones) of its depth-related distribution. With only minor exclusions these regularities were confirmed later on by other scientists.

Ottoman (Osman) Empire (Great Porta) – this was the official name of the Sultanate of Turkey (by the name of Osman I—the founder of the Osman dynasty). It was formed in the fifteenth to sixteenth centuries as a result of Turkish conquering in Asia, Europe and Africa. At the peak of its development in the second half of the sixteenth century to the mid 1670s it incorporated, apart from Turkey proper, the whole Balkan Peninsula, vast expanses in Northern Africa, the Mesopotamia and other territories. From the late seventeenth century it gradually lost some of its conquered territories and in 1918 after defeat in World War I it disintegrated eventually.

Ovchinnikov Ivan Mikhailovich (1931–2000) – outstanding Soviet physical oceanographer, professor, who plaied a key role in the investigation of the Black and Mediterranean seas. Since 1960 he worked at the Black Sea experimental research station in Gelendzhik (from 1967—Southern Branch of Institute of Oceanology, USSR Academy of Sciences). From 1965 to 1970 he headed that Southern Branch of the Institute. Since 1973 he is a head of Hydro-Physical Laboratory, which in 1981 was converted to a Hydrophysical Department. In the early 1980s, he was involved in the study of dynamic processes of the Black Sea, which largely determine the functioning of the entire Black Sea ecosystem. These studies have led to a revision of the traditional understanding of the mechanisms of formation of cold intermediate layer (CIL), which is one of the most important elements of the hydrological structure. It was found that large amounts of CIL water are formed in the centers of cyclonic gyres of the Black Sea as a result of the winter convective mixing. He established the presence of coastal anticyclonic eddies carrying water exchange between the shelf and deepwater areas of the Black Sea. These results led to a completely new approach to the assessment of the Black Sea ecosystem, the forecast of its variability and mechanisms of biological production. Ovchinnikov participated in 27 major research expeditions, in 13 of them as a chief of the cruise. He was an author of over 200 scientific papers, which addressed the most important hydrophysical problems of the Black and Mediterranean seas. Many of the ideas put forward by him, are developed at present.

Ovchinnikov I.M. (http://www.ocean.ru/content/view/1342/41/)

Overgrowths – colonies of aqueous organisms formed on natural and artificial solid surfaces, such as rocks, stones, underwater parts of ships, hydraulic structures, oil platforms, etc. They are made by bacteria, algae and invertebrates.

Ovidipol – a city that was initially a fort on the bank of the Dniester Liman. Its construction was started on 15 May 1793. On its territory there were two Scythian colonies (in fourth to third centuries B.C.) and two Sarmatian colonies (in the second to third centuries A.D.). The place for the fortress was chosen by A. Suvorov. In the eighteenth century the Turkish fort of Khadji-Dere (Khadjider) was built on the Liman bank. In 1769 it was conquered by the Zaporozhie Cossacks and in 1789 it was stormed and seized by the Russian troops. The new fort was of great strategic significance—it guarded the entrance into the Dniester mouth and it was an outpost protecting Ochakov and Nikolaev from the Ottomans. In 1795 while constructing the settlement there was discovered the Antique burial—the funeral urn with bones and several Roman amphoras. This was considered to be a burial place of Ovidius, the Ancient Roman poet, who was banished here by Emperor Augustus in the first century A.D. Later it was proved that he lived in the Roman fort of Tomis (presently Constanta in Romania). But the name of the poet was commemorated in the name of the city in 1795 by order of Russian Empress Catherine II.

P

Paliastomi Lake (in the Greek "*palaios*" meaning "ancient" and "*stome*" meaning "mouth") – a shallow relict lake located in the southern part of the Caucasian coast of B.S near the city of Poti, Georgia. Its area is 17.3 sq. km, mean depth is 2.6 m. This is the former lagoon of the Rioni River which mouth shifted northward. The shores of the lake are waterlogged. It is fed by several rivers and the water seeping from Rioni. It is connected with B.S. via the Kaparcha River. The winds from the sea drive the saline water into the lake, so the average water salinity there is about 5‰. Construction of the drainage system on the Kolkhida Lowland also contributed to the salinity growth. A wide sand beach separates the lake and the sea. P.L. is used for fishing and also as a water sport center.

Pallas Peter Simon (1741–1811) – the Russian who was born in Berlin. In Russia he became the world renowned natural scientist and became one of the most prominent scientists-travelers of the eighteenth century, academician. In 1767 P. arrived to Saint-Petersburg where in 1768 he became Professor at the St. Petersburg Academy of Sciences. He led the academic expedition for exploration of the central regions of Russia, Lower Povolzhie, Caspian Lowland, Middle and Southern Urals, Western and Eastern Siberia. In 1793–1794 he visited Middle and Lower Povolzhie, Caspian Lowland, Northern Caucasus, Crimea. P. collected and generalized the enormous materials on geography, geology, fauna, flora, mineral deposits, ethnography for Southern Russia and Crimea where he settled down and lived his last years (1795–1810). His explorations of Crimea were very careful and in-depth. His work about the Tavricheskiy District (1795) was the most successful theoretical work on regional geography of Russia in that time. P. identified the main physiographical regions of Crimea and described them in detail specifying geographical peculiarities of each of them. P. was the first explorer of the B.S. fauna. The results of his investigations were generalized in the book "Zoographia Russo-Asiatica" (three volumes) published in 1811 in St. Petersburg. Thanks to originality and a manner of material presentation that was quite unusual for that time, the information about numerous new species, orderly classification this work opened a new stage of development of the world zoological science.

S.R. Grinevetsky et al., *The Black Sea Encyclopedia*,
DOI 10.1007/978-3-642-55227-4_17, © Springer-Verlag Berlin Heidelberg 2015

Travels of P. were very important for revealing the past of the south Russian seas. He was the first to suggest an idea on connection in the past of the Caspian Sea with the Azov-Black Sea basin which was confirmed later by many facts.

In 1795 the scientist was granted two estates in Crimea, including the vineyards in Sudak. P. described in detail about 40 local grape varieties and delved into secrets of champagne making. P. did much for development of wine-making in Sudak. In 1804 on an order from the government P. organized here the Russia's first vine-growing and wine-making college and in 1804–1810 was its supervisor (this college existed till 1847). In Crimea P. lived for 15 years till 1810. From Sudak he went to his native country to Berlin where he died. His most significant geographical works in the Russian language are: "Traveling over Various Provinces of the Russian Empire" (in 5 vols., 1773–1778), "Observations Made During Traveling Over Southern Provinces of the Russian State in 1793–1794" (in 2 vols., 1799–1801) (the first full publication in Russia was published in Moscow in 1999), "Brief Physical and Topographic Description of the Tavricheskiy District" (1795) and others. In Sudak the memorial plate was installed on the house where P. lived. It says: "Peter Simon Pallas, the well-known Russian natural scientist, academician, lived in this house from 1795 to 1810".

Pallas P.S. (http://upload.wikimedia.org/wikipedia/commons/2/2c/Pallas_PS_by_Tardier_grey.jpg)

Panfilov Alexander Ivanovich (1808–1874) – Russian Admiral (1866). He finished the Naval Cadet Corps in 1824 in the rank of a warrant officer. He navigated on different ships in the Baltic and North seas. In 1834 he was appointed to serve on the Black Sea Fleet. He commanded the tender "Luch" and from 1837 he was the envoy in Constantinople where he fulfilled very important tasks. In 1839–1848 he was the on-duty office at the Chief of the Black Sea coastline. From 1849 to 1853 he commanded the ship "Twelve Apostles", cruised over B.S. and took part in the battle at Sinop with a unit of frigates. In 1854 he commanded the frigates and later the third defense line of Sevastopol. In 1855 for repelling the storm of Sevastopol he was awarded the Order of St. Anna of category I. From June to October 1855 he acted as the commander of the Sevastopol port, was the assistant chief of the Sevastopol garrison. After the city was left he commanded the troops and fortifications of the Northern Side. In 1856 he was in charge of marine affairs in Nikolaev and was the military governor of Nikolaev and Sevastopol.

Panticapeum – the ancient capital of a state on the territory of Crimea, the capital of the Bosporus Kingdom. The city was founded in the first half of the sixth century B.C. by the Milesians, supposedly in place of the Ionian trading point. It was quickly developing and outpaced other Greek colonies in this region. In the second half of the sixth century B.C. the city started minting silver coins and from the fourth century B.C.—the gold and copper coins. In the first half of the fifth century B.C. P. became the capital of the Bosporus Kingdom that was formed after uniting of the Greek cities located on the coasts of the Kerch Strait. At the peak of its development P. covered an area about 100 ha. The city had a very convenient harbor; already in the sixth century B.C. it was surrounded by a fortification wall. It situated on slopes and at the foot of the Mithridates Mountain. The Acropolis was rising on the mountain top with temples and public buildings. In the first to second centuries A.D. P. was the large handicraft and trade center. However in the third to fourth centuries A.D. the handcrafts started declining. In the late fourth century P. was ruined by the Huns. In the early Middle Ages in its place there was a small town of Bosporus. The archeological excavations of P. and its necropolis were initiated in the first half of the nineteenth century.

Panticapeum (http://en.wikipedia.org/wiki/Panticapaeum)

Paphlagonia – an ancient seaside area on the B.S. southern coast situated between Bithynia and Pontus. It supplied timber and metals. The seaside cities were populated mostly by the Greeks (Sinope). In the sixth century B.C. P. was conquered by the Lydians and later Persians. In the late fourth to early third centuries B.C. it was ruled by Alexander the Great and his successors. From 281 B.C. it was an independent state. From first century B.C. the coastal part of P. was incorporated into the Roman province of Pontus and Bithynia, while the interior—into Galatia.

Parliamentary Assembly of the Black Sea Economic Cooperation (PABSEC) – was established in June 1993 for promotion on a parliamentary level of development of the multilateral economic cooperation of the Black Sea states and consolidation of the political stability in the region. It represents the national parliaments of the states being members of the Black Sea Economic Cooperation and ensures permanent support of the Black Sea region development by providing consultation services. Close interaction among the national parliaments services to intensify and to make more efficient the cooperation of the countries in the Black Sea region.

Parus Rock – see Dzhankhot.

Paspalev Georgi (1895–1967) – Bulgarian Professor, Corresponding Member of the Bulgarian Academy of Sciences. For a long time he worked at the Marine Biological Station in Varna as its director. He conducted hydrological and hydrochemical investigations in the Varna Lake and also coastal fauna in B.S.

Patash (from French "*patache*") – a small boat used in France as a customs vessel. It appeared in the late sixteenth century in Spain as a military-transport vessel and in the early seventeenth century it was used for the same purposes in Turkey.

Paustovskiy Konstantin Georgievich (1892–1968) – a Soviet writer. In 1912 he finished a gymnasium and for 2 years studied in the Kiev University. During World War I he served as a paramedic in a hospital train carrying the wounded from Moscow to the cities in the rear. After the October Revolution of 1917 he went to Kiev and served in the Red Army in the sentinel unit. After Kiev he moved to Odessa where he worked in the newspaper "Moryak" (Sailor). Then he returned to Moscow. His first book was the collection of stories entitled "Oncoming Ships" (1928). After his visit to Poti P. wrote "Kolkhida"; after a trip along the B.S. coast—"Black Sea" (1935). During World War II P. worked as a military correspondent on the Southern Front. He was the participant of the Odessa defense.

Paustovskiy K.G. (http://img12.nnm.ru/c/5/a/1/f/02b4d8023147005b4f1e706c96f.jpg)

Pazar – a village on the coast of the Black Sea, Rize Province, Turkey. Population: 15,900 (2012). The first recorded occupation is the trading colony Αθήνα established here by the Ancient Greeks of Miletos in the eighth century B.C. Along with the rest of what is now Rize Province, Athina then became part of the Roman Empire and its successors the Byzantine Empire and the Empire of Trebizond until it was brought into the Ottoman Empire by Mehmet II in 1461, although this coast was always vulnerable to pirates and threats of invaders from across the nearby Caucasus.

Peace Treaty of Paris of 1856 – it was signed on 18(30) March 1856 by the representatives of Russia, Austria, France, Great Britain, Prussia, Sardinia and

Turkey. This Treaty put an end to the Crimean War and marked a severe setback to the Russian influence in the region. The Treaty consisted of 34 clauses and one "additional and temporary provision". This Treaty was very unfavorable for Russia as it lost the lands along the Danube River and lost its influence over the Danube principalities of Moldavia, Wallachia and Serbia. The most serious defeat for Russia was giving a neutral status to B.S. Clause XI of this Treaty stated that B.S. was open for navigation of merchant ships, but it was closed officially and forever to all warships, both coastal and all others and prohibited fortifications and presence of armaments on its shores. Russia was prohibited to have the fleet on B. S., shipyards, sea berths and to build coastal fortifications. However, Russia kept the right to construct fortresses on the Sea of Azov and in the Kerch Strait. Three conventions were attached to this treaty that had the same force as the treaty itself. The Convention on the Dardanelles and Bosporus declared about strong resolution of the Sultan to follow the "ancient rule" of closing the Straits for passage of foreign warships till Porta existed in the world. The Convention on warships in B.S. stated that each of the Black Sea states could have for coastal service 6 steamers 50 m long in their waterline and with a carrying capacity to 80 tons and 4 light steamer or sailing ships with a capacity to 200 tons. This was an exclusion from the neutralization principle. The impossibility to have warships on B.S. and armaments and fortifications on its shores undermined the positions of Russia in this region. Formally the neutralization principle put both Black Sea states in equal condition, but, in fact, Russia was affected much heavier. It could not have any more the naval forces in the region, except some light ships, while Turkey maintained the right to keep its ships in the Straits.

Pechenegs – the community of the Turk and other tribes populating the Zavolzhie steppes in the eighth to ninth centuries. They roamed between the Aral Sea and the Volga. In the ninth century they occupied the territory of the Northern Prichernomorie extending from the Don to Danube and in Crimea. Their main occupation was cattle growing. They regularly organized raid to the lands of the Kievan Rus.

Perekop Bay – a shallow bay washing in the west the Perekop Isthmus, Crimea. In the west several sand spits cut into its water area.

Perekop Isthmus, Perekop – a land strip connecting the Crimean Peninsula with the mainland, Ukraine. It separates the Karkinit Bay of B.S. and Sivash of the Sea of Azov. It has got its name from "perekop" meaning "cross ditch" that was dug in 1540 after formation of the Crimean Khanate for defense purposes. This Perekop was not the first one: the ancient Greek scholars Strabo (64/63 B.C. to 24 A.D.) and Ptolemy (90 A.D. to 168 A.D.) mentioned "ditch" under the name "Taphros". Its length from northwest to southeast is 30 km, width 8–23 km, height to 20 m. It is composed of clays and loams. Its shores are steep (to 5 m). Its surface is flat overgrown with steppe and semi-steppe vegetation. Several saline lakes are found in the south of P.I. The North-Crimean irrigation canal crosses P.I. During the Civil War of 1918–1920 in Russia and Great Patriotic War of 1941–1945 in the USSR the fierce battles were waged here. More than once the Perekop lands became the battlefield where Zaporozhie Cossacks fought against Tatar feudals and Turkish invaders. During the Russian-Turkish wars the Russian troops three times seized the P. fortifications.

Perekop Storming – (1) Russian-Turkish War of 1735–1739. The Perekop Isthmus had a system of potent defense structures that included the 8-km ditch that ran from B.S. to the Sea of Azov and the earth embankment. On 21 May 1736 the city of Perekop was first stormed and seized by the Russian Army (58,000 people) under command of General—Field Marshall B.K. Munnich. The Perekop fortifications were defended by the 3,000 Turkish garrison. After seizure of Perekop on 16 June 1736 the Munnich Army moved to Crimea and for the first time in the history of the Russian-Crimean Wars captured the capital of the Crimean Khanate—Bakhchisarai. However, due to diseases and shortage of water the Russians had to leave Crimea in August 1736.

(2) Russian-Turkish War of 1768–1774. Seizure of Perekop on 14 June 1771 by the Second Russian Army commanded by General V.M. Dolgorukov (35,000 people). The Perekop fortifications were defended by the Crimean-Turkish troops commanded by Khan Selim-Gyrei (57,000 people). The Russians seized the fortifications practically without casualties. Having conquered Perekop, later on Dolgorukov occupied the whole of Crimea overthrowing the Turkish ruling and in 1783 Crimea was incorporated into Russia. For the Crimean operation Prince Dolgorukov was awarded the Order of St. George of 1st Degree and the honorable attachment to his surname—Crimean. It's remarkable that in his youth Dolgorukov servicing as a private in the Russian Army distinguished himself during Perekop storming in 1736 and became the first Russian soldier who climbed on the Perekop fortification.

(3) Civil War in Russia, seizure of Perekop on 7–17 November 1920 by Red Army commanded by M.V. Frunze. Crimea was defended White Army commanded by General P.N. Vrangel. As a result, Red Army took cities of Simferopol (13 November), Sevastopol and Feodosiya (15 November), Kerch (16 November) and Yalta (17 November). White Army was evacuated on ships and left Russia.

Perekop-Chongar Opeartion – the seizure during the Civil War in Russia of the Perekop-Chongar fortifications on 8–12 November 1920 by the Red Army of the Southern Front under command of M.F. Frunze. These fortifications that obstructed the way for the Red Army to the Crimean Peninsula were defended by the White Army commanded by General P.N. Vrangel. The storm began on the night of 8 November at −11 °C frost when the Red Army crossed on foot the shallow lake of Sivash and occupied the Litovsky Peninsula. As a result, the Red Army was threatening to penetrate into the rear of the Perekop defenders. Simultaneously, they started frontal attack of the Perekop fortifications. The first attack was defeated. On 9 November the Red Army units took another attempt and seized the key position—the Turkish rampart by striking frontally and in bypass over the Sivash Lake. After this the White forces retreated to the second line of defense—the

Ischunskiy positions that were breached and the Red Army rushed out into the Crimean Peninsula. The White troops headed for the ports for evacuation abroad. Over 145,000 people managed to leave Russia on 126 ships. On 15–17 November the Red Army without any fighting occupied such seaside cities as Sevastopol, Kerch and Yalta. After conquering Crimea the Red Army liquidated the last large front of the Civil War (1918–1922).

Periple (from the ancient Greek meaning "sailing around") – a navigation guide, the so-called "coastal pilots", practical recommendations for seafarers that contain descriptions of travels and instructions to a master. Several texts of P. have survived to the present. For B.S. there is known "Periple of Arrian" from Nikomedia (second century) for which there is also anonymous P. These "coastal pilots" represent standard descriptions of travels as we know nothing about the maps attached to them. The Greek texts mentioned "drawn P." or sea maps. It was assumed that the origin place of sea maps was Byzantine.

Pervomaisk – a city, regional center, Nikolaev Region, Ukraine. It is located on the Southern Bug River which bisects the city. Its population in 2001 was 70,170. In 1744 in Tatar-Orel ("Tatar corner") shanets (field fortification) was built and it was given the name Orlovskiy or Orlik using the Turk name of this area. The settlement that was built near this fortification in 1773 turned into a town given a pseudo-Greek name—Olviopol. The basis of the name is Olvia (in Greek meaning "happy") as some ancient Greek colonies located in Prichernomorie were called. In 1920 it was renamed into Pervomaisk. The industries of the city include production of diesels and diesel generators, agricultural facilties, sugar, canned milk and garments.

Peter I the Great (1672–1725) – Russian Tsar (from 1682). From 1721 he became the Russian Emperor. He created the regular Russian Army and the Navy. The lifework of Peter I was buildup of the military might of Russia and consolidation of its role internationally. Thanks to the successful foreign policy of P. and his persistent efforts to get firm footing in the area of the Sea of Azov and Prichernomorie Russia turned into the great power. First P. decided to take hold of the seaside city of Azov being the Turkish fortress (1695). The first storming of Azov ended in failure. For the new Azov campaign P. decided to build the row fleet on the Don River and chose for this purpose the shipyards in Voronezh. The Admiral of the fleet was appointed Lefort and Vice Admiral Lima. The tsar himself was on the galley "Principium". The first victory of the Russian Tsar on sea was won at Azov in May 1696 during the Azov campaign and in June of the same year the Turks were driven out from the Don mouth.

Peter I the Great (http://www.sedmitza.ru/data/004/465/1234/10331.jpg)

Petkov S. – Bulgarian Academician. In 1904–1905 he conducted the first studies of the flora near the Bulgarian coast of B.S. In 1919 and 1932 he published the detailed lists of the lower and higher algae found near the coast, in coastal lagoons and river mouths. Before 1943 he studied the bottom algae.

Petrov Ivan Efimovich (1896–1958) – Soviet Army General, the Chief of Staff of the First Ukrainian Front during the Great Patriotic War (1941–1945). In 1917 he finished the military college. He was an active participant of the Civil War in and World War I in Russia. P. was one of those who commanded the defense of Odessa and Sevastopol. In 1941–1944 he was the commander of the Primorsky Army, Black Sea Group of the North-Caucasian Front. The troops under command of P. took part in the Kerch-Eltingen landing operation, in the battles for liberation of the Taman Peninsula and the cities of Maikop, Krasnodar and Novorossiysk. From August 1944 he was the commander of the Fourth Ukrainian Front. On 26 October 1944 P. was conferred the highest military rank Army General. From April 1945 to the end of the war P. was the Chief of Staff of the First Ukrainian Front. For skilful guidance of the troops in the Berlin and Prague assault operations and for initiative and dedication Army General P. was awarded in 1945 the Title of the Hero of the Soviet Union.

Phyllophora (*Phyllophora*) – unique red algae. Discovered in 1908 by S.A. Zernov in the north-western part of the Black Sea. P. genus comprises 3 species, of which most wide-spread is ribbed P. A large number of invertebrates and fishes dwell in the thickets of P. It plays an important role in their propagation, fattening and wintering. The algae is extracted for the purpose of obtaining agaroid at the Odessa Agar Plant; agaroid is a product used in food-processing industry and in many branches of economy. P. grows attached to rocky ground at the depth of 1–2 to 10–15 m.

"Phyllophora field", "Zernov's Phyllophora Sea" – unique phenomenon in the World Ocean formed by unattached phyllophora. Sited in the center of the north-western shelf of the Black Sea, looks like a flat sheet up to 1 m in height and lies above the sea bottom between the depths 20 and 60 m. The existing data indicate that reserves of 3 species of P. exceed 5.6 mln tons, of this, 75 % falls on ribbed phyllophora. Phyllophora field occupies an area of around 11,000 km^2. The area in question was named "phyllophora field" or "Zernov's phyllophora sea". The causes for such accumulation of algae have not been established finally.

Phytobenthos of the Black Sea – large algae growing on rocks, stones and other hard subsea objects. It is just as crucial to the life of the sea as phytoplankton, but is confined to the shallow-water coastal zone, where the algae are capable of growing. Numerous animals feed on the algae, fishes build their nests and spawn in the algae, fishes also find shelter from enemies in algae thicket. Over 300 species of Black Sea large bottom algae have been described. Besides, small unicellular algae dwell on the sea bottom: these cover with a thin film not only underwater objects, but the surface of the soils, too. Around 400 species of such fine bottom algae have been described. Large algae are divided into three large groups (departments) that can be distinguished by color, namely: green, brown and red. The bulk of large algae grows at the depth of 5–10 m, but sometimes these may be encountered at the depth of 80 m. Some species of the Black Sea large algae have industrial uses. These are Cystoseira or bluestem (*Cystoseira barbata*) of brown algae and four phylloflora species—gelatinizing agent agaroid. In addition to algae, Black Sea phytobenthos includes 7 species of higher (flowering) plants. Foremost in terms of occurrence and the size of reserves are two species of seaweed: zostera or Zostera marina and Zostera nana). These feature a rather thick rootstock buried in the sand at a rather considerable depth and long leaves (the former species—1.5 to 1.8 m). In shallow bays and blights as well as in brackish maritime lagoons, zostera forms real underwater meadows. This thicket is a habitat of numerous animals that form a biocoenosis or zostera community. Both marine organisms and some waterfowl feed on zostera. Dried out seaweed leaves are used as a packaging or padding material.

Phytoplankton of the Black Sea – there are over 700 unuicellular species of algae in the Black Sea phytoplankton. The principal systematic groups of the Black Sea phytoplankton are these: diatomous, dinophytic (Peridinaceae class species are particularly numerous) coccolithophora, green, Euglenaceae, blue-greens. Despite the tiny size (usually, under 1 mm), phytoplanktyon organisms are crucial in the seas, above all because, using inorganic matter of the sea water, carbon dioxide and solar energy, they produce, in the course of photosyntehsis, organic matter, disengaging oxygen. As they do, microscopic phytoplankton algae contained in the water mass produce conditions vital to the life of animals that are in need of ready-to-use organic matter, yet they are unable to produce from inorganic matter. Besides, phytoplankton is the source of oxygen that not only is dissolved in sea water, but is disengaged into the Earth's atmosphere.

Pilots of the Black and Azov Seas – the first Russian printed pilot of B. and A. seas was prepared by Lieutenant Budischev in 1808. It was titled "The Pilot or Sea Guide Describing Seaways and Entrances into Ports, Bays of the Azov and Black Seas, in the Bosporus and Byzantine Straits with Speculations on Winds and Currents". Only its first part devoted to the Sea of Azov was published. The first pilot of B.S. that became classical was published in 1851. It was based on the descriptions of the B.S. coasts made in 1825 by the expedition of G.P. Manganari, in 1847 by Lieutenants G.I. Butakov and I.A. Shestakov (later they were promoted to Admirals) who commanded the vessels "Pospeshnyi" and "Skoryi". The pilots also used the results of previous hydrographic investigations. The description of the Caucasian coast was based on the handwritten "The Pilot of the Eastern Coast of the Black Sea" prepared by navigator Tarashkin. In 1854 navigator A. Sukhomlinov after 2 years of coast investigations on the schooner "Astrolyabia" published "The Pilot of the Sea of Azov and Kerch—Enikale Strait". These two Pilots were ample enough to meet the needs of the developing shipping and remained in use for many years. Only 16 years later the second edition of the B.S. Pilot was published, revised and enlarged, by Lieutenant Pavlovskiy. The next third edition of the B.S. Pilot appeared in 1892 and it incorporated the Sea of Azov Pilot. The preface to this edition said that it was a preparatory one for the next version that would be published after completion of hydrographic investigations conducted that time. The fourth edition significantly revised and enlarged was published in 1903. It was the last pilot combining B. and A. seas. Beginning from 1915 all follow-on versions contained description of only one of these seas. A rich collection of the pilots of B. and A. seas may be found in the Sevastopol regional museum. It contains nearly all printed pilots and atlases. Some of these publications are placed in the exposition of the history of oceanographic investigations of B. and A. seas.

Pipeline transport – is the transportation of liquid, gaseous and dry (crushed) goods via tubes due to the pressure difference created by pumping stations. The carrying capacity of pipelines depends on the tube diameter and capacity and quantity of pumping stations. By the purpose and parameters the pipelines may be classified into three groups: main; local (on-field, collecting, distribution); internal (communication lines of oil refining works, oil bases). P.T. is the most economically effective transit means. Pipelines are usually installed along the shortest route. P.T. ensures regular and uninterrupted delivery of goods and this is most important for plants with a continuous flow process (oil refining and petrochemical works). T.P. is most widely used for transit of fuel: oil, its products, gas and recently coal. Main oil product lines connect the oil refinery centers with the bases supplying users. On the Caucasus the oil pipelines lead from oil production fields to the oil refinery centers and to large sea ports and also to users. In the Black Sea there are "Blue Stream" and "Dzhubga-Lazarevskoe-Sochi" offshore gas pipelines.

Piracy – attack on the sea of a private ship belonging to one or several owners by some other ship, usually merchant or passenger, for robbery. P. may be offshore. It also includes privateering and raiding. In B.S. all kinds of P. were used.

Piri Reis – the real name of Haci Ahmed Muhiddin Piri (between 1465 to 1470—c. 1553). The Turkish Fleet commander, geographer and cartographer. He prepared the world maps, the atlas and pilots of the Mediterranean basin. He is thought to be of the Greek origin. In the early sixteenth century he took part in numerous campaigns of the Turkish Fleet in the Mediterranean. Preparing his famous world map (1513) P.R. used 14 earlier maps, including Venetian, Portugese and the map of Christopher Columbus that did not survive to the present. The survived part of the P.R. maps (the western part of the Atlantic and the western part of the Old World with the coasts of America and Europe) was published in 1935 by the Turkish Historical Society. It is interesting that the map has a partial contour of an Antarctica coast, which was discovered 300 years later. The Mediterranean Atlas that included B.S. "Bakhrie" (Poem about the Sea) that was prepared by P.R. in 1523 was a reference book for navigation over the Mediterranean basin. It represented the full collection of maps (Portolan Charts) that were made European style and a guideline in verse over terrains shown on maps with description of coasts, marking of currents, shallows, harbors, raids, bays, etc. P.R. was appointed the "kapudan" (Admiral) of Egypt. He commanded the marine forces in the Red Sea and the Indian Ocean where his fleet was defeated by the Portugese squadron and for this P.R. was tried and executed following the court judgment. Several warships and submarines of the Turkish Navy have been named after Piri Reis.

Piri Reis (http://annoyzview.files.wordpress.com/2011/12/piri-reis-111.jpg)

Pirogov Nikolay Ivanovich (1810–1881) – prominent Russian doctor and scientist; one of the founders of the surgery anatomy, military field surgery, organization of medical support of troops. In 1824 at the age of 14 P. entered the Medical Department of the Moscow University. After graduation from it in 1828 he entered the Dorpatensis Professor Institute (today Tartu University in Estonia), the surgery department. In 1841 P. organized and headed the surgical hospital at the Petersburg Military Medicine Academy. In 1856 he left the academy and accepted the appointment of the trustee of the Odessa Educational District. From this time his 10-year period of educational activity started. In military medicine P. developed the scientific foundations of the Russian field surgery and the new field of military medicine—medical service organization and tactics. When in 1853 the Crimean War broke out P. decided that his place was not in the capital, but in the besieged city. He took much effort to get appointment to the acting army. In 1854–1855 P. twice went to battlefields and took part in organization of medical services to the troops and treatment of the wounded. He encouraged female volunteers ("Sisters of Mercy") for taking care of the wounded on the front. The most important doing of P. during the Crimean War was organization of the orderly performing field medicine service. P. proposed the elaborate system of evacuation of the wounded from a battlefield.

The last years of his life (from 1866) P. spent in his estate in the village of Vishnya near Vinnitsa from where he went as a field medicine consultant to the front during the Franco-Prussian (1870–1871) and Russian-Turkish (1877–1878) wars. The results of his observations P. generalized in his work "The Fundamentals of the General Field Surgery Developed on the Basis of Field Hospital Practice and Recollections about the Crimean War and Caucasian Expedition" (1865–1866) and "Field Medicine and Private Assistance on the War Theatre in Bulgaria and in the Rare of Acting Army in 1877–1878" (1879). In 1897 the monument to N.I. Pirogov was installed in the Tsaritsynskaya Street (from 1919—Bolshaya Pirogovskaya Street in Moscow) in front of the surgical clinic. In 1947 by the decision of the Soviet Government the memorial estate-museum was opened in the village of Pirogovo (former Vishnya) where the burial vault with the embalmed body of the great doctor still exists.

Pirogov N.I. (http://www.medical-enc.ru/topographic_anatomy/images/pirogov.jpg)

Pitsunda – a cape on the B.S. coast near the Bzyb River mouth; a seaside climatic resort, Republic of Abkhazia. Population: 4,200 (2011). It was first mentioned in the second century B.C. as Pitius, in the first century A.D. as Pitiunt. Both forms are from the Greek *Pithyus* meaning "pine" which is connected with the famous grove of relict pines ("Pitsunda pine"). The Greek name came to us in the form of Pitsunda. Presently "Asmara ("pine grove") is in use in Abkhazia. This urban-type settlement is situated 25 km southward of Gagra, 90 km to the northwest of Sukhumi and 15 km to the southwest of the railway station "Bzyb". The territory of this resort is a slightly undulating plane on a marine terrace surrounded in the north and northeast by the Gagra and Bzyb ridges of the Greater Caucasus. The vegetation here is represented by the box grove (63 ha in area) and the unique grove of Pitsunda pines (about 200 ha in area) running for 7 km along the seashore. The climate is humid subtropical. The winter is very mild, without snow. The mean temperature of January is 6 °C. The dull cloudy weather in this season is quite frequent here. The spring is early with the prevailing dry and sunny weather. The summer is very warm—the mean temperature of August being 24 °C. The autumn is long, sunny, and warm. First frosts may occur only in December. The average annual precipitation are 1,500 mm. The number of sunshine hours is 1,800 a year. The bathing season lasts from May through late October (in some years even to mid-November). The sea water temperature is from 22 to 26 °C. Tender climate, wonderful beaches of sand and small pebble, ionized air and warm sea created conditions for opening here in 1960 of the premium-class climatic resort. The main curative methods here are aero-, helio- and thalassotherapy. The resort provides treatment for the people suffering from respiratory (non-TB), cardiovascular and nervous diseases.

On the P. cape you can see the city and port of Pitiunt that was founded by the Greeks in Antiquity and Middle Ages. It was also a Roman fortress, one of the Christianity centers on the Caucasus and an important outpost of the Byzantine. In the fourteenth to fifteenth centuries the Genoan trading post of Petsonda located here. In the 1950s the excavations conducted here uncovered the ruins of the ancient settlement dated to the second half of the first millennium B.C., churches (including Basilica of the fourth to fifth centuries with the mosaic floor and the church of the tenth to eleventh centuries), fortifications and living houses, baths. To the northeast of the ancient town the cross-domed church of the tenth century with the frescoes of the sixteenth century is preserved. Presently it is a museum that is the venue of the traditional festivals and concerts of organ music. In the P. vicinity there are ruins of the Bzyb fortress of the ninth to tenth centuries.

Pitsunda (http://photos.wikimapia.org/p/00/02/93/75/59_full.jpg)

Pitsunda Pine Reserve – locates in Pitsunda on the coast of B.S., Abkhazia. Its area is about 200 ha. Approximately 27,000 trees are growing here. The rest houses and resorts are found along the margins of this famous protected pine grove. This place is favorable for treatment of lung, nervous diseases and anemia.

Pitsunda-Myussera Nature Reserve – in Abkhazia. It is situated on the B.S. coast of the Caucasus extending from the Pitsunda Cape as far as the Myussera Mountains. Its area is 3,800 ha. From 1926 the relict Pitsunda pines are under protection in the forests along the Myussera River. In the same year the Pitsunda site was established and in 1946—the Myussera Nature Reserve. In 1966 they were united forming the P.-M. nature reserve. It comprises the grove of the relict endemic Pitsunda pine (over 100 ha in area) and of box (60 ha in area). In the nature reserve you can find the Georgian oak, beech, hornbeam, strawberry tree, arboraceous heather. Among the fauna the roe, boar, badger, marten, squirrel and other are quite common.

Plaka Cape – it is situated in Crimea between Alushta and the resort town of Partenit on the shore of the Kuchuk-Lambat Bay of B.S. The height of P.C. is 40–45 m and its width is 80 m. It is stretching in the northeastern direction for 250 m. The smooth hard-rock slope dips abruptly into the sea. The cape is a massif of volcanic rocks where porphyrite is prevailing. Nearby the cape there is the town of Lampada that was mentioned by Ancient Greek historian Flavi Arman (95–175 A. D.). Later in the sixteenth to fifteenth centuries the colonists from Genoa settled down here and called this place Lambat. Then the Crimean Tatar settlement called Kuchuk-Lambat (Smaller Lambat) appeared and in 1958 it was renamed into "Utes" (Rock). Here the well-known sanatorium "Utes" is located. Its administration is placed in the palace built in 1907 for Princess A.D. Gagarina. The park around the sanatorium is the monument of the garden-park architecture of the nineteenth century. It was founded in 1814. About 220 decorative varieties of trees and bushes from tropical America and Africa are growing here.

Planerskoe – see Koktebel.

Plavni – floodplains in the lower reaches of the southern rivers of Ukraine and Russia (Dnieper, Dniester, Danube, Don, Kuban and others). They are usually waterlogged or overgrown with hydrophilic grass vegetation, mostly cane, reed, cattail. Sometimes these are wide valleys in the lower reaches of rivers occupied by meadows and covered with forests. The meadows are used for hay-making. Cane and reed are used as building materials (for making reed-fiber mats) and also as raw material for the chemical industry.

"Pobeda" (Victory) – a passenger liner of the Black Sea Shipping Company. Its construction was ordered by the German Company Hamburg-Amerikanische Packetfahrt-Actien-Gesellshaft (Hapag) and it was set afloat in 1928. The ship was given the name of "Magdalena". It was capable to accommodate 123 passengers in the first-class cabins, 102—in the second-class and 106 in the third-class cabins. Its speed was 15 knots. Its sailing range was about 28,000 km. The ship navigated along Europe, West Indies islands and Central America. In 1934 it went aground near the Curacao Island. After it was brought afloat again the ship was repaired and refurbished. It became single-funnel, its displacement and speed were increased and it got a new name—"Iberia". During World War II the liner was used as a floating base in Kiel. In 1946 as a result of division of the German merchant fleet this ship was passed to the USSR and was named "Pobeda". After its rehabilitation it was capable to carry 600 passengers. This motor vessel was put in operation in 1947 and navigated along the line Crimea-Caucasus. It sailed from Odessa to New-York also. In September 1948 while sailing from Batumi to Odessa there was a fire in which 40 passengers died. The ship was repaired in Germany and in 1952 it again returned to B.S. In late 1977 it was sent for cutting.

"Pobeda" (http://et.wikipedia.org/wiki/Pobeda_(reisilaev))

Polovets (Kipchaks) – the Turk people. In the eleventh century they roamed in the steppes in the south of Russia. Their main occupations were cattle breeding and handicraft. In the twelfth century they populated vast expanses. In the west the Polovets roamers reached Ingulets. But the great many of them concentrated on the left bank of the Dnieper and over the shores of Sivash. In the east they went as far as the Volga. But still they preferred living on the Donets River and its tributaries. The northern border of their living space came close to the borders of Rus, while in the south it ran along the Sea of Azov coast. In the early thirteenth century they were defeated and went under control of the Mongol-Tatars first on the Northern Caucasus and later on the Kalka River where they were allies of Russian Princes. Some P. joined the Golden Horde, while the other crossed the border and went to Hungary.

Pomorie (before 1934—Anchialo) – the seaside climatic and mud resort in Bulgaria (the Burgas Region), to the northeast of the city of Burgas. It is situated on the B.S. coast, on the small Pomoriysky Peninsula. The climate here is marine. The mean annual air temperature is about 13 °C. The winter is mild, practically without snow. The mean temperature of January is 2.6 °C. The autumn is long, warm and dry. Precipitation are 469 mm a year. The number of sunshine hours is 2,200–2,360 a year. Mists are rare. The tender climate, warm sea (the water temperature in summer is 24–25 °C, in August—to 28–30 °C) and wide sandy beaches (the sand is dark colored due to the presence of titanomagnetite) surrounded by sand dunes allow good opportunities for climatotherapy and thalassotherapy. Apart from this special fruit and vine diets are applied here (the surroundings of P. are famous for orchards and vineyards). To the northeast of P. the resorts Sunny Beach and Nesebr are found. In the vicinity one can see the Antique Roman dome tomb (third century). P. is one of the most ancient cities of Bulgaria. In the fourth century B.C. the Ancient Greek colony of Anchialo was founded here by the people of the other Greek colony on the Bulgarian coast—Apollonia-Sozopol. Still in the Antiquity the sea salt was produced here. In the Roman times the city of Anchialo was a very important economic and cultural center on the B.S. coast. Many prosperous Thraceans moved here from nearby places. Population: 14,200 (2010).

Pomorie Lake – the saline lagoon lake situated 2 km northward of Pomorie, Bulgaria. Its area is 7.6 sq. km, depth—to 1.4 m. It is separated from B.S. with a low-elevated sandy spit about 6 km long. Its bottom is covered with a thick layer of sulfide mud that is used for treatment purposes (this is one of the best curative mud deposits in the country). The brine of the lake is also used for treatment. It is also the source of table salt.

Pontic Mountains (from the Greek "*Pontos*" meaning "Black Sea") – a range of mountains in Turkey that stretches along the southern shore of B.S. They make the northern folded margins of the Asia Minor and partially Armenian plateaus. Its length is about 1,000 km, width—to 130 km. The ranges of P.M. running in parallel are broken by longitudinal valleys and chains of hollows and they are cut through by narrows worked by the Kyzyl-Irmak and Ishil-Irmak rivers. Eastern P.M. feature the Alpine relief. Their highest point is the Kachkar Mount (3,937 m). Western P.M. are of medium height. This area is subject to high seismic activity. Significant coal deposits are found in Zonguldak. The northern slopes are overgrown with coniferous forests, while the southern slopes are covered with steppes and semi-deserts and the higher belts—with silver fir forests and mountain meadows.

Pontus – (1) An area in Asia Minor, a part of Kappadokia (Pontic Kappadokia) on the northeastern coast of Asia Minor. In 280 B.C. after the wars waged by Diadochi it became an independent Kingdom and during the ruling of Mithridates VI (132–63 B.C.) it reached the apex of its economic and political development (from 63 B.C. it was the Roman province and from 40 to 64 A.D. it was again the independent Kingdom and then again the Roman province). (2) Name of B.S.—Pontus Euxinus.

Pontus Euxinus (from the Greek "*Pontos Euxenos*" meaning "hospitable sea") – the ancient name of B.S. Before the sixth century A.D. the ancient Greek name originated, in all probability, from Pontos Axeinos meaning "Inhospitable sea", i.e. the names changed for the opposite ones. It appears the word "Axeinos" is based on the ancient Persian word "*akshaina*" meaning "dark, black" that was perceived by the Greeks as "inhospitable". The change of the name was connected with widening of regular economic and cultural relationships of the Greeks with the local population and establishment of numerous Greek colonies on the coast of Pontos Euxenos. This sea was a very important trade route linking Northern Prichernomorie via the Thracian Bosporus with the Greek states and via the

Bosporus Cimmerian (Kerch Strait) with the tribes of Southeastern Europe. The Greeks knew about P.E. for a long time, which was proved by the legend about Argonauts. In the seventh century B.C. it was colonized by the Greeks and acquired economic importance (grain, fish, slaves). The detailed description of the P.E. coast that first appeared in the second century A.D. was made by Arrian in his work "Periplus of Ponti Euxeni" (131 A.D.) from which later on the anonymous shorted version appeared.

Pontus, Pontus Kingdom – the Hellinistic state in Asia Minor (302/301—64 B. C.). It appeared on the southern coast of B.S. as a result of disintegration of the empire ruled by Alexander the Great. The Kingdom reached its greatest height under Mithridates VI Eupator who conquered the Bosporus Kingdom (Northern Prichernomorie) and other territories and who for many years carried on the wars with Rome (89–84, 83–81 and 74–64 B.C.). As a result of these wars P.K. fell under control of Rome.

Popov Andrey Aleksandrovich (1821–1898) – Russian shipbuilder, Admiral (1891). He finished the Naval Cadet Corps. From 1838 he served on the Black Sea Fleet, commanded the support cruiser "Meteor". He participated in the Crimean War of 1853–1856. In 1854 P. was the commander of the ship "Taman" that broke through the line of enemy ships from Sevastopol to Odessa and destroyed 6 Turkish merchant ships. In September 1854 he organized the Sevastopol defense from the sea using cannons taken off from ships. Till the end of the Sevastopol defense he was in charge of artillery support of the fortress. He implemented very important shipbuilding programs—construction of the first Russian armored ships, including the ship "Peter the Great". During the Russian-Turkish War of 1877–1878 he supervised retrofitting of the merchant ships for military purposes.

Port Kavkaz – is a small port and harbour on the Kerch Strait in Krasnodar Kray, Russia. The port is able to handle vessels up to 130 m in length, 14.5 m in breadth and with draft up to 5 m. It is the eastern (Russian) terminal of the railroad and car ferry line connecting Krasnodar Kray with Crimea (Ukraine). The Ukrainian terminal of the ferry line is Port Krym.

Port Kavkaz (http://bellona.ru/imagearchive/port_Kavkaz.jpg)

Port Krym – is a small port in Crimea, Ukraine. It is located on the western coast of Kerch Strait, in the north-eastern part of Kerch city. Next to the port is located a train station "Krym" as well as a customs and border checkpoint. Port Krym has a ferry connection with Port Kavkaz on the eastern (Russian) coast of the strait. The port is also a base for pilot service and boats which guide navigation through the Kerch Strait.

Poskochin Fedor Vasilievich (?–1804) – Russian Rear Admiral (1799). He finished the Naval Cadet Corps for Nobility. Every year he took part in campaigns in the Gulf of Finland and on the Baltic Sea. In 1782 P. was transferred to the Azov Fleet commanding the ship "Patmos", cruised along the Crimean coast (1782), cruised over B.S. (1783), commanded a pink in A. and B. seas (1784–1786), the frigate "Strela" (1787). P. took part in the Russian-Turkish War of 1787–1791, fought the enemy near the Phidonisi Island on the ship "St. George Pobedonosets". Commanding the same ship he cruised with the fleet over B.S., escorted the row fleet from Sevastopol to the Danube mouth (1789), sailed with the fleet from Sevastopol to Sinop. Later P. participated in battles with the Turkish Fleet in the Kerch Strait (1790) and in the Gadzhibei Lagoon. In 1791 P. was transferred to the Danube rowing fleet and distinguished himself in the battles against onshore enemy fortifications. In 1796 he was appointed the treasurer of the Black Sea Admiralty Board.

Poskochin Ivan Stepanovich (?–1803) – Russian Captain-Commander (1803). He finished the Naval Cadet Corps for Nobility. Commanding the war galley "Iput" on which Catherine II was going he navigated over the Dnieper from Kiev to Kaidakov and then to Kherson. He was the participant of the Russian-Turkish War of 1787–1791. Commanding Floating Battery No. 1 in the Fleet of Prince Nassau-Siegen who sailed on it from Glubokaya Pristan to Ochakov where he took part in battles with the Turkish ships. He cruised between Ochakov and Gadzhibei, participated in storming of the Sulinskiy fortifications, seizure of Tulcha, storming of Izmail. Commanding the first unit of galleys he in 1791 participated in seizure of the Brailov shore batteries. In 1796 he was appointed to serve on the Black Sea Fleet. Commanding the frigate "Fedor Stratilat", ship "St. Trinity" in the squadron of Vice-Admiral F.F. Ushakov he navigated from Sevastopol to Arkhipelag where P. took part in the battles for Tserigo, Zante and Cephalonia islands in the Ionian Sea. Reaching the Corfu Island P. fought the enemy vessel with 80 cannons onboard. He continued commanding the ship "St. Trinity" and stayed for defending the Corfu Island. In 1800 on the same ship he sailed from this island to Sevastopol. In 1801–1803 he commanded the ship "St. Paul".

Potemkin Grigoriy Aleksandrovich (1739–1791) – Russian statesman and military figure, diplomat. His titles are "Prince, President of the State Military Collegium, General-Field Marshal from 1784, Grand Hetman of the Cossack, Yekaterinoslav and Black Sea Troops, Commander-in-Chief of the Yekaterinoslav army, light cavalry, Black Sea Fleet and other land and marine forces; Senator, General-Governor of Yekaterinoslav, Tavria and Kharkov; general-inspector, general-aide-de-camp, chamberlain of the troops of Her Emperor Highness; Colonel of Lifeguard of the Preobrazhensky regiment; chief of the cavalry guard corps. In 1756–1760 he studied in the gymnasium of the Moscow University. He took part in the palace coup of 1762 as a result of which Catherine II came on the throne. During the Russo-Turkish War of 1768–1774 he distinguished himself in the battles at Khotin, Focsani, Large and Kagul for which he was promoted to General. In 1774 he organized punitive actions against Ye.I. Pugachev. Being the favorite and the closest assistant of Catherine II he took a high position at the court and in the state administration. He showed himself a gifted administrator and army and fleet organizer. In 1776 he was appointed General-Governor of the Novorossiysk, Azov and Astrakhan provinces. In 1783 he carried out the project of annexing Crimea to Russia, for which he received the victory title of His Serene Highness Knyaz Tavrichesky. In 1783–1791 he was the Commander-in-Chief of the Black Sea Fleet. He did much for founding and construction of Kherson, Nikolaev, Sevastopol that received privileges in marine trade along with Petersburg and Arkhangelsk. He created the military and trade fleet on B.S. and organized construction of fortifications along the sea coast. During the Russo-Turkish War of 1787–1791 he commanded the Yekaterinoslav army. He was buried at the Catherine Cathedral of Kherson, Ukraine. In honor of P. the squadron armored ship was named "Prince Potyomkin of Tauris" (1904) (see).

Potemkin G.A. (http://upload.wikimedia.org/wikipedia/commons/thumb/1/1c/Princepotemkin. jpg/428px-Princepotemkin.jpg)

Potemkin Ladder – was constructed in Odessa. It was a present of Prince M.S. Vorontsov to his wife Elizabeth. The ladder was built by the design of architect F.K. Boffo in 1837–1841. First it was called "Boulevard" and later "Giant". Originally 200 steps led to the sea. But after construction of the Primorsky road 8 of the steps happened to be under it and were covered with asphalt. After shooting of the world renowned film "Bronenosets Potemkin" by Sergey Eisenstein this symbol of Odessa (along with the monument to Duc de Richelieu—"Duc") came into the history not as the Vorontsov, but as Potemkin ladder. Its length is 142 m, height 30 m. It has 10 flights, 10 landings and 182 steps. The 2-m thick parapets on both sides seem running parallel. This visual effect is achieved by gradual widening of the ladder downward—on top the steps are 12.5 m wide, while at the bottom—21.6 m. According to public opinion poll in Europe, P.L. is the 6th famous feature on the continent. More well-known are only the ladders in the Spain Square in Rome, Montmartre in Paris, Acropolis of Lindos on the Rhodes Island, Fillgraderstiege in Vienna and Cavalier ladder in Prague.

Potemkin ladder (http://spox.ru/cache/imagemanager/2705116.jpg)

"Potemkin" ("Prince Potemkin of Tauris") – Russian squadron armored ship of the Black Sea Fleet. It was built in the yards of Admiralteiskiy and "Naval" in Nikolaev. The construction was started in 1898 and in 1904 the ship was put afloat. This was the first ship on which during revolution of 1905–1907 there was the mass sailor revolt (1905) in which G.N. Vakulenchuk, Bolshevik leader of sailors, was deadly wounded. The ship sailed to Constanta (Romania) for replenishment of fuel and food. Having failed to receive these it went to Feodosiya where the coal and food were loaded on the ship. It had to sail to Constanta again where on 25 June 1905 the sailors surrendered the ship to the Romanian powers and called themselves political emigrants. The members of the crew who returned to Russia were arrested and convicted. The Romanian authorities returned the ship to the Russian government. In October 1905 it was renamed into "Panteleimon". In April 1917 the ship was returned its previous name, while in May it got the name "Borets za svobodu" (Fighter for Freedom). In April 1919 the ship was blasted in Sevastopol and sank.

After the Civil War in Russia it was lifted and cut to scrap. Its displacement was 12,500 tons, length—113 m, width—22.2 m, speed—16 knots, crew—730 people. Armaments included 2 turrets with two 305-mm cannon in each, in vaults—sixteen 152-mm, fourteen 75-mm and ten small-size cannons, 5 torpedo tubes. In 1925 Sergey Eisenstein shot his famous film "Bronenosets Potemkin".

"Potemkin" (http://academic.ru/pictures/wiki/files/80/Potemkin.jpg)

Poti – a town and port in Georgia on the B.S. coast, in the mouth of the Rioni River. In the ancient times the town of Phasis was in this place, it was mentioned by Ancient Greek historians in the fifth to sixth centuries B.C. In the medieval times the town was called Fasso and in 1578 the Ottomans built the fortress of Fash here. The present-day Georgian version—Poti was formed from these two names. It was incorporated into Russia as a result of the Russian-Turkish War of 1828–1829. The industries here include plants of hydraulic mechanisms, electrical devices, tar paper and roofing felt, peat-processing, bakeries and fish processing plants, and repair shipyards. This is an important fishery center. The port in the city handles manganese ores, bauxites, coal, metal, equipment. Population: 53,400 (2008).

Poti Marine Port – locates on the eastern coast of B.S. (Western Georgia). It was founded in 1858. In 1991 after Georgia became an independent state the port was retrofitted applying the newest technologies and now it is capable to handle all kinds of cargo. It has 14 berths. The total length of berths locating in three basins is over 2,800 m, the water depth at berths is 6.1 to 12.5 m. Six cargo terminals are operating here: container, railway ferry, complex for handling Ro-Ro vessels, oil, for bulk cargo and for general cargo. In the cargo turnover of the port the transit operations are prevailing (48 % of all cargo). The share of export and import is 33 % and 19 %, respectively. By the cargo structure there are distinguished: general

(44 %), bulk (40 %) and liquid (16 %). The general cargo in the port turnover includes grain (about 50 %), container freight (about 30 %) as well as cotton, sugar, wood and timber. The bulk cargo includes metal scrap (about 48 %), metals and metal products (approximately 23 %), coke, ores and bauxites. Among the liquid cargo prevailing are oil (over 55 %), chemicals and oil products (mainly diesel fuel). P. services the ferry carriages from Bulgaria, Ukraine, Romania and Turkey. Three Ukrainian, two Bulgarian and one Georgian ferries run along the ferry line Ilyichevsk-Poti-Batumi. In March 2005 a new ferry line Port Kavkaz (Russia)— Poti (Georgia) started operating. For further development of the port it is planned to construct two new complexes for handling oil tankers.

Poti Marine Port (http://commons.wikimedia.org/wiki/File:POTI.JPG?uselang=ru)

Prichernomorie – a territory around the Black Sea. It includes Nortern, Southern, Western, and Eastern Prichernomorie.

Prichernomorie archaeology – the historical and archaeological explorations in the Prichernomorie areas of Russia, Romania, Bulgaria and Turkey. They were carried out influenced by excavations in the Mediterranean Sea in the nineteenth century. They were preceded by excavations of the Scythian mounds that provided the most valuable artifacts—jars, precious things and decorations. These findings made to wonder about the relationships between the Scythians and the Greeks. And there was quite logic reasoning on the need to study in detail the colonization by the Greeks of the B.S. coast and to conduct archeological excavations in Greek cities.

The targeted objects of these efforts that have not been completed so far were Tira (Belgorod-Dnestrovsky), Olvia on the Bug, Chersonesos (Sevastopol), Panticapaeum (Kerch), Tanais-on-Don, Istria and Kallatida in Romania, Mesembria (Nesebr) and Apollonia (Sozopol) in Bulgaria, Heracleum Pontic, Sinop and Amis in Turkey. Detailed explorations of Northern Prichernomorie were conducted most intensively in the USSR when the interest to the historical past of the country was most great. In the recent decades the investigations have been conducted not only in cities, but in villages and trade centers. Therefore, the new factual material is collected for taking new look at and detailed elaboration of such issues as the Scythian-Greek relationships (excavations in Neapol-Scythian in Simferopol), the Cimmerian issue, the relationships between the Thracian and Greek history and culture (excavations in Bulgaria and Romania).

Prichernomorie Lowland – a flat plain slightly inclined southward, adjoining B. and A. seas between the Danube Delta in the west and the Kalmius River in the east, Ukraine. Its elevations vary from 10 to 150 m. It is composed of the thick Paleogene and Neogene marine deposits (limestone, sand, clay) overlain with Quaternary loess and loess-like loams. It is cut through by wide (with numerous terraces) valleys of the Dnieper, Southern Bug, Dniester and other rivers. The water divides are flat and characterized by the presence of hollows—pods. The coast is mostly abrupt, often subject to landslides. There are many deeply incising lagoons (Dnieper-Bug, Dniester and others); the ancient Dnieper Delta is well developed. P.L. is situated in the steppe zone composed of chernozem soils replaced in the south with chestnut soils.

Primorsko (before 1934—Kyupria) – a seaside climatic resort in Bulgaria (Burgass Region) to the southeast of Burgas. It is situated on the B.S. coast at the inflow of the Dyavolska River. The climate is marine, warm and humid with a great number of sunny days. The mean temperature of January is about +3 °C, of July—about 23 °C. Precipitation are 550–600 mm a year. In the west and southwest the resort is protected by the hilly uplands of Bosna and Khasekiyata (outspurs of the Strandzha range) which slopes are overgrown with coniferous forests. Tender climate, warm sea and sandy beaches are favorable for climatotherapy and thalassotherapy here. Numerous rest centers are found in the vicinity of P. The well-known international recreational center for youth also locates here. Primorsko was declared a national sea resort in 1953. In 1981, it was merged with Kiten and became a town, and in 1998 it split from Tsarevo municipality. Population: 3,700 (2010).

Primorsko-Akhtarsk – a regional center, Krasnodar Territory, Russian Federation. In 1774 during the Russian-Turkish War the Russian troops seized the Akhtar-Bakhtar fortress (a distorted Turk name "seaside white cliff") in place of which in 1778 the Akhtar redoubt was built. In 1829 the farmstead of Akhtarskiy appeared here. In 1900 it was turned into the Cossack village of Primorsko-Akhtarsk and in 1949—into the city of Primorsko-Akhtarsk. The railway station (Akhtari) locates 151 km to the northwest of Krasnodar. P.-A. is the natural climatic resort area in the

Kuban-Priazov Lowland on the coast of the Sea of Azov. It has a sandy-pebble beach stretching for 8 km. The winter is mild and rainy, without a permanent snow cover. The mean temperature of January is about −3 °C. The summer is warm, dry and sunny. The mean temperature of July is 22 °C. Precipitation fall mostly in the form of rare heavy rainfalls with thunderstorms. The autumn is also warm and sunny. Population: 32,150 (2010).

Primorskoe (Primorskiy resort, formerly Budaki) – a seaside climatic resort in Ukraine (Odessa Region), 30 km southward of Belgorod-Dnestrovskiy. It is a part of the Odessa resort zone. It is situated on the shore of the Budakskiy Liman on B.S. The fine-sand beach is about 3 m long and to 50 m wide. A large park is growing on the resort territory. The climatotherapy, including thalassotherapy, is practiced here for curative purposes. Large territories near the resort are used for vine growing, thus, the vine treatment may be conducted here. The resort was founded in 1946.

Prince Igor (?–945) – the Prince of Kiev since 912. In 913–914 on 500 vessels he went to the Transcaucasia to fight the Khazars. Having descended by the Dnieper to B.S. the P.I. fleet sailed around Crimea and via the Bosporus Kimmerian (Kerch Strait) got into the Sea of Azov, went up the Don as far as Kachalinskiy village where they portaged the ships to the Volga. Having descended by the Volga into the Caspian Sea the Russian sailed along the coast of Gilan, Dailem, Tabaristan and Azerbaijan and defeated the Shirvan-Shakh Fleet near the Shirvan coasts. On their way home they were attacked by Khazars and were defeated near Ityl. In 941 P.-I. navigated to B.S. The Russian Fleet attacked the enemies on the coast of the Bosporus Thracian (Bosporus Strait), but was beaten near Constantinople. In 944 P.I. made another march to Byzantine. The Russian ships approached the Danube mouth. The Byzantines weakened in the struggle against the Arabs and Bulgarians paid to the Russian a great tribute and in the same year they signed the peace treaty in Kiev under which Byzantine and Rus' agreed to render military assistance to each other, in particular, the Prince of Kiev had to defend the Byzantine lands in Crimea from the nomad raids.

Prince Svyatoslav (?–972) – the Great Prince of Kiev. He was the son of Prince Igor. In 964–967 he was famous for his incessant campaigns against Khazaria. He also subdued the Volga Bulgars and Vyatichs who paid tribute to the Khazar Khan. He defeated the Khazar troops and conquered such Khazar cities as Sarkel (in Slavic–Belaya Vezha), Ityl and Semender. He attacked the Yass (Ossetians) and Kasogors (Cherkess) and forced them to subservience. Having destroyed the Khazar main city—Ityl the Russian warriors went out to the Caspian Sea, established complete control of the Volga trade route and extended relations with Iran and Central Asia. During the P.S. ruling three seas—Black, Caspian and Baltic were free for sailing of the Russian ships. S. was seeking to dominate on the shores of the Danube. In 968 the Russian Fleet entered the Danube. Having defeated the Bulgarian Army he seized about 80 Bulgarian cities. P.S. was killed in 972 by the Pechenegs on the Dnieper.

Privateering (piracy) – a private trade vessel authorized by a country's government by the so-called "letters of marque" to attack the enemy trade ships or ships of neutral states carrying cargo for a hostile state. P. was widely developed on B.S. In the late eighteenth century the Russian Squadron of F.F. Ushakov comprised the permanently operating privateering ships that were referred to as corsair in the documents. Among the most well-known corsairs operating on B.S. there were Paul Jones and Lambro Cazzoni. P. was prohibited and abolished by the Declaration of Paris of 1856.

Program EROS-2000—Black Sea (European River Ocean System) – the international interdepartmental research program. Its purpose is study of the interaction of the Danube River with the northwestern part of B.S. in the area from Burgas (Bulgaria) to Sevastopol (Ukraine) for elaboration of the integrated actions on control of eutrophication, pollution, bank destruction and other anthropogenic impacts. The main directions of research include revealing the main sources of pollutants brought to the sea with the river flow and atmospheric precipitations; prediction of the impacts of river flow regulation on stratification of sea waters and lithodynamic processes, especially on bank erosion; assessment of the role of gas emissions from the research area into the atmosphere in terms of the global greenhouse effect and forecasts of large-scale changes of climate occurring in some regions.

Prokofiev Foma Yakovlevich – Russian Captain, General of the major rank (1797). He finished the Naval Cadet Corps for Nobility. On the frigate "Africa" he participated in the Russian-Turkish War of 1768–1774 in the battles at Napoli-di-Romania, in the Chios Strait and at Chesma. He was sent to the Don Fleet and served at the Novokhoper shipyard (1779–1783), in the port of Sevastopol (1783–1785), was Captain of the Kherson port (1786–1787), commander of the frigate "Pospeshnyi" (1787–1788) staying at the entrance into the Sevastopol Bay as a floating battery.

"Prut" – a training ship of the Russian Black Sea Fleet. Its displacement was 5,460 tons, speed—to 14 knots, sailing range—12,600 km, armament: 10 cannons and 3 machine gun, the crew—278 people. On 19 June (2 July) 1905 when the ship moored near the Tendrovskaya Spit the revolt broke out on it (following the armored ship "Potemkin"). Under the red flag the ship navigated to Sevastopol, but at the approaches to the city it was met by the torpedo boats and escorted to the naval base. Then there was a trial and the revolt participants were arrested, while the ship was used as a prison for several years. In 1909 it was retrofitted into the mine layer. During World War I on 30 October (12 November) 1914 the German cruiser "Geben" tried to seize the ship, but "P." was sent to the bottom by its crew.

Psou River – flows into B.S. The state border between Russia (Krasnodar Territory) and Abkhazia runs over it. This is a hydronym from the Adygean *"psy"* meaning "water, river".

Publius Ovidius Naso (43 B.C.–c. 18 A.D.) – also known as Ovid, the Roman poet. Emperor Augustus ordered to banish the unwanted poet to Tomis (presently Constanta) on the Black Sea. Staying there far from Rome the poet wrote his poems. In the mid-century O. was ranked the second after Virgil. Dante put O. alongside such poets as Homer and Horace. Already the first poem of O.— "*Amores*" in three books brought him fame. But the views of O. on love run counter to the efforts of Emperor Augustus to fight against moral degradation. Therefore, it is quite likely that his love poems were the cause of his exile. While in exile O. wrote in 12 A.D. the collection of poems "*Tristia*" in five books and in 13–16 A.D. four books of "*Epistulae ex Ponto*" (messages from the Black Sea). The citizens of Constanta esteemed high the memory of O. and gave his name to one of the squares in the city. The statute of this great poet was made in 1887 by Italian sculptor Ettore Ferrari and installed in Tomis. In 1957 the Archological Museum was opened in Constanta where one of the expositions was devoted to Publius Ovidius Naso.

Publius Ovidius Naso (http://upload.wikimedia.org/wikipedia/commons/0/0d/Latin_Poet_Ovid.jpg)

Pushkin Aleksandr Sergeevich (1799–1837) – the greatest Russian poet. In 1820 he was banished from Petersburg to the south of Russia for his freedom-loving poems. The first years of his political exile in the south the poet spent in Kishinev. In August 1820 he made a 3-week trip over Crimea making a start in Kerch (in 1999 the commemorative sign was put on the embankment where P. first put his foot on the land of Tavrida), then traveled to Feodosiya, Karadag, Gurzuf, Yalta, Bakhchisaray and Simferopol. He left Crimea via Perekop. During these 3 weeks P. made also a 1-day voyage over the sea onboard the war brig "Mingrelia". In 1823 he was transferred to Odessa where he took the position of the collegiate secretary in the chancellery of M. Vorontsov, General-Governor of Novorossiya. Here the exiled poet met future Decembrists and many friends of his from Petersburg and Moscow. The 13 months (from 3 July 1823 through 1 August 1824) spent in Odessa were very fruitful for P. and significant for his creative life: he finished the poems "The Fountain of Bakhchisaray" and "Robber Brothers", wrote a part of the lyrical poem "The Gypsies" and nearly three chapters of the verse novel "Eugene Onegin" and 28 poems.

Pushkin A.S. (http://sanke.chupakabr.ru/wp-content/uploads/2012/03/A.S.Pushkin-.jpg)

Pustoshkin Pavel Vasilievich (1749–1829) – Russian Vice-Admiral (1799). He finished the Naval Cadet Corps for Nobility. In 1768 he was sent to serve on the Black Sea Fleet. From 1771 to 1780 he served on the Sea of Azov. In 1772–1774 commanding the ship "Khotin" on B.S. he made description of the Dnieper and Southern Bug rivers. In 1777 he was promoted to Captain-Lieutenant. Commanding the frigate "Pochtalion" and transport mail vessels he was sent from Kerch to Taman from where he carried Crimean Khan Shagin-Gyrei to Yenikale and later navigated B.S. In 1780 he was transferred to the Baltic Fleet. In 1782 he again returned to the Black Sea Fleet and commanded the frigate "Pochtalion" at the port of Taganrog. In the rank of Flag-Captain P. navigated from Taganrog to Sevastopol. After this he made measurements and descriptions of the Dniester Liman and later commanded a ship at the port of Taganrog. In 1784–1787 he was the chief of the Gnilovskiy shipyard. He was appointed the chief of the Taganrog port. In 1789 he was promoted to Captain of the brigadier rank. In 1790 commanding the squadron he sailed from Taganrog to Sevastopol where taking the command of the ship "Knyaz Vladimir" he navigated to the Danube mouth and near the Rumeli coast. In 1791 he took part in the battle at Kaliakria. In autumn 1791 P. was appointed the commander of the Danube Fleet and the Black Sea rowing fleet. In 1794 P. commanded the second squadron on the Sevastopol roads. In 1795 he was appointed the Commander of the Nikolaev port and the member of the Black Sea Admiralty Board. From 1797 he commanded the rowing fleet on B.S. and in 1798—the Dniester squadron.

Puzanov Ivan Ivanovich (1885–1971) – Soviet zoologist, zoogeographer. He was the Doctor of Biological Sciences (1938), Honored Worker of Science of Ukrainian SSR (1965), professor, head of the vertebrate zoology chair of the Odessa State University. He studied the nature of Crimea and Caucasus. He conducted investigations of the B.S. coast, Red Sea, in Eastern Sudan, on Ceylon, etc. Among his major works there are "The Great Canyon of Crimea" (1954) and "Over Untrod Crimea" (1960).

R

Raevskiy Nikolay Nikolaevich, Junior (1801–1843) – Russian General, founder of Novorossiysk City (Russia) together with and Vice-Admiral M.P. Lazarev and L.M. Serebryakov (since 1856 Admiral). The son of a war hero in 1812, General Nikolay Raevskiy. Nikolay Raevskiy soon began to make a military career: in 10 years, has been identified with ensign in Orel Infantry. Received a baptism of fire in 11 years during a war with Napoleon. In a rear-guard battle near the village Saltanovka 29 June (11 July) 1812, along with his father was the head of the Smolensk regiment during the counterattack. For the case in Saltanovka Nikolay Jr. was promoted to Lieutenant and was awarded the Order of Vladimir 4th degree. He participated in the foreign campaigns of the Russian Army in the war against Napoleon. In 1814 transferred to the Life Guards Hussar Regiment and was appointed adjutant I.V. Vasilchikova. In 1817—the Staff-Captain, in 1819—Captain, in 1821—appointed adjutant Baron I.I. Dibich, in 1823—Colonel Sumy Hussars. In 1826 he was transferred to Nizhny Novgorod Dragoon Regiment and appointed him commander. In 1828–1829 he participated the Russian-Turkish War, was promoted to Lieutenant General. In 1838 he was appointed Head of the Black Sea coastline defence. In 1841 he was dismissed from the cavalry and the service.

In 1837, it was decided to set up the coastline fortification from Taman to Poti. Russian Tsar appointed a chief and responsible for the construction of coastal fortifications General N.N. Raevskiy Junior. Admiral M.P. Lazarev, Chief commander of the Black Sea Fleet, received instructions to study carefully the Caucasian coast and find a comfortable place for landing and building forts. Military Ministry did not plan to construct a fort in the Sudzhuk Bay (Tsemess Bay) because of local conditions, but Raevskiy Jr. proposed to build it. In the beginning of September 1838 he got a permission to start a construction of a fort. On 12 September 1838 ships of Russian Black Sea Fleet eneterd Tsemess Bay and 5,816 people commanded by General N.N. Raevskiy Jr. and Vice-Admiral M.P. Lazarev landed near ruins of the Turkish fortress. Now this day is celebrated as a birthday of the city. On 14 January 1839 Russian Military Minister issued an order to name Novorossiysk the constructed fortifications in the Tsemess Bay.

S.R. Grinevetsky et al., *The Black Sea Encyclopedia*,
DOI 10.1007/978-3-642-55227-4_18, © Springer-Verlag Berlin Heidelberg 2015

Today a mountain peak at the western edge of the village Verkhnebakanskiy near Novorossiysk and village Raevskaya village located a few km west of this peak have a name of N.N.Raevskiy Jr.

Raevskiy N.N., Junior (http://www.polyarka.ru/pages/images/g25-1.jpg)

Raiders – independent military actions on the enemy communications using warships, often at great distances from the main base of the fleet. In 1914–1918 the raiders on B.S. were German cruisers "Goeben" and "Breslau".

Railway ferry – a structure designed for transportation of the railway rolling stock in the operational condition over water bodies. The ferryline "Crimea-Caucasus" in the Kerch Strait (about 5 km wide) between Russia and Ukraine makes much shorter the railway route between Crimea and the Caucasus.

Razelm, Razim – the liman lake on the B.S. coast in Romania south of the Danube Delta. A narrow sandy spit separates it from the sea. Its area is 394 km^2. It incises into land for 30 km. The lake is shallow. The water in R. is desalted thanks to the canal built from the Georgievsky gyrlo. Fishing is practiced here. R. is connected with B.S. via the narrow strait of Portica.

"Red tide", "Red water" – "coloring" of the sea observed usually on the surface of coastal waters caused by very intensive propagation of certain kinds of phytoplankton, mostly of the red (more seldom yellow, green or brown) color, as a result, the water acquired their coloring. During such periods the quantity of organisms may reach hundred million specimens in one liter of water. Most often these were peridinean algae which waste products may be toxic and cause mass death of marine animals, including fish. In B.S. such R.T. is caused by exuviella, thus, fish and other mobile organisms in B.S. try to leave the places of exuviella blooming.

Redox-zone – an intermediate zone between the oxygen and hydrogen sulfide zones in the water layers of oxygen–hydrogen sulfide coexistence locating at the lower border of the chemocline. This zone is the place of most intensive processes of chemical and biological oxidation of the reduced compounds formed in the hydrogen sulfide zone due to bacterial anaerobic decomposition of the organic remnants. Therefore, R.Z., especially its top part, features a very high gradient of the oxidation–reduction potential.

P.P. Regir Shipping Company – was founded in 1889 in Petersburg by Regir Petr Petrovich (1851–1919) for coal transportation on the vessel "Progress" from Mariupol to the B.S. ports. In the early twentieth century the company had at its disposal nine cargo vessels—"Petr Regir", "Maria Regir", "Engineer Avdakov", "Rossia", "Bessarabia", "Velikorossia", "Belorossia", "Novorossia" and others. They carried coal, wood and grain over the Black, Mediterranean and Black seas. For further business extension the Russian Commercial Shipping Company was founded in 1912 by P.P. Regir, Jr. The company had four ships. In 1913 the capital of the company was 1 mln Rubles.

Repnin Nikolay Vasilievich (1734–1801) – Russian prince, military commander and diplomat. During the war with Turkey in 1768–1774 he commanded the corps and in 1770 seized Izmail and Kilia. In 1774 he took part in preparation of the Kuchuk-Kainarji Peace Treaty. In 1775–1776 R. was assigned the Ambassador to Turkey. In 1791 he was appointed the supreme commander of the Russian Army in the war with Turkey (for the time of absence of G.A. Potemkin). After the victory at Machin Repnin forced Turkey to accept the preliminary truce of Galati. In 1798 he negotiated with Prussia and Austria the creation of a new coalition against France. After return to Russia he fell in disfavor and in November 1798 he had to resign.

Repnin N.V. (http://maloarhangelsk.ru/wp-content/uploads/2010/08/repnin.jpg)

Republic of Turkey, Turkey (Turkiye Cumhuriyeti) – one of a few states located simultaneously on two continents—Europe (Eastern Thrace or Rumelia) and Asia (Anatolia, Anadolu cover over 97 % of the country's territory). The area of the country is 783,562 km^2; the population—75.8 mln people (2013). The border separating these two parts of T. runs over the Bosporus (Istanbul) and Dardanelles (Canakkale) Straits and the Sea of Marmara. The coastline is 7,200 km long. The greater Asian part (97 % of the total territory) lies within the Asia Minor Peninsula. The European part occupies the southeast of the Balkan Peninsula. Administratively T. is divided into 81 provinces (ils) that, in their turn, are subdivided into ilche (districts). The capital of T. is Ankara (it was transferred from Istanbul in 1923). The territory of the country is surrounded by the B.S., Sea of Marmara, Mediterranean and Aegean seas. The Bosporus and Dardanelles forming the natural outlet from B.S. into the Marmara and Mediterranean seas belong to T. The position of T. at the juncture of Europe and Asia and on important international routes facilitated mutual penetration of cultures, rapid development of trade, mass migration of the peoples. T. borders on Bulgaria, Greece, Georgia, Armenia, Azerbaijan, Iran, Iraq and Syria.

Republic of Turkey (http://suzideanne.files.wordpress.com/2011/03/turkey_map_02.jpg)

T. is the country of plateaus and mountains. The central part of the country is occupied by the vast Anatolian Plateau (800–1,500 m high, on the average). The volcanic area Kappadokia is situated eastward of the Tuz Lake. The northern areas of the plateau reveal high seismicity: devastating earthquakes occur frequently along the so-called Anatolian fault. The eastern part is covered by the Armenian Plateau—the highlands with volcanoes and picturesque valleys. The highest point of the country is dead volcano Greater Ararat (or Buyuk Argydak, 5,137 m) near the border with Armenia. There is a belief that during the Flood the Noah's Ark landed at the southern slopes of this mountain. Western (Aegean) Anatolia consists of mountains, plateaus and valleys dividing them. The largest of them is the valley of the Greater Menderes. The mountains are extending latitudinally (the highest point is Mount Uludag or Smaller Olympus, 2,543 m). They are composed of granites and gneiss. The rugged coast of the Aegean Sea abounds in convenient harbors. In the north Asia Minor is confined by the Pontic Mountains. They consist of several parallel ranges with the wide valleys in between. The Pontic Mountains (to 3,428 m high) rise as high as the permanent snow line. In the southern margins of Asia Minor the Tauri Mountains (Toros) are found. More compact and high ridges of Central Tauri (to 3,734 m high) are cut by through valleys. In the east the Tauri stretches in several parallel ridges (Northern, Internal and Armenian Tauri) 2,000 to 3,000 m high. The Mediterranean coast of Asia Minor is mountainous, in some places limestone cliffs drops abruptly into the sea. Wonderful sand beaches are found in the Antalia Bay.

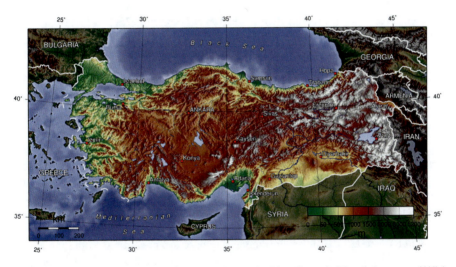

Republic of Turkey, topography map (http://upload.wikimedia.org/wikipedia/commons/d/db/ Turkey_topo.jpg)

The country is rich in ore deposits, such as iron, lead, zinc, copper, tungsten, mercury, antimony and molybdenum as well as coal which reserves are estimated at 590 mln tons and lignite—7,700 mln tons. T. possesses the world's largest chrome iron ore deposits. The Zonguldak-Eregli coal basin on the B.S. coast is very important for development of ferrous metallurgy. The oil reserves are not large (66 mln tons) and they satisfy less than 15 % of the country needs. In the 1980s the development of natural gas fields was started. There are also deposits of uranium ores.

The climate is subtropical, on the coasts and seaside mountain slopes—Mediterranean, in Anatolia—continental with continentality growing eastward. In Anatolia the summer is hot (+30 °C) and in winter the air temperatures may drop to −20 °C. The annual precipitation varies from 200 mm in the center to 400–500 mm towards the plateau periphery. They fall mostly in spring. Relative humidity is very high on the B.S. coast; the precipitation here vary from 1,000 to 3,000 mm a year; they are most copious in summer and autumn. On the coast of the Aegean and Mediterranean seas and also on the southern slopes of the Tauri the summer is hot (+28 °C) and without clouds, while the winter is warm (+10 °C) and rainy. On the windy slopes of the Tauri the precipitation amounts to 500–1,500 mm a year. Such major rivers of Asia as Euphrates and Tigris take their origin in T. as well as large rivers of Kizilirmak, Sakarya, Greater Menderes and Seyhan. The rivers in the country are not, in fact, navigable, but the country has the significant hydropower resources. Saline lakes, the largest of which are Tuz and Van, locate in the drainless depressions of the Anatolian Plateau.

In the central and eastern regions of Anatolia the dry (grass-sagebrush) steppes cover wide expanses composed of the gray-brown soils. The stony terrain is overgrown with astragals and prickly thrifts. At present no forests are left on this territory, although still in the fifteenth to seventeenth centuries the oak, pine and juniper forests were growing here. The forests covering 26 % of the country's territory can be found only in the mountains, but they have changed greatly as a result of cattle grazing, tree felling and fires. The hard-leaved oak or pine forests are substituted with secondary thickets of maquis and phrygana. The most valuable oak and beech forests growing on brown soils have survived on the northern slopes of the Pontic Mountains at altitudes higher than 400–700 m, while at altitudes 1,200–2,400 m the coniferous fir and silver fir forests are growing. Some mountains have Alpine meadows. The seaside slopes of the Eastern Pontic Mountains receiving excessive moisture are covered with Kolkhida-type forests of beech, hornbeam, linden growing on mountain red and yellow soils. On the slopes facing the inner regions the dry forests of black pine are dominating. The lower slopes of the Tauri and the Mediterranean coast are occupied by maquis vegetation (madrona, pistachio, oleander). The middle-mountain belts are covered by oak and hornbeam forests, the drier slopes—by pine forests. At altitudes of 1,700–2,200 m the coniferous forests of Cilician fir, pine and juniper are prevailing. In the mountains the stone barren lands overgrown with phrygana vegetation are found in addition to forests. These are more often found on the northern slopes of the Tauri featuring highest aridity. The mountain forests are inhabited by fallow deer, row deer, leopard, jackal, bear and badger. In the high mountains the bezoar goats and moufflons are met. The steppe lynx—caracal lives in the Anatolia steppes. There are also many species of rodents and reptiles (Central Asian vipers, small lizards and agamas). The country has 49 protected territories (1.4 % of the country's total territory).

The Asia Minor Peninsula is one of the most ancient cultural centers. The Neolithic settlements found here are considered to be among the earliest human settlements in the world. Beginning from the second millennium B.C. the Hititis states (eighteenth to twelfth centuries B.C.), Phrygian (tenth to eighth centuries B.C.) and Lydian states (seventh to fourth centuries B.C.) succeeded each other on the territory of T. Then Asia Minor was conquered by the Persians, and later fell to Alexander the Great. From the fourth to eleventh centuries Asia Minor was the Byzantine province. In the eleventh century the Turk nomad tribes (Oguz and

Turkmen) settled down here and formed the Seljuk Empire. In the thirteenth century the Seljuk armies were defeated by the Mongols and the empire slowly disintegrated into several principalities, the largest of which was the Ottoman principality that in the fourteenth to sixteenth centuries as a result of the wars of conquest turned into the powerful Ottoman Empire expanding throughout Asia Minor, Balkans, some Arab countries of Western Asia and North Africa. By the late nineteenth century the Ottoman Empire was on decline. In 1918 after defeat in World War I the Turkish territory was divided among the Allied Powers. As a result of the struggle for liberation during 1919–1923 that was led by Kemal Ataturk ("*Father Turk*"), T. became an independent state. The 1920s–1930s was the period of reforms: the Islamic leaders were alienated from government of the state, the Moslem law (Shariat) was replaced with the modern law system, foreign concessions were annulled, and the state sector in economics was established. They say that Ataturk is the author of the following words: "We have the only way—to Europe". During World War II T. was neutral. In 1952 it joined NATO. In 1960 the coup in T. resulted in adoption of a new constitution. In 1980–1983 the military government was in power that banned and later disbanded all political parties. In 1982 as a result of the national referendum, the new constitution was adopted and since 1983 T. had the civil government.

T. is a candidate to accession the EU. T. is the parliamentary republic. Its supreme state power is the unicameral parliament (Mejlis), the Grand National Assembly of Turkey elected for 5 years. The head of the state is President. The executive power is exercised by the Prime Minister and the Council of Ministers. The official language is Turkish. The basic ethnic community in the country is the Turks (70–75 %), but the country has also Kurd population (18 %) living mostly in the mountains in so-called Kurdistan in the southeast of the country. The remaining is distributed among the Arabs (in the southeast), Greeks (about 100,000 people), Armenians (60,000), Jews (25,000), Assyrians (25,000), and others. The territories of the B.S. coast near the state border with Georgia are populated by the Lazs, Adjarians and Georgians. The Turkish population also includes the groups of nomad Turks who did not merge with the local population: Uryuks, Takhtadzhi, Abdaly and others. The Bulgarians, Bosnians and Albanians live in the west of the country. The male population is prevailing. The most densely populated regions are seaside lowlands, while the most sparsely populated are the inner regions of Anatolia. The rural population, in particular males, tends to migrate to large cities. More than 1.2 million Turks worked in the countries of EU (Germany) and Near East. T. is an urban country. 75 % of the population lives in cities. There are 8 cities with the population over one million. These are Istanbul (13.7 mln), Ankara (4.6), Izmir (3.4), Bursa (1.98), Adana (1.6), Gaziantep (1.4), Konya (1.1) and Antalya (1.07). Most people speak the Turkish language belonging to the Turk group of the Altai family. Two dialects are in use—Anatolian and Rumelian. Since the thirteenth century the written language had used the Arab alphabet, while in 1928 T. changed over to the Latin alphabet. The literary language is based on the Istanbul dialect. The written language of the Kurds uses the Latin graphics. The prevailing religion (but not official) is Sunni Islam (98 %). In 1924 the law on separation of religion and state was adopted. About 200,00 people of T. are Christians. Some pre-Moslem beliefs are also preserved. The currency is Turkish Lira.

T. is an industrial-agrarian country. The share of industry in GDP is 30 %, which is more than of agriculture (15 %). In the 1980s–1990s the country adopted a new strategy of economic development aimed at reduction of state regulation and amplification of market mechanisms. The country became open for free traffic of goods and capitals. The private sector takes the leading positions in industry, banking business, transport and communication. The power engineering sector is developing rapidly and in the 1990s the country started exporting electricity to the East European countries and Georgia. T. has thermal and hydropower plants that are the largest in Western Asia and that generate one-third of the country's power production. The cascade of power plants was constructed on the Euphrates River the major of them "Ataturk" was commissioned in the late 1980s. The processing sector prevails in the industry. Developed are also the ferrous and nonferrous metallurgy, chemistry, oil refinery, machine building, automobile, aircraft and ship building. The steel and milled products are the second largest items in the Turkish export (after textile industry). The country has three large metallurgical full-cycle plants in Eregli, Karabyuk and Iskanderun and 15 privately-owned mini-plants. The processing industries maintain their importance as before. T. is a major producer of coal and lignite as well as chromium, iron, manganese, copper and other ores. Hi-tech industries, such as electronics, precision machine-building are developed rather weakly. The key industry in T. is, traditionally, the textile manufacturing. It includes manufacturing of tailored clothing (25 % of the export earnings) and knitwear. In this industry T. is the third in the world after Italy and Hong Kong. Traditionally strong is the leather industry. It manufactures 2 % of the world leather products and is among the major importers of leather materials.

T. has diverse agriculture that meets the internal needs of the country. The rain-fed farming dominates. More than 80 % of the arable lands are used for growing wheat, barley, corn, oat and millet. The production of industrial crops that are exported and used as raw material by the local industry is significant. Among them most important are tobacco (T. is one of the world leaders by tobacco production), cotton (export— 750,000 tons a year, one of the world leader), sugar beet, oil crops and rose. In the coastal areas the growing of grapes and fruit trees (citrus, olive, apricot, peach) is developed. The export goods include hazel nuts (50 % of the world supply), resin and figs. In the agricultural sector small private farms operate side-by-side with large mechanized farms applying the most advanced agricultural practices. The cattle breeding was always an extensively developing sector that used new and new pasturelands. The sheep and goat husbandry (45 % of the total cattle population) is the key occupation. T. is a major supplier of the Angora wool to the world market because the Angora sheep makes the quarter of all ship population.

The tourism sector has experienced rapid growth in the recent years and consti-tutes an important element of the country's economics. The number of tourists annually visiting T. has reached 31.5 million (2011), ranking as the 6th most popular tourist destination in the world. The country has 60 airports. T. is the major sea power. Its merchant fleet has at its disposal over 500 vessels. The major sea ports are Istanbul, Mersin, Izmir, Trabzon, Samsun, and Iskenderun (the Turkish naval base). The ports on B.S. include Bartin, Erdemir, Giresun, Hopa, Ordu, Rize, Samsun, Trabzon, Zonguldak.

At present the role of transit carriages, including oil pumping through the country's territory, is growing. The motor transport takes the greatest load in internal carriages. The total length of automobile roads is 382,000 km, of railroads is 8,600 km. The route of the famous "Oriental Express" connecting Europe with Iran and India goes over the Turkish territory. There are plans on further development of railroads so that the cargo and passengers can reach via railways any part of the country and also the EU countries. To spur this process some major projects were developed and a part of them is already in implementation. In 2003 the construction of the high-speed railway Istanbul-Ankara was started. It is currently running between Ankara and Eskişehir, with the Eskişehir-Istanbul part under construction. One of the major projects "Sea of Marmara" launched in 2004 is also connected with the railroad transport. It envisaged construction of a 14-km tunnel under the Bosporus for linking two continents. It was open on 29 October 2013.

The development of transport communication between Asia and Europe becomes most urgent. Among the ongoing EU projects of international transport corridors the most important for T. is TRASECA the connection to which diversified the transport routes in T. Within the TRASECA framework the railroad from the Turkish city of Kars to Tbilisi is being constructed. It is planned to open a ferry line from Samsun port to the Mediterranean ports of Mersin and Iskenderun. The cargo traffic via the Fourth Pan-European Transport Corridor is growing. One of its branches runs from Bulgaria over the Turkish territory to Greece. For improvement of the delivery conditions T. signed an agreement with Greece on electrification of the railroads making a part of this corridor.

From the 1990s T. has witnessed a rapid development of the pipeline transport. In 2005 the oil pipeline Baku-Tbilisi-Ceyhan was constructed that in the future can become the main pipeline for transit of the Caspian oil to Europe. Still earlier the gas pipeline Dzhubga (Russia)—Samsun (Turkey) known as "Blue Flow" was constructed for export of the Russian gas. The export of the country includes textile products (more than 25 %), foodstuffs (15 %), fabric (12 %), leather, metal and transport equipment. T. imports machines, semi-finished products, chemicals and fuel. The basic foreign trade partners of T. are Germany, the USA, Italy, France, Great Britain and Russia.

Turkey has very diverse culture, customs and rites that are a blend of the Turkish and Moslem traditions with European innovations. T. can boast of a very rich folklore that includes the heroic epos "Oguz-name" (eleventh to twelfth centuries), epic legends (dastans) and lyric songs. The ornamental art uses widely the stylized presentations of plants and arabesques that can be seen in decorations of carpets, metal and ceramic dishware, clothing. The painting and sculpture started development only in the newest time. The architectural masterpieces and monuments of various epochs and peoples—Hellenes, Romans, Byzantines, Seljuks and Ottoman Turks survived in the country. The Ataturk's reforms, including secularization of education and substitution of the Arab alphabet with the Latin alphabet easier for writing, facilitated further development of education and science. At present the population literacy in the country is high. Istanbul and Ankara have large universities and many libraries. The Academy of Economics and Trade is located in Izmir. During the time of the Ataturk's reforms the country became more European preserving at the same time its national attributes.

Research vessels (R/V) of the Black Sea – in the Soviet Union the research institutes and organizations located in Sevastopol and having at their disposal well equipped R/V played the leading role in B.S. studies. These are the Marine Hydrophysical Institute (R/V "Mikhail Lomonosov", R/V "Akademik Vernadskiy", R/V "Professor Kolesnikov") and the Institute of Biology of the Southern Seas (IBSS) (R/V "Akademik Kovalevskiy", R/V "Professor Vodyanitskiy") of the USSR Academy of Sciences (presently the National Academy of Sciences of Ukraine). At the same time this was the base for research vessels of the Sevastopol Branch of the State Oceanographic Institute (SB SOI) and the Hydrographic Service of the Black Sea Fleet. Sevastopol was also the home's port for the research vessels of the M.V. Lomonosov Moscow State University ("Akademik Petrovskiy", "Moskovskiy Universitet") on which in summer the research voyages became at the same time the study practical course for students. For nearly 30 years (1961–1989) the Odessa Branch of IBSS successfully used R/V "Miklukho-Maklai" for detailed exploration of the environmental situation in the northwestern part of the sea, including its deterioration from the early 1970s. Beginning from 1949 the P.P. Shishov Institute of Oceanology replenished and replaced its research fleet permanently. Among the most remarkable vessels that studied B. and A. seas there were "Geolog" and "Truzhenik" (1950–1957), "Akademik Shirshov" (1953–1978; in 1967 it was renamed into "Akademik Obruchev"), "Akademik S. Vavilov" (1954–1975), "Professor Dobrynin" (1980–1982), "Akademik "Orbeli" (1975–1985), "Priboy" (1974–1989), "Akvanavt" (1978—to the present), "Gidrobiolog" (1983–1995), "Akvanavt-2" (1989–1991).

Research-Production Center "EKOSIGIDROFIZIKA" – the self-supporting division of the Marine Hydrophysical Institute (MHI) of the National Academy of Sciences of Ukraine (NASU) was established for environmental studies and expertise, interpretation of satellite information, preparation and publication of research works of MHI and other research establishments of the NASU, popular-science publications, methodological materials. It is located in Sevastopol.

Research-Technological Center "SHELF" – the isolated self-supporting division. In 1996 following the order of Presidium of the NASU it was incorporated into the Marine Hydrophysical Institute to address the following tasks: control of the condition and functioning of experimental mussel and rapan growing farms, improvement of fishery, fishing tools and tactics, search for new forms of mariculture organization, organization of farms in the offshore zone. It is located in Sevastopol.

Resort complex "Druzhba" – is known as the oldest, qiuet and romantic area for rest and recreation on the coast to the north of Varna, Bulgaria. More intensive resort construction started in 1956. During a short time 18 hotels were built. In 1977

Grand Hotel "Varna" was opened. It was a modern, by that time parameters, recreational and balneological center. The resort was given back its old name "Sveti Konstantin and Elena".

Rezovo – a village on the coast of the Black Sea, Bulgaria. It is a seaside resort in southeastern part of Bulgaria, part of Tsarevo municipality in Burgas Province. Lying at the mouth of the Rezovo River, it is the southernmost point of the Bulgarian Black Sea Coast and the southeasternmost inhabited place in Bulgaria. Population: 77 (2008).

Richelieu Duc Armand du Plessis, Emmanuel Iosifovich (1766–1822) – Duke, first General-Governor of Odessa, General-Governor of Novorossiya (1803–1804). Having left Odessa he twice was Prime-Minister of France. He distinguished himself during the Izmail storming (1790) and was awarded the golden sword. He became the prototype of a character in the Byron's poem "Don Juan". He initiated planting of the Duke Garden. The statute of Duc de Richelieu built in 1826 in the a-la Antique style was one of the last works of well-known sculptor I.P. Martos, the author of many masterpieces (including the monument to Minin and Pozharskiy in Moscow, to M.V. Lomonosov in Arkhangelsk and others). The base for the monument was designed by architect A. Melnikov. The monument to Duc de Richelieu and the Potemkin Steps make a single ensemble.

Richelieu Duc Armand du Plessis (Photo by Elena Kostyanaya)

Richelieu Lyceum – was founded in 1817 on the basis of the Noble Educational Institute that existed at the first General-Governor of Odessa—E. Richelieu. The first director of L. was Jesuit Abbot Nicol and the whole teaching was practically in French. At the Richelieu's successor the Noble Educational Institute was turned into the advanced education establishment—"Lyceum named after Duc de Richelieu". The lyceum located in the former house of Sobanskiy at the crossing of the Deribasovskiy and Ekaterininskiy streets. From 1829 the status of the secondary educational establishment (gymnasium) had only the first four classes. The next five classes were lyceum and they were divided into two branches: one with the philosophy orientation and the other—legal. The lyceum included the Pedagogical Institute. In 1838 the new statute of the lyceum was adopted and it practically enacted separation of lyceum from the gymnasium classes. At the same time the Institute of Oriental Languages was incorporated into lyceum. In 1839 there was established the agricultural department and in 1841—the geodetic department. In the same year the "laboratory division" was organized that, apart from physics, physical geography, chemistry and natural history that were also studied at the physico-mathematical division, taught political economy, sciences on finance, trade and agriculture.

Rioni (perhaps in the Svanetian "rien" meaning "large river") – a river in Georgia. Its length is 327 km. It originates in the glaciers on the southern slope of the water divide ridge. As far as the city of Kutaisi it flows in the mountains, first in a narrow gorge, then over a rather wide valley and then again enters the narrow canyon-like valley where the Gumatsky and Rioni hydropower plants are constructed. Downstream Kutaisi the river flows over the waterlogged Kolkhida lowland in the meandering channel that breaks into arms forming an island, bends and oxbows. Not reaching Poti it again runs in one channel and then with two arms it flows into B.S. It has mostly the snow and glacial recharge; in the downstream it is fed with rains. The waters of R. in its middle reaches are used for irrigation. Beginning from the city of Oni it is used for rafting. Such cities as Oni, Kutaisi and Poti locate in the R. valley. The Voenno-Ossetian road runs over the upstream valley of the river.

Rioni Lowland – a lowland on the Caucasus, in the lower reaches of Rioni, Kolkhida lowland.

Ritsa Lake – a beautiful lake located in Abkhazia. R.L. is surrounded by the Caucasian Mountains with heights of 2,200 to 3,500 m with forests and subalpine meadows. It is one of the deepest lakes in Georgia (116 m). Surface elevation— 950 m, lake surface—1.49 km^2. R.L. is fed by six rivers and drained by one, the Iupshara River. The average annual temperature in the area is 7.8 °C (in January −1.1 °C, in August 17.8 °C). The mean annual precipitation is 2,000–2,200 mm. Winters are sometimes snowy, summers warm. Its water is cold and clear. The region around R.L. is a part of the Euxine-Colchic deciduous forests ecoregion with a high concentration of evergreen boxwood groves. Many specimens of the Nordmann Fir, which reach heights of 70 m are found around the lake. In 1930 the Ritsa Nature Reserve (162.89 km^2) was established to protect the natural state of the lake and the surrounding land. The road from the Black Sea coast was built in 1936. The lake was an important tourist attraction during the Soviet period. It is still visited by Russian tourists. The resort Avadhara lies to the north of the lake. The Soviet leader Joseph Stalin had one of his summer-houses (dacha) at the lake. Today this dacha belongs to the Abkhazian Government.

Ritsa Lake (http://upload.wikimedia.org/wikipedia/commons/thumb/5/5c/RitsaMounts.jpg/ 1024px-RitsaMounts.jpg)

River basin – a territory from which the water flows into a river over the surface and via underground waterways (from mountain rocks). The largest river basins of B.S. are the Danube and Dnieper rivers.

River sediments – the suspended matter transported with river waters and settling down in the riverbed. They are formed as a result of scouring of the watershed surface with rain and melt waters as well as scouring of the riverbed proper.

Rize – a city in the northeast of Lazistan, Turkey, situated on the coast of the Rize Bay; the administrative center of the Rize Il (Province). Its population is 104,500 (2012), predominantly the Lazys—these are the Georgians who were converted into Moslems in the fifteenth century. The citrus, tobacco, rice and corn are grown here. The port on B.S. The major tea producer—about 350 tea factories may be found in the city and its surroundings. The Institute of Tea locates in R. The factories manufacturing cod-liver oil, canvas ("docum") and shawls are working in the city. It has a small beach and a botanical garden. The mineral water spring Dzjamaly-Khemshi locates 5 km from R. and the baths are constructed here. A not large port may receive small sea craft and "river-sea" vessels with the draft to 3 m. Rize University was founded in 2006.

Rize   (http://upload.wikimedia.org/wikipedia/commons/b/b6/Rize%27den_bir_g%C3%B6r%C3%BCn%C3%BCm.jpg)

Rize Bay – locates on the Anatolian coast of B.S., Turkey. It incises into the shore between the Askoros Cape and the Pyrios Cape located 6.5 km westward of it. Many mountain rivers bring their waters into the bay. The mountain range of Charshidag 350–380 m high runs along the eastern coast of R.B. Its slopes do not reach the sea and in the shore area between the Askoros Cape and the city of Rize there is a wide low sandy strip that is densely built up. On the western coast of the bay the low coastal strip is much narrower and it is stony.

Rize Il – the province on the B.S. coast, Turkey. Its area is 3,920 km^2, population—319,600 (2011). Its capital is the city of Rize.

Romania – the B.S. state in the southeast of Europe, in the basin of the Lower Danube. In the first centuries B.C. the territory of present-day Romania was at the outskirts of the Roman Empire where the local population mixing with the Roman colonists was called "Romanus" (from the Latin "*romanus*" meaning "Roman"). In 1862 the state was formed here called Romania by the name of the people living here. The territory of this country is 238,400 km^2; the population over 20.1 mln people (2011). The capital is Bucharest with the population of 1.9 mln (2011). Other major cities of the country are Cluj-Napoca (325,000), Timisoara (319,000), Iasi (290,000) and Constanta (284,000. The official language is Romanian; the prevailing religion—Christianity (Orthodox). Local currency is leu. The form of government is presidential parliamentary republic. The head of the state is President elected for a term of 4 years. The legislative body is Parliament consisting of the Chamber of Deputies and Senate elected for 4 years, too. R. is divided into 41 counties plus the municipality of Bucharest which has the equal rank. The southern state border of the country runs over the Danube River and the eastern—over the Prut River. In the north R. borders on Ukraine, in the east—Moldova, in the southwest—Federation of Serbia and Montenegro, on the south—Bulgaria (along the Danube), on the west—Hungary and on the southeast it is washed by B.S. (the Romanian section of the B.S. coast is 225 km long).

Romania (http://education.randmcnally.com/images/edpub/Romania_Political.png)

In the first century B.C. to third century A.D. the territory of present-day R. was populated by the Getae-Dacian tribes (Dacians). In the early second century the area populated by the Dacians was turned into the province of Roman Dacia. After the Romans left this place (in 271) this territory was invaded by the Goths, Gepids and Avars. In the sixth to seventh centuries this place was inhabited by the Slavs. In the mid-fourteenth century the feudal states of Wallachia and Moldova appeared here that later on (in the sixteenth century) got under the Ottoman ruling. In 1829 with the help of Russia these principalities attained the autonomous status and in January 1862 they united forming the Romanian Kingdom that was under suzerainty of the Ottoman Empire. The independence was proclaimed in May 1877 after the Russian-Ottoman War. In 1881 R. became the Kingdom. In 1918 after the defeat of Austria-Hungary the principalities of Bessarabia, Bukovina and Transylvania proclaimed unions with the Kingdom of R. The country fought in World War II on the side of Germany, but a short time before its end it changed the sides and joined the Allies. In 1940 Bessarabia and the northern part of Bukovina were ceded to the USSR, while Northern Transylvania to Hungary. In 1944 the fascist dictatorship of Ion Antonescu was toppled. After liberation of R. the coalition government came to power. The Treaty of Prague (1947) restored the country in its former borders as of 1 January 1938. The monarchy was liquidated and King Michael I abdicated. As a result, the people's republic was proclaimed. In 1948–1989 the Romanian Communist Party was in power. In 1990 and 1992 the first democratic elections were conducted. The policy of the new government was oriented to democratic development and transition to the market economy. In 2004 R. acceded to NATO.

The third of the country's territory is occupied by the mountains; the central part of the country—by the Transylvania basin. R. is surrounded by mountains: the Carpathian in the north and east, the Transylvania Alps (the highest peak is Mont Moldovyanu—to 2,544 m) in the south, the Bikhor (the highest peak is Mont Kukurbeta to 1,848 m high) with the dead volcanoes and karst—in the west. The rest of the territory is flat—the large Lower Danube Plain is extending in the south of R. The mountains are cut by deep river valleys. The major river is Danube. Its tributaries running over the country's territory—Olt, Siret, Muresh and other take their origin in the Carpathian. The country has many lakes being limans of B.S., the largest of them is the Razelm Lake. The climate is temperate continental; draughts may occur on the Lower Danube Plain. The mean temperature of January on the plains varies from 0 to -5 °C, of July—from 20 to 23 °C; in the mountains the temperatures are somewhat lower. The precipitation is 300–700 mm a year (in the mountains—1,500 mm). About the half of the country's territory is cultivated and occupied by orchards, farmlands and vines. The broad-leaved forests cover 27 % of lands in mountains and river plains. The steppe and forest-steppe vegetation is growing on plains. The resorts with abundant artificially grown vegetation situate on the sandy coasts of B.S. Coniferous forests are growing in mountains in the northern areas. The wetlands in the Danube Delta remote from the sea are overgrown with shrubs and are a habitat for many bird species. The animal world is not rich, only roe deer and fox are preserved here. In the Carpathian the mountain goat, bear and lynx may be met. There are also found rodents, lizards and snakes.

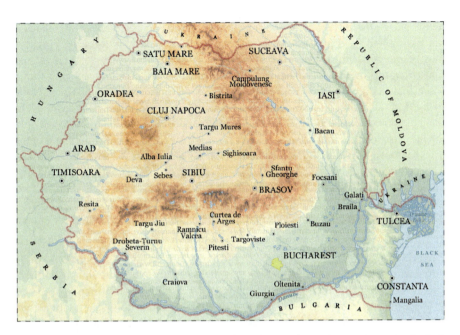

Romania, topography map (http://www.romaniatourism.com/romania-maps/romania-physical-map.gif)

The Romanians make nearly 89 % of the country's population. There are also large diaspores of the Hungarians (6.5 %), Roma (3.2 %), Germans (0.5 %) living in compact groups in some regions of Transylvania, the Ukrainians an others. A considerable number of gypsies is populating here. The Romanian language refers to the Roman group, but it has many words borrowed from the Slavs and from the Dacians and Thracians who populated this territory in the ancient times. The architecture of Romanian cities was influenced by the nearby Orthodox countries: Byzantine, Serbia and Bulgaria. The medieval buildings, castles and fortresses survived in the mountains. Bucharest has many buildings that remind of the Ottoman ruling. Of greatest interest are Turkish baths. Modern architecture follows the best examples of the world urban construction. The resorts on the B.S. coast and in the mountains have comfortable hotels and cultural centers, such as Mamaia, Kostinesti, Sinaia and others. Well developed are the decorative art and handcrafts, including making of ceramics, lace and leather products, weaving of wickers and straw, decorative embroidery, wood carving, etc. Among the most interesting museums there are the Museum of Arts possessing a large collection of paintings and household items, the National Museum of Antiquities, the Museum of Folk Art in Bucharest, the Museum or Art and Ethnography in Yassy and others.

At present the economics of R. is facing great difficulties. The industrial products are competitive. R. has the gas and oil fields. It is one of the European leaders in production of coal, copper, iron ores, lead, zinc, gold, silver and other deposits. The industry includes mining, metallurgical, chemical, forest, woodworking enterprises, tractor construction, production of building materials, furniture, textile and footwear factories, light machine building and assembly of vehicles, oil processing and food production. In agriculture the key sectors are crop growing, mostly grains—corn, wheat and barley as well as sugar beet, potato, sunflower and vine growing. In animal husbandry the sheep and cattle breeding are prevailing. More than 8,000 joint ventures with foreign companies from South Korea, Germany, Italy and others have been established. Great attention is paid to development of small and medium business and foreign tourism (6.5 mln tourists a year). In the infrastructure the importance of river ports is growing due to development of the trans-European system Rhine–Danube which length within R. exceeds 1,000 km. The economic policy is targeted to overcoming of crisis, market reforms and integration into the economic structures of Europe. The country exports tractors, automobiles, petroleum products, equipment, mill products, fertilizers, cement, furniture, footwear and fabric; imports machines and equipment, foodstuffs, consumer goods. The main trade partners are Germany, Italy and Russia.

Romanian Black Sea coast – the coast of Romania stretching for about 260 km. In terms of resort and tourist attraction it is divided into two subregions—northern (from the Ukrainian border to the Midia Cape) and southern (from the Midia Cape to the Bulgarian border). They differ by the size, nature, tourist attractions and tourism development level. The northern subregion (about 200 km long) includes the Danube Delta and large seaside lakes of Razelm (310 km^2), Golovica (120 km^2), Sinoye (180 km^2) and others. Extensive sandy sandbanks are stretching between the lakes and the seashore. The southern subregion is much smaller in size—its length is about 60 km, but it is more convenient for construction of tourist centers. Sand beaches are running along the whole coast. Several lakes (Syutgel, Tekirgel, etc.) are found in the coastal zone. Some of them are saline and contain significant peloid resources. The climate here is temperate. Cold winds in winter are often responsible for a long freeze-up period in the Danube Delta and seaside lakes. The summer, in particular in the southern subregion, is rather hot. The mean annual temperature on the coast is 11.3 °C. The precipitation is approximately 500–600 mm a year, and in the Danube Delta 300–400 mm. As a result, the vegetation here is not rich. The broad-leaved forests are few in number. Steppe shrubs and grasses are prevailing here. Very good asphalt-paved highways connect the coast with Bucharest and other large cities in Romania and in neighboring states. The water transport includes marine and river vessels that serve major resorts. In the northern subregion the marine transport is the most convenient means connecting seaside towns. The river transport is well developed in the Danube Delta. From airports in Constanta and Tulcha one can easily reach any part of the country and other world countries as well.

"Romanian Riviera" – the B.S. coastal strip 50 km wide where 16 best resorts are situated. The greatest attraction among them is Mamaia, Euphoria Nord (north), Costinesti, Neptun, Venus Olymp, 2 Mai.

Romanians – "Romanus" is the nation, the indigenous population of Romania (over 85 % of the country's inhabitants). They originate from the Dacian tribes that were under control of the Roman Empire in the second to third centuries A.D. The language and culture of R. were influences, to a large extent, by Southern Slavs. The Romanian language belongs to the Roman group of the Indo-European family of languages. R. include the ethnographic group of the people—semi-sedentary Makans living in Southern Carpathian whose main occupation is sheep breeding.

Roman-Kosh – the highest point on the Crimean Peninsula, Ukraine. This name appeared from the original Turk name "Orman-Kosh" where "*orman*" meant "forest", "*kosh*"—"temporary living place of cattlemen" or "a place where sheep are driven for night", i.e. "forest pasture". According to other versions this name originated from the Turk name "*orman*". It is located in the Babugan-Yaila area, in the southern range of the Crimean Mountains. Its height is 1,545 m. It is composed of limestone.

Romanovskoe – see Krasnaya Polyana.

Root-mouthed jelly (*Pilema pulmo*) – the largest (scyphozian) species of the Coelenterates (*Coelenterata*) in B.S. The jelly semi-transparent dome of R.M.J. has the form of a high bell to 40 cm in diameter with 8 complex structured tentacles hanging from the dome center. Eight sensory organs locate over the wavy edge of the jelly dome. In summer one can often see small fish sizing to 2 mm among the tentacles that find there shelter and protection from the enemies; the stinging capsules of the jelly do not touch them.

Rose pelican (*Pelecanus onocrotalus*) – nests in the Romanian part of the Danube delta and feeds mostly in the Ukrainian part. This species is put on the Red Book of Ukraine (1994). It is referred to the category of endangered species. It is thought that the reproduction of the R.P. population is reduced greatly due to high levels of DDE (derivative of DDT).

Rostov-on-Don – is a portand city, the administrative center of Rostov Oblast and the Southern Federal District of Russia. It lies on the Don River, 32 km from the Sea of Azov. Population: 1.1 mln (2010). Since ancient times, the area around the mouth of the Don River has been of cultural and commercial importance. Ancient indigenous inhabitants include the Scythian, Sarmat, and Savromat tribes. Later it was the site of an ancient Greek colony Tanais, Fort Tana under the Genoese and Fort Azak in the time of the Ottoman Empire. On 15 December 1749 a custom house was established on the Temernik River, a tributary of the Don River, by edict of Empress Elizabeth, the daughter of Peter the Great, in order to control trade with Turkey. It was located near a fortress named for Dmitriy of Rostov, a metropolitan bishop of the old northern town of Rostov the Great. Azov town, located closer to the Sea of Azov on the Don River, gradually lost its commercial importance in the region to the new fortress. In 1756, the "Russian commercial and trading company of Constantinople" was founded at the "merchants' settlement" (Kupecheskaya Sloboda) on the high bank of the Don River. Towards the end of the eighteenth century, with the incorporation of previously Ottoman territories into the Russian Empire, Rostov lost much of its militarily strategic importance as a frontier point. In

1796, the settlement became an administrative centre for the newly Russified region. In 1806, the name of this founding site was changed to Rostov and later to Rostov-on-Don, thus, the modern day town was established.

Rostov's favorable geographical position at trading crossroads promotes economic development. The Don River is a major shipping lane connecting southwestern Russia with the north. Rostov-on-Don is a trading port for Russian, Italian, Greek and Turkish merchants selling of wool, wheat and oil. It is also an important river port for passengers. The Rostov-on-Don agricultural region produces one-third of Russia's vegetable oil from sunflowers. Rostov-on-Don has experienced economic growth. Numerous start-up companies have established headquarters in the city, the median income is increasing, and the city is being transformed into a modern, industrial and technology-rich hub. Rostov-on-Don is a center for helicopter and farm machinery manufacturing. With the construction of the Volga-Don Shipping Canal in 1952, Rostov-on-Don became a port of five seas: the Black, Azov, Caspian, White and Baltic seas.

Rostov-on-Don hosts higher educational establishments, including universities, academies, secondary schools, colleges, technical schools, and schools of general education. The largest educational establishments of the city is Southern Federal University.

Rostov-on-Don (http://upload.wikimedia.org/wikipedia/commons/6/6f/View_Rostov.jpg)

Rotten Sea – see Sivash.

Roubaud Franz Alekseevich (1856–1928) – the outstanding Russian painter who depicted historical battles. He was born in Odessa where he received his initial education attending the art school (1865). In 1877 he went to Munich where he studied at the Bavarian Academy of Arts and where he stayed to live. Here for 2 years (1902–1904) in a special pavilion R. together with a team of painters from the Bavarian Academy of Arts created the canvas "The Sevastopol Defense of 1854–1855". The sketches for the future canvas R. made having visited the place of battles. After this canvas R. painted also "The Storm of Achulgo" and "The Battle of Borodino" (1911). He was Academician of Arts; from 1910—the member of the Petersburg Academy of Arts where he taught from 1904 to 1912. In 1913 he went to live and work to Germany. He painted over 20 canvases showing mostly battles, such as "Darjal Gorge", "Storm of the Geok-Tepe Fortress", "Battle at Kushka", "Pereprava" (Crossing).

Rules of navigation in the Bosporus 1982 – they were enforced by Turkey on 1 May 1982. They envisaged: waive of the rule "to keep left" that was the cause of many collisions; ban on movement at a speed more than 10 knots (18.5 km/h), except in some emergency situations; sending of at least once-week notice about passage of Turkish, foreign and tourist vessels with displacement over 10,160 tons; demand to take onboard a Turkish pilot by all foreign merchant ships (this applied to specific transit voyages and routes with moorages); temporary stopping, if necessary, of the passage of ships via the strait enforced by the Istanbul port authorities. These rules were adopted for better organization and creation of conditions for prevention of accidents in the strait, including by attraction by the ship captains of pilots and tow boats. Unilateral enforcement by Turkey of these rules caused displeasure among the interested parties. New rules were not applicable to warships.

Rumeli Cape – on the European coast of the Bosporus Strait, 3.7 km to the northwest of the Anadolu Cape at the northern entrance into the Bosporus. The cape is elevated and stony. The lighthouse is constructed here.

Rumeli coast – the section of the Turkish coast of B.S. westward of the Bosporus Strait to the Igneada Cape.

Rumyantsev Aleksandr Ivanovich (1680–1749) – a count, the notable Russian military figure and diplomat. In 1703–1712 he was on the military service. In 1712 he was sent to Turkey for ratification of the Treaty of Prut of 1711. In summer 1724 together with I.I. Neplyuev, the Russian resident in Constantinople, he forced Porta to agree to concessions concerning the borders among Russia, Turkey and Iran. In 1740–1742 he was the Ambassador in Constantinople where in summer 1741 he put his signature under the convention obliging Russia to liquidate the fortifications on Azov, while Turkey acknowledged the Emperor title of the Russian Tsar.

Rumyantsev-Zadunaiskiy Petr Aleksandrovich (1725–1796) – a count, Russian General-Field Marshal (1770). During the Russian-Turkish War of 1768–1774 in the campaign of 1768 he commanded the Second Army which task was to defend the southern borders of Russia and the area of action of the First Army commanded by A.M. Golitsyn from attacks of the Crimean Tatars. For distraction of the Turkish forces from the main theater of military actions the simultaneous attacks of the Russian Fleet in Archipelago were planned. In November 1768 Rear Admiral A.N. Senyavin initiated restoration of the Azov Fleet. From 1770 and onward its base was in Taganrog. In June 1770 the First Army led by R.-Z. defeated the troops of Crimean Khan Kapsan-Girey and in the battle near the Kagul River—the Turkish Army of Khalil-Pasha as a result of which first the fortress of Izmail and later Kilia had to surrender. The Russian troops came on the left bank of the Danube. On 26 July 1770 Catherine II conferred to R.-Z. the rank of General-Field Marshal. Preparing for military actions across the Danube he started creation of the Danube Fleet that included 67 armed vessels. In spring 1772 Admiral Charles Knowles was appointed the commander of the Fleet. The plan of the 1774 campaign developed by R.-Z. included Danube crossing by the Russian troops. In June the corps of A.V. Suvorov defeated the Turkish troops at Kozludzhey. The main forces of R.-Z. crossed the Danube and blockaded the fortresses of Silistria and Ruschuk. The efficient actions of the Azov Fleet influenced greatly the outcome of the war. On 10 July 1774 the Peace Treaty between Turkey and Russia was signed in the village of Kuchuk-Kainarji under which after long-time efforts Russia got access to B.S. and was able to create the Black Sea Fleet. For these dazzling victories over the Turks in 1775 Rumyantsev was awarded the victory title Zadunaiskiy. At the beginning of the Russian-Turkish War of 1787–1791 R.-Z. commanded the Second Army. In 1799 on the Marsovo Field in Saint-Petersburg the monument "To Victories of Rumyantsev" was constructed. In 1818 it was moved to a new place and installed between the buildings of the Academy of Arts and the First Cadet Corps.

Russia's Black Sea Navy – the commencement of establishment of Russia's regular navy is dated the end of the seventeenth century, when by order of Peter I the Azov Fleet was built on the Voronezh River in 1695–1696 by way of struggling for access to the Balck Sea and the Sea of Azov. In 1696, in Peter's second Azov campaign two Russian battle ships, 4 branders, 23 galleys and 1,300 strougs took part. On Ocober 20 (30), 1696, based on the report of Peter I on the construction of the Azov Fleet, the Boyar Duma decreed: "Let there be sea-going vessels". That day is regarded the birthday of Russia's regular navy.

R.B.S.N. was established by Prince G.A.Potemkin during the reign of Catherine II. Permanent reinforcement of the navy was necessitated by the need to inhibit Turkey's aggressive intentions. Rapid growth of the Russian Navy early in the eighteenth century and its brilliant victories made Russia one of the most mighty naval powers of the world. In 1770, Admiral G.A.Spiridov's squadron defeated the Turkish Fleet in Chesma Bay and became dominant in the Aegean Sea. In 1771–1773, after the Russian Army reached the Danube and the Sea of Azov, the Danube Flotilla and the Azov Military Flotilla were established.

The Emperor's decree of December 11, 1775 determined he main directions of naval forces development in the Black Sea and of the ship-building production facilities. In 1778, the new port Kherson was founded—the first ship-building base on the Black Sea. The construction of the Black Sea Navy was conducive to the growth of the Black Sea region itself. In 1783, thanks to skilful diplomacy backed by the might of its navy, Russia asserted its superiority in the Crimea finally. The date of May 2(13), 1783, when the Azov Flotilla entered Akhtiar (Sevastopol) Harbor, is generally regarded as the day of Black Sea Navy establishment. In 1788, the shipyard was built in Nikolaev. In 1804, Sevastopol was declared the main naval port on the Black Sea. The young B.S.N. won its first victory in the Russian-Turkish war of 1788 near Fidonisi Island, were it defeated the Turkish Fleet. In 1790, the Black Sea Navy had 22 battle ships, 12 frigates, nearly 80 rowing and sail vessels. That same year, "the sea Suvorov" Rear-Admiral F.F. Ushakov gained a brilliant victory first near Fidonisi Island (1788), in Kerch Strait (1790), near Tendra Island (1790) and near Kaliakria Cape (1791). The latter expedited the conclusion of a peace treaty with the Turks. In 1799, under the command of F.F. Ushakov during the liberation of seven Ionic islands the Corfu Fortress was seized which hitherto had been regarded impregnable. The Black Sea Navy thanks to F.F. Ushakov's untiring efforts turned into a mighty force: not only Turkey, but also Britain and France had to reckon with it.

In 1804–1804, several battle ships of the Black Sea Navy took part in operations on the Mediterranean Sea under the command of Admiral D.N.Senyavin against France. A few major victories over the Turkish Fleet were scored in the battle of Dardanelles (1807). In 1827, the Russian squadron led by Admiral L.P.Heiden in conjunction with the British-French Fleet destroyed the Turkish Fleet in the sea battle of Navarin. During next Russian-Turkish war of 1828–1829, a heroic feat was performed by the Russian brig "Mercury". In 1831, at the initiative of Vice-Admiral A.S.Greig the management of the Black Sea department was reorganized. The Main Black Sea Department was instituted. The Chief Commander was put in charge of the Navy; attached to him were the front office, headquarters and a commissariat. In 1832, the first chief of staff of the Black Sea Navy became Rear Admiral M.P.Lazarev.

By the time the Crimean war commenced (1853–1856), the Russian Navy essentially comprised sailing ships (40 battle ships, 15 frigates, 24 corvettes and a brig) an 15 steam-powered frigates only. The Black Sea Navy was remarkable for its training: the leading posts were occupied by the outstanding Russian Admirals M.P.Lazarev, P.S.Nakhimov, V.A.Kornilov, V.I.Istomin. The Crimean (Eastern) war that began in 1853 engaged the Black Sea Navy as the main sea battles were fought essentially on the Black Sea. During the war, the Black Sea Navy scored brilliant victories under the command of Vice-Admiral P.S. Nakhimov in the battle of Sinop (1853) over the Turks. That was the last battle of sailing ships.

The defense of Sevastopol became the central event of the Crimean war. To prevent penetration of the British and French battle ships that had a numerical superiority to the city from the sea most of the Black Sea Navy sailing ships were sunk. During the Sevastopol defense of 1854–1855, the Russian sailors demonstrated an example of active defense of the naval base from land and sea. Despite the heroism and self-denial of the Russians, Russia was defeated. The outdated sailing fleet of the Russian Black Sea Navy was unable to confront the newest steam –powered Navy of Britain and France. The war ended in a prohibition for Russia to have its Navy on the Black Sea (over 100 Black Sea naval vessels were sunk), Russia was denied access to the Black Sea straits.

Russia entered the twentieth century with iron-clad steam-powered fleet, which called for the development of new sea action tactics, new methods of ship armament application. The Admirals G.I.Butakov and S.O.Makarov made a great contribution to solving these problems. By the time WWI broke out, major sea powers built huge fleets. During WWI, the Black Sea Navy supported maritime flanks of the fronts, especially the Caucasian. The Navy managed to curb substantially the sea shipments of Germany and Turkey. The first engagement of the Russian and German navies took place on the Black Sea. Here, in November of 1914, near Sarych Cape a battle was fought between the Black Sea Squadron and the newest German battle ships "Goeben" and "Breslau". The former was badly damaged. In 1915, "Breslau" tripped a mine barrier that had been placed by the Russian subsea mine layer "Crab". Until the end of 1916, the Black Sea Navy destroyed several battle ships, seriously damaged three cruisers, sank over 60 carriers and 3,000 various sailing ships of the enemy. The activity of the Black Sea Navy became particularly successful in 1916 after it succeeded in blocking the enemy fleet in Bosporus. From July of 1916 to July of 1917, the Black Sea Navy carried out 17 mine-laying operations near Bosporus and planted 8,000 mines near Varna. As a result of such activity the German High Command stopped sending the cruisers "Goeben" and "Breslau" to the Black Sea, and after the 'UB-46" submarine was destroyed by mine tripping, no more submarines were sent to the Black Sea.

The October Revolution marked the start of the history of the Soviet Navy. After the Civil War in Russia, construction of the Black Sea Navy began. In 1923, the navy included the cruiser "Komintern" (former "Pamyat Merkuria" ('Memory of Mercury')), the squadron destroyers "Nezamozhnik" (former 'Zante'), two submarines and some other ships. In November of 1926, a 6-year program (1925–1932) of naval ship-building was approved, it provided, among other, for the completion of construction of the cruisers "Chervona Ukraina", "Krasny Kavkaz" and others. By 1928, the Black Sea Navy had 2 cruisers, 5 squadron destroyers, 8 patrol craft, 5 submarines, 4 battle-boats and other ships.

By the time the Great Patriotic War broke out. There were 4 task forces set up: one of these—Black Sea Task Force (headed by the Commander-in-Chief Vice-Admiral F.S. Oktyabrskiy). From the first day of the war, the Black Sea Navy took part in the defense of crucial ports and imparting stability to the southern wing of the land front. The Black Sea Navy provided for the heroic defense from the sea of Odessa, Sevastopol. The Navy played an important role in the battle for the Caucasus. The Black Sea Navy also assisted land troops by landing amphibious parties. Among the major amphibious landing operations—Kerch-Feodosiya (1941–1942), Kerch-Eltigen (1943), Novorossiisk operation (1943). During the war, in line with the old combat tradition, marine guard was re-established to become successor of the feats of heroic ships of the Russian naval ships that had been decorated with St.George's flags. The cruisers "Red Crimea", "Red Caucasus" and other ships became part of the guard.

At present, the surface-to-surface missile cruiser "Moskva" (former "Slava"), flagship of the Black Sea Navy, has a sonorous attribute of a guard ship. In 1965, the Black Sea Navy was awarded the Order of Red Banner. After the war, the Black Sea Navy kept building up its combat potential, was upgraded, enlarged, emphasis being laid on promoting military cooperation with the friendly states on the Black Sea shores. Early in the 1970s, unified Black Sea Navy was established on the basis of the forces of the Soviet Black Sea Navy, naval forces of Bulgaria and Romania. For four decades the Black Sea Navy provided the security of the Black Sea boundaries. During the years of the "cold war", the Black Sea naval ships participated in localizing and settling the armed conflicts in Angola, in the Arab-Israeli wars, during the Cyprus crisis of 1964. Besides the Black Sea naval ships guarded Soviet fishing activity near the shores of West Africa in 1979–1991. In 1964, a group of battler ships of the Black Sea Navy for the first time went into the Mediterranean Sea to keep an eye on the aircraft carrier battle group of the US Navy. In 1967, the Black Sea Naval ships arrived in the Indian Ocean.

The collapse of the Soviet Union put the Black Sea Navy in an extremely complex situation. Part of the naval ships was lost irretrievably, the fleets' infrastructure underwent considerable changes. The "Black Sea Navy Agreement" was signed in 1995, whereby Russian Black Sea Navy and the Navy of Ukraine were instituted, each having its own base. After the division, Russia's Black Sea Navy took part in armed conflicts in the countries of near abroad. At the end of the twentieth to early in the twenty-first century, the Black Sea Navy remains one of the most effective tools of Russia's state policy geared to provide for long-term economic, foreign-policy and military-strategic interests in the system of Russia's national security priorities.

Russian Federation, Russia – the state covering the east of Europe and the north of Asia. In the northwest it borders on Norway and Finland, in the west—on Ukraine, Poland, Estonia, Latvia, Lithuania and Belorussia, in the southwest—on Georgia, Southern Ossetia, Abkhazia, Azerbaijan and Kazakhstan, in the southeast—on China, Mongolia and Democratic Republic of Korea. The area of the country is 17.075 mln km^2. The length of the land border is 20,322 km, sea border—about 38,000 km. The population is 143.6 mln people (2013). The average population density is 8.4 men/km^2. The Russians make 81 % of the population. In general, over 100 peoples live in this country. Nearly all confessions are represented here—Christians, mostly Orthodox, Moslems, Judaists, Buddhists and others. The capital is Moscow (11.5 mln in 2009).

The northernmost point of the mainland is Cape Chelyuskin on the Taimyr Peninsula; the southernmost point is in Daghestan on the border with Azerbaijan. The distance between the western and eastern borders of R. is about 9,000 km. The territory of R. is broken into 9 time zones (since 2011). The territory of the country is washed by 12 seas belonging to the basins of three oceans: Atlantic Ocean—Baltic, Black seas (the southern part of the Krasnodar Territory goes out to the sea) and Sea of Azov (the Rostov Region and the northern part of the Krasnodar Territory go out to the sea); Arctic Ocean—Barents, White, Kara, Laptev, East-

Siberian and Chukchi seas and Pacific Ocean—Bering Sea, the Seas of Okhotsk and Japan as well as the landlocked Caspian Sea.

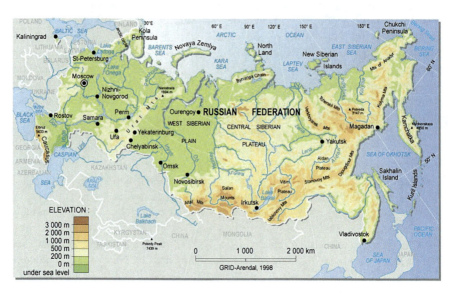

Russian Federation (http://maps.grida.no/go/graphic/russian_federation_topographic_map)

The greater part of the R. territory has a flat relief. In the west there is the East-European plain within which not high uplands and lowlands (pre-Caspian and others) are found. It is confined in the east by the Ural Mountains. To the east of the Urals the West-Siberian plain is stretching. The Middle Siberian Plateau locates between the Enisey and Lena rivers. The flat relief is broken in some places by mountains. In the east it passes into the Central Yakutian Plain. Mountains are mostly found in the east and south of the country. In the European part these are the ranges of the Northern slope of the Greater Caucasus (the highest mountain—Elbrus, 5,642 m); in Southern Siberia—Altai, Salairskiy Range, Kuznetskiy Alatau, Western Sarjan, Eastern Sarjan, the mountains of Tuva, Pribaikalie and Zabaikalie. In the northeast of Siberia and in the Far East the mountain ranges of medium height are prevailing. The mountains of Kamchatka and Kuril Islands with the active volcanoes are stretching along the Pacific coast.

The greater part of R. locates in the temperate zone; the islands of the Arctic Ocean and northern regions of the mainland—in the Arctic and sub-Arctic zones; the Black Sea coast of the Caucasus—in the subtropical zone. Nearly everywhere the climate is continental. The mean temperatures of January vary from 0 °C in pre-Caucasus to −50 °C in Yakutia and of July from −1 °C on the northern coast of Siberia to 26 °C in the pre-Caspian Lowland. The greatest amount of precipitation is recorded in the Caucasus Mountains—to 3,200 mm a year, the minimum amount in the semideserts of the pre-Caspian Lowland—170 mm a year. The snow cover lays for 60–80 days in the south and for 260–280 days in the Far North.

The country has about 120,000 rivers with a length over 10 km. The largest of them are Volga, Ob and Irtysh, Lena, Enisey, Kolyma, Amur, Don, Kuban and Neva. The Volga and Ural flowing into the Caspian Sea refer to the inland water basin. There are about 2 mln freshwater and saline lakes in R. The major are Baikal, Ladoga, Onega and Taimyr lakes.

The territories of R. with flat relief feature quite vividly the natural zonal differences. Seven natural zones are discernible in the Arctic, sub-Arctic and temperate belts (from north to south): Arctic deserts, tundra, forest-tundra, forest, forest-steppe, steppe and semidesert. The narrow strip of the Black Sea coast of the Caucasus refers to the forest zone of the subtropical belt.

R. possesses the most copious mineral resource potential. The share of the country in the world reserves of coal, iron ores, potassium salts, and phosphorous deposits reaches or exceeds 30 %. R. has to 30 % of the world production of natural gas, 10–20 % of ores of rare, nonferrous and precious metals, 15–17 % of oil, to 14 % of iron ores and 5–6 % of coal. Oil and natural gas are the key elements in the fuel and power budget of the country and in the raw material export.

On the territory of R. there are over 1,080 cities, 2,000 urban-type settlements and 24,000 villages. The largest cities with the population over 1 mln people are Moscow (11.5 mln), Saint-Petersburg (5.2), Novosibirsk (1.5), Ekaterinburg (1.4). Nizhny Novgorod (1.3), Samara (1.2), Omsk (1.15), Kazan (1.14), Chelyabinsk (1.13), Rostov-on-Don (1.10), Ufa (1.10), Volgograd (1.09), Perm (1.09), Krasno-yarsk (1.0), Voronezh (1.0).

R. is a democratic federative state with the republican form of government. The country is managed on the basis of the Constitution adopted in 1993. R. incorporates 83 subjects of the Federation exercising equal rights: 21 republics, 9 kray (territories), 46 oblast (provinces), 2 cities of the federal significance (Moscow and Saint-Petersburg), 1 autonomous oblast and 4 autonomous districts. The official (state) language over the whole country is Russian.

The head of the state is President who is also the Supreme Commander of the armed forces. It is elected by popular vote. The executive power is exercised by the RF government. The chairman of the government is appointed by President upon consent of State Duma. The representative and legislative powers are vested in the two chambers of the Federal Assembly (Federation Council and State Duma). The membership of the Federation Council (166 members) incorporates two represen-tatives from each Federation subject: one from the representative and executive powers each. State Duma comprises 450 deputies. The deputies of Federation Council are elected for a term of 4 years.

In the 1st millennium B.C. some part of the present territory of R. was occupied by the antique cities (city-states), the Bosporus State and the Scythian State both in the Northern Prichernomorie. In the mid-seventh to late tenth centuries the Khazar Khanate located on the territory covering the Lower Volga, Northern Caucasus and circum-Azov areas. As a result of incorporation in the sixteenth to nineteenth centuries of the territories of the North, Povolzhie, Ural, Siberia, Far East as well as Ukraine, Caucasus and Transcaucasus coming out to B. and A. seas the multi-national state was formed called Russian Empire that existed till the February

Revolution of 1917. The October Revolution that occurred on 25 October 1917 brought to power the soviets of the workers', soldiers' and peasant deputies. On January 1918 the Russian Soviet Socialist Republic (RSFSR) was formed. RSFSR together with the Ukrainian, Belorussian and Transcaucasian Soviet Socialist Republics formed the Union of Soviet Socialist Republics (USSR) or Soviet Union, on 30 December 1922. Fifteen republics make up the USSR, the largest in size and over half of the total USSR population was the RSFSR, which came to dominate the union for its entire 69-year history.

On 12 June 1990 the First Congress of the RSFSR Deputies adopted the Declaration on State Sovereignty of RSFSR. In March 1991 the post of the RSFSR President was constituted. In December 1991, the heads of RSFSR, Ukrainian SSR. Byelorussian SSR signed the Belovezhskiy Treaty stating dissolution of the USSR and formation of the Commonwealth of Independent States. In September 1993 following the President's order the system of Soviets was liquidated. In December 1993 the Constitution of the Russian Federation was adopted and elections of the members of the Federal Council were conducted.

R. is an industrial and agricultural country. All kinds of mineral fuel are produced in RF in which the largest is the share of oil (with gas condensate) and natural gas. The process of transition to the market economy is underway in R. About 70 % of GNP is generated in the non-federal sector of economy. About 60 % of territories in agricultural use are arable lands the four-fifth of which is found in the Central and Central black-earth areas, Povolzhie, Northern Caucasus, Ural and Western Siberia. Farming provides over 50 % of the gross agricultural product, animal husbandry (meat-milk and meat-wool)—about 50 %. RF is the largest transport power exploiting all kinds of transport—railway, aviation, river, sea and pipeline. In December 2011 Russia joind World Trade Organization.

RF numbers about 2,000 higher educational and academic institutions. The oldest and largest Russian Universities are Moscow State University and Saint Petersburg State University. In the 2000s the government launched a program of establishing "federal universities", mostly by merging existing large regional universities and research institutes and providing them with a special funding. These new institutions include the Southern Federal University, Siberian Federal University, Kazan Volga Federal University, North-Eastern Federal University, and Far Eastern Federal University.

"Russian Riviera" – in this way the Black Sea coast of Crimea has been referred to for long and presently the mountain Black Sea coast of the Caucasus. In 1915–1916 this was the title of the journal publishing articles about the Black Sea coast of the Caucasus and Crimea.

Russian Shipping and Trade Company – this joint stock company was founded in Petersburg after the end of Crimean War of 1856. It was formed for revival of the sea trade in the Black Sea ports with foreign countries and also for support of the navy in the wartime. The capital of the company was 6 mln Rubles. RS&TC should purchase and build such freight and passenger vessels that in the wartime could be retrofitted into cruisers. The tsar government permitted 126 officers and several hundreds of seamen to retire and went to work with the company. In addition, RS&TC with the central office in Odessa was given the right to establish the tow shipping company for navigation over the Dnieper, Bug and Don. It was given on the non-return basis the Admiralty in Sevastopol (but, in fact, it was exploded by the British when in 1855–1856 they occupied the Korabelnaya side). RS&TC was capable to construct mechanical, ship-repair and shipbuilding workshops, warehouses and chambers as well as to open colleges for education of pilots, machine operators and other specialists needed by the Company.

Initially the company had 5 ships handed over by the Russian government. However, with time on it turned into the prosperous company possessing its own shipyards and ports. It opened the freight and passenger traffic over B.S. In 1858 the company had already 17 ships and a year later—35. In 1865 its capital had grown to 9 mln Rbls. After restoration of the plant in Sevastopol in 1868 the company started construction of its own ships. By the mid-1870s it had already 97 ships that navigated over B.S. and Mediterranean.

During the Russian-Turkish War of 1877–1878 some ships of RS&TC were chartered by the Naval Department. The armaments were installed on them andthe boat flotilla was created. The RS&TC ships were now called the active defense vessels and on the initiative of Admiral S.O. Makarov the vessel "Knyaz Velikiy Konstantin" (Grand Duke Konstantin") was retrofitted into the world's first floating base for mine launch boats. At the beginning of World War I the company had 73 ships with the total capacity of 120,900 tons. However, from this time onward the company's activities slacked off due to closing of the Bosporus and Dardanelles. Because of the Russian-Turkish War of 1877–1878 and industrial crisis in the first half of the 1880s the position of RS&TC worsened significantly. Seeking for new, more profitable marine lines in 1893 the company's ships navigated for the first time from B.S. to the Baltic. In 1904 it established its own Admiralty in Odessa (with a shiphouse and a floating dock). In the early twentieth century RS&TC was the major company in marine carriages over B.S. and Mediterranean (including on the most profitable lines Crimea-Caucasus and to Alexandria). In 1914 the company possessed 43 mail-passenger vessels, 3 freight-passenger, 23 freight and 7 tow ships. Its capital that time was equal to 10 mln Rbls. In 1918 RS&TC as well as voluntary fleet and other private shipping companies were nationalized.

Russian Squadron – In November 1920 following the order of Commander Vice-Admiral M.A. Kedrov the Black Sea Fleet (the remaining part of the Black Sea Emperor's Fleet) was reorganized into R.S. That time 120 vessels were taken off and about 150,000 people were evacuated. The fleet was reorganized into a squadron after the fleet reached Constantinople and after initiating the demobilization of ships, auxiliary war ships and reformation of the services and establishments that became needless. The word "Fleet" no longer matched the notion of armed forces by the quantity, class and composition of ships. R.S. left Constantinople on 8 December 1920 and called on Bizerta (Tunis) on 22–23 December 1920. After recognition of the USSR by France on 30 October 1924 the Andreevskiy flags were hauled down on R.S. ships. In 1950 the marble plate was installed in the memorial church in Bizerta constructed by Russian seamen. The names of all ships that sailed from Crimea were carved on it.

Russian Transport and Insurance Company – founded in 1844 under the name "Russian Marine and River Insurance Company". From 1877 it opened the traffic line Crimea-Caucasus employing four ships. In the 1880s the traffic line Odessa-Nikolaev was organized. By the early 1910s it exploited nine sea vessels with the displacement of 12,500 tons. Four of these were passenger vessels "Chernomor", "Svyatogor", "Ruslan" and "Vityaz". During a year the company transported to 240,000 tons. By 1914 the fixed assets of the company was 4.1 mln Rbls.

"Russian Troy" – this was the name of Sevastopol mostly in memory of the Crimean war.

Russian-Turkish Union Treaty of 1798 – fixed the accession of Turkey to the Second Anti-France Coalition. The treaty was signed on 23 December 1798 (3 January) 1799 in Constantinople for a term of 8 years. Turkey decided to sign this treaty beware of aggression from France that occupied Egypt in 1798. Russia was seeking to consolidate its positions in the war against France as a member of the Second Coalition of European powers as well as its influence on the Balkans. The treaty guaranteed the entirety of the ally holdings and counteraction of the French designs. In the secret clauses of this treaty Russia was obliged to render military assistance to Turkey in the struggle with France and for the first time Russia was granted the official right for its warships to pass through the Black Sea Straits.

Russian-Turkish Union Treaty of 1805 – it confirmed the Russian-Turkish Treaty of 1798 and contained 25 clauses, including 10 secret ones. It was signed in Constantinople on 11(23) September for a term of 10 years. The parties undertook in the wartime to assist each other with weapons, ships, troops or money as well as to abstain from entering into treaties that could be to the detriment of any party. Concrete actions in case of the likely French aggression were provided for in the secret clauses: Turkey had to agree its actions with Russia in the interests of the whole anti-French coalition, during the war with France to facilitate passage of the Russian warships and transport vessels via the Bosporus and Dardanelles to the Mediterranean blocking the entrance into B.S. for warships of other states. Turkey was also obliged to maintain self-government and other privileges of the Christian nationals of Turkey in the coastal areas of Albania. Russia guaranteed the entirety of Ionic Islands being under its ruling. The parties confirmed commitment to the provisions of the Treaty of Yassy (1791) and the Treaty on the Republic of Seven (Ionic) Islands of 21 March (2 April) 1800 and all other treaties. But this treaty was short-lived. After the victory of Napoleon at Austerlitz on 20 November (2 December) 1805 the rapprochement between Turkey and France was marked. Turkey started breaching the treaty which led to Russian-Turkish War of 1806–1812.

Russian-Turkish War of 1676–1681 – the war was caused by the spreading Turkish aggression against Ukraine after its unification (in 1653) with Russia. After having captured the region of Podolia the Ottoman government strived to spread its rule over all of Right-Bank Ukraine with the support of its vassal Hetman Doroshenko. In 1674 Ivan Samoilovich, Hetman of Left-Bank Ukraine was elected the sole Hetman of all Ukraine. Doroshenko decided to fight back and in 1676 his army of 12,000 men seized the city of Chigirin, the Hetman capital, hoping with the support of the Ottoman troops to be restored in his rights. In order to destroy these plans in spring 1676 the Russian and Ukrainian forces crossed the Dnieper and besieged Chigirin and on 23 July (2 August) seized it. The struggle for Chigirin between the Russian-Ukrainian troops and the Turkish-Tatar Army continued during the Chigirin campaigns of 1677 and 1678. In 1679–1680 the Russian-Ukrainian troops repelled successfully the attacks of the Crimean Tatars. On 13(23) January 1681 having failed to attain its goals Turkey had to sign the Bakhchisaray Peace Treaty that recognized the unification of Left-Bank Ukraine with Russia.

Russian-Turkish War of 1686–1700 – it was started after Russia's accession to the anti-Turkish "Sacred League". During the war, the Russian Army organized the Crimean campaigns of 1687 and 1689 and the Azov campaigns of 1695 and 1696. In view of Russia's preparations for the war with Sweden and signing by other countries of the coalition of peace treaties with Turkey the Russian government signed the Treaty of Constantinople in 1700, thus, ending the war. As a result of this war Russia got the town of Azov and the A.S. coast as far as the Mius River.

Russian-Turkish War of 1710–1713 – the war waged in the period of the Great Northern War of 1700–1721. It was unsuccessful for Russia. As a result of this war Russia had to abandon the town of Azov and to liquidate fortifications on the Azov coast.

Russian-Turkish War of 1735–1739 – this war was caused by intensified contradictions over the results of the Russian-Polish War of 1733–1735. In 1736 The Russian Dnieper Army under the command of General—Field Marshal B.C. Munnich took by storm the fortifications at Perekop on 20(31) May and then occupied Bakhchisaray, the capital of the Crimean Khanate. But the lack of supplies coupled with the outbreak of epidemic forced the Russian troops to retreat to Ukraine. On 19(30) June 1736, the Russian Don Army supported by the Don Military Flotilla seized the fortress of Azov. In June 1737 the Don Army backed by the flotilla forced a crossing of Sivash and defeated the Army of Crimean Khan Fetkhi-Girey in the battle at Salgir, while the Dnieper Army on 2(13) July took by storm the Turkish fortress of Ochakov. In the same year Austria entered the war. In August 1739 the Dnieper Army defeated the Turkish forces in the battle at Stavuchany. The imminent threat of the Swedish invasion and signing by Austria of the peace treaty with the Ottoman Empire forced Russia to sign in 1739 the Belgrade Peace Treaty with Turkey. In this war the Russian troops gained experience on crossing complicated water bodies (Sivash and Chongar crossings in Crimea).

Russian-Turkish War of 1768–1774 – on 25 September (6 October) 1768 Turkey supported by France and Austria declared the war on Russia. After this Russia moved the First Army of General A.M. Golitsyn (80,000–90,000 men) from Kiev to Khotin and the Second Army of General P.A. Rumyantsev (about 35,000 men) for fighting the enemy between the Dnieper and Don. Having repelled the attacks of the Crimean Khan Kaplan-Girey troops (70,000–80,000 men) the Second Russian Army went out to A.S. and blocked Crimea. At the Danube theatre of operations the army of Golitsyn after two failures to storm Khotin had to retreat beyond the Dniester. In September due to short supplies the Turkish garrison left Khotin. Rumyantsev who became the commander of the First Army organized the march to Yassy and on 26 September (7 October) seized the city. The Second Army commanded by General P.I. Panin conducted warfare along the Southern Bug. In June 1769 the naval squadron under the command of Admiral G.A. Spiridov proceeded from the Baltic to the Aegean Sea to take part in the war. In the military campaign of 1770 the First Army defeated the enemy in the battles at Larga and Kagul. After destroying totally the Turkish Navy in the battle at Chesma the Russian squadron gained naval supremacy in the Aegean Sea and blocked the Dardanelles. In September the Russian Army seized the fortress of Vendory, in July–November—Izmail, Kilia, Vrailov and Akkerman (Belgorod-Dnestrovskiy). In 1771 the Russian First Army supported by the Danube Military Flotilla seized Jurja (Djurdja) and in March blocked the fortresses of Tulcha and Isakcha. On 14 (25) June the Second Army of General V.M. Dolgorukov with the support of the Azov Military Flotilla took by storm Perekop and occupied Crimea. In the face of such spectacular victories of Russia on land and on sea Turkey on 19 (30) May 1772 had to sign the truce in Jurja and on 1(12) November Russia signed the peace treaty with Crimean Khan Sakhib-Girey under which Crimea formally got independence of Turkey and went under protection of Russia.

During the campaigns of 1773 the military actions were conducted on the Balkans. In June 1774 the Russian army (52,000 men) under the command of P.A. Rumyantsev crossed the Danube. On 9(20) June the Russian troops (25,000 men) of the A.V. Suvorov corps defeated the Turkish troops (40,000 men) at Kozludzja (presently Suvorovo), while the detachment of I.P. Saltykov (10,000 men) defeated the enemies (15,000 men) at Turtukai. The Rumyantsev's Army besieged the fortresses of Shumla, Ruschuk (Ruse) and Silistria, while the vanguard detachment moved beyond the Balkans. With such status quo the Peace Treaty was signed in Kuchuk Kainarji on 10(21) June 1774. This Treaty fixed the victory of Russia in this war and ensured the free access of Russia to B.S. During this war the Russian military art gained the experience in strategic interaction of the army and fleet, crossing of large water bodies (Bug, Dniester, Danube, Sivash), offensives in mountains, seizure of fortresses. During this war Rumyantsev created the Liman Rowing Flotilla (see Danube Military Flotilla) for joint military operations with the land forces. Instead of the linear arrangement of forces there were widely practiced the new arrangements, such as division and regiment squares as well as columns in combination with spread formation of gunners (hunters). The warfare experience gained in this war was included into instructions developed by Rumyantsev and Suvorov.

Russian-Turkish War of 1787–1791 – it was caused by the claims of the Ottoman Empire to regain Crimea, to acknowledge Turkish ruling over Georgia and to permit Turkey to inspect the Russian merchant ships on their way through the Straits. Russia rejected such claims. Turkey started the war using strong Army (20,000 men) and Fleet (19 battleships, 16 frigates and 5 corvettes). Russia used two armies—Ekateringubrg Army (82,000 men) commanded by General-Field Marshal G.A. Potemkin and the Ukrainian Army (37,000 men) commanded by General-Field Marshal P.A. Rumyantsev. The Black Sea Fleet (24 ships) was also involved. On 21 August (1 September) 1787 the Turkish ships attacked the Russian guard ships at Kinburn. On 1(12) October the Ottomans (5,000 men) landed on the Kinburg spit, but as a result of the sudden attack of the corps under command of A.V. Suvorov the Turkish troops were completely destroyed. In 1788 Austria entered the war on the side of Russia. During this war the fortresses of Khotin and Ochakov were stormed and captured and here the Fleet played its important role.

In the campaign of 1789 the Ukrainian Army and the united Southern Army commanded by G.A. Potemkin fought for Bendery and other fortresses in Bessarabia. On 21 July (1 August) 1789 under Fokshany the Russian detachment (5,000 men) commanded by A.V. Suvorov and the Austrian corps (12,000 men) commanded by Prince Frederick Josias of Saxe-Coburg-Saalfeld defeated the corps (30,000 men) of Osman-Pasha. On 11(22) September the Russian and Austrian troops (25,000 men) under command of Suvorov destroyed the troops (100,000 men) of Great Vizir Yusuf-Pasha in the battle at the Rymnik River. But Potemkin did not use these victories for advancing across the Danube, but only seized the fortresses of Vendory, Khadzhibey (presently Odessa) and Akkerman.

On B.S. and Mediterranean the Russian ships operated on the routes of the Ottoman Fleet. In the 1790 campaign Potemkin concentrated the main forces for storming the fortresses and not destroying the enemy in field battles. The Ottomans directed their main assault to the Caucasian coast of B.S. having sent to the fortress of Anapa the Batal-Pasha Army (40,000 men) for attacking Kuban. They also designed the landing in Crimea. The Black Sea Fleet under command of Rear-Admiral F.F. Ushakov (from March 1790) destroyed the Turkish Fleet in the battles at Sinop, near the Kerch Strait (see Kerch Marine Battle of 1790) and at the Tendra Island. The victory at Tendra on 28–29 August (8–9 September) ensured the Russian Fleet supremacy on B.S. Supported by the fleet the Southern Army seized the fortresses of Kilia, Tulcha and Isakcha. The corps of Batal-Pasha moving to Kabarda was defeated. In September Austria signed the separate peace treaty with Turkey which aggravated the position of Russia. But despite this it launched an assault on the Danube. On 11(22) December 1790, the Suvorov troops captured the allegedly impermeable Turkish fortress of Izmail. In the 1791 campaign the Russian Army won great victories. On 4(15) June the detachment of General M.I. Kutuzov crossed the Danube and beat down the Turkish corps (23,000 men) at Babadag. On 28 June (8 July) 1791 the main forces of the Russian Army commanded by General-Field Marshal N.V. Repnin won victory over the Turkish Army in the battle at Machin. On 22 June (3 July) on the Western Caucasus the detachment of General I.V. Gudovich took by storm Anapa. The victories of the Russian Army on land and on sea and also defeat of the Turkish Fleet by F.F. Ushakov at Kaliakria on 31 July (11 August) accelerated the conclusion in 1791 of the Treaty of Yassy.

Russian-Turkish War of 1806–1812 – it was unleashed by Turkey that was supported by France. Turkey hoped for revenge, because that time Russia was at war with France (1805–1807) and Persia (1804–1813). The *casus belli* was breaching by Turkey of the Russian-Turkish Union Treaty of 1805 about free passage of the Russian vessels through the Bosporus and Dardanelles and deposition by Turkey of Russophile hospodars (lords) of Moldavia and Wallachia. In response in November–December 1806 a 40,000-strong Russian Army contingent led by General I.I. Mikhelson advanced into Moldavia and Wallachia. The Danube Cossack troops took the side of Russia. On 18(30) December Britain also joined Russia and its Fleet attempted, but without success, to seize Dardanelles fortifications and the coast of Egypt. In February 1807 the Russian squadron of Vice-Admiral D.N. Senyavin based on the Tenedos Island, blockaded the Dardanelles and in 1807 defeated the Turkish Fleet in the sea battles in the Dardanelles (1807) and at Athos (1807). On the Balkan and Caucasian theaters of military operations the Russians were also victorious. In August 1807 after breakup of the British-Russian Alliance against Turkey Russia signed truce with Turkey. In spring 1809 the hostilities were resumed. On the Caucasus the Russian troops together with the Azerbaijan and Georgian home guard forced the Ottomans out of Poti (1809) and Sukhum-Kale (presently Sukhumi) (1810) and in 1811 they seized the fortress of Akhalkalaki. The Russian Army (80,000 men) under command of General-Field Marshal A.A. Prozorovskiy (from August 1809—General P.I. Bagration) with the support of 140 vessels of the Danube Fleet captured the fortresses of Isakcha, Tulcha, Babadag, Machin, Izmail, Brailov and others. In May 1810 headed by new supreme commander—General N.M. Kamenskiy (it was appointed in February) seized the fortresses of Pazardzhik, Selistria and Razgrad. In the battle at Batin on 26-27 August (7–8 September) the 100,000 Ottoman Army was defeated. In September the fortresses of Ruschuk and Jurja capitulated. On 17 (29) October 1810 the Russian Army entered the city of Lovchu, but soon left it. On 28 January (10 February) 1811 it took this city by storm the second time. In early 1811 due to a threat of the Napoleon Army invasion into Russia a part of the Russian Danube Army was moved to the western border. In March 1811 M.I. Kutuzon was appointed the commander. On 22 June (4 July) 1811 in the battle of Ruschuk and on 23 November (5 December) in the battle at Slobodzen Kutuzon forced the Turkish troops to capitulate. According to the Treaty of Bucharest signed by Kutuzov in 1812 the Turks ceded Bessarabia and western Georgia to Russia. On the eve of the 1812 War Napoleon lost one ally—Turkey.

Russian-Turkish War of 1828–1829 – it was caused by the struggle of the European countries for division of the lands of the Ottoman Empire that was suffering that time a very acute crisis that was spurred by the national-liberation revolution of 1821–1829 in Greece. In 1827 the ally fleet of Russia, Britain and

France defending the Greece independence defeated the Turkish-Egyptian Fleet at Navarino. Having learned about this and about the growing conflicts among the allies the Turkish Sultan revoked all earlier Russian-Turkish treaties and in December 1827 declared the "sacred war" (jihad) on Russia. On 14(26) April 1828 Russia declared war on Turkey. The Army of Field-Marshal P.K. Wittgenstein (95,000 men) was sent against the Army of Khusein-Pasha (150,000 men) to the Danube war theater and the corps of General I.F. Paskevich was sent against the Turkish troops (50,000 men) to the Caucasian war theater. During a month the Wittgenstein Army succeeded to conquer the Danube Principalities and crossed the Danube. On 29 September (11 October) Varna was seized as a result of joint assault from land and from sea. On 23 June (5 July) the corps of Paskevich took by storm the fortress of Kars and in July–August the fortresses of Ardagan, Akhaltsih, Poti and Bayazet. In the 1829 campaign the hostilities were more active. In the battle at Kulevicha the Russian troops (18,000 men) under command of General I. Dibich (Count Hans Karl von Diebitsch) defeated the twice as large enemy army and on 18(30) June they took Silistria. In July Dibich led the 17,000-strong contingent across the Balkans and advanced to Adrianopol (Edirne) which garrison capitulated on 8(20) August. Proceeding further to the south, the Russian Army was on the way to Istanbul (Constantinople). On the Caucasus on 27 June (9 July) the Russians captured Erzurum and came out to Trapezund. Victories of the Russian Army on both war theatres, advancing of the main forces to Istanbul, blockade by the Russian Fleet of the Bosporus and Dardanelles and its cruising operations near the Turkish coast of B.S. urged Turkey to sue peace that was signed on 2(14) September 1829 in Adrianople. Under this treaty Turkey recognized Russian sovereignty over South Ukraine, Crimea, Bessarabia and a part of the Caucasus and Russia established firmly on the B.S. coasts. As a result of this war the management of troops and interaction of land forces with the navy were improved. Some new experience in storming fortresses and conducting field battles was gained.

Russian-Turkish War of 1853–1856 – see Crimean War of 1853–1856.

Russian-Turkish War of 1877–1878 – has its origins in the rise of the national-liberation movement against the Turkish ruling on the Balkans and aggravation of international conflicts in Near East. Russia supported the national-liberation movements of the Balkan peoples and was also seeking to recover its prestige and influence on the Balkans that were undermined as a result of Crimean War of 1853–1856. It also pursued the goal of settling the problem of free navigation through the Bosporus that was needed for further economic development of the country. First the Russian government decided to support the Balkan peoples diplomatically. But its proposal to organize the collective defense of the Slav people was rejected by Britain and Austria-Hungary. In April 1877 Turkey, following the advice from Britain, turned down the new draft proposal on granting independence to Bosnia, Herzegovina and Bulgaria elaborated by the European states on the Russia's initiative. Then on 12(24) April Russia declared war on Turkey. On 9(21) May Romania and later on Serbia and Montenegro joined Russia.

By the beginning of the war Russia organized two armies: the Danube Army (185,000 men, 810 cannons) commanded by Grand Duke Nikolay Nikolaevich and the Caucasian Army (75,000 men, 276 cannons) commanded by Grand Duke Mikhail Nikolaevich. The Danube Army was joined by the Bulgarian home guards (about 8,500 men) led by Russian General N.G. Stoletov, while the Caucasian Army—by militia forces from Georgia, Armenia, Azerbaijan, Ossetia and others. The Danube Army was confronted by the Turkish troops headed by Abdul-Kerim-Nadir-Pasha (more than 206,000 men, 400 cannons), the Caucasian Army—the troops of Akhmet-Mukhtar-Pasha (65,000–75,000 men, about 100 cannons). On 10 (22) June the Lower Danube detachment crossed the Danube near Galati and Brailov (Brail). On 15(27)—20 June (2 July) the 14th infantry division of General M.I. Dragomirov and the main forces of the Danube Army (4 corpses) crossed the Danube near Zimnitsa. From Sistovo the Danube Army was divided into three parts: the Western Detachment (about 35,000 men, 108 cannons) was assigned to capture Nikopol, Plevna; the Eastern Detachment (75,000 men, 216 cannons)—Ruschuk; the Advance Detachment (to 12,000 men, 32 cannons)—to Balkan passes. About 70,000 men were on their way on the left bank of the Danube and in reserve. Insufficiency of forces, their improper arrangement (only 12,000 men were concentrated on the main direction of attack) and incorrect management of troops resulted in dragging out the war. The Western Detachment took Nikopol, but failed to seize Plevna before approach of the 15,000-strong Ottoman corps of Osman-Pasah from Vidin. As a result, the communications of the Danube Army were threatened. The Eastern Detachment only organized surveillance of the Turkish fortresses. Meanwhile, the Advance Detachment of General I.V. Gourko seized as a result of a sudden attack on 25 June (7 July) the capital of Bulgaria—Tyrnovo and on 2(14) July crossed the Balkans and occupied the Shipka Pass. The way to the Bosporus and Constantinople was open, but Russians had not enough forces to launch a new offensive. Having captured the city of Eski-Zagra (Old Zagora) the Advance Detachment was attacked by the forces of Suleiman-Pasha (37,000 men) that came from Montenegro, so the Russians had to retreat to Shipka. The newly established Southern Detachment was also moved here (20,000 men).

In April-May the Russian Caucasian Army seized the fortresses of Bayazet (presently Dogubayazit), Ardagan and blocked Kars. But attacked by the Army of Akhmet-Mukhtar-Pasha in June it retreated to the border and passed over to the defensive. In August the attempt of Suleiman-Pasha with his troops (to 27,000 men) to seize Shipka Pass failed due to staunch defense of this place by the Russians. The Turkish assault against the Eastern Detachment was also repelled. The Russian troops made three attempts to storm Plevna, but only after they seized Lovcha on 22 August (3 September) and blockaded Plevna they forced its garrison to capitulate on 28 November (10 December). On the Caucasus the Russian troops (56,000 men, 220 cannons) led by M.T. Loris-Melikov in September-October 1877 defeated the Turkish troops (about 37,000 men, 74 cannons) in the battle at Avliyar-Aladzhi. On 5-6(17–18) November the Russian troops commanded by I.D. Lazarev took by storm the fortress of Kars (the storm was preceded by 8 days of bombing of the fortress) and proceeded further to Erzerum. The seizure of Plevna was the turning point of the whole war. The Danube Army that number reached 377,000 men (1,343 cannons) unrolled hostilities against the Turkish troops (275,000 men, 441 cannons). Serbia that entered the war against Turkey on 1(13) December sent its army (81,500 men, 232 cannons) to the city of Niche. Regardless of the hardships of severe winter the Western Detachment of General Gourko (over 71,000 men) crossed the Balkan Mountains, defeated the 42,000-strong Turkish contingent and on 23 December 1877 (4 January 1878) occupied Sofia. The Central (former Southern) Detachment of General F.F. Radetskiy (over 54,000 men, 83 cannons) crushed the troops of Suleiman-Pasha in the battle at Filippole. The Danube Army advancing beyond the Balkans seized on 3–5(15–17) January Adrianople and a month later moved to Constantinople. On the Caucasus the Russian troops blocked Erzerum and captured Batum. Russian victories stirred alarm in the ruling circles of Britain and Austria-Hungary. The British sent a Fleet of battleships to the Sea of Marmara to intimidate Russia from entering Constantinople. In January Russia accepted the truce offered by the Ottoman Empire. On 19 February (3 March) 1878 the Treaty of San-Stefano was signed on the beneficial for Russia terms. The Russian Empire regained the southern part of Bessarabia lost as a result of Crimean War of 1853–1856, annexed Batum and the Kars region. In the course of this war the comradership-in-arms of the Balkan peoples with the Russian people that brought them freedom and independence was consolidated. The war revealed a tendency to extension and changed nature of the armed struggle due to appearance of mass armies equipped with rifled weapons, use of railway and field telegraph. More than one million of people were involved in the war.

Rzhevskiy Campaign – during Russian-Turkish wars in the sixteenth century. The march of the Russian detachment under command of M.I. Rzhevskiy (djak) to the Crimean-Turkish lands in spring-summer 1556. It was organized by Tsar Ivan the Terrible for drawing off the forces of Crimean Khan Devlet-Giray who was preparing a raid to Moscow. At the same time Tsar sent troops against the Astra-khan Khanate (see Astrakhan Campaign II). In May 1556 the detachment of Rzhevskiy started off from Putivl and went down the Dnieper where 300 Ukrainian Cossacks joined them. After reaching the Turkish fortress of Ochakov on B.S. the Rzehvsky troops seized and destroyed the jail in the fortress and then turned back having fought off the attempts of the Ottomans to pursue them. Near the fortress of Islam-Kermen on the Dnieper the Russian-Ukrainian troops repelled for six days the assaults of the Crimean troops led by the senior sun of Devlet-Giray. In this battle Rzhevskiy succeeded to beat off from the enemy the horse herds and successfully returned to Russia with rich loot. This was the first appearance of the Moscow troops on the lower Danube. This impressed greatly the Ukrainian Cos-sacks who from this time on perceived the Russian Tsar as a mighty and real ally in the struggle against the Crimean-Turkish aggression. The campaign of Rzhevskiy urged Khan Devlet-Giray to give up the idea of new assaults and to turn his troops for defense of Crimea.

S

Sablin Mikhail Pavlovich (1869–1920) – Russian Rear-Admiral (1915), Vice-Admiral (1919). In 1890 he finished the Naval Corps. He participated in the Russian-Japanese War of 1904–1905: an officer on the armored ship "Oslyabya"; participant of the sea battle at Tsusima where he was wounded. After the war with Japan S. served on B.S. commanding different ships (1906–1907). He was the participant of World War I: chief of mine defense on B.S. (1915–1917); Chief-of-Staff and later Commander of B.S. Fleet (1917–1918). In May 1918 S. received an order from Lenin on sinking the Fleet, but he refused to fulfill this order and went to Moscow to report personally to Lenin. However, in Moscow S. was arrested and put to prison, but he escaped. In 1918 S. joined the White Movement—the Voluntary Army and Armed Forces of the South of Russia (AFSR). From October 1918 to April 1919 he served at the Naval Department of the Headquarters of Supreme Commander General A.I. Denikin. From March 1919 he was the Commander of the Sevastopol port (commander-in-chief of vessels and ports on B.S. and A.S.). In April 1919 he was evacuated from Sevastopol to Novorossyisk on cruiser "Katull" ("General Kornilov"). From February to October 1920 he was the commander of the B.S. Fleet at General P.N. Vrangel. He evacuated from Sevastopol onboard the ship "Velikiy Knyaz Aleksandr Mikhailovich" (Grand Duke Alexander Mikhailovich).

Sagaidachnyi Petr (Konashevich-Sagaidachnyi) (c. 1570–1622) (later on the patronymic became a part or attachment to the surname) – Hetman of the Zaporozhie Cossacks. He was born the village of Vishenike near Sambor (in Precarpathian Rus') in an aristocratic family. He received very good education in the famous Slavonic-Greek-Latin Academy of Ostrozhsk and was one of the most well-educated figures of his time. The exact time of his joining the Zaporozhie troops is not known. It happened approximately in the late sixteenth century. Rather quickly he became prominent among Cossacks because of his intellect, education, courage, initiative and political foresight. Soon after 1601 when the Cossack troops were returned their rights and privileges the Cossacks community elected him their Hetman.

S.R. Grinevetsky et al., *The Black Sea Encyclopedia*,
DOI 10.1007/978-3-642-55227-4_19, © Springer-Verlag Berlin Heidelberg 2015

As the head of the Cossacks, S. took part in all military events of that time. In 1616 he led troops to the Turkish lands and seized Kafu (Feodosiya). With 2,000-strong Cossack troops he crushed the Turkish detachment of 14,000 men and the Turkish Fleet in the Dnieper Liman. He burnt Kafu and liberated many Christians who were kept in captivity, then took by storm Sinop and Trabizond. One of the most successful military actions of S. was the famous Khotin campaign. From May 1618 Poland and Turkey were at war after the Turks attacked Podol. This was a critical period for the Polish state and its government applied to S. who that time was not in power. And S. used this situation to procure privileges for his troops from the government. He demanded annulment of the laws that restricted the Cossacks independence. He also claimed that the power of hetman in Ukraine was officially recognized. In August 1621 having collected 40,000-strong troops S. came up to Khotin, the fortress on the Dnieper, where hostilities against the Turkish troops had been waged for already 5 weeks. And the Cossack troops made the key contribution into the victorious outcome of the Khotin campaign. S. was deadly wounded in the Khotin campaign that became his last military exploit. The Polish government awarded generously hetman and his troops. Soon on 10 April 1622 S. died in the Kiev-Bratskiy Monastery. No traces of his grave were found. Most likely it was destroyed during refurbishment of the monastery. There are some data proving that before his death S. took monastic vow. His name was given to the flagship of the Ukrainian B.S. Fleet—frigate "Hetman Sagaidachnyi".

Saint Clement of Rome – the fourth pope of Rome [from 92(?) to 101(?)]. According to a legend S.C. was banished by Emperor Trajan from Rome to Tauric Chersonesos (Crimea) where he was martyred. He was the author of "The First Letter to the Catholic Church in Corinth" (c. 96). Saint Clement is considered one of the Apostolic Fathers.

Saint Konstantin and Elena (former "Druzhba") – the most long-living resort in Bulgaria on the B.S. coast to the north of Varna. It got its name from the name of the monastery of the fourteenth century located here. Vast parks with a great number of different trees and clear air allow opportunities for good rest here. The beach is narrow, but it abounds in hot spouting mineral water springs. More than 15 hotels can be found here.

Sakalin Island – it is situated in the southern part of the sea edge of the Danube Delta. Its northern part locates opposite the northern arm, while the southern— 5 miles to the south of the lighthouse "Sfyntul-George". It was formed by sediments of the Georgievskiy arm of the Danube Delta. The island is low, sandy, cut in some places with channels. Because of the permanent sedimentation process its outlines and position are changing constantly. In 1972–2002 its southern end became nearly 5 km longer, while the island itself shifted westward for more than 600 m.

Sakarya – a river in the northwest of Turkey. Its length is 790 km. it originates in the Emirdag Mountains of the Anatolian Plateau and flows into B.S. The river is full-flowing and meandering. Its main tributaries are: left Parsuk-chai (208 km long), Ankara-chai (140 km), Chubuk-chai and Gek-su. The volume of the annual flow into B.S. is 6.4 km^3. The greatest water flow is observed in the winter and spring seasons (72 % of the annual flow). The waters of the lower reaches are used for irrigation of fertile lands in the depressions of Geive and Adapazary. The Anatolian railway runs in the gorge of Lower S.

Sakarya Il – the B.S. province of Turkey. It locates in the northwest of Turkey. Its adjacent province is Kocaeli Il. Its area us 4,895 km^2, population—872,900 people (2011). The capital is Adapazari (239,300 in 2011).

Saken von Der Osten Johann Reinhold (Christopher Ivanovich) (1755–1788) – Russian naval officer. Finished the Naval Cadet Corps. In 1772 he sailed from Kronstadt to the Greek Archipelago onboard the ship "Chesma" in the squadron of Rear-Admiral V.Ya. Chichagov. In 1772–1775 he served on the Mediterranean Fleet and in 1775–1786—on the Baltic Fleet. In 1786 he was transferred to the B.S. Fleet. In 1787 when Turkey started hostilities against Russia S. was conferred the rank of Captain 2nd Rank. He served as the master of a galley and freight ship in the Dnieper Liman; supply officer and inspector on the B.S. Fleet. He took part in the Russian-Turkish War of 1787–1791. In May 1788 near the Bug mouth the boat commanded by S. was attacked by the Turkish squadron of 30 ships. To avoid surrender to the enemy S. blasted his ship together with 4 Turkish galleys that grappled it. S. with his crew (43 men) died in this explosion. (The first similar exploit known in the history of the Russian Fleet was made by Captain Petr DeFremery, Russian officer of the French origin. He blasted his ship practically in the same place and under similar circumstances escaping from the Turkish galleys.) The actions of S. had very important consequences: further onward the Turks did not take risks of fighting the Russian ships by grappling even if they were numerically superior. The marble plaque was installed in the church of the Naval Cadet Corps in Saint-Petersburg to commemorate the exploit of Captain Sacken. In 1888 the mine cruiser "Captain Sacken" was built and put afloat on the Baltic Sea.

Saki (allegedly from the Turk "*saki*" meaning "sack"; the Persians called the Scythians "sacks") – from 1952 this was the balneological and climatic resort, the regional center in Crimea, Ukraine. Its population is 23,700 (2012) people. It is situated in the western part of Crimea, 45 km to the northwest of Simferopol and 20 km eastward of Evpatoria, 4 km from the Kalamita Bay on the northeastern shore of the drainless saline Saki Lake. In the first century A.D. Pliny the Senior, the Roman scholar, mentioned in his works about the curative properties of the Saki mud and about its use by the local population for curing diseases long before the new era. In Russia the first who described the use in folk medicine of the mud treatment in S. was Russian explorer Academician P.I. Sumarokov (1799). In 1828 the first Russian mud-treatment center was constructed here on the money collected in the province. In 1843 doctor N.A. Ozhe who practiced in Evpatoria published his work "About Curative Mud of Saki" in which on the basis of his investigations he described the physical and chemical properties of the mud from the Saki Lake and the concept and method of treatment. Near S. the thermal (+37 to +45 °C) hydrocarbonate-chloride-sodium waters (salinity 2.2–2.3 g/l) were channeled to the surface and were used for curative baths and drinking. It was also bottled as the curative-table water named "Krymskaya". In 1837 a branch of the Novorossiya Military Hospital was opened here. In the early twentieth century Academicians S. Nalbandov and N. Burdenko developed in S. the scientific fundamentals of the peloidotherapy (treatment of various diseases with mud). Beginning from 1938 the medical centers in S. started receiving people for treatment not only on the seasonal, but year-round basis. S. was the first resort in the USSR with the developed infrastructure for the invalids (on wheelchairs). The resort deservedly boasts of its park where one can see the rare combination of the forest, forest-steppe and Mediterranean vegetation that creates a very picturesque and unique green land-scape. The resort has a railway station of the same name. It has the chemical and food industries. In the Soviet time the aviation garrison located here that had the world's largest ski-ramp for training deck-based aircraft. In 1944 the planes of G3—Stalin, Roosevelt and Churchill landed here.

Saki Lake – a drainless saline lake in Crimea, Ukraine. It is situated 20 km from Eupatoria. Its area is 8.1 km^2, length—5 km, width—0.5 to 2 km, the maximum depth—about 1 m. It locates at 2 m below the mean sea level. It was formed from a liman that was separated from B.S. with a sand barrier 2 km long and 400–600 m wide. The lake is divided with an earthen dike into two basins: eastern—curative and western—for industrial use, being a source of raw materials for the chemical plant in Saki. The lake receives sea water via a canal. The bed of S.L. is covered with a layer of silty mineral mud 1–3 m thick and its reserves are evaluated at 4.7 million m^3. The mud is of the sulfide type. Brine or lake water is a strong solution of sea salts. In summer its salinity level reaches 250–280 g/l. The town and resort of Saki are situated on the lake shore.

Saki mud – curative mud, velvet by touch, oily, viscous, black, smelling of hydrogen sulfide. It represents the intricate organic mineral complex. It contains iron, potassium, magnesium and calcium oxides, the sodium, potassium salts and many other elements as well as live microorganisms, bacteria and fungi. The content of vitamins in S.M. is three to ten times higher, of lipids and their derivative fat acids two to three times higher than in the curative mud of the Dead Sea. The silty sulfide mineral mud of Saki is well-known over the world and is recognized as very effective in treatment of infertility.

Salgir (from the Turk "*salgur*" where "*sala*" means "river tributary, mountain outspur, intermountain hollow, ravine") – a river on the Crimean Peninsula flowing into Sivash of A.S., Ukraine. Its length is 232 km, the watershed area—4,000 km^2. It takes origin with two streams—Kizil-Koba and Angara on the northern slopes of the Main Range of the Crimean Mountains where it runs in a deep and narrow gorge. Having taken in its main tributary—Biyuk-Karasu the river flows into the Sivash. The reservoir is constructed upstream the city of Simferopol. In summer Lower S. often dries out. The river has a mixed recharge, mostly from rainfalls.

Salgir Battle – the battle got its name by the Salgir River where in July 1737 the Russian Don Army commanded by General-Field Marshal P.P. Lassi fought against the troops of Crimean Khan Fetkhi-Girey. The Russian troops invaded Crimea to stop the frequented raids of the Crimean Tatars to the southern lands of Russia and to put an end for the Ottoman Empire to use the troops of the Crimean Khan as its vassal in the rear of the Russian Army. On 3(14) May the Don Army started off from the town of Azov in bypass of the Perekop fortifications. On 27 June (8 July) it landed on the Arabat Spit having crossed the Genichesk Strait and moved to the town of Arabat. Fetkhi-Girey directed his 60,000-strong army to the same place. On learning about this, Lassi sent to Arabat a cavalry regiment imitating attack, while the main forces crossed the Sivash near the Salgir mouth and supported by the Don military flotilla commanded by Vice-Admiral P.P. Bredal entered Crimea. Believing that the main Russian forces are on the way to Arabat, Fetkhi-Girey sent against them only the 15,000-strong cavalry regiment that on 12(23) July attacked the Russian troops on the Salgir River near Karasubazar. Soon the main forces of Lassi (to ten cavalry and infantry regimes) came to the battlefield and crushed the enemy. Russian troops occupied Karasubazar and chased the enemy as far as Bakhchisaray, the capital of the Crimean Khanate. But remoteness from the bases, short supplies of food, fodder, water and epidemic diseases forced Lassi to stop hostilities and retreat from Crimea.

Saline "zasukhi" – in this way the coastal parts drying out in summer of saline lakes, limans and bays are called in Crimea. These are depressions with a rather well delineated coastline on the land side with the gradually sloping bottom towards the water body where the depression has no coastline. During the off-water periods and, in general, during low-water months S.Z. became exposed, their bottom is usually smooth and bare. During summer heat the S.Z. surface cracks down forming a kind of polyhedra and is overgrown with rare thistle shrubs. In winter during high water and sometimes in summer during rainstorms or water surge from a water body S.Z. are covered with water. The example of such S.Z. is the so-called "Alekseevskiy zasukha" in Sivash.

Samsun (from Ancient Greek "*Amisos*") – a city on the Turkish coast of B.S., the largest city in Northern Turkey, the main port on the western coast of the Samsun Bay of B.S., the administrative center of the Samsun Il. It was founded by the Greek colonists in 562 B.C. Its population is 538,100 (2012). This is a railway station and the crossing point of automobile roads. The city has the tobacco plant (processing of the local tobacco "samsun"), light industry and copper works. The fishery is also developed here. The port is used for export of tobacco, grain, oil, raw hide and fruit. The well-known places of attraction are the monument to Ataturk (he came here in 1919) being the symbol of the city, the library and museum of Ataturk (Gallery of the 19th May), the park, the annual fair, Bazar Mosque of the thirteenth century and many others. It is the generally recognized motherland of fabulous Amazons.

Samsun Bay – a wide bay westward of the Dzjiva Cape to the Kaljon Cape, Turkey. The Samsun harbor incises into the western coast of the bay. Samsun is one of the major ports on the B.S. coast of Turkey, is situated on the northwestern shore of the harbor that is limited in the south by the abrupt Cinekoglu or Dervend that is overgrown with forests.

Samsun Il – a B.S. province of Turkey. Its area is 9,579 km², population—1.25 million people. The capital is the city of Samsun. Its adjacent provinces are Sinop on the northwest, Corum on the west, Amasya on the south, Tokat on the southeast, and Ordu on the east. Mustafa Kemal Ataturk, the founder of the Turkish Republic, on 19 May 1919 started here the Turkish War for independence.

Samsun Port – a large port on B.S. in the mouth of the Mert River, Turkey. It has 15 jetties with a total length of 1.7 km. The maximum depth in the ship channel is 12.2 m and near the freight jetties—to 10 m. It is capable to handle ships over 150 m long. The port has a grain warehouse. One of the promising directions of port development is commissioning of the rail ferry that together with other B.S. ports with the similar infrastructure (e.g. Burgas, Batumi, Ilyichevsk, Constanta, Poti) will contribute to further development of rail ferry lines. The ferry length is 184.5 m, width—26.5 m. The complex is capable to receive the ships with the deadweight to 12,000 tons. It is designed with a possibility of further retrofitting for servicing five regular ferry lines. Tobacco, grain, chrome ores, cement, metal and vehicles are imported via this port. It can also handle containers.

Sapun-Gora (from the Tatar "*sabun*" meaning "soap, soapy") – the famous mountain located eastward of Sevastopol on the way to Yalta. Its height is 231.7 m. In the past the soapy clay was extracted here. During the Crimean War on 13 October 1854 the Balaklava battle took place at the foot of S.-G. During the Sevastopol Defense of 1941–1942 one of the lines of resistance of defenders was organized on the mountain slopes. The pillboxes, trenches and dugout shelters are still found here. In 1944 this mountain was the most important position of the German fortifications the storming of which caused great casualties among the Soviet troops. The final episode of this storming is represented in the diorama "Storm of Sapun-Gora on 7 May 1944" in Sevastopol. The memorial complex and the Obelisk of Glory were constructed here.

Sarmatia – the territory in Northern Prichernomorie in the third century B.C. inhabited by cattle breeding tribes of Sarmatians who having expelled the Scythians in the first to early second centuries settled down over whole Crimea. In the fourth century they were crushed by the Huns.

Sarp – a Turkey village on coastal border with Georgia.

Sarpi – a Georgian village on coastal border with Turkey, southward of Batumi. Population—858 (2002).

Sarych Cape (from the Turkish meaning "bird of prey, vulture") – the south-ernmost geographical tip of Sevastopol, Crimean Peninsula, Ukraine. The cape was formed by the hogged slope of the offspurs of the Baidarskiy yaila dropping southward. In 1888 the navigation lighthouse was built on the top of the cape and the Mayachnyi (lighthouse) township appeared nearby. Administratively S.C. refers to Sevastopol, but traditionally it is considered a part of the resort of Foros. On this cape there was the state dacha "Object Zarya" where during the August coup of 1991 first and last USSR President M.S. Gorbachev was detained. In late 1991 Ukraine transferred this state dacha to the ownership of the administration of the Ukrainian President. The famous resort zone of the Southern Coast of Crimea is extending from Foros to the east and onward to the northeast.

Sarych Sea Battle – the cape (Sarych) near which on 5(18) November 1914 there was the first sea battle between the B.S. squadron consisting of five battleships commanded by Admiral A.A. Ebergard and the new German cruisers "Goeben" and "Breslau" commanded by Rear Admiral Wilhelm Anton Souchon during World War I. In fact, the battle resulted in the artillery duel between "Goeben" and Russian head battleship "Evstafiy". Thanks to the precision fire of Russian gunmen "Goeben" received 14 accurate hits. The German ship was set on fire, and Souchon not waiting for other Russian ships to join the battle ordered to retreat to Constantinople. "Evstafiy" was hit only four times and disengaged without serious damages. Having tested the strength of the B.S. defense line of Russia the German and Turkish Fleet stopped active actions near the Russian shores. On the contrary, the Russian Fleet became more active on the sea communications.

Sasyk Reservoir – it was built in 1978 when the Sasyk Lake was separated from B.S. with a dam. Its length is about 30 km, its width in the northern part is 4.5 km, in the southern—to 12 km. The water surface area is about 210 km^2, the water volume is over 425 km^3. The depths in the reservoir vary evenly from 0.5 to 3.2 m. The water body floor is flat and covered with silt deposits. The coastline is rugged, but slightly. The water regime of the reservoir is regulated by a pumping station discharging waters into the sea depending on the available schedule of water inflow from the Danube via the Danube-Sasyk Canal and withdrawal of water for irrigation purposes. The Kogilnik and Sarata rivers flow into S.R. Their total flow is 30–60 million m^3. In the period from 1978 to 1986 the hydrochemical regime was formed. The water salinity in the water body dropped from 18–20‰ to 1–2‰ and during inflow of the Danube waters to 0.6–0.8‰. At present the freshwater ecosystem is taking shape in S.R. that substitutes the marine ecosystem typical of the Sasyk Liman—the populations of new freshwater plant varieties, new species of aqueous animals and new freshwater biocenoses appear. The freshwater fish fauna is formed from the young fish migrating here via the Danube-Sasyk Canal.

Sasyk, Sasyk-Sivash – a saline lake (liman) in Crimea, Ukraine. This is the Crimea's largest lake located in the west of the peninsula. The lake is shallow and very saline. Its area is 75.3 km^2, length—14 km, width—5.5–9 km, depth— 0.5–1.2 m. The lake was formed as a result of flooding by the B.S. waters of wide mouth areas of five large ravines: Bogaiskiy, Mamaiskiy, Tyumenskiy, Aidarskiy and Temeshskiy. In the north and northeastern parts it receives four rivers drying out in summer when it shrinks significantly. Its depth is to 1.2 m. It is isolated from the sea with a barrier spit that was formed about the second millennium B.C. The lake level is 0.6 m lower than the sea level. It is recharged by sea waters filtered through the barrier and by ground waters. The main economic use of S. is connected with salt production.

"Saving the Black Sea" – the official newsletter of the Black Sea Environmental Program (BSEP) and the Global Environmental Facility (GEF). It started circulation in September 1994. It is published twice a year in Istanbul by the Coordinating Center of the GEF/BSEP Program.

Savitskiy Evgeniy Yakovlevich (1910–1990) – Marshal of Aviation of the Soviet Union, war pilot, double Hero of the Soviet Union (1944 and 1945). He was born in Novorossiysk. In winter 1924 he was transferred to the first orphanage house in Novorossiysk. Later he entered the college at the cement plant "Proletariy" where he got two specialities: foreman on excavation of raw materials for cement plants and mechanics on diesel engines. He worked for 2 years as a driver in a petroleum company. In 1929 the Komsomol organization of Novorossiysk sent him to the Seventh Stalingrad Military School of aircraft pilot commanders which he finished in 1932. In 1937 in the rank of Senior Lieutenant he commanded the aircraft unit and in 1941 being the Major he commanded the division of fighter aircraft. From March to May 1942 he commanded the Air Force of the 25th Army. From May 1942 to the end of the war S. served in operating army: commanded the fighter aircraft division and from December 1942 to the end of the war—the fighter aircraft corps. He participated in the battles for liberation of Kuban, Donbass, Ukraine, Crimea, Belorussia, Baltic Republics and Poland, took part in storming Berlin. He supported from air the landing troops on Malaya Zemlya near Novorossiysk. During the war S. made 216 combat flights, shot down personally 22 enemy planes. In 1950 the monument to him was installed in Novorossiysk. In 1970 he was made the Honorary Citizen of the city of Novorossiysk. In 1980 the Krasnodar publishers made his book of reminiscences entitled "In the Sky over Malaya Zemlya" in the series "Feat of Arms of Novorossiysk".

Scampavia (from the Italian *"sampare"* meaning "disappear" and "via" meaning "away") – a long, low war galley, a small rowing vessel of the Russian skerry fleet of the eighteenth century. Such vessels were used by Italians. They were intended for reconnaissance, transportation of troops, landing of troops and grappling battles in skerries. The first Russian S. was built in 1703. Its length was about 22 m, width—about 3 m, draft—0.7 m. it had 12–18 pairs of oars and 1–2 masts with gaff and stay sails. It was able to carry to 150 people.

"Scythia Minor" – a Scythian state. The Scythians appeared in Crimean in the seventh century B.C. "M.S." as a state was formed only after breakdown in the second century B.C. of the Scythian Kingdom into tribal communities: agricultural Scythians, nomad Scythians and Royal Scythians. They established three independent Scythian states—Balkan, Dnieper and Crimea forming S.M. that incorporated the Crimean Peninsula without Kerch Peninsula and Chersonesus. The capital of S.M. was Scythian Neapolis (the territory of Simferopol). From Greek sources some other Scythian towns of that time became known—Khabei and Palakion. The latter, according to some authors, was on the territory of present-day Balaklava. The Scythians permanently fought Chersonesus and the latter had to apply for military aid first to Mithridates VI Eupator of Pontus and later to the Romans. In the sixth century S.M. was destroyed by the Huns.

Scythian State, Scythia – it was first mentioned in the 670s B.C. Having defeated the Cimmerians on the Crimean Peninsula the Scythians formed a new state that stretched from the Don to the Danube. The Scythians were divided into four tribes: pastoralists living in the Bug River basin; farmers living between the Bug and the Dnieper; nomads living on the territories more southward; Royal Scythians living between the Dnieper and the Don and in Crimea. The territory in Crimea inhabited by Royal Scythians was called in antiquity the Scythia. The Scythian tribal community was based on the principles of military democracy with the popular assembly of personally free nomads. The Scythians had the cult of sword. The supreme male God was always depicted on a horse, while the female Goddess was Great Goddess or the Mother of Gods. The Scythians had a very large army using the best iron arms and famous bows.

Scythia reached its highest flourishing and might during the ruling of legendary Tsar Ateas in the fourth century B.C. The period of rapid development at the turn of the fourth and third centuries B.C. was followed by a deep crisis caused by some unfavorable factors, such as frequent draughts, degraded grass cover in steppes as a result of overgrazing and others. Thus, in the third century B.C. the Scythians were forced out of the steppe territories lying between the Don and the Dnieper. Having moved from vast steppe lands to a relatively small territory mostly in Tauri the Scythians had to adjust to new conditions. They gradually turned into sedentary farmers and pastoralists living in permanent settlements. Such radical economic changes shaped quite different way of life, material culture, social relations and religious beliefs that significantly affected the further history of the Scythians. Unless the previous period the Scythians, having found themselves confined to a "limited space", started fundamental development of the Crimean Peninsula. They started construction of settlements in mountain Crimea, in river valleys, on fertile lands with water sources. Archeologists uncovered the Scythian settlements in the valleys of the Kacha, Belbek, Alma and Salgir rivers. In the third century B.C. the Sarmatians continued their pressing on S. and, finally, they overwhelmed the Scythians and S. was turned into Sarmatia. In the antique sources the population of Crimea was referred to as the Tauri Scythians.

Scythians – the general name of the tribes living in ancient times in the B.S. area, on the Don, Dnieper and Danube. They were described by Herodotus. Some of them settled down in Southern Russia, others continued their roaming and invaded Asia Minor. Already in the seventh to sixth centuries B.C. S. established very stable economic and cultural relations with the Greeks on the Pontus coasts.

Sea Battle at Kaliakria (the present name Kaliakra) – the cape on the B.S. coast of Northeastern Bulgaria where on 31 July (11 August) 1791 there was a sea battle that ended the Russian-Turkish War of 1787–1791. The Russian squadron of ships commanded by Rear Admiral F.F. Ushakov came across the Turkish Fleet under command of Kapudan-Pasha Hussein that was anchored near Kaliakria under defense of land batteries. Trying to gain time, take the advantageous windward position and take the enemy unawares Ushakov did not rearrange the ships. Despite the fire from the Turkish land batteries he led three columns of ships between the shore and Turkish ships, took the windward position and quite unexpectedly attacked the enemy. Turkish ships having cut the anchor ropes in commotion took the leeward position and started to line up. Following them Ushakov arranged the ships in a stern line and directed them parallel to the Turkish Fleet. The vanguard ships of the enemy commanded by Algerian Pasha Seyit Ali tried to take the windward position, but Ushakov on the battleship "Rozhdestvo Christovo" (Christmas) left the line and resolutely attacked the flagship of the enemy and forced it to go out of action. Other Russian ships started firing the Turkish ships that stopped fighting and under the guise of darkness moved to the Bosporus. In the battle at Kaliakria Ushakov used a new tactic—attacking of the enemy ships in a march disposition. The key factor in the victory was a bold maneuver—passage of the Russian ships between the shore and the enemy squadron for taking the advantageous windward position before the attack and also leaving by the Ushakov's ship of the line for attacking the enemy's flagship. The victory at Kaliakria brought closer the conclusion of the 1791 Peace Treaty of Yassy.

Sea Battle of Chesma – night attack on June 26, 1770 in Chesma harbor of the vanguard of Russian ships under the command of Rear-Admirasl S.K. Greig against the Turkish Fleet commanded by Kapudan-Pacha Hasan Bei. After the battle in Khioss Strait, the Turkish Fleet found refuge in Chesma Harbor. At the war council of the Russian squadron it was decided to attack the Turks in the harbor and set their fleet on fire with the aid of branders. At night, the Russian ships entered the harbor. They opened fire on the Turkish Fleet with incendiary shells and set some vessels on fire. At midnight, the Russians set four branders afloat in order to set the other ships on fire. The fourth brander commanded by Lieutenant D.S. Ilyin was lucky. It was he who was to praise for the destruction of the Turkish Fleet. Having chosen the lager ship, Ilyin reached its side using the shortest way, attached his brander quickly and lit the taper leading to the explosive. A fire that followed the explosion reached the other ships, too. The Turkish Fleet burned down completely except one battle ship and five galleys that were seized by the Russian squadron. The Turks lost 10,000 men in the battle of Chesma, while only 11 Russians were killed in action. After Chesma, the Russian Fleet had absolute superiority in the Aegean Sea. It blocked Dardanelles and engaged in subversion attacks on the Turkish shore. A special medal was minted for the participants of the Sea Battle of Chesma: it was the most brilliant victory of the Russian Fleet from the day of its establishment. The medal featured this laconic inscription: "Was". Prince Orlov was decorated with the order of St. George 1st Degree and an honorable prefix to his sirname—Chesmenskiy. In memory of Lt. Ilyin's feat one of the cruisers of the Russian Fleet was subsequently named after Ilyin.

Sea Charter – see UN Convention on the Law of the Sea.

"Sea Coast" – the national program of Ukraine aimed at conservation and development of the Crimean coastal area.

Sea of Azov – See Azov (Sea of Azov).

Sea of Azov—Black Sea Basin – encompasses the Black Sea and Sea of Azov linked by Kerch Strait as well as the surrounding coastal areas, including the Crimean Peninsula. The A.-B.S.B. is enclosed. It is connected with the Atlantic and Indian Oceans by the only route—via Bosporus Strait, Sea of Marmara and Dardanelles Strait, the Aegean and Mediterranean Seas and on via Gibraltar Strait and Suez Canal. Besides, the basin is connected with other European basins by domestic waterways. Thanks to its lucrative geographical position, the basin is very important economically to the countries on its shores. The A.-B.S.B.is a mighty base for the fleets of the circum-Black Sea countries. There are major deep-water ports on their coasts participating in substantial export and transit flows of goods. The basin directly adjoins advanced economic areas: industrial and agricultural whose products are in high demand on the world and domestic markets. The favorable climatic conditions are conducive to the development of health resort and sanatorium construction, tourism and maritime passenger traffic. International transport corridors (ITC) pass through the basin's northern ports. Out of ten crucial transport arteries of Europe, the third, fifth, seventh, ninth ITC as well as the corridors Europe-Caucasus-Asia (TRACECA) and "North-South" (the Baltic Sea—Black Sea) are connected with the A.-B.S.B. Using these corridors for transporting transit cargoes is crucial the economies of these countries. In the basin, major sea ports feature a number of shipbuilding and ship-repair facilities with floating drydocks and drydocks, workshops furnished with up-to-date equipment, where all kinds of repair can be done. These are Odessa, Izmail, Ilyichevsk, Nikolaev, Novorossiysk, Tuapse, Mariupol, Kerch, etc.

Sea of Marmara – (Propontis in antiquity) is an inland sea entirely within the borders of Turkey that connects the Black Sea and the Aegean Sea, thus separating Turkey's Asian and European parts. The Bosporus Strait connects it to the Black Sea and the Dardanelles Strait to the Aegean Sea. The sea takes its name from the Island of Marmara, which is rich in sources of marble. The sea's ancient Greek name *Propontis* derives from *pro* (before) and *pont* (sea), deriving from the fact that the Greeks sailed through it to reach the Black Sea. The Sea has an area of 11,350 km^2 (280 km × 80 km) with the greatest depth reaching 1,370 m. The surface salinity is about 22‰, which is slightly greater than that of the Black Sea significantly less than in the Mediterranean Sea. However, the water is much more saline at the deep layers (38‰)—similar to that of the Mediterranean Sea. Water from the Susurluk, Biga and Gonen Rivers bring freshwater to the sea. There are two major island groups known as the Princes Islands and Marmara Islands (including Avsa and Pasalimani). The southern coast has the Gulf of Izmit, the Gulf of Gemlik, the Gulf of Bandirma and the Gulf of Erdek.

Sea of Marmara (http://b.static.trunity.net/files/122701_122800/122706/620px-Sea_of_Marmara_map.png)

Sea Trade Port of Tuapse – is situated on the Caucasian coast of B.S. in the Tuapse Bay, to the southeast of the steep cape of Kadosh, Russia. An approach channel 400 m long, 120 m wide and 13.5 m deep leads to the port water area. First port facilities (Beregovoy and Reidovyi piers) were constructed in 1896 making the basis of the Old Port. On 26 December 1898 the first trade ship called on the port. This date is considered the birth date of the Tuapse port. In 1917 after construction of the southwestern breakwater and the southern pier the New Port appeared. It happened to locate inside the city of Tuapse, not far from its historical center. During the Great Patriotic War the city and port were damaged severely as a result of bombardments and artillery fire. Despite serious damage and great number of victims among the personnel the work in the port did not stop even for a day. In postwar years the city and port were restored rather quickly. In these years the oil-loading part of the port was developing most speedily. It was refurbished and provided with the modern technological equipment that made it possible to receive and handle large-tonnage tankers at its berths. A high-capacity plant for treatment of ballast waters was constructed and it served the needs not only of the oil harbor, but also of the refinery works in Tuapse. The following establishments are operating on the port territory: OJSC "Tuapse Marine Trade Port", loading-unloading works are conducted on the berth of CJSC "Tuapse Fish Port", OJSC "Tuapse Ship Repair Yard" and CJSC "Tuapse Ship Mechanical Yard". OJSC "Tuapse Marine Trade Port" provides transportation of oil and petroleum products as well as bulk (coal, ore) and general cargo to other countries. The port incorporates three special areas: dry cargo, bulk cargo and passenger. The port is capable to handle annually about 17 million tons of cargo, including 13 million tons of bulk cargo. The port cooperates with the countries in Europe, Near East, Africa, Southern, Southeastern and Eastern Asia, North and South America and other countries. At present the port is the second largest on B.S. in Russia by the cargo turnover.

Sea Trade Port "Yuzhnyi" – the universal non-freezing sea port, 30 km eastward from Odessa, Ukraine. By the turnover it takes the second line on the list of Ukrainian ports. Most favorable geographical location makes this port highly competitive among other Ukrainian ports. The strategy of the port development envisages the growing handling of bulk solid and liquid cargo and increase of general cargo turnover.

Seaside edge of delta – a transitional zone between delta and offshore mouth areas. Its dynamics is indicative of the direction and intensity of the delta formation processes, interaction between the river and sea factors in delta formation. In the non-tidal river mouths the main factors contributing to dynamics of S.E.D. are water flow, sediment transport of a river and its distribution by arms and also sea waves.

Sebastopolice – see Sukhumi.

Seismicity – the likelihood and recurrence of earthquakes of certain intensity. B.S. and its nearby territories refer to a high seismicity zone.

Senyavin Aleksey Naumovich (1716–1797) – Russian Admiral (1775). He was the son of Vice-Admiral N.Ya. Senyavin. Aleksey Senyavin began his career in the Navy in 1734 as a warrant officer. In 1737–1738 serving with his father on the Dnieper Military Flotilla he took part in the Ochakov defense. In 1739, A. Senyavin was transferred to the Baltic Fleet. He took part in the Seven Years' War (1756–1763). During the Russian-Turkish War of 1768–1774 he commanded the Don (Azov) Military Flotilla. Together with G.A. Spiridonov he designed new ships with shallow draft that could be used in battles on rivers and in A.S. He organized construction of the flotilla base in Azov, construction of ships in shipyards of Kherson and other cities. In 1769 he was promoted to the rank of Vice-Admiral. In 1771 commanding the Azov Flotilla he navigated over A.S. and gained the foothold in the Enikale Strait. He took part in creation of the B.S. Fleet, organized cruising over B.S. and the naval base in Kerch, defense of the Kerch Strait. In 1773 he was put in charge of the squadron and navigated over B.S. In 1774 he rebuffed the attack of the superior forces of the Turkish Fleet in the Strait of Kerch, making them retire with losses. After the war S. continued working in the shipbuilding area. During 5 years more than 130 ships, including 30 large ones, were built under his supervision.

Senyavin Dmitriy Nikolaevich (1763–1831) – Russian naval commander, Admiral (1826). In 1780 he finished the Naval Cadet Corps and served on the Azov Military Flotilla. In 1783 he was transferred to the B.S. Fleet. During the Russian-Turkish War of 1787–1791 he took part in the battle at the Kaliakria (1791). He assumed command of the flagship "Saint Peter" in the Mediterranean campaign of the Russian Fleet during the war against France in 1798–1800. He also commanded the squadron of ships in the Adriatic campaign of 1806–1807, prevented seizure of the Ionian Islands by the French troops and captured some fortresses. His talent as a commander was demonstrated most brilliantly during the 2nd Archipelago expedition in 1807 when the Russian Fleet under his command blockaded the Dardanelles and defeated the Turkish Fleet in the battles of Dardanelles and Athos (1807). Both engagements were Russian victories and ensured Russia's ascendancy in the

Aegean Sea. In 1813 he had to resign. In 1825 after aggravation of the Russian-Turkish relations S. was recalled to active service and was appointed the commander of the Baltic Fleet. The Senyavin Islands in the archipelago of the Caroline Islands, the Senyavin Cape in the Bristol Bay of the Bering Sea and the Senyavin Cape in the southeast of the Sakhalin Island still commemorate his name. Some battleships of the Russian and Soviet Fleet bear his name.

Senyavin D.N. (http://upload.wikimedia.org/wikipedia/commons/5/5b/Senyavin_D_N.jpg)

Serebryakov Lazar Markovich (Artsatagortyan Ghazar Markosovich) (1792–1862) – Russian Admiral (1856). Founder of Novorossiysk together with General N.N. Raevskiy Jr. and Vice-Admiral M.P. Lazarev He started his naval service in 1810. Until 1837 he served, first, on the Baltic Sea and later on B.S. In 1837 he was sent to the Caucasus where he served for two decades, first, as the on-duty field-officer on movement and action of naval units on the eastern coast of B.S. In 1839 he was appointed the chief of the First Division of the Black Sea Coastline, Novorossiysk fortification and the under-construction port. Under his supervision in 1840–1850 N.N. Raevskiy, the founder of Novorossiysk, realized his plans and projects on construction and improvement of the city. Being the chief of the First Division of the Black Sea Coastline (this division had in its control 13 fortifications from Anapa to Golovinka) he had to address a wide scope of military, political, administrative and economic issues. For a short time he managed to create the effective fighting detachments, to organize normal supply of troops. He initiated and directly supervised a wide-scale construction of military and civil objects in this territory. In 1851 he was appointed the Chief of the whole Black Sea Coastline. Till 1920 the central street in Novorossiysk was called Serebryakov Street. Today the city has Admiral Serebryakov Seafront.

Serebryakov L.M. (http://www.khachkar.ru/images/enc/0000000454.JPG)

Sergeevka – the seaside climatic and mud resort in the Odessa Region, Ukraine. This is a small town 19 km from the railway station "Shabo" and 18 km from Belgorod-Dnestrovskiy. It is situated on the shore of the Budakskiy Liman. The summer here is mild with the unsteady snow cover; the mean temperature in January is +2 °C. The summer is warm, dry and largely cloudless. The mean temperature of July is +22 °C. Very hot weather is quite often here. The precipitation is about 250 mm a year, falling mainly in April-October. The main curative factors here are climate and sulfide mud from the Budakskiy (Shabolatskiy) Liman that may be used in climatotherapy and mud treatment. The Budakskiy Liman 17 km long, 1.5 km wide and 0.2–1.0 m deep is separated from B.S. with a narrow (to 150 m) sand barrier being an excellent beach used in thalassotherapy. The resort also offers artificial sulfide, radon and other curative baths prepared on the basis of sea water. This resort is very good for treatment of locomoter, respiratory (other than tuberculosis), nervous, gynecological diseases. It is also intended for the patients suffering from the osteoarticular tuberculosis. This resort was constructed in 1940.

Sergeev-Tsenskiy Sergey Nikolaevich (1875–1958) – Russian/Soviet novelist who wrote much about the sea, Academician of the USSR Academy of Sciences (1943). From 1905 he lived in Crimea. He travelled much over the country. His first works were published in 1898. His main work was the unfinished epic "Transfiguration of Russia" in several volumes that included 12 novels, 3 stories, 2 essays. He wrote it for 45 years. It depicted the life of the Russian society in the pre-revolutionary time, the events of World War I, February Revolution of 1917, the Civil War, the first years of the young Soviet State. Most popular was his marine epic book "Sevastopol Campaign" (1937–1939) that described the heroic battles of the Russian sailors and soldiers during the Sevastopol defense in the Crimean War of 1853–1856. He presented the whole gallery of the famous Russian Fleet commanders, such as V.A. Kornilov, P.S. Nakhimov, V.I. Istomin and others. He was awarded many orders and medals. He was the Laureate of the USSR State Award (1941). His house in Alushta is now the literary and memorial museum.

Sevastopol (from Greek "*sebatos*" meaning "sacred" and "*polis*" meaning "city"—"sacred city") – the city-port, Crimea, Ukraine. S. as a fortress was twice deservedly glorified in the world history for its brave and staunch defense. It is located on the shore of the Sevastopol Harbor of B.S. The city extends for nearly 25 km along the Heraclea Peninsula. The notion "Sevastopol" includes urban built-up territories on the coasts of the Sevastopol, Karantinnyi, Streletskiy, Kruglyi, Kamyshovyi and Kazachia harbors as well as the administrative municipalities subordinated to the Sevastopol City Council with an area of 86,400 ha. The Ukrainian Constitution refers S. together with Crimea, 24 oblasts (provinces) and the city of Kiev to the basic administrative-territorial units of the country.

A month after Crimea was incorporated into Russia as a result of the victorious war with Turkey in 1768–1774—in April 1783 Capitan 2nd Rank I.M. Bersenev on the frigate "Ostorozhny" (Cautious) inspected the Akhtiar Harbor where in the Antiquity the city of Chersonesos located and decided to organize here the naval base. In May of the same year five frigates and eight battleships of the Azov Fleet commanded by Vice Admiral F. Klokachev who was later appointed the Commander of the A. and B.S. Fleet as well as a part of ships of the Dnieper Fleet commanded by Sidor Bilyi who headed the Black Sea Cossack troops entered the harbor. In 1788 Catherine II ordered to build there a fortress and to give it the name of S. The entrance of the ships into this harbor was the birth date of the Black Sea Fleet (the Black Sea rowing (liman) fleet was also operating on B.S.). On 3 June 1783 the sailors came ashore and started construction of a city and port. The new city appeared not only nearby ancient Chersonesos, but for its construction the cut stones and marble of this antique city were used. In 1797 under the edict of Emperor Paul I the city was renamed into Akhtiar, but in 1826 after his death the city was returned its former name—Sevastopol. In 1804 it was declared the main port of the B.S. Fleet. The Fleet and the city were developing in parallel, but after in 1804 the main naval base was moved here from Kherson it spurred further the city growth and expansion of trade via the port of S. In the mid-nineteenth century it became the largest port in Crimea. Its population was 42,000 people. A great contribution into the city development was made by Admiral M.P. Lazarev who between 1834 and 1851 was the Chief Commander of the Fleet and ports of B.S. and also the military governor. In this period the new beautiful buildings appeared in the city, the fleet was renovated, the admiralty with dry docks, barracks and defense structures were built. Two of five stone forts—Konstantinovskiy and Mikhailovskiy that were built that time had survived to the present. However, by the beginning of the Crimean war the city had no fortifications on the land side.

The young B.S. Fleet based in S. won brilliant victories led by such outstanding Russian commanders as F.F. Ushakov, D.N. Senyavin, A.S. Greig and others in B.

S., Mediterranean and on the Danube. The famous seafarers of the Russian Fleet—
P.S. Nakhimov, V.A. Kornilov, G.I. Butakov and others served in S. The defenders
of S. during the Crimean War of 1854–1855 heroically stood the whole siege lasted
for 349 days. This heroic siege of Sevastopol was represented by artist
F.A. Roubaud as a huge panorama. In 1875 the railway connecting S. with
Lozovaya was constructed. S. became a large reloading center mostly of the
agricultural products exported from Russia.

During the Civil War of 1918–1920 the city was occupied by the German,
French and White Guard troops. The exploit of S. defenders in the Crimean War
was repeated nearly a century later. During the Great Patriotic War the city was
heroically defended for 250 days.

S. is the major naval port as well as the industrial, scientific and cultural center of
Ukraine. Its population is 344,100 (2013). The city has fishery and canned fish, food
industries, repair shipyards, instrument-making, woodworking enterprises and production of flux and construction limestone. It is the crossing point of automobile
roads. It has the airport "Belbek". Some design and research institutions are also
found here, among them there are Marine Hydrophysical Institute of National
Academy of Sciences of Ukraine and the A.O. Kovalevskiy Institute of Biology
of the Southern Seas. The latter has the aquarium containing a rich collection of
marine inhabitants.

Such famous Russian scientists as F.F. Bellinsgausen, G.A. Sarychev,
S.O. Makarov and others worked in S. The radio transmitters developed by
A.S. Popov were tested on the Sevastopol roadsteads. In 1905 S. witnessed the
mutiny on the ship "Potemkin", the armed rebellion, including on the cruiser
"Ochakov" under command of P.P. Shmidt. In 1983 in commemoration of
200 years of the city existence research vessels "Admiral Vladimirskiy" and
"Faddey Bellinsgausen" left the port of S. and headed to Antarctica following the
pioneer route of its explorers. They enriched the world science with new discover-
ies. The Panorama "Sevastopol Siege of 1854–1855" by F.A. Roubaud, the diorama
"Storming of Sapun-gora on 7 May 1944" and the exposition in the Black Sea Fleet
Museum tell about heroic exploits of the Sevastopol people.

More than 100 archeological monuments, including the ancient Greek colony of
Chersonesos, are found on the territory of S. The city dates back to the ancient
times. The Tauric settlements had already existed here in the early first millennium.
In the first quarter of the fifth century B.C. the Greek colony of Strabo Chersonesos
was founded on the Mayachnyi Peninsula and it was given the name of Heraclea
Taurica. In 422–423 B.C. the Greek colonists constructed the town of Chersonesos
Taurica on the western coast of the Karantinnyi Bay. Later on it was named
Chersonesos, Kherson and in Russian chronicles—Korsun. Under this name it
existed for nearly 1,800 years—till the end of the fourteenth century when it was
destroyed and burnt by the Mongol-Tatar Hordes of Murza Edigey. In the fifteenth
century the Turks referred to the ruins of Chersonesos as Sary-Kermen ("yellow
(orange) fortress"). In the eighteenth century by the time of Sevastopol foundation
two Tatar villages—Inkerman and Akht-Yar located on the coast of the
Sevastopol Bay.

On 8 May 1965 for outstanding war and labor exploits S. was given the title of the Hero City. In 1954 it was awarded the Red Banner Order, in 1983—the October Revolution Order and in 1965—the Order of Lenin. S. is the Crimean base of the Ukrainian and Russian Black Sea Fleet. The status of the Russian Black Sea Fleet in Crimea is regulated by the Framework Treaty of 1996 and the USSR-Ukrainian Treaty on Status and Conditions of Staying of the Black Sea Fleet of the Russian Federation on the Ukrainian Territory of 18 May 1997.

Sevastopol (Photo by Elena Kostyanaya)

"Sevastopol" – a ship of the Russian and Soviet Navy of the dreadnought type, the leader in a group of seven ships. It was built on the Baltic yards in Petersburg by the design of Professor I.G. Bubnov and Academician A.N. Krylov. It was put afloat in November 1914. In engineering terms the ships of the "S." type were the best in their time. It took part in battles on the Baltic Sea during World War I. It was an active participant of the Great October Socialist Revolution and Civil War. In 1918 together with other ships of the Baltic Fleet "S." made a heroic "ice" march from Helsingfors (Helsinki) to Kronstadt. It took part in the Petrograd Defense of 1919. In 1925 it was renamed into "Parizhskaya Kommuna" (Paris Commune). In late 1929—early 1930 "S." together with cruiser "Profintern" navigated to B.S. and became the flag ship of the Black Sea Fleet. In 1933–1937 it was capitally repaired and updated. During the Great Patriotic War it took part in the Sevastopol defense and in other battles. In 1943 in honor of the Hero City it was returned its former name—"Sevastopol". For orderly fulfillment of its military tasks in July 1945 it was awarded the Red Banner Order. It was in service till 1956. Its original deadweight was 23,000 tons, length– 181.2 m, width—26.8 m, speed—23 knots, sailing range—to 1,625 miles, crew—1,126 people. It was constructed with the following armaments onboard: four turrets of the major caliber with three 305-mm cannons in each, sixteen 120-mm antimine cannons (in vaults), four 47-mm antiaircraft guns, four subsea torpedo tubes.

"Sevastopol" battleship (http://upload.wikimedia.org/wikipedia/commons/4/48/Sevastopol_bat tleship.jpg?uselang = ru)

Sevastopol Biological Station (SBS) – one of the world's first station. First it appeared in Odessa, but later, in 1871–1872, it was transferred to Sevastopol (the world's second was built in Naples in 1872–1874). The decision on its construction was taken by the Union of Natural Scientists and the University of Novorossiya on the initiative of young anthropologist and ethnographer N.N. Miklukho-Maklay who later became the outstanding scientist and explorer. Such prominent scientists as A.O. Kovalevskiy, N.I. Mechnikov, I.M. Sechenov and others were involved in its creation. The first director of SBS was Professor of the Novorossiya University A.O. Kovalevskiy who later was conferred the title of Academician. In 1889 he proposed to transfer it to the management of the Academy of Sciences which was done. A building with the aquarium was specially built for it. After appearance of SBS the explorations of the B.S. flora and fauna became regular. A considerable contribution to these studies was made by Honorary Academician N.M. Knipovich, Academician S.A. Zernov, N.V. Nasonov, Professor V.N. Nikitin, V.A. Vodyanitskiy, N.M. Morozova-Vodyanitskaya and others. In 1934 after expansion the SBS building was capable to accommodate the physiological and biochemical laboratories. In 1963 SBS together with the Karadag Biological Station and the Odessa Biological Station were united to form the A.O. Kovalevskiy Institute of Biology of Southern Seas with the center in Sevastopol.

Sevastopol Biological Station (Photo by Elena Kostyanaya)

Sevastopol Defense Museum – was opened in Sevastopol on 14 September 1869 on the initiative of the participants of the Defense of 1854–1855. It occupies five halls in the house that belonged to General E.I. Totleben. In 1895 the Naval Department decided to construct a special building for the museum that was called already the War Historical Museum of the Black Sea Fleet. The building was designed by architect Academician A.M. Kochetov. The building is made in a classical style and is distinguished by splendid and abundant decoration.

Sevastopol Defense of 1854–1855 – defense by the Russian troops of the main B.S. Fleet base in Sevastopol from 13(25) September 1854 to 27 August (8 September) 1855 during the Crimean War of 1853–1856. After defeat on 8(20) September 1854 in the first field battle on the Alma River the Russian troops under command of Admiral A.S. Menshikov retreated to Sevastopol and then to Bakhchisaray. The Sevastopol garrison numbering about 7,000 men was open to the fire of the British-French troops (67,000 men). The allies conquered Balaklava and Kamyshovaya Bay turning them into the main base of their fleet and the main point for supply of their troops. The situation of Sevastopol was aggravated still more due to the lack of adequate defense fortifications of the city on the land side. On the sea side the city was defended by 13 coastal batteries (611 cannons). The main forces of the Black Sea Fleet (16 sailing battleships, 6 steam and 4 sailing frigates and other ships and 24,500 men) were concentrated on the Sevastopol sea roads. The enemy had three times more the battleships and nine times more the steamers compared to the Russian forces. Vice-Admiral V.A. Kornilov, the Chief Commander of the B.S. Fleet, and Admiral P.S. Nakhimov, Commander of the squadron of battleships, headed the defense and took urgent actions on its strengthening. At the entrance into the Severnaya Bay 5 old linear battleships and 2 frigates were sank, some cannons were moved from the ships to the land and 22 battalions were formed out of the ship crews. Under supervision of Engineer-Colonel E.I. Totleben only several weeks were used for construction of 20 fortifications and the number of cannons on the southern side was increased to 341, including 118 big guns, against 144 enemy cannons. So, within a very short time the in-depth defense was created that permitted to use the forces and facilities most effectively, including naval and coastal artillery. The main elements of the defense were bastions. Three lines of defense and entrenchments were built near Sevastopol. The dugouts were used here for the first time in Russia. The British-French command planned to storm Sevastopol from the south after intensive bombardment from land and from sea. By this time it amassed near Sevastopol 67,000 men (the Sevastopol garrison numbered 36,600 soldiers and sailors). But forestalling the enemy, in the morning of 5(17) October that was the day appointed by the British-French command for the first bombardment the Russian troops opened heavy fire on the enemy batteries and destroyed them. On the same day Kornilov was deadly wounded and the command of the Sevastopol defense passed to Nakhimov.

Repelling the enemy offences the defenders of the city successfully attacked the enemy batteries, laid mines and organized night raids by small units. But the balance of forces was in the allies favor. By May 1855 the enemy had 175,000-strong army against 85,000-strong Crimean Army (of which 43,000 defended Sevastopol). This enabled the British-French troops to launch more active offensive. In late May they seized the front redoubt at the approaches to the key place in Sevastopol—Malakhov Mound which defense was headed by Rear-Admiral V.I. Istomin. In May the casualties of the Sevastopol garrison were about 17,000 men. On 28 June (10 July) Nakhimov was deadly wounded. After hard battles the British-French troops seized the Malakhov Mound. The Russian troops left the southern side and at night of 28 August (8 September) moved to the northern side to unite with the army of Menshikov. The well-organized retreat of the Russian Army with artillery and logistic support units only during one night was the unprecedented doing having no analogs in the history of wars. In the battles for Sevastopol the enemy lost about 73,000 men dead and wounded not including all those who got sick and died of diseases, while the Russian casualties were about 102,000 men. The ally troops sent 1,356 million shells on the city.

The defense participants were awarded the special Medal "For Sevastopol Defense". The Russia's first Sisters of Mercy also took part in the battles for Sevastopol. They were awarded the Silver Medal "Crimea 1854–1855". The Sevastopol defense was the climax of the Crimean War. Thanks to persistence demonstrated by the city defenders the offensive of the allies fizzed out. Soon the parties started peaceful negotiations in Paris. The Sevastopol defense that lasted for 349 days demonstrated expert organization of land and sea forces joint actions in defending a seaside fortress. Many local people took part in this defense. The names of such commanders as V.A. Kornilov, P.S. Nakhimov, V.I. Istomin, S.A. Khrulev, E.I. Totleben, of such officers as A.V. Melnikov, N.A. Birilev, F.M. Novosilskiy, P.L. Zherve, A.I. Panfilov, of such sailors as P.M. Koshka, I. Dimchenko, F. Zaika, A. Rybakov, of such soldiers as A. Eliseev, I. Shevchenko, Ya. Makhov; of doctor N.I. Pirogov and of many other defenders of Sevastopol were put forever in the military history of Russia. The panorama "Sevastopol Defense", its affiliate museum on the Malakhov Mound, the monuments to V.A. Kornilov, P.S. Nakhimov, E.I. Totleben, sailor P.M. Koshka, to the sunk ships were constructed to commemorate the heroic deeds of the Sevastopol defenders.

"Sevastopol Defense of 1854–1855" Panorama – the amazing piece of the battle art and a monument to the heroic defenders of Sevastopol during the Crimean War. The picture shows battles for Sevastopol fought on 6(18) June 1855 giving rebuff to the enemy storming fortifications on the Malakhov mound and the third bastion. The author of this panorama is outstanding battle-painter F.A. Roubaud. The canvas of the panorama sizes 115 m × 14 m, while the total area of the panorama is 1,610 m^2. The space between the viewing site and the canvas is occupied by the life-size plan 900 m^2 in an area that gradually merges with the canvas giving the 3D perspective to the battle. A special building for the panorama was designed by architect S.V. Bedyaev and military engineer O.I. Eiberg with participation of V. Feldman. The panorama was opened for the visitors on 14 May 1905—the 50th anniversary of the first Sevastopol defense. It was not closed even during the defense of the city in 1941–1942. But on 25 June 1942 during heavy bombardment and shelling of the city the building of the panorama was set on fire and suffered serious damage. Thanks to the rescue works 86 parts of the canvas were saved. After the war the panorama was, in fact, created a new by a group of artists directed by Academicians of Art V.N. Yakovlev and P.P. Sokolov-Skal. The panorama was re-opened in 1954 by the 100th anniversary of the first Sevastopol defense.

Sevastopol Defense of 1941–1942 – the defense of Sevastopol by the Soviet troops that lasted from 30 October 1941 through 4 July 1942. The city was besieged by the Eleventh German Army commanded by General E. von Manstein. The attempt of the Germans in October-November 1941 to seize the city on a rush failed due to staunch resistance of the local garrison (23,000 men). In these battles the defenders of the city were joined by the detachments of the Primorskiy Army commanded by General I.E. Petrov that retreated from the north. On 7 November the Sevastopol Defense Area (SDA) was headed by Vice-Admiral F.S. Okrtyabrskiy and his deputy was I.E. Petrov. On 17 December 1941 the Germans launched a new storming of Sevastopol. Manstein sent 100,000-strong army against 60,000 defenders of SDA. They managed to break through the defense line in the north, near the Makkenzy Mountains, but the Sevastopol defenders just that time were reinforced with the troops that were delivered on the B.S. ships and by joint forces they beat back the Germans. The support of the B.S. Navy was very vital and without it the defenders were unlikely to repel the German attacks for such a long time. During the Sevastopol siege the ships delivered to the city more than 100,000 fighters, dozen thousand tons of ammunition and food. With the beginning of the Kerch-Feodosiya operation Manstein had to stop the siege of Sevastopol. In January-March 1942 as a result of counterattacks the defenders returned to their former positions. After the German victory in the Kerch battles they concentrated the forces of the Eleventh Army (more than 200,000 men) against Sevastopol. By this time the number of defenders was 106,000 men who had at their disposal 600 cannons, 38 tanks and 53 aircraft. On 2 June 1942 the Germans started massive fire of the city. Even the famous German cannon "Dora" was brought here. The battles were fierce on both sides. Because of German domination in air the supply of fresh forces and ammunition became impossible. Regardless of permanent gunfire and bombardment the combatants of SDA resisted stiffly the fierce attacks of the German Eleventh Army for more than a month. On 30 June the Germans rushed at last into the southern part of the city that was turned into ruins and burnt sites. The Russian troops had to retreat to the Khersones Peninsula where they fought hero-ically till 4 July. The greater part of the Khersones defenders (30,000 men) were taken prisoners because only a few boats and submarines were available for evacuation of the troops. However, the 250-day defense of Sevastopol kept a considerable part of the German troops. It demonstrated that having a solid system of fortifications the Soviet troops were able to effectively repel the enemy attacks for a long time. The German casualties under Sevastopol were 300,000 men, out of which 150,000 died in the last storming. For seizure of Sevastopol Manstein was promoted to Field Marshal. For the Soviet defenders the Medal "For Sevastopol Defense" was instituted.

"Sevastopol Defense" – the first Russian full-length historical film made in 1911 by V.M. Goncharov, a film director (1861–1915), and A.A. Khanzhonkov (1877–1945), Russian cinema producer and cinematographer. This film is about the heroism of the Russian soldiers and seamen who defended Sevastopol in 1854–1855. It was first demonstrated in 1911 to the Emperor's family at the Livadia Palace.

Sevastopol Harbor – prior to the 1970s such name was used in respect of the B.S. bay situated eastward of the line connecting the Konstantinovskiy and Aleksandrovsky capes. After construction of the Southern Protective Pier in the Sevastopol Harbour the SB area included also Martynova Bay that was earlier an external one. The notion "SB" includes 19 large and small harbors formed in meanders of its coast: Konstantinovskiy, Matyushenko, Mikhailovskiy, Staro-severnyi, Severnyi, Inzhenernyi, Dokovyi, Gollandia, Sukharnyi, Mayachnyi, Neftegavan, Kilen, Appolonova, Korabelnyi, Yuzhnyi, Artilleryiskiy, Khrustalnyi, Aleksandrovskiy, Matyushenko. Moreover, S.B. also includes one more "bucket"-like basin of the Chernaya River into which the river channel is running. S.B. is the most convenient bay on B.S. for port construction. The name "S.B." appeared simultaneously with the foundation of the city of Sevastopol. But there are also known its other names: Ktenus/Ktenunt, Kalamita-liman, Korsun Sivash, Akhtiar, Inkerman, Bolshaya bays, Bolshoy roads, Glavny roads, Sevastopol roads, Khersoness liman. These names are connected wither with the names of coastal settlements or with the bay size.

Sevastopol Harbor (Photo by Dmytro Solovyov)

Sevastopol Harbors – large and small harbors in the Sevastopol administrative territory from the Sarych Cape in the south to the Lukull Cape in the north. There are 36 of them, including Sevastopolskiy, Konstantinovskiy, Matyushenko, Mikhailovskiy, Staro-Severnyi, Severnyi, Inzhenernyi, Dokovyi, Gollandia, Sukharnyi, Mayachnyi, Neftegavan, Kilen, Apollonova, Korabelnyi, Yuzhnyi, Artillery, Khrustalnyi, Aleksandrovskiy, Karantinnyi, Pesochnyi, Streletskiy, Kruglyi, Abramova, Kamyshovyi, Lebyazhya, Dvoinaya-Troinaya, Solenyi, Golubaya, Aleksandra, Mramornyi, Balaklavskiy, Ershinaya, Laspinskiy. Some of these harbors are used for navigation, others—only for recreation.

Sevastopol Minor – due the Sevastopol defense of 1941–1942 in this way Balaklava was referred to.

Sevastopol Sea Port – locates in the unique non-freezing Sevastopol Bay protected from winds and storms and situated in the southwestern part of the Crimean Peninsula, Ukraine. The port in the Sevastopol Harbor was constructed and developed together with the city. On 22 February 1784 Catherine II signed the Manifesto "On Free Trade in the Cities of Kherson, Sevastopol and Feodosiya" that declared the ports in the mentioned cities open for international trade. From now on the ships of all friendly states were allowed to call on Sevastopol. On 7 March 1789 the port office was instituted and on 22 March it started operating. It existed till 1804 (with the interval for military actions) when Sevastopol was called the main port of the B.S. Navy. In 1863 the customs house was opened in the city. In 1867 the naval port in Sevastopol was liquidated. In 1875 after commissioning of the railway Sevastopol was officially declared the export commercial port and from 1880 it was developing rather intensively. The regular ship routes to Constantinople were organized. The cargo and passenger traffic had increased enormously. In 1884 the western coast of the Yuzhny Harbor was given for construction of a commercial port ("Tsar Wharf").

The Black Sea Fleet developed in the conditions of aggravating international relations and once again the problem of impossibility of side-by-side location of a commercial port and a naval base in Sevastopol was brought to the fore. On 11 May 1890 the Council of Ministers decided to transfer the commercial port from Sevastopol to Feodosiya. In 1905 the foreign trade port was closed. During World War I and the Civil War the port facilities were destroyed completely. The Soviet period of the Sevastopol trade port began in the 1920s. By 1925 the Sevastopol port became one of the main points of domestic and foreign trade in the south. During the Great Patriotic War the port was not operating and only in May 1944 it resumed its work. In 1950 the port organized the local passenger traffic. In 1954 the Sevastopol Shipping Company was transferred to the USSR Ministry of Marine Fleet. In 1962 the Sevastopol Agency was reorganized into the Sevastopol Marine Trade Port of Rank III. In 1968 a new building of the marine terminal was put into operation. In December 1975 the Ministry of Marine Fleet conferred Rank II to the port. In 1992 the Sevastopol Marine Trade Port was once again transferred to the management of the Ukrainian Ministry of Transport. In mid-1993 the port resumed its international passenger and cargo traffic that was closed in the Soviet time. The advantageous geographical location, developed network of automobile roads provide opportunities for good servicing of the cruise vessels and organization of sightseeing tours over the whole Crimean Peninsula. The port has two passenger wharfs 200 m and 135 m long, respectively, that are capable to receive passenger liners with the draft of 8.6 m and 4 m, respectively, while the cargo wharf receives vessels with the draft to 7.8 m. The port has a well equipped water area where all operations on preparation of the vessels for voyages can be done. The port possesses all facilities for repair of vessels of various deadweight and purpose, including heavy aircraft carrying cruisers, whale boats, vessels with the deadweight of about 100,000 tons. There is also a large fishery port in the Kamyshovyi Bay.

Seventh Army – it was formed in July 1914 in the Odessa Military District. At the beginning of the war it was tasked to defend the B.S. coast and the state border with Romania. In October 1915 it was moved to the Terebovlya (Trembovlya) Region, Chortkov and incorporated into the Southwestern Front. It took part in the offensives in December 1915, in 1916 and in June 1917. In July 1917 it was attacked by the enemy troops and retreated eastward together with the Eleventh Army which led to disintegration of the whole Southwestern Front. And only on 15(28) July the enemy was stopped. The troops of the Seventh Army took on the defensive on the line westward of the Zbruch River. It was disbanded in April 1918.

"Severnaya Tavria" (Civil War of 1918–1922) – the offensive operation of the Red Army against the White Army of General P.N. Vrangel in the south of Ukraine (eastward of the Dnieper lower reaches to the Berda River) in the period from 28 October to 3 November 1920. The southern front under command of M.V. Frunze took part in it. It was opposed by the First Army of General A.P. Kutepov and the Second Army of General F.F. Abramov. In the center of "S.T." (Lower Seragoz area), the assault unit was deployed commanded directly by Vrangel that incorporated the infantry and cavalry detachments. It had to make counterstrikes from the center to prevent the likely breakthroughs of the Red Army. The Vrangel unit in "S.T." numbered 41,000 bayonets and 17,000 swords.

By this offensive operation the Soviet command was seeking to take the Vrangel troops in "S.T." into a circle cutting them off from the ways of retreat to the Crimean Peninsula. The main offensive was launched from the Kakhovka base by the troops of the Sixth Army of A.I. Kork and the First Cavalry Army of S.M. Budennyi. Having broken through the defense lines of the White Army, on 29 October the Red Army detachments came out to the approach avenues to Crimea. The 51st Rifle Division under command of V.K. Blyukher blockaded the Perekop Isthmus, while the advance forces of the First Cavalry Army having made a 100-km raid conquered by the end of 30 October the "Salkovo" station cutting off the escape routes for the Vrangel troops to Crimea via the Chongar Peninsula.

However, the Red troops failed to encircle the Vrangel Army in "S.T." The assault of the Red Army in other front sectors was not dashing enough. Thus, in the north the offensive of the Second Cavalry Army under command of F.K. Mironov was stopped for some time by the counterattacks of the Don cavalry units. The attacks of the 13th Army under command of I.P. Uborevich from the east were developing rather slowly, they had to storm the defensive lines of the Whites near Melitopol. In this situation Vrangel taking advantage of the central position of its assault unit used its forces (the Cavalry Corps of General Bobrovich) for counterattack of the detachments of the First Cavalry Army that broke through into the Salkovo-Rozhdestvenskoye-Otradnoe region. After the raid the Buddenyi's cavalry had overstretched communications and it was unable to withstand the sudden assault and was knocked off from the ways to Crimea. This permitted the Vrangel unit to pass to the peninsula over the Chongar bridge destroyed later on by the retreated troops. On 3 November the Red troops again conquered the stations "Salkovo" and "Chongar", but their attempt to rush into Crimea at once was not successful. During this operation the Soviet troops took 20,000 men prisoners, but the main Vrangel forces managed to retreat to Crimea. Regardless of the victory in "S.T." the Red Army failed to liquidate the main Vrangel forces that organized defense on the Crimean Peninsula. Severe frosts that occurred in the Prichernomorie steppes threatened the Red Army with disaster (remember here the insufficiency of lodging, water and fuel). This forced the Soviet command to start, practically without any pause, storming of the Perekop-Chongar fortifications.

Shabla, Cape – easternmost point of Bulgaria, sited on the Black Sea shore on the border with Romania. The cape is precipitous and is almost imperceptible. Can be identified by the light-house standing on the cape.

Shabo – climatic health-resort location in Odessa Region, Ukraine. Sited on the western shore of Dniester Liman, to the north-west of Karolino-Bugaz-Zatoka health-resort. A beach is available. In the environs of Shabo—vineyards, gardens.

Shakalin Bottleneck – location where Arabat Arrow comes nearest to the Crimean shore. It divides Eastern Sivash into two parts: Northern and Southern.

Shakhe River – river in Lazarev region of Sochi city, Krasnodar Territory, Russia. It reaches the Black Sea in Golovinka district of Sochi. There are 33 waterfalls on the river.

Shalanda (scow) – (French *"chaland"*, from late Greek *"chelandion"*) (1) Small shoal- draft, usually not-self-propelled craft, used in ports for loading and offloading large vessels berthed on the roads or for soil transportation from dredgers (the so-called soil-carrying Sh.). Such Sh. are equipped with devices for offloading via the bottom. (2) Fishing sail flat-bottomed boat with a dagger keel, is in common use in the Black and Azov Seas and is designed for fishing using nets or a trot line. Sometimes was used for transporting small cargoes (the so-called Ochakov Sh.). Was furnished with spritsail harness, had good sea-going characteristics. Length up to 7.5–8.5 m, width around 2.5 m, height of the side 0.8–0.9 m, draft 0.6–0.7 m, tonnage around 3–5 tons.

Shapsukho River – interesting because it's hardly the only river or on the coast of Tuapse and Gelendzhik region, the mouth of which is open all year round in the sea, and never closes, the other river pebble shafts, which are formed by storms. The river flows between the mountains in the beautiful valley. Above the village of River Valley Tenginka practically unaffected by urbanization and river channel retains its natural origin. The air here year round remarkably clean and fresh, and from early spring to late fall is filled with the heady aromas of great variety of flowers and herbs.

Shepsi – a small village at the Black Sea coast of Caucasus, at the Shepsi River mouth, 20 km (by road) southeastward from Tuapse. Population—3,100 (2002).

Sheskharis – Russia's largest oil terminal sited in Novorossiysk. Commissioned in 1964. Deep-water pier capable of handling tankers of up to 250,000 tons.

Sheskharis (http://www.nvr.transneft.ru/index.php?sub=1&about=2&podrazdel=4)

Shestakov Ivan Alekseevich (1820–1888) – Russian Admiral. In 1830–1836, studied at the Naval Cadette Corps. In 1837–1838 and 1840–1842, took part in cruising near the Caucasian shore of the Black Sea, and in 1838–1840 and in 1842–1843—in the Mediterranean Sea. From April of 1843 promoted to Lieutenants with appointment to the position of the aide de camp of the Chief Commander of the Black Sea Navy and Ports Admiral M.P. Lazarev. In 1845–1846 served on the steamer "Bessarabia". In 1847–1850, concurrent with G.I. Butakov, was making a hydrographic description of the Black Sea and in association with him drafted the first sailing directions of the Black Sea. In 1860–1862 commanded a squadron of ships in the Mediterranean Sea. From 1864—in civil service. From 1872—again at the naval department: marine agent in Austria-Hungary and Italy. From 1881—head of the Sea Ministry. With Shestakov in office, the Black Sea Navy was re-established up-tempo and the Siberian Flotilla grew rapidly. An island in the Barents Sea (near Novaya Zemlya) is named after him.

Shevchenkovo – (formerly known as Alupka-Sara) maritime health-resort location in Ukraine. Adjoins Alupka on the west. Sited on the high shore of the Black Sea (southern slopes of the main Range of the Crimean Mountains). The main curative natural factor—subtropical climate of Mediterranean type, extremely congenial to climotherapy of pulmonary TBC patients, of locomotor and support organs. Sanatoria are available.

Shilder Carl Andreevich (1786–1854) – Russian military engineer, Engineer-General (1852). Served in engineer troops, participated in the 1812 Patriotic War, in the Russian-Turkish War of 1828–1829, in the Crimean War of 1853–1856.

Developed an effective countermining system, designed stone and case-shot fougasses, invented an ingenious design of a suspended pontoon bridge. In association with B.S. Yakobi, designed galvanic and galvano-impact-detonated sea mines (1838–1848), the world's first all-metal missile-carrying submarine (1834), and naval steamer "Otvazhnost" ('Bravery') (which was viewed as a prototype destroyer) were built to Sh.'s designs. Sh. died of grave wounds he received in the battle of Silistria.

Shilling Nikolay Nikolaevich (1870–1946) – Russian Colonel (1909), Major-General (1915). Lieutenant-General (1919). Finished the Nikolaev Cadette Corps (1888) and the 1st Pavlov Military Training School (1890). Participant of WWI. General at the Staff of the Commander-in-Chief of Hetman Skoropadskiy's Army (March-August, 1918). In the White Movement: Officer at the Staff of Volunteers' Army Representative in Kiev from September of 1918. In 1919—Commander of the 5th Infantry Division, the Crimea (January-May, 1919). Wounded, from 1919— Commander of the 3rd Army Corps (reorganized former Crimean-Tavrian Army), simultaneously Governor-General and Commander of the troops of Tavrian and Kherson Provinces. Commander of troops of Novorossiya. While remaining in that position, he assumed command over part of the troops of Kiev Region that retreated from Kiev to Tavria (November-December, 1919). Commander of Novorossiya troops and in the Crimea (December, 1919–March, 1920). After Odessa was occupied by the Red Army (by Kotovskiy's Detachment), was evacuated on the ships of the Black Sea Navy from Odessa to the Crimea, where he continued to act as the commander of Novorossiya troops. After General Wrangel's Army was established in the Crimea in March of 1920, was transferred to the reserve in view of Novorossiya troops being disbanded. In emigration from 1920 in Czechoslovakia. After the Soviet Army occupied Prague in May of 1945, Sh. was arrested. Released for reasons of health and age, and soon died in Prague.

Shipbuilding Plant named after 61 Kommunar – one of the oldest enterprises on B.S. It was constructed in 1788 in the mouth of the Ingul River (Nikolaev) as a shipbuilding yard. In 1789 it was the Nikolaev Admiralty. From 1930 it bore the name of 61 Kommunars given in memory of 61 hostages executed in 1919. In 1790 the first 50-gun frigate "St. Nikolay" was put afloat. 287 sailing ships were built here until 1865. In subsequent years till 1910 the armored whole-metal vessels were constructed here. In September 1900 the armored vessel "Potemkin" was sent afloat here. In 1911 the Russian Shipbuilding Company rehabilitated the Admiralty for the second time and gave it a new name—"Russud" Plant. In pre-revolutionary years two battleships, cruisers "Admiral Nakhimov" and "Admiral Lazarev" were built here. In the 1930s the plant built tankers, dry-cargo ships, ore carriers, all-welded floating docks; after the Great Patriotic War—self-propelled barges, whaleboats and refrigerator ships. In 1967 it was awarded the Order of Lenin.

Shipbuilding Production Association named after 60-letie Leninskogo Komsomola (60 years of Lenin's Komsomol) – it was founded in 1976 after merging of the Kherson Shipbuilding Plant (1951) and the "Pallada" Plant (1931).

It builds dry-cargo vessels, icebreakers, transports, expedition, research-fishing vessels, towboats as well as ship facilities. It also fulfills agricultural orders. In 1970 it was awarded the Order of Lenin and in 1980 the International Award "Golden Mercury".

Shipping Company – an integrated water transport organization having at its disposal the fleet, sea (river, lake) ports, ship repair yards, special management department and other production and economic divisions as well as special educational institutions. It undertakes carriage of passengers and cargo via waterways. In the USSR such companies were organized on the geographical principles: by sea basins and sections of sea coasts, by river basins, etc. The Marine Fleet Ministry had 16 shipping companies. The Azov Shipping Company (ASC) was founded in 1871 in Mariupol as a joint stock company. From 1924 to 1953 ASC located in Rostov-on-Don. Due to its small size it was reorganized into the Azov Department of the Black Sea Marine Shipping Company and transferred into Mariupol (then Zhdanov). In 1967 on the basis of this department the independent AMSC was established. It had four ports (Zhdanov, Kerch, Taganrog and Berdyansk), the railway ferry in Kerch, the repair-construction department, and the naval school. The fleet comprised more than 200 different ships. AMSC was awarded the Order of the Labor Red Banner.

The Georgian Marine Shipping Company (GMSC) was founded in 1967 in Batumi. It comprised: three ports (Batumi, Poti and Sukhumi), three port-points (Novy Afon, Gagra, Gantiadi). The fleet had 46 crude oil tankers and dry-cargo vessels with the total carrying capacity over 700,000 tons.

The Novorossiysk Marine Shipping Company (NMSC) was organized in 1967 on the basis of the oil fleet department of the Black Sea Marine Shipping Company (BSMSC) that was preceded by established in 1932 in Tuapse of the "Sovtanker" S.C. It comprised: three ports (Novorossiysk, Sochi, Tuapse), port-points, two repair shipyards (Novorossiysk, Tuapse) and the naval school (Novorossiysk). The fleet included 138 transport vessels with the total carriage capacity over 5 million tons. It was awarded the Order of the October Revolution (1971).

The Soviet Danube Shipping Company (SDSC) was founded in 1944 in Izmail. It comprised: four ports (Izmail, Kilia, Renni, Ust-Dunaisk), two repair shipyards (Izmail, Kilia), three fleet maintenance bases (Izmail, Vilkovo, Renni) and a naval school (Izmail). The transport fleet numbered 62 marine dry-cargo vessels with the total capacity of 236,300 tons, 2 lighter carriers, over 600 river self-propelled and non-self-propelled ships and also marine and river passenger ships.

The Black Sea Marine Shipping Company (BSMSC) was established in 1967 in Odessa on the basis of the state Black-Azov Shipping Company that was founded in 1922 the predecessor of which was the shipping company founded on B.S. in 1833. BSMSC included: 11 ports (Odessa, Ilyichevsk, Kherson, Yalta, Yuzhnyi, Belgorod-Dnestrovskiy, Nikolaev, Skadovsk, Evpatoria, Sevastopol, Feodosiya), three repair shipyards, the fleet maintenance department "Cherazmorput". The fleet numbered 260 dry-cargo vessels with the deadweight over 3 million tons and more than 30 passenger ships. It was awarded the Order of Lenin (1971).

Shipping Company for Navigation over the Dnieper and its Tributaries – it was founded in 1858 in Petersburg for cargo and passenger carriages over the Dnieper. In 1903 the fixed assets of the company were evaluated at 1 million Rbls.

Shirokaya Balka – a resort area 10 km from Novorossiysk, pebble beach, holiday houses and hotels. Local population—105 (2010).

Shirokaya Balka (http://static.tonkosti.ru/images/)

Shirokiy Inlet – sited on the northern shore of the eastern part of Karkinite Bay, Crimea, Ukraine. On the east, the inlet is confined by Domuzgla Peninsula, on the west—by Gorkiy Kut or Khorlovskiy Peninsula with Khorly port arranged on it.

Shishkov Georgi (1865–1943) – professor, founding father of zoological research in Bulgaria. Late in the nineteenth century, began scientific excursions; the findings of studies he conducted during the excursions he synthesized in two scientific publications (1907, 1911). The first dealt with saltwater mites he had discovered in the Black Sea, the second one contains a list of 280 animal species.

Shmidt Petr Petrovich (1867–1906) – Russian seaman, Lieutenant of the Black Sea Navy who headed the uprising on the cruiser "Ochakov" in 1905. Finished the Naval Training School in St. Petersburg in 1886. Served in the Baltic and Pacific Navies. In 1898, retired in the rank of Lieutenant. Used to sail on merchant ships. In 1904, mobilized. From 1905—commander of destroyer No 253 in the Black Sea Navy. In 1905, organized "Union of Officers—People's Friends", took part in establishing the "Odessa Society for Mutual Assistance of Merchant Marine Seamen". Was arrested on November 2, 1905 for addressing the rallies of workers, soldiers and sailors. The workers elected Sh. by absentee ballot a lifelong deputy of the Sevastopol Soviet of workers' deputies and secured his release from prison. When Sevastopol uprising began, the military organization of social democrats offered him the position of the military leader of the uprising. Sh. arrived on "Ochakov" cruiser, where the red flag had been hoisted on November 27. After the uprising was quelled on November 28, Sh. was arrested and on March 2, 1906 was sentenced by the military tribunal to capital punishment. Together with other leaders of the uprising, was executed by firing squad on Berezan Island. From 1926 has been posthumously elected honorable member of the Sevastopol Soviet of people's deputies. In 1962, the museum named after Sh. was commissioned in Ochakov. The monument to Sh. was set up on Berezan Island in 1972. A bridge across the Neva River in St.-Petersburg was named after P.P. Shmidt.

Shmidt Vladimir Petrovich (1827–after 1898) – Russian Admiral (1898). In 1841 promoted to midshipmen of the Black Sea Navy. In 1845–1848, sailed on various ships in the Baltic and North Seas. Transferred to the Black Sea Navy. During 1849–1850, sailed on various ships near the eastern shores of the Black Sea, then served in a separate task force under the flag of Rear Admiral V.A. Kornilov to supervise Black Sea ports and Caucasus fortifications. In 1851, took part in the campaign on Odessa roads. In 1851–1852, sailed to the Mediterranean Sea and back. In 1853, participated in te battle of Frigate "Flora" against the Turkish steamers at the height of the Pitsunda fortification. In 1854, commanded frigate "Flora", "Rostislav" ship on Sevastopol roads, then was in charge of a marine battalion on Malakhov Mound. In 1854–1855, was assistant commander of Bastion No 5, then, successively commanded Bastion No 2, battery in the gully of Ushakov's balka and Rostislav redoubt. In 1856, sailed between St. Petersburg and Kronstadt, whereupon was transferred from the Baltic Navy to the Black Sea Flotilla and commanded steamer "Skromnyi" ('Modest') in Nikolaev roads. In 1857, was posted to Tulon to superpose the construction of screw-propelled scooner. In 1858–1862, sailed on it near the Caucasian shores and in the Mediterranean Sea. In 1872–1876, sailed in the Gulf of Finland and on the Baltic Sea. In 1877, posted to the Danube Army. Promoted to Rear-Admirals for distinction demonstrated when he was ferrying troops across the Danube near Zemnitsa. From 1878 to 1884, in the Baltic Navy. In 1886 promoted to Vice-Admirals with approval in the filled position. In 1887, appointed Chief of the Squadron on the Pacific. In 1892 appointed member of the Admiralty-Council.

Shokalskiy Juliy Mikhailovich (1856–1940) – outstanding Russian geographer, oceanographer, cartographer, Corresponding Member of the USSR Academy of Sciences (1939), professor of the Naval Academy (1910–1930). In 1873–1877, studied at the Marine Training School. Having finished it, sailed in the Baltic Sea on the iron-clad battle ship "Petr Velikiy" and the cruiser "Krechet" ('Arctic Falcon'). In 1878, promoted to midshipman and enrolled as a student of the hydrography department of the Naval Academy. Upon leaving the academy as a first-rate graduate, from 1880 to 1882 worked as the chief of the meteorology department in the Main Physical Observatory. From 1882 to 1908, lectured at the Marine Training School, worked as the head of the Main Marine Library (1891–1907). In 1907–1912, Chief of the Hydrometeorological section of the Main Hydrography

Department. His activity in that position is above all related to the growing significance of hydrometeorological support of naval operations, to organizing more efficient performance of the network of marine hydrometeorological stations and to publishing the results of observations. At his initiative, the Tidal Yearbook began to be published, containing detailed data on the tides regime in the seas near the Russian shores. Took an active part in activities on commencement of hydrographic research with a view to ensuring the safety of seafaring, introduction of general rules in both marine cartography and in preparing and publishing sailing directions. From 1912 to 1930, lectures at the Marine Training School, where he established a chair of oceanography. Besides the Naval Academy, he lectured at the teacher-training and Geographic Institutes. From 1925 and until his death he was a professor of the Leningrad State University. In 1883–1908, took an active part in the development of the Northern Sea Route, in the study of rivers, Ladoga Lake and the Caspian Sea. In the field of mapping, Sh. devised a new hypsometric map of Russia to the scale of M 1:12000000 (1914), initiated the development of the mapping course in higher school, edited a number of specialized maps and atlases. In 1923–1927, was in charge of the oceanographic expedition for integrated study of the Black Sea. He substantiated the causal relationship between the phenomena in the World Ocean as well as the need for studying as a whole marine hydrology and meteorology. Participant of numerous international conferences and congresses in geography, oceanography, hydrometeorology and other disciplines. In 1930, dismissed from the Navy for the reason of age. Among his 130 works—monograph "Oceanography" (1917, 1959), which has become part of the world science as a classical research work as well as "Hydrography and Physiography" (1900), "Essay on Oceanography Development" (1900), "A View of the Current State of Oceanography" (1911), "Physical Oceanography" (1933). 12 geographic locations and a vessel are named after him.

Shokalskiy J.M. (http://spb.rgo.ru/)

Shores of the Sea of Azov – are largely low. All along their length, steppe alluvial and loess plains approach the sea. The plains are worked out on the structures of the ancient (Russian pre-Cambrian) and more recent (Scyth Hercynic) platforms. It is only at the south-eastern edge of the sea that the wide Azov-Kuban piedmont foredeep of the alpine system is laid bare, to which the extensive (area of 4,300 km^2) multi-arm delta of the Kuban River is confined. The characteristic feature of this delta formation is the availability of numerous infradeltaic lakes of 1,500 km^2 combined area. Typical of S.A. are accumulative shore forms of varying types, sometimes unique. In the west, it is the vast (nearly 110 km long, with average width of 0.5 km) Arabat Arrow structured almost entirely by shell material and genetically presenting a barrier beach—a strip of silted dryland; in the north—classical Azov-type spits, structured by quartz sands and shelly ground; in the east—large barrier beach hemming the sea margin of the Kuban Delta, and a series of sand spits and bay-bars of varying size and shape. The most ancient of these is Arabat Arow: it began to take shape in early Holocene, when the sea level was low. As the sea level started to rise, the Arabat Arrow was shifting increasingly shorewise, separating Sivash Lagoon. Due to small depths, poor connection with the sea (narrow Genichesk Strait) and intensive evaporation Sivash water is rather saline and has large reserves of salt resources. The present-day Sivash is a vast base of raw materials for the development of mineral-salt production, using bitterns.

The current state of the Azov Sea shores is characterized by the domination of abrasion processes. Natural development of the shores of late has been severely disturbed by economic and recreational activity in the river basins and on the coast. Most dynamic are the shores in the eastern part of the sea, structured by Quaternary loams and clays and exhibiting significant existing tectonic warping. Here, the wave resultant is directed almost normal towards the shore which conditions manifestations of strong positive surges and a prevalent development of abrasion processes with a mean rate of washout of 0.4–3.0 m/year (maximum—near Primorsko-Akhtarsk City—up to 6.0 m/year). On the shores of Taganrog Bay, that tend to rise and are characterized by outcrops of sand-and-clay deposits, abrasion is attended with the development of landslides most active around the city of Mariupol. Miocene and Pliocene limestones, protecting the shore precipices, feature wide occurrence within the Kerch-Taman coast. For this reason, the mean rate of abrasion here is notably lower (maximum 0.5–1.5 m/year).

The peculiar feature of the Azov Sea accumulative forms is the high content of shelly ground in them (from 60–70 % in Belosarai Spit to 90–99 % in Dolgaya Spit and Arabat Arrow, which ground is brought out by the waves from the seabed. Therefore, the accumulative forms are rather sensitive to variations of biogenic material in the continental slope. Working out the deposits for construction and for agricultural needs that had been on for almost 20 years manifested itself in the spits being washed out. The volume of shelly ground extraction from Dolgaya Spit in 1966–1975 was more than 1 million tons, which reduced accretion of the pit considerably and caused its washout. This was also assisted by an abrupt change of hydrochemical conditions in Taganrog Bay in connection with the Don flow regulation which resulted in reduced productivity of biocenosis of *Cardium edule*—the primary component of biogenic material. The gradual stabilization of conditions that augured well for vital activity of this mollusk coupled with four to fivefold reduction of biogenic material from the seabed conditioned the resumption after 1976 of the accumulative processes and restoration of part of the islands in spit continuation.

"Shota Rustaveli" – cruise liner on the Black Sea. Built in 1968 by Mathias-Thesen-Werft, Wismar, East Germany. Dimensions: $175.77 \times 23.55 \times 8.10$ m, Brt/Dwt. 19,567/5,696, capacity: 15,666 kW, speed: 21 knots, passengers: 750 people (after the reconstruction—650 people). It was considered the fourth in a series of five similar ships. The lead ship was "Ivan Franko" built in 1964, "Alexander Pushkin" in 1965, "Taras Shevchenko" in 1966, he was succeeded by "Shota Rustaveli" and completes the series "Mikhail Lermontov" in 1971. On 30 June 1968 "Shota Rustaveli" came and passed to cruise shipping company in Odessa, USSR. In 1995 transferred to an offshore company "Blasco UK", Monrovia, Liberia. In 1997 sold to "Ocean Agencies", Odessa, Ukraine. In 2000 sold to "Kaalbye Shipping International", Kingstown, St. Vincent, and renamed "Assedo" (the same name "Odessa", but written in Latin letters back to front). In November 2003 sold to the Indian company for scrap. On 28 November 2003 arrival at Alang, India and the dropping of the beach for scrapping.

On board the ship, lovely interior: paintings, tapestries and ornaments in the Georgian style, bright, spacious cabins with showers, most of them with a safe. For the comfort of a vessel equipped with stabilizers pitching. There were two swimming pools and a sauna, gym, massage, sports ground on the top deck, cinema, disco, seven bars with a menu to suit all tastes, casino, amusement arcade, kids club, shops, hair salon and medical center. Since July 1968 the passenger liner "Shota Rustaveli" started to make cruises between ports of Odessa and Batumi, and 2 months later came to the shores of the UK. In the port of Southampton has taken on board the ship tourists and went to a 3-month world cruise. Flight ran through the ports of Las Palmas, Sydney, Auckland, Papeete and Panama. For such a long journey the ship left nearly 26,000 nautical miles and crossed the Atlantic, Indian and Pacific Oceans, as well as the Tasman Sea and the Caribbean. Passenger ship chartering of various foreign firms' "Grande Viaggio", "Italnord","Orienturist","Trans-stage" and others

"Shota Rustaveli" (http://korabley.net/_fr/1/4027886.jpg)

Shuleikin Vasiliy Vladimirovich (1895–1979) – Professor, Academician of the USSR Academy of Sciences (1946), prominent scientist, oceanologist and geophysicist, one of the founding fathers of the Russian school of marine physics, Captain 1st rank. In 1912 entered Moscow Higher Technical School, Department of Hydropower. Upon graduation in 1916, lectured at the same institution, and in 1918 remained there for training to become a full professor. In 1923–1929—professor. Simultaneously, worked as a researcher at the Institute of Physics and Biophysics (1920–1931), from 1927, head of the department of geophysics of the same institute. In 1929, at the initiative and under the guidance of Sh. the world's first stationary hydrophysical station was established at Katsiveli near Simeiz (Crimea) on the shore of the Black Sea; the station was intended for the investigation on a regular basis of the offshore and atmospheric processes and phenomena. Sh. was the station's director until 1941. In 1942–1945, served at the Department of Hydrography of the Navy. Took direct part in building ice crossings on Ladoga Lake ("Life Road"), in offshore areas of the Baltic, White and Barents Seas. From 1943 to 1945, headed the Chair of Marine Physics he organized at the Moscow State University. In 1945–1947—Professor, Head of the Chair of Marine Hydrometeorology of the Krylov Naval Academy.

In 1948, Marine Hydrophysical Institute (MHI) of the USSR Academy of Sciences was established on the basis of the Black Sea Hydrophysical Station of the USSR Academy of Sciences and Marine Hydrophysical Laboratory (previously, a section of the Institute of Theoretical Physics of the USSR Academy of Sciences). Sh. Worked as Director of the Institute until 1957, then was in charge of the Marine Hydrophysical Institute. The most crucial results of MHI activity include a study of heat phenomena in the ocean, heat interaction between the ocean, atmosphere and continents, impact of the ocean on climate and weather. Sh. was the first to introduce in geophysics a thermodynamic view of thermal engines of four different types, working in the system ocean-atmosphere-continent. Experimental work carried out in the storm basin devised by him made it possible to develop a sea wave theory.

From 1947 to 1950—Chief of the Main Department of Hydrometservice of the USSR Council of Ministers. In 1949–1950—deputy executive editor of the Marine Atlas (v. 2). Participated in several research expeditions: 1922—sailed in the White, Barents and Kara Seas, doing research that laid the foundation of marine optics; in 1926–1930—took part in the Spitsbergen expedition, made hydrological and meteorological observations in the Barents and Greenland Seas; in 1927, did oceanographic research on the steamer "Transbalt" that was sailing from Evpatoria to Vladivistok via Suez Canal; in 1932—assistant Chief of Expedition for science at the research vessel "Taimyr" which for the first time in world practice traveled via Shokalskiy Starit to Severnaya Zemlya; in 1957–1959, headed an expedition to the Atlantic Ocean.

The main areas of scientific research: development of a general theory of interaction between the World Ocean, atmosphere and continents in shaping up the climate-forming and weather factors as well as development of the theories of wind-induced waves, sea currents, tropical hurricanes, etc. Fundamental theoretical and experimental research of Sh. and his disciples in the field of hydrooptics, hydroacoustics, electromagnetic phenomena in the ocean, molecular marine physics and biology of marine environment constituted the basis of a new area of geophysics—marine physics. Sh. established the Russian school of physicists-marine scientists. Awarded a Semenov–Tyanshanskiy medal for his research in marine physics. In 1942,—a Stalin Prize winner, decorated with orders and medals. Published around 400 books, papers and training aids (the book "Marine Physics" won international acclaim and was reissued four times). Besides science, Sh. was fond of music (composed romances and symphonic works), painting and poetry, wrote a book of memoirs "The Days I Lived". A submerged mountain in the Pacific and a research vessel were named after him. Major works: "On Chromacity of the Sea" (1922), "Essays on Marine Physics" (1927), "Marine Physics" (1941), "Brief Course on Marine Physics" (1959).

Shuleikin V.V. (http://antarctic.su/books/item/f00/s00/z0000020/pic/000006.jpg)

Sile – town and port on the Asian coast of Turkey, health-resort settlement. Frequented by the dwellers of Istanbul who come here when summer heat becomes unbearable. Beaches, restaurants, hotels. Famous for the fabrication of cotton shirts.

Silistar Beach – a recreation area on the Black Sea, Bulgaria.

Simeiz – the seaside climatic resort, Crimea, Ukraine. A small town 21 km to the southwest of Yalta and 68 km to the southeast of the railway station "Sevastopol". It is situated on the Southern Coast of Crimea at an altitude of 260 m over the mean sea level. It has good beach covered with small pebble. The center of the town is occupied by a large park with various species of evergreen trees and shrubs. The climate is Mediterranean subtropical. S. is one of the warmest places on the Crimean coast. It is characterized by high aridity and insignificant daily fluctuations of air temperatures. The mountains sloping down to the sea protect the resort from cold winds. The winter is very mild; the mean temperature in February is +4 °C. The summer is warm; the mean temperature in August is +24 °C. The relative humidity in summer is about 60 %. Dry air and sea breezes help to endure easier the summer heat. The autumn is also warm and dry. The precipitation is about 400 mm a year. The sunshine hours are 2,360 in a year. The bathing season lasts from June through October (the water temperature is above +18 °C). The natural climatic conditions in S. and availability of many sanatoriums provide opportunities for climatotherapy of the adults, children and teenagers suffering of TB (mostly of lungs). The resort has the winery, the astrophysical observatory. The Black Sea Branch of the Marine Hydrophysical Institute of NAS of Ukraine locates in the village of Katsiveli 3 km from S. Population—3,900 (2012).

Simferopol (from the Greek *"symphero"* **meaning "useful, helpful" and** *"polis"* **meaning "city", i.c. "city of common use")** – a city on the Salgir River, regional center of Autonomous Republic of Crimea, Ukraine. It was founded in 1784 in place of the town of Aqmescit ("White Mosque") as the administrative center of the Tavria Region. This is the railway station, the crossing place of automobile roads and air routes. The Simferopol reservoir was constructed near the city. The machine-building enterprises manufacture complicated automatic machines for the canning, fish, meat, milk and other industries. The city also has light industry factories. Of great importance is the food industry. Many institutes, vocational educational institutions, museums, picture gallery also locate in the city. Population—362,400 (2013), among them—Russians (67 %), Ukrainians (22 %), Crimean Tatars (7 %), and others. Simferopol is connected with Alushta and Yalta by the world longest tramway line (86.5 km).

Simferopol (http://image.goodvin.info/images/original/Aug2012/19925.jpg)

Sinemorets – a village on the B.S. coast of Bulgaria, situated 13 km from the state border with Turkey. This place is famous for its wonderful sandy beaches.

Sinemorets Cape – locates on the Bulgarian coast of B.S. 15 km to the northwest of the Rezovo Cape. It is the northeastern tip of a small peninsula. The cape is rocky and devoid of vegetation. The village of Sinemorets is situated westward of the cape. To the south of the cape there are two beautiful sandy beaches divided by an abrupt cape.

Single Mooring Point (SMP) – used for oil loading into tankers. SMP is a circular floating buoy attached to the sea bottom over an underwater oil pipeline. Oil is delivered to the tankers via flexible floating hoses. A buoy may drift vertically and horizontally and is capable to withstand rough sea and storm winds. SMP floating buoys are installed at a considerable distance from the shore which ensures the minimum environmental impact, enables to avoid calling on overcrowded ports and diminishes the danger for a tanker to come aground. The pipeline system of the Caspian Pipeline Consortium (CPC) ends with SMP near Yuzhnaya Ozereika. This SMP is installed at a depth of 65 m. It is capable to load 60,000-ton tankers at a rate of 12,700 m³/h. More than 500 types of SMP are used in the world.

Sinoe – a liman lake on the B.S. coast to the south of the Danube Delta, Romania. Its area is 164 km², depth—2–2.5 m. It is separated from the sea with low sand barriers cut by narrow channels.

Sinop (in the Ancient Greek "Sinope") – the city in the north of Turkey; the port in the central part of the Anatolian coast of B.S.; the administrative center of the Sinop Province. It was the birthplace of two famous figures of the ancient times—Mithridates Eupator, the ruler of the Pontus Kingdom and the Bosporus Kingdom, and Greek philosopher Diogenes (412 B.C.). From this place Apostle St. Andrew the First Called, the first preacher of the Sacred Gospel in the Slav Land, came to Rus'. This is one of the most ancient cities in Asia Minor. The story says that it was founded by Amazons as the Greek colony in the seventh century B.C. That time its population reached 60,000. It was situated in a convenient bay. In the Middle Ages it was one of the major military ports. S. was the capital of the Pontus Kingdom (183–64 B.C.). During the Caesar ruling it was the Roman colony. Thanks to a convenient harbor S. flourished for a long time as the major trade center on B.S. From the fifteenth century it was used by the Turks mostly as a naval base which they called Sinub. On 18(30) November 1853 at the start of the Crimean War, in the battle at Sinop, the Russians under the command of Admiral Nakhimov destroyed the Ottoman Fleet.

At present the population of S. is 38,600 (2012). This is the trade center, the crossing point of automobile roads. The construction of small vessels and the forestry industry are developed here. Tobacco, fruits and timber are exported via the port. The city may boast of numerous archeological monuments: the foundation of the Roman Temple of Serapis, the citadel, the armory, the Alaeddin Mosque (1267), the Alaiye Medresse (thirteenth century), the library of Riza Nur, the ruins of the Byzantine Fortress of Balatlar and other places of interest. One of the oldest archeological museums (1921) is also found here. The Sinop sea roads are considered one of the best and safest on the Anatolian coast of B.S. The port has two wharfs that are able to receive large- and small-tonnage vessels. Hamsilos Fjord located not far from the city center is the only fjord in the country.

Sinop (http://www.netgazete.com/files//sinop_854162104.jpg)

Sinop Bay – locates southward of the Boztepe Peninsula on the Turkish coast of B.S. The coast of the bay drops steeply into water, except the strip near the city of Sinop where the remnants of ancient piers formed a reef. The water depth in the bay is 10–60 m. The sea roads in S.B. provide the only large comfortable anchorage place near the Anatolian coast. This place is commemorated in the history because of the battle at Sinop that occurred here in 1853.

Sinop Peninsula – a peninsula located in the central part of the Turkish coast of the Black Sea.

Sinop Province – the Black Sea II of Turkey. Its area is 5,862 km^2, population— 201,100 (2011). The center is Sinop city. Its adjacent provinces are Kastamonu on the west, Corum on the south, and Samsun on the southeast.

Sirko Ivan Dmitrievich (c. 1610-1620–1680) – the Cossack military leader. In September-October 1663 the united forces of the Cossacks headed by S. and dragoon of Russian voivode Grigoriy Kosagov stormed and seized the fortress of Perekop. In 1667 S. taking the occasion that the main Khan troops together with the detachments of Petro Doroshenko, Hetman of Right-Bank Ukraine, were fighting that time Rzech Pospolita, defeated in autumn 1667 the garrisons of Perekop and headed deeper into the Crimean Khanate. For 20 years the Cossacks led by S. made numerous raids to the Crimean Khanate. In spring 1672 S. claimed to be the Hetman of Left-Bank Ukraine and instead he was sent to Siberia by Tsar Aleksey Mikhailovich. A year later the Moscow state together with its ally Rzech Pospolita faced great difficulties. In spring 1672 the united forces of Sultan of Turkey Mehmed IV, Crimean Khan Selim Girey and Hetman Doroshenko with their 300,000-strong army initiated hostilities against Poland and Moskovy. In this situation Tsar Aleksey Mikhailovich had to recall former Ataman S. from exile. In 1675 Mehmed IV was ready to start a new campaign against Zaporozhie Secha, but before starting off he sent to the Cossacks a letter demanding to vow voluntarily loyalty to him as "the invincible warrior". And the Cossacks sent their famous reply to Sultan. But this was not an end to this. In revenge for an attempt to seize Sech, S. with his 20,000-strong detachment in summer of 1675 invaded Crimea, defeated the 50,000-strong Crimean troops at Sivash, took hold of rich loot, sent free many prisoners among which there were about 7,000 Christians.

Siutghiol Lake – one of the largest freshwater lakes in Romania. It extends over 20 km^2 and has a maximum depth of 18 m. In winter, up to 90 % of its surface may be covered by ice.

Sivash (in the Turk meaning "rotten sea") – a system of salty bays of the Sea of Azov; a vast, but very shallow liman near the western coast of A.S., Ukraine. Geographically it is a part of the Crimean Peninsula. It was mentioned by Ancient Greek geographer Strabo (first century A.D.) as Sapro Lake or "rotten" lake called so due to the rottening smell of water. It stretches from north to south for 115 km and from west to east for 160 km. Its area is approximately 2,560 km^2. It is separated from A.S. with a narrow (from 270 m to 7 km) sand bar—Arabat Spit. It is connected with the sea via the Genichesk (Thin) Strait. In this wetland territory there are many small and three large islands, they are Russkiy and Kuyuk-Tyup in Western Sivash and Kiyanly in the north of Eastern Sivash. The Chongar Peninsula and the Tyup-Dzhankoy Peninsula coming here from the Crimean side, with the bridges and dikes constructed here divide S. into Western and Eastern. In the south of Eastern S. there is a small spit and behind it a small, often drying out shallow bay called Alekseevskiy "zasukha" (drought).

The coasts are meandering, low, flat, swampy, forming numerous bays, peninsulas ("tyups") and solonchaks ("droughts"). They are covered with salt. Sea winds drive water on mainland. During strong western winds and severe heat the water level in S. drops significantly revealing barren solonets soil covered with viscous and impassable silt. The local people call this soil "sivashes". When the western winds drift water from S. to A.S. the S. floor gets exposed, at times so significantly that the coastline recedes for 5–6 km and even more. The eastern winds drive water from A.S. into the lagoon, so S. receives much water and floods vast expanses of low-lying coastal lands. In such cases its area increases 10 %.

S. is not subject to the mitigating impact of B.S. on Crimea. The mean air temperature in the S. area in July is +24 °C, in January from −2 to −4 °C. But the heat in summer may reach +44 °C and the winter frosts −33 °C. The snow cover in the Sivash area is observed every winter, but it is not deep and stays for a month and a half. Short-time temperature rise may involve melt of snow. The water temperature in summer in S. (keeping in mind its small depths) may be as high as 30–35 °C and in winter it may drop to − 1 or 2 °C, while in the southern part—even to −3 °C (due to high salinity). Therefore, the northern part of S. gets frozen, but only for 2 weeks. The precipitation in the S. area is about 300 mm. Every year S. receives to 1.5 km^3 of water from A.S. Due to high evaporation the water in S. turns into the saturated salty solution—brine with the salinity to 170‰. S. possesses various chemical resources: table salt, magnesium sulfate, sodium sulfate, bromine, etc. Since old times the saltworks had existed here. The Sivash brines are also used for recovery of mirabilite by the sedimentation method.

S. abounds in aqueous vegetation. In the eastern part the sea grass (damask) dominates, while in the western part—cladophora. These plants concentrate near coasts and while decaying release great quantities of hydrogen sulfide. Through the whole summer S. and its nearby territory are enveloped in heavy rotten odor vindicating the name of this lake—"rotten". In 1947 some islands in S. were made nature reserves. They are Kitai, Kuyuk-Tuk and Russkiy islands. Numerous nestling and migrating birds stop here for rest or live here. Especially plentiful are silver gulls, gadwalls and shelducks. High air temperatures in summer, partial contamination of water with hydrogen sulfide and constantly changing salt concentrations are responsible for inhibition of the flora and fauna development. The flora of S. is represented by slightly saline-aqueous (20–30‰) vegetation—sea grass, zoostera, algae (red, green and blue-green); saline (30–60‰) vegetation—zoostera, algae, but in less quantities; excessively saline (50–240‰) vegetation—green algae cladophora.

In the fauna fish species are prevailing. They may be divided into three groups: permanent inhabitants of S.—flatfish (glossa) and grass bullhead; fish running to S. for rather a long period for fattening—gray millet, black millet, Azov anchovy, herring aterina, garfish, monkey goby, horse mackerel; fish accidentally running into S. areas with the high salt concentration—Artemia salina and Chironomus salinaris. S. provides only a small fraction in the total fish product of A.S.

Skadovsk – the town and port on the coast of the Karkinit Bay; the center of the Skadovskiy District, Kherson Region, Ukraine. The city was founded in 1894 on the lands of landholder Skadovskiy and was named after him. This is the climatic seaside resort for children. The commercial-scale farms for growing Black Sea oysters are organized near S. in the Dzharylgach Bay and in the Egorlyk Bay. The resort has very good beaches surrounded by sanatoriums and rest houses. The port of S. locates southward of S. on the northern shore of the Dzharylgach Bay. The port water area is "bucket"-shaped the entrance into which goes between two jetties. The port has four berths, including for bulk oil and dry cargo. Population—19,100 (2011).

Skalovskiy Ivan Semenovich (1776–1836) – Russian Rear Admiral. In 1791 he started his service on the B.S. Fleet as a midshipman. In 1798–1799 with the squadron of F.F. Ushakov he sailed to the Mediterranean Sea. He took part in storming the fortresses on the Vido and Corfu islands. Commanding brig "Alexander" he distinguished himself in the war with France (1804–1807). In the

Russian-Turkish War of 1828–1829 he showed himself well during storming of the fortresses of Anapa and Varna. In 1829 he successfully fought enemy transports near the Anatolian coast. S. was promoted to Rear Admiral. In 1830 he was transferred to the Baltic Fleet and in 1834 from the Baltic to the Black Sea Fleet.

Skates (*Rajidae*) – the suborder of the *Selachii* order. These are mostly marine, rarelym freshwater species. They have a flattened body shaped like a disk or rhomb. Many S. prefer living in the coastal areas of the tropical and subtropical seas. They are native to the Barents, White, Black and Far-Eastern seas. Some S. possesses electrically charged organs. By striking with its tail having barbs it can make wounds where poison is spit. Such wounds are very painful. Some of these species are fished in commercial scales.

Skate (http://photo.planetakrim.com/big/906.jpg)

Skopintsev Boris Aleksandrovich (1902–1983) – Soviet chemical hydrologist, professor. He studied marine chemistry of organic matter, oxygen and anaerobic zones of the World Ocean. He was one of the first to investigate the water chemistry of B.S. and its hydrological conditions. His main works are "Formation of Modern Chemical Composition of Black Sea Waters" (1975), "Organic Matter in the Ocean Waters" (1979).

Skrudzh Bay – it incises into the coast between the Iskuria Cape in the west and the Tamysh Cape in the east, Georgia. This is a small bay open for winds and waves from the sea. The Tsurgili River flows into it. The village of Akhali-Kindgi locates on the open plateau on the eastern coast of the bay.

Skrydlov Nikolay Illarionovich (1844–1918) – Russian Admiral. In 1864 he finished the Naval Cadet Corps. As a member of the Guard he took part in the

Russian-Turkish War of 1877–1878. He commanded the mine boat of the Danube Military Flotilla. In 1877 he commanded the ship "Karabia". From 1893 he served as junior flagman on the Baltic Fleet. In 1898 he commanded the squadron in the Mediterranean and in 1900–1902—the Pacific squadron. In 1903–1904 he was appointed the chief commanded of the B.S. Fleet and ports. After death of S.O. Makarov he went to serve as the fleet commanded in the Pacific Ocean. In 1906–1907 he was called back to the position of the chief commanded of the B.S. Fleet and ports. In 1909 he was promoted to Admiral.

Slaschev Yakov Aleksandrovich – Russian Colonel (1916); Major General (1919); Lieutenant General (1920). In 1903 he finished the nonclassical college in Saint-Petersburg, in 1905 the Pavlovo Military College and in 1911 the General Staff Academy in Nikolaev. He took part in World War I where he was wounded five times. In the White Movement he was in charge of formation of the Voluntary Army on an order from General Alekseev in the Mineralnye Vody area (January-May 1918). He commanded the First Kuban Infantry Brigade and was the Chief of Staff of the Second Kuban Cossack Division (April-August 1919), the Third Army Corps (December 1919–February 1920). He participated in defense of the Perekop Isthmus on 27 December 1918 having anticipated the invasion of the Red Army into Crimea. He defended Perekop and Chongar which permitted to prolong the existence of White Russia for nearly a year. He commanded the Crimean (former Third) Corps (February-April 1920), the Second Army Corps (former Crimean, renamed by General Vrangel) (April-August 1920). On 18 August 1920 by order of General Vrangel S. as a hero of the Crimea defense was named from now on Slaschev-Crimean. In the Russian history of Crimea he was the second man [the first was Prince V.M. Dolgorukiy (1722–1782)] who was honored to bear this title. On 18 August 1920 Vrangel dismissed him and dropped from command of the corps. In November 1920 he evacuated from Crimea. From November 1920 to November 1921 he was in emigration. In November 1921 he returned to Russia. In June 1922–January 1929 he lectured at the courses "Vystrel" (Shot). On 11 February 1929 he was killed in his room in Lefortovo in Moscow. In 1990 his book "White Crimea. 1920. Reminiscence and Documents" was published in Moscow.

"Small London" – in this way the British based during the Crimean War in Balaklava called it.

Snezhinskiy Vladimir Apollinarievich (1896–1978) – Soviet engineer, Rear Admiral, Doctor of Military Sciences, hydrometeorologist and oceanographer, explorer of the White, Black and Azov seas. In 1918 he finished the Special Midshipman School in Petrograd. In 1918 he volunteered into the Navy and till 1921 he served as the navigation officer, senior captains' mate, commander of the destroyer "Deyatelnyi" (Active), took part in the battles on the Volga and Caspian Sea. From mid-1928 and for 10 years onward he headed the Hydrometeorological Division of the Department on Safe Shipping over B. and A.S. ("Ubekochernaz"), the Sevastopol Marine Observatory (SMO) and the Seismic Station "Sevastopol" of the USSR AS. Pursuant to Resolution of the Central Executive Committee of the

USSR Soviet of the People's Commissars of 1929 "On Integration of the USSR Gidromet Service", in 1931–1933 he was a military representative in the Committee on Organization of Gidromet Service of Crimea and Gidromet Committee of Crimea. By 1936 the network of basic hydrometeorological stations on B. and A.S. were provided with modern facilities, the methods of synoptic analysis, typization of atmospheric processes was applied in the practical activities of the weather forecast service, regular aerological observations were organized. In 1932 the synchronous pilot-balloon sounding of the atmosphere on land (Sevastopol, Yalta) and on sea was conducted for the first time in the history. In 1928 he led the multi-purpose expeditions of SMO conducted after the disastrous earthquake in Crimea in 1927. In these expeditions using the large core sampler they managed to receive the record long ground sample—485 m. In 1929–1938 he headed the annual oceanographic expeditions to B.S. and A.S. In 1930–1931 he studied the relief of the B.S. floor for placement of the underwater cable communication lines. In his work "Relief of the Black Sea Floor" (1936) he pioneered the development and application of the concepts on geomorphological study of the sea floor (in 1946 he used it for defending the candidate dissertation). He organized regular year-round hydrological profiling that laid the basis for stationary oceanographic observations on B.S. The observations of the currents started still in 1923 were continued. Using the obtained results he prepared the first schematic map of permanent currents that was included into the B.S. Pilot of 1931–1932. The first Physiographic Atlas of B.S. was prepared that became a prototype for presentation of data on other seas. In 1936 he participated in preparation of the new bathymetric map of B.S. published by the State Hydrological Institute. His activities were closely connected with the practical needs. He initiated creation of the hydrometeorological service on B.S. that survived the ordeals of the Great Patriotic War. In 1938–1941 S. served in the Naval Hydrographic Department. From 1943 to 1964 he headed the Hydrometeorology Chair (later the Chair of Military Hydrography and Navigation) in the Naval Academy. In 1954 he became the USSR's first doctor of naval sciences. In 1963 he was conferred the rank of Engineer Rear Admiral. S. published over 100 scientific works and wrote many physiographical essays and sections in pilots and other specialty publications. In 1954 he was awarded the F.P. Litke's Gold Medal. He also had other orders and medals. Among his major works there are "Relief of the Black Sea Floor" (1936), "Practical Oceanography" (1951, 1954), "Marine Hydrometeorology" (1962), "Oceanographic Devices" (in co-authorship, 1975).

Sochi (in the Adygean "*shacha*" meaning one of local tribes) – the seaside balneological and climatic resort of the federal significance, Krasnodar Territory, Russia. This is the southernmost point of the northernmost subtropics in the world. Its area is 3,506 km^2. Population is 368,000 (2013). Greater Sochi sprawls along the shores of B.S. in a narrow strip for nearly 150 km stretching from the Shepsi River in the northwest to the Psou River on the border with Abkhazia and this against the background of the snow-capped peaks of the Caucasus Mountains. It includes the

Adler, Khosta, Central and Lazarevskoe regions with their numerous resort towns from Magri in the northwest to Adler and Krasnaya Polyana in the southeast (Greater S. within the borders of 1961). It has a sea passenger port and a large airport (in Adler).

The resort territory is surrounded in the northwest, east and southeast by the Greater Caucasus ridges: Alek, Mamaiskiy, Bykhta, Mount Akhun and others (from 300 to 1,100 m over the sea level) that run down to the sea like an amphitheatre, protecting the coastal strip from cold northern and eastern winds blowing in winter. In the central area the coastline forms the so-called Sochi Cape occupied by the city center. To the northwest of the center the New Sochi with a group of sanatoriums is located. The lower part of the city is situated in the Sochi River valley. To the southeast of the center, on the southwestern slopes of the Vereschaginskiy and Bykhta ridges there are complexes of large sanatoriums, parks that are integrated into an entire system by the main highway—Kurortnyi Avenue. Matsesta district situated in the valley of the river with the same name is southward of the Bykhta Ridge. Numerous mountain rivers and springs flow over the resort territory into the sea.

The resort territory is overgrown with the evergreen varieties (to 70 % of the park area) of trees and shrubs typical of subtropics, they are cactus, cedar, cypress, magnolia, palm, eucalypt and others. In total, there are about 900 (in the parks—to 2,000) varieties of trees and shrubs many of which are brought here from Australia, Japan, Mexico and other countries. The dendrarium, Park "Riviera" may boast of the richest flora and very good landscaping. In the mountain-forest belt on the slopes of the Greater Caucasus the forests of fir, silver fir, beech, hornbeam trees as well as wild fruit and nut trees are growing. Still at higher altitudes there are sub-Alpine and Alpine meadows.

The climate here is humid subtropical. The solar radiation (about 2,000 sunshine hours a year) is very intensive, especially in summer. The precipitation is about 1,400 mm a year. The winter is mild, with changing weather and frequent long rains; the average January temperature is +6 °C, on a sunny day at noon the temperature may reach +15 to +17 °C and even more. The most dry and sunny month in winter is December: the number of sunshine hours is 90, the average relative humidity is 69 %. In the period from November through March over 700 mm of precipitation fall. The spring is early with unstable rainy weather. In March the average daily temperature is +10 °C. But in general the spring is much colder than autumn. The summer is very warm with the prevailing clear and nearly cloudless weather. The precipitation is the minimum in late spring and early summer. Quite often are the days with the hot and dry weather. Beginning from June the daytime temperature does not drop below +20 °C; in July and August it is often as high as +25 to +28 °C and even more. But the effect of summer heat is smoothed by sea breezes.

Apart from climate the main natural curative factor of this resort is sulfide sodium-chloride natural waters of Matesta that are used for baths, sprinkling, inhalations, etc. About 30 wells are drilled on the resort territory. In addition to the sulfide natural springs the resort has also the carbonate mineral water springs in Krasnaya Polyana. Very healthful is the carbonate waters "Sochinskiy narzan" (its spring is in Chvizhepse) containing arsenic that are used in curative baths and for drinking.

S. is nearly 150 years old. On 23 April 1838 not far from the present-day lighthouse the marine unit landed in the Sochi River mouth and founded the fort of Alexandria named after Empress Alexandra. A year and a half later N.N. Raevskiy Jr., who commanded that time the Black Sea Coastal Line, applied to the Military Minister with a request to rename the fort of Alexandria into Navaginskiy fortifications in honor of the Navaginskiy Regiment that glorified itself in the battles with the Turks. After the end of the Caucasian War the Navaginskiy fortifications and the B.S. fortresses have lost their military significance. A civil settlement was constructed around the former fortress in which after institution of the Black Sea Governorate the simplified city management was practiced. This place was named Sochi. In 1896 Sochi was made the district center of the Black Sea Governorate.

The basis for the resort development in Sochi was laid at the turn of the nineteenth and twentieth centuries. Doctor V.F. Podgurskiy who came to S. in 1898 found that the local population quite successfully used the primitive Matsesta baths (the pits dug in ground were filled with mineral water warmed by the sun and heated stones) in treatment of the locomotor diseases. In 1898 the task commission comprising Professors A.I. Voeikov, F.I. Pasternatskiy and mine engineer M.V. Sergeev examined the mineral water sources, climate peculiarities and geological structures of the B.S. coast in the vicinity of S. The meeting of the Russian Geographical Society considered the materials of Professor A.N. Krylov who called the B.S. coast near Sochi the "Russian Riviera". From this time on the resort boom was on the rise. Nobility, large industrialists and prominent merchants having learnt about prospects of this area rushed purchasing land sites there, started construction of country houses and set up commercial enterprises. From the early twentieth century S. was developed as a resort city. In 1902 the development of the sulfide sodium-chloride mineral waters of Matsesta was initiated for treatment purposes. 14 June 1909 is usually considered the birth date of the S. resort when the complex "Caucasian Riviera" was opened. It included two 4-storey hotels with a theatre, café and restaurant. In 1911 the bath building comprising 18 booths was put into use. The first years of the "Caucasian Riviera" operation brought enormous profits to its owner—Moscow businessman A.G. Ternopolskiy who without any delay constructed one more 43-storey dwelling building and a building for water-, light- and electricity-based therapy. Surrounded by luxury country houses of the celebrities S. as it is looked rather miserably.

The apex of S. development was in the Soviet period. After establishment of the Soviet power and nationalization of all resorts in the country S. was developing at a quick pace. After the Civil War the first sanatoriums were opened in the nationalized country houses, villas and rest homes. In 1925 S. acquired the status of the resort of state significance. In 3 years many treatment and health-improvement establishments were operating here. The first balneological building was constructed in Matsesta. In 1933 the government approved the first Master Plan on reconstruction of the Sochi-Matsesta resort area. In 1936 the construction of the Stalin Avenue, the main thoroughfare of this resort, was completed. It connected the sanatoriums and living quarters of the city. The research base of the resort was continuously improved. In 1936 the Research Institute of Balneotherapy and Physiotherapy was opened. In the years of the Great Patriotic War S. turned into the major hospital base of the country. About 11 hospitals functioned here.

In the postwar years the attention to the resort development was not declined. Many sanatoriums here specialized on treatment of the people with blood circulation and locomotor, nervous, skin, gynecological and other diseases. In 1967 the second Master Plan of the Greater Sochi development was elaborated. It included sanatorium complexes in Central, Adler, Khosta and Lazarevskoye districts. This master plan envisaged creation of eight resort groups on the territory of Greater Sochi: Magri, Ashe, Lazarevskoe, Golovinka, Loo, Dagomys, Sochi, Adler. And each group should have balneo- and climatotherapy centers, sanatoriums as well as boarding homes, holiday inns, hotels, tourist bases, resort polyclinics, bath complexes, cabinets for electricity- and light-based treatment, swimming pools, cinemas, winter theatre, library, etc. It was planned to develop seaside parks on the basis of the existing ones and to lay promenade avenues, paths, terrain cure routes.

On 10 February 1961 a part of the Adler and Lazarevskoye districts was included into S. The industry of S. serviced the needs of the resort. In 1980 in commemoration of 35 Years of the Great Victory S. was awarded the Order of Patriotic War 1st degree.

About 30 tourist routes cross S. Among the places of attraction for tourism and excursions there are the Akhun Mount (663 m high over the sea level) with the 30-m viewing tower, the Agurskiy waterfalls, Orlinyi rocks, yew-boxwood grove in Khosta, the park with subtropical vegetation in Adler, Vorontsov caves, Krasnaya Polyana, Ritsa Lake. In S. you can find the Historical Museum of Sochi, the Literary and Memorial Museum of novelist N.A. Ostrovskiy, the Art Museum, the Historical Museum of the Adler Area, the Ethnographic Museum (in Lazarevskoe), the Museum-Dacha of Singer V.V. Barsova, the Garden-Museum "Tree of Friendship". Very popular among the visitors is the S. dendrarium. The sanatorium and resort complex in S. is the Russia's largest. The city concentrates over 50 % of the whole resort potential of the Krasnodar Territory. About two million people visit Greater Sochi each summer. In early 2000 the work on elaboration of the Master Plan "Mountain-Marine Complex "Krasnaya Polyana" was started. Sochi will host the XXII Olympic Winter Games and XI Paralympic Winter Games in February 2014, as well as the Russian Formula 1 Grand Prix from 2014 until at least 2020. It is also one of the host cities for the 2018 FIFA World Cup.

Sochi (Photo by Evgenia Kostyanaya)

Sochi Nature National Park – it is located northward of the city of Sochi, on the northwestern slope of the Greater Caucasus, Krasnodar Territory, Russia. The park covers 190,000 ha. It was founded in 1983 for preservation of the natural complexes of higher-elevated zones in Western Precaucasia, the typical flora and fauna of the mountain-forest and mountain-meadow landscapes, the monuments of nature, history and culture, for organization of cognitive tourism.

Sochi (Sochinka) River – a small river which flows into the Black Sea in the Central District of Sochi, where the banks are taken into cement. Other settlements on its course are Baranovka, Plastunka, Orekhovka and Azhek. The length of the river—45 km, drainage area—296 km^2, average river runoff at the village Plastunka is of 15.6 m^3/s.

Sogdaya – see Sudak.

"Solemn League" (1684–1698) – the union of some European states against Turkey. It was based on the anti-Turkish defensive treaty of 1683 between Austria (that headed the Holy Roman Empire) and Rzech Pospolita (Poland). In 1684 Venice joined them. The league members thought it appropriate to drag in Russia that was also interested in this union. Russia had to defend itself from Turkey and Crimean Khanate that claimed the lands of Left-Bank Ukraine, Zaporozhie, Lower Don and Kabarda being a part of the Russian state. By joining this league Russia wanted to force Poland to abandon its claims to the Ukrainian lands and Kiev. In 1686 the Russian government succeeded to force signing in Moscow of the Eternal Peace Treaty that settled the relations between Russia and Rzech Pospolita. In 1686 Russia joined the league and pledged to break the relations with Turkey and Crimean Khanate and to organize a military campaign against them together with its allies. During the war of the anti-Turkish coalition the troops of Austria and Venice defeated the enemy in Venice, Dalmatia and on B.S. Being faithful to its ally commitments Russia sent its armies in 1687 and 1689 against the Crimean Khanate and in 1695 and 1696 to Azov drawing off the Turkish forces from other fronts. The league proved to be unstable and in 1698–1699 during the Karlowitz Congress it broke down due to the contradictions among its members.

Solkhat (in the Tartar "*sal*" meaning "left" and "*khat*" meaning "side") – a Crimean town first mentioned in 1263. This town was populated with Kipchaks, Russians and Alans. In the Genoese sources the town was called Solkhat, in the Tatar—Crimea. While speaking about economics, first of all, about trade, the name Solkhat was used, but while speaking about politics—Crimea. Some researchers connected the name Crimea with the word "ditch". After appearance of Crimean Yurt (Crimean province of the Golden Horde) S. became the home for the Emir of Crimea from where he ruled the territories of Yurt. Later it was decided to construct the residence for the vicegerent of the Golden Horde near the trade town of S. As a result, the residence together with S. was surrounded by ditches. Perhaps after in this way the Crimea appeared. Afterwards the name Crimea was applied to the whole peninsula. Initially S. was the political and administrative center of Crimean Yurt.

Southern Bug – (ancient Greek name—*Hypanis*) river in the south-west of Ukraine. Length is 792 km. Catchment basin area 63,700 km^2. Originates in Volyn-Podolsk upland, falls into the Dnieper-Bugskiy Liman of the Black Sea. In the upper reaches, it flows within a wide valley with low gently-sloping sides and a water-logged flood plain. There are many dikes and mill dams. Downstream of Novo-Konstantinovo, the river cuts into granites, its valley becomes narrow, sills and rapids being formed in the river channel. Where granites occur at a greater depth, the flow becomes quiet. Within the segment from Pervomaisk City to Aleksandrovka Village (over 70 km), the banks are high (up to 90 m) almost its entire length they are rocky and steep, and the river not infrequently forms sills (Migei, Bogdanov, etc.). Downstream of Aleksandrovka Village, S.B. flows along Circum-Black-Sea Lowland, its valley and channel widen. The alimentation is mainly snow and rains. The river breaks up in mid-March, freezes from early November to Mid-January. In some years, freeze-up is unstable. The main tributaries: Volk, Zgar, Rov, Kodyma, Chichikleya—on the right; Sob, Sonyukha, Mertvovod, Ingul—on the left. In the S.B. basin there are over 4,000 lakes and storage reservoirs of around 270 km^2 total area. Regular navigation begins from Voznesensk City. Steamer traffic exists with Kherson City (on the Dnieper) via the Dnieper-Bug Liman. The following cities are sited on S.B.: Nikolaev, Pervomaisk, Gaivoron, Khmelnitskiy, Khmelnitsk, Voznesensk. In the estuary, they fish pike perch, pike, goby, roach and other fishes. Used for irrigation. There is a hydropower plant.

Southern Coast of Crimea – a term (YuBK—Yuzhny Bereg Kryma in Russian) which combines the southern coastal zone of Crimea from Sevastopol on the west to Feodosiya on the east. SCC is one of the crucial and most popular maritime zones of health-resort therapy, holiday-making and tourism, sited between the Black Sea coast and the Main Range of the Caucasus Mountains. Includes health-resort cities of Yalta, Alushta and numerous villages and locations. The length of SCC is around 150 km: from Aiya Cape to Karadag Massif in the east. Narrow—from 2 to 8 km— gently-sloping and hilly terrain structured by shales and limestones is limited in the north by Yaily Ledge—Main Ridge of the Crimean Mountains, ending abruptly toward the sea in the segment from Aiya Cape to Alushta (this area is usually called SCC proper). In the vicinity of Gurzuf, there is a peculiar dome-shaped laccolith mountain Ayudag (Bear-mountain), near to Planerskoe—ancient volcanic massif Karadag. Quite common are landslides, heaps of large debris of crystalline rocks— "chaoses" (e.g. Alupka chaos). Despite the diverse natural vegetation of Mediter- ranean type (oak and juniper woods with a lower storey of evergreen and deciduous shrubs; on the Yaila slope—woods of beech, oak, Crimean pipe), the SCC land- scape is determined in large measure by ornamental, including exotic, garden-park plantations: fountain-tree and cedar of Lebanon, cypress, cherry laurel, magnolia, ivy; cultivated are pears, apricots, peaches, fig trees, persimmon; here are large vineyards; there is also Californian giant sequoia, Spanish broom and other tree species. East of Alushta, the Main Ridge of the Crimean Mountains grows lower and starts retreating from the shoreline. The space between Alushta and Sudak is a cross-country, featuring alternation of small plateau-shaped mountain massifs and basins going down to the sea as pebble beaches. The short mountain rivers: Ulu-Uzen with Golovin waterfall, Biyuk-Uzen with Jur-Jur waterfall, Sudak and others. The coastal ranges and hills are covered with shrub vegetation. The wide potholes are used for vineyards.

The primary natural curative factor—subtropical climate of Mediterranean type. The winter is mild, rainy. Mean temperature of January 4 °C, daytime—7 °C; during any winter month, days with a temperature of 15–20 °C are not uncommon. The bulk of annual precipitation falls in November-March (400 mm; in April- October—under 240 mm). In winter, dull weather prevails. Days with strong (15 m/ s) easterly wind, storms are likely. In winter, mostly unstable windy weather is observed. Light frosts are possible until the end of March. Mean daytime temper- ature of April is 14 °C. The summer is long, sunny and dry. Mean temperature during July-August is around 24 °C, during daytime hours 28 °C (absolute maxi- mum 39 °C). The heat is mitigated by sea breezes. Precipitation falls mostly in the form of showers; the mean number of days with rainfall in July is 7, with thunder storms—5. Sometimes, northerly winds drive warm surface water away, which results in an abrupt cooling of water (coastal upwelling) near the shore (by 10 °C and more) with subsequent slow temperature rise. Dry and sunny autumn is the best season of the year. Mean temperature in October is 14 °C, daytime temperature is around 19 °C. In September, like in August, rainfall is minimum. Mean temperature in October is 14 °C, daytime temperature is around 19 °C. In September, like in August, precipitation is minimum. The first light frosts occur at the end of November. The bathing season lasts from June to October, the temperature of sea water in July-August is up to 23–26 °C. The number of sunshine hours equals

2,200–2,360 per year. The special features of SSC eastern part are hotter and dry weather in summer, lower relative air humidity in autumn (53 % at Sudak in September).

Treatment mainly of the diseases of respiratory organs, cardio-vascular and nervous systems, disturbances of metabolism. Along with climotherpy and thalassotherapy, diseases of the digestive organs are treated with drinking mineral water from a source in the Melass health-resort location, treatment with grapes (September-November).

Southern Federal District – originally called North-Caucasus District, one of the eight federal districts in Russia. Instituted on the territory of the Russian Federation pursuant to Decree of the RF President No 849 of May 13, 2000. The SFD comprised the Republic of Daghestan, Republic of Kalmykiya, Astrakhan Region, Republic of Adygeya, Kabardino-Balkariya, Ingushetiya, Karachaevo-Cherkessiya, North Osetia-Alaniya, Chechnya; territories: Krasnodar, Stavropol, Regions: Volgograd, Rostov. Center—Rostov-on-Don. On 19 January 2010 North-Caucasus federal district (with center in Pyatigorsk) was split from the S.F.D. Since 2010 S.F.D. has sea and land borders with Ukraine on the west, with Kazakhstan on the east, with Abkhazia and North-Caucasus F.D. on the south, and with Central and Privolzhie F.D. on the north. On the west it is washed by the Black and Azov seas, on the east by the Caspian Sea. Thus, today S.F.D. includes Republic of Adygeya, Astrakhan Oblast, Volgograd Oblast, Republic of Kalmykiya, Krasnodar Territory and Rostov Oblast. Territory—416,840 km^2, population—13.9 million (2013).

Southern Fortresses Building Expedition – established in 1792. The fortresses under its purview included Kinburn, Ochakov, Simferopol, Feodosiya, Sevastopol and others. Under the decree of Catherine II of November 10, 1792, the expedition was headed by Prince A.V. Suvorov. Emphasis was laid on the construction of Sevastopol fortifications; the plan of construction had been developed by Lieutenant-Colonel F.P. De Wollant. By 1796, eight coastal batteries were built in Sevastopol—Konstantinovskaya, Alexandrovskaya, Nikolaevskaya, Pavlovskaya as well as Batteries No 1, 2, 4 and 5 sited on the capes of the same name. On January 10, 1798, pursuant to the decree of Pavel I, SFBE was dissolved.

Southern Smaller Stickleback (*Pungitius platygaster platygaster*) – of the sticklebacks family (*Gasterosteidae*) small fish up to 7 cm in length. Common in the basins of the Azov and Black Seas. The male builds a nest of underwater plants and protects it until the larvae pip out.

SOVCOMFLOT – is the largest Russian shipping company offering a full range of crude oil, refined petroleum products and liquefied gas transportation services. It is one of the world's leading tanker owners. The company actively participates in key Russian oil and gas development projects including operations in harsh ice environment, as well as successfully competes in the international maritime shipping markets. Its fleet is amongst the five leading tanker companies in the world: 158 vessels with 12,54 million dwt. Sovcomflot provides a full spectrum of services for the safe and reliable seaborne transportation of energy to its customers:

(1) Operating crude oil tankers in Suezmax (120–200,000 dwt) and Aframax (80–120,000 dwt) segments; (2) Product tankers (45–47,000 dwt), chemical carriers (5–20,000 dwt); (3) Liquefied natural gas and petroleum gas-carriers; (4) Ice-class ships; (5) Logistical support for offshore development (shuttle oil deliveries in ice conditions, Floating Storage and Offloading units (FSO) services); (6) Rendering port-related services including management of oil terminals and tugs operations; (7) Technical management of the company's and third party vessels. It includes "Novoship" Company located in Novorossiysk.

Soviet Socialist Republic of Tavrida – it was formed on 21 March 1918. It was included into RSFSR. On 7–10 March 1918 the Foundation Congress of Soviets, Revolutionary Committees and Land Committees of the Tavria Province was held in Simferopol. The Congress proclaimed the Tavria Central Executive Committee to be the supreme revolutionary power body. It also passed the resolution on implementation of the decrees of the Soviet power. On 21 March 1918 the Tavria CEC adopted the Decree on SSRT establishment. Its government was headed by Polish communists Naftali (Anatoliy Iosifovich) G. Slutskiy ("Anton"), the SSRT CEC member, and Yan Yu. Tavratskiy, chairman of the Communist Party Province Committee, sent from Petrograd by the Communist Party Central Committee. But SSRT existed for only a month. On 18 April 1918 the German troops invaded Crimea. People's Commissars were seized by the rebellious Crimean Tatars and the White Guard and on 24 April 1918 they were executed. On 30 April 1918 SSRT was collapsed. The monument to the members of the SSRT government was constructed in Alushta.

Sozopol – the seaside climatic resort in the Bourgass Region, Bulgaria. It is situated on the rocky Kiril Peninsula. Climate is marine here; the mean annual temperature is +13.3 °C. The winter is mild, practically without snow; the mean temperature in January is about +3 °C. The summer is very warm, dry and sunny; the mean temperature of July is +24 °C. The precipitation is over 500 mm a year. The number of sunshine hours is 2,350 a year. Mild climate, warm sea and beautiful sandy beaches provide opportunities for climato- and thalassotherapy here. Rich history of S. attracts here many tourists. S. is the center of summer rest and tourism. This is the starting place for many interesting sightseeing tours over the Bulgarian coast. Approximately in 610 B.C. the Greeks from Miletus founded a colony here and called it Apollonia on account of a temple dedicated to Apollo and containing a famous colossal statute of the God of Apollo (its height is about 14 m). The archeologists uncovered the necropolis of the fifth to third centuries B.C., the gravestone of Anaximander (about sixth century B.C.) and ceramic pottery. The ruins of some fortress walls also survived. The town has an archeological museum. Not far from S. one can see small, but very picturesque islands—St. Ivan, St. Peter, St. Cyricus (the latter is connected to the mainland via a jetty). Here there is a museum containing the Europe's riches collection of stone anchors.

Sozopol (http://carsburgas.com/wp-content/uploads/2012/03/sozopol1.jpg)

Sozopol Bay – is situated between the narrow Kiril Peninsula 3 km westward of the bay and the Chrisosotira Cape; a part of the Bourgass Bay, Bulgaria. The southeastern coast of the bay is abrupt; the southern and western coasts are low, flat and overgrown with not dense forests. The Bay shores are slightly incised and drop deeply into the sea. The average water depths in the bay is 11–14 m. The small island of Kiril or Milos is found near the northern tip of the Kiril Peninsula in the northeastern part of the bay. The St. Ivan Island locates to the northeast of the bay entrance.

Special Design and Technological Bureau (SDTB) – it is a part of Marine Hydrophysical Institute of Ukrainian NAS. This is an independent self-supporting organization targeted to intensification of works on automation of hydrophysical and meteorological researches, equipment of research vessels with the modern facilities for collection, processing and transfer of the information on measured parameters of the marine environment and also to development of special technical facilities for measurement of hydrophysical and hydrometeorological parameters from a ship, aerospace and submersible vehicles. It has its own trial production. It is located in Sevastopol.

Special Sea Port "Oktyabrsk" (Ukraine) – it was founded in 1965. It is situated 25 km from the city of Nikolaev on the left bank of the Dnieper-Bug Liman. It is one of the best ports in Ukraine. It is capable to handle to 1 million tons of general cargo. The port has seven wharfs 1.9 km long in total, the railway and automobile access roads. It is capable to receive vessels with the draft of 10.4 m in freshwater and 215 m long.

Spit – (1) a narrow section of land extending into a water body, mostly, perpendicular to the shore. It is formed in meeting places of two sediment flows moved by waves along shores. (2) a strip of deposited ground that often appears at inflow of a tributary into main river, e.g. Arabat Spit in A.S.

Sprat (*Sprattus sulinus*) – small fish of the herrings family (*Clupeidae*). Reaches 6–8 cm in length, hardly ever 10 cm. Mass 8–10 g. Omnipresent in the Black Sea. Schooling pelagic fish, cold-water—in warm months (June-September) keeps away from the shore, sticking closer to deep waters over 30–100 m. During the winter-spring period dwells mostly in waters closer to the shore where it fattens and propagates. Lives 4 years, maximum. S. galores in the Black Sea, but is not fished as extensively as it should be. It is fished in large quantities along the shores of Bulgaria and Romania as well as in the north-western part of the Black Sea. S. has tender, tasty meat, comes on the market fresh, salted. Used for canning. Local name—kilka, sardel. Black Sea kilka contains much fat and for this reason is not suitable for smoking. Therefore, local canning factories fabricate not "sprats in oil", but canned "Kilka in tomato sauce".

St. George Monastery (Balaklavskiy) – locates in Crimea, on the B.S. coast, not far from Sevastopol. It was founded in 891. This is the oldest monastery in Crimea. There is a legend about S.G.M. The Greek merchants sailing over B.S. were caught in a very strong storm. Seeing the inevitable death they appealed to Sacred Great Martyr George the Victorious and they were saved. On a cliff they found an icon of Saint George, took it to the shore and in a rock wall near the place where they had nearly died they built a monastery as a token of gratitude. This legend was first published in the book "About the History of Christianity in Crimea" of A.L. Bertie-Delagard who wrote that this legend appeared in 1862 and was concocted by monastery prior Archimandrite Nikon. But the first official mention of the monastery dates back to 1578. It appeared in recollections of M. Bronewskiy, Polish Ambassador in the Crimean Khanate. Before the time when Crimea was incorporated into Russia the St. George Monastery subordinated to the Constantinople Patriarch and was known only to a few. After 1783 it was often visited by the officers of the Black Sea Fleet among whom there were many Greeks as well as military and civil functionaries visiting Crimea. In 1794 this convent was put under subordination of the Holy Synod and from this time on it was officially referred to as the Balaklava St. George Monastery. On March 23, 1806 the Holy Synod reported to the Emperor about the need to turn this monastery into the basic church for the fleet monk-priests. The staff of the monastery was approved. It included a prior, 13 monk-priests and 14 monks for the fleet. The growing Black Sea fleet needed also more clergymen. That is why from December 1, 1813 their number was doubled, and the St. George Monastery was called from now the fleet monastery. In 1850 the monastery was transferred to class one.

During the Crimean war the monastery was in the hands of the British and French troops. They did not do any harm either to a prior or monks, and the monastery itself was brought under protection of the allied forces. But having spared the monastery, the enemy troops destroyed the monastery farmyard, cut off orchards, vines and forests. Before leaving Crimea the British took with them some most ancient icons from the monastery. After the end of the Crimean war the fleet hieromonks and the brotherhood of the St. George Monastery were awarded the medals "For Sevastopol Defense" and "In Memory of 1853–1856 War" as well as the Pectoral Cross on the Vladimir ribbon "In Memory of 1853–1856 War" that was specially instituted on August 26, 1856 for the clergymen. In early 1860 the Tavria Ecclesiastical Consistory was established.

In different years these places were visited by Pushkin and Griboedov, Emperors Alexander II and Nikolay II. On November 29, 1929 the Balaklava St. George Monastery was liquidated and handed over to the resort administration for the Sanatorium "OSAVIAHIM". From late 1939 to November 1941 the former monastery was given to the military-political courses of the Black Sea Fleet. On July 22, 1993 the Council of Religious Affairs at the Cabinet of Ministers of Ukraine registered the statute of the St. George Friary Monastery of the Crimean Eparchy of the Ukrainian Orthodox Church of the Moscow Patriarchy.

St. George Monastery (http://data.sobory.ru/pic/02900/02946_20090517_210848.jpg)

Stakhiev Aleksandr Stakhievich (1724–1796) – Russian diplomat. In 1745–1748 he served in the Russian mission in Turkey. In 1775–1781 he compelled Turkey to abide by the Kuchuk-Kainarji Treaty of 1774. In 1779 he signed with Turkey the Ainali-Kawah Convention confirming the independence of Crimea from Turkey. Upon return to Russia he abandoned the diplomatic career.

"Standard Wind and Wave Fields of the Black Sea" – the guidebook on the wind and wave regime in B.S. containing tables and maps of atmospheric pressure, wind and waves for five areas of B.S. as well as the information about the maximum wave heights of different probability (Sevastopol, 1987).

Stankevich (Sudak) Pine (*Pithyusa v. Stankevicusi*) – it was named so in 1906 by Academician V.N. Sukachev in honor of Crimean botanist, forester V.N. Stankevich who first described this tree in the early twentieth century. This is a large sprawling tree with the brownish bark and a semi-circular crown, with intricately bent snake-like branches, dense and long needles and large cones growing strictly upright. Branches of some trees are stretching horizontally. The stem is branched. The tree grows well on not easily accessible rocks. The pine is salt resistant. It is found only in two areas on the Southern Coast of Crimea—from the Aiya Cape to Balaklava and near Sudak in Novy Svet. This pine is referred to rare varieties and is put on the Red Book of Ukraine.

Stankevich (Sudak) Pine (http://static.panoramio.com/photos/original/28906956.jpg)

Stanyukovich Mikhail Nikolaevich (1786–1869) – Russian Admiral. In 1797 he entered the Naval Cadet Corps. From 1810 to 1814 he served on the Baltic Fleet. In 1817 he was transferred to the B.S. Fleet as a brig commander. In 1826–1829 commanding the boat "Moller", he made round-the-globe voyage to the shores of Kamchatka and to Russian America. From 1837 to 1851 in the rank of Rear Admiral (1837) and later Vice-Admiral (1848) he commanded the ship squadron of the B.S. Fleet. In 1852 he was appointed the commander of the Sevastopol port and the military governor of the city. In 1855 he became a member of the Admiralty Board and a year later he was promoted to Admiral. His name is given to the mountain in the Bristol Bay and the cape in the Taunskiy Gulf in the Sea of Okhotsk.

Staryi Krym (Old Crimea) – a city in Crimea, Ukraine. In place of the town of Solkhat founded in the fourth century there was built a fortress (from the Mongolian, Crimean Tartar "*kerem*"). This name was given to the city of Krym that was constructed near the fortress. It became the administrative center of the whole peninsula. After losing its administrative significance the city from the late fifteenth century was called Eski-Krym ("Old Krym"). In 1783 after accession to Russia it was given the pseudo-Greek name of Levkopol ("White City"), but from 1787 it took the Russian name "Staryi Krym". S.K. locates 26 km westward of Feodosiya, in the valley of the Churuk-Su River limited in the south with the Dzhady-Kai ridges. Climatotherapy and other methods of complex treatment are applied here for treatment of the lung TB. Some architectural and historical monuments survived in the city: ruins of mosques, including the mosque built by Khan Uzbeg (thirteenth to fourteenth centuries). Medrese, mint house and caravan-sarai (fourteenth century), the house-museum of A. Grin who ended his life here, the ruins of the Armenian monastery Surb-Khach, founded in the fourteenth century, with the survived frescos of the late fourteenth to early fifteenth centuries that may be found 5 km to the southwest of the city. Population—9,500 (2012).

Stellate sturgeon (*Acipenser stellatus*) – the fish of the sturgeon family (*Acipenseridae*). Its head is elongated and sword shaped. Its length reaches 200 cm, weight—3.2 kg. They are native to B.S., A.S., Caspian Sea, the Bosporus, and the Sea of Marmara. Quite seldom they are met in the Adriatic Sea. The males reach maturity at the age of 4–6 years and females—at 7–8 years. They choose to propagate in the Danube in May-June. The female fish lays from 35,000 to 350,000 eggs. They feed on shellfish, snail, bivalve mollusks and sometimes small fish. They are fished mainly in the Caspian Sea and A.S. The sturgeon meat is very tasty and is used mostly fresh or canned.

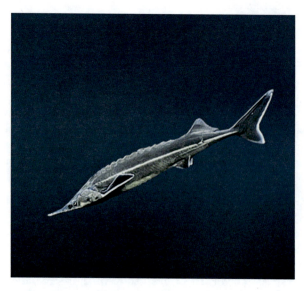

Stellate sturgeon (http://www.theanimalworld.ru/img/encycl/fish/big/sevrjuga.jpg)

Stetsenko (Stetsenkov) Vasiliy Alekseevich (1822–after 1898) – Russian Admiral. In 1830 he entered the naval unit of the Alexander Corps as a cadet; in 1833 he was transferred to the Naval Corps. In 1837 he was promoted to midshipman. In 1837–1842 on different vessels he navigated over the Baltic Sea. In 1842 he was transferred to the B.S. Fleet. In 1843–1846 S. navigated onboard various vessels, cruised near the Caucasian coasts. In 1847 he was appointed aid-de-camp of Vice-Admiral L.M. Serebryakov, Chief of the Black Sea Coast Line. On the frigate "Midia" S. navigated from Sevastopol to Novorossiysk from where with the landing troops he crossed the Muzemel Ridge and took part in reconnaissance of the Baku, Attikai and Neberdzhay gorges. In 1848 he cruised near the eastern coast of B.S., took part in the land march from Novorossiysk to the Attikai gorge in pursuit of the hostile mountaineers. In 1849–1850 on different vessels S. cruised near the Caucasian coasts; in 1849–1852 he participated in expeditions against mountaineers. In 1853 he was appointed assistant commander of the Naval Junker School in Nikolaev.

In 1854 S. took part in the battles against the British-French troops at Evpatoria, in the battles at Alma and Inkerman. He was promoted to Captain-Lieutenant and appointed aide-de-camp to the Chief of Naval Staff, Supreme Commander of Military Land and Naval Forces of Crimea Aide-de-Camp General Prince A.S. Menshikov. In 1855 he was conferred the rank of Captain 2nd Rank. He was aide-de-camp at Chief Commander of troops in Crimea, Aide-de-Camp Genera Prince M.D. Gorchakov. S. was sent to Ochakov to observe the movement of the enemy fleet at Kinburn. He was appointed aide-de-camp to General-Admiral Grand Prince Konstantin Nikolaevich who sent S. to the besieged fortress at Kinburn. In 1856 he was directed to Nikolaev and to the Caucasus to take part in the operation on conquering by the Russian troops of some points on the northeastern coast of B.S. in 1857 S. was appointed a special envoy in the rank of commander of the naval staff at Supreme Commander of the Caucasian Army. He studied the sailing range of ships over the Kuban River. In 1858–1859 the frigate under his command sailed from Kronstadt to the Mediterranean Sea calling on different ports. From 1859 to 1882 S. commanded the vessels, squadrons of vessels on the Baltic Sea. In 1893 he was promoted to Admiral.

Steven Christian Christianovich (1781–1863) – Russian prominent botanist, academician, of the Finnish origin. In 1799 he graduated from the Medical-Surgical Academy in Petersburg and was awarded the academic doctoral degree for his special botanical researches. In 1800–1805 he traveled over the Caucasus and gathered herbaria and collections. In 1807 S. visited P. Pallas in Sudak and continued his researches. In mid-1807 he arrived to Odessa. He was the assistant of the agricultural inspector for the South of Russia who was a prominent botanist and after his death he worked as assistant of the inspector's deputy. In 1807 he settled down in Crimea and lived in Simferopol and Sudak. In Sudak he spent the last years of his life. For nearly 60 years of his life he devoted to studies of the vegetation of Crimea and development of agriculture there. In 1812 he was appointed the first director of the Nikitskiy Botanical Garden and occupied this post for 14 years. The Nikitskiy Garden became the prominent center of introduction and acclimatization of fruit and decorative crops, vines, tobacco.

S. was widely known due to his long-lasting studies of flora and ethnofauna of the Caucasus and Crimea. In 1826 he went to live to Simferopol and in 1835 he decided to naturalize in Russia. S. published some papers devoted to the Crimean flora and establishment of the Nikitsky Botanical Garden. In 1846 he suggested a bold idea—to direct the Dnieper waters to Crimea. In 1849 he was made the Honorary Member of the Petersburg Academy of Sciences. In 1855 he completed his work "List of Wild Plants Growing on the Crimean Peninsula" (1855–1857) that took together all results of researches of the Crimean plants conducted during five decades. This was the first and fullest catalog of the Crimean flora that incorporated 1654 plant varieties growing in Crimea which was 735 varieties more than described by Pallas. During his lifetime many plants brought from the countries which climatic conditions were similar to SCC were adapted here. His very rich herbarium S. decided to present to the Helsingfors University (Helsinki) and A. Nordman came to Sudak in 1860 specially to take it. According to the contemporaries, the Crimean Peninsula was well known in Europe because S. lived there. In 1940 S.S. Stankov published the biography of S.

Still spots, Sliks – areas of the sea, where the sea surface is smooth as a mirror, glossy. Produced as a result of accumulation of a significant quantity of dead organic matter or when oil products or other surfactants find their way into the sea, whereby small waves are damped.

Stomoplo marsh – it is situated on the western coast of B.S. northward of the city of Primorsko, near the Stomopolskiy Bay, Bulgaria. The jetty breaks it into two parts: northern and southern. The water is exchanged between them via tubes made in the jetty. The water salinity here varies with seasons from 0.3 to 13‰. The bog is recharged from ground waters, precipitations and sea waters. It is overgrown with different varieties of reeds and sedge, while its western part with cane.

Stone pine – Italian pine (*Pinus pinea*) of the pine family. The crown of the young trees is tapering and of old trees—flat, umbrella-shaped. Its stem is to 25 m high. It is native of the Mediterranean area. It is cultivated in Crimea and Caucasus.

Storm of Izmail (Russian-Turkish War of 1787–1791) – storming and conquering by the Russian troops commanded by General A.V. Suvorov in 1790 of the Turkish fortress—the robust citadel constructed by Turkey to control the Danube Delta. Suvorov sent to the fortress governor a proposal to surrender: "24 h for consideration—freedom. Our first shot—captivity. Storming—death." The legend said that Mehmet-Pasha haughtily declined the ultimatum saying that sooner the sky falls on the land and the Danube flow backward than Izmail fall down. After 2 days of preliminary cannon fire nine columns of the Russian troops stormed the fortress on all sides, including on the river side so that to cut off the Turks from their reserves. The sixth column under command of General M.I. Kutuzov had a severe battle. The soldiers were unable to break through the dense fire and they entrenched. The Turks started attacking. They Suvorov sent to Kutuzov the order on appointing him the governor of Izmail. Inspired by such trust the general personally led the infantry into attack and conquered the Izmail fortifications. At the same time the assault troops led by General O.M. De Ribas landed in the city on the southern side. By 4 p.m. the battle finished and the Izmail fell. This was the most severe battle in all Russian-Turkish wars. The Russian lost 4,000 men and 6,000 were wounded. Out of 650 officers who took part in the storm more than the half were either wounded or killed. The Turk losses were 26,000 and among them Mehmet-Pasha. The remaining 9,000 people became prisoners. The Russians got hold of all Turkish cannons, 400 banners, large stocks of food and ammunition, as well as gold, silver and precious stones ten million piastre worth. In honor of conquering Izmail the gold cross "For Great Bravery" was specially established. The lower-rank officers were awarded with the Silver Medal "For Great Bravery in Izmail Storming".

Storms in the Black Sea – north-easterly (essentially) hazardous winds that constitute a winter continental monsoon. More often than not observed in December and March. Storm duration averages 10–30 h, in some cases reaching 100 h and more. The storms are accompanied by severe air temperature decrease, snowfall and sea steaming. In spring, there occur north-westerly storms with dull unstable weather. Southerly storms are observed more seldom than others. On hot days, spouts occur sometimes. In spring and autumn, cyclones break through into the sea; these bring with them severe and protracted storms from the west, south, sometimes, from the south-west. Black Sea storms are observed 50–60 days a year. Some blights and bays are characterized by their own peculiarities of storm regime. These include, for example, the autumn-winter bora in Novorossiisk, Gelendzhik and Tuapse. In Batumi harbor, during westerly winds, harbor oscillation is observed, when south-westerly winds blow—there is swell. On the southern sea-shore, the most hazardous areas are Mapavri and Kerempe Cape. If the mountains are covered by clouds, it means that during next few days rain will start with westerly winds. In Rise-Limany Harbor (between Askoros and Fener Capes), north-westerly wind induces large waves, while the south-westerly wind succeeding it causes sea wave. Maximum recurrence is exhibited by storms during winter here. In

Trabzon Bay (between Khupsi and Guzelhisar Capes), storms are noted during all seasons of the year. Particularly hazardous are north-easterly and north-westerly winds producing strong sea wave. An easterly wind produces swell called "kalash". In Samsun harbor, north-easterly winds in winter may reach the force of a storm and cause strong wave. These are accompanied by rain, sometimes hail, and last several days. In Sinop Bay, storms are usually observed in winter induced by south-easterly. North-westerly or westerly winds.

In Bosporus Strait at daytime, north-easterly winds usually blow from the Black Sea, these die down by night. In winter, north-easterly wind blows here in a stable manner, the sky being cloudless; when the wind breaks through to the west, it brings haze and clouds with rain, not infrequently with snow and even hail. Frost may reach -10 °C. North-westerly winds blow seldom, but are gusty and are often accompanied by rain. Southerly winds bring fair weather.

Near Odessa bay, gusty storms arrive largely when north-easterly winds blow. In winter, they bring fogs. At Ochakov Port, when a strong south-westerly wind blows, there begins harbor oscillation. Strong easterly Ochakov gusts are also observed here: sudden gusts of wind, sometimes reaching the force of a hurricane. These are produced on the northern periphery of the Balck Sea cyclone and are related to the atmospheric front arranged latitudinally. In Nikolaev, during the warm part of the year, sand storms are observed. On the Southern Coast of the Crimea (Alushta, Yalta), strong winds blow as a rue from the mountain passes.

Storms in the Sea of Azov – northerly and north-easterly hazardous winds of monsoon type that encompass vast expanses and are accompanied by severe frosts. Occur 20–30 times during a year and reach high speed. In the Sea of Azov dangerous to ships. These winds drive waves to the distance of 300–400 km from the apex of Taganrog, conditioning a strong wash. The waves break up the ice, heap up ice reefs along Arabat Arrow, roll over it, produce alluvial fans of sand and shell. In spring, southerly and westerly winds blow more often. In summer, maximum number of stormy days is 1–2 a month. These are mostly gusts of wind on cold atmospheric fronts. Once and again, there occur spouts. One of the main peculiarities of the sea wind regime is the significant fluctuation of the water level induced by winds causing positive and negative surges. The wind-induced wave involves the entire water mass down to the bottom (due to prevalence of shallow depths). As the wind weakens, waves are damped quickly. Wind-induced waves are short and quite steep.

Strategic Plan of Action for the Rehabilitation and Protection of the Black Sea (Istanbul, 1996) – preparation of this Strategic Plan of Action (SPA) was included into Resolution 3 adopted at the Diplomatic Conference on Protection of the Black Sea, Bucharest, 21–22 April 1992, and the Final Act of the Ministerial Meeting on the Declaration on Protection of the Black Sea, Odessa, 6–7 April 1993. The document outlined the principles, policy and actions of the Black Sea coastal states relevant to attainment of the environmentally sustainable development applying wise principles of the Black Sea environment management, including live organisms and water quality, ensuring long-time benefits to the population of the Black Sea coastal states with simultaneous creation of the prerequisites for health and environment safety as well as economic sustainability and environment stability for

future generations. Herewith it was stressed that the state of the Black Sea environment continues to be a matter of concern due to ongoing degradation of its ecosystem
and the unsustainable use of its natural resources. Among the causes of such
degradation there are disposal of pollutants, in particular agricultural wastewaters,
inadequately treated industrial domestic wastewaters, oil products, appearance of
invaders, unwise natural resource management. Elaborating SPA, the Black Sea
coastal states considered and took into account some principles, such as precaution-
ary principle, anticipatory actions, use of clean technologies and economic instruments, environmental and health considerations, wider public involvement and
transparency, closer cooperation among Black Sea coastal states, cooperation
among all Black Sea basin states. According to SPA, each Black Sea coastal state
shall elaborate the national SPA (by the end of 1997) or a similar document that
would detail on the actions taken at the national level for implementation of this Plan
of Action. The Conference of Donor States shall be convened once in 5 years. The
national SPA shall be realized using both their internal reserves and outsourcing (the
finance of regional and international foundations and organizations).

Sturgeons *(Acipenseridae)* – fish of the sturgeon order. Its body has five rows of
bony scales; its skeleton is mostly cartilaginous. Its sizes may reach 3 m and weight—
1,000 kg. It is mostly anadromous fish, spawning in rivers and fattening in the coastal
areas of the sea. There are also freshwater species (sterlet). They feed on water
animals. S. refers to the valuable game fish. It includes sturgeon, beluga, starred
sturgeon, bastard sturgeon, shovel-nosed sturgeon and others. The representatives of
this bottom fish species populate extensively the coastal waters of B.S., giving
preference to its northwestern part. In the 1950s their regular fishing reached 1,000
ton. Spawning and early (3–4 months after spawning) stages of development of the
Russian sturgeon occur in the Dnieper (May-July) and Danube (February-September).
The starred sturgeon is spawning in the Danube from March through December with
interruption in July-September, while beluga—from January through June.

Sturgeon (www.planeta-neptun.ru/)

Sudak – a seaside resort in Crimea, Ukraine. It is situated on the coast of a large B.S. bay in a wide Sudak valley, 60 km from the railway station "Feodosiya". Its population is 15,400 (2012). It is protected from winds by the rocky Megan Cape in the east and outspurs of the Main Range of the Crimean Mountains in the west. It is one of the oldest settlements on the Crimean Peninsula. It first appeared in 212 as a settlement of Sogdia (in the Ossetian "*sugdoeg*" meaning "sacred"), later it was a Greek colony. In the eighth century this was the prominent trade center. In the early thirteenth century it was the trading center of Venetians. The city is famous for the ruins of the Genoese (Sudak) fortress of the fourteenth to fifteenth centuries, the outstanding medieval architectural monument of world significance. In Russian sources of the tenth to fourteenth centuries it was referred to as Surozh. After accession of Crimea to Russia S. became a resort and agricultural area near Feodosiya of the Tavria Province. In 1922 S. was assigned the status of an urban-type settlement and in 1929—the status of a city. In the Soviet time S. developed as a vine-growing and wine-making center. The plants of vintage wines, champagne ("Novyi Svet" Plant) and "Dolina Roz" (Rose Valley) Factory manufacturing rose oil locate in the city. In the middle ages the Great Silk Road passed through S. Painters Aivazovskiy, Lagorno, Bogaevskiy, Latri and others made famous pictures here.

Sudak (http://odissey.net.ua/images/stories/sudak/sudak2.jpg)

Sudak Bay – it incises into the coast between the Peschernyi and Rybachiy capes. The elevated cape of Alchak-Kai projecting from the mid of the bay's northern coast divide the bay into the eastern and western parts. On the western part of the coast 1.5 km to the northeast of the Ploskiy Cape the Sokol Mount is rising to 472 m over the sea level. The Sudak Rock with two Genoese fortresses on its top is found 2.5 km from the mountain. The coast between the rock and the Alchak-Kai Cape is low and here the Suuk-Su River flows into the sea. The city of Sudak locates in its wide valley.

Sudak Fortress – see Genoa fortress.

Sudzhuk Bar – it projects for 6.5 km to the northeast of the Myskhako Cape from the southwestern shore of the Novorossiysk (Tsemess) Bay at its entrance. The bar limits in the east a small and shallow bay known as Solenoe Lake. A small low-elevated island of Sudzhuk covered with pebble is found 500 m from the bar.

Sudzhuk Liman – it is situated on the low Sudzhuk Cape at the entrance into the Novorossiysk Bay. The liman is separated from B.S. with a spit having a small strait. The water salinity in the liman ranges from 16 to 22‰.

Sudzhuk-Kale (Russian-Turkish War of 1768–1774) – the Turkish fortress on the eastern coast of B.S. near which on 23 August 1773 the Russian ships commanded by Captain 2nd Rank I. Kingsbergen fought against the Turkish ships. Using the numerical superiority the Turks resolutely attacked the Kingsbergen's squadron. After a fierce battle the enemy "unable to resist any more the heavy fire of the Russians and having sustained serious damage of ships" escaped under protection of the S.-K. fortress. The battle at S.-K. became one of the first victories of the Russian fleet on B.S.

Sukhotin Yakov Filippovich (1730–1790) – Russian Vice-Admiral (1783). He finished the Naval Academy (1743–1745). In 1745–1751 and 1756–1766 he took part in the military campaigns on the Baltic Sea. In 1796 he served in Tavrov, near Voronezh, where the shipyard and the admiralty located. He commanded 29 newly built military boats, sailed from the shipyard in Ikorets, located at the inflow of the Ikorets River into Don, to Cherkassk. He conducted measurements and prepared description of A.S. as far as the Berdyansk Spit. In the same year he was appointed the commander of the A.S. squadron of 8 "newly-invented" vessels and a boat. In 1771, commanding the vessel "Khotin" he navigated in the A.N. Senyavin squadron from Taganrog to Enikale (Kerch) Strait. In 1772 he commanded the squadron on A.S. In 1773 commanding the squadron of four vessels he cruised over B.S. near the Enikale Strait. On 29 May 1773 sailing with the squadron between Kafa and Sudzhuk-Kale he was victorious in the battle against six Turkish ships on the Kuban River. In February 1779 he was the commander of the Don Military Flotilla. In 1783 he was promoted to Vice-Admiral and appointed the supreme commander of the B.S. Fleet. In 1785 he was transferred to the Baltic Sea Fleet. In 1790 on the flagship "Twelve Apostles" he commanded the vanguard of the squadron of S.K. Greig in the battle at Krasnogorsk (23–24 May 1790) with the Swedish ships in the Gulf of Finland.

Sukhoy Liman – incises into the shore for 31.5 km to the northeast of the Dnieper-Tsargrad gyrlo, Ukraine. The liman shores are steep, in some places even abrupt, cut with gullies. The dikes break the liman into three parts that are called basins. The depths here vary from 5 to 11.5 m. The entrance into the liman is between two sandbars from the tips of which two parallel protective jetties project seaward—northern and southern. The city and port of Ilyichevsk locate on the southwestern shore of the liman.

Sukhum Landing Unit (Russian-Turkish War of 1806–1812) – storming of Sukhum on 11 July 1810 by the landing troops from the B.S. Fleet ships under command of Lieutenant-Captain P.A. Dodt. The city was defended by the troops of Arslan-Bey who having overthrown his father Kelesh-Bey who took Russian nationality came to power in Abkhazia. For suppression of the rebellion the detachment of the B.S. ships with a battalion of infantry was sent from Sevastopol to Sukhum. On 9 July it reached Sukhum and started gunning of the fortress. After 2-day bombardment the troops landed onshore. After fierce battles the landing troops seized Sukhum where they took hold of enormous stocks of gunpowder (more than 1,000 poods). During Sukhum storming the Russians lost 109 men. Having suppressed the rebellion the Dodt detachment headed to Turkish coast fortification Sudzhuk-Kale and captured it in early August. As a result, the whole eastern coast of B.S. from Anapa to Poti came under Russian control.

Sukhumi, Sukhum – a town and port on the coast of the Black Sea, the capital of the Republic of Abkhazia. Population—64,500 (2011). This is a railway station and a port on B.S. It is located on a wide and very beautiful Sukhumi Bay and in the valleys of the mountain rivers of Besletka, Sukhumka and others. In the fourth century B.C. the antique colony of Dioscurias was founded here and it had been so named for the Dioscuri, the "Zeus twin sons"—immortal Pollux and mortal Castor who took part of the Argonauts travel to Colchis. Then the Roman fortress of Sebastopolis appeared here that was named in honor of the Roman Emperor Octavius (27 B.C.–14 A.D.) who had the title of Augustus-Sebastos meaning "glorified by Gods, sacred". In the Middle Ages the city of Tskhumi (in the Svanetian "*tskhumi*" meaning "hornbeam") entered the Georgian Kingdom. In the sixteenth to early nineteenth centuries it was under Turkish control and was called Sukhum-Kale. The Russians called it Sukhum. From 1939 it was referred to as Sukhumi.

The vegetation of S. is rich and diverse. It is represented mainly by cultured plants. Prevailing are subtropical varieties, both deciduous (palm, magnolia, eucalyptus, laurel) and coniferous (cypress, arborvitae, cedar, pine). In numerous gardens the tangerine, persimmon, fig trees, grapes and also apple, bear, cherry and other trees are growing. The center of S. is given to the Botanical Garden of the Academy of Sciences being one of the oldest gardens (it was founded in 1840). Its collection numbers over 4,500 varieties and life forms of plants. Subtropical fruit (avocado, papaw, podocarpus and others), technical and medicine (pilocarpus, boldo, tung, varnish trees, etc.), aqueous (lotus, water hyacinth, etc.) plants are growing here in open ground. Widely represented are evergreen trees of the Japanese, Chinese and North American origin. The Botanical Garden possesses a unique collection of fossil plants.

The climate of this resort is humid subtropical. The winter is very mild, without snow. The mean temperature of January is about +6 °C; the interchange of the cloudy, rainy and clear, sunny weather is quite typical of this place. The spring is not long, cooler than autumn, with the prevailing sunny, moderately wet weather. The summer is very warm and, mostly, sunny. The mean temperature of August is +24 °C. The autumn is long, warm and sunny. The non-frost period is about 300 days. The precipitation is to 1,500 mm a year falling mainly in April-October;

the minimum rainfall is in May, the maximum—in September. The greatest number of rainy days is recorded in autumn and winter. The number of sunshine hours is 2,120 a year. The temperature of sea water in summer is +26 to +27 °C. Sukhumi is the seaside climatic and balneological resort. The mild climate, being the main curative factor here, provides opportunities for climatotherapy, including in combination with sea bathing (from May through November). Good climate of S. had been used for treatment of people in sanatoriums still in the late nineteenth century when Russian doctors A.A. Ostroumov and S.P. Botkin sent to S. patients with lung disorders. That time more and more sanatoriums and holiday hotels were constructed there. Apart from climate, one more curative factor in S. is mineral water springs. These are thermal waters (+35.5 to 42 °C) of different salinity. Nearby S. there are many sightseeing attractions for tourists, among which the Novyi Afon Mount with its friary and the unique natural feature—Novo-Afon cave. Many tourists always visit the monkey nursery and botanical garden.

The fruit canning, tobacco, leather and footwear industries are developed here as well as electrical instrument-making, casting and mechanical, shipbuilding industries. The Sukhumi HPP was constructed on the Gumista River. The Research Institute of Abkhazian Language, the Research Institute of Tea and Subtropical Plants, the Institute of Experimental Pathology and Therapy of the Medical Academy of Sciences and musical colleges locate in the city. The Sukhumi Physico-Engineering Institute was established in 1945. It produced unstable isotopes and referred to secret enterprises of the USSR. After the Georgia-Abkhazia military conflict of 1992–1993 the Georgian specialists left the institute and S. Now it is known under the name Research & Production Association "Sukhumi Physico-Engineering Institute" (SPEI).

Sukhumi (http://media-cdn.tripadvisor.com/media/photo-s/01/71/03/9c/sokhumi.jpg)

Sulina – a village and port in the mouth of the Danube middle arm. The easternmost point in Romania. Its population is 3,540 (2011). The main industries here are fish processing and metal-working. It is also a tourist center. It is mentioned in the written documents of the tenth century. The old lighthouse "S." 2 km from B.S. was constructed in 1802. Its location may be indicative of the growth rate of the Danube delta.

Sulina Gyrlo, Sulina Arm – the left channel of the Tulcha Arm of the Danube Delta in Romania; one of the main delta channels. This arm has an artificially straightened and canalized bed. The straightening and bottom dredging works were conducted in 1868–1902. As a result, the arm length became 21.2 km shorter (from 83.8 to 62.6 km). At present its length is 62.6 km (the bed from the origin— Georgievskiy Chatala to the port of Sulina) plus 13 km (paired piers projecting into the sea), making in total about 76 km. The arm channel width is 130–160 m.

Sulkevich Magomet Suleimanovich (1864–1920) – Russian Lieutenant-General (1915). He finished the Voronezh Cadet Corps, the Mikhailovskiy Artillery School and the Nikolaev General Staff Academy (1894). He was Moslem. S. took part in the Russian-Japanese War of 1904–1905 and in World War I. He commanded the Moslem Military Corps (from 1916 to February 1918). After the revolution and the Brest-Lithuanian Peace Treaty he served in Crimea. After the Soviet Socialist Republic of Tavria was broken down by the Tatar nationalists supported by the German troops that invaded Crimea S. headed the Crimean Government (25 June-18 November 1918). He put into use the blue flag with the Tavria coat of arms. He started formation of the army. S. visited Germany and conducted separatist negotiations to attain recognition of Crimean independence and even proposed establishment of the Crimean Khanate under control of Turkey and Germany. After disagreements that had arisen among the government members and after landing of the allies in Sevastopol he assigned the authorities of the government chairman to socialist-revolutionary S.S. Krym who on 15 November 1918 was officially appointed to this position at the meeting of country and urban representatives of Crimea.

Sunny Beach – the major seaside climatic resort in the Bourgass Region, Bulgaria. It is situated on the B.S. coast occupying a flat plain along the bay between the Nesebr Peninsula and the Emine Cape. On the north it is protected by the southern slopes of the Eminska Planina (about 620 m over the sea level) that is partially overgrown with oak forests and shrubs. The resort territory is very green. Many tree varieties are growing here (Canadian poplar, pine, cypress, sycamore, acacia) and amongst them the hotels are rising. Climate here is marine. The winter is mild, practically without snow; the mean temperature in January is about +2.5 °C. The summer is very warm, dry and sunny; the mean temperature of July is about +24 °C. The precipitation is over 430 mm a year. The number of sunshine hours is 2,300 a year (in July-August to 11 h a day). In summer the sea breezes smooth the summer heat. The beach 300–400 m wide is stretching for 6 km along the bay. It is covered with fine gold sand and in some places it is rimmed with sand dunes. In the south it closes up with the beach of Nesebr resort. Mild climate, warm sea, beautiful sandy beaches and flat gently sloping sea floor provide opportunities for climato- and thalassotherapy here. The bathing season here lasts from mid-June through

October. The resort has many hotels, often distinguished by specific architectural design. S.B. is the major resort complex in Bulgaria of international significance. It is connected by the automobile roads with Bourgas and Varna. It is also linked practically with all seaside resorts of Bulgaria via sea. This resort abounds in restaurants, cafes, tennis courts, sport centers, casino, etc.

Supsa – a river in Georgia. The area of its basin is 1,100 km^2. Its average many-year flow (at the village of Khidmagala 6 km from the mouth) is about 1.5 km^3 a year. It varied from 2.4 km^3 in 1985 (164 % of the average) to 0.9 km^3 in 1947 (64 % of the average). Within a year the flow is distributed rather evenly between the warm and cold seasons. In spring the flow tends to grow (32 %), while in summer—to decrease (18 %).

Surmene – a village on the coast of the Black Sea, Turkey. It is situated 40 km east of Trabzon. Population—15,500 (2012).

Suvorov Aleksandr Vasilievich (1730–1800) – Russian military commander, one of the founders of the Russian military art of war, Generalissimo (1799), Count of Rymnik (1789), Prince of Italy (1799). From his childhood he studied artillery, fortification, and military history with his father—General, Senator and the author of the first Russian military dictionary. In 1742 he was enrolled as a musketeer into the Life Guard Semenovskiy Regiment in which in 1748 he started his active military service as corporal. S. received his first military experience during the Seven Years' War of 1756–1763. He took part in the battle at Kunensdorf and in storming Berlin (1760). In 1773 he was sent to the Balkan theatre of war operation (during the Russian-Turkish War of 1768–1774) to the First Army of Field Marshal P.A. Rumyantsev. In May-June S. made two successful raids and crushed the Turks at Turtukan. In September, defending Girsovo, he repelled the attack of the Turkish troops. On 9(20) June 1774 at Kozludzha, he defeated the 40,000-strong Turkish Army. In August 1774 he was sent to repress the peasants uprising led by E.I. Pugachev. In 1776–1779 he commanded the troops in Crimea and Kuban and played a very important role in preparation of accession of Crimea to Russia. During the Russian-Turkish War of 1787–1791, that time in the rank of general (1786), he commanded the defense of the Kherson-Kinburn region. On 1(12) October 1787 at Kinburn, his troops crushed the Turkish landing troops. In 1788 he took part in the Ochakov siege; in 1789 he won two great victories at Focsani and by the Rymnik River; in 1790 he successfully stormed the fortress of Izmail. In 1792–1794 he commanded the troops in the south of Russia. In 1798 when Russia joined the second anti-French coalition (Britain, Austria, Turkey, Kingdom of Two Sicilies) on the insistence of the allies Paul I had to summon S. to military service as the supreme commander of the Russian troops in Northern Italy. Within a short time Northern Italy was liberated from the French control. In August 1799 the Switzerland march of S. began. It made the invaluable contribution into the art of war. In October 1799 Russia broke off the union with Austria and S. returned to Russia where he was dismissed in disgrace and the old veteran died a few days afterwards unable to survive this. He was buried in the Aleksandr-Nevskiy Monastery in St. Petersburg. The simple inscription on his grave states: "Here lies

Suvorov". S. elaborated and applied in practice more advanced forms and methods of waging of war and battle and in this he was much ahead of his time. He took part in more than 60 battles and all were victorious.

Suvorov A.V. (http://zhitejnik.ru/uploads/posts/2011-01/1295620663_suvorov.jpg)

"Svobodnyi Plyazh" (free beach) – the name of artificial sand beaches washed on without any auxiliary systems, but simply as a result of construction of dikes or underwater breakwaters with cross bars. Such beaches may appear if there are underwater quarries or other accessible soil sources, in some cases—the so-called "bypassing" (artificial drift of loose material from the place of its buildup). Their efficiency is usually the highest only at regular replenishment, "feeding" as a result of additional aggradation. The slopes of artificial beaches shall correspond to the natural slopes of the local beach, while the coarseness of the material (sand) shall be greater or equal to that of a natural particular beach. The best conditions for S.P. formation are considered to be bays and deeply incised harbors (coasts protected from the wave effect). On B.S. such beaches were formed in the Gelendzhik Bay, between Novorossiysk and Tuapse, in the Gagra Region of Georgia, nearby Sochi, Yalta, etc.; on A.S.—near the Taganrog Cape (Adreeva Bay).

"Swallow's nest" – a symbol of Crimea. Roman warriors founded a castle on the western spur of the steep cliff of Ai-Todor. This flat site (10 m wide, 20 m long and 12 m high) is on the eastern tip of the Limen-Burun Cape ("Cape of harbor") that was called the Captain Bridge. This cliff was given the name of Aurora, the Roman goddess of the dawn. The first structure on the Aurora Cliff was a wooden cottage that was named "S.N." by Moscow merchant woman Rakhmanina. In 1911 Baron von Steinheil who made his fortune by extracting oil in Baku purchased this cottage and in a year a stone palace was built here designed by well-known Russian architect A.V. Sherwood. In 1912 the amazing Neo-Gothic castle of a step-like design appeared on the site of Limen-Burun. A building 12 m high rose on the foundation 10 m wide and 20 m long. Its original design envisioned a foyer, guest room, stairway to the tower, and two bedrooms on two different levels within the tower. In 1914 this castle was purchased by Moscow trader P.G. Shelaputin who opened here a restaurant. During the Soviet time the building was used as a tourist base and later some rooms of the building were given to the library of the nearby "Zhemchuzhina" ("Pearl") sanatorium. In 1927 "S.N." survived a serious earthquake. A portion of the cliff fell into the sea and the observation deck was menacingly hanging over the abyss. The building was not, in fact, damaged. However, the cliff itself developed a huge crack. It was decided that the palace was in danger and it was closed to the public for a certain period. During this time it started dilapidating. The restoration and renovation of the building was started only in 1968. The project involved reinforcement of the foundation, making some changes in the façade and internal rooms. Constructed of the gray stone "S.N." with its graceful Gothic towers is well visible from the road running along the coast and from boats cruising between Yalta and Simeiz. The building is crowned with a round tower embraced with light balconies. After restoration this castle was officially recognized the monument of architecture of the twentieth century.

The castle-palace "Swallow's nest" (http://sergek.org.ua/images/dsc_4388.jpg)

Symplegades – in the Greek mythology S. also known as the Clashing Rocks, were a pair of rocks that clashed together randomly. In different myths they were placed either at the entrance into Pont Euxinian (B.S.) or near Sicily or near Pillars of Hercules (Gibraltar Strait). They were defeated by the Argonauts who would have been lost and killed by the rocks except for the Phineus' advice. They let a dove fly between the rocks and rowed mightily after it so that their ship "Argo" passed successfully through the rocks with only a part of its stern ornament being lost. After that S. lost their magic, stopped moving and, thus, obstructing the free passage into Pont.

Sysoev Viktor Sergeevich (1915–1994) – Soviet Admiral. He served in the Navy from 1937. In 1939 he finished the M.V. Frunze Higher Naval School. He took part in the Great Patriotic War. In 1941–1944 he served on destroyers on B.S. In 1952 he graduated from the Naval Academy. In 1955 he was appointed the Chief of Staff of the B.S. Fleet; in 1960—Head of the Chair of Surface Craft Tactics in the Naval Academy; in 1965—First Deputy Commander and in 1968—Commander of the B.S. Fleet. In 1970 he was promoted to Admiral. In 1974–1981 he was the Head of the Naval Academy, professor; in 1981 he was included into the General Inspector Group of the USSR Ministry of Defense. In 1987 he resigned. He was awarded many orders.

T

Taganrog – a city in the Rostov Region, Russia; a port in the Taganrog Bay of A.S.; a railway station. It was founded in 1698 by Peter I as a military port and a fortress on the Taganyi Rog Cape. In the eighteenth century this was the trade port specializing mainly in grain export. In the 1890s large plants and factories were built here. In the twentieth century the local industry included new plants, such as metallurgical, boiler, self-propelled combine, press-forging equipment, ship repair, furniture. The city has the dramatic theatre, A.P. Chekhov Regional Museum located in the writer's house, the literary museum. Population—254,800 (2013).

Taganrog (http://upload.wikimedia.org/wikipedia/commons/d/d1/Port_of_Taganrog.jpg?uselang=ru)

S.R. Grinevetsky et al., *The Black Sea Encyclopedia*,
DOI 10.1007/978-3-642-55227-4_20, © Springer-Verlag Berlin Heidelberg 2015

Taganrog Bay – a large bay in the Sea of Azov. It is separated from the sea by Dolgaya and Belosarayskaya Spits. T.B. is 140 km long, 31 km wide at the entrance, 5 m deep in average. It's surface is 5,600 km², and volume—25 km³. Yearly T.B. is covered by ice from December to March, but in mild winters it is ice free. The Don, Kalmius, Mius and Eya rivers flow to the bay. There are several large ports in the Bay: Taganrog and Eisk in Russia, and Mariupol in Ukraine.

Taganrog Port – locates on the northern shore of the Taganrog Bay 15 km westward of the Don River mouth. This is the oldest port in the south of Russia. The Azov-Don canal 4.2 m deep and the Taganrog canal 3.9 m deep lead to the port area. The total length of five wharfs is 1.1 km. The water depth at port wharfs varies from 3.9 to 4.2 m. The port is operating year-round. The port complex includes OJSC "Taganrog Sea Trade Port" (TSTP), cargo wharfs of JSC "Priazovie" and JSC "Taganrog Ship Repair Yard". TSTP is capable to handle 1.2 million tons a year. In the handled cargo the greatest share is taken by metal products (33 %), coal (16 %), chemical products (7 %) as well as metal scrap, tubes, forest products. The port is adequately equipped for loading bulk, general and container freights. In winter when the port water area freezes for 3.5 months the ships are navigated to the port by icebreakers.

Taman – a village in Temryuk Region of Krasnodar Territoru, Russia. Located in the west of the Taman Peninsula (often referred to simply Taman) on the coast of the Taman Bay, located in the Kerch Strait. There is a version of the origin of the word "Taman" fron the Adyghe Language "*temen*"—a swamp, marshes, or Kabardian "*taman*"—a lake. Population—10,000 (2010). About 592 B.C. the ancient Greeks founded the city Hermonassa at this place. On 4 September 2008 a village celebrated 2,600 years anniversary of Hermonassa-Slavyanskaya-Tmutarakan—the capital of ancient principality Tmutarakanskiy (second half of X–XI). After the defeat of the Khazar Khanate in 965 by Kievan Prince Svyatoslav Igorevich, the city came under the rule of Russia. At this time, it was known as a major market town with a good harbor. Political and economic links between the Russian principalities, peoples of the North Caucasus and Byzantium have been supported via Tmutarakan. The town was populated by Kasogs (Adyghe people), Greeks, Alans, ancient Russians and Armenians. In the tenth century Tmutarakan was surrounded by a strong wall of brick. In 1023 Prince Mstislav Vladimirovich, who ruled in Tmutarakan from 988 to 1036, built the Church of the Virgin. In 1068 Prince Gleb measured the distance from Tmutarakan to Korchev (Kerch) on the sea ice and wrote on the historical "Tmutarakan Stone" the value of 14,000 sazhen (30.24 km). The Stone was found on Taman Peninsula in 1792 by Russian Admiral P. Pustoshkin.

A few kilometers south of Taman (Volna village) it is planned to establish a new international cargo seaport on the Black Sea coast—Taman Port. The construction of the terminal for the shipment of ammonia "ToliattiAzot" is going on.

Taman Bay – a large bay deeply incising into the eastern coast of the Kerch Strait. In Antiquity it was called the Korokondam Lake. The Markitanskiy Spit projects from the southern shore of T.B. and the Rubanova Spit—from its northern shore. These spits divided the bay into two parts: western (external) and eastern (internal). In the northwest the bay is limited by the Chushka Spit. The northern part of T.B. to the east of the Chushka Spit base has its own name—the Dinskiy Bay. The shores of the bay are mostly steep. The bay is shallow, the depths here are no more than 5–5.5 m. The village of Taman is situated on the southern coast, in the external part of the bay, while the settlement of Sennaya—on the eastern coast, in the internal part of the bay.

Taman Peninsula (in the Adygean "temen" meaning "marsh") – the western end of the Caucasus. It is separated from the last ridges of the Greater Caucasus by a low alluvial band of the old Black Sea arm of the Kuban River. In the south, north and west T.P. is washed by the Black and Azov seas and the Kerch Strait. The western B.S. coast of the peninsula from the Tuzla Cape to Taman is elevated (15–30 m) and abrupt. The northern part of T.P. is called the Fontalovskiy Peninsula, sometimes—Cimmerian Island. Eastward they are replaced with long sandy bars of the old (Black Sea) and recent (Azov) mouths of the Kuban. The peninsula stretches for 40 km from south to north and for 66 km from east to west. Geologically T.P. is quite a young formation. According to ancient Greek geographer Hecataeus of Miletus, still in the fifth century B.C. there were five isolated islands among the arms of Hipanis (Kuban) that later joined together and formed a peninsula. The channels and straits among them were filled with the products erupted from mud volcanoes, deposits of ancient Kuban and wave and tidal action of the sea. In the sixth century B.C. the Pontic Greek colonies of Hermonassa, Phanagoria, Kepa and others located on the shores of T.P. In the fifth century B.C. the territory of this peninsula belonged to the Bosporus Kingdom.

The area of T.P. is 1,700 km^2, of which slightly more than the half is land, other territory is covered by limans and plavni (flooded lowlands). The shores are low, cut with bays, the largest of which is Taman Bay closed with the sand spit of Chushka. The relief of the peninsula is represented by not high, elongated (mostly latitudinal) anticlinal ridges composed of the Neogene rocks, predominantly clays. The anticlinal domes are "crowned" with the cones of active mud volcanoes (maximum height being 164 m). Their number may reach 30. The most well-known among them are Akhtanizovskiy Mound, Karabetova Mount, Miska, Gnilaya Mount, Azovskiy Peklo. The depressions among ridges are occupied by shallow, mostly closed basins and limans with the bitter-saline water that are surrounded by silty, often solonchak lands. The outcrops of oil and combustible gases as well as iron ores are found here. The climate is arid (about 400 mm of precipitation). The greater part of T.P. is used for cultivation of wheat and corn. There are also many vines and large wine-making plants, such as "Phanagoria" (village of Sennoy), "Kubanvino" (Staratitorovskiy) and others.

Taman Peninsula (http://www.google.com/earth)

Tamara (Tomara) Vasiliy Stepanovich (1740–1813) – Russian diplomat. During the Russian-Turkish War of 1768–1774 he was actually the Russian envoy in Venice and was obliged to support expeditions of the Russian Fleet to the Aegean Sea. In 1783 he was sent to Georgia for participation in the negotiations that ended in signing of the Georgievskiy Treaty of 1783. From May 1797 to July 1802 T. was Envoy Extraordinary and Minister Plenipotentiary of Russia in Turkey. In January 1799 he signed the Russian-Turkish Treaty of Alliance; in 1800 he concluded with Porta the Convention of Constantinople. In 1802 T. resigned.

Tanfiliev Gavriil Ivanovich (1857–1928) – Russian geobotanist, soil scientist and geographer. From 1905 T. was Professor of the Novorossiya University in Odessa. He conducted research in the European part of Russia, Transcaucasia, and Western Siberia. He was the author of the comprehensive work "Main Features of Vegetation in Russia" (1902); he also developed a university course on geography of Russia and prepared the summary report on the USSR seas. Among his other works we can mentioned "Essay on Geography and History of Main Cultural Plants" (1923), "Caspian, Black, Baltic, Arctic, Siberian Seas and Eastern Ocean" (1931).

Tanker – a cargo ship designed for transport of liquids in bulk in special tanks for liquid and semi-liquid cargo (oil, benzene, alcohol, lubricating and food oil, wine, chemicals, bitumen, liquefied gases, etc.). The cargo is loaded on T. via pipelines and unloaded with the help of onboard pumps. Water displacement of T. is up to 500,000 tons and more.

Tarkhankut Cape – the westernmost point of the Crimean Peninsula, Ukraine. The sea near this cape is very clear, thus, the divers prefer training here. But T.C. had always been in bad repute among the seafarers. In the past many shipwrecks occurred near it during the misty stormy weather. In 1837 the lighthouse was constructed on the cape and in 1947 a meteorological station was founded here.

Tarkhankut Peninsula – the western end of the Crimean Peninsula between the Karkinit and Kalamit bays of B.S. Its length is about 80 km. The terrain is flat here. Its elevation is to 180 m. It is composed mostly of limestone. The shores are steep. The peninsula is overgrown with different grasses, wormwood-grass steppe vegetation, while the gullies—with shrub thickets. Many saline lakes are found here. In the southwest the Tarkhankut Cape is located.

Tarle Evgeniy Viktorovich (1875–1965) – Soviet historian who contributed much into military studies and naval history of Russia; correspondent member of the Russian Academy of Sciences (1922); academician of the USSR Academy of Sciences (1927). In 1896 he graduated from the History & Philosophy Department of the Kiev University. He started researches in the naval history in the late 1930s. In his fundamental work "Crimean War" (2 volumes, 1941, 1943) using the archive documents T. presented a detailed picture of the origin, course and outcome of the Crimean War of 1853–1856, described the military actions on sea and on land, the episodes of the heroic defense of Sevastopol, many prominent and ordinary participants of the events. He focused much attention on the hero of the Sevastopol defense—Admiral P.S. Nakhimov. During the Great Patriotic War T. published some patriotic articles, such as "Heroic Past of the Russian People", "Heroic Past of the Russian Fleet" and others. In the 1940s his fundamental works devoted to the naval history were published. Among them there were: "The Battle at Chesma and First Russian Expedition to Archipelago of 1769–1774", "The Role of the Russian Navy in Foreign Policy of Russia at Peter I", "Admiral Ushakov in the Mediterranean Sea" (1798–1800)", "Expedition of Admiral D.N. Senyavin in the Mediterranean Sea (1805–1807)". He is also the author of many papers on the history of France, Britain, Russia and some other countries, the history of international relations and wars.

Taukliman – the beach strip stretching 15 km northward of the Kaliakria Cape, Bulgaria. It has small islets and mineral water springs. The large resort complex "Rusalka" (Marmaid) was built here.

Tauri – the ancient tribes that lived in the south of Crimea. According to Herodotus, T. lived "entirely from war and plundering" sacrificing to the Virgin Goddess "the shipwrecked travelers and all Greeks waylaid in open sea". Perhaps T. received their name from Mountain Crimea that was called Taurika by analogy with mountain ridges of Asia Minor and other regions denoted as "Tauri". Herodotus wrote that T. covered "the mountain country" stretching from the Greek city of Kerkinitis (within the borders of present-day Evpatoria) to the Rocky (Kerch) Peninsula. It was thought that T. included the so-called Kizil-Kobin villages. They were great in number and were arranged in a band going from Belogorsk as far as Sevastopol surroundings. The villages usually located along rivers. The mountain villages were protected by two walls of natural stone with backfilling of the space between them. The villages were not large and had a thin cultural layer indicative of not long staying in a place. Archeological findings proved that unlike ancient Greek authors T. were quite peaceful people. The main occupations of T. who inhabited the territories at some distance from the sea shores were farming and animal husbandry as well as weaving and bronze casting. Only seaside T. (their villages did not survive because of thriving seaside resorts development) could be engaged in piracy, but even this they practiced in combination with sea gathering and fishing. T. did not create their own state. Later they were called Taurioscythae. Perhaps they mixed with the Scythians inhabiting Taurika and dissolved in the conglomerate of tribes and peoples who settled down in Crimea in different times. By the fifth century B.C. the ethnic composition of the Crimean population had changed significantly. That time the Greeks founded Tauric Chersonesos and many Scythians went to live here although they started appearing in this area still in the seventh century B.C.

Tauri Mountains – see Crimean Mountains.

Taurida (in the Greek "*Tauris*") – in antiquity the southern part of the Crimean Peninsula or whole Crimea that received its name from the Tauri, the ancient tribes populating this territory. The reference to T. may be found in the works of many authors, including "Iliad" by Homer, "*Epistulae ex Ponto*" (Letters from the Black Sea) by Ovid. More often this place was referred to by its Greek name "*Taurike chersonesos*" or "*Skythike chersonesos*" or by its Latin name "*Chersonesos Taurica*". According to Russian writer and diplomat A.S. Griboedov, "Taurida is a natural museum keeping millennia-old mysteries". A.S. Pushkin also wrote: "Beautiful are the shores of Taurida" or "Among the green waves kissing Taurida". The name of T. came into use again after Crimea was incorporated into Russia.

Taurika – in the ancient times this referred to mountain Crimea by analogy with mountain ridges of Asia Minor included into the notion "Tauri". The antique authors gave this name to Crimea by the name of the Tauri tribes that inhabited this territory.

Tauroscythae – in this way the antique settlement of Taurika was referred to in the Greek sources. This place was also called "Scythotauri" or "Tauri Scythae" which was indicative of the friendly relations between kin peoples.

Tavria Province – the administrative unit formed in October 1802 that incorporated the Crimean Peninsula, Dnieper, Melitopol and Tmutarakan uezds (districts). In 1820 the Tmutarakan uezd was excluded from T.P. and in this composition the province existed till 1917. It received its name by the Tavrida Peninsula.

Tavria Region – it was established after Crimea (Crimean Peninsula and Taman) was acceded to Russia following the Decree of Catherine II of 2 February 1784. On 22 February 1784 Sevastopol and Feodosiya were declared open cities for all peoples friendly to the Russian Empire. Foreigners could freely come to live in these cities. That time Crimea had 1474 villages; its population was about 60,000. It existed till 1802 when as a result of reforms of Pavel I the Tavria Province was set up.

Techirghiol – the largest saline liman lake in Romania located on the B.S. coast 18 km southward of Constanta. Its area is 11.7 km^2, max depth—9 m. T. is surrounded by the hillocky terrain overgrown with steppe vegetation. Once in the past it was a bay of B.S. But the sandy bar was gradually extending and finally separated the lake from the sea. Water was evaporating, the salt concentration was growing and, finally the curative muds were formed there. The winter is mild here; the mean temperature in January is minus 0.5 °C. The summer is warm with the mean temperature being +23 °C. The balneological seaside climatic resorts were constructed on the northwestern shore of the lake. Near T. one can find the famous vines "Murfatlar". The resorts use such curative procedures as mud baths (brines with salinity 70–80 g/l) and climatotherapy.

Temryuk – a town in the Anapa District, Krasnodar Territory, Russia. Established in 1860. It has a berth on the right bank of the Kuban River, not far from its inflow into A.S. This is also a sea port (4 km from the town) and a railway station. The food industry (canning, fish processing, wine-making and other plants) and the engine-repair plant are found here. The town has a regional historical museum. Population—38,400 (2013).

Temryuk Bay – a shallow bay in the southeastern shore of A.S. incising into mainland for 27 km, in entrance width is 60 km. The shores of the bay are low, overgrown with reeds. The Kuban River flows into T.B. and near its mouth the Kurchanskiy and Akhtanizovskiy limans are situated. The bay freezes completely by mid-January and in March the river breaks up. The port of Temryuk locates here.

Temryuk Port – it is located in the Temryuk Bay in the southeastern part of A.S., on the left bank of the Kuban River, distanced 52 km from the Kerch Strait. In August 1994 it was put on the list of sea ports of Russia. The port development is oriented to export of bulk freight, including coal (to 2 million tons a year). For this purpose the protective embankments were restored, the bottom dredging works in the canal were conducted to increase its depth to 5–7 m. The length of wharfs of OJSC "Temryuk Marine Transport Department" is 300 m. The Department has at its disposal five vessels of the "river-sea" type. In winter the navigation is supported by icebreakers.

"Temryuk Surge" – it occurred in late October 1969 when water of A.S. rushed inside the mainland for many kilometers, flooded many villages; many people died.

Tender – a small sail military seacraft designed for reconnaissance, patrol and dispatch services. T. were equipped with 10–12 smooth-bore light guns. They had one masthead and a large bowsprit. The sail rigging included a main gaff, a topsail, a staysail and 1–2 jib sails. Its length was 22–28 m. width 3.5–5 m and deadweight to 200 t. Tenders were used in the Russian Fleet from 1817.

Tendra Bay – locates southward of the Dniester Liman. The entrance into the bay lies between the northwestern end of the Kinburn Spit and the northern end of the Tendra Spit. In the northeastern coast of T.B. there is a large shallow Egorlyk Bay separated from T.B. by the Dolgiy and Kruglyi islands. The northwestern and southeastern parts may be distinguished in this bay. The northwestern part is deep and is limited by the Dolgiy Island, Egorlyk Kut Peninsula and the western part of the Tendra Spit. The southeastern part is shallow and unsuitable for navigation. It cuts into the coast between the southern shore of the Egorlyk Kut Peninsula and the eastern part of the Tendra Spit. The islands of Smolenye and Babin are found in this part of the bay. The shores of T.B. are low and sandy. Many shallows with depths less than 2–m are found here. The seafloor is muddy, covered with shells. The water depth is 4–5 m.

Tendra Sea Battle – an island in the northwest of B.S., presently the Tendra Spit, where on 28–29 August 1790 the battle was fought between the Russian Fleet commanded by Rear Admiral F.F. Ushakov and the Ottoman Fleet commanded by Hussein Pasha. Ushakov came upon the Ottoman vessels near the Island of Tendra and attacked them not putting the Russian ships in a line. After fighting for 2 h the Ottoman vessels started retreating. In pursuit the Russians seized one tail linear ship and two more ships were sunk. The Ottoman Fleet left this area and retreated to the Bosporus. As a result, the Danube Mouth came under control of the Russian Fleet and this complicated significantly the supply of the Turkish fortresses on the Danube.

Tendra Spit, Tendra – a sand bar separated from the mainland by a narrow passage. In fact, this is a narrow sandy island near the northern coast of B.S., to the southeast of Odessa. Its length is about 65 km, width—to 1.8 km. The lighthouse was constructed on the western coast of the spit. In 1790 the sea battle occurred near T.S. in which F.F. Ushakov victoriously defeated the Turkish Fleet. In 1824 the sea explorations for lighthouse construction were conducted here by naval officer N.D. Kritskiy. The riot of sailors on the battleship "Potemkin" happened near T.

Tendra Spit (http://fishing-ua.com/attachments/22824/)

Terme – a town on the coast of the Black Sea, Turkey. Terme (formerly spelled *Termeh*; Ancient greek: *Thèrmae*) is a town, the headquarters of Terme District, Samsun Province, Turkey. It is named after the Terme River. Scholars have identified Terme or its environs as the site of the ancient city of Themiscyra. Terme District is the site of an annual festival celebrating the Amazons, an ancient nation of all-female warriors who are believed to have lived in the Samsun region. Population—31,200 (2012).

Tethys – a system of ancient sea basins in the Mediterranean geosynclinals area. In the Paleozoic, Mesozoic and early Cenozoic time they stretched along the north-west of Africa, Western and Southern Europe, Asia Minor, Caucasus, Iran, Afghanistan, the Himalayas, Indochina and Indonesia. In the Neogene as a result the Alpine folding processes the high ridges of the Alpine-Himalayan mountain belt (Atlas, Pyrenees, Alps, Caucasus, Hindu Kush, Himalayas), mountain structures of Indochina and Indonesia were formed in place of T. The Mediterranean, B.S., Caspian Sea, Persian Gulf and seas of the Malay Archipelago are, in fact, the remnants of T. The name *"tethys"* was suggested in the late nineteenth century by Austrian geologist Eduard Suess by the name of ancient Greek Sea Goddess Thetis.

Thalassotherapy (in the Greek *"thalassa"* meaning "sea") – the medical use of sea climate, bathing and solar baths. It is widely practiced in seaside resorts on B. and A. seas. In a broad sense of the word T. is close to climatotherapy as it applies the same curative methods, although the treatment on the seaside has its specifics, while in the narrow sense T. is reduced to the use of sea bathing for health improvement and better fitness. The physiological effect of sea bathing is attained by a combination of factors. For example, the lower (compared to the body temperature) water temperature produces a thermal effect on an organism. Sea waves act as a mechanical factor ("water massage"). During bathing the salts dissolved in sea water deposit on the skin causing chemical irrigation of its receptors. The combined effect of these factors maintains for some time the response of an organism developed during bathing. Certain effect is also produced by the marine bacterial flora and phytoncides released by sea algae. During bathing a man inhales the ionized sea air, his body is exposed to solar radiation because UV rays are capable to penetrate through water to a depth to 1 m. As a result, T. trains the thermal regulation mechanisms; regulates the functioning of heart, breathing organs, muscles; helps to increase oxygen intake by an organism; improves metabolism and fitness of an organism. Inclusion of sea bathing into a health treatment complex improves its efficiency.

The Gyreis – the ruling dynasty in the Crimean Khanate that stayed in power the longest: 340 years (from 1443 through 1783). Mengli-Gyrei I, the son of Khadzhi Devlet-Gyrei who founded the state, ruled the Crimean Khanate for the longest time: from 1467 to 1475 and from 1478 to 1515, i.e. 45 years in total. Unlike other Gyreis, this outstanding state and political ruler was in power till his natural death date. During the ruling of Mengli-Gyrei I the very important events occurred in the Crimean Khanate. Thus, in 1474 the first Russian ambassador visited Crimea; in 1475 the Khanate got into vassal dependence on Turkey; in 1489 the Crimean Tatars made their first raid on the Ukrainian lands and burnt Kiev; in 1502 the vassal dependence on the Blue (Desht-i-Kipchak) Horde was broken. Mengli-Gyrei

I was buried in the mausoleum of his father built at the foot of the Lufut-Kale near Bakhchisarai.

"The journey beyond three seas" – a narrative of Afanasiy Nikitin about his trip over Persia and India in 1466–1472. This is a literary and geographical monument of the fifteenth century. In 1466 Ivan III after receiving the embassy from Shemakha sent there his envoy Vasiliy Panin with a team that included merchants from Moscow and Tver who were going under protection of the embassy to establish trade relations with the eastern coast of the Caspian. One of the merchants was Afanasiy Nikitin. After death of A. Nikitin in Smolensk on his return home his travel notes were delivered to the Moscow Embassy Board where they were studied very carefully. At the beginning of his notes Nikitin wrote the following: "I described here my travel beyond three seas: first sea—Derbenskiy (Caspian), second sea—Indian and third sea—Black". The notes may be divided into three parts. The first part describes his travel from Tver to Shemakha and then via the Persian lands to the shores of the Persian Gulf; the second part—his travel over India and the third part—his traveling back via Persia and Turkey, from the shores of the Persian Gulf to the Black Sea and Crimean coast.

The struggle against the Wite Guard on the Black Sea – it is customary to regard its starting moment the departure of the Black Sea Fleet under the command of Read-Admiral M.P. Sablin from Sevastopol on April 29, 1918, after the Germans occupied Ukraine and the Crimea.

Theodoro – principality. A medieval principality that existed in the Crimea, a feudal state in the south-west of the Crimea around the twelfth–fifteenth centuries. The territory of T. (in the north and north-east as far as the Kacha River, in the south—up to the Black Sea coast, from the contemporary Balaklava to Alushta). The principality was established by Constantine, a Byzanthian émigré from the rich aristrocratic Armenian ancestry of the Gavrases. Constantine had been exiled from Trapezund in 1140. T. was engaged in foreign trade on a large scale. The stronghold of T. principality in their struggle against the Genoese on the Black Sea was Calamita Fortress built in 1427, which defended the only sea port of Avlit principality in the Chernaya River mouth. The capital of the principality sited on Mt. Mangup was the last citadel in the fight against the Turkish invaders in the Crimea.

The capital Mangup, a former Got fortress, belongs to the group of "cave cities"; it used to be sited 20 km from Bakhchisarai, near to Sevastopol. Mangup Plateau is 250–300 m higher than three picturesque adjoining valleys: Karalez, Jan-Dere and Ai-Todor. Its elevation above the mean sea level is nearly 600 m. The city area used to be quite significant: around 90 ha, and was one of the largest fortresses of medieval Tavrika. Funa Fortress built at the foot of Mountain Demerdzhi was the second largest city of the principality.

The period of Prince Alexis rule (middle of the fifteenth century) was particularly successful to the Principality. The Prince managed to secure T.'s exit to the sea trade routes. With this in mind, he remodeled and fortified Kalamita Port which, growing lucratively as it did, began competing with the ports of Genoa. Economy of the principality was, above all, based on agriculture that developed in the fertile valleys of the south-western Crimea. The principality was the center of domestic craft industry which advanced to a rather high level, which was, for the most part, true of pottery, ironmongery and weaving.

The year 1475 proved fatal to the Principality of T. The Turks approached the T. capital on a huge squadron, destroyed the city and killed most of the dwellers. Prince Alexander was taken prisoner, sent to Constantinople and executed there. The only person representing the aristocratic ancestry in the male line, who survived, was Alexander's minor son. He was subsequently converted to Islam and became the founding father of a noble Turkish clan. After the fortress was seized, the former capital of Feodorites was turned by the Turks into a fort and became the center of a kadalyk (district) that included lands hitherto possessed by the T. principality. After the signing in 1774 of the Kuchuk-Kainardzhi Treaty, the Crimean Khanate got rid of Turkey's vassalage, and the Turkish garrison left Mangup for good. The last dwellers of T. were the Crimean Karaims.

Tikhmenev Aleksandr Ivanovich (1878–1959) – Russian Captain first Rank (1917), Rear Admiral (1920). He finished the Naval Corps (1901) and the Mine Officer School (1904). He took part in World War I commanding destroyers "Zhutkiy" and "Nepokornyi". He was commander of the battleship "Volya" (till February 1917 its name was "Emperor Alexander III") on B.S. With the detachment of ships under general command of Admiral Sablin in February 1918 he went from Sevastopol to Novorossiysk saving the ships of the B.S. Fleet from the German troops that intruded Crimea. When in May 1918 Sablin received the V.I. Lenin's order to sink the ships in the Novorossiysk Bay as the Voluntary Army of General Denikin was approaching Novorossiysk, Sablin refused to fulfill the order and went to Moscow for personal reporting to V.I. Lenin. Sablin transferred his commander powers to T. who was that time captain first rank. T. also refused to sink the ships, moreover, regardless of this order on 18 June 1918 he led some ships (battleship "Volya" and six destroyers "Bespokoinyi", "Derzkiy", "Zhivoy", "Pylkiy", "Zharkiy" and "Pospeshnyi") back to Crimea. He transferred the ships led from Novorossiysk to Commander of B.S. Fleet in Crimea Admiral Kanin (who came on 18 June 1918 from Moscow to Novorossiysk fulfilled the Lenin's order and sank the ships that remained in Novorossiysk). After coming of foreign ships and occupation of Crimea by enemy troops (November-December 1918), T. was enlisted to the reserve of the Marine Department of Headquarters first of Denikin and later of Vrangel. On 17 October 1920 he was appointed by Vrangel the chief of staff of the Marine Department of the Russian Army. In November 1920 he organized evacuation of the Russian Army from Crimea. After coming of the B.S. Fleet to Constantinople and navigation to Bizerta (Tunisia) in December 1920 Admiral T. was appointed the chief of staff of the Russian squadron (in emigration). On 29 October 1924, after France recognized Bolshevik Russia, T. remained as emigrant in Tunis where he died in Bizerta.

Tiligul Liman – one of the largest limans in the northwestern part of B.S., Ukraine It was formed as a result of water salinization in the Tigul River mouth. For a long time its natural connection with the sea was not regular and long. Disturbance of the liman water balance led to shrinking by nearly a half of the water surface area, but after heavy rainfalls and spring floods the liman restored its connection with the sea. Such periodical connection of the liman with the sea and from 1959 the functioning of the sea canal facilitated penetration into T.L. of the saline-water and sea marine organisms. At present the water regime and the surface water area depend on how long the liman is connected with the sea. The mean water level in the liman is 40–50 cm lower than the sea level. The average annual volume of water exchange with the sea is about 3 km^3. The year-to-year variations of the water salinity in T.L. are rather large. The maximum salinity recorded in 1870 was 40‰, while the minimum was 5‰ in 1945. In 1960–1964 the quantity of fish species inhabiting T.L. was the largest and reached 45–49. At present among the fish species having commercial significance there are atherine, flatfish (gloss), B.S. gray mullet, bullheads.

Tira – see Belgorod-Dniestrovskiy.

Tirebolu – a small town on the coast of the Black Sea, Turkey. It is located on the hill named Ayana which rises from the Black Sea shore just to the west of the Harşit River mouth. Tirebolu has a little harbour and a fishing fleet but the mainstay of the local economy is growing hazelnuts. Population—14,800 (2012).

Tmutarakan Principality – the accurate time of its foundation is unknown due to lack of documentary proofs. The city of Tmutarakan (Tmutorokan, Matarkha) being the capital of the principality is identified with present-day Taman on the Taman Peninsula. The fortified city of Tmutarakan appeared in place of the ancient colony about 965 after the southern military campaigns of Svyatoslav Igorevich. T.P. was first mentioned in 988 in the chronicle "Povesti vremennykh let" (Tales of Ancient Times) when Vladimir Svyatoslavovich founded there a principality and gave it to his son Mstislav, thus, acceding these lands to Old Rus'. Although the borders of T.P. are not known exactly it is thought that it extended over the half of the Kerch Peninsula and territories on both banks of the Kuban River spreading into mainland for 200 km. In 1054 T.P. was included into Northern land. In the early 1060 T.P. became the core of dispute between Kiev and Chernigov being transferred to control of one or another. In the middle of the twelfth century T.P. was conquered by the Polovets.

Tolstoy Lev Nikolaevich (1828–1910) – Russian count, great Russian novelist who stayed in besieged Sevastopol from April 1855 till raising the siege. That time he was a young writer who was transferred from the Danube Army to Sevastopol. Second Lieutenant T. was assigned to the battery of the Yazon redoubt (fourth bastion) ("Yazon" is the name of a brig which crew built this redoubt). On 4 (16) August T. took part in the battle on Chernaya Rechka and later in the battles in the city. "For Composure and Quick-Mindedness" in the battles he was conferred the rank of second lieutenant and awarded the Order of St. Anna, fourth degree "For Bravery". Being here he wrote the first of his famous "Sevastopol

Sketches"—"Sevastopol at Day and at Night" that was followed by other stories. The obelisk was constructed in the place of this redoubt. In 1901–1902 T. lived in B.S. resort Gaspri. He is best known for two long novels—"War and Peace" (1869) and "Anna Karenina" (1877).

Tolstoy L.N. (http://www.tolstoy.ru/way/art/images2/739.JPG)

Tolstoy Petr Andreevich (1645–1729) – count (1724), Russian statesman and diplomat. He was the first count in the Tolstoys family and the ancestor of three famous novelists and many famous figures of Russian culture and state. He took part in the siege of Azov (1696). In 1697–1698 he studied in Italy. In 1707–1714 he was the Russia's first regularly accredited ambassador to the Ottoman Empire. He influenced significantly the policy of Porte keeping the Turks from starting the war on Russia. He took part in negotiations about extradition of Charles XII and Mazepa who took refuge on the Turkish soil after the battle at Poltava. In 1710 during the war with Russia the Ottoman government had thrown T. into the Seven Towers castle and after the Peace Treaty of Prut (1711) he was kept in Constantinople as a hostage. In 1716–1717 he accompanied Peter the Great on his voyage over Europe. Many people thought him to the "cleverest brains" in Russia. T. made the best description of the Ottoman Empire focusing on Prichernomorie and the situation in the Ottoman Fleet. In 1725 he assisted in raising Catherine I to the throne, but after a conflict with A.D. Menshikov he was arrested and exiled. In 2006 in Moscow his diary and messages of the time when he was the ambassador "Description of the Black Sea, Aegean Archipelago and Ottoman Fleet" were published for the first time.

Tolstoy P.A. (http://www.tula.rodgor.ru/news_images/tolstoy2.jpg)

Tomy – a city-colony founded in the mid-seventh century B.C. on the western coast of B.S. where the modern city of Constanta is situated at present in Romania. Despite the fact that soon after its foundation T. became economically dependent on Histria and Callatis, already in the Hellenistic time this colony was in the focus of attention and with time onward it turned into a major economic center of Western Prichernomorie. Ovidius Publius was sent in exile here. The archeological excavations in the area of the Antique port (buildings decorated with mosaic, storehouses, terms) proved that in the late Antiquity T. was a large city.

Toprak-Kaya Cape (in the Turkic meaning "clay rock") – has also another name—"Chameleon" that was received thanks to its ability to change color depending on the position of the Sun. This visual effect is created by specific spreading of clay shales composing the cape and reflecting sunrays differently. This cape closes in the northeast the Koktebel Bay, Crimea, Ukraine.

Tortoise (Shell) Ridges – name of ancient sea bars of the Sea of Azov that are at the distance of 20–30 m from the seashore.

Totleben Eduard Ivanovich (1818–1884) – a prominent Russian military engineer, Aide-de-Camp General (1855), Engineer-General (1869), count (1878). T. developed further the ideas of gifted Russian engineer A.Z. Telyakovskiy who first suggested considering the fortification systems in conjunction with the tactics and strategy of war. In 1832 he entered the Chief Engineering College, however, the heart problems prevented T. from finishing the full course of studies. In 1836 he was promoted to engineer-warrant officer. In 1838 he went to serve to the Riga fortress team and in 1840 in the rank of lieutenant he was transferred to the training sapper battalion. Here K.A. Shilder noticed him and ordered to study the mine system. For further studies T. was sent with a team of sappers to Kiev where he headed experiments on waging underground war. In 1845 for his work T. was promoted to staff-captain. In 1848 he went to the Caucasus where he took part in several expeditions. He contributed much into success of the siege of Gergebil and after seizure of Gergebil T. was promoted to captain. In 1849 he was in charge of all works on siege of the fortress of Chokh. After return from the Caucasus T. was appointed aide-de-camp to General Shilder and in 1851 he was appointed the guard engineer and settled down in Petersburg where he commanded the guards sapper battalion during encampment exercises. In early 1854 T. was called to the head-quarters of the Danube Army and here he fulfilled the orders of General Shilder having fulfilled some brilliant reconnaissances under the Turkish fire and developed the plan of offensive against fortifications at Calafat. With the beginning of preparatory works for siege of Silistria, T. was appointed the trench-major. After General Shilder was wounded T. headed all works and on 7 June blasted the whole frontline of fortifications of Arab-Tabi. When the siege of Silistria was raised he was commanded to Sevastopol where enemy landing was expected. First Chief Commander Prince Menshikov thought that because it was already autumn the allies would not dare to land in Crimea and declined the initiative of T. to start immediately the defensive works. But they landed and only then the defensive works were started. T. widened the frontline position on the line of Northern fortification and created practically anew the defense line on the southern side. The defense line was constructed as follows: first, the fortified place nearest to the city was found and powerful cannons were installed in its main points that were connected with trenches for rifle defense and for shelter, some batteries were organized between the main points. In this way the approaches to the city were strongly defended with gun and rifle fire in front and on flanks. The works were conducted day and night. During a short time the terrain where previous enemy reconnaissance detected only weak fortifications with large unprotected spacing was covered with a continuous defense line. The allies had to abandon the idea on seizure of Sevastopol by direct offensive and on 28 September they started the siege of the city. The first bombardment on 5 October demonstrated the strength of Sevastopol fortifications and well-chosen direction of gunning. After this the enemy decided to change over to the underground war and to blast the fourth bastion, but here T. again was ahead—the enemies quite unexpectedly came across very skillfully made network of mine galleries. On 8 June T. was wounded in the foot, but despite this he continued commanding the defense works until he felt so badly that had to leave Sevastopol. According to T. estimates, during the Sevasto-pol siege the enemy rushed on the city about 1,356,000 artillery shells.

The efforts of T. during the Sevastopol defense were duly appraised by the Tsar. In October 1854 he was promoted to lieutenant-colonel, in April 1855—to major general and was appointed to the escort of Tsar. On 25 April it was ordered to engrave his name on the marble plaque of the Engineering College in Nikolaev. After the fall of Sevastopol, in September 1855 T. was called to Nikolaev, which strategic importance after the fall of Sevastopol had grown enormously, for con-struction of city defense. The message of T. concerning fortification of Nikolaev may be considered one of his most valuable research works. His ideas that took shape as a result of the fresh experience gained in the recent battles opened a new era in the fortification art. They departed sharply from traditional views dominating that time even in France regardless of the experience of the Napoleon wars. T. pointed to the need to have a system of forts with in-between artillery positions to which railroads should be constructed. He considered the forts as the key points of struggle. T. focused much attention on distribution of all kinds of armaments and the role of each of them.

On his return to Petersburg T. was employed in strengthening the fortifications of Kronstadt. After this for 2 years he studied the fortresses in Germany and France and engineering art. In 1859 he was appointed Director of the Engineering Department of the Military Ministry (1859–1877); in 1861—the head of the General-Inspector Staff Office; in 1863—the assistant of general-inspector on engineering. In 1863 in view of the expected political complications T. conducted improvement of some fortresses—Sveaborg, Dinaburg, Nikolaev, Vyborg, construction of fortifications in the mouth of the Neva and Zapadnaya Dvina rivers. Kronstadt was fortified against sea attacks. In 1869 T. prepared the project on fortification of Kiev. Being the Chairman of the Artillery Engineering Commission, T. took an active part in equipment of Russian fortresses with rifled guns. At the same time, T. initiated reorganization of engineering troops in accordance with the most recent achievements in the military art.

From 1871 to 1875 T. developed a new system of defensive lines with the basic fortresses. For this purpose he conducted investigations near Brest-Litovsk, Kovna, Belostok, Goniondza, Grodno, Dubna and Proskurovo. In 1873 the meeting devoted to strategic situation in Russia chaired by Alexander II adopted the T. plan that included the following main concepts: (1) to improve outposts at Novogeorgievsk, Ivangorod and Warsaw and to construct outposts around Brest for covering railroads; (2) to fortify the cities of Grodno, Kovna and positions near Vilno, to construct fortifications near Osovets and a crossing over the Zapadnaya Dvina River near Riga; (3) to construct fortifications in front of Dubna and Proskurovo; (4) to construct outposts in Bendery and fortifications at Ochakov and Yampol. The implementation of this plan was interrupted by the Russian-Turkish War of 1877–1878.

In 1876 T. was called to Livadia and appointed the Head of the Black Sea Defense with control of naval forces. On the T. orders at approaches to Kerch, Ochakov, Odessa and Sevastopol there were installed mines, constructed new batteries and their armament was improved. In late 1876 T. returned to Petersburg and it was not until 2 September 1877 when the Plevna siege was dragging on that T. was called to the front where he undertook command of siege works at Plevna. After seizure of Plevna on 29 November T. was appointed the Commander of the Eastern Detachment, but on 8 February he was called to Petersburg to take part in the meeting on conquering the Bosporus and on its closing for the British Fleet anchored at the Princes Islands. T. was appointed the Chief Commander of the Army. Arriving to the army he found that seizure of the Bosporus near Buyuk-Dere, keeping in mind impossibility to obstruct the Strait with minefields and to maintain communication with the Russian B.S. ports, would be useless and that in the event of successful storming of Constantinople the benefits would be only temporary and in the event of a failure the results of the previous campaign could be lost. So, the main task of T. as the Chief Commander was to support the Russian diplomacy during negotiations on signing of the final peace treaty forcing the Turkish government to the soonest and accurate satisfaction of the Russian demands and control the return of the Russian troops to Russia. At the same time T. proposed some measures on self-defense of Bulgaria after departure of the Russian troops.

In 1879 T. was appointed the Governor-General of Odessa and Commander of the Odessa Military District; in 1880 T. held the post of Governor-General of Vilno, Koven and Grodno and Commander of the Vilno Military District. In commemoration of 25 years from the first bombardment of Sevastopol he was made a hereditary count. T. was awarded many highest orders of Russia and foreign states. In 1909 the monument to T. was installed in Sevastopol.

Totleben E.I. (http://historydoc.edu.ru/attach.asp?a_no=1204)

"Tovarisch" – (1) Four-mast barque, sail vessel for training of the staff of the USSR Marine Fleet. It was built in 1892 in Belfast (Northern Ireland) as a cargo vessel that sailed under the name of "Lauriston" and was provided with the sailing equipment (all masts had only square sails). The hull and part of masting of the vessel were steel. In 1908–1909 the sail vessel was re-equipped into barque (the jigger was installed on the mizzen mast). After 1917 the vessel was captured by invaders and brought to Britain. Only in 1921 the barque was returned to the Soviet Republic. In 1922 in Petrograd it was retrofitted into a training sail vessel. Soon the vessel was given a new name—"Tovarishch" (Friend) and was transferred to the marine college. Every year the students of marine schools and colleges passed their practical course on this vessel navigating over B.S. In autumn 1941 in Mariupol (Zhdanov) the vessel was captured by the Germans and sank. At present the anchor of this vessel that was raised from the sea floor is installed in the park of this city.

The vessel's displacement was 4,750 tons, maximum length—over 100 m, total area of sails—3,000 sq.m, crew—32 men, number of students—to 120 persons.

(2) Three-mast barque, the training vessel of the Kherson Nautical College named after Lieutenant Shmidt. It was built in 1933 in Hamburg and was given the name "Gorch Fok". It was included into the German Navy as a training vessel. It was handed over to the USSR in 1945 as a result of division of the German fleet among allies following the decisions of the Potsdam Conference and was renamed into "Tovarishch". The Soviet flag was hoisted on this vessel in 1950. Its displacement was 1,800 tons, length—89 m, speed—8 knots, total area of sails—1,860 sq. m, crew—45 persons, number of students—to 140 persons.

"Tovarisch" (http://windgammers.narod.ru/Korabli/Tovaris002.jpg)

Trabzon (Trapezund) (in ancient Greek "*Trapezius*", in the Turkic "*Trabzon*") – a city and merchant port on the southeastern coast of B.S., on the western coast of the Trabzon Bay, westward of the Guzelhizar Cape; situated in the northeast of Turkey; administrative center of the Trabzon Province (Il). In the first century Strabo mentioned it as *"Trapezus"* from the ancient Greek *"trapeze"* meaning "table" which may indicate to the flat tops of the surrounding mountains. Possibly, this name was brought by Greek colonists. Europeans refer to it as Trapezund (*"zund"* meaning "strait"). T. was founded by the Greek traders from Miletus in the second half of the seventh century B.C. In the second century B.C. it was conquered by the Romans and for several centuries it was one of the main ports in the east of the Roman Empire. T. situated at the starting point of an important trade route to the East and reached the top of its development during the Roman ruling. In 257 it was destroyed by the Goths. In the fifth—early thirteenth centuries when it was a part of Byzantine it was again on the rise and became the capital of the Trapezund Empire (1204–1261) that existed till 1461. T. was founded with the support of the Georgian Kingdom by Alexey Komnin, the grandson of Byzantine Emperor Andronik I Komnin. This dwarf empire stretched as a narrow strip along the northeastern coast of Asia Minor and from 1244 it was under control of the Mongols. T. was an important center on the trade routes from Chernomorie to Asia Minor. The city turned into a flourishing trade center under the ruling of Khulagids and was the most important harbor at the beginning of the transit route from B.S. to the inner regions of Iran and Transcaucasia. T. was connected by seaways with Kafa, Soldaier and Pero and traded with the merchants on the western coast of the Caucasus. In the transit trade of T. the key role was played by Genoese who in 1306 became the owners of Leonastro castle, the highest point in the city, and in 1314 they took hold of the armory and shipyard. Via T. the Genoese merchants exported salt, grain, furs, fish, cheese and wine to Armenia and Iran.

In 1461 T. was conquered by the Ottomans. In the seventeenth century it was frequently raided by the Don Cossacks. The Armenians and Greeks contributed much to T. flourishing. In 1920–1921 it was the center of the so-called "autonomous Armenia". T. was the place of birth of prominent sultans of the Ottoman Empire— Selim the Terrible and Suleiman the Beautiful. The population of T. was 243,700 (2012). This is the starting point of automobile road Trabzon-Erzurum-Iran. The city has an airport. The food, cement, shipbuilding (large vessels), leather industries and fishing are well developed here. The port is used for export of hazelnuts, tobacco, wool and timber. The Black Sea Technical University and architectural monuments of the seventh–fifteenth centuries are situated here. At present T. gradually turns into a business center. The city locates on three hills gently sloping to the sea. The hills are divided by valleys over which bridges are built. The middle hill is surrounded by walls and here the administrative institutions are located. In the east of the city, at the turn of the Erzurum highway to the south there is a large suburb town of Degirmendere. More eastward of this town the industrial region begins. The port of T. is a transit point for Erzurum and Diyarbakir. T. has many historical monuments among which there are: the citadel, the Hagia Sophia, a stunning Byzantine church of the thirteenth century, old Christian churches turned into mosques, old houses. T. is put on the UNESCO World Heritage List.

Trabzon Bay – a narrow bay that incises not deeply into the mainland. The shores are sandy. The depths in its middle part are 5–7 m and more seaward they grow and reach 15 m. The city-port of Trabzon is situated on its western shore.

Trabzon Province (II) – the Black Sea province in Turkey. Its area is 6,685 sq. km; population—763,700 (2011). The capital is the city of Trabzon. Neighbouring provinces are Giresun to the west, Gümüşhane to the southwest, Bayburt to the southeast and Rize to the east. Trabzon province is divided into 18 districts. The province was a site of major fighting between Ottoman and Russian forces during the Caucasus Campaign of Worl War I, which resulted in the capture of the city of Trabzon by the Russian Army under command of Grand Duke Nikolay and General Nikolay N. Yudenich in April 1916. The province was restored to Turkish control in early 1918 following Russia's exit from WWI with the Treaty of Brest-Litovsk.

Trachean Peninsula (in Greek meaning "stone peninsula") – one of the names of the Heraclean Peninsula.

Transport Corridor Europe-Caucasus-Asia (TRASECA) – a project of a new Eurasian transport corridor crossing the Caucasus region. The restoration of the Great Silk Road was the main idea of the program on technical promotion of corridor development that was considered in 1993 at the conference in Brussels (Belgium) attended by the leaders of eight countries of Transcaucasus and Central Asia. The Brussels Declaration on transport corridor Europe-Caucasus-Asia (TRASECA) was signed by member states of the European Union. It was assumed that the transit flows of the South Caucasian corridor will be divided in the ports on B.S. and Caspian Sea: Poti and Batumi (Georgia)—to East and Central Europe via B.S. and Mediterranean ports of European countries; Baku (Azerbaijan)—to Central and Far East Asia via the ports of Aktau (Kazakhstan) and Turkmenbashi (Turkmenistan).

Trapezund – see Trabzon.

Trapezund seizure – a landing operation on seizure of the Turkish port of Trapezund from the ships of the B.S. Fleet commanded by Rear Admiral A.A. Sarychev in October 1810. On 6 October the squadron of ships carrying 4,000-strong landing troops was sent for seizure of Trapezund. Having approached the port Sarychev decided to immediately attack the city. But quite unexpectedly the troops that landed near the port were attacked by the Turkish troops and after a fierce and bloody battle they were forced back to the sea. Only half of the landing troops could reach the ships. After such defeat Sarychev did not venture to make a new landing attempt. After bombardment of Trapezund the ships returned to Sevastopol.

TRASECA – see Transport Corridor Europe-Caucasus-Asia.

Traversay Ivan Ivanovich (Jean-Baptiste Prevost de Sansac, marquis de Traversay) (1754–1831) – a French seaman, admiral, Navy Minister of Russia. In 1765 he joined the Navy College in Rochefort and when this college was closed, his class resumed training in the Naval School in Brest. At the age of 13 he took part in his first military battle. As after the Revolution the French Fleet was falling apart, T. went abroad and soon received an invitation to join the Russian service. In 1791 being in the rank of the captain of the French Navy he was appointed to the rowing fleet as Captain of the General-Major rank under the name Ivan Ivanovich. In 1801 he was promoted to Admiral; in 1802 he was appointed the Chief Commander of the B.S. Fleet and ports, the military governor of Sevastopol and Nikolaev. During the Russian-Turkish War of 1806–1812 he commanded in 1807 the operation on storming Anapa and the Ottoman fortress in the North Caucasus. In 1809 in the absence of Navy Minister P.V. Chichagov he was entrusted the management of the Navy Ministry. After resignation of P.V. Chichagov in 1811 he was appointed the Minister of the Navy. He declined the offers from Napoleon I to take the highest positions in the French Fleet on any conditions. He did much for development of shipbuilding, sponsored long-range expeditions on boats "Vostok" and "Mirny" as well as "Otkrytie" and "Blagonamerennyi" for exploration of Polar seas. In 1828 he retired.

Traversay I.I. (http://upload.wikimedia.org/wikipedia/commons/3/39/Traverse-%D1%81.jpg)

Treaty between the Russian Federation and the Ukraine on cooperation in the use of the Sea of Azov and Kerch Strait – was signed in Kerch on December 24, 2003 by the Presidents of Russia and Ukraine. This Treaty defines that A.S. and Kerch Strait are the internal waters of Ukraine and Russia. A.S. is delineated by the state border as stipulated in the bilateral treaty. The non-commercial vessels under the flags of the parties are entitled to free sailing in A.S. and Kerch Strait. The vessels of other states may sail in A.S. and pass through the Kerch Strait only on an invitation of one party having received permission from the other party. The signing of this Treaty was preceded by a conflict between these two states in mid-2003 in connection with construction by the Russian Party in late September of a dam from the Taman Peninsula to the Tuzla Island. Supreme Rada of Ukraine ratified this Treaty on April 20, 2004 and the Federal Assembly of the Russian Federation—on April 22, 2004.

Treaty between the Russian Federation and the Ukraine on the Russian-Ukraine state border – was signed by the Russian and Ukrainian Presidents on January 28, 2003 in Kiev. Its signing became possible after complete coordination of the delimitation maps of the land section of the Russian-Ukrainian border. This Treaty fixed the land border between these two states. Its length is 2,063 km.

Trebaca, Trebaccolo (in Italian "*trebaccolo*") – a sailing cargo or fishing coaster with two masts most often used in the Mediterranean, Black and Azov seas. The length of cargo T. is about 28 m, width about 6 m, the board height about 2 m; the length of fishing T. is 10–20 m, width 3–5 m, board height 1–2 m. It has lugsails and two jib sails on the bowsprit.

Truce of Andrusovo, Andrusovo Truce Treaty (1667) – between Russia and Rzeczpospolita for 13.5 years. Signed on January 30 (February 9) at Anrdrusovo Village near Smolensk. The truce was the outcome of the Russian-Polish War of 1654–1667. Russia was represented by the Grand and Plenipotentiary Ambassador A.L. Ordin-Nashchokin, Rzeczpospolita—by Commissar Jerzy Chlebowicz. Rzechpospolita returned Smolensk and Chernigov Provinces to Russia, recognized reunification of Left-bank Ukraine with Russia. Kiev with a small district was handed over to Russia for 2 years. Zaporizhian Sich was declared to be managed jointly by Russia and Rzeczpospolita. The truce made provisions for joint action against Crimean and Osman attacks. T. was reaffirmed by "Eternal Peace" of 1686.

Tsarevo – one of the largest settlement on the Bulgarian Black Sea coast. The village is sited on the shore of a picturesque harbor at the foot of Mount Pipiya. Legends have it that Byzanthian emperors (Vasilevs) used to like it very much, therefore its original name was "Vasoliko"—"Imperial".

Tsargrad – see Constantinople, Istanbul.

Tsemes (Novorossiysk) Bay – deep-water, non-freezing harbor at the north-eastern coast of the Black Sea, Krasnodar Territory, Russia. In terms of location and depths, one of the best Black Sea harbors. Named after the Tsemes River falling into the harbor; often is also called after the city-port of Novorossiisk at the top of the harbor. The harbor is in the northern part of the Caucasian coast. It is formed by Sudjuk Spit and Doob Cape. Cuts into the shore 4.5 km to the north-east-north of Myskhako Cape. Length 15 km, the width at the entrance is 9 km, in the northern part—4.6 km. The shores in the south-west are low-lying, in the north-east— elevated, little-dissected. The harbor shores are hemmed by a reef stretching almost everywhere parallel to the shoreline. The bay has flat topography of the bottom and depths ranging from 21 to 27 m; near the shores the depths decrease abruptly. In this middle of the entrance, there are large Penai shoalbanks. During the autumn-winter seasons of the year, the north-easterly wind in bay gains the force of a hurricane and spreads 30–50 km from the shore. This wind has a local name "Novorossiysk bora" and presents a real hazard to navigation. In 1918, most of the Black Sea Navy was sunk in the harbor lest Germany should seize a single ship.

Tsemes Bay (http://dic.academic.ru/pictures/wiki/files/78/Novo_bay.jpg)

Tsemes River – narrow and shallow river in the city of Novorossiysk. The river originates in the north-eastern slope of Mount Gudzeva (height of 425.6 m). Riverbed passes in the city center of its industrial parts. From the river name occured the name of Tsemess Bay into which it flows from Tsemess groves.

Tsikhisdziri – maritime climate health-resort in Georgia (Adjaria, Cobuleti District); railway station 19 km south-east of Batumi. Sited on the Black Sea coast of the Caucasus. Vegetation is represented by subtropical ornamental and fruit trees and shrubs; in the environs—tangerine, orange and lemon groves, tea plantations. The climate is subtropical, characterized by abundant heat and moisture (mean annual relative humidity 78 %). The winter is very mild, with no snow; mean temperature of January is 6 °C. The summer is very warm. Mean temperature of August is 23 °C. Rainfall is uniform, its average annual total is around 2,750 mm. Mild climate is good for the treatment of cardio-vascular diseases and the nervous system, respiratory organs.

Tsygan Pier, Harbor or Chehgene-Skelya – cuts into the shore between Chukalya and Foros Capes. Sited on the Bulgarian Black Sea coast. The shores of the harbor are mountainous, mostly upright, with rare trees; in places, there occur small beaches. The eastern shore of the harbor has steep slopes, while the south-western shore is more gently-sloping. The Marinka and Otmanlii Rivers fall into the southern part of the harbor. Kraimorie Settlement is near a beach on the western shore of the Harbor. The depths at the entrance to the harbor are 11–13.4 m. There are the best protected and safest anchorage sites in the harbor not only in Burgas Bay, but all along the coast from Bosporus Strait to the Danube River; even when storms come from the north-east, wave is almost imperceptible here.

Tsygane (Gypsies) – (selfname—*"Roma"*) people, apparently associated by its origin with some castes of the peoples of North India. Language—Tsygan, belongs to the Indo-Aryan group of the Indo-European Language family, is divided into a number of dialects. In mid-first millennium, Ts. left India for good and settled first in the countries of the Middle East. In the eleventh century, they emerged on the Balkan Peninsula, and in the fifteenth–sixteenth century proliferated all over Europe. The most significant groups of Ts. dwell in Bulgaria, Romania, Ukraine and Moldavia. Population in the world is estimated as 8–10 million.

Tuapse – a city, center of the Tuapse District (Krasnodar Territory), Russia; a port on B.S.; a railway station. It is situated on the B.S. coast, in the mouth of two rivers—Tuapse and Pauk to the southwest of Krasnodar. Its population was 63,000 (2013). This is the sea cargo port handling oil products and dry cargo. The oil refinery, ship-repair, processing, food and construction industries are developed here. Resort areas are located around T. Fruit, vegetable and tobacco growing are widely practiced here. It was founded in 1838 as the fort of Veljaminovskiy by Admiral Mikhail Lazarev in memory of General Veljaminov who commanded the B.S. coastal line. As many other Russian fortresses on B.S., it was destroyed during the Crimean War. It was rebuilt in 1864 and in 1870 a civil settlement named Tuapsinskiy appeared here. In 1913 the merchant port of Tuapse started functioning. In 1916 Tuapse received the municipality rights.

The name "Tuapse" may be translated from the Adygean as "two waters" (*"tua"* meaning "two" and *"pse"*—"water"). And this name may be traced to the Tuapse River that is formed by confluence of two mountain rivers—Chilipsi and Psynacho. For some distance the two flows are easily discernible in this river. The construction of a railroad connecting T. with the Armavir railway junction and an oil pipeline from the oilfields in Maikop spurred quick development of T. as an industrial center and a port. One cement and two brick plants were constructed near the city and food factories and handicraft undertakings appeared in the city. Construction of the Armavir-Tuapse and Black Sea railroads, new trade port and other plants attracted here qualified workers. The population of the city was growing. Many new workers' settlements appeared in the suburbs. In 1929 the oil refinery works was commissioned and in later decades its output reached 4 million tons of oil.

In the prewar (before 1941) years the oil pipelines Khadyzhensk-Tuapse and Grozny-Tuapse, the oil jetty, ship-repair yards were constructed; the sea port and the machine-building plant were retrofitted. During the Great Patriotic War (1941–1945) practically everything of the above was destroyed. In the war years T. was the main base, main port and main naval base supporting the whole B.S. group of troops fighting against the Germans at Novorossiysk. The city was bombarded from air and from the sea. There were days when the enemy aircraft made to 11 raids a day. There were also the so-called "star" raids in which 70–90 aircraft was involved simultaneously. The German aircraft had thrown 10,000 multi-ton ground bombs, the enormous quantity of flame bombs and containers with inflammable mix. For persistence and bravery showed by the people of T. during these war years in 1981 the city was awarded the Order of Great Patriotic War Degree I. T. is called the gates of Greater Sochi. Many sanatoriums, rest houses and rest hotels are found in the Tuapse area. At the same time T. (like Novorossiysk) with its numerous industrial enterprises can be hardly referred to resorts. The resort area includes the picturesque villages and towns located nearby—Dzhubga, Novomikhailovskiy, Nebug, Agria, Olginka, Gizel-Dere, Shepsi. The city has branches of three higher education institutes: Kuban State Technological University, Rostov State Railways University, and Russian State Hydrometeorological University.

Tuapse Bay – is limited in the west by the Kodosh Cape, in the east—by the high steep cape with white cliff debris. The high mountains overgrown with forests slope down to the bay shores; they are cut by small valleys of the Pauk and Tuapse rivers. The port of Tuapse locates between the mouths of these rivers.

Tuapse bora – a cold northeastern wind; the dropping of air from not high (300 m) and narrow (6 km) Goitkhskiy Mountain pass along the Tuapse River valley to the B.S. coast. T.B. differs from simply strong winds observed in Tuapse (brought with southeastern storms) by its direction, low temperature and great gustiness.

Tuapse resort area – Krasnodar Territory, Russia. It includes the following seaside climatic resorts (from northwest to southeast): Dzhubga, Novomikhailovskiy, Nebug, Agria, Olginka, Gizel-Dere, and Shepsi. This resort zone stretches for about 100 km along the B.S. coast of the Caucasus to the northwest and southeast of Tuapse. In the nearby forests not only deciduous varieties of trees typical of the moderate climate zone are growing, but also subtropical trees and shrubs. This resort area abounds in small mountain rivers with narrow valleys. The beaches here are covered with small pebble or sand. The climate is subtropical Mediterranean. The climatic conditions are most favorable in the Dzhubga-Tuapse zone where the summer heat is not wearisome, the refreshing sea breezes are blowing and the air humidity is lower. The climate in Tuapse proper is characterized by the warm winter with the mean temperature in February being -5 °C, with long copious rainfalls and by the humid hot summer with the mean temperature in August being $+23$ °C. The coastal zone here is protected from dry steppe winds, and only the northeastern winds penetrate sometimes from the Goitkhsky mountain pass. The precipitations are over 1,200 mm a year. The average annual relative humidity reaches 70 %. The number of sunshine hours is 2,300 a year. The main natural curative factor here is climate, but the physiotherapy, remedial gymnastics and others are also applied for health improvement. The people suffering from disorders of the breathing organs and functional diseases of the nervous systems come here for treatment.

Tuapse Sea Trade Port – the second largest Russian port in the Black Sea by the volume of handled cargo (12 million tons in 2013). The port has 13 wharfs with the total length of 2.3 km. It is divided into three zones by the handled cargo: bulk oil, dry cargo and passenger. The depths at six oil-loading and three dry-cargo wharfs are 13 m. The port is capable to receive vessels with the draft to 12 m.

Tuapse Sea Trade Port (http://www.morport.com/files/225/tuapse.jpg)

Tulcea – the major city-port in the Romanian part of the Danube Delta, 80 km from B.S., Dobrudzha (Dobrogea) Region, Romania. Its population is 92,400 (2007). This is also a railway station. The city was built on seven hills. The canning (fish, fruit, vegetable) industry, cane cutting and processing (building plates, cardboard) and a small shipyard are available here. This is the center of fishing and fish industry. The sea vessels came here via the Sulinsky Canal. The Delta Museum is located here.

Tulcea Mountains – this is the northern part of the Dobrudzha (Dobrogea) upland in Romania. It is situated between B.S. and the Danube lower reaches. They are over 400 m high. Their surface is mostly plateau-like.

Tuna (*Thunnus thynnus*) – large schooling fish of the mackerel family (*Scombridae*) preferring warm waters. Its length may be 1.5–2.5 m. Known are also species to 4.5 m long and weighing over 600 kg. This fish runs into B.S. via the Bosporus in summer for fattening and at times for spawning. The main spawning grounds of T. are in the Mediterranean Sea. In summer it may be found near the shores of Crimea and Northern Caucasus, while in autumn the large schools numbering to 10–15 specimens may accumulate near the Kerch Strait in the zone of autumn migration of anchovy from the Sea of Azov. Turkey and Bulgaria are basic producers of T. This fish is considered most valuable in the world fishing.

Tunbas – a sailing cargo vessel that existed in Turkey in the seventeenth–eighteenth centuries. It was often used in landing operations.

Turk Kaganate – the state that existed from the fourth to the eighth centuries. It was founded by Ashina, which was a tribe and the ruling dynasty of the ancient Turks who rose to prominence in the mid-sixth century when their leader, Bumin

Khan, revolted against the Rouran. In the second half of the sixth century T.K. together with Byzantine launched a war against Persia for control over the Great Silk Road. In the early seventh century T.K. was broken into the Eastern and Western Kaganate which control was extended to Crimea. In 740 Western Kaganate ceased to exist.

Turkeli – a village on the coast of the Black Sea, Turkey. Türkeli is also a district of Sinop Province. The town has a year-round population of about 7,000 people, but in summer the population exceeds 10,000 (2012).

Turkish Black Sea coast – it stretches for about 1,500 km from the Bulgaria-Turkey state border (the Rezovsky River mouth) in the southwest to the state border between Georgia and Turkey (the Chorokh River mouth) in the south. Different climatic conditions in various regions provide different tourist opportunities. The western B.S. tourist region covers the European B.S. coast of Turkey and the westernmost part of the Asian coast (from the Bosporus to the Sakarya River mouth). It is divided into two subregions—European and Asian. The coastline in this region is broken, but slightly. There are no large bays and still lagoons. In the north of the European subregion the shores are steep and cliffy. The hillocky-flat southeastern part of this subregion abounds in sandy beaches. The same relief is observed in the Asian subregion. The climate in the coastal region is moderate. The winter is mild and wet. The maximum precipitations fall in the north of the coast—about 800 mm a year. The vegetation is rich—the oak and beech forests grow on coastal slopes of mountains and in some coastal areas near the Bosporus. The central B.S. tourist region stretches from the Sakarya River mouth as far as Ordu. The shores are hillocky, slightly broken; there are many beaches here. The Pontic mountain ranges (to 3,900 m high) run parallel to the coast and protect it from cold continental winds, summer heat and droughts coming from the south, while the sea alleviates the effect of northern winds. The climate in this region is moderate: the winter is mild and the summer is rather humid and chilly. In some areas the precipitations vary from 700 to 1,300 mm a year. This favors development of rich and variegated vegetation represented by broad-leaved trees, shrubs and grasses. The transport is developed here better than on the western B.S. coast. The railroad connects the ports of Zonguldak and Samsun. A very convenient automobile road runs parallel to the coast. The passenger vessels make regular voyages connecting large port cities (Zonguldak, Sinop, Samsun, Ordu) and smaller settlements. The eastern B.S. tourist region extends over the coastal zone from the city of Ordu to the state border with Georgia. The coast is broken insignificantly; some areas are covered with extensive sandy beaches. The climate on the greater part of the coast (in particular, in the east) is humid subtropical. About 1,300–2,100 mm of precipitations fall here during a year which favors development of rich vegetation. The coastal slopes are overgrown with broad-leaved forests and shrubs. The subtropical plants, such as citrus, orange, tea, laurel, myrtle and others are met here.

"Turkish Coastal Law" – passed in 1984. However, in 1986, the Constitutional Court established that the law did not comply with the Constitution (in substance). In 1990, it came into effect in the form of a modified version. But in January of

1992, some articles of this new version of the law were again declined by the Constitutional Court. Finally, the laws was adopted with an amendment in July of 1992. However, the law is not exhaustive as far as the problems of offshore zone regulation are concerned, and deals essentially with economic activity in the vicinity of the shoreline.

Turkish Naval Museum – the country's largest national museum showing the history of the Turkish Fleet beginning from the fifteenth–sixteenth centuries. It is situated in Istanbul, on the European side of the Bosporus Strait. It possesses a rich collection of exhibits on the history of shipbuilding and seafaring: navigation instruments, sea maps, arms. Quite unique is the multi-color world map prepared by Turkish seafarer and geographer Piri Reis that was made on white leather in 1462. The other map dates back to the seventeenth century. The museum also stores the golden-yellow flag woven with silver threads from the Turkish Admiral vessel that took part in 1571 in the battle at Lepanto—the last major battle of the Mediterranean rowing fleet when the allied fleet of Venice, Spain and Roman Pontiff defeated Turkey and also the flag of Sultan Selim III dated back to the eighteenth century. Very interesting are also the models of vessels, drawings, arms and maps. The museum also has the aquarium.

Turkish Navy – (Turkish *Türk Deniz Kuvvetleri*) is the naval warfare service branch of the Turkish Armed Forces. The Navy can trace its history back to the establishment of the Aegean Fleet of the Seljuk Navy under the command of Çaka Bey in 1081, and later to the Ottoman Navy. However, the modern naval traditions and customs of the Turkish Navy can be traced back to 10 July 1920, when it was established as the Directorate of Naval Affairs during the Turkish War for independence led by Mustafa Kemal Ataturk. Since July 1949, the service has been officially known as the Turkish Naval Forces. In 2008, the Turkish Navy reported 48,600 active personnel which included an Amphibious Marines Brigade as well as several Special Forces and Commando detachments. The Turkish Naval Forces were represented under the title of the Naval Undersecreteriat at the Turkish General Staff Headquarters in Ankara from 1928 to 1949. The decree of the Higher Military Council on 15 August 1949 led to the foundation of the Turkish Naval Forces Command (*Deniz Kuvvetleri Komutanlığı*). After Turkey joined NATO on 18 February 1952, the Turkish Naval Forces were integrated into the organizational branches of the alliance In 1961, the Turkish Naval Forces Command was organized into four main subordinate commands: The Turkish Fleet Command, the Turkish Northern Sea Area Command, the Turkish Southern Sea Area Command, and the Turkish Naval Training Command. In 1995, the Turkish Naval Training Command was renamed as the Turkish Naval Training and Education Command. The Turkish Navy maintains several Marines and Special Operations units. These include the Amphibious Marines Brigade (Amfibi Deniz Piyade Tugayı), several Commando detachments and two special operations forces. As of 2013, there are approximately 112 commissioned ships in the Navy (excluding minor auxiliary vessels), including; 16 frigates, 8 corvettes, 14 submatines, 27 missile boats, 22 patrol boats, 20 mine countermeasures vessels, 5 landing ships, and various auxiliary ships. In 2002, the total displacement of the Turkish Navy was approximately 259,000 tons. The Turkish Navy also operates a total of 50 aircrafts, including 14 fixed-wing aircrafts and 36 helicopters.

Frigate TCG "Turgutreis" (F-241) in Novorossiysk (Photo by Evgenia Kostyanaya)

Turks (self-name "Turk") – the people, the basic (about 70–75 %) population of Turkey. The Turk population in Turkey is equal to 55–60 million people (70 million in the world). Turks are also living in Germany (3.5–4 million), France (0.5–1 million), Bulgaria (0.6–0.8 million), UK (0.5 million), USA (0.5 million), the Netherlands (0.4–0.5 million), Austria (0.4–0.5 million), Northern Cyprus (0.3 million), Australia (0.3 million), Belgium (0.2 million), Greece (0.15 million), etc. Several millions of Turkish people live in Arab world. In Turkey T. make the majority of the population in all regions except Kurdistan and Khatai vilajat. The Turkish nomad tribes—Seljuks (mostly Oguzes and Turkmens) appeared in Asia Minor in the early eleventh century. During the twelfth–thirteenth centuries the resettlement of other Turk tribes from the east was ongoing. During many centuries they contacted the local population (Kurds, Armenians, Georgians, Greeks, etc.) and gradually the Turk tribes mixed with them and adopted the sedentary life. The Balkan and Arab ethnic groups also made their contribution into formation of T. ethnogenesis. As a result of Ottoman conquering wars T. appeared on the Balkans in the fourteenth–fifteenth centuries and on Cyprus in the sixteenth century. The Turk language refers to the southwestern subgroup of the Turkic group of the Altai group of languages. The modern Turkic language is represented by the literary and folk spoken language having several dialects. The written language uses the Latin alphabet (until 1928—the Arabic alphabet). The Sunnite Moslems left many works in the Turkic language. The church is separated from the state, but the primary school has retained the teaching of religion. Several ethnic groups still exist among T.

Tuzla Island – it was formed in 1925 when the Tuzla Spit suffered from massive erosion during a major storm in Kerch Strait. Natural processes of sediment deposition on both sides from the Tuzla spit helped to save the remnants of the spit in the form of an island. In 2006 its length depending on the water level in the straight varied from 6.5 to 7 km, the maximum width was 500 m. By 1950 the width of the Tuzla eroded channel (300 m) increased to 3 km and in the late 1970s it was 4 km. By Decree of the Russian Senate of 28 November 1869 the Middle Tuzla Spit was included into the Kuban Region. In 1922 Tuzla was incorporated into the Crimean Region. In 1941 the Decree of the Presidium of the RSFSR Supreme Council "On Transfer of the Middle Spit (Tuzla) Island from the Temryuk District of the Kranodar Territory to the Crimean ASSR" was issued. In the early 1970s after transfer of the continental part of the Crimean Region to administrative control of the Ukrainian SSR the authorities of the Krasnodar Territory and Crimean Region agreed the border between these administrative units of RSFSR and Ukrainian SSR. And the sea border was delineated along the Kerch Strait and a part of T.I. was included into the Crimean Region. After USSR disintegration Ukraine unilaterally declared this administrative border as the state border. Regardless of the Treaty on Friendship, Cooperation and Partnership signed between the Ukraine and the Russian Federation, until December 2003 the Russian side did not recognize such border in the Kerch Strait. In September 2003 the Russia attempted to extend the spit by construction of an embankment (1.5 km × 800 m) towards T.I. that belonged to Ukraine. After the Note of the Ukrainian Foreign Ministry, the representations of the Russian Foreign Ministry and meeting of representatives of two states in late October 2003 the further construction of embankment was stopped. Now the sides initiated negotiations on construction of a bridge over the strait. The state appurtenance of the island proper will be defined in a separate treaty.

Tuzla Island in Kerch Strait (http://commons.wikimedia.org/wiki/File:Tuzla,_Kerch_Strait,_near_natural_colors_satellite_image,_LandSat-5,_2011-08-30.jpg?uselang=ru)

Tuzla Lake – a system of lakes 5 km to the northeast of Balchik, Bulgaria. The lakes cover an area of 80 ha. This waterlogged terrain is turned into the bird nature reserve. Many herons, pewits and diving ducks are nestling on the banks of these lakes. The mud curative center is organized here for treatment of the peripheral nervous system, locomotive organs and skin diseases.

Tuzla Spit (local name "Middle Spit") – it is located in the eastern part of the Kerch Strait. It runs for 7.5 km to the northwest of the point located 5.5 km to the northwest of the Tuzla Cape. The spit is sandy, low-elevated and shaped like a sword. The shallow Tuzla washway was formed in 1925 during a strong storm between the southeastern tip of T.S. and the mainland.

Tyapkin Vasiliy Mikhailovich (the years of birth and death are unknown) – Russian diplomat. In 1664 he was sent as a courier to Poland, in 1666—to Turkey. In 1668 he took part in negotiations with the Hetman of Right-Bank Ukraine. In 1673–1677 he worked in Warsaw and was the first permanent Russian resident abroad. He advocated active actions against Turkey. After defeat of Sweden in the war with Brandenburg he suggested starting war for Pribaltika. In 1680 he headed the peace delegation to Crimea. In 1681 T. signed the Peace Treaty of Bakhchisaray on recognizing by Turkey of reunion of Kiev and Left-Bank Ukraine with Russia.

U

Uch-Dere – maritime climatic health-resort location (Lazarevskiy District, Sochi, Russia), 28 km north-west of Sochi center. Sited on the Caucasus Black Sea coast. The beach is pebble, stow Uch-Dere ("Three Gorges") is covered with woods and gardens with ornamental subtropical vegetation. On the slope of one of the mountain spurs, at the height of 100 m, N.A. Semashko sanatorium was built—for children with skin diseases.

Uch-Dere Cape – flat, gently-sloping ledge overgrown with high trees. Sited on the Black Sea coast of Krasnodar Kray, Russia. To the south-east of the cape are the gorges Dagomys and Mamaika. The wide gorge Dagomys which is at the distance of 4.2 km to the south-east of Uch-Dere Cape is divided in the middle by a hill in the shape of a triangle. The Dagomys River flows at the foot of the hill down the gorge.

"Ukraina" – passenger liner that began operating on the line Odessa—Crimea—Caucasus in June of 1945. Built in Copenhagen on the shipyard "Burmeister & Wain" in 1938. Had been part of the Romanian Navy as "Basarabia". After Romania terminated its participation in the war, the liner was handed over to the USSR by reparation and was given a new name. Tonnage 6,970 tons, maximum speed 25 knots. "U." Operated for nearly 20 years without overhaul. In 1967, the ship was refurbished and was capable of taking 76 passengers accommodated in the lux and first class suites, 314 passengers in the second and third class suites. Besides, 330 passengers could be accommodated on deck. "U." was overhauled for the second time early in the 1970s in Yugoslavia: the number of first class suites was doubled. From 1975, the liner began cruising the Mediterranean Sea and sailed as far as Cuba. After the shipwreck of "Admiral Nakhimov", the vessel was written off.

Ukraine – state in East Europe. Territory—603,550 km². Population—45.5 million people. Capital—Kiev. Major cities: Kharkov, Dnepropetrovsk, Donetsk, Odessa, Simferopol, Sevastopol. The highest point—Goverla (2,061 m). Official language—Ukrainian. Dominant religion: Christianity (Orthodoxy, autocephalous Greek-Catholic

S.R. Grinevetsky et al., *The Black Sea Encyclopedia*,
DOI 10.1007/978-3-642-55227-4_21, © Springer-Verlag Berlin Heidelberg 2015

Church). Monetary unit: Grivna. State system: parliamentary-presidential republic. The president is elected for 5 years. Legislative body: single-chamber parliament (Supreme Rada) made up of 450 deputies.

Ukraine (http://cooler-online.com/lt/map_ukraine.jpg)

Borders: in the north with Belorussia, in the north-east and east with Russia, in the west with Poland, Slovakia, Hungary, Romania, Moldova, in the south is washed by the Black Sea. U. is a member of GUAM.

Around 3,500 years ago this territory was inhabited by cattle-raising tribes. In the ninth century A.D. the Slavic state of Kievan Rus came into being around Kiev. In the fourteenth century, Ukraine lost its independence and first was subjugated by Poland and Lithuania, then after the victorious culmination of the liberation struggle of the Ukrainian people of 1648–1654 against the Polish gentry, U. reunified with Russia (Pereyaslavskaya Rada). The left-bank U. gained autonomy as part of Russia. In the second half of the eighteenth century, the southern Ukrainian lands shed the Turkish yoke. The right-bank Ukraine was reunified with Russia at the end of the eighteenth century.

The Soviet power in Ukraine was established in late 1917—early 1918. As a result of the Russian-Polish War of 1920 Western Ukraine was under the rule of Poland and was reunited with the UkrSSR in November of 1939. In August of 1940, the UkrSSR was enlarged with North Bukovina and part of Besarabia. In 1941–1945, the territory of Ukraine was under the German occupation, and partisan liberation movement spread everywhere. U. was liberated by the Soviet Army by early 1945. From June of 1945, Transcarpathian U. becomes part of the UkrSSR. From 1945, U. is a member state of the United Nations. During the years of Soviet power national culture kept developing, a program of intense industrialization was under way in the country. In 1954, Russia transferred the Crimea to U.

The country was proclaimed independent in August of 1991, after the collapse of the Soviet Union.

Most of U. territory is within the south-western outskirt of East-European Plain. In the extreme south—on the Crimean Peninsula—the Crimean Mountains tower (height up to 1,545 m, Mt. Roman-Kosh), in the west—Ukranian Carpathians (height up to 2,060 m, Mt. Goverla). Major rivers fall into the Black and Azov Seas: the Dnieper with tributaries (divides the country into right-bank and left-bank Ukraine). The Severskiy Donets, South Boug, Dniester, Danube. There is a cascade of storage reservoirs built on the Dnieper: Kiev, Kanev, Kremenchug, Dneprodzerzhinsk, Kakhovka.

The climate is temperate, largely continental, on the South Shore of the Crimea—subtropical, Mediterranean. Mean temperatures of January vary from—8 °C in the north-east to +2–4 °C in the south of the Crimea, temperature of July—from 18–19 °C to 23–24 °C. The amount of precipitation falling in the north-west is 600–700 mm per annum, in the south-east—up to 300 mm, in the Crimean Mountains—1,000–1,200 mm, in the Carpathians—under 1,600 mm. The snow cover on most of the territory stays for 4–5 months.

Vegetation is, for the most part, strongly modified by man. In the conserved natural areas of the north (the zone of mixed forests) there prevail pine woods, in the forest-steppe zone, oak woods alternate with woodless spaces. In the steppe zone, forb-mat grass and sagebrush—sheep's fescue steppes on chernozem soils have survived in reserves only. The slopes of the Crimean mountains feature woods of oak, beech, pine, juniper. The Carpathian slopes are covered with woods of oak, beech, spruce, silver fir, cypress, still higher are alpine meadows—polonines. The Black Sea coast features cultivated vegetation of the subtropical belt. Natural vegetation and wildlife are under protection in 11 reserves, the largest of which are—Black Sea, Polesskiy, Yalta, Carpathian, Ascania-Nova. There are sanctuaries in each region, where rare species of plants and animals as well as natural places of interest are protected. The wildlife of Ukraine does not differ from the adjacent regions of Russia or Belorussia: dwelling in the woods are roe deers, deers, elks, boars, foxes, hares, squirrels, wolves, bears, numerous birds, lizards. In the steppes and in the forest-steppe, there dwell steppe birds (partridge, quail and others), lizards and snakes are also available.

Ukrainians account for most of the population. Besides, there live Russians, Belorussians, Jews, Tatars, Armenians and other peoples in Ukraine. There lives an ethnic variety of Ukrainians in the circum-Carpathian area—Gutsuls. The Crimean Tatars (around 250,000 people in Crimea) claim to be "indigenous".

U. is endowed with reserves of mineral resources: coal, iron and manganese ore, uranium, graphite, building raw materials, etc. Reserves of oil, gas and timber are insufficient. Of the branches of manufacturing industry, prominent are ferrous and non-ferrous metallurgy. The primary branches of machine-building are the manufacture of plant and equipment, power engineering industry, electrical engineering industry, machine-tool industry, instrumentation, radioelectronics, transport machine-building, aircraft industry, ship building, tractor industry, agricultural machinery industry. Chemical industry is at an advanced stage of development.

Agriculture is specialized in grain and cash-crop growing as well as in stock-raising. The main agricultural crops are—grain crops: winter wheat, corn, rice, barley, millet, buckwheat, leguminous crops and others. Also cultivated are sugar beet, sunflower, flax, potato, hem, tobacco, fodder crops. Besides, vegetable-growing, melon-growing and grape-growing table, dessert wines and port wines are produced. The main branches of animal husbandry are cattle-breeding, sheep-raising and pig-breeding.

U. exports industrial raw materials and foodstuffs. It imports energy carriers, equipment, consumer items. Major trading partners: Russia, Turkmenistan, Germany, China, Belorussia.

There are still traces of several cultural-and-historic periods on the territory of Ukraine: primitive communal system (Stone Grave, Zaporizhye Region), stone sculptures (images), mounds and Scythian sites of ancient settlements; on the Crimean Peninsula, there are still ruins of Greek and Genoese settlements. Ancient Russian culture is portrayed in architectural monuments of Kiev (Temple of Sofia, Golden Gates and others), Chernigov. The culture of the unique Ukrainian ethnos was forged during the fourteenth–fifteenth centuries. Uzhgorod and Chernovtsy resemble medieval European cities. Of interest are buildings and churches built in a style of the so-called "Ukrainian baroque" (Poltava, Kharkov, Kiev and other cities). Quite peculiar are Ukrainian handicraft items: embroidered towels, pottery items, tiles, woven carpets, carved wood, castings, etc. In literature, the beginning was made by the epic "Slovo o Polku Igoreve" of the eleventh century unknown author. The founding father of the literary Ukrainian Language was I.P Kotlyarevskiy (early nineteenth century). The poetry of Taras Shchevchenko and Lesya Ukrainka. Prose of M. Kotsubinskiy, Panas Mirnyi, philosophical works of Ivan Franko and other writers have become part of the treasury of the world culture.

U. is rich in architectural places of interest, reserves. There are world-famous museums in Kiev, including "Museum of Sofia"—architectural-historic reserve; Vydubetskiy Monastery and Temple of Sofia of the eleventh century; the Saviour on Berestov Church of the twelfth century and others. In Kharkov—University Museum of Nature, Historic and Cultural Museums, Pokrovskiy Temple of the seventeenth century and the Temple of Assumption of the eighteenth century in Odessa—ensemble of buildings in the city center with Potemkin's stairs (1826–1841) and a monument to Duke Richelieu; Archeological, Marine and other museums. The architecture of Lvov evokes associations with Vienna: Cathedral Temple of the fourteenth–fifteenth centuries, "Black Kamenitsa" of the fourteenth–seventeenth in the style of Renaissance, the St. Juri Cathedral of the eighteenth century, etc.

There are museums and interesting features in small and medium cities of Ukraine. Natural monuments: Karadag, Martyan Cape and Yalta Reserves in the Crimea, eighteenth century garden-and-park art "Sofievka" near Uman, the 22 km long Crystal Cave (Ternopol Region), Askania-Nova Reserve, Nikitinskiy Botanical Gardens. The territory of the Autonomous Republic of Crimea is extremely rich in interesting features. These include museums and memorials: ruins of the classical cities Mirmekii and Nimphei, Demetra's Tomb—the first century A. D., the Church of the tenth–thirteenth centuries—in Kerch; Genoese fortresses and towers of the fourteenth–fifteenth centuries—in Sudak and Feodosiya; picture gallery with paintings by I.K. Aivazovskiy and Museum of A.S. Grin, a carpet museum, Mufti-Jamie Mosque of the seventeenth century—in Feodosiya; Literary Museum of M. Voloshin—in Planernoe (Koktebel); House-Museum of A.P. Chekhov, Great Livadia Palace (1910), Museum of A.S. Pushkin in Gurzuf, Alupka Palace-Museum of the nineteenth century with a park—in Greater Yalta; in Bakhchisaray—Museum of Architecture and History: Khan Palace of the sixteenth century; Cave Monastery of Assumption (sixteenth century), Cave Assumption Monastery of the fourteenth century; cave cities Chuft-Kale, Tepe-Kermen, Eski-Kermen. In Sevastopol—State Reserve of History and Archeology "Classical Khersones" of the fifth–first centuries B.C.; Panorama "Defense of Sevastopol in 1854–1855" (1902–1904); Grafskaya Pristan and Petropavlovskiy Cathedral of the nineteenth century; Aquarium in the Institute of Biology of the Southern Seas.

U. has attractive recreational locations (the Crimea, Carpathians, south of Ukraine) and is capable of promoting the tourism infrastructure. Health resorts of the Crimea, balneal health resorts of Odessa, circum-Carpathian area (Truskavets, Kvasov Meadow), Central Ukraine make use of medicinal mineral water sources and muds.

U. is a member of the Commonwealth of Independent States, GUAM, from June of 1992, has been taking an active part in the BSEC organization. U. Has been member of the UN ever since it's was established in 1945 (initially,—as part of the USSR; in 2000–2001, U. was elected a non-permanent member of the UN Security Council), is a member of the Council of Europe and has been participating actively in the NATO Program "Partnership for Progress"—since 1995.

Ukrainian Navy – on August 24, 1991 the Supreme Council of the Ukrainian Soviet Republic declared the independence of the Ukraine. On the same day the Parliament passed the resolution "On Military Formations in Ukraine". On April 5, 1992 the President issued Order "On Urgent Actions on Construction of Military Forces of Ukraine" where it was ordered to form the Ukrainian Navy on the basis of the Black Sea Fleet dislocated on its territory. Pursuant to Yalta Agreement of July 3, 1992, the Black Sea Fleet became the united Fleet of Russian and Ukraine and it was subordinated to presidents of both countries. Signing by Presidents of Ukraine and Russia on June 17, 1993 of the historical Treaty "On Urgent Actions on Formation of Russian Navy and Ukrainian Navy on the Basis of the Black Sea Fleet" laid the basis for the legal division of the Black Sea Fleet.

In June the Ukrainian Navy included the assault landing air-cushion ship "Donetsk" and in June 1993 the Ukrainian flag was hoisted on the patrol ship of the Ukrainian Navy "Getman Sagaidachnyi". In June-August 1994 the frigate "Getman Sagaidachnyi" made a far cruise to the shores of France. This was the first in the history of independent Ukraine international visit of the Ukrainian military ship to a foreign port.

On June 9 1995 in Sochi the Treaty between Presidents of Ukraine and Russia on the Black Sea Fleet was signed. According to these documents, 81.7 % of ships and vessels were assigned to the Black Sea Fleet of the Russian Federation and 18.3 %—to Ukraine; the property, armaments and technical stocks were divided equally 50 × 50. In January 1996 the first stage in the Black Sea Fleet division was started. From this date the naval flags of Ukraine were hoisted on four guided missile boats, in a separate missile-artillery brigade, on 17 ships of the former Crimean Naval Base of the Black Sea Fleet. The Ukrainian side was officially passed the naval bases and rear logistic bases located in Ochakov, Odessa, Izmail, Donuzlav, Balaklava and others. In August 1996 on B.S. there were organized the first in the Ukrainian Fleet history wide-scale military exercise "More-96" under the flag of the Supreme Commander-in-Chief of the Ukraine Military Forces— President of Ukraine. The Decree was signed on annual celebration of August 1 as the Day of the Ukrainian Navy.

On May 31, 1997 in Kiev the Presidents of Ukraine and Russia signed the Treaty of Friendship, Cooperation and Partnership between Ukraine and the Russian Federation and the Treaty on the Status and Conditions of Staying of the Black Sea Fleet of the Russian Federation on the Territory of Ukraine. In fact, this was the beginning of the second and final stage in division of the Black Sea Fleet of the former USSR. The Ukrainian Navy includes the surface ships and submarines, naval aviation, coastal missile-artillery troops, marines, special-mission units, rear logistic and technical support units. The headquarters and the main naval base of the Ukrainian Fleet locate in Sevastopol. In addition, the Ukrainian Naval bases are deployed in Odessa, Ochakov, Chernomorsk, Novoozernyi, Nikolaev, Evpatoria and Feodosiya. The staff of the Ukrainian Navy comprises 11,000 sailors and officers. Among the Black Sea countries Ukraine is the fourth by the Fleet potential after Russia, Turkey and Romania.

Ukrainians – Ukraine's basic population (37.5 million). 45 million Ukrainians live in the world. Just like Russians and Belorussians, U. evolved from a single ancient Russian people that had been forged from congeneric East Slavic tribes. The Ukrainian nationality as an independent ethnic community came into being in the fourteenth—fifteenth century. From the seventeenth century Ukraine began to take shape as a bourgeois nation. This process was intensified after the reunification of Ukraine with Russia in 1654 and ended in the nineteenth century. After the collapse of the USSR in 1991, U. accounted for the bulk of Ukraine's population—78 %. Ukraine is also home for Russians (around 17 %), Belorussians (0.6 %), Tatars (0.5 %) and a number of other peoples. There dwells an ethnic variety of Ukrainians in the circum-Carpathian area—Gutsuls. Major settlement areas are the western, central and northern areas of Ukraine. In the southern and eastern regions, the percentage of U. is lower. In the west and to a lesser extent in the central part of Ukraine, the Ukrainian Language is dominant, in the south and east the Russian Language prevails.

Ukrainian Danube Shipping Company – established in 1944 in Izmail. At present, the largest shipping company in the Ukraine. Today, this is a fleet with versatile possibilities, deadweight of more than 1 million tons. The ships operate on ocean and river lines, transport cargoes between the ports of the Danube countries; between the ports of the Black Sea, the Sea of Marmara, Agaean Sea, Adriatic, Mediterranean and

Red Seas; in the combined sailing "Danube-sea-Danube". The fleet includes all-purpose and specialized river, sea, "river-sea" ships, passenger fleet.

Ulagay Sergey Georgievich (1875–1947) – Russian Colonel (1917), Major-General (1918), Lieutenant-General (1919). Finished Voronezh Cadette Corps (12895) and Nikolaev Cavalry School (1897). Participant of the 1904–1905 Russian-Japanese War. Participant of WWI: commander of Kuban Cossack Regiment (1917). Supported Kornilov's attempted coup, was arrested, fled to Kuban (August-November of 1917). In the White Movement: an officer in the Kuban Cossack troops, head of cavalry in the Kuban volunteers' unit, Commander of Cherkessk Regiment. Commander-in-Chief of the cavalry task force of the Caucasus Army near Tsaritsyn (June-August of 1919). Commander-in-Chief of the Kuban Army, successor to General Shkuro in that position. Was evacuated to the Crimea; being in reserve of the Russian Army, was acting troop chieftain of the Kuban Cossack Troops. Elaborated and was preparing a plan of landing troops on Taman Peninsula of the Kuban. Commander of a task force of the Kuban Cossack troops that landed on Taman Peninsula of the Kuban through Kerch Strait. The landing party of U. (around 14,000 men) that was dropped on Taman Peninsula was defeated by the superior forces of the Red Army and evacuated back to the Crimea (July-September of 1920). U. was dismissed from the Russian Army and in October was evacuated from the Crimea on board the ship "Constantine". In emigration from October of 1920. Died in Marseilles, France on March 20, 1947.

UN Convention on the law of the sea (1982) – was developed by the Third UN Conference on the Law of the Sea, signed on 10 December 1982 and enforced on 16 November 1984. It contains a complex of mutually acceptable arrangements on the basic issues of the World Ocean management, including external borders, the regimes of territorial waters and continental shelf, the right to unobstructed passage of ships via straits, the regime of the international sea floor area and others. This Convention sets the new legal status for the seas and oceans, the rules governing the environment protection standards as well as provisions related to the sea environment pollution. This Convention is ratified by 135 countries, including all the Black Sea states.

Underwater archeology – hydroarcheology, a section of archeology studying the monuments of history and culture located on the floor of water bodies or buried beneath bottom sediments. They include immobile objects—settlements, towns and separate structures found in the coastal zone and happened to be under water due to the water level rise or on the bottom; sank ships, boats, rafts, their parts and cargo; items having no relation to ships, but happened to be on the bottom for some reasons. These monuments define the main directions of U.A. explorations. Submerged settlements and structures are usually found in the narrow coastal zone of seas and landlocked water bodies at depths not exceeding 10 m. The most ancient of the known settlements that had been covered by water as a result of the rising level of the World Ocean were Neolithic and they once located on the shores of lakes in Europe. The greatest number of the known submerged settlements and towns was in the Antiquity (seventh century B.C.–fifth century A.D.) and they situated on the shores of the Mediterranean, Black and Azov seas. The Medieval settlements (eighth–fourteenth centuries A.D.) are known on the Mediterranean, Baltic and Caspian seas.

Studies of shipwrecks, their equipment and cargo helped to delve into the history of shipbuilding and maritime trade, the everyday life of seafarers. Examination of the submerged settlements provides for the historians the most valuable material for plotting the graph of the sea level fluctuations, reconstruction of the ancient topography of the coastal settlements and towns. The facts of rising cargo and equipment from wrecked ships and other artifacts have been known for long time. U.A. started shaping as a research area only in the late nineteenth century. In 1894 during the bottom dredging works in the port of Feodosiya headed by Russian archeologist A.L. Bertie-Delagade about 4,000 pine piles arranged in rows forming a right angle were lifted from the sea floor. Perhaps this was the ancient defense structure, something like a jetty. In 1905 L.P. Kolli, the keeper of the Antiquity Museum in Feodosiya, organized the first purposeful search of ancient artifacts in the Feodosiya Bay. Fifteen large amphorae were taken up from the seafloor.

The first underwater archeological expedition in the USSR headed by K.E. Grinevich was organized in 1930 near the Khersones Cape. The divers explored the ruins of Chersonesos on the bottom of the Karantinnyi Bay described in "Geography" of Strabo in the first century A.D. The film "The City on the Seafloor" was also shot in 1930. The great contribution into U.A. development was made by R.A. Orbeli who proposed to make the hydroarcheological map of the country. The intensive development of U.A. started after invention of aqualung that provided the archeologists a possibility to participate personally in underwater investigations. This also enabled to expand the investigation site. Jacques-Yves Cousteau and Philippe Tailliez pioneered the use of aqualung in underwater archeological surveys. The most ancient ship that was lifted from the seafloor dates back to the thirteenth century B.C. It was found in 1958 near the Helidonja Cape at the southwestern coast of Turkey. This operation was supervised by Doctor George Fletcher Bass, the U.S. underwater archeologist. Together with four other ships lifted in this area this ancient vessel is demonstrated in the underwater archeological museum in Bodrum, Turkey.

In the USSR Professor V.D. Blavatskiy pioneered the underwater archeological investigations. In 1957 he organized the underwater expedition of the Institute of Archeology of USSR AS that worked for 11 seasons near the shores of Black and Azov seas. In 1971–1976 the expedition of this institute explored the submerged part of the ancient Greek city of Olvia in the Bug Liman of B.S. In different years it also took part in explorations of the port site and nearby harbors of ancient Chersonesos-Korsun. Explorations are also underway in the water area of the ancient Bosporus Kingdom covering the shores of the Black and Azov seas and the Kerch Strait. In 1983 and 1984 on the floor of the strait they found the ruins of fortification walls and other structures of the antique city of Akra which was mentioned by Strabo, Plinius the Senior and other Antique authors in their works. In 1983 the International Federation for Underwater Activities and UNESCO initiated preparation of a list of underwater archeological explorations in all world countries having settled some methodological, organizational and other issues of U.A.

Underwater laboratory – a stationary or self-propelled underwater manned complex of structures for long-time underwater explorations and works connected with letting out divers into the sea. The stationary laboratories are mounted on a pontoon or a frame towed (or carried onboard) to a working site, submerged into water and installed on the floor at depth to 150–200 m. Stationary U.L. has the living compartment and air lock, research laboratories, the systems of life-support, heating, moisture absorption, control of the complex. The crew of stationary laboratories may be delivered, changed and evacuated with the help of transportation chambers or rescue submersible vehicles. Stationary U.L. designed for installation at depth of 40 m has the normal atmosphere and are sometimes called the subsea houses. The atmosphere of stationary U.L. lowering to more than 40 m represents a breathing gas mix which composition and pressure match the sea depth in the site of complex installation. During staying in this U.L. the crew is working in the regime of "saturated" diving. In view of high heat conductivity of helium contained in the U.L. atmosphere the temperature inside it is maintained at about 30–34 °C. Decompression of the crew is conducted once, directly after completion of works and this is usually done in a special chamber on board a support ship. The crew goes there from a transportation chamber after rising to the surface. The support ship is anchored near the place of U.L. installation. The laboratories are connected with this ship or with the shore via cables and hoses that supply them with electricity, breathing gas mixes, compressed air and drinking water. The cables also ensure communication. The food, devices, spare parts are delivered to U.L. from the surface in containers. Storage batteries are used as emergency power supply sources. Emergency stocks of breathing gas mixes are kept in cylinders, fresh water—in tanks. The equipment and some mechanisms used in U.L. are of special modifications because of the higher atmospheric pressure in a laboratory and presence of helium in the air mix.

The stationary laboratory weight is 200–500 tons; it is capable to operate autonomously for 20–30 days; the crew is 4–16 people. For supporting purposes the subsea shelters may be installed on the site of stationary U.L. The self-propelled underwater manned complexes are designed as subsea vehicles with great displacement or they are made of retrofitted submarines. They are provided with research facilities and air locks for letting the divers out into the sea. For letting the divers out into the sea the self-propelled U.L. are fixed to the seafloor with supports or guide ropes. Operation of U.L. requires most careful preparation of the laboratory itself, technical devices, experimental facilities and training of the personnel. In 1966–1968 in B.S. near the coast of Crimea the Soviet researchers started experimental works using subsea "houses" "Ikhtiandr-67" and "Ikhtiandr-68". Then there were works at a depth of 25 m in U.L. "Sadko-1" (1966), "Sadko-2" (1967) and "Sadko-3" (1969). P.P. Shirshov Institute of Oceanology of USSR AS conducted works near the Caucasian coast in subsea vehicles "Chernomor-68" and "Chernomor-69". In 1967 Bulgaria conducted the first experiment on population of the continental shelf with the help of the subsea "house" "Gebros-67" working at a depth of 10 m in the Varna Lake. In 1973 the Soviet-Bulgarian experiment named "Shelf-Chernomor-73"using the Soviet manned U.L. "Chernomor" was conducted near the Maslen Nos Cape southward of Sozopol.

Union of Soviet Socialist Republics (USSR, Soviet Union) – the state that existed in 1922–1991 covering the greater part of the territory of the former Russian Empire. Under the USSR Establishment Treaty of 30 December 1922, it comprised the Russian Soviet Federative Socialist Republic (RSFSR), Transcaucasian Soviet Federative Socialist Republic (TSFSR), from 1936—the Azerbaijan SSR and others. Later on the Turkmen SSR (1925), Kazakh SSR (1938) and other union republics were formed. In December 1991 the leaders of Belorussian SSR, RSFSR and Ukrainian SSR signed the Belovezhsky treaties stating termination of the USSR existence. On 8 December 1991 in Minsk they also signed the Commonwealth of Independent States (CIS) Establishment Treaty. In the Declaration of 21 December 1991 in Alma-Ata, Azerbaijan, Kazakhstan, Turkmenistan and others declared about their commitment to the principles of the CIS Treaty of 25 December 1991. From this time the USSR as a subject of international law ceased to exist.

Unkiar Skelessi Treaty (or the Treaty of Hünkâr İskelesi, 1833) – signed on June 26 (July 8), 1833 between Russia and Turkey. The treaty was of defensive nature and was concluded for 8 years. Under the treaty, Russia undertook, if necessary, to come to Turkey's rescue "overland and by sea". The crucial to Russia condition was contained in a "separate and secret article" that said that the Russian side, "willing to relieve the Resplendent Ottoman Porta from the hardship and inconveniences it would have had as a result of delivering substantial aid", would be satisfied by Turkey meeting just one commitment. Under the terms of the secret article, Turkey was to restrict its own actions in respect of the Russian Imperial Court by closing down Dardanelles Strait, i.e. "not to allow any foreign combat ships to enter the strait on whatever pretext". The treaty thus concluded in a way filled the gaps of Adrianopol Peace Treaty (1829) that had ignored the straits regime. In this respect, the Treaty of Hünkâr İskelesi was an unconditional success of Russian policy even though it did not state specifically that the Russian ships were entitled to sail in the Straits.

Unye – a town on the coast of the Black Sea, Turkey. This is a town and district of Ordu Province in the Black Sea region of Turkey, 76 km west of the city of Ordu. In 2009 it had 74,800 inhabitants.

Unye Bay – cuts slightly into the shore to the west-north-west of Fatsa Bay, Turkey. Tashkhane Cape delimits U.B. from the west. The bay is open to the winds from the north-west via north to the east. Mountain spurs come close to the bay shores. To the west of the bay these mountains give way to the hills and retreat far from the shoreline, emerging near the shore only near Samsun Bay. On the western shore of the bay, the city of Unie is arranged as an amphitheater. There are many vineyards in its environs.

Upwelling – (from English *up*—go up и *well*—rise, rush)—process of vertical movement of water in the ocean (sea) as a result of which from the depth of 100–200 m colder waters rise to the surface. U. occurs when steady winds blow parallel to the coastline so that the coast remains on the left in the Northern Hemisphere and on the right in the Southern Hemisphere (due to Ekman transport). U. may occur also as a result of a wind surge of surface waters from the shore or ice edge, flows diverging from the land, in the cyclonic eddies and cyclonic gyres. Permanent upwelling zones are observed near the western shores of U.S.A. (Californian and Oregon U.), Peru (Peruvian U.), North- and South-West Africa (Canary and Benguela U.), Australia. The coastal upwelling zones, thanks to the greater amount of biogenic elements brought up to the surface, are the most productive fishing areas of the World Ocean, accounting for over 50 % of the world yield.

In the Black Sea, U. Is most often observed in the north-western and western parts of the sea, near the southern coast of the Crimean Peninsula and along the Turkish coast. In these coastal areas, U. is most intensive, characterized by a sharp difference in the sea surface temperature between the upwelling zone and surrounding waters, which may reach 12–14 °C. The width of the upwelling zone is 10–30 km. In most cases, the process takes place from May to September. Thus, in summer, the waters of 22–24 °C as a result of U. may be replaced with waters with temperatures 10–12 °C for a period from several days to several weeks in spite of hot weather. For example, at Kaliakra Cape on July 15, 2000, temperature of water in U. was as high as 7.6 °C, while the temperature of ambient waters equaled 21.5 °C, whereas U. lasted 3 weeks. In the summer of 2003 (from mid-June to the end of July) near the southern coast of the Crimea there developed so powerful and long-lasting U. that it actually destroyed all bathing plans of the holiday-makers. For example, on July 10, on all beaches from Balaklava to Feodosiya, the temperature of water did not exceed 10–12 °C, while the air temperature in Yalta was 27 °C. In the north-western part of the Black Sea, the shoreline changes its orientation dramatically (actually by 180°), at the same time, U. may be observed in various segments of the shoreline simultaneously, which occurs both as a result of the wind surge from the coast, and as a consequence of Ekman transport. Here, the area of the most frequent rise of coastal waters is Tendrovskaya Spit, which is related to maximum recurrence of north-westerly wind for U. in this segment. U. in the north-western part of the Black Sea is accompanied by the formation of short-lived (one to several days) transverse jets of 4–14 km in width. These transfer the upwelling-generated water over a distance of 50–160 km from the shore with a speed of 30–40 cm/s as well as by cyclonic eddies (e.g. near Khersones Cape) of 10–15 km in diameter, which spread beyond the shelf zone. The upwelling jets are typical of U. of the Anatolian coast, too. The time between the onset of the wind favorable for U. and the surfacing of cold waters in most of the cases does not exceed 1 day. U. relaxation occurs just as quickly.

Ushakov Fedor Fedorovich (1745–1817) – Russian naval commander, Admiral (1799), one of the founding fathers of the Black Sea Navy and its Commander-in-Chief from 1790. In 1766, finished the Naval Cadette Corps. Served in the Baltic Fleet and in the Don (Azov) Flotilla (from 1769). In 1769 promoted to Lieutenants, sailed down the Don River on pram No 5. Took part in the Russian-Turkish War of 1768–1774. In 1770, commanded the same pram on the Don River, then on pram "Defeb" reached the mouth of the Kutyurma River. In 1771, stayed on board the "Pervyi" frigate during its pilotage down the Don River from Pyatiizbyannaya Cossack Village to Taganrog. In 1772–1773, commanded the boat "Courier" on the Azov and Black Seas, in 1774—commanded the ship "Modon" in the Sea of

Azov. In 1775, transferred from Taganrog to Kronstadt (Gulf of Finland, the Baltic Sea), promoted to Captain-Lieutenants. In 1783, served on the Black Sea, supervised ship building at Kherson, participated in the construction of the naval base in Sevastopol. Promoted to Captain first rank (1784). In 1785, steered his ship to Sevastopol. In 1786, commanded the ship "St. Pavel". In the battle of Fidonisi (1788), being at the head of the squadron vanguard, defeated the Turkish squadron. As the commander of the Blacvk Sea Navy, scored spectacular victories in the sea battle of Kerch near Tendra Island (1790). In 1791, had another victory at Kaliakria Cape. In 1793–1797, stayed at Nikolaev. In 1797, being in possession of the flag on the ship "St. Pavel", commanded practical squadron and sailed between Sevastopol and Odessa. In 1798–1800, headed the squadron in the Mediterranean Sea in the war against France. Was promoted to Admirals. In 1800, returned to Sevastopol with the squadron. The Emperor Alexander I failed to appreciate U.'s successes. In 1807, U. was dismissed to resignation. U. had been decorated with numerous Russian and foreign orders. A harbor in the south-eastern part of the Barents Sea, a cape on the northern shore on the Sea of Okhotsk are named after U. His name was also given to some ships of the Russian Navy. On March 3, 1944, by Decree of the Presidium of the USSR Supreme Soviet, some of the highest orders were established: Ushakov's order of 2° and a medal. On November 30, 2000, the Canonization Commission of the Russian Orthodox included the naval commander in the community of saints of Saransk Episcopacy. On August 4–5, 2001, during canonization, his ashes were reburied at the Blessed Virgin Nativity Cathedral of the Sanaksar Monastery of Blessed Virgin Nativity.

Ushakov F.F. (http://sfw.org.ua/uploads/posts/2010-07/thumbs/1278788775_ushakov.jpg)

Ushkol – light sail-and-rowing vessel in seventeenth century Turkey. Used mainly on the Black and Azov Seas to guard trade caravans. Had one mast with a fore-and-aft sail.

USSR-Turkey Treaty on delineation of the continental shelf of the Black Sea between the USSR and Turkey of 23 June 1978 – it was signed in Moscow and included four clauses. According to Clause 1, the border of the continental shelf between the USSR and Turkey shall start from the terminal point of the sea border dividing the territorial waters of the USSR and Turkey in the Black Sea fixed by the Protocol of 17 April 1973 signed by the governments of the USSR and Turkey and defining the sea border between the territorial waters of the USSR and Turkey in the Black Sea. This clause also defines the geographical coordinates via which the border shall go.

Ust-Dunaisk – port built in Zhebriansk Harbor in 1977 as the base of lighter-carriage fleet, Ukraine. To accommodate large-tonnage ships, the southern part of the harbor of around 0.5 km^2 in area was dredged to the depth of 10 m and connected with the sea by means of a 7 km long approach channel, its width across the bottom 10 m and depth 13.2 m. Besides, for process considerations, a channel was dug in 1979 that linked U.-D. port water area with the Arm Prorva.

Utluk Liman – bay in the Sea of Azov. Formed near its north-western shore by a hydraulically-filed strip of land: by Fedotova and Biryuchyi Island spits. Length 46 km, width at the entrance around 12 km, maximum width 17 km. Depth 5–7 m. The shores are precipitous, in the east—low-lying. The city of Genichesk is on the western shore.

Utochkin Sergey Isaevich (1876–1916) – Russian aviator. Made his name as a racing cyclict. During 17 year stayed on track and scored numerous victories in Russia and abroad. Used to set speed records in motor racing. In 1910 in Odessa ventured a flight without any instruction: a rare case in the history of aviation. Became the second Russian pilot. In 1911, had an accident during the Petersburg-Moscow flight.

Utrish Cape – sited 16.5 km to the south-south-east of Anapa Cape, Krasnodar Territory, Russia. Low cape, hemmed by a shoal featuring above-water, underwater rocks, shoalbanks and sunk vessels. Outrish Islet is on the same shoal at a distance of 200 m to the north-west of the southern tip of U.C. At the distance of 3.5 km north of U.C. is Mountain Sukko or Ekonomicheskaya, its height 120 m, and 4 km east of the cape—Mountain Kobyla (531 m). South of Mt. Sukko is gorge Sukko.

Utrishenok Cape – sited 7.5 km south-east of Utrish Cape and resembles it very much. Sakharnaya Golova Mountain (544 m) is 5.5 km to the east-north-east of the cape.

Uzunlar Liman – brackish lagoon on Kerch Peninsula between Chauda and Opuk Capes. Its length around 10 km, width of the maritime part—5.5 km, area 21.2 km^2, maximum depth 0.1 m. A gray mullet feeding farm is established on Uzunlar Lake. In spring, young gray mullets enter from the Black Sea via special sluices into the lake, where warm shallow water enables them to feed well; this done, year-old mullet returns to the sea. The lake is adjoined by Uzunlar trench and levee.

Uzun-Syrt, Klementiev's Mountain – (Turk. *"long range, back"*) sited 13 km from Feodosiya, the Crimean Peninsula. A mountain elongated almost 8 km with an ideally square, 270 m above the sea level. The place is the homeland of Soviet glider flying. The local relief provides for excellent conditions for the circulation of the rising air flows. From 1923 to 1935, there were held 11 forums of the Society of the air fleet friends. At the proposal of K.K. Artseulov, one of the founding fathers of glider flying, November 1, 1923 has been regarded the birthday of glider flying. In 1924, Peter Klementiev, a young glider flier, was making his 23rd flight on his glider, and he died in a crash. Since then, U.-S. has also been referred to as "K.M." The outstanding Soviet plane builders and future scholars of the outer space S. Ilyushin, A. Yakovlev and S. Korolev took part in forums on K.M. There is a glider flying museum here.

In 1925, even German pilots who visited the Soviet Union recognized the advantages of Uzun-Syrt, stating that Koktebel Mountain is best place in Europe to fly gliders. Among them was the future commander of the Luftwaffe in Third Reich Hermann Goering. When, during the Great Patriotic War (1941–1945), the Germans took the Crimea, Goering issued a special order not to bomb Klementiev downhill, and not to damage the unique qualities of the ridge. And not a bomb fell on the Mount.

Klementiev's Mountain (http://silvar.livejournal.com/196014.html)

Uzunye Cape – sited 3.5 km north-west of Rumeli Cape which is distanced most from the shoreline in the segment from Bosporus Strait to Dalyan Cape. The cape is high, precipitous.

V

Vakfikebir – is a small town and district of Trabzon Province in the Black Sea region of Turkey. Population—13,400 (2012).

Vakulenchuk Grigoriy Nikitich (1877–1905) – a noncommissioned Russian officer, one of organizers of the social-democratic unit on the Black Sea Fleet. He was a member of the underground Central Naval Executive Committee of the Russian Social-Democratic Workers' Party (RSDRP). In spring-summer 1905 the revolutionary rebelion engulfed the Army and the Navy. Near Odessa the battleship "Prince Potemkin the Tavrian" came for military exercises. V. was the commander at this ship. On June 14 the crew refused to take dinner prepared from rotten meat. The commander ordered the crew to line up on the deck and called the guard. Suddenly some sailors cried out: "Pals! We can't bear it anymore!" The moment this revolt started that was led by V., one of the officers fired at him and wounded him deadly. The sailors dealt violently with the officers. The rebels seized power on the ship.

Vama-Veke – a terrain in Romania on the B.S. where major Romanian seaside resorts locate: from the Danube's arm Kilia in the north to V.-V. in the south near the Bulgarian border. V.-V. may be translated as "old customs." For a long time V.-V. in Romania has been known as a remote and solitary place, although, in fact, it is rather easily accessible. Northward of V.-V., along the B.S. coast the infrastructure of the Romania's second largest seaside city—Mangalia is found. The distance to the main Romanian port Constanta is 50 km, to Bulgarian Varna—about 100 km. At all times big Black-Sea resorts have been found to the south and north of V.-V. The beach is covered with wonderful fine-grained sand, the sea is calm here. Such conditions favor development of diving and water sports here. The holiday season lasts here from the second half of May through October.

Vardane – a small village on the coast of the Northeastern Black Sea, 21 km northwestward of Sochi, Russia. Recreation area, beach.

Varna – the major city-port on B.S. in Bulgaria, the administrative center of the Varna Region. This is a large industrial, resort and cultural center extending as an

S.R. Grinevetsky et al., *The Black Sea Encyclopedia*,
DOI 10.1007/978-3-642-55227-4_22, © Springer-Verlag Berlin Heidelberg 2015

amphitheater on the coast of the harbor with the same name. The population here is 334,700 people (2012). The first settlement here called Odessili, Odessos was founded in the sixth century B.C. by Greek colonists from the city of Mileta of Near Asia. Not long ago the traces of Odessos were found during construction of dockage facilities near the shipbuilding yard in Varna. Odessos played a significant role in the life of Thrace in the Antiquity. Approximately in year 50 A.D. Varna was captured and destroyed by Goths. When beginning from year 15 A.D. the Romans organized on this captured territory the Province Nizhnyaya Moesia, Odessus Varna was annexed to this province, but as before it remained closely linked economically and culturally with the life of the Thraceans. Although the city had lost its political independence, the Romans nevertheless allowed them to have the local self-government, the right to mint coins and to use the Greek language and not Latin as in other centers of the Roman Empire. Comfortable and well protected by the fortress wall Odessos during the Roman reign grew and the second protective wall was constructed encircling the new quarters. The center of the city was intensively built up and there in the second century A.D. the public baths Thermae were constructed and their ruins survived to the present. Once the public baths took an area of 10,000 m^2 (about 7,000 m^2 had survived).

In 681 the city was conquered by Bulgarian Khan Asparukh. That time the city received its present name—Varna by the name of the Varna River (this name is associated with the Bulgarian variant "mineral spring"). In the time of the Second Bulgarian Empire, during the ruling of Ivan Asen II, Varna turned into a thriving commercial hub that, according to historical documents, traded with large ports of that time—Tsargrad (Constantinople), Genoa, Venice and Dubrovnik. Since that time this port had played and still plays an important role in the economics of the country. During the Russian-Turkish Wars of the eighteenth to nineteenth centuries Varna taking the strategically important position, was besieged by the Russian troops in 1773, 1810 and 1828. The last siege was most difficult and ended in capture of the city. In 1949–1956 the city was named Stalin in honor of I.V. Stalin.

About the half of the country's marine cargo traffic passes through Varna. The industries here include textile, food and fishery. The port has the shipbuilding and ship-repair yards with the dry dock being the largest on the Black Sea. Varna is crossed by several operational intermodal lines, including the ferry line Varna-Ilyichevsk-Poti/Batumi (the TRASECA main branch). The economic, electrical engineering and medicine institutes, the Higher Marine College, the Fishery Institute, the Institute of Marine Research and Oceanography of the Bulgarian Academy of Sciences are found in Varna. The Opera Theatre and the Dramatic Theatre. The ruins of the Antique and Medieval buildings have survived to the present. The Archeological Museum, the Black Sea Museum and marine aquarium.

Varna is a well-known center of musical and theatrical art. The city often becomes the venue of international festivals and competitions. Varna is also the world's renowned place for holding congresses and meetings.

The climate here is moderate and if you add here the sea, sand beaches and warm mineral springs this place becomes most favorable for rest and health improvement. The well-known seaside resorts, such as "Druzhba", "Golden Sands" and "Albena", are found near Varna. 50 km to the northeast of Varna the Kaliakria Cape projects far into the sea where on July 31, 1791 there was a sea battle that was the last in the Russian-Turkish War of 1787–1791.

Varna (http://www.varna.bg/en/index.html)

Varna Bay – a bay of B.S. in Bulgaria located between the Franga Plateau and the Avren Plateau. The bay origin is connected with the tectonic Varna depression. The bay protrudes into land for 3 km, its width at an inlet is 7 km, depth 10–18 m. During severe winters it gets frozen. The coasts in the north and south are elevated, while in the west—low-lying. Via the navigable canal it is connected with the Varna Lake. The city and port of Varna locates on V.B. coast.

Varna Lake – the largest and deepest fresh-water lake on the Bulgarian coast of B.S. that in 1907 was connected with the sea by a canal. It originated as a liman in the mouth of the Provadiya River. The average depth of the lake is 9.5 m. It is separated from the sea by a constantly windening sandy strip about 2 km wide. V.L. has an elongated form. Its southern shore is high and steep, the northern is flat. The bottom of the depression enclosing the lake is covered by a silt layer from 10 to 30 cm thick. In the deepest places the bottom is covered by black hydrogen sulfide mud (fango). V.L. is of a tectonic origin. It was formed as a result of the sea level rise in the Late Pleistocene. The lake is surrounded by numerous valleys 30 to 120 m wide, but this terrain does not abound in large water sources. The lake is replenished largely from the Devnya sources via the Beloslav Lake and in the western part—from underwater sources. The major rivers are Devnya and

Provadiya that, before correction of the Provadiya River in 1945, flowed into the same place in the western part of the Beloslav Lake. At high water levels these sources aggraded quartz sands that formed numerous beaches.

Before the twentieth century the water flowed from the lake into B.S. via the full-flowing, but shallow river Devnya (Varna River). In 1906–1909 after port construction and drainage of the river the navigation canal was excavated between the lake and the sea. As a result, the water level in V.L. dropped to 1.4 m and the water from the sea started inflowing into the lake. With the time the waterlogged shores of the lake dried out providing spaces for agriculture development. The water level in V.L. depends largely on fluctuations of the sea level and, to some extent, on the flows of the Provadiya and Devnya rivers. The water temperature and salinity depend on the incoming sea waters. On the lake surface the water temperature varies greatly depending on a season: at times such variations may reach 25°. The mean annual temperature near the surface is 14 °C and near the bottom—about 8 °C.

In spring the surface water salinity decreases due to the increased flow of the Provadiya River in this time of a year. The growing salinity in the summer–autumn months may be attributed to the decreased inflow of river waters and evaporation from the lake surface. Sometimes in summer the sea water inflow increases due to the wind effect, but this water is of less density (because of its higher temperature) and it spreads over colder and heavier deep lake waters. Apart from vertical stratification the salinity variations are observed in different areas of the lake: in the areas close to B.S. it is approximately 12‰, while in the western part—approximately 7‰. Intruding sea waters killed the fresh-water fish, so only marine species survived in the lake. On the Genoa hill near the lake shore one can find the ruins of the ancient basilica reminding of the Genoa colony once existed here.

Varna Province – an administrative territorial unit in the northeast of Bulgaria. Its area is 3,800 km^2, the population—475,000 (2011). The administrative center—the city of Varna. The territory of the province had a flat and hillocky relief. In the south V.P. is confined by the eastern spurs of Staro-Planina—Kamchiya-Planina, in the north it includes a part of the Dobrudja Plateau—Franga Plateau, in the west—Provadiya Plateau, in the east is it washed by the Black Sea. The main rivers are Kamchiya and Provadiya; there are also large karst springs (Devnenskiy). The central flat part of V.P. has the deposits of rock salt and manganese ores. About one-third of the province territory is covered by forests, mostly oak.

V.P. combines the developed industry and intensive agriculture; transport and seaside resorts being very popular in Bulgaria and over the world. The main industries here include machine-building (ships, engines, etc.), chemical (production of soda, chlorine, plastics), cement, glass and food. A new coal port is constructed on the shore of the Varna Lake. The major industrial centers are Varna and Devnya. Varna is the major shipbuilding and ship-repair center in Bulgaria. The greater part of the constructed ships is exported. Agriculture is practiced on nearly the half of the province area. Here such crops as corn, wheat, sunflower and sugar beet are grown; vine-growing is also well developed. As concerns the animal husbandry, the sheep and pigs are grown here. Varna is an important Black Sea port. The territory of V.P. is crossed by railroads Rousse—Varna, Dobrich—Varna, Varna—Gorna-Oryakhovica—Sofia and Karnobat—Varna. Such international seaside resorts as "Golden Sands" and "Druzhba" are the world renowned places.

Varna Sea Port, Port of Varna – due to a restricted water area the port is being developed by construction of a new region on the Varna Lake shores. For this purpose the canal 2,830 m long, 40 m wide and 10.5 m deep is being built that will link the lake with the Varna Bay. The new region will specialize in handling of bulk cargo (in the plans the handling of coal only will reach 3 mln tons a year). In the future this canal may be deepened to 14 m and widened to 60 m which will ensure the working of the PANAMAX-type vessels. The Varna Port includes three cargo regions (ports)—Eastern, Western and Balchik. Varna Port—East is in the center of the city. Here grain is loaded for export. There is also a container terminal. Commissioning of a new container terminal in Varna Port—East will include Bulgaria into the EU transport structure. Port Varna—West is the most up-to-date port in Bulgaria; it was constructed in 1974. The port located 30 km from the city services five nearby major industrial enterprises—reloading of fertilizers, soda ash, cement, quartz sand and sugar. Port Balchik locates 50 km northward of Varna.

Varna siege (1828) – siege and capture of the Ottoman citadel—port Varna in Bulgaria by the Russian troops commanded by Rear-Admiral A.S. Menshikov and General M.S. Vorontsov in July–September 1828 in the course of the Russian-Turkish War of 1828–1829. In late July the assault unit of Menshikov (6,000 people) landed near Varna. The citadel garrison numbered to 18,000 people. In the battle on July 26 the Russian ships destroyed the Turkish ships that protected the citadel from the sea. Once this obstacle was removed the Black Sea Navy could start firing of Varna. On August 8 during beating off the attack Menshikov was wounded and the command was taken up by General Vorontsov. By September, after arrival of the guard detachment the number of siege forces reached 32,000 people. This made the siege more "dense" and helped the troops on September 16 to repel successfully the attack of 30,000 strong Turkish corps commanded by Omar-Vrione-Pasha who tried to de-block the besieged garrison. On September 25–26 the general assault of Varna was launched. The garrison managed to repel this assault, but being completely isolated the situation of the garrison was very difficult. The Turkish troops suffered large casualties from the fire of Russian cannons from the sea and land that sent more than 107,000 shells to the city, including combat and flame missiles. It's worth mentioning that the first missile unit commanded by Colonel V.M. Vnukov operated near Varna. It was actively involved in firing the citadel. Shortage of food, grave casualties and lack of hope for breaking the siege urged the Turkish command to enter into negotiations that were completed on September 29 by capitulation of the Varna garrison. The capture of Varna became the most significant success of the Russian troops in the military campaign of 1828 on the Balkan theatre of military operations.

Varvara – is a small village in southeastern Bulgaria, located in the Tsarevo Municipality of the Burgas Province. This seaside resort is situated on the Black

Sea coast, between the towns of Tsarevo and Ahtopol, near the border with Turkey. Population—250 (2005).

Vasilchikov Viktor Illarionovich (1820–1878) – Russian Prince, military figure, one of the leaders of the Sevastopol defense of 1854–1855, Lieutenant-General (1857), and Aide-de-Camp General. In 1839 he finished the Corps of Pages. In 1842 he excellently showed himself in the Caucasian War. He took part in the Hungarian campaign of 1849, in the Crimean War of 1853–1856. From November 1854 he acted as the Headquarters Chief of the Sevastopol garrison. He revealed the extraordinary organizational capabilities. He did much for improvement of the sanitary unit and supply of the garrison, he managed to attract local people who helped the Sevastopol defenders with money and personal participation; he put in order the ammunition use. He was not only the key figure in organization of the march of the Sevastopol garrison to the northern side, but also a person responsible for this operation because M.D. Gorchakov, the Commander-in-Chief in Crimea, refused to sign the disposition for leaving the city. From late 1855 he was appointed the Headquarters Chief of the Southern Army. From 1858—the manager in the Military Ministry. In 1861 he fell ill and was given leave and in 1867 he resigned. A year before the death he prepared the notes "Why the Russian Arms were constantly defeated on the Danube and in Crimea in 1853–1855". And the main cause of the Russian Army failure, as he saw it, was inadequate strategic preparation of military operations.

Veleka River – flows into B.S. It locates one kilometer from the Veleka village (sometimes on the maps it is given as Velika), Bulgaria. V.R. is a small river 15 m wide originating near the Strandzha Mountains. A sandy beach running for about two kilometers was formed in the river valley. In winter the mouth of the river disappears as the river filters through a sandy swash bar and in summer the river carves its way over the surface. Before a beach the river forms several oxbow lakes and the beach looks like a wide sandy bar. The sea bed near the shore has sandy and stony areas. The flora and fauna here is rich enough. The water is transparent. In July–August the plankton proliferates in B.S. and the water transparency becomes worse. The best months for diving are May, June, September and October (in early May and in late October the water is cool here—about 14–18 °C).

Verkhush (Gorishnyak) – a northern or northeastern wind on the Azov Sea that forces water via the Kerch Strait to B.S. There are beliefs that if V. does not die out during a day, it will blow for 3 days. If it does not stop on the third day—it will blow for a whole week.

Vienna Conferences (1854–1855) – diplomatic negotiations in the course of the Crimean War of 1853–1856. The USA, Belgium, Prussia and Austria proposed their agency services for peaceful settlement of the war. The Austrian Cabinet after negotiating its proposals with Paris and London submitted in July 1854 to A.M. Gorchakov, the Russian Ambassador in Vienna, the program of "4 items" that included: (1) revision of the position of the Danube Principalities and Serbia and their relations with Russia; (2) defining of the navigation conditions over the Danube; (3) revision of the 1841 London Convention on the Black Sea Straits;

(4) defining of the stand of Russia towards the Christian subjects of Turkey. After receiving the consent from the Russian Government (in late 1854) these "items" made the basis for peaceful negotiations. The preliminary negotiations were conducted in December 1854 to January 1885, in March–June 1855 in Vienna. There were 14 official meetings that were referred to as conferences. The Russian delegation by means of compromises in relation to items 1 and 2 important for Austria succeeded to get its support in relation to item 3 being of principal importance for Russia. Russia proceeded from the failure of the Sevastopol siege, loss of allies in Crimea and also reluctance of Austria to initiate military actions. After discussion and coming to an agreement on items 1 and 2 about establishment of general protectorate of France, Britain, Austria, Russia and Prussia over the Danube principalities and temporary occupation of these territories by the Austrian troops and also on forming of a multilateral commission on elaboration of the navigation codes and on collective control of the Danube River mouth, the Parties started discussion of the issue of the Black Sea straits. This discussion revealed a lack of unity in the approaches taken by France, Britain and Austria to settlement of this issue, and this fact helped Gorchakov to come most close to the Austrian proposals being most acceptable for Russia. However, the possibility of consensus was frustrated by the categorical demand of the British-French diplomats on disarmament of Sevastopol and consent of Russia to significant reduction of its naval fleet on B.S. In his turn Gorchakov resolutely waived any attempts to restrict the Russia's sovereign rights to B.S. and named as acceptable the "counterbalance" principle suggested by Austria (it envisaged the right of allies to have any quantity of ships on B.S. not restricting at the same time the number of ships in the Russian Black Sea Fleet). In the course of debates the French party showed some hesitations, but pressed by the British delegation it waived the Austrian initiative and the negotiations had no continuation. Item 4 was not discussed at these conferences. The Turkish delegation, although trying to protect the interests of its Government, had not demonstrated a clear-cut independent position at these conferences.

Vienna Union Treaty of 1697 – a treaty concluded by Russia, Austria and Venice on the offensive alliance against the Ottoman Empire and Crimea Khanate for a term of 3 years. It was signed on January 29 (February 8) 1697. The Treaty consisted of three articles. The allies undertook to carry out coordinated military actions on the land and on the sea, to inform each other in due time about their military plans and to abstain from entering into a truce. This Treaty became an important step in diplomatic actions taken by Tsar Peter I for preparation for the war with Ottoman. In 1698 Austria getting ready for the war with France concluded a separate peace with the Ottoman Empire and the Republic of Venice followed it. Being left by its allies Russia had to conclude a truce with the Ottoman which fact was confirmed in the Karlowitz Peace Treaty of 1699.

Vifian Peninsula, Kocaeli – the peninsula in Turkey lying among B.S., the Bosporus Strait and the Izmir Bay of the Marmara Sea. Its length is about 75 km, width—to 50 km. The plateaus and hills (the northwestern terminal ranges of the Pontic Mountains) up to 537 m high are composed of quartzite, limestone and sandstone. It is covered by deciduous forests, maquis; on the southern coast there are vines, plantations of mulberry trees, seaside resorts (the so-called "Anatolian Riviera"). The main cities—Izmir and Uskudar.

Vilkovo (former village Lipovanskoe) – a city and port on the left shore of the Kiliyskiy Channel of the Danube where it ramifies forming a kind of a "fork" ("vilka" in Russian means a fork), Odessa Region, Ukraine. It locates 18 km from B.S. It was founded in 1746 by the Don Cossacks. In 1775 the Zaporozhie Cossacks settled here who gave the present name to this place. V. locates on the islands separated by numerous arms and canals making "water streets" that are used for transportation and as a place from where excursions over the river to nearby lakes start. All the above justifies its second name "Danube Venice". V. has fish processing factory, shipbuilding yard, fishery.

Vitkovskiy Vladimir Konstantinovich (1885–1978) – Russian Colonel (1916), Major-General (1918), Lieutenant-General (1920). In 1903 he finished the First Cadet Corps and in 1905—the Pavlovo Military School. He started his military service in the life guard of the Keksgolm regiment. He took part in World War I. On October 2, 1917 he was appointed the commanded of the 199th Kronshtadt Infantry Regiment. He struggled on the side of the Belaya (White) Guard: battalion commander in the Colonel Drozdovskiy unit during the march from Yassy to Novocherkassk in March–July 1918; later on he was the commander of the Second General Drozdovskiy Officer Unit and the brigade commander. In February 1919 he was promoted to the position of the Chief of the Third Infantry (from October 1919—"Drozdovskiy") Division (Donbass—Orel—Novorossiysk—Crimea). After evacuation from Novorossiysk to the Crimea he served in the Russian Army of General Vrangel. He distinguished himself in the battles at landing near Khorol (the Sea of Azov) in Northern Tavria and on the Perekop Isthmus (Crimea). In August 1920 he was the commander of the Second Army (instead of General Dratsenko). In November 1920 he evacuated from the Crimea to Gallipoly in Turkey. His emigration geography was as follows: from late 1920—Turkey, Bulgaria, from 1928—France, from 1945—the USA. He was an active figure in the Russian Combined Arms Union (1924); deputy of General Kutepov. He died in Palo-Alto (San-Francisco, USA).

"Vityaz" – an oceanographic research vessel of the P.P. Shirshov Institute of Oceanology of the USSR Academy of Sciences. In the 1950s–1960s it was the flagship of the Soviet expedition fleet. It was designed for carrying out integrated oceanic studies. It was constructed in 1939 in the dockyard "Schichau Seebeckwerft" (often abbreviated as SSW), a German Shipbuilding company, headquartered in Bremerhaven, Germany, as a banana-carrying vessel called "Mars". In 1945 this ship as a military trophy appeared in Britain where it was given the name "Empire Forth". Later on by reparation this ship was transferred to the Soviet Union, but this time bearing the name "Ekvator" (equator). In 1946 it was renamed into "Admiral Makarov". In the same year it was given its final name "Vityaz" in honor of two Russian famous corvettes—on one corvette N.N. Miklukho-Maklay went to the coasts of New Guinea and on the second Admiral Makarov conducted investigations in the Pacific Ocean. The name of the latter corvette is carved on the fronton of the Oceanographic Museum in Monaco among 10 other outstanding ships that contributed much into study of the World Ocean.

In 1948 this ship was refurbished into a research vessel in Wismar (GDR). The deep-sea winch with a large cargo-lifting capacity, the equipment for ocean bottom sampling and other modern mechanisms and instruments were installed on the ship. In 1949 "V." was transferred to Odessa and made the second trip over B.S. "V." made 65 expeditions (the last one in 1979) in the Far-Eastern seas, the Pacific and Indian oceans during which new oceanographic, meteorological, biological and geological data were collected, some deep troughs in the Pacific were investigated. In 1958 the maximum depth of the World Ocean in the Mariana Trench—11,022 m was measured from board of this ship. During its service life the ship covered more than 600,000 miles of which in over 500,000 miles depth sounding, thousands of deep hydrographic stations, sometimes at depths over 6,000 m, were carried out. "V." took part in international expeditions on ocean studies, in investigations within the frame of the Second International Geophysical Year, worked in the Indian Ocean under the program of the International Indian Ocean Expedition, participated in the Soviet-American expedition for investigations in the "Bermuda triangle". It was excluded from the fleet in 1980. The ship was put on its eternal anchorage in Kaliningrad and is the key exhibit in the Museum of the World Ocean. Its displacement was 5,710 tons, speed 13.5 knots, autonomous cruising range—to 18,500 miles, main engine capacity—2,206 kW. This ship accommodated 12 research laboratories, the crew—72 people and the team of scientists—63 people.

R/V "Vityaz" in the Museum of the World Ocean, Kaliningrad (http://commons.wikimedia. org/)

In 1981 instead of the above veteran the Soviet expedition fleet received the new "Vityaz-IV" built in Szczecin, Poland. This was an eight-deck steel vessel with the unlimited cruising range. It was designed for comprehensive oceanic researches, primarily, hydrobiological, and for works with the hyperbaric complex. It was provided with the equipment for underwater investigations: a marine hyperbaric complex with a diving bell for operation at depths to 165 m, submarine units— manned "Argus" and unmanned "Zvuk". Its displacement was 6,358 tons, length— 111 m, width—16.6 m, speed—to 18 knots. It had onboard 21 research laboratories, the crew of 66 people and the research team of 59 scientists. The operating regions were the Atlantic and Indian oceans, Black and Mediterranean seas. Having started its operation in 1982 this ship already in November 1993 made its last 27th cruise. In May 1996 following the resolution of the Presidium of the Russian Academy of Sciences "Vityaz-IV" was removed from the research fleet.

R/V "Vityaz-IV" in the Black Sea (1984) (http://s017.radikal.ru/i430/1202/b9/80da00eb2e29.jpg)

Vityazevo – is a village in Vityazevskiy Rural Okrug under the administrative jurisdiction of the town of Anapa, Krasnodar Kray, Russia. It is located 11 km north of Anapa proper on a bank of the Vityazevskiy Liman, at the coast of the Black Sea. Population—7,900 (2010).

Vladimir (Admiral) Cathedral – In 1825 Commander-in-Chief of the Black Sea Fleet and ports Admiral A.S. Greig applied for the Highest permission to install a monument on the Chersonesos ruins where Prince Vladimir was baptized and from where Christianity started spreading over Rus'. Later on the initiative of Admiral M.P. Lazarev it was decided to construct a cathedral in the center of Sevastopol. After death of M.P. Lazarev in 1851 he was buried in a tomb in place of the future cathedral. The foundation of the cathedral was laid in 1854, but the Crimean War interrupted the construction. During defense of Sevastopol the followers and companions-in-arms of M.P. Lazarev—V.A. Kornilov, V.I. Istomin and P.S. Nakhimov were buried in the same tomb. Thus, the unfinished cathedral became the burial place of admirals of the Black Sea Fleet and heroes of the Sevastopol defense of 1854–1855. In 1858 the construction of V.C. was resumed and finished in 1888. The height of the cathedral with the cross is 32.5 m; the foundation is made in the form of a cross. On the outside of the southern and northern façade walls four black marble plates with the names of Admirals M.P. Lazarev, V.A. Kornilov, P.S. Nakhimov and V.I. Istomin were inserted. In addition, marble plates with the names of admirals and officers of the Marine Department awarded the Order of St. George for bravery were installed in the walls of the cathedral. In the period from 1869 to 1920 there were buried here nine marine Admirals P.A. Karpov, P.A. Pereleshin, M.I. Defabr, V.P. Schmidt and I.M. Dikov—participants of the Sevastopol defense of 1854–1855; S.P. Tyrtov, G.P. Chukhnin, M.P. Sablin—commanders of the Black Sea Fleet in the late nineteenth to early twentieth centuries; I.A. Shestakov, Marine Minister in the 1880s. Till 1917 V.C. was maintained on the money of the Marine Ministry. In 1932 it was closed. During the Great Patriotic War (1941–1945) it was damaged. In 1966 it was decided to restore V.C. In 1972 it became a branch of the Museum of Defense and Liberation of Sevastopol. In late 1991 The cathedral was sanctified once more and the sacred services were resumed.

Vladimir Cathedral (Photo by Andrey Kostianoy)

Marble plate in memory of Admiral M.P. Lazarev (Photo by Andrey Kostianoy)

Marble plate in memory of Vice-Admiral V.A. Kornilov (Photo by Andrey Kostianoy)

"Vladimir" – a paddle steamer-frigate of the Black Sea Fleet that showed itself well in the 1853–1856 Crimean War. It was built in Britain in 1848. Its displacement was 1,713 tons, speed—about 12 knots (22 km/h). It had 7 cannons onboard. On November 5(17), 1853 "V." under command of Captain-Lieutenant G.I. Butakov sailing under the flag of the Chief of the Black Sea Fleet Headquarters Vice-Admiral V.A. Kornilov detected to the north of Eregli (Anatolian coast of Turkey) the Turkish 10-cannon ship-frigate "Pervaz-Bakhri", started fighting it and after 2 hours of the cannon duel forced it to lower the flag. After this the frigate "Pervaz-Bakhri" was led to Sevastopol where she was renamed into "Kornilov" and included into the Black Sea Fleet. This was the first battle of steamers in the history. During the Sevastopol defense of 1854–1855 "V." supported the defenders of the city with its cannon fire. On the night of August 31, 1855 the frigate was drowned by the crew on the northern side of the Sevastopol Bay. In 1860 she was lifted and passed to breaking.

Vladimirskiy Lev Anatolievich (1903–1973) – Soviet Admiral, Commander of the Black Sea Fleet. In 1925 he finished the M.V. Fruzen Naval School and was appointed the watch officer on the torpedo boat "Lieutenant Schmidt" of the Black Sea Fleet. In 1926 he was the senior watch officer, assistant commander of the cruiser "Chervona Ukraina" which construction was nearing completion. In 1927 he finished the special courses for the Navy command staff; he was the senior gunner, senior assistant commander of the destroyer "Shaumyan". In 1928–1930 he sailed to Turkey, Italy and Greece. In 1932 he was appointed the commander of the patrol ship "Shkval" which construction was nearing completion, in 1935— destroyer "Petrovskiy", in 1937—leader ship "Kharkov". In 1938 he escorted the cargo destined to Spanish republicans; being the captain of the 2nd rank he led to Vladivostok the hydrographic vessels "Polyarnyi" and "Partizan". In 1939 after all tests he accepted the leader ship "Tashkent" that was built in Italy and led it to Odessa. In 1939 he was the commander of the cruisers brigade, the commander of the squadron of the Black Sea Fleet. In November 1939 he successfully led three low-speed tankers and the icebreaker "Mikoyan" to the Bosporus.

During the Great Patriotic War (1941–1945) on his proposal the ships were dispersed so that they suffered less from air attacks. Some ships were transferred to the Caucasian ports. V. commanded the landing operation under Grigorievka, was wounded on the destroyer "Frunze" that was later drowned, but he continued to command the operation. He successfully guided the evacuation of the Maritime Army from Odessa. He commanded the squadron during gunning of enemy positions near Sevastopol and during the Kerch-Feodosiya landing operation that was launched on the night of December 29, 1941. On the night of January 16, 1942 he commanded the landing operation in Sudak. Regardless of the fact that he was not allowed to command personally the escorts he went into the sea many times. During the war he took part in 20 military and 4 supporting operations. He commanded the Black Sea Fleet from April 1943 to March 1944. On his proposal on the night of September 9, 1943 there was a successful assault landing under Novorossiysk. He commanded the fleet in the Novorossiysk (1943) and Kerch-Eltigen (1944) operations. After great casualties during landing in the Crimea in spring 1944 he was removed from this position and appointed the commander of the Baltic Sea Fleet squadron. In 1944 he took part in the Vyborg operation.

In late 1946 he was moved to the Army Inspectorate to the position of the Naval Deputy Chief Inspector. He advocated construction of aircraft carriers. In spring 1948 he was appointed the chief of the Naval educational establishments. From 1951 he headed the Military Training Department and in 1952 he graduated from the Headquarters Military Academy. In 1954 he was promoted to Admirals. From April 1955 he was the deputy on shipbuilding of the Navy Commander-in-Chief. In 1956–1959 he was the chairman of the Marine Research Committee. Beginning from 1959 he was engaged in research and pedagogical activities. In 1960 he was appointed the Deputy Chief of the Navy Academy. In the period from 1962 to 1967 V. headed the Academic Courses for the Officer Staff at the Navy Academy. In 1967–1970 he was professor-consultant. In the 1960s he organized several hydrographic expeditions and led the most important and significant of them: in 1965—in the Mediterranean Sea and in 1968–1969—in the Indian Ocean on the oceanographic vessel "Polyus" (pole). This was the first world cruise of the Soviet surface ship sailing under the Navy flag. Upon return he defended the thesis and was conferred the academic degree—Candidate of sciences. He published his articles in the journal "Marine Newsletters". Some of his articles were devoted to the war on B.S. In 1970 he resigned. He was awarded many orders and medals. His name was given to one of the streets in Sevastopol and the research oceanographic vessel "Admiral Vladimirskiy".

Vladimirskiy L.A. (http://upload.wikimedia.org/wikipedia/ru/c/cf/Vladimirskiy.jpg)

Vodyanitskiy Vladimir Alekseevich (1893–1971) – the outstanding Soviet scientist in hydrobiology and oceanography. He devoted many years to creation of the Novorossiysk Biological Station and later the Institute of Biology of the Southern Seas (IBSS). After finishing the Nonclassical College he entered the Kharkov Technological University. But a year later he changed over to the natural science chair of the physico-mathematical faculty. In 1920 V. together with his wife N.V. Morozova-Vodyanitskaya organized in Novorossiysk the Biological Station at which till 1924 they were the only permanent research workers. In 1930 V. was transferred as a senior zoologist to the Sevastopol Biological Station. In 1934 V. was conferred the academic degree—Doctor of Biology. In 1939 he became the head of the Novorossiysk Biological Station and at the same time Professor of the Hydrobiological Chair at the Rostov University. Here he held two optional courses: "Oceanography" and "General Hydrobiology". In 1944 V. returned to Sevastopol and got down energetically to restoration of the Sevastopol Biological Station. In summer 1946 V. went on the first postwar expedition. From board of a minesweeper he made profiles from Yalta to Batumi and laid the basis for investigations in a new field—deep-water microbiology. After establishment of the Ichthyological Commission at the USSR Academy of Sciences that was headed by Academician L.S. Berg he became his deputy. In 1957 V. was elected the Corresponding Member of the Ukrainian Academy of Sciences. He was conferred the title of the Honored Worker of Science and the Honorary Member of the All-Union Hydrobiological Society. In 1957 he took part in the Mediterranean expedition—the USSR's first hydrobiological expedition to the Mediterranean Sea. Next year a new expedition to the Mediterranean was organized on the vessel "Akademik A. Kovalevskiy" devoted to the International Geophysical Year. In 1960 he worked in the Red Sea and in the Gulf of Aden. He was among those who organized the Soviet-Cuban marine expedition in 1964. In 1975 after his death his autobiographical book "Notes of the Natural Scientist" were published under editorship of V.V. Shuleikin.

V. started a large cycle of integrated researches of B.S. with observations over the eggs and larvae of the Black Sea fish. The obtained results enabled him to refute the erroneous idea that the fauna of B.S. in its open part, far from the coast was poor. He assumed that there was a rather prominent mutual exchange between the deep waters rich in biogenic substances and the surface waters of B.S. Seeking an answer to the question "How such water exchange is possible at the pronounced density stratification in B.S.?" V., apart from hydrobiology and hydrochemistry, turned to physical oceanography. On the basis of the results obtained by various scientists in numerous expeditions he demonstrated that in the open part of B.S., far from its coasts, the upwelling of water should exist and these waters should entrain upward the bottom waters. The correctness of this statement was fully confirmed, but, unfortunately, after his death. Only quite recently it was proved the dominating role of the advection transfer of substances in sharply stratified seas where no strong turbidity could be generated. Relying upon the research results of V. the fishing activities were moved to the new areas of B.S. V. permanently underlined that hydrobiology and ecology were the theoretical basis for fishery development. V. energetically opposed the West-European projects on burial of nuclear wastes in the depth of B.S. His estimates had shown that the water flowing out of B.S. had not zero salinity, but a salinity level of 18 ‰ because the fresh waters of the surface flow mixed with the deep waters and that the time of the B.S. water renewal was measured not by thousands, but hundreds of year.

Vodyanitskiy V.A. (http://s005.radikal.ru/i210/1307/6e/fdeae20c81a9t.jpg)

Voeikov Aleksandr Ivanovich (1842–1916) – the outstanding Russian climatologist and geographer. In 1910 he was elected the Corresponding Member of the Petersburg Academy of Sciences. In 1865 the Gottingen University awarded him the degree of Doctor of Philosophy. In 1880 the Moscow University conferred him the academic degree of Doctor in Physical Geography. On the V. initiative the Meteorological Commission was set up at the Russian Geographical Society (RGS). In 1891 he founded the Russia's first meteorological journal "Meteorological Newsletters". In 1872–1876 he traveled over Western Europe, North, Central and South America, India, visited Central Asia, China, Ceylon, Java and also Japan. In 1884 he published his fundamental work "Climates on the Globe, Especially in Russia" for which he was awarded the RGS Great Gold Medal. More than once he visited Caucasus and the Black Sea coast. As a result of these visits he published such works as "Black Sea Coast" (1899) (together with F.I. Pasternatskiy and M.V. Sergeev), "Black Sea railroad and Black Sea coast of the Caucasus" (1911), "Climatic places on the Black Sea coast and in Middle and Southern Trancaucasus" (1912), "Northern part of the Black Sea coast, its climate and colonization" (1914). V. may be rightly considered the pioneer in study of the Caspian Sea water balance. V. evolved the idea on dependence of the sea level variations on the climate. He focused special attention on rivers. Depending on climatic conditions he identified 8 basic types of rivers differing by their hydrological regime. He supported the idea of construction of the Dnieper-Don canal for connecting the North and Baltic seas in the north with the Black and Mediterranean seas in the south and also of the Sea of Azov with the Caspian Sea (Black Sea—Caspian Canal). For the first time in the geographical science he applied the balance method to study of geographical events (water balance in glaciers, water balance in air, etc.). He laid the foundation of such sciences as paleoclimatology, agricultural meteorology and phenology. V. was a member of many Russian and foreign scientific societies. His name was given to the Chief Geographical Observatory in Saint Petersburg. In 1948–1957 his "Selected Treatises" in 4 volumes were published. He was a member of the representative Commission on Study of the Black Sea Coast (together with A.N. Krasnov, F.I. Pasternatskiy and others) that was engaged in investigation of the climate peculiarities, curative waters, in particular in the coastal zone for construction of harbors.

Voeikov A.I. (http://upload.wikimedia.org/wikipedia/commons/5/55/Brockhaus_and_Efron_ Encyclopedic_Dictionary_B82_11-4.jpg?uselang=ru)

Voice of the sea – the infrasound effects caused by storm waves. This is a natural phenomenon discovered by the Black Sea Branch of the Marine Hydrophysical Institute of the NASU in Katsiveli, Crimea, Ukraine.

Voinovich Mark Ivanovich (1750–1807) – Russian Admiral, hydrographic specialist, count. He was born in Montenegro. In 1770 he started his service in the Russian Fleet as a midshipman. Later he was promoted to Lieutenant and on the ship "St. George the Victorious" he went from Kronshtadt to the Mediterranean Sea where he stayed till 1776 and took part in the Russian-Turkish War as a commander of ships. During this time he was promoted to Captain-Lieutenant. In 1777–1779 he commanded the ships of the Baltic Fleet and in 1780–1783 being the Captain of the 2nd and 1st ranks he commanded the squadron of ships on the Caspian Sea. Under his guidance the reconnaissance marine inventories were conducted near the Apsheron Peninsula, in the Astrabad and Krasnovodsk bays. Comparing the obtained materials with the available data he revealed changes in the coast configuration. In 1783 he was transferred to the Black Sea Fleet where he commanded the first battleship "Slava Ekaterine". In 1787–1789 during the Russian-Turkish War in the rank of the Rear Admiral he commanded, at first, the Sevastopol squadron and then temporary acted at the Commander-in-Chief of the Black Sea Fleet. At the same time he was a member of the Admiralty Department. In 1790 he was again transferred to the Caspian Sea and in 1791 he was dismissed and went to his native country. In 1796 he again returned to the Russian service and in 1797 was appointed a member of the Black Sea Admiralty Department. In 1797 he was Vice-Admiral and 1801—Admiral. In 1802 he became the director of the Black Sea Navigation School. He was awarded three Russian orders. In 1805 he resigned.

Volga-Don Navigation Canal (in the past the canal named after V.I. Lenin) – a lock shipping canal. Its construction was started before the Great Patriotic War (1941–1945) and finished in 1952. This canal connects the basins of the Volga and Don Rivers in the place where they come most close—near Volgograd-Kalach. It opens the way for ships to the Caspian, Black, Azov and Baltic seas. Its length is 101 km. This canal begins from the Volga near Krasnoarmeisk. The ships go up to the water divide between the Volga and Don via nine locks (the total lifting height—88 m). Several reservoirs are constructed on the Don slope of the canal. They are filled with water pumped by three pumping plants from the Don. The water from these reservoirs is fed to the water divide stretch of the canal and to the locks. The ships go down from the water divide via four locks (total height—44 m). Having passed Lock No. 13 the ships go out into a large reservoir provided with backup walls of the Tsimlyanskiy waterworks. The reservoir width in some places is as large as 20 km, the maximum wave height is 3 m. Through two locks of the Tsimlyanskiy waterworks the ships go down to the Don along which they sail to the river mouth.

Voloshin (Kirienko-Voloshin) Maksimilian Aleksandrovich (1877–1932) – Russian poet, artist, translator, one of the most bright individualities of the Silver Age. He finished the college in Feodosiya, entered the Law Faculty of the Moscow University. In 1900 his involvement in resurgence of the radical student movement results in his expulsion from the 1st course of the University. In 1901 V. left for Paris to attend lectures in Sorbonne. In 1903 he returned to Russia and became a correspondent of the journal "Vesy" for which he wrote articles about art, art exhibitions and book reviews. In 1905 V. got acquainted with the outstanding Germany philosopher Rudolf Steiner (1861–1925) who was the founder of the spiritual science Anthroposophy. This acquaintance influenced greatly the further creative life of B. He became the member of the General Anthroposophy Society. In 1908 V. started cooperating with the journal "Apollon". His first book "Poems of 1900–1910" was published in 1910. In the periods between his travels abroad he often visited both Russian capitals and lived in Koktebel, in the Crimea, in his "house of a poet" that became a kind of a cultural center, a hangout and place of rest for the writing elite. G. Shengeli, a poet and translator, called this place "Cimmerian Athens". At different times this place was visited by V. Bryusov, Andrey Belyi, M. Gorkiy, A. Tolstoy, N. Gumilev, M. Tsvetaeva, O. Mandelshtam, G. Ivanov, E. Zamyatin, V. Khodasevich, M. Bulgakov, K. Chukovskiy and many other novelists, artists, actors and scientists. In 1914 V. went to Switzerland and in 1915—to Paris. That time he published his book of poems "Anno mundri ardentis". After the war his collection of articles titled "Paris and War" was published.

In 1916 V. returned to Russia for good. He lived in Feodosiya where he delivered lectures about literature and art. In spring 1917 he settled down in Koktebel. His house welcomed all newcomers during the Civil War: "and a red leader and a white officer" found hospitality and even a hideout here, as he wrote in his poem "House of a Poet" (1926) and "a red leader" was assumed to be Bela Kun. In the Soviet times V. was allowed to retain his house and he lived rather secure. In the USSR in the period from 1928 to 1961 his name was in oblivion and not a single line of his verses was published. V. was buried, as he wished, on top of the Kuchuk-Enishar Mountain from where a wonderful panoramic view opens on the Koktebel Bay.

"Volunteer Fleet" Society – a marine shipping company founded in 1879 on private donations with a view to develop the merchant navigation in Russia and to create a reserve for the Navy, first of all, on B.S. where the military threat from Britain and Austro-Hungary was pending and for assistance to the Russian Government with cargo transportation over B.S. and Mediterranean, the Indian and Pacific Oceans (the concept of this society was developed in the years of the Russian-Turkish War of 1877–1878). During the wartime these ships could be converted into auxiliary cruisers capable to operate on sea routes of the enemy or into military transport ships. Created in April 1878 the Committee on Establishment of "V.F." consisted of three divisions: operating (financial-business activities), naval (elaboration of assignments and requirements to selection of the type and ammunition of ships) and organizational (development of an organization structure) as well as a secretariat. The Chairman of this committee was heir apparent, future Emperor Alexander III, Vice-Chairman—Secretary of State K.P. Pobedonostsev; the naval division was headed by Admiral K.N. Posiete. In June 1878 "V.F." spent the collected money (8 mln Rubles during 1878–1902) on purchase of the first three ships in Germany. They were given the names of "Russia", "Moscow" and "Petersburg". Already in July these ships equipped with canons and stuffed with the naval crews took part alongside other Navy ships in the parade on the Kronshtadt roads that was attended by the Tsar. After the 1878 Berlin Congress that removed the threat of war the ships were disarmed and directed to B.S. for evacuation of the Russian troops from San-Stefano to Russia.

Later on this society was granted the exclusive right to maintaining the regular cargo-passenger traffic between Odessa and the Pacific ports which happened for the first time in the Russian history. And a very important role here was played by regular transportation of migrants (with their families and belongings) from Odessa to Vladivostok that was realized for nearly two decades beginning from 1883. From 1888 special speedy ships of the cruiser type capable, if necessary, to cope successfully with the military tasks were constructed for the society. For this purpose it was possible to install cannons on decks. Construction of the Trans-Siberian road and the Russian-Japanese War led to decline of "V.F." activities. During the Russo-Japanese War of 1904–1905 four ships from "V.F." were used as auxiliary cruisers and seven ships—as military transport and hospital vessels. From 1907 the attempts were made to revive its activity, primarily, by increasing the number of local voyages in the Far East (over the Sea of Japan, Okhotsk and Bering seas). By this time the fleet comprised 21 ships. During World War I it comprised already over 45 ships that were used for transportation of military and commercial cargo among ports of the Far East, Russian North and ports of the allied states.

Vorontsov Mikhail Semenovich (1782–1856) – Russian outstanding military commander and state figure, Aide-de-Camp General (1815), a member of the State Council (1826), the honorary Member of the Petersburg Academy of Sciences (1826), His Highness Prince (1852) and the General-Field Marshal (1856) being the highest military rank in Russia. He descended from one of the most ancient nobility clans. He was educated at home. V. started his military service from 1801 in the Life Guard of the Preobrazhenskiy Regiment. He distinguished himself in the battles on the Caucasus where he was directed as a volunteer (1803), in the wars with France of 1805 and 1806–1807, and the Russian-Ottoman War of 1806–1812. In the rank of the Major-General he commanded the unified Grenadier Division in the 1812 War and later on in the rank of Lieutenant-General he took part in foreign campaigns of the Russian Army in 1813–1814. In 1815–1818 he was a commander of the Russian Occupation Corps in France. In 1823–1854 V. was the General-Governor of Novorossia (from 1828—the General-Governor of Bessarabia). It was he who promoted development of vine-growing in the Crimea. He patronized A.S. Pushkin exiled to the south (1823–1824). During the Russian-Ottoman War of 1828–1829 he commanded the siege and capture of Varna. In 1844–1854 V. was the vicegerent and, at the same time, the Commander-in-Chief of the troops on the Caucasus. In the areas being in his control he did much for development of education, farming, industry, trade and shipping. He paid much attention to buildup and improvement of the living environment in Odessa, in particular, he guided the construction of the famous Primorsky Boulevard. V. turned Odessa into the main trade city in the south of Russia, succeeded to prolong the porto-franco for 10 more years. In 1839 he created in Odessa the Society of History and Antiquity. The merits of V. were duly praised in Russia. His name could be found on marble plates in the Georgian Hall of the Moscow Kremlin in a sacred list of the dedicated sons of the Motherland; on the monument "Millennium of Russia" in Novgorod the figure of V. could be seen among the outstanding people; the portrait of V. was placed in the famous "Military Gallery" in the Winter Palace. His monuments were installed in Tiflis and Odessa.

Vorontsov M.S. (Photo by Evgenia Kostyanaya)

Vorontsov Caves – the largest caves known on the Black Sea Coast of the Caucasus making the so-called Vorontsov system. The Vorontsov, Labyrinth, Dolgaya caves and Kabany gap being 4,092 m, 3,827 m, 2,325 m and 1,476 m long, respectively, were connected. The total length of the galleries in this cave system was as large as 11,720 m, the height drop—about 300 m. The system represents a chain of specific labyrinths that has 14 entrances at an elevation of 417–720 m over the sea surface. V.C. are of a karst origin and were formed in the zone composed of limestones. The atmospheric precipitation seeping into the rocks erode them forming grottos and halls. Water decorates them with very picturesque formations—stalactites, stalagmites, stone fountains, waterfalls, flower beds and flowers. The caves abound in water bodies, such as mineral springs, underground rivers and lakes. Now these are the places of sightseeing.

Vorontsovskiy (Odessa) Lighthouse – locates on the tip of the Reid pier in
Odessa. In 1862 a wooden lighthouse was built here at first, then, in 1872 it was
built anew in stone. It existed till 1941. During the war it was blown up. The present
lighthouse as we know it was built in 1954. Its height from the water surface to the
watch light is 27.2 m. At night the red gleam of the lighthouse is well visible from
the distance of 12 miles. It sends both light and sound signals. V.L. became one of
the Odessa emblems.

Vorontsovskiy Lighthouse (Photo by Elena Kostyanaya)

Vorontsovskiy (Alupka) Palace – one of the main showplaces in the Crimea.
V.P. was built in 1828–1848 by M.S. Vorontsov by the design of Edward Blore, the
court architect of British Queen Victoria. The southern side looking on the sea made
in the Moresque style somehow reminds of the famous Alhambra Palace of the
Arab rulers of Spain in Granada built in the late fourteenth century. In front of the
southern façade of the palace there is made the "lion's terrace" from which grand
stairs with three pairs of lions made of the white Carrara marble descend in several
flights to the sea. In front of the palace two parterres are made with a marble
fountain in the center of each. The "Fountain of Tears" reminding of the famous
Bakhchisaray Fountain is built not far from the palace. The palace has about

150 rooms. The lobby, gala dining-room and billiard room are decorated with wood in the Gothic style. Of special interest are the winter garden, a typical attribute of palaces in the northern European countries, and a blue guest room with fretted ceiling and walls. The northern façade of the palace and its western part are the architectural variations of the "Tudor style" of the sixteenth to early seventeenth century.

On the outside, the palace looks like a patterned Gothic frame against the grand view of Ai-Petri Mountain. Terraces go down from the palace to the sea. The upper terrace is called Lvinaya (lion's). The park around the palace also deserves special mention. Its total area is 40 ha. The Alupka Park is a remarkable feature because it has no traces of man's interference. In the park you can come across mountain lakes, waterfall cascades, natural heaps of stones, beautiful sunny meadows. The Alupka Park can boast of the richest collection of rare plants. It has over 200 species of trees and shrubs from various world countries: Japanese padoga trees, cedars of Lebanon, Italian pines and Chilean parana pines.

The fame of M.S. Vorontsov Palace was spread over the world by scientists, writers and artists. Deserved reminiscences about this palace one can find in the works of L. Tolstoy, M. Gorkiy, Lesya Ukrainka, I. Bunin, V. Bryusov, A. Kuprin, V. Nabokov, in drawings and pictures of N. Chernetsov, K. Bossoli, I. Aivazovskiy, I. Levitan, V. Surikov, A. Lentulov, Z. Serebryakova and many other well-known figures in literature and art.

Vorontsovskiy Palace (Photo by Evgenia Kostyanaya)

Vortex structures of the Black Sea – analysis of numerous satellite images of B.S. in the infra-red and optical bands of spectrum, hydrological data and information about drifter transport has revealed availability of various mesoscale vortex structures with the spatial scales from 10 to 100 km: coastal anticyclonic vortices (CAV), anticyclones and cyclones in the open sea, dipole vortex structures and jets of various origin. The vortical movements may penetrate as deep as 300–400 m (maximum to 1,000 m). Their drift velocity is about 5–20 cm/s, the orbital velocity—20–40 cm/s (maximum being 45 cm/s). Their lifetime is from several days (dipole) to 6 months (large vortices in the open sea). Merging of anticyclones prolongs their lifetime.

To the west of Sevastopol the so-called "Sevastopol" anticyclone is often observed. Its diameter is to 100 km and it originates over the continental slope and either remains between this slope and the main Black Sea Current (Rim Current) during 12 months or moves with the Rim Current to the southwest. This process may recur several times generating a chain of vortices moving from the Crimea to the Bulgarian coast. They may be alive for 3 months. Large quasistationary anticyclonic vortices are recorded also along the eastern part of the Turkish coast. In the southeastern part of the sea near the so-called "Batumi anticyclonic circulation" the anticyclonic and cyclonic vortices to 60 km in diameter and also more complicated structures both over the continental slope (coastal) and in the open sea may be registered in all seasons. The dipole vortex structures (or "mushroom" currents) consisting of a cyclone-anticyclone pair to 100 m in diameter and to 160 km long may be often observed in all regions of B.S. A chain of CAV 40 to 100 km in diameter may be often witnessed along the Russian coast. It slowly migrates from Sochi to Anapa; their lifetime may reach 5 months.

The vortices are generated due to instability of the Rim Current, interaction of vortex structures with each other and the wind action. This process is facilitated by specific features of the bottom relief and a coastline, a wind regime and water circulation. The mesoscale vortices are dominating in B.S. at weak Rim Current and, on the contrary, are practically absent at intensive Rim Current. Vortices, dipoles, streamlines and nonstationary dynamic formations connected with their evolution play a key role in the horizontal mixing of waters, in particular, in the water exchange between the shelf and the deep open sea, influence the coastal current regimes, horizontal and vertical distribution of hydrological, hydrochemical and hydrobiological parameters (e.g. the vertical shift of the border of a hydrogen sulfide zone) and also fishery.

Vrangel Ferdinand Ferdinandovich (1844–1919) – Russian 2nd rank Captain, hydrographic specialist, explorer of B.S. He was born in Saint-Petersburg in the family of Rear Admiral F. Vrangel. In 1857 he entered the Naval Cadets Corps which, among the first graduates, he finished in 1860. In 1870 he finished the Naval Academy in Nikolaev. As a naval cadet he sailed on frigates "Svetlana" and "Oleg" in the Mediterranean Sea. In 1862 he was promoted to a Midshipman. During two years he attended lectures at the Derpt University. After this he went on frigate "Peresvet" over the Baltic Sea and in the Gulf of Finland. In 1869 he was the commander of the training boat "Priboy". In 1871–1876 V. and F. Mandel were the first to measure the temperature and density of the surface waters in the northwestern part of B.S. and also near the Crimean and Caucasian coast. The explorers defined that the density of the B.S. waters is lower than in oceans. In 1873–1876 he headed the physical explorations during the Caucasus surveys within the frame of the Black Sea hydrographic expedition, took part in the Vienna Meteorological Congress (1873) and the London Scientific Exhibition (1876). During the Russian-Turkish War of 1877–1878 in the rank of Captain-Lieutenant he was the flag-officer at the commander of the coast defense of the Ochakov distance, later he headed the hydrographic team of the separate surveys of the B.S. northern coast. From 1883 he held the post of an inspector–tutor at the Alexandrovskiy Lyceum and taught cosmography there. In 1885 he was promoted to 2nd rank Captain. In 1885–1888 he delivered lectures on oceanography, hydrology and meteorology at the Nikolaev Naval Academy, was elected a member of the Naval Academy Confederation and approved as a member of the Naval Scientific Committee (1886). In 1890–1891 he took part as a hydrologist in the first multi-purpose depth-measuring oceanographic expedition to B.S. on gunboats "Chernomorets" (1890), "Donets" and "Zaporozhets" (1891). During these expeditions the first conclusions were made about specific features of the hydrological regime, flow structure in the Kerch Strait and the factors were found out that governed the water transport in this strait. In 1892–1896—Director of the Alexandrovskiy Lyceum. Editor of "Notes on Hydrography" and the "Pilot for Sailing from Kronshtadt to Vladivostok and Back". The author of many articles on hydrography and meteorology published in the "Naval Newsletters". In April 1896 he was dismissed from the Naval Department, resigned and left for Switzerland where he lived in Ascona. His name was given to a cape in the Taimyr Bay (the Kara Sea).

Vrangel Petr Nikolaevich (1878–1928) – Russian Baron, Colonel, Major-General (1917), Lieutenant-General (1918). He graduated from the Mining Institute (1901), finished the General Headquarters Academy in Nikolaev (1910) and the course at the Officer Cavalry School (1911). He took part in the Russian-Japanese War of 1904–1905 with the Second Verkhneudinskiy and Second Argunskiy Cossack troops. He was a participant of World War I: squadron commander in the Life Guard of the Horsemen Regiment (1914); Chief of Staff of the United Cavalry Division (1914); at the court (aide-de-camp) of Emperor Nikolay II (1915); commander of the First Nerchinsk Regiment (1916); commander of the Second Brigade of the Ussuryisk Cavalry Regiment (1916–1917); commander of the Seventh Cavalry Division (1917); from 1917 commander of the United Cavalry Corps. In August 1918 he joined the White Movement—commander of the First Brigade of the First Cavalry Division, later the commander of this division; commander of the First Cavalry Corps (from November 1918 to January 1919). Upon arrangement between General Denikin and General Krasnov, on December 26, 1918 the single command of the Armed Forces of Southern Russia (AFSR) was created that united the Volunteer Army and the Don Army under the common command of General Denikin. At the same time V. was appointed the commander of the Volunteer Army (from December 1919 through January 1920). In mid-January 1920 he left Crimea for Constantinople (Turkey). In the emigration in Turkey he stayed till March 20, 1920 because of disputes with Denikin. From March 23 to May 11, 1920—AFSR Commander. On April 28, 1920 he transformed the former AFSR into the Russian Army. Commander of the Russian Army (Crimea, Novorossia, Northern Tavria; from April 28 to November 17, 1920). On November 17, 1920 he evacuated from the Crimea. In emigration: from Novebmer 1920—in Turkey, from 1922—in Yugoslavia, from September 1927—in Belgium. In September 1924 he created the Russian All-Military Union (RAMU) that united the former Russian military men from all branches of troops of the White and Russian armies. He died on April 25, 1928 in Brussels (Belgium) and was buried in Belgrade, Serbia.

Vrevskiy Pavel Aleksandrovich (1809–1855) – one of the commanders of the Sevastopol defense during the Crimean War of 1853–1856, Russian Baron, Major-General (1848), Aide-de-Camp General (1854). After finishing the School of Guards Ensign-Bearers and Cavalry Junkers (1828) he served as an ensign in the Life Guard of the Izmailovskiy Regiment. He participated in the Russian-Turkish War of 1828–1829. For his perfect services during the Varna siege (1828) he was awarded the "Golden Sword". In 1830 after being wounded he resigned. But a year later he returned to the military service. During the Polish campaign of 1830–1831 he fought near Prague, Grokhov and Ostrolenkoy, took part in seizure of Warsaw. In 1833 he was transferred to the Life Guard of the Grodno Hussars Regiment. In 1834–1838 he served on the Caucasus fulfilling various administrative and political orders. During military operations he showed great personal courage. On November 4, 1837 together with others he signed the Code of the first meeting of the Chess Players Society. Beginning from 1838 V. was the Chief of the First Department of the Ministry for Military Affairs. In 1841 he was promoted to Colonel, in 1848—to Major-General and took the position of the Secretariat Chief in the Ministry for Military Affairs. In 1855 during the Crimean War he submitted to the military minister a note where he analyzed the position of Sevastopol besieged by the enemies and the enemy troops. In this note V. proved the need for the troops besieged in Sevastopol to change over to offensive actions and expressed his readiness to lead them. In the final run, the Emperor agreed to such reasoning and V. was sent to Sevastopol. On August 4, 1855 he received a deadly wound in a battle on the Chernaya River near Sevastopol. He was buried in the Assumption Monastery in Bakhchisaray. He had many Russian and foreign orders and medal and was awarded the "Golden Sword" with the inscription "For courage" for his participation in the Varna siege.

Vulf Evgeniy Vladimirovich (1885–1941) – Russian botanist and geographer, florist. In 1909 he graduated from the Vienna University. In 1914–1926 he worked at the Nikitskiy Botanical Garden and at the same time held the position of Professor of the Crimean University (1921–1926). From 1926—the worker of the Leningrad All-Union Institute of Plant Breeding. V. was the author of the book "Historical Geography of Plants. History of the Earth's Flora". It was published in 1994 after his death. In this book he detailed on the history of flora development in the major regions of the Earth. V. specially studied the flora and vegetation of Crimea and wrote generalizing monographs "The Flora of Crimea" and "The Vegetation of Eastern Yail of Crimea" (1925). In his numerous other works V. investigated the useful plants in the Russian and world flora (ethereal-oil, tanning, medicine and others), history of botanic and also plant systematics (the family of scrofula, beech and others). He was the author of summary works about ethereal-oil, tanning, tinctorial and medicine plants.

"Vulnerability Arch" – a geopolitical conflict-forming belt crossing the Mediterranean, Black Sea, Northern Caucasus and Transcaucasus as far as Central Asia and involving the interests of Russia and other countries, too.

Vylkanov A. (1904–1971) – Professor, Corresponding Member of the Bulgarian Academy of Sciences (BAS). In 1940 he was appointed the Director of the Marine Biological Station where he conducted systematic investigations of the sea fauna. His series of works about single-cell animals deserves special mention.

W

Water balance of the Black Sea – the input part of this balance is made by the river flow, atmospheric precipitation and sea waters coming through the Bosporus and Kerch straits. The groundwater inflow is of minor importance. The output part of the balance is made by evaporation from the water surface and the outflow of the B.S. waters through the Bosporus and Kerch straits. The average value (with some allowances) of the input-output components of the balance is about 800 km^3/year, i.e. only 0.15 % of the B.S. volume. Every year B.S. receives approximately 338 km^3 of river waters in which the share of the Danube reaches 200 km^3. With the atmospheric precipitation in the form of rains and snow the sea receives 237 km^3 of water. Via the Bosporus Strait with the bottom current about 175 km^3 on the average of saline water from the Sea of Marmara are brought into B.S., while via the Kerch Strait from the Azov Sea—to 50 km^3. The water losses to evaporation make, on the average, about 396 km^3 a year. About 370 km^3 of water are evacuated from B.S. via the Bosporus Strait with the surface current to the Sea of Marmara and 34 km^3 to the Azov Sea. Therefore, in the input part of the balance the key role is played by the river flow which takes about 42 % of the whole inflow. Its variability is great. In the output part of the balance the main components are evaporation (49 %) and water outflow via the Bosporus (46 %). However, the evaporation varies insignificantly, thus, it cannot produce a perceptible effect on the variations of the sea regimes.

In the water area the incoming waters spread rather unevenly. The river flow concentrates mainly in the northwestern part of B.S. (to 80 %) and to a less degree in the southeast. Desalted water of the Sea of Azov (with salinity of 10–14‰) come via the Kerch Strait into the northeastern part of B.S. Featuring not high density these waters are spread by the currents in the surface layer of the sea. Saline waters (about 30‰) from the Sea of Marmara are brought with the bottom currents via the Bosporus Strait. The desalinization degree of the sea surface waters during a year depends largely on the volume of the river flow input and its distribution over the sea. In the autumn-summer period the sea receives approximately 60 % of the annual river flow. The greatest river flow is recorded in May, while the minimum—in September. The seasonal and annual variations of the water

S.R. Grinevetsky et al., *The Black Sea Encyclopedia*,
DOI 10.1007/978-3-642-55227-4_23, © Springer-Verlag Berlin Heidelberg 2015

balance components influence, finally, the sea water level and the water exchange via the Bosporus.

Watershed basin of the Black Sea – B.S. has a large watershed area that is crossed by more than 300 large and small rivers. It covers about one-third of the Europe land area. It exceeds 2.3 million km^2 and comprises, in full or in part, the territories of more than 20 countries of Europe and Asia Minor. These are coastal states—Abkhazia, Bulgaria, Georgia, Russia, Romania, Turkey and Ukraine and countries of Central and Eastern Europe—Austria, Albania, Belarus, Bosnia-Herzegovina, Hungary, Germany, Italy, Macedonia, Moldova, Poland, Slovakia, Slovenia, Croatia, Czech Republic, Switzerland, Serbia and Montenegro. The share of each of the above countries in formation of the watershed basin is not identical. Thus, Albania, Poland and Italy cover no more than 100–300 km^2, Switzerland—1,700 km^2, Moldova—33,700 km^2.

Western Asia, Front Asia – the southwestern part of Asia comprising the Asia Minor Peninsula, Arabian, Oman and Sinai peninsulas together with the nearby territories: Caucasian isthmus, Armenian and Iranian highlands, Mesopotamian Lowland and Levant states. Its area is about 7.5 million km^2. The northern part of W.A. (Caucasus, the northern margins of the Iranian highlands) lies in Russia, Georgia, Armenia, Azerbaijan, and Turkmenistan (the southernmost part). The remaining territory of W.A. covers Turkey, Iran, Afghanistan, Iraq, Jordan, Israel, Saudi Arabia, Yemen, Lebanon, Kuwait, Qatar, Oman, a part of Egypt, Syria, Bahrain and UAE.

Western Black Sea Coast of Turkey – the western part of Prichernomorie in Turkey stretching from the border with Bulgaria to the delta of the Kyzyl-Irmak River. In the ancient times this part of Turkey was called Paphlagoria. The western third of Turkey's Black Sea coast is its most remote and beautiful (having been spared the indignities of the coastal highway), home to the ancient fortified port city of Sinop, and the beautiful resort town of Amasra.

Western Pontic Mountains – the western part of the Pontic Mountains in the north of Turkey. They extend for about 400 km. They comprise 2–3 chains of mid-height ranges broken by densely populated valleys and depressions (with lakes) and also plateaus 900–2,000 m high. Their height is to 2,565 m (the city of Delitepe). They are composed of sandstone, limestone and metamorphic rocks. Coal deposits (Zonguldak basin) are found here. The northern slopes are overgrown with forests ranging from the Mediterranean in the piedmont areas to coniferous with the meadows occurring near the tops. The southern slopes are represented by phrygana vegetation and mountain steppes.

Wetlands – according to the Convention on Wetlands of International Importance Especially as Waterfowl Habitat, the wetlands are understood as areas occupied by marshlands, peat lands or water bodies, natural or artificial, permanent or temporary, drainless or through-flow, fresh, slightly saline or saline, including marine water areas, which depth at a low tide does not exceed 6 m.

Wilson Robert – Vice-Admiral (1801). In 1770 he left the British Navy and joined as a volunteer the Russian Fleet staying in Archipelago (Aegean Sea). In 1773 he was admitted in the rank of a Lieutenant to the crew of frigate "Venera" on which he sailed in 1774 between the Archipelago and port Leghorn. Upon return to Russia in 1776 he was a commander on a frigate, in 1777—on yacht "Duchess Kenston", in 1781—on the frigate "Paros". In 1783 he was transferred to the Black Sea Fleet and in 1784 he commanded the frigate "Saint Andrew" on which he took part in the Russian-Turkish War of 1787–1791 in the battle in the Kerch Strait as a part of the squadron commanded by Rear-Admiral F.F. Ushakov against the Turkish Fleet (1790). In 1791–1798 he commanded the ship "Bogoyavlenie Gospodne" (God's theophany). In 1798–1804 he commanded the squadron of the Black Sea Fleet in Sevastopol. In January 1805 he was dismissed. During his service life he was awarded numerous medals. In 1796 he was given the rank of Brigade Captain and in 1797—Rear-Admiral.

Wind and waves on the Black Sea – The variations of the main wind directions over B.S. are a function of the seasonal variability in distribution of the atmospheric pressure. In winter the northeastern, northern and northwestern winds prevail over the sea and in the southeastern part of the sea—the eastern winds. In summer the northwestern, western and southwestern winds are more frequent. In the cold season the average wind speed in the western part of the sea reaches 7–8 m/s, in the coastal zone it is less than 7 m/s. In the southeastern part of the sea the wind speed is 5–6 m/s, while in the northeastern—6–7 m/s. In the warm season of a year the average wind speeds are 1–1.5 m/s lower, but they tend to grow from east to west. Throughout a year the least wind speeds are recorded near the Southern Coast of Crimea and in the southeastern part of the sea. Apart from this, in the coastal zone of B.S. breezes are often blowing. They are generated due to dissimilarity in heating and cooling of the land surface and the sea. In the daytime they usually blow onshore and at night—seaward. In Yalta there are recorded to 190 days with the breeze circulation. Breeze winds may penetrate into land or sea to a distance of 30 km.

The wind speedup over the sea is most often connected with the passage of atmospheric cyclones. The number of days in a year with strong winds (over 15 m/s) is approximately 35–40 that are observed mostly on the northeastern and northwestern coasts of the sea. The strongest wind on B.S. is Novorossiysk bora (or "nord-ost") that occurs several times a year in the autumn and winter months (40–50 days). The speed of the northeastern wind may be as high as 40 m/s with some gusts to 80 m/s. Bora is often accompanied by temperature lowering by 5–10 °C. Bora may blow for 1–3 days to one week. Bora is usually observed in an area from Anapa to Tuapse, quite rare it is witnessed on the Southern Coast of Crimea.

Every year several cases of tornadoes are registered in the northeastern part of B.S. Their coming on the shore entails at times the disastrous consequences. Thus, in June 1991 the people of Tuapse were affected seriously and in August 2002—the people living in Novorossiysk and its residential suburbs. Houses, power transmission lines, motor roads and railroad were damaged, several nearby settlements were destroyed. Dozens of people died or were recorded missing.

The wave regime of B.S. is studied inadequately because practically no systematic instrumental observations over waves in the open sea have been conducted. The main characteristics of waves were determined by estimations by which the average wave height was 0.5–1 m and the maximum height (in winter)—7 m. However, on November 14, 1854 during the Crimean War near Balaklava the strongest storm caused wreck of the united British-French squadron: 34 military ships sank and the casualties were 1,500 people. Later on such extreme storms were registered in late October 1969, in November 1981 and 1992. During the 1981 storm, near the Tarkhankut Cape (the western end of the Crimea) the wave slaps were so strong that they knocked down the deck of an oil and gas platform located at a height of 14 m over the sea level. The maximum wave height in the open sea could reach 16–17 m.

According to various sources, in last 2000 years 25 events of tsunami were registered on B.S. the greater part of which was associated with the earthquakes. Some of them, such as Dioskuria (in place of modern Sukhumi, first century B.C.), Sevastopol (109), Varna (543), Bosporus (557), Evpatoriya (1341), Foros (1427), Turkey (1598, 1939), Crimea (1869, 1875, 1902, 1908, 1919, 1927, 1941), Anapa (1905, 1966), were accompanied by tsunami waves 1–3 m high and they did not result in disastrous consequences (except destruction of Dioskuria).

World Sea Day – one of international days celebrated within the UN frame. It has been celebrated since 1978 following the decision of the tenth Session of the Assembly of the Inter-Governmental Maritime Consultative Organization (IMCO, since 1982—International Maritime Organization) in the last week of September. The purpose of W.S.D. is to draw the attention of the international community to the irreparable damage caused to the seas and oceans by overfishing, water pollution and global warming. Two most important targets—to ensure higher security on the sea and prevention of marine environment pollution, in particular with oil.

Y

Yaila – (Cimean Tatar *"yai"*—*"location"*, *"yaila"*—*"summer rangeland on a mountain plateau"*). Main range of the Crimean Mountains with steep southern and more gently-sloping northern slopes. Extends 110 km along the Black Sea coast from Balaklava Heights in the west to Ilya Cape in the east; the main water divide of the Crimean Peninsula. The highest summits of the Mountainous Crimea are here: Mt. Romano-Kosh (1,545 m), Demir-Kapu (1,540 m), Zeitin-Kosh (1,534 m) and others. The high part has no ridge and is plateau-shaped massifs, structured largely by limestones. Karst is very much in evidence; numerous landslides and rockfalls; ravines, canyon-shaped gorges (Grand Canyon of the Crimea). Mountain-forest landscapes (in the south—of Mediterranean type), on the summits—meadows and steppes. Some parts of Y., are called (from the west to the east): Baidar Yaila, Ai Petri Yaila, Yalta Yaila Nikitin Yaila, Babugan Yaila, Demerdzhi Yaila, Dolgorukov Yaila, Karabi Yaila The total area of the Crimean Yaila is around 320 km^2.

Yalta – town, maritime climate health resort, port on the Black Sea, 79 km south of the Railway station Simferopol (Crimean Peninsula, Ukraine). Arranged in the form of an amphitheater on the southern slopes of the Main Range of the Crimean Mountains, covered with coniferous and broad-leaved woods, on the shore of a wide and deep harbor delimited by Ai-Todor Cape and Nikitin Cape. To the north and north-west of Y. there towers Ai-Petri Yaila—part of the Main Range of the Crimean Mountains with the Ai-Petri Mountain top (1,234 m). From the early centuries A.D. in place of the present-day city there used to be the ancient Greek colony Yalita, called by the name of the location where it came into being, and the name of the location stemmed from the name of the Turk genus Yaltan or Yaltalar. The exact time of Y. emergence is not known. The name of Y., in Greek "Yalita" (from Greek *"yalos"*—*"shore"*) indicates that the city was apparently founded by the Greeks.

In the fifteenth century, Y. becomes a Turkish possession as part of Mangup Kandalyk. The Christians fled to the mountains seeking to escape rom the Turkish pursuers. As Russia grew stronger, the authority and importance Turkey in the

S.R. Grinevetsky et al., *The Black Sea Encyclopedia*,
DOI 10.1007/978-3-642-55227-4_24, © Springer-Verlag Berlin Heidelberg 2015

Crimea waned gradually. After the first Turkish War, in 1778, despite the protests of Tatars and Christians, nearly all Christian population was moved to Russia. Y. ceased to exist. After the second Turkish campaign, the Crimea was annexed to Russia. Thanks to the enthusiasm of Prince Vorontsov, re-birth of the Southern Coast commenced and Y. became a district center on September 17, 1837. During the same year, Y. was connected with Sevastopol by a road and the original Yalta mall was built, which was destroyed by a storm a year later.

Prof. Sergey Botkin, a famous Russian clinician, therapist, one of the founders of modern Russian medical science and education, was the first to note the outstanding climatic conditions of the Southern Coast of Crimeae. On his advice, the wife of Alexander II began spending the autumn months at Livadia from 1866. From that time on, Y. began to grow rapidly. The war with Turkey of 1877–1878 had an extremely adverse impact on Y. that only began taking a proper shape. As a health-resort city, Y. develops from the end of the nineteenth century, after Livadia becomes the Tsar's summer residence. Following the final establishment of Soviet power in the Crimea (1920), Y. becomes a major health resort. In 1921, there were already 18 sanatorias (for 1,000 visitors) in the city and its environs. At present, Y. is the center of a health-resort location known as the Southern Coast of the Crimea.

The main natural curative factor—mild, warm, moderately humid climate of Mediterranean type. The bathing season lasts from early June to early October. Clean air, abundant sunny days, picturesque landscapes are conducive to air-therapy, helio-therapy, thalasso-therapy procedures by persons who have chronic diseases of the lings and the upper respiratory organs, diseases of the cardio-vascular system and nervous system as well as with a view to toughening up. Mineral sources of sulfate water with a small yield are available; for the treatment of digestive organs, sulfate-hydrocarbonaceous calcium-sodium-magnesium water is used. Besides, during the warm period of the year, kidney patients are treated.

In Y., there are wine-making factories of the "Massandra" winery, fish and meat-processing factories, factoring fabricating souvenirs and other manufacturing facilities. The Institute of Viniculture and Wine-making and Sechenov Institute of Climotherapy are also in Y., along with several other institutes.

Y. is a major center of tourism, one of the main attractions for those cruising the Black Sea. Of interest to excursions are architectural and historic monuments of the city: ruins of the medieval fortification Isar and remains of a temple in the Iograph Cave in the environs of Y. There is a museum of local lore, along with literature-memorial museums, including those of A.P.Chekhov who spent the last days of his life in Y. Nikitinskiy Botanical Gardens. Y.is the cradle of Russian cinematograph. In 1911, the first in history A.A. Khanzhonkov's film studio was built here, in 1918, a division of Joseph Ermoliev's Moscow film studio was set up in Y. The head-quarters of the Ukrainian scouts is also in Y.

Yalta (Photo by Evgenia Kostyanaya)

Yalta – a small village on the northern coast of the Sea of Azov, Ukraine.

Yalta Bay – cuts into the shore of the Crimean Peninsula between Aitodor Cape and Nikitin Cape. The shore of the bay are largely precipitous, with pebble beaches. Near the shore of the bay apex are Yalta Port and Yalta City.

Yalta Sea Port – cruise gates of the Crimea. The territory of the Yalta Merchant Marine Port includes: passenger complex in the city center—5.5 ha; freight-and-passenger complex 2 km east of the city center at Massandra—3.3 ha; territory of the Yalta port-stations. The passenger complex of the port is capable of handling motor vessels as long as 215 m (up to 240 m under suitable weather conditions) and a draft of 8.6 m. To receive tourists arriving by larger vessels, raids services are provided. The freight-and-passenger complex includes a freight wharf. In 1986, a modern transshipment complex capable of handing up to 2 million tons of cargoes per year was put into operation at Massandra. The port stations and wharfs—Foros, Simeiz, Alupka, Miskhor, Lastochkino Gnezdo, Zolotoi Plyazh, Livadia, Intiourist, Nikitinskiy Botanical Gardens, Gurzuf, Frunzenskoe, Rabochiy Ugolok, Alushta, Morskoe, Sudak and others are purposed to perform passenger operations by the craft of the port. Y.S.P. is a member of the International Mediterranean Sea Cruise Ports Association "Medcruise". Services foreign cruise ships.

Yalta Sea Port (Photo by Evgenia Kostyanaya)

Yarylgach Blight – cuts into the shore 30 km to the south-west of the base of Bakal Spit between Chernyi Cape and a cape lying 3 km of it. The shores are low-lying, sandy. Mezhvodnoe Settlement is on the north-eastern shore of the blight. The depths at the entrance to the blight are 12 m.

Yasen Reserve – sited 30 km south of Eisk on the Sea of Azov, Krasnodar Kray, Russia. Unique circum-Azov plavni (seasonally flooded land) has lived to this day here. One can see pelicans, flamingoes and swans in a natural setting here. Marvelous beaches are available, too.

Yasun (Jason) Cape – famous "Jason's Cape". Sited not far from Fats City in the central; part of the Turkish Black Sea coast, 8.5 km west of Cham (Vona) Cape. The cape is pointed, rocky. There are ruins of an ancient monastery and eight towers.

Yazykov Nikolay L'vovich (prior to 1773–1824) – Russian Vice-Admiral (1805). Graduated from the Naval Cadette Corps as a Midshipman in 1773. From 1783, served in the Black Sea Navy. In 1784–1785, participated in the description of the Dnieper. In 1786—adviser of the state-financed expedition at Nikolaev and commander of a navigator's company. In 1787, while commanding the ship "St. Alexander Nevskiy", took part in the battle against the Turkish rowing vessels in the vicinity of Ocvhakov. In 1788, sailed between the Glubokaya ('Deep') wharf and Ochakov on the same ship and participated in the battle against the Turkish Fleet. In 1790, took part in the battles near Kerch Strait and Khadzhibey, at Kaliakra. In 1792–1798, was Oberster Kriegs-Kommissar at the Black Sea Admiralty Board. In 1798–1802, served at the office of the Chief Commander of the Black Sea Navy and Ports as an adviser to the Commissariat Department. In 1802, was the squadron commander. In 1805, appointed Director of the Black Sea School of Navigation. In 1808, appointed the Chief of naval crews in Sevastopol. From 1809, was in charge of the Black Sea Department, in 1811, he became the

Chief Commander of the Black Sea Navy and Military Governor of Nikolaev and Sevastopol. In 1816, surrendered his position to Vice-Admiral A.S.Greig.

Yesilirmak – a river in the north of Turkey, 416 km long. Its origin is near the origin of the Chorokh River on the western margins of the Armenian highlands. It crosses the Pontic Mountains and flows into B.S. The main tributaries of this river are Cherek-Chai and Charum-Su. It is not full-flowing and not navigable. The annual flow of Y. is about 5 km^3. Sharp fluctuations of the water level are registered here.

Yuznyi (Southern) Research Center of the National Academy of Sciences of Ukraine – established in 1971 г. as a center of the Academy of Science of UkrSSR, one of the five existing in Ukraine. Called upon to coordinate the activity of higher educational establishments and research institutes of Odessa, Nikolaev, Kherson and Crimean regions on crucial problems of science. The Center has a number of academic divisions in the field.

Yukharin Pavel Matveevich (prior to 1816–1876) – Russian Admiral (1871). Finished the Naval Cadette Corps in 1816 as a midshipman. In 1816–1821, sailed in the Baltic Sea, twice sailed from Archangelsk to Kronstadt. In 1837–1838, took part in the expeditions against the highlanders. In 1838–1840, while commanding the corvette "Ithygeniya", sailed in the Mediterranean Sea. In 1845, commanding the frigate "Flora", sailed in the Black Sea. In 1855, was the commander of the first and third brigades of the fourth naval divisions on the roads of the besieged Sevastopol. In 1859 promoted to Vice-Admirals and dismissed from service. In 1870, readmitted for service with the Black Sea Rowing Flotilla.

YUZHMORGEO – Russian State Research Center on marine geology, located in Gelendzhik, Krasnodar Territory, Russia. Operates in the field of geological problems related to the study and development of mineral resources on the continental shelf, inland seas and oceans. It is one of the leading research organizations in Russia with more than 55 years of practical experience, is involved in federal and international programs and agreements; presents its services on the marine geophysical industry. It has 14 departments, three special research vessels, 15 special research boats, and 110 units of marine geophysical equipment.

Yuzhnaya (Southern) Ozereika (Ozereevka) – a small village on the Black Sea coast, 12 km northwest of Novorossiysk City, Krasnodar Territory, Russia. This is an end point of the Caspian Pipeline Consortium oil pipeline, where tankers are loaded by oil in several km from the coast. Population—1,100 (2010).

Yuzhnaya Ozereika Operation – amphibious operation of Soviet troops north of Novorossiisk on February 4–15, 1943 during the Great Patriotic War (1941–1945). The main amphibious part was on land near Yuzhnaya Ozereika Village. However, it was only possible to land the assault groups of the first exchelon (around 1,500 men and 10 tanks). Due to stormy weather and violent fire of the Germans who discovered Soviet ships at sea, the rest of the amphibious party returned to their bases at Gelendzhik and Tuapse. The troopers that were left on the shore after the bitter combat did manage to seize Yuzhnaya Ozereika and then began moving toward Glebovka Village, where they were engaged in a new battle with the Germans who had received large additional reinforcements. In that combat, the troopers used up the entire combat stock, and nearly all of them were shot dead. A small group of the troopers did manage to infiltrate into the area of Myskhako Peninsula (closer to Novorossiysk), where an auxiliary amphibious party landed under the command of Major Ts.L. Kunikov. Only a few men left for the mountains. Another 25 men joined the paratroopers who had landed the moment the operation began. And then broke through to the shore, from where they were taken out on the ships. The failure of the operation did not allow Soviet troops to seize Novorossiisk and cut off escape routes of the seventeenth German Army via Taman Peninsula. This rough luck was the cause for Vice-Admiral F.S. Oktyabrskiy's removal from office of Commander-in-Chief of the Black Sea Navy.

Yuzhnyi (Southern) Front – by resolution of the Central Committee of RCP(b) of September 21, 1920, the Crimean Segment of the South-Western Front was transformed into an independent Y.F. M.V.Frunze was appointed the Commander-in-Chief of the Y.F; S.I.Gusev and Bela Kun—members of the Revvoensoviet (Revolutionary Military Board). Late in October, the Y.F. troops began assault and destroyed General Vrangel's main forces (White Army) in North Tavria.

Yuzhnyi (Southern) Research Institute for Fishery and Oceanography (YugNIRO) – the institute was started as Kerch Ichthyological Laboratory. In 1923–1924, the Azov-Black-Sea Research-and-Fisheries Expedition headed by N. M.Knipovich was organized to study the fish resources. It marked the commencement of integrated fishing research in the Southern Basin. In 1933, the Azovo-Chernomorskiy Research Institute for Fishery and Oceanography—AzCherNIRO—was established, renamed in 1988 as YugNIRO. YugNIRO is the only research institute in Ukraine dealing with multidisciplinary research, design and consultancy-expert-appraisal works in the field of sea fishery, fisheries oceanography and mariculture. The geography of the institute's research operations is quite extensive: the Azov-Black Sea area, the Mediterranean Sea as well as the water areas of the Atlantic, Indian and Pacific oceans, the waters of the Antarctic regions. The primary object of YugNIRO is to provide scientific support of Ukraine's current activity in the field of sea fisheries through the development and implementation of integrated measures for long-term and stable use of marine living resources. Along with the search for new fishing areas, objects of fishing and development of scientific foundations of rational fishing, efficient technologies of hydrobiont processing are devised, the problems of disposing of production waste and obtaining valuable fodder and medical-veterinary products are resolved.

Ichthyologists of YiugNIRIO have developed a scientific basis for rational use of major game fishes of the Black and Azov Seas, which has been conducive to the commencement of developing the resources of the hitherto unused objects and has made it possible to increase the total volume of seafood production by 80–100,000 tons each year. From 1961 to 1991, the institute made a detailed study of the shelf waters of the western part of the Indian Ocean, Gulf of Bengal, and the eastern part of the Andaman Sea. As a result, the coastal waters of Somalia, Oman, Pakistan and West India were recommended for fisheries development. The most promising object of such development were determined in the open waters of the Indian Ocean: tuna, sharks and oceanic squids. The institute researchers have played a significant role in the development of the resources of Antarctic fishes, such as notothenoide, white-blood spikes, Patagonian toothfish as well as Antarctic krill.

After the collapse of the USSR, the region of institute's fisheries research has broadened and now also includes the areas of the Atlantic and Pacific Oceans, where the Ukrainian fishery fleet operates. During the last decades, in view of the degradation of natural conditions for natural reproduction and decrease of the populations of some valuable species of fish and other hydrobionts in the Azovo-Black Sea Basin, the institute's scholars have developed processes of obtaining viable young fishes of Black Sea gray mullets and plaices, cultivation of live foods; the spawning schools of young steelhead salmon and Far Eastern gray mullet acclimatizer—Mugil-so-iuy; biotechnology of industrial cultivation of commercial mussels and oysters has been improved.

The institute is developing a new area of process research: obtaining biologically active agents from hydrobionts and fabrication on this basis of products and pharmaceutical drugs that cure and prevent diseases. A preparation from muscles-BIPOLAN—unique in its antitumoral properties, has been developed at YugNIRO: it has no analogs in the world. The drug is capable of bringing back to full-fledged life many people affected by the Chernobyl disaster. A group of carrageenans also exhibiting curative-and-prophylactic as well as radio-protective effect has also been isolated from Black-Sea phylloflora. Besides, a valuable product called philagar—substitute of the microbiological agar imported by Ukraine—has been obtained. The process of isolating a biopolymer called spirulin, from algae, has been devised. Spirulin exhibits antioxidant, antiviral (in respect of the Herpes virus) activity and exerting hormonal effect in the reproductive functions of the organism.

In 1996, two independent research organizations—in Sevastopol and Berdyansk—were affiliated with YugNIRO. YugNIRIO is part of the Ukraine's National Academy of Sciences. YugNIRO consists of the chief institute in Kerch City (Crimea) and four branches: in Odessa with laboratories for raw material resources, for other than fish object processes, lacustrine fish farming (Izmail City), Azov (Berdyansk City) with laboratories for ecology and forecasts of bioresources for commercial sturgeon breeding and aquaculture, Sevastopol Branch with laboratories for submarine sound detectors and quantification of fish reserves, heat-conserving equipment. From 1995, YugNIRO has been national partner of ASFA (Aquatic Sciences and Fisheries Abstracts is an International Cooperative Information System which comprises an abstracting and indexing service covering the world's literature on the science, technology, management, and conservation of marine, brackish water, and freshwater resources and environments, including their socio-economic and legal aspects).

"Yuzhnyi" Sea Merchant Port – versatile, non-freezing, deepest and young, one of the three leading port of Ukraine (Ilyichevsk, Odessa, Yuzhnyi). The construction of the port began in 1973 30 km from Odessa, in the non-freezing Maly Adzhalyk (Grigoryev) Liman. The liman is linked with the sea by an approach channel, 3 km long and 15.5 m deep, the depth of the liman is 14 m. The new port's date of birth is generally believed to be July 27, 1978. The port specializes in the transshipment of dry and liquid cargoes, among these coal, metal, carbamide, building materials, liquid ammonia and others. Here comes the 230 km- long ammonia pipeline laid from Togliatti. Most transit cargoes (90 %) comes from Russia and neighboring CIS countries: Moldova, Kazakhstan, Belorussia. The total length of the wharfs if around 2.3 km. The anchorage accommodating 24 vessels is on the outer water area. Three railway stations adjoin the port: Beregovaya, Khimicheskaya and Promyshlennaya of the Odessa Railway Line. The stations are interconnected via Chernomorskaya Railway Station. The main shipment destinations are: the Black Sea and Mediterranean Basins; USA, Latin America; Middle East; South-East Asia. When the liman's shore line is developed finally, the combined length of the wharf will total 9 km, the freight turnover will increase to 35–45 million tons per year; it will be possible to receive up to 40 million tons of oil per year. Strategic plans of the port include participation in the common European transport corridors. There is a large port-attached liquid ammonia and carbamide plant and a large oil terminal.

Z

Zabache Sea – presently the Sea of Azov. Probably, the word "Zabache" conceals the name of fish—"*chabak*" that still means bream in the lower Don. A.S. and mouths of the rivers flowing into this sea abound in fish (sturgeon, stellate sturgeon, great sturgeon, pike perch, common carp, sabrefish). Oriental authors called it Tanu (fourteenth–fifteenth centuries), Azak, from here the name of the Sea of Azov—Ozachskoye used in "The Journey of Ignatiy Smolnyanin": "we passed through the mouth, i.e. Bosporus Kimmerian, Kerch Strait, of the Ozachie Sea and went out into the Great, i.e. Black Sea." Sometimes A.S. was referred to as Frankish. It was called in this way by Persian author Abd-ar-Razzak Samarkandi (1413–1482): "They (tsareviches and Emirs of Timur) devastated and plundered this district (Cherkess ulus) stretching as far as the Thracean Sea shores called the Azakskiy Sea". Also there is a well known and exact map of the Sea of Azov engraved by Nicolas Visscher in Amsterdam in the end of seventeenth century, which has a title in French: "Nouvelle Carte Geographique de la Mer d'Asof ou de Zabache & des Palus Meotides. Exactement dessignee & mise en lumiere par Nicolas Visscher".

Zagreba – a name of reefs in B.S.

Zaionchkovskiy Andrey Medardovich (1862–1926) – Russian well-known military commander, military historian and theorist, General of infantry (1916). He finished the Engineering College in Nikolaev (1883) and General Staff Academy (1888). He started his service as a Second Lieutenant in the sapper battalion. From 1888 he served in different headquarters in Petersburg. During the Russian-Japanese War of 1904–1905 he commanded the infantry regiment and then division. He took part in World War I. Simultaneously from 1908 he was the assistant of Great Prince Alexander Mikhailovich on management of the Sevastopol Museum. From March 1915 he commanded the 30th Army Corps that took part in offensive in the southwestern front in 1916. Then he was appointed the commander of the Army corps in Dobrudja. After failure of Army operations he vacated command and later, in March 1917, he was dismissed. From 1918 during the Civil War in Russia he served in the Red Army and was the Chief of Staff of the 13th Army. Then he lectured in military educational institutions. He was the author of the works on the history of World War I. In 1899 he published the work "The Sevastopol Defense. Heroic Deeds of Its Defenders. Brief Historical Review". His most fundamental work was "The Eastern War of 1853–1856 in the Context of the Political Situation of That Time" (vols. 1–2 and two volumes of appendices, 1908–1913) prepared by the 50th anniversary of the Eastern War. In 2002 this work was reprinted in Moscow in three volumes.

Zapadnyi Cape – (1) one of the names of the western entrance cape of the Balaklava Bay. The cape was given this name by its geographical location in relation to the entrance into the bay (*"zapadnyi"* meaning "western"). There are other names of this cape: Batareinyi, Deli-Christo and Kurona. (2) Name of the western entrance cape of the westernmost Dvoinoy (Troinoy) Bay in Sevastopol. It got its name by the geographical location in relation to the bay entrance.

Zaporozhie Region, Zaporozhie – locates in the southeast of Ukraine. It borders on the Kherson, Dneptropetrovsk and Donetsk Regions. In the south it borders on the Sea of Azov, which coastline stretches for over 300 km within this region. This region was established in 1939. Its area is 27,200 km^2 (4.5 % of the Ukraine territory). It stretches from north to south for 208 km, from east to west—for 235 km.

The climate here is moderately continental with pronounced aridity. The mean annual temperature in summer is +22 °C, in winter −4.5 °C. There are about 225 sunny days in a year; the precipitation is about 500 mm a year. The region has a flat topography. The chernozem soils prevail here. Natural resources are abundant and variegated. The region has considerable resources of iron and manganese ores, granites. The share of the region in the total mineral reserves in Ukraine is: pegmatite—88.06 %, apatite—63.42 %, manganese ore—69.1 %, secondary kaolin—22.9 % and fireclay—8.6 %.

The population of the region is 1.8 million (2013) or 4.0 % of the whole population of Ukraine, of which the urban population is 1.4 million and the rural—448,000. The administrative center is the city of Zaporozhie (767,000 people in 2013). The region incorporates 20 districts, 14 towns, 23 urban-type settlements, 918 villages.

After the Tatar-Mongol invasion of 1237–1240 the Z.R. territory was a part of the Golden Horde for nearly two centuries. In 1445 the Zaporozhie steppes on the left bank of the Dnieper were included into the Crimean Khanate. From the late fifteenth to mid-nineteenth centuries these territories were inhabited by nomadic and semi-sedentary Nogaians. The natural-geographic and historical conditions in the fifteenth–sixteenth centuries made the South Ukrainian lands one of the centers of Zaporozhie Cossack formation and the Khortitsa Island, well-known since the times of Ancient Rus', became one of their main strongholds and symbols. The event that became very important for the whole of Ukraine was formation of the social, political and military organizations from separate Cossack units and production teams that was called Zaporozhie Secha ("Zhaporozhie Troops").

Zaporozhie Secha became the first political association in Ukraine. It enjoyed independence for rather a long time and played an outstanding role in international relations—the European states established diplomatic relations with this organization and sought military alliance with it. In the late eighteenth century the lands of Southern Ukraine were incorporated into the Russian Empire. The process of settlement and development of the lands of modern Z.R. was not easy. It involved people of many nationalities which made this region a multinational formation. Among the population of the present-day Z.R. there are Germans, Bulgarians, Jews, Gagauzes and others.

Z.R. became one of the territories where the USSR strategic plans of industrialization were tested. In 1927 the construction of Dnieper HPP, the most high-power electrical plant in Europe that time, was initiated. Simultaneously a complex of new power-consuming enterprises was built speedily. The economic development of Z.R. was interrupted by World War II. And one of the most significant events in the postwar years was revival of the Zaporozhie industrial complex— Dnieper HPP and enterprises of ferrous and nonferrous metallurgy. Commissioning of Dnieper HPP-2 was accompanied by construction of the Zaporozhie Regional Power Plant and Zaporozhie NPP being the Europe's largest. Its installed capacity is 600 MW.

Z.R. is one of the basic industrial and agricultural regions in Ukraine. It takes the ninth place in Ukraine by the population and the second place after Kiev by the volume of the gross added value per one inhabitant. The principal industries here are metallurgy and power engineering that provide 12.1 % of the total volume of cast iron, 14 % of steel, 15.6 % of ready rolled stock, 11.9 % of coke, and 26.5 % of electricity in Ukraine. Such enterprises in the region as Zaporozhie Aluminum Works and Zaporozhie Transformer Works are the only in Ukraine that produce aluminum and transformers complying with the world quality requirements. The metallurgical complex includes such world renowned ferrous and nonferrous plants as OJSC "Zaporozhstal", the leader in steel and cast iron production; "Dneprospetsstal" producing special steels; "Ukrainskiy Grafit", the leader in production of graphite electrodes; "Zaporozhie Aluminum Combine", the Ukraine's only plant producing aluminum and alumina being the key component for aluminum production; "Titanium-Manganese Combine", the Ukraine's only plant producing spongy titanium and the leader in production of germanium and crystalline silicon. The machine-building plants with hi-tech production well known over the world are also found here and among them—OJSC "Motor Sich" producing engines for aircraft and helicopters for the leading aviation companies (Design Bureaus of Antonov, Yakovlev, Tupolev, Beriev, Kamov and Mil'), OJSC "Zhaporozhie Transformer Works, the Ukraine's only plant producing power transformers; "Zhaporozhabraziv", the leader in production of abrasive materials and tools. The Joint Venture "AvtoZAZ-Daewoo" manufacturing cars is the leader on the Ukrainian market. The aviation industry develops quite dynamically.

Z.R. is one of the major centers of agriculture and food industry in Ukraine. The area of agricultural lands is 2.25 million ha or 5.4 % of the Ukraine's agricultural lands. Crop growing dominates in agricultural production, but of no less significance are vegetable growing and horticulture. Z.R. is the Ukraine's major producer of sunflower seeds. The share of this region in the total sunflower output exceeds 10 %. In all agroclimatic zones there are favorable conditions for vegetable and melon crop growing. The priority direction in animal husbandry is poultry and pig rearing.

A dense railway network connects all major industrial centers of the regions with other cities in Ukraine and CIS countries. The automobile roads with the total length of 6,690 km link all settlements in the region. The automobile and railway transport ensures the greater volume of cargo and passenger traffic. A very important role in the transport system of Z.R. is given to the Dnieper, the Europe's third navigable waterway that is also a very important transport waterway in Ukraine. The waters of the Dnieper with its reservoirs support the industry development. The river cargo port in Zaporozhie handles the greater part of the total cargo traffic in the region. It specializes on handling of industrial, including metallurgical, cargo. The sea port in Berdyansk (on the Sea of Azov) is the sea gates into Z.R.

In the National Nature Reserve "Khortitsa" that received its status in 1993 you can find the museum of the Zaporozhie Cossack history.

Zarudnyi Viktor Ivanovich (1828–1897) – Russian Vice-Admiral, nautical surveyor, explorer of B. and A.S. In 1847 he finished the Naval Cadet Corps among its first graduates. From 1847 to 1850 he served on B.S. where he took part in hydrographic surveys for preparation of the B.S. pilot. From 1850 to 1852 he worked together with well-known nautical surveyor M.F. Reineke on the Baltic Sea where they made surveys and depth measurements in the Gulf of Finland. In 1852 he was again transferred to the Black Sea. Till 1860 he taught mathematics and astronomy in Nikolaev. In 1860 he was appointed the assistant Chief of the Hydrographic Unit of the Nikolaev Port. From 1863 he was the Chief of the Hydrographic Unit and at the same time the Director of lighthouses and pilots of B. and A.S. He contributed much into improvement of the navigation facilities installed on the coasts of B. and A.S. From 1872 keeping the mentioned posts he also supervised the Black Sea hydrographic expedition on exploration of the B. and A.S. coasts and the Dniester Liman. In 1882 "for successes in hydrographic and mapping works" he was conferred the rank of the Rear Admiral. In 1888 he was promoted to Vice-Admiral, then he resigned. His contemporaries believed that he was the best nautical surveyor of the Russian Fleet. For his great achievements in development of the hydrographic science he was awarded several orders.

Zatoka – a resort located 60 km to the southwest of Odessa on a very picturesque sand bar between B.S., Dniester Liman and saline lakes, Ukraine. Zatoka is a quickly developing tourist and health improvement center in Ukraine. The sandy strip is running for two dozens of kilometers. One can find here warm sea and a microclimate peculiar of a liman. The bar has a beach strip more than 100 m of fine clean sand. The favorable geographic position, remoteness from industrial centers and absence of industrial enterprises make this region one of the ecologically most pure places on the Black Sea coast. A smoothly sloping sea floor creates very good conditions for bathing. Many recreation centers locate directly on the seashore, just several meters from the coastline. The holiday season lasts from mid-May to mid-September. The mean daily air temperature in summer is 24–28 °C, while the water temperature is 19–23 °C.

Zekhti-Burun Cape – more widely known as Maslenyi (oily) Nos, Bulgaria. The legend says that in the ancient times a ship that carried olive oil was wrecked in this place and from here is the name of the cape.

Zelinskiy Nikolay Dmitrievich (1861–1953) – was born in Tiraspol of the Kherson Province. Specialist in organic chemistry, Academician of the USSR Academy of Sciences (1929). He graduated from the natural science faculty of the Physico-Mathematical Department of the Novorossiya University in Odessa (1884). In 1885 he went abroad for continuation of his education. In 1888 he returned to Odessa and was enrolled as out-of-staff private-docent to the Novorossiya University. In 1889 he defended the master's and in 1891 the doctor's thesis. In 1893 Z. discovered the bacterial origin of hydrogen sulfide contained in waters of B.S. Studying the formation of hydrogen sulfide in bottom sediments he found out that this was a result of fixation of the oxygen from sulfates by the specific bacteria living in bottom silt in B.S. He isolated the culture of these microorganisms having the form of spirillas and vibrios and called them *Microspira desulfuricans*. In 1893 he was invited for work to the Analytical Chemistry Chair of the Moscow University. In 1911 he went to live to Petersburg. In 1915 after application by the German troops of the warfare chemical agents Z. made a carbon respirator. In 1917 he was again with the Moscow University. He was the founder of the organic chemistry scientific school. He took part in establishment of the Organic Chemistry Institute of the USSR AS (1934). In 1953 this Institute was given his name.

Zelinskiy N.D. (http://www.tonnel.ru/calendar/kniga/611270213_tonnel.gif)

Zenkevich Lev Aleksandrovich (1889–1970) – the outstanding Soviet oceanologist and biologist, the Honorary Member of the USSR Geographical Society (1962), Academician of USSR AS (1968). In 1908 he entered the Law Department of the Moscow University. For his involvement in the revolutionary student's movement in 1911 he was dismissed from the University and forced to move to town of Tula. In 1912 he returned to Moscow, graduated as a non-resident student from the Law Department and entered the natural science faculty of the Physico-Mathematical Department. Still in his student years having visited as a member of the expedition the Barents Sea he wrote a number of scientific articles about the sea fauna. In 1916 he graduated from the University and stayed to work there as an assistant. In 1921 he took part in the scientific expedition to the north. Together with I.I. Mesyatsev he founded and guided the work of the Floating Marine Research Institute ("Plavmornin") where he took the position of the Deputy Director on research. In 1921 he took part in the first expedition of the Institute on the icebreaker "Malygin" to the Barents and Kara seas. On the research vessel "Persey" he took part and headed many multi-purpose research expeditions. He studied the biological productivity of the Caspian, Black and Azov seas. From 1948 he headed the Benthos Laboratory of the P.P. Shirshov Institute of Oceanology in Moscow; in 1949–1952 he was the head of the integrated oceanographic expedition. Z. initiated investigations of the deepwater ocean fauna. In 1949 he headed the First Pacific Expedition on the vessel "Vyatka" during which the trawl was lowered to a depth of 8 km and it brought to the surface many bottom organisms. The quantitative methods of study of the bottom fauna developed by Z. enabled preparation of the benthos biomass distribution map of the World Ocean and revealing of some general regularities in the life of bottom inhabitants of the ocean. For organization of this expedition, in 1951 he was conferred the USSR State Award. He was the author of many fundamental works on biogeography of the USSR seas. In 1968 he organized and headed the deepwater expedition on R/V "Akademik Kurchatov" to the Pacific region about which the knowledge was scarce. He was the Chairman of the Oceanographic Commission at the Presidium of USSR AS, the representative of the All-Union Hydrobiological Society, the Vice-President of the Moscow Natural Scientists Society, the Editor-in-Chief of the journal "Oceanology" organized by him, a member of editorial boards of some Russian and foreign scientific journals. He was the Honorary Doctor of some foreign universities, a member of the Scientific Council of the Oceanographic Institute in France that conferred him its highest award—the Medal "In Memory of Albert I of Monaco". His principal works are "Fauna and Biological Productivity of the Sea", vols. 1–2, (1947–1951); "The Seas of the USSR, Their Fauna and Flora" (1951); "The Biology of the USSR Seas" (1963) and others.

Zenkevich L.A. (http://biologylib.ru/books/item/f00/s00/z0000008/pic/000002.jpg)

Zenkovich Vsevolod Pavlovich (1910–1994) – the outstanding Soviet oceanologist and geomorphologist. In 1927 he entered the Geological Faculty of the Moscow University and soon he went to the sea on the research motor-sailing vessel "Persey" belonging to "Plavmornin" Institute. In 1928 he was a member of the Kerch multi-purpose expedition. In 1929 he was admitted to the laboratory of the sea as a laboratory assistant. In 1932–1935 Z. studied the processes of sedimentation in the coastal zone of the White and Barents seas on the Kola Peninsula. In 1936 the Dissertation Board thought it possible to confer him the academic degree of the candidate of geographical sciences without defending the theses, but on the basis of all his published scientific articles. In the same year Z. went to work to the Institute of Geography of USSR AS. At the same time, from 1937 he taught as an Assistant Professor at the Moscow Hydrometeorological Institute. In summer 1939 he studied the underwater coastal slope of Eastern Caspian using the light diving breathing apparatus. In 1943 Z. defended successfully his doctoral theses "Coasts and Floor of the Seas" in 3 volumes. In 1944 he was offered to take part in establishment of the Institute of Oceanology of the USSR AS (1946) and to head its Department on sea coasts. In 1946 Z. published under the auspices of the already established Institute of Oceanology his fundamental work "Dynamics and Morphology of the Sea Coasts". In 1951 Z. together with a group of cartographers was conferred the Stalin (State) Award for preparation of the USSR's first Gypsometric Map at scale 1:25,000,000 (marine part). In 1952 he headed the Coastal Section of the Interdepartmental Oceanographic Commission of USSR AS. Then he was elected the President of the International Coastal Oceanography Committee. In 1958 he was awarded the Gold Medal of F. Litke of the USSR Geographical Society for his fundamental work about the coasts of the Black Sea (cadastre). In 1958 he

published the book "Coasts of the Black and Azov Seas" and in 1958–1960—the
book "Morphology and Dynamics of the Soviet Coast of the Black Sea" (in two
volumes). In 1962 his book "Fundamentals of the Teaching on Sea Coast Development" was published for which in 1964 he was conferred the Lenin Award.
Z. was a self-made artist, learned to play a guitar by himself; he was an excellent
dancer; he swam and skied very well. In the P.P. Shirshov Institute of Oceanology
one of the laboratories is given his name.

Zenkovich V.P. (http://images.myshared.ru/469958/slide_4.jpg)

Zernov Sergey Alekseevich (1871–1945) – the outstanding Soviet hydrobiologist,
Academician (1931), the founder of the Russian hydrobiological science. In 1895
he graduated from the university and stayed there as an assistant in the Zoological
Museum. In 1897 he was arrested for "criminal agitation among the workers" and
was exiled to the Vyatka Province. In 1899 at the end of his exile period he received
an invitation to become the keeper of the Natural Science and Historical Museum of
the Tavria Province Zemstvo (country council) in Simferopol. Here he started
investigations of plankton in B. and A.S. At the same time he studied fishery
possibilities near the Crimean coast. In late 1901 Academician V.V. Zelenskiy,
Director of the Sevastopol Biological Station, offered Z. a position of the senior
zoologist and head of the Station. In March 1902 Z. moved to Sevastopol. Here he
prepared his master's thesis and organized an expedition to study the coastal zone
of B.S. from Batumi to the southern border of Bulgaria. Using a dredger and a sea
gauge he gathered the materials on the bottom fauna distribution depending on the
depth and ground nature from the water edge to the border of the hydrogen sulfide
zone. He studied the seasonal events in the life of benthos, plankton and nekton.
During the expedition on the trawler "Fedya" (1909) to the northwestern part of the
sea he discovered a large community of the red algae of the Phyllophora gender in a

depth interval of 34–60 m. This discovery was important both in scientific and practical terms. He called this community a Phyllophora Field, later on it was referred to as the "Zernov Phyllophora Field". Originally it was found that the Phyllophora community covered about 10,000 km^2 and its total stock was evaluated at approximately 10 million tons. During World War I these algae were used for iodine production and after the 1917 October Revolution in Russia they became a source for production of the valuable agar-agar. The name of Z. is closely connected with transition to the ecologobiocenotic researches. Z. laid the foundation of a new ecological stage in studies of the B.S. flora and fauna.

Z. lectured at the Fishery Faculty of the Agricultural Institute in Moscow. He was the author of the Russia' first study course on hydrobiology. Using the accumulated data on the species composition of the sea life, Z. divided the benthos into communities (biocenoses) by the nature of deposits, depth and domination of this or that animal species in a given region. His methodology is still in use in the present-day investigations. The expeditions of Z. during 1908–1911 in the Black Sea coastal areas were most fruitful. During 3 years he managed to collect enormous materials from 200 stations from various depths along the Russian as well as Romanian and Bulgarian coasts. And on the basis of these materials he was able to make certain generalizations. All subsequent investigations of benthos in B.S. confirmed the basic conclusions of Z. in relation to biocenosis distribution and composition in B.S. For comparison of the life in B.S. and in other seas Z. conducted observations in the Sea of Marmara in 1904 and in the Adriatic Sea in 1906. In the same years he visited the marine biological stations in Naples, Villefranche-sur-mer, Marseilles, and on the Turkish coast (1912). The obtained results he generalized in his books "On Study of the Life in the Black Sea" (1913), "General Hydrobiology" (1934). In 1902–1914 he was the Chief of the Sevastopol Biological Station. In 1914 he moved to live to Moscow where he organized the first Hydrobiological Chair in the Moscow Agricultural Institute. He was elected the Russia's first Professor of hydrobiology of the Fishery Department of the Agricultural Institute (later on the Timiryazev Agricultural Academy). After 1917 Z. organized the Hydrobiological Chair in the Moscow University and was elected the Academician. From 1931 to 1942 he headed the Zoological Institute of the USSR AS.

Zernov S.A. (http://www.ras.ru/nappelbaum/924bf3ec-5760-4926-b111-997b2dfc5c33.aspx)

Zhandr Aleksandr Pavlovich (1825–1895) – Russian Admiral, outstanding naval historian. After finishing of the Naval Corps (1844) he was conferred the rank of midshipman and served in the Black Sea Navy. In 1846–1847 he navigated in the Mediterranean. In 1849–1852 he was the flag-officer at Aid-de-Camp General V.A. Kornilov. In 1854 he supervised the on-land telegraphs in the vicinity of Sevastopol; was the officer-at-large at V.A. Kornilov; took part in the Sevastopol defense (1854–1855). In 1857 he was appointed the chief of the shipbuilding expedition of the Black Sea intendance and in 1858—Vice-Director of the Ship-building Department. In 1859 Z. published his fundamental work "The Materials on the History of the Sevastopol Defense and Biography of Vladimir Alekseevich Kornilov". In 1863 he became General-Controller of the Naval Report Department. In 1867 he was promoted to Major-General, in 1877—to Lieutenant-General. In 1878 he became a member of the State Control Board and in 1879—a senator later on renamed into Vice-Admiral. In 1855 he retired. Z. was awarded many orders.

Zhebrianskiy Bay – cuts into the coast to the north of the Kilian girlo delta on the Danube, Ukraine. The southern shore of the bay has many low-elevated islands of the Ochakov girlo, while the northwestern shore represents a low-lying barrier beach confining the vast Sasyk Liman. From time to time this barrier beach is scoured and the liman gets connected with the bay via a narrow passage. Settlement Primorskoe is situated on the northwestern shore of the bay.

Zheleznyakov Anatoliy Grigorievich (1895–1919) – a sailor of the Baltic Fleet, a participant of the October Revolution in Petrograde and Moscow, a hero of the Civil War in Ukraine. In 1918 he was a commissar of the Danube Navy, a chief of the Birzulskiy fortified region, a commander of a regiment and armored train. While breaking through the enemy's entrapment he was deadly wounded in the battle near Verkhovtsevo village. His name was given to a monitor of the Danube Fleet, to an armored train constructed in 1941 in Sevastopol and also to a destroyer.

"Zheleznyakov" – a flat-bottomed river monitor which crew became famous for its heroic deeds during the Great Patriotic War (1941–1945). It was built in 1934–1936 and incorporated into the Dnieper Navy. In 1940 it was transferred to the Danube Navy. It took part in the battles on the Danube, Southern Bug and Dnieper, defended Nikolaev and Kherson; in autumn 1941 it fought in Sevastopol. In 1941 "Z." was transferred to the Azov Navy. From May 1942 it fought on the Don and Kuban, near the Taman Peninsula and near the Black Sea coast of the Caucasus. In 1942 after severe battles it broke through to Poti where it stayed for repair till early 1943. In 1944 the monitor was incorporated once again into the Danube Navy and moved fighting as far as Budapest. In 1967 this veteran ship found its eternal moorage near the Dnieper embankment in Kiev. Its water displacement was 240 tons, length—50 m, width—8 m, draft—0.75 m, speed—20 km/h; it had 221 kW engines and a crew of 70 people. Its armament included two 102-mm and four 45-mm cannons and 4 stationary machineguns; during the Great Patriotic War it was additionally equipped with antiaircraft guns.

Zhomini Aleksandr Genrikhovich (1814–1888) – Russian Baron, diplomat, secret counselor (1875), state secretary (1882). He started his service in 1835 at the Russian Foreign Ministry, later on he took various posts in the Internal Relations Department and then a foreign relations officer; from 1845—at the Department for Asia. In 1856–1888 he was the senior counselor at the Foreign Ministry; he was the first to study all political papers of the Ministry, diplomatic notes, conventions; the editor of all most important documents prepared in the Ministry. He worked together with A.M. Gorchakov. After the end of the 1853–1856 Crimean War Z. was sent with a special diplomatic mission to Berlin (1856) and to Paris (1862). In 1867 he escorted Emperor Alexander II to the Paris World Exhibition. In April-November 1875 in the absence of Gorchakov he managed the Foreign Ministry. During the Russian-Turkish War of 1877–1878 he was in Bucharest with Gorchakov. In 1879 and 1880 he acted as the Minister's deputy. Z. was one of the founders of the Russian Historical Society. In 1863 he wrote a diplomatic history of the Crimean War enlightening on many factors which history had not been known. He wrote this work in French. It was translated into Russian under the title "Russia and Europe in the Epoch of the Crimean War" and published in 1886.

Zhovtnevoe (in the Ukrainian "*zhovten*" means "October") – a town, from 1961 the center of the Zhovtnevskiy district of the Nikolaev Region, Ukraine. It was situated in the mouth of the Southern Bug River, 12 km southward of Nikolaev. A railroad station of Pribugskaya. In 1973 it was included in the territory of the City of Nikolaev.

Zhukov Gavriil Vasilievich (1899–1957) – the Soviet military commander, Vice-Aadmiral (1944). From 1918 served in the Navy. He finished the Naval Commander School (1925). From 1928 he was the commander and commissar of the cannon boat, later the commander of a cruiser and a team of training ships of the Odessa Naval Base. During the Great Patriotic War he commanded the Naval Base in Odessa, the Odessa defense region; later on he was appointed the deputy commander of the Black Sea Fleet on defense, the commander of the naval base in Tuapse. After the war he took command positions in the Navy. In 1948–1951 he headed the Black Sea Higher Naval College.

Zhukov Georgiy Konstantinovich (1896–1974) – the major Soviet military commander, Marshal of the Soviet Union, four times Hero of the Soviet Union. In January-July 1941—Chief of Staff of the Deputy Commissar on Defense. From June 23—a member of the General Headquarters of the Military State Committee. From August 1942—First Deputy Commissar on Defense and Supreme Commander-in-Chief. A direct participant of elaboration and realization of the major military operations during the Great Patriotic War (1941–1945)—Moscow, Leningrad, Stalingrad, Kursk, Belorussia and Berlin. On May 8, 1945 on an order from the Soviet leadership he accepted the unconditional capitulation of fascist Germany in Berlin. On June 24, 1945 he took the salute of troops at the Moscow Parade. After the war he commanded various military districts. At the time of heading the Odessa military district he conducted significant organizational reforms.

Zhukov G.K. (http://topwar.ru/uploads/posts/2010-11/1290067544_zhuk.gif)

Zmeinyi Island (former Achilles, Beg Achilles, Levka, Belyi, in the Antiquity— "Fidonisi") – one of a few islands in B.S. located 37 km from the Kilian arm in the Danube Delta, thus, being of strategic significance. It belongs to Ukraine. Its area is 1.5 km^2, elevation is to 40 m over the mean sea level. In the late 1940s Romania handed over this island to the USSR, but after the Soviet Union disintegration it laid claims to this island again. The surface of the island is leveled, the shores are steep. This island was mentioned in the works of Antique authors, such as Aristotle, Pseudo-Skimn, Pavsanium, Maxim Tirskiy, Strabo and others. The very first mention of this island may be found in the poem "Ephopide" by Arkin Miletian, the Greek poet of the eighth century B.C. According to the ancient authors, after the death of Achilles, the hero of the Trojan War, who was hit, as is well-known, in the heel by the Paris arrow his remains were buried on the Levke Island.

The name of Fidonisi remained in the history in connection with the brilliant victory of F.F. Ushakov over the Turkish Fleet here. In the early twentieth century in honor of this battle at Fidonisi his name was given to one of the Russia's best torpedo boats of the Black Sea Navy that in 1918 raised the signal "Die, but not surrender!" The boat was sunk in the Tsemess Bay to prevent seizure by Kaiser Germany.

Here a cathedral was built in honor of the B.S. ruler—Achilles Pontarch. In the Middle Ages the cathedral was more than once robbed by pirates and in 1837 its ruins were used in construction of a lighthouse that was put into service in 1843. Now this lighthouse is technically the most advanced one. In 1841 the island was visited by the first research expedition of the Odessa Society of History and Antiquities led by Professor A.D. Nordman. After this the whole scientific community learned that this island seemed white because of the abundance of seagulls and their manure. In the Soviet time the island was declared a closed area. At present a small hotel is built on the island. The frontier guard and lighthouse services are stationed on the island and the research workers (archeologists, ornithologists, meteorologists) locate here, too. Two Soviet submarines that perished here during the Great Patriotic War are lying on the sea floor near the island. According to the international law, Z.I. has all features of an island, not a rock (Romania believes that it is a rock and cannot have any other status). The island has a 12-mile zone of territorial waters which existence is not questioned by Romania. The subject of the disputes with Ukraine was division of the continental shelf belonging to Ukraine. And this issue become ever more important in view of considerable oil and gas reserves in it the exploration of which was conducted in the 1980s–1990s near Z.I. located beyond these waters.

On February 3, 2009, the International Court of Justice delivered its judgment, which divided the sea area of the Black Sea along a line which was between the claims of the Ukraine and Romania, and Zmeinyi Island became a territory of the Ukraine.

Zonguldak – a town, the administrative center of the Zonguldak Il, Turkey. Industrial and trade center. Railway terminal on the western Black Sea branch of the railroad from Ankara. One of the major ports in Western Prichernomorie exporting coal. The port developed here thanks to the coal deposits. The center of the country's largest Eregli-Zonguldak coal basin. This is the main port on import of the Ukrainian and Russian coal via the Black Sea ports. It can accept vessels with the draft to 9 m. The cargo turnover of the port exceeds 3 million tons. The city ascends in terraces around a small harbor. Somewhat to the east of Z. there is a small city of Thylios being a resort and a tourist center. The population is 109,000 (2012). Three times a week the train "Black Diamond" transporting coal runs over the railroad.

Zonguldak Province – the Black Sea il in Turkey. It is located in the west of the country. Its area is 3,480 km², the population—620,000 (2011). The administrative center of the il is Zonguldak City.

Zuev Vasiliy Fedorovich (1754–1794) – the Russian traveler, natural scientist, ethnographer, teacher and translator. In 1779 he was elected the Adjunct and in 1787—Academician of the Petersburg Academy of Sciences. In 1707 he was a member of the Siberian Expedition headed by P.S. Pallas. He provided a detailed description of the nature of the Polar Ural. The data gathered by Z. and his traveling notes were used in his work "The Journey Over Various Provinces of the Russian State" (1783–1788). Taking into consideration the opinion of Pallas and the works presented by Z. to the Academy of Sciences, he was admitted to the Academic University. For further advancement of his knowledge he was sent abroad to the Leiden and Strasbourg universities. In 1779 he returned to Petersburg. He was conferred the degree of adjunct and worked under the guidance of Academician Pallas. In 1781 on an assignment of Academy of Sciences he went on an expedition to investigate the southwestern belt of European Russia. During this expedition he crossed the Central Russian and Dniester uplands, went over Dniester rapids. In February-March 1782 he made a voyage to Istanbul. In the same year he crossed the Crimean steppes from Perekop to Karasu-Bazar. His first published work about the nature of Crimea was entitled "The Extracts from the Traveling Notes Concerning the Crimean Peninsula" (1782). The expedition was sent there "for observations and discoveries in natural science history", for investigation of lands, waters and study of the natural conditions for growing of cereals, fruit trees, tobacco and forest management. Z. was the first to notice and describe the geographical peculiarities of the peninsula, in particular possible geological links of the Mountain Crimea with the Balkan and Caucasian mountains, focused his attention on limestones composing the mountains, their crushing and subsidence (sliding), on convenience of the Akhtiar (Sevastopol) Bay and other things. In the flat part of Crimea he found chernozem soils. In 1783 Z. taught natural science at the Main Public College, natural science history—at the Teacher's School in Petersburg. In 1784 he prepared the textbook "Natural Science History for Public Colleges of the Russian Empire". In 1785–1787 he was the editor in the journal "Vine Growing". In 1787 he published his "Traveling Notes from Saint-Petersburg to Kherson in 1781 and 1782". He also translated scientific treatises, in particular of Pallas—the book "Flora Rossica" (description of the flora of the Russian state), and others.

Chronology of the Key Historical Events on the Black and Azov Seas

Eighteenth Century

1700
- The Treaty of Constantinople was signed on 13 July 1700 between Russia and the Ottoman Empire. It ended the Russian-Turkish War of 1686–1700.

1701
- Engraver from Holland A. Schoonebeek engraved a map called "Eastern part of the Sea Palus Maeotis, which is now called the Sea of Azov." On the map there were shown: the coastline of the sea, the grid of parallels and meridians, the compass grid, depths, anchor positions, cities. The map scale is about 1:700,000, the size is 52 × 63 cm, published in Moscow.

1702–1704
- P. Picart issued in Moscow a map called "Direct drawing of the Black Sea from the town of Kerch to Tsargrad." On the map there were shown the towns of Bendery, Ochakov, Taman, Trapezund, Tsargrad and there was an inset of the Bosphorus Strait with depths depicted on the fairway.

1703
- Issue of the Atlas of the Black and Azov seas with the navigation chart of the track from Kerch to Constantinople (observation was performed by the ship "Krepost").

1703–1704
- The "Atlas of the Don River, Azov and Black Seas" compiled by Admiral C. Cruys was printed in Amsterdam; it consisted of a description and 17 maps.

1706
- A lot of new ships were added to the Russian Azov Fleet.

S.R. Grinevetsky et al., *The Black Sea Encyclopedia*,
DOI 10.1007/978-3-642-55227-4, © Springer-Verlag Berlin Heidelberg 2015

1710–1713
- Russian–Turkish War.

1711
- Treaty of the Pruth was signed between Russia and Turkey according to which Russia was to hand over the fortress of Azov to Turkey.

1713
- Treaty of Adrianople was signed between Russia and Turkey.

1720
- "Perpetual Peace" between Russia and Turkey was signed in Constantinople; it replaced the temporary Treaty of Adrianople of 1713.

1722
- The Turks founded the fortress of Sudjuk-Kale (later Novorossiysk)

1733
- The Don Military Flotilla was formed; formation of the Russian Fleet to assist Russian troops in repelling attacks of the Crimean Tatars in the basin of the Don River and fighting with the Turkish Fleet in the Sea of Azov.

1735
- War with Turkey in 1735–1739 with the aim to return the Azov coast, Crimea to Russia, and to gain access to the Black Sea

1735–1739
- Russia in alliance with Austria is fighting against Turkey and takes Crimea twice.

1736
- The Russian Fleet of 50 ships under the command of Rear Admiral Bredal took part in the siege and taking of the Turkish fortress of Azov.

1737
- Peace negotiations between Russia and Austria on the one hand and Turkey on the other hand during the Russian-Turkish War of 1735–1739. The negotiations were called The Congress of Nemirov after a Ukrainian place named Nemirov (now an urban-type settlement in the Vinnytsia Oblast).
- Formation of the Dnieper War Flotilla to assist the Russian Army which was fighting on the coast of the Dnieper Liman during the Russian-Turkish War of 1735–1739.
- Russian boat under the command of Captain De Frémery sent to the Azov was met at the Fedotov Spit by the Turkish squadron of 31 ships. De Frémery sent the team to the shore, set afire the boat, with which he died.

1739

- The Treaty of Belgrade between Russia and the Ottoman Empire. It ended the war of Russia and Austria against the Ottoman Empire in 1735–1739. Russia was prohibited from keeping its ships in the Black and Azov Seas. The Azov district now belonged to Russia.
- Special navigation atlas was published by L. Renard; a map of the Black Sea was included in the atlas.

1768

- The Azov Military Flotilla started to form in Russia.

1768–1774

- Russian-Turkish War, which resulted in the proclamation of the Crimean Khanate independent of Turkey, Kerch becoming a Russian city, and forming of Russian garrisons in all ports.

1768–1796

- Some reconnaissance marine surveys of the Black and Azov seas under the leadership of Senyavin D., Odintsov I., Bersenev I. et al. were carried out.

1769

- Building of the Azov Military Flotilla in Russia.

1770

- Night attack in Chesma Bay during the Russian-Turkish War of 1768–1774 by the avant-garde of the Russian ships under the command of Rear-Admiral S.K. Greig (The Naval Battle of Chesma).

1771

- Admiral Senyavin introduced the map of the Sea of Azov with the Strait of Kerch to the Collegium of Admiralties.
- Creation of the Danube Military Flotilla in Russia
- Proclamation of the independence of the Crimean State.

1772

- Seizure of the fortress Chesma by assault.
- The map of the Black Sea compiled by Jacques Nicolas Bellin was published in Paris. In the Russian Fleet it was used in copies and had significant errors: coastlines of the Black Sea were applied with an accuracy of up to 65 miles, coastlines of the Azov Sea—with an accuracy of 40 miles, and the mouth of the Don River was shown 100 miles further eastward.

1773

- Sea battle near Balaklava between two Russian ships "Koron" and "Taganrog" and the Turkish squadron of four ships.

– Topographical survey and measurement ("... with measurements and tools") of the Sevastopol Bay were performed by the team under the command of navigator I.V. Baturin. The first map of the bay was made—"Map of the harbor of Akhtiar with measurements" and "Plan of the harbor of Akhtiar with measurements and the position of the town of Ackerman", but the plan didn't receive widespread use.

1774
– Battle in the Kerch Strait between a Russian squadron under the command of Rear Admiral V.Ya. Tchichagov and a Turkish squadron.
– Treaty of Kuchuk-Kainardji (Küçük Kaynarca) between Russia and Turkey was signed, which ended the Russian-Turkish War of 1768–1774. Russia gained free access to the Black Sea.

1775
– Publication of the description of southern ports and seas, compiled by Academic Johann Anton Güldenstädt

1778
– Beginning of the building of the town and port of Kherson near a small fortress of Aleksanderschanz, which will be home of ship-building at the Black Sea.
– Foundation of the Kherson commercial seaport.

1779
– Beginning of the construction of the first Russian Black Sea Fleet battleship "Saint Catherine" (60 guns) in Kherson. In 1784–1785 years it was dismantled on the stocks.

1782
– For the first time the Russian ships "Ostorozhnyi" and "Otvazhnyi" entered the Bay of Akhtiar and wintered there. During the winter their teams compiled the first sea map of the Sevastopol Bay.

1783
– Foundation of Sevastopol—the main base of the Russian Black Sea Fleet.
– Foundation of the Russian Black Sea Navy.
– Foundation of the Admiralty of Sevastopol—predecessor of the Sevastopol sea plant.
– The ships of the Azov Flotilla under the command of Vice-Admiral F.A. Klokachev came into the Bay of Akhtiar (the Bay of Sevastopol).
– G.A. Potemkin was appointed Commander-in-Chief of the Russian Black Sea Fleet. Building of the marine and commercial fleets at the Black Sea was performed under his command and upon his initiative.

– Compilation of the "Map of the harbor of Akhtiar from the mouth of the Belbek River to the former town of Kherson is compiled upon demand of His Excellency F.A. Klokachev in 1783".

1784
– The name "Sevastopol" was given to the town.

1785
– By a decree of Catherine II, all Crimean ports were exempt from customs duties for a period of 5 years, and customs guards were transferred to Perekop.
– The Russian government approved the first sea staff of the Black Sea Fleet.
– Turkish cartographer Mustafa Rezmi painted by hand on silk a map of the Black Sea.

1786
– The Bay of Sevastopol became the main base for the Russian Black Sea Fleet.

1787
– Joint actions of the Russian Army and the flotilla in the Dnieper Liman.
– Severe storm in the Black Sea, destroying the Sevastopol squadron.

1787–1791
– Russian-Turkish War; Turkey acknowledges the incorporation of the Crimea into Russia.

1788
– During the Russian-Turkish War of 1787–1791 a naval battle took place in the Dnieper Liman between the Turkish Fleet under the command of Hassan al-Gasi (43 ships) and the Russian rowing Liman Flotilla under the command of Rear-Admiral Karl Heinrich de Nassau-Siegen (50 ships).
– Expedition of the Russian squadron and its victory over the Turkish Fleet at Sinop.
– The defeat of the Turkish Fleet under Ochakov.
– The Black Sea Fleet squadron under the command of F.F. Ushakov defeated the Turkish Fleet at the Island of Fidonisi.
– Construction of shipyards in Nikolaev.
– A ship under the command of Captain II Rank Johann Reinhold von der Osten-Sacken was attacked by 30 Turkish ships near the mouth of the Southern Bug River. Osten-Saken blew up his vessel along with Turkish janissaries.
– John Paul Jones, the first officer of the US Navy, Rear-Admiral of the Russian Black Sea Fleet, the famous privateer of the Atlantic Ocean, came to serve at the Black Sea.

1789

- Russian victory over the Turkish Army at Rymnik.
- Foundation of the city of Nikolaev—the rear base of the Russian Black Sea Fleet.
- F.F. Ushakov was appointed Commander of the Sevastopol naval squadron.

1790

- Rear-Admiral F.F. Ushakov was appointed Commander of the Russian Black Sea Fleet.
- Sea fight of the Liman rowing flotilla under the command of Major-General de Ribas with the Turkish river flotilla.
- Renaming of the Liman (Danube) rowing flotilla into the Black Sea rowing fleet.
- Battle in the Kerch Strait at Cape Takil between the Russian squadron under the command of Rear-Admiral F.F. Ushakov and the Turkish squadron of Kapudan Pasha Hussein.

1791

- Treaty of Jassy between Russia and Turkey was signed, which ended the Russian-Turkish War of 1787–1791.
- The Black Sea Fleet squadron under the command of F.F. Ushakov defeated the Turkish Fleet at Cape Kaliakra.

1793

- A fortress was built on the bank of the Gulf of Khadzhibey, future Odessa.
- A.V. Suvorov signed a plan for the construction of Odessa, prepared by the engineer-colonel of the Russian army François Sainte de Wollant.

1794

- Foundation of Odessa. De Ribas directs the construction of Odessa.
- On the stocks of the Sevastopol Admiralty there were laid the first two schooners of the rowing fleet, marking the beginning of sailing shipbuilding in the Sevastopol Admiralty.
- Catherine II issued a rescript on founding a new port city at the site of Khadzhibey.

1795

- Khadzhibey was renamed into Odessa.

1796

- Sevastopol was renamed into Akhtiar by Emperor Paul I, who was against everything what had been done during the reign of Empress Catherine II.

1798

- Foundation of the school of ship architecture in Nikolaev for the Black Sea Fleet. Dissolved in 1803.

1799
- Publication of the first "Atlas of maps and plans of the northern coast of the Black Sea from the Dniester River to the Kuban River" compiled by Captain I Rank Joseph Billings.

Nineteenth Century

1801
- Foundation of the first Black Sea hydrometeorological station in Nikolaev.

1801–1803
- Joseph Billings' description of the Black Sea was continued with inclusion of the west coast of the Black Sea from Odessa to the Bosphorus Strait, the south coast from the Bosphorus Strait to Cape Samsun under the leadership of Lieutenant I.M. Budischev, warrant officer N.D. Kritskiy, Capitan Lieutenants A.E. Vlito and P. Adamopulo; besides description of the Sea of Azov was performed under the leadership of A.E. Vlito and N.D. Kritskiy.

1803
- Depot of the Black Sea Fleet maps was created in Nikolaev in order to collect and store drawings of ships, marine surveys books and navigation maps. Later functions of the maps depot were expanded and it became the leading agency of the fleet, which was engaged in hydrographic research, compilation and publication of maps, as well as in supplying maps to ships.

1804
- Map of the Sea of Azov compiled by Capitan Lieutenant A.E. Vlito and warrant officer N.D. Kritskiy with the use of the description they had made earlier was published in Russian, French and Greek.
- Sevastopol regained its name, and it became the main naval port of the Russian Black Sea Fleet.

1805
- Treaty of alliance between Russia and Turkey was signed in Constantinople, Turkey, according to which Russia was permitted to have free passage through the Black Sea straits.

1806–1812
- Russian-Turkish War.

1806
- Turkey terminated the Russian-Turkish treaty of alliance.
- Supervisor of the Depot of the Black Sea Fleet Lieutenant I.M. Budischev compiled the "Flat Map of the port of Akhtiar".

1807
- Opening of the hydrometeorological station in Kherson.
- The Admiralty published the Atlas of maps and plans of the Black and Azov Seas, as well as a map of the western part of the Black Sea in the Mercator projection.
- Battle of Athos. The defeat of the Turkish Fleet by the Russian squadron of D.N. Senyavin off Cape Athos.

1808
- Laying of the corvette "Crimea", which marked the beginning of a systematic construction of sailing ships in the Sevastopol Admiralty.
- Publication of the "Pilot chart, or a marine guide of the Azov and Black Seas", compiled by Lieutenant I. Budischev.

1809
- The Russian rowing flotilla took part in the siege by the Army of the Turkish fortress of Silistra.

1811
- Academician Peter Simon Pallas in the book "Zoographia Rosso-Asiatica" published the first scientific information on the fish fauna of the Black Sea, which he had collected during a trip to the Black Sea in 1793–1794.

1812
- Signing of the Bucharest Peace Treaty that ended the Russian-Turkish War of 1806–1812.

1816
- The first stone lighthouses of Tarkhankut and Chersonesus were built on the Black Sea.
- Reni Port was founded on the Danube.
- Fabian Gottlieb Thaddeus von Bellingshausen (Faddey Faddeevich Bellingshausen) described the Black Sea coast.

1817
- The general map of the Black and Azov Seas from the Atlas of I.M. Budischev was republished with data from supplementary inventories.
- Foundation of the Directorate of the lighthouses and pilot chart of the Black and Azov Seas, responsible for the operation of lighthouses and other means of navigation equipment. Capitan Lieutenant M.B. Berkh was appointed the first director, later he was a full Admiral.

1820
- Russian-French hydrographic expedition on the ship "La-Chevrette", led by French Captain Gauthier, with the participation of Russian Captain Lieutenant

M.B. Berkh, describes the shores of the Black Sea. The map of the sea was published 2 years later in France.
– Building of the Enikalskiy lighthouse.
– Launching of the brig "Mercuriy" in Sevastopol, which became a legendary ship, awarded St. George's Ensign.

1821
– Opening of the hydrometeorological station in Odessa.
– In order to provide a safe entrance to the Sevastopol Bay Inkerman Leading lights were built with the visibility of the front one—20 miles, of the one in the back—28 miles.
– Foundation of the Kerch Sea Commercial Port.

1822
– Foundation of the Marine astronomical observatory in Nikolaev.
– Publication of the map of the Black and Azov Seas in France.
– The first pilots (pilot service) appeared in Kerch.

1824
– Opening of the hydrometeorological station in Sevastopol.

1826–1836
– Hydrographic expedition under the command of Captain I Rank E.P. Manganari with the aim to make a survey of the Black and Azov Seas.

1826
– Akkerman Convention was signed between Russia and Turkey, which confirmed the Treaty of Bucharest of 1812.
– Dzharylgach lighthouse built on the same name island in the Karkinit Bay of the Black Sea.

1827
– Statute and staff of the Corps of naval navigators were approved. The head of the Corps became General—hydrograph G.A. Sarychev.
– The first small sailing ships were built—pilot boats #1 and #2 for hydrographic work at the Black Sea.
– Odessa (Vorontsov) lighthouse was built.

1828
– Russian-Turkish War of 1828–1829.
– Shipping began at the Black Sea; the first steamer "Odessa".

1829
– Peace treaty of Adrianople was concluded between Russia and Turkey. Russia gained the mouth of the Danube with islands and the coast of the Caucasus.

- Publication of the atlas and the map of I.M. Budischev.
- Depot of the Black Sea Fleet maps made a new general map of the Black and Azov Seas on the basis of the map of Gauthier—Berkh.
- Battle of the brig "Mercuriy" with two Turkish line ships "Selimie" and "Real-Bei"—a naval battle with the inequality of forces.

1831
- Russian authorities founded fortresses of Gelendzhik and Anapa.

1832
- Depot of the Black Sea Fleet maps was abolished.
- On the 1st of September within the headquarters of the Black Sea Fleet the Hydrographic Office was established, which included the hydrographic depot. That day was the date of foundation of the Hydrographic Department of the Russian Black Sea Fleet.

1833
- During the stay of the squadron under the command of M.P. Lazarev in Turkey there was made a survey of the Bosporus and the Dardanelles Straits, which was conducted by Lieutenants E.V. Putyatin and V.A. Kornilov.
- Establishment of the first steamship company at the Black Sea.

1834
- Opening of the Merchant Shipping School for preparation of skippers, navigators and shipbuilders in Kherson.

1835
- Laying of the first sluice of the five-chamber-dry dock in Sevastopol.

1836
- Captain Lieutenant E.P. Manganari published a map of a part of the northern shore of the Black Sea from Chersonesus Cape to Taman.

1836–1837
- E.E. Sabler, A.N. Savich and E.N. Fus conducted leveling of the Black and Caspian Seas.

1837
- Beginning of the building of a lighthouse on the island of Fidonisi (Zmeinyi Island) from the ruins of the temple of Achilles Pontarches, which means "Master of the Black Sea".

1838
- Admiral Mikhail Lazarev founded Fort Velyaminovskiy, future Tuapse.
- Amphibious landing at the mouth of the Sochi River; laying of the fortification of Aleksandria.
- Foundation of the city of Novorossiysk.
- In the roads of Tuapse during the storm of incredible force there were lost: frigate "Varna", corvette "Mesembria", brig "Themistocles", tenders "Luch" and "Skoryi", steamship "Jason", several merchant ships—a total of 13 vessels, of which five military ones.
- Two warships and seven merchant vessels were lost in the roads of Sochi.

1840
- London Convention, signed by Great Britain, Russia, Austria and Prussia on one side and Turkey on the other on support for the Turkish Sultan against Egyptian Pasha Muhammad Ali.

1841
- London Straits Convention, signed by Russia, Great Britain, Austria, Prussia and Turkey; it regulated the regime of the Black Sea straits.

1841–1842
- Depot of the Black Sea Fleet maps published the first copies of the Atlas of maps of the Black and Azov Seas. The publication of it was completed in 1844. On the maps of the Atlas for the first time in the Russian marine cartography submarine relief was depicted with isobaths.

1842
- Publication of the "Atlas of the shores of the Black and Azov Seas", complied by Major General E.P. Manganari.

1843
- Permanent steamship service was established between Odessa and Constantinople on state steam frigates "Odessa" and "Crimea".
- Completion of the construction of the lighthouse on Zmeinyi Island (Fidonisi).

1846
- Beginning of steamship service between Odessa and Russian Danube ports up to Galati.
- Discovery of the largest natural gas field Golitsynskoe on the Crimean Black Sea shelf in the Karkinit Bay, about 70 km from the Tarkhankut coast.

1847–1849
- Lieutenants G.I. Butakov and I.A. Shestakov (commanders of the tenders "Pospeshnyi" and "Skoryi") carried out marine survey of the shores of the Crimea and the Caucasus (1847), the Bug River, the Dniester Liman and a

part of the coast between Ochakov and the mouth of the Danube River (1848), the coast of Anatolia (1849) (with the assistance of two Turkish military brigs) and the coast from the Strait of Constantinople to the mouth of the Danube River. On the basis of the materials of the expedition there was compiled the "Pilot chart of the Black Sea".

1848
– Novorossiysk was granted status of a city.

1850–1851
– Depot of the Black Sea Fleet maps published the first pilot chart of the Azov Sea.

1851
– Depot of the Black Sea Fleet maps in Nikolaev published the first "Pilot chart of the Black Sea" by G.I. Butakov and I.A. Shestakov.
– Beginning of the description of lighthouses of the Black and Azov Seas in the Black Sea hydrographic depot.

1853–1856
– The Crimean War. Sevastopol becomes a place of heroic battles on land and sea; Russia is fighting against England, France and the Kingdom of Sardinia, which are protecting the influence of Turkey in the Black Sea.

1853
– The first in history battle of steamers: steam frigate "Vladimir" (of the Black Sea Fleet) with the Turkish ship of the similar class "Pervaz Bahri".
– Sea Battle of Sinop (Sinop is a port on the Black Sea coast of Turkey) between the Russian squadron under the command of Vice Admiral P.S. Nakhimov and the Turkish squadron under the command of Osman Pasha. All the Turkish ships were destroyed, except for one.
– An Anglo-French squadron enters the Black Sea in order to block the Russian Fleet.

1854
– Russia declares war on Great Britain and France.
– Joint Fleet of England, France and Turkey dropped anchor in the roads of Evpatoria.
– Landing of the "allies" at Yalta in the Crimea.
– Military action near the mouth of the Danube.
– Landing of the Anglo-French troops in Varna, blocking the Russian Black Sea Fleet in the Sevastopol harbor.
– Flooding of a part of the Russian squadron (seven sailing ships) at the entrance to the bay of Sevastopol in order to prevent the entrance of the enemy ships.
– The first bombardment of Sevastopol, which led to considerable destruction of defensive fortifications.

– Bombardment of Odessa by the Anglo-French squadron started.
– Publishing of the first pilot chart of the Azov Sea, compiled by Second Lieutenant A. Sukhomlin on the basis of the work that he had performed on the pilot schooner "Astrolabe" in 1850–1851,—"Pilot chart of the Kerch-Enikalskiy Strait".
– Admiral A.A. Popov organized the defense of Sevastopol from the sea, using artillery taken from the ships.
– Battle of Balaklava.
– "The great storm of Balaklava"—a storm that hit the Crimea destroying much of the Anglo-French squadron that participated in the Crimean War, and resulted in the deaths of 1,500 people.

1854–1855
– Diplomatic negotiations during the Crimean War of 1853–1856 (Vienna Conferences).
– Defense of Sevastopol.

1856
– Signing of the Treaty of Paris between Russia and Great Britain, France, Turkey and Sardinia that were at war with Russia. The Black Sea was declared neutral.
– The Russian Steam Navigation and Trading Company was established (ROPiT, also referred to as Russian S.N.Co.).

1857
– The Russian Steam Navigation and Trading Company (ROPiT) opened the cargo-passenger service in the Black Sea.

1858
– Foundation of the sea port Poti.

1859–1860
– Capitan Lieutenant I.M. Dikov did geomagnetic measurements in 41 locations in the Black Sea. On the basis of these data and the coordinates of the ship he built maps of isogons and isoclines of the Black Sea.

1860
– Hydrographic Office of the Staff of the Black Sea Fleet was transformed into the Hydrographic part of the port of Nikolaev with the existing functions. The hydrographic part was directly subordinate to the chief commander of the port, and on special issues—to the Hydrographic Department.

1860–1862
– Magnetic survey was performed under the direction of warrant officers I.M. Dikov and M.I. Samoilovych on the shores of the Black Sea that belonged to Russia, and in some locations of the Anatolian coast. The Odessa magnetic anomaly was identified and examined; 41 locations were identified. On the basis

of the results of the work a magnetic map of the Black Sea was compiled under the direction of warrant officer I.M. Dikov. I.M. Dikov continued observations in 1871 and 1875.

1861
- The Hydrographic part of the port of Nikolaev and the Directorate of the Black and Azov Sea lighthouses were merged. In 1863 V.I. Zarudnyi was appointed Head of the organization.
- The Workshop of nautical instruments in Nikolaev became subject to the Hydrographic Department (till 1861 it was part of the Admiralty of the Black Sea Fleet).

1862–1863
- Military Geodesist I.I. Stebnitskiy performed geodetic measurements of the Black and Azov Seas' levels.

1864
- Creation of the pilot assistance in Nikolaev.

1865
- Foundation of the Hydrometeorological center of the Black and Azov Seas in the University of Novorossiya (now University of Odessa).

1867
- The Pilot chart of the Black Sea was republished with amendments and changes.
- The first electric lighthouse in Russia—the Odessa lighthouse—came into operation.

1868
- The Society of pilots of Nikolaev in the Black Sea was founded.
- A comprehensive expedition in the Black Sea on the corvette "Lvitsa" ("Lioness") under the command of Capitan Lieutenant F.N. Kumani. The aim of the research was to study the route for laying a telephone cable in the area of Feodosiya—Sukhumi. Depths (up to 800–1,000 m), temperature and salinity of water were measured, soil samples were taken.

1869
- Foundation of the Military History Museum of the Russian Black Sea Fleet.

1871–1914
- The first detailed, based on triangulation, marine survey of the Black and Azov Seas under the direction of V.I. Zarudnyi, K.A. Myakishev, Yu.K. Ivanovskiy, P.E. Belyavskiy, K.P. Andreev, L.V. Antonov, N.V. Malina.

1871

- Foundation of the Sevastopol Biological Station, later A.O. Kovalevskiy Institute of Biology of the Southern Seas of the National Academy of Sciences of Ukraine.
- Establishment of the hydrographic expedition of the Black Sea.
- Organization of the Black Sea hydrographic expedition led by Captain I Rank V.I. Zarudnyi. During 16 years by the means of plane-table survey, boat and sometimes ship soundings the most important areas of the Black and Azov Seas were examined.
- Signing of the agreement on the amendment of some articles of the Treaty of Paris of 1856. The document was named the London Convention. It canceled the "neutralization" of the Black Sea.
- On the right riverbank of the Southern Bug near the village Bogdanovka the signs of the first on the Black Sea dimensional line were set.
- Establishment of the compass observatory in Nikolaev.
- Foundation of the Azov Sea Shipping in Mariupol.

1872

- Beginning of the construction of armored ships in the port of Nikolaev.

1873

- The Department of the Black Sea Fleet and ports allocated to the jurisdiction of the Directorate of lighthouses and pilot charts an iron schooner "Ingul" with a mobile lighthouse workshop on it.
- Lieutenant F.F. Vrangel began hydrological studies of the Black and Azov Seas.
- Publication of a paper named "The Black Sea and its significance for Russia: historical and geographical essays".

1877

- Russian-Turkish War of 1877–1878.
- Battle between the Russian armed steamer "Vesta" and the Turkish ironclad warship "Feth-i Bülend".
- The seizure of the Turkish steamer "Mersina" by the Russian cruiser "Russia". The biggest prize in the war history in the Black Sea (silver ore, valuables, government papers, a lot of "pocket money", etc.).
- The roads of Batumi, attacks of Russian mine boats against Turkish ships in the port of Batumi during the Russian-Turkish War of 1877–1878.
- Foundation of the seaport of Batumi.

1878

- Attack of Russian boats against Turkish ships and running down of the guard vessel "Intibah" in the roads of Batumi.
- In the Admiralty of Sevastopol the Russian Steam Navigation and Trading Company (ROPiT) built the first Russian Navy torpedo boats (12) for the Black Sea, intended for use against enemy ships at the coast and for defense of access to ports.

- The "Russian sea and river insurance company" opened service along the Crimea-Caucasus line.
- End of the Russian-Turkish War and military operations in the Black Sea in the nineteenth century.
- The preliminary Treaty of San Stefano, which ended the Russian-Turkish War of 1877–1878.

1879

- Signing of the Peace Treaty of Constantinople—an agreement between Russia and Turkey.
- Meeting in Livadia, chaired by Alexander II, who discussed the fate of the Black Sea straits in case of disintegration of the Ottoman Empire.
- Foundation of the society "Voluntary Fleet" to assist the Government in the transport of goods across the Black Sea.

1880

- The "Russian sea and river insurance company" established the connection between Odessa and Nikolayev.

1880–1883

- E.V. Maydell built a new map of isogons of the north-western part of the Black Sea, additionally making measurements of magnetic declination in 23 locations on land and in 14—at sea.

1881

- Signing of a secret Austro-Russian-German treaty, known as the Three Emperors' Alliance. The third article of the treaty was dedicated to the Black Sea straits. It provided Russia with a guarantee of diplomatic support from Germany and Austria-Hungary to prevent the Sultan from letting the English Navy entry the Black Sea.
- A special meeting on a long-term program of building of the Russian Black Sea Fleet.
- Republication of the Pilot chart of the Black Sea, with account of the data of the hydrographic expedition of V.I. Zarudnyi.

1881–1882

- S.O. Makarov, being Commander of the ship "Taman", conducted observations of currents in the Strait of Bosporus: about 4,000 measurements of temperature and specific gravity of water and 1,000 measurements of velocity of currents. He discovered deep and surface currents and, thus, determined the pattern of water exchange in the Black and Marmara Seas.

1882

- The Black Sea Company "Portlandcement" built in Novorossiysk the first cement factory "Zvezda" ("Star") (now "Proletariy" ("Proletariat")).

– Approval of a large-scale shipbuilding program of the Russian armored fleet for 20 years for the Baltic Sea, Black Sea and Pacific Fleets consisting of 24 battleships, 15 cruisers, 19 gunboats and 127 torpedo vessels.

1883
– Beginning of construction of the West dock in Sevastopol.

1884
– Laying of the first in the world multitowered ironclads "Chesma" and "Sinop", which laid foundation for the armored Black Sea Fleet.
– Introduction of the first pilots (pilot service) in Batumi.

1885
– S.O. Makarov published a paper named "On water exchange in the Black and Mediterranean Seas," which received the Prize of the Russian Academy of Sciences.
– Completion of construction of the West dock in Sevastopol; the first ship—the steamer "Nakhimov"—was put in the dock.

1887
– Hydrographic expedition of the Black Sea due to the completion of major work was transformed into a Separate survey of the Black Sea, reduced to a Separate team in 1908.
– The "Reinsurance Treaty" –an informal name of a secret Russian-German treaty signed in Berlin. The agreement recognized the European importance of the straits of Bosporus and Dardanelles being closed to warships of all nations.

1888
– Beginning of the transition to lighthouses with kerosene in the Black Sea.

1889
– Formation of the pilot society in Kerch.

1889–1890
– At the 8th Congress of Russian naturalists and physicians A.V. Klossovskiy and N.I. Andrusov reported their project on physiographic and biological studies of the Black Sea.

1890
– Creation of the Black Sea pilot district.
– Russian Navy Ministry (Principal Hydrographic Department) at the request of the Russian Geographical Society organized an expedition on the gunboat "Chernomorets", led by the Head of the meteorological office of the Department I.B. Spindler, with the participation of F.F. Vrangel and N.I. Andrusov, with the aim to study the Black Sea in four main directions: (1) along the axis of the

western basin, (2) along the middle parallel of the sea between Varna and Pitsunda, (3) along the shortest line between the Crimea and the Anatolian coast, (4) along the axis of the eastern basin.
– Creation of the pilot society in Novorossiysk.
– Foundation of the Constanţa shipyard—the largest and most modern shipyard in Romania.
– N.I. Andrusov published a paper named "On the need for deep-sea research in the Black Sea".

1891
– Lieutenant M.E. Zhdanko collected all measurements of magnetic declination made on the shores of the Black Sea and compiled a map of isogons of the sea.
– Expedition led by I.B. Spindler on the gunboats "Donets" and "Zaporozhets" with the aim to study the content of hydrogen sulfide in the deep waters of the Black Sea.
– The Sevastopol Biological Station came under the jurisdiction of the Academy of Sciences.

1892
– Beginning of the construction of the sea commercial port of Feodosiya.
– Foundation of the "Bulgarian Trade Steamship Company" (later "Bulgarian Sea Fleet Shipping Company").

1894
– Emperor Alexander III died in the Smaller Livadia Palace in Yalta.

1895
– Sevastopol became the main base and administrative center of the Russian Black Sea Fleet.
– Putting into operation of the Feodosiya sea commercial port.
– The first revolutionary armed uprising of the Russian Black Sea Fleet sailors.
– Beginning of the shipbuilding facility construction in Nikolaev (future "Naval").

1896
– Doctor of Zoology A.A. Ostroumov published the "Indicator of fish of the Black and Azov Seas".
– Completion of the construction of the Novorossiysk sea port.

1897
– Creation of a public aquarium at the Sevastopol Biological Station.
– Opening of the "Naval" plant in Nikolayev, later the Black Sea Shipyard.

1898
– Completion of the Eastern dock construction in Sevastopol; the first ironclad "Georgiy Pobedonosets" ("Saint George the Victorious") was put into the dock.

- The second cement plant "Tsep" ("Chain") [now "Octyabr" ("October")] was put into operation in Novorossiysk.

1899
- The Navy Ministry published "Materials on the hydrology of the Black and Azov Seas in the expeditions of 1890–1891" (processed by I.B. Spindler and F.F. Vrangel).
- Closing of the Sevastopol commercial port.

1900
- Creation of the Simeiz Observatory.

Twentieth Century

1901
- Laying of the cruiser "Ochakov"—the legendary ship of the first Russian revolution—on the first stocks of the Lazarevsky Admiralty in Sevastopol.
- L.N. Tolstoy visited Sevastopol.

1902
- Completion of the construction of the Massandra Palace.

1902–1904
- Franz Alekseevich Roubaud with a team of artists and students of the Bavarian Academy of Fine Arts created the panorama "Defense of Sevastopol in 1854–1855".

1903
- Republication of the "Pilot chart of the Black Sea" with the pilot chart of the Sea of Azov as its part.
- Emergence of the first revolutionary groups of the Russian Social Democratic Labour Party (RSDLP) on ships and in crews of the Black Sea Fleet.

1903–1905
- under the leadership of Captain II Rank A.M. Bukhteev triangulation of the Black and Azov Seas was recalculated and equalized.

1904
- Construction of the building of the panorama "Defense of Sevastopol in 1854–1855" (destroyed in 1942, restored in 1953, repaired in 1974).
- The first revolutionary actions of the Black Sea Fleet crews in Sevastopol.

1905
- The Sunken Ships Monument was erected in Sevastopol on an artificial island.
- Opening of the panorama of F.A. Roubaud "Defense of Sevastopol", dedicated to the heroism of the defenders of Sevastopol during the Crimean War, on the occasion of the 50th anniversary of the first defense of Sevastopol.
- Creation of the "Novorossiysk Republic", that lasted from the 12th till the 25th December.
- The world's first reinforced concrete lighthouse—Ozharskiy was built in the Black Sea; height—36.7 m (from the bottom).
- Uprising on the battleship "Kniaz Potemkin Tavritcheskiy" ("Prince Potemkin of Tauris") supported by the battleship "Georgiy Pobedonosets" ("Saint George the Victorious"), hydrographic vessel "Vekha" ("Milestone") and torpedo vessel "#267".
- Uprising of the Sevastopol garrison, naval crews and a part of the Black Sea Fleet led by the cruiser "Ochakov" under the leadership of Lieutenant P.P. Schmidt. The cruiser "Ochakov" was joined by the battleship "Panteleimon", mine cruiser "Griden", training ships "Dnestr" and "Prut", torpedo boat destroyers "Svirepyi", "Zavetnyi", torpedo vessel "# 270" and others.

1906
- Leaders of the Black Sea Fleet uprising in Sebastopol—Lieutenant P.P. Schmidt, conductor Chastnik, sailors Antonenko and Gladkiy—were shot on Berezan Island.

1908
- Creation of the pilot society in the port of Odessa.
- The Principal Hydrographic Department published the "Atlas of the winds and fogs of the Black and Azov Seas".
- In connection with the completion of hydrographic surveys of the Black and Azov Seas the Separate survey was transferred into the Separate hydrographic team.
- Oral Russian-Austrian agreement was reached in the castle of Buchlov (Czech Republic). It was envisaged that Austria-Hungary would not make difficulties for changing the regime of the Black Sea straits.

1909
- Huge cluster of the red algae Phyllophora, known in the literature as "Zernov's Phyllophora field", was discovered in the north-western part of the Black Sea.
- Foundation of the Sevastopol Naval Observatory—one of the oldest naval observatories of Russia. It consisted of three parts: the compass, astronomical and hydrometeorological ones.
- Foundation of the first trawling teams of the Black Sea Fleet, which included specially built trawlers of 100–150 tons and adapted for trawling out-of-date torpedo vessels and harbor vessels.

1911
- Laying of three Black Sea Fleet battleships of the dreadnought type in Nikolaev: "Imperatritsa Mariya" ("Empress Maria") (came into commission in July 1915), "Imperatritsa Ekaterina II" ("Empress Catherine II") (came into commission in November 1915), "Imperator Aleksandr III" ("Emperor Alexander III") (came into commission in July 1917).
- Development of the shipbuilding program for 1912–1916 for the Baltic and Black Sea fleets consisting of 4 battle cruisers, 8 cruisers, 44 torpedo boat destroyers, 24 submarines.
- The Sevastopol Naval Observatory was made responsible for the management of the entire network of hydrometeorological stations of the Russian Black Sea Fleet.
- The Greater Livadia Palace was built.

1912
- Yu.M. Shokalskiy founded the Department of Oceanography in the Russian Naval Academy.

1913
- Future academician S.A. Zernov (1922) published a fundamental monograph "On the study of the life of the Black Sea".
- The Sevastopol Naval Observatory began compiling synoptic maps on a daily basis.

1914–1917
- Blockade of the Bosphorus, military operations of the Russian Black Sea Fleet during the First World War in order to interdict the passage through the Bosphorus for enemy warships and to disrupt the enemy's shipping operations in the southern and south-western parts of the Black Sea.

1914
- Yu.M. Shokalskiy developed a detailed plan of the Black Sea study for a special comprehensive expedition.
- Opening of the Karadag Biological Research Station, which had been built for a long time on very modest means of Moscow University Privat-docent T.I. Viazemskiy.
- L.S. Bagrov published "The list of ancient maps of the Black Sea" in Petrograd (St.-Petersburg).

1914–1915
- The Russian Navy was active in the sea links of Turkey in the region between the Bosporus and the ports of the southern coast of the Black Sea (mines' placing, firing ports).

1915

- Bosporus campaign of a Black Sea Fleet squadron going to the Bosporus.
- Publication of the "Physical-geographical survey of the Black and Azov Seas".
- The world's first underwater minelayer "Crab" (sunk in 1919) became part of the Black Sea Fleet.
- A secret Anglo-French-Russian agreement on the Black Sea straits (cancelled in 1917).

1916

- Opening of the meteorological department of the Headquarters of the Russian Black and Azov Seas Fleet.
- At the request of the Black Sea Fleet Commander Admiral Andrey Augostovich Eberhardt there was formed a hydrographic expedition of the Black Sea, in technical aspects subordinate to the Principal Hydrographic Department, and in other aspects during the war—to the Black Sea Fleet Commander. Its main task was to perform hydrographic survey near the Anatolian coast of Turkey occupied by the Russian troops (Head—Major General A.M. Bukhteev).
- The newest battleship "Imperatritsa Mariya" ("Empress Maria") was blown up at Apollonov Pier in Sevastopol; it had come into commission 1 year before.
- A.V. Kolchak was made Rear Admiral and appointed Commander of the Russian Black Sea Fleet.

1917

- Pilot societies, partnerships, associations ceased to exist; pilots were now employees at the Hydrographic Department.
- Foundation of the Ukrainian General Secretariat of Maritime Affairs in Kiev.
- Publication of the classical scientific work by Yu.M. Shokalskiy "Oceanography", which became part of the world history of science.
- The Black Sea after the armistice with Germany was divided by the demarcation line into two parts—the Russian and the Turkish one. The line ran from the Georgievsky delta arm of the Danube to Trapezund which was occupied by the Russian troops.
- Creation of the Ukrainian Soviet Socialist Republic.

1918

- The main part of the Russian Black Sea Fleet was intentionally sunk in the Tsemess Bay, so that not a single ship could be taken by Germany.
- Central Rada adopted the "Temporary Law on the Fleet of the Ukrainian People's Republic".
- The Soviet power was established all over the Crimea.
- German troops took the Crimea, then they were replaced by Anglo-French forces.
- Bessarabia, Bukovina and Transylvania became part of Romania after the fall of the Austria-Hungary.
- Nationalization of ROPiT and the Voluntary Fleet.

- Establishment of the Ukrainian People's Republic.
- Creation of the Black Sea Soviet Republic on the territory of the Black Sea Governorate.
- The first landing troops of the Entente in Odessa.

1918–1921
- The Crimea became the scene of brutal battles of the Civil War and interventions of Imperial Germany, that led to the formation of the Crimean Autonomous Soviet Socialist Republic within the Russian Soviet Federative Socialist Republic.

1919
- Creation of the Black and Azov Sea fishery research station as a public service.
- The Kuban region was occupied by the White Army.
- Creation of the "Armed Forces of South Russia".
- Order of the People's Commissar for Military and Naval Affairs of Ukraine established the Hydrographic Department of the Black and Azov Seas, which was located in Odessa. The Department was an independent body and was not subordinate to the Principal Hydrographic Department as it was difficult to maintain communication with it.
- The Red Army liberated the whole Crimea, except for the Kerch Peninsula.
- Army of General A. I. Denikin occupied the Crimea.
- Creation of the Black and Azov Sea White Fleet, that played an important role in the Crimean saga of 1919–1920.
- Liquidation of the Ukrainian People's Republic.

1920
- Taking of Perekop-Chongar fortifications by the South Front troops of the Red Army under the command of M.F. Frunze (Perekop-Chongar operation).
- The Black Sea Fleet upon arrival in Constantinople after the evacuation from the Crimea was reorganized into a Russian squadron.
- Creation of the Department of Safe Navigation in the Black and Azov Seas (Ubekochernaz).
- Creation of the Black Sea hydrographic unit, which consisted of three teams, with the base in Odessa.
- Creation of the Novorossiysk Naval Base.
- Formation of the government of the South of Russia in April; in November it escaped with the remaining parts of the Russian Army to Constantinople.
- The Russian squadron left Constantinople and arrived in Bizerte, Tunisia.
- The Red Army completely liberated the Crimea from the White Army.
- Foundation of the Admiral S.A. Makarov Nikolaev Shipbuilding Institute.
- Creation of the Novorossiysk Biological Station.

1921
- Signing of the Treaty of Friendship and Brotherhood between the RSFSR and Turkey.

- Opening of the Kerch Ichthyology Laboratory.
- Commander of the Naval Forces of the RSFSR adopted the 1st program of hydrographic research of Ubekochernaz in the Black and Azov Seas, which included a range of activities related to installing navigational facilities on the shore, hydrographic and hydrological studies.
- Publication of the first Black Sea Fleet (after 1917) issue of "Notices to Mariners".

1922

- Creation of the Commission composed of Lev Trotskiy, Karl Radek and Georgiy Chicherin, following the decision of the Political Bureau of the Central Committee of the Russian Communist Party of Bolsheviks for development of the Soviet platform regarding the Black Sea Straits at the conference in Lausanne. Georgiy Vasilyevich Chicherin sent a letter with the theses regarding the Straits to Joseph Stalin and all the members of the Political Bureau.
- International Conference in Lausanne, Switzerland, on the Middle East, with the issue of the Black Sea Straits being included in the agenda.
- All hydrographic expeditions were dissolved; conducting of surveys was now under the supervision of the Department of navigation security in the Black and Azov Seas (Ubekochernaz) that had been formed earlier.
- Creation of the State Black Sea—Azov Sea Shipping Company.
- Azov Sea—Black Sea scientific-commercial expedition led by Nikolay Mikhailovich Knipovich.

1923

- The second stage of the Conference of Lausanne. The participants signed the Peace Treaty between the Allied Powers and Turkey as well as a number of documents attached thereto, including the Straits Convention.
- Creation of the Special-Purpose Underwater Rescue Party ("EPRON") in Sevastopol by order of F.E. Dzerzhinskiy for salvaging of sunken ships and rescue operations.
- Black Sea hydrographic team performed a magnetic survey of the whole sea coast from the mouth of the Dnieper to Batumi, including the Sea of Azov.
- Proclamation of the Turkish Republic.
- The Sevastopol Biological Station of the Academy of Sciences of the USSR started to conduct oceanographic transects along the line of Sarych (Crimea) to Cape Inebolu (Turkey).
- Opening of the "Maritime Museum" in Varna, Bulgaria (now the Naval Museum).

1923–1927

- The Black Sea oceanographic expedition led by Yu.M. Shokalskiy in order to conduct a comprehensive study of the Black Sea. By 1928, the principal research of the sea regime had already been made, but the work continued until 1935.

1924
- I.V. Kurchatov, the future most prominent atomic scientist published his work on the tides of the Black Sea.
- Creation of the Weather Bureau of the Black and Azov Seas in the USSR.

1925
- The Tuzla Spit was broken through during a violent storm.
- Foundation of the pioneer camp "Artek", the most famous children's recreation center of the Crimea and USSR.

1926
- Creation of the marine zoological station in Adjija (Mamaia, Romania) at the initiative of Borca Jon.

1927
- Crimean earthquake completely destroyed Yalta.
- The Kerch Ichthyologic Laboratory was renamed into the Kerch Scientific Fishery Station.

1928
- Creation of the seismic station of the Academy of Sciences of the USSR within the Sevastopol Naval Observatory.
- The Black Sea hydrographic team was moved from Odessa to Sevastopol.
- The Kerch Scientific Fishery Station was named the Azov and Black Sea Research Institute of Raw Materials.

1929
- By the initiative and under the leadership of V.V. Shuleykin in the village of Katsiveli, Crimea, on the Black Sea coast there was created the first in the world stationary marine hydrophysical station for performing research of processes and phenomena in the coastal area.

1930
- Creation of the Odessa Institute of Marine Fleet Engineers.
- The scientific fishery and biological station in Batumi was put into operation.
- The first in the USSR underwater archaeological expedition was conducted under the supervision of Prof. K.E. Grinevich in the area of Chersonesus Cape.

1931
- Signing of the protocol that envisaged responsibilities of the USSR and Turkey to increase their naval forces in the Black Sea without notice to the other party.
- Formation of the Dnieper Military Flotilla (disbanded in 1940) in the USSR.

1932

- Signing of the Treaty on economic and cultural cooperation between the USSR and Turkey.
- The first aerial photographic survey of the Black Sea coast from Cape Chauda to Anapa, the Kerch Strait and the Taman Bay.
- Creation of the "Sovtanker" enterprise in Tuapse—predecessor of the Novorossiysk Shipping Company.
- N.M. Knipovich published two works: "Hydrological research in the Azov Sea" and "Hydrological research in the Black Sea".
- Foundation of the Odessa Hydrometeorological Institute (in Kharkov, then in 1944 it was moved to Odessa).

1933

- The Black Sea oceanographic expedition by order of the Academy of Sciences of the USSR conducted a study of topography and bottom sediments of the Black Sea in connection with the earthquake in the Crimea in 1927.
- The Azov and Black Sea Research Institute of raw materials was transformed into the Azov and Black Sea Institute for Fishery and Oceanography (AzCherNIRO).
- Creation of the experimental ichthyological station in Varna, Bulgaria.

1934

- The scientific fishery station in Odessa was put into operation.

1936

- The heads of the delegations of Bulgaria, France, Great Britain, Greece, Japan, Romania, Turkey, the USSR and Yugoslavia signed the Convention regarding the Regime of the Straits (the Montreux Convention).
- Formation of the system of standard hydrological sections on the Azov Sea.

1937

- The "Paris Commune" battleship was given to the Navy after its modernization—it was a powerful gunship which added bright pages into the heroic annals of the USSR Navy.
- In the Dnieper-Bug mouth there was discovered a crab species which was new for the Black Sea—the Dutch crab.
- Formation of the Krasnodar Krai (territory).

1938

- The first conference on the study of the Black Sea took place in Sevastopol, which was attended by Yu.M. Shokalskiy, N.M. Knipovich, A.D. Archangelskiy.
- Publication of the book by A.D. Archangelskiy, N.M. Strakhov "Geological structure and history of the development of the Black Sea".
- Part of the Danube Delta in Romania was declared a nature reserve.

1939
- Catastrophic earthquake in Turkey that claimed more than 23,000 lives was accompanied by a tsunami.

1940
- Council of People's Commissars of the USSR adopted a decree "On the organization of special marine secondary schools". One of them was opened in Odessa. Such schools were created to form reserves for naval schools.

1941
- Off the western coast of the Crimea there took place large training maneuvers of the Black Sea Fleet.
- The beginning of the Great Patriotic War on 22 June.
- The 73-day-long defense of Odessa by the forces of the Odessa defense area.
- The first and the largest in the history of the Great Patriotic War Kerch—Feodosiya landing operation.
- The first military assault of Sevastopol by the Germans.
- The beginning of the defense of Sevastopol by the Soviet forces. The defense lasted until July 4, 1942.
- At the very beginning of the war there was conducted a raid of a light forces squadron of the Black Sea Fleet at Constanta. The destroyer leader "Moskva" ("Moscow") was sunk during the operation.
- I.V. Kurchatov and P.A. Alexandrov, working in Sevastopol, theoretically substantiated a method of protecting ships from noncontact electromagnetic mines and performed the first test of degaussing ships.
- Almost the whole Crimea was occupied by the Nazis.
- Publication of the fundamental monograph by V.V. Shuleykin "Physics of the Sea".
- Special-Purpose Underwater Rescue Party ("EPRON") joined the emergency rescue service of the USSR Navy.
- The first in the history of the Great Patriotic War amphibious landing operation took place near the fishing village of Grigorievka near Odessa.

1942
- Evpatoria landing operation.
- The cruiser "Krasny Kavkaz" ("Red Caucasus") came into operation after the plant had removed emergency and battle damages.
- The Novorossiysk defense operation—August 19–September 26.
- A raid was made at Feodosiya. During the operation the cruiser "Molotov" was made inoperative for almost a year.
- The biggest weapon of World War II—cannon "Dora" (caliber of 800 mm) was delivered near Bakhchisarai.
- The Kerch operation "Bustard Hunt"—military advance of the 11th German Army under the command of General Erich von Manstein on the Kerch Peninsula, taking of Kerch.

- Liquidation of the Crimean group of German-Fascist troops was finished on Chersonesus Cape.
- The start and the end of the operation "Sturgeon Catch" on taking Sevastopol by the 11th Army of Erich von Manstein.
- Beginning of the transfer of German small submarines from Germany to the Black Sea.
- Italian Admiral Arturo Riccardi signed an agreement with the Germans, according to which "Italian light forces" would be used to help the German Air Force in the Black Sea.
- Novorossiysk fell as a result of the Nazi military advance in the direction of the Caucasus.

1943
- On the basis of the marine department of the State Hydrological Institute there was founded the State Oceanographic Institute (GOIN)—the central research, scientific and methodological institution of the Main Department of the Hydrometeorological Service of the USSR in the field of oceanography and marine forecasts.
- Additionally the map-making Southern institution was created in order to provide maps for the Black Sea Fleet.
- Leader "Kharkov" and destroyers "Sposobnyi" ("Able") and "Besposhadnyi" ("Merciless") raided Feodosiya and Yalta. The operation resulted in sinking of three large ships and death of 692 people.
- Conducting of a combat operation, which was unique in the history of the Black Sea Fleet—simultaneous destruction of two ships by one salvo from the "C-33" submarine under the command of Soviet Captain II Rank B.A. Alekseev.
- Kerch—Eltigen amphibious operation.
- The Black Sea landing operation by Tsesar Kunikov at "Malaya Zemlya" near Novorossiysk.
- Failed landing operation of Soviet troops in Yuznhaya Ozereika north of Novorossiysk.
- V.V. Shuleykin opened the Department of Physics of the Sea at the Moscow State University and was its head until 1965.
- Novorossiysk was freed after the defense that lasted for 225 days.

1944
- Creation of the Danube Military Flotilla.
- The Soviet Union declared war on Bulgaria.
- The Black Sea Fleet landed troops in the area of Varna.
- A squadron of Soviet marine troopers seized a bridgehead in the Black Sea port of Nikolaev (Nikolayev bridgehead).
- Operation of the NKVD (People's Commissariat for Internal Affairs) of mass deportation of Crimean Tatars, Bulgarians, Greeks, Gypsies into the regions of Central Asia and Siberia. Political rehabilitation took place in 1967 and massive return of deported peoples—since 1989.

- An official armistice was signed between the USSR, USA and Great Britain on the one hand, and Romania—on the other.
- September 9—the official date of the termination of armed hostilities of the Black Sea Fleet.
- V.P. Zenkovich developed a new method to depict underwater topography, based on drawing isobaths not according to the method of linear interpolation, but on the basis of geomorphological understanding of the bottom.
- Odessa was freed from Nazi invaders.
- The Higher Marine Engineering School (later Odessa National Marine Academy) was founded in Odessa.

1945
- The Yalta Conference (The Crimea Conference) of the Heads of the governments of the USSR, USA and Great Britain defined the postwar world: there were adopted decisions on the division of Germany into occupation zones and on reparations, on the USSR participation in the war with Japan, on the postwar system of international security and on the establishment of the United Nations.
- Decree of the Presidium of the Supreme Soviet of the USSR on the transformation of the Crimean Autonomous Soviet Socialist Republic into the Crimean region within the RSFSR.
- The beginning of the inventory of the Black Sea under the leadership of V.P. Zenkovich.
- Liberation of Ukraine by the Soviet Army.
- The issue of the Black Sea Straits was on the agenda at the Potsdam Conference of Heads of Government.

1946
- Creation of the P.P. Shirshov Institute of Oceanology of the USSR Academy of Sciences.
- The Hydrometeorological Service of the Black Sea Fleet was renamed into the Marine Observatory; regional hydrometeorological stations were subject to it.
- Functions of the pilot service of the Azov and Black Sea basin by a joint decision of the People's Commissars of the Navy Fleet and the Ministry of the Sea Fleet were transferred to the administration of commercial ports (except for military pilot sites of the ports of Bulgaria, Romania and Sevastopol).
- Foundation of the shipyard "Okean" ("Ocean") in Nikolaev.
- An unknown large gastropod—Rapana—was found near Novorossiysk.

1947
- Paying tribute to the scientific and organizational merits of the academician A.O. Kovalevskiy, his name was given to the Sevastopol Biological Station.
- In Romania the Monarchy was brought down and the People's Republic was proclaimed.

1948

- Commencement of work of the Black Sea Scientific and Industrial VNIRO expedition.
- The Marine Hydrophysical Institute (MHI) was founded in Moscow on the basis of the Black Sea hydrophysical station of the Academy of Sciences of the USSR and Marine hydrophysical laboratory.
- Reconstruction of the Sevastopol Biological Station after the war.
- The Danube Commission was established instead of the European Danube Commission.

1949

- Creation of the Black Sea experimental research station (CHENIS) of the Institute of Oceanology of the USSR Academy of Sciences in the Golubaya ("Blue") (Rybatskaya, "Fisherman's") Bay near Gelendzhik. One of the first research vessels of the Institute of Oceanology "Forel" ("Trout") was obtained in the framework of reparations from Romania and was transferred to the ownership of the CHENIS.
- In the Black Sea the research ship "Vityaz" made its first scientific voyage.

1950

- In the memorial church erected in the memory of Russian sailors in Bizerte, Tunisia, there was made a marble plaque on which there were carved the names of all the ships that came from the Crimea.
- Nazim Hikmet, a famous poet, escaped from Turkey by boat across the Black Sea to the USSR.

1951

- Japan under the San Francisco Peace Treaty left the Montreux Convention.

1952

- Turkey joined NATO.
- 12 Soviet torpedo boats of the Black Sea Fleet were given to the Navy of the Romanian People's Republic.
- Foundation of the port city of Ilyichevsk in the Odessa region.

1953

- Creation of the Department of Oceanology in the Moscow State University.
- Opening of the Kerch ferry service.
- Foundation of the Institute of Fishery Research in Istanbul, Turkey.

1954

- Creation of the Odessa Marine School.
- Decree of the Presidium of the Supreme Soviet of the USSR "On the transfer of the Crimean region from the RSFSR to the Ukrainian Soviet Socialist Republic" during the celebration of the 300th anniversary of the reunification of Ukraine with Russia. Crimea became a region within Ukraine.

- Compilation of the inventory of the Black Sea was finished.
- The "Atlas of morphology and dynamics of the Black Sea shores" was created manually.
- Creation of the Lower Danube River Administration (RAND) for hydrotechnical works and regulation of shipping.
- Beginning of the submarine fleet of the People's Republic of Bulgaria (the USSR gave three small submarines of the XV series to Bulgaria).

1955
- Sinking of the battleship "Novorossiysk" in the Sevastopol Bay.

1956
- Publication of the "Climatic and hydrological atlas of the Black and Azov Seas".

1957
- Prof. V.D. Blavatskiy organized an underwater expedition of the Institute of Archaeology of the Academy of Sciences of the USSR at the Black Sea coasts.
- Beginning of the construction of a top-secret facility #825 at Balaklava for basing submarines.

1959
- Trilateral agreement on fisheries in the Black Sea.

1960
- At the 11th Session of the UNESCO General Conference it was decided to establish the Intergovernmental Oceanographic Commission (IOC). The proposal to establish the IOC was made by the USSR delegates at the Copenhagen Conference on Oceanographic Research.
- Creation of the Danube Hydrometeorological Observatory in Izmail.

1961
- The Marine Hydrophysical Institute was transferred into the system of the Academy of Sciences of the Ukrainian Soviet Socialist Republic.

1962
- Publication of the monograph by Yu.I. Sorokin "Black Sea: nature, resources".
- Publication of the "Ice atlas of the Black and Azov Seas".

1963
- The Marine Hydrophysical Institute of the Academy of Sciences of the Ukrainian Soviet Socialist Republic was relocated to Sevastopol.
- The Presidium of the Academy of Sciences of the USSR approved the establishment of the Institute of Biology of the Southern Seas of the Academy of Sciences of the Ukrainian Soviet Socialist Republic (IBSS) with the center in the city of Sevastopol.

1964
- The terminal—oil depot "Sheskharis", built on the outskirts of Novorossiysk, came into operation.
- Archaeological research on Zmeinyi Island ("Snake Island") was conducted under the direction of N.V. Pyatysheva.
- A squadron of warships of the Black Sea Fleet for the first time came into the Mediterranean Sea to monitor the aircraft carrier strike group of the US Navy.

1965
- Foundation of the museum of the Navy of Ukraine.
- Foundation of the specialized seaport "Oktyabrsk" on the bank of the Dnieper-Bug mouth.
- Hero-City Sevastopol was awarded the Order of Lenin and the medal "Gold Star".
- Romania was declared a Socialist Republic.
- Hero-City Odessa was awarded the Order of Lenin and the medal "Gold Star".
- The "Catalogue of level observations in the Black and Azov Seas" was compiled.

1966–1967
- Research work of the underwater laboratories of the Leningrad Hydrometeorological Institute "Sadko-I" and "Sadko-II" at the depth of 25 m near Sukhumi.

1966–1968
- Experiments near the Crimean coast in the underwater houses "Ichthyandr-67" and "Ichthyandr-68".

1966
- A sand gaper, or Mya (Mya arenaria), was found in the Black Sea.
- The "Atlas of hydrological characteristics of the northwestern part of the Black Sea" was compiled.

1967
- Two scuba divers during 24 h were in the underwater house "Gebros-67" at the depth of 10 m in the waters of the Gulf of Varna. They performed different physiological and psychological research.
- Foundation of the Georgian Shipping Company in Batumi.
- Foundation of the Novorossiysk Shipping Company.
- Foundation of the Black Sea Shipping Company in Odessa.
- CHENIS (The Black Sea experimental research station) was transformed into the Southern Branch of the Institute of Oceanology of the USSR Academy of Sciences.
- For the first time the blue crab was found at the Bulgarian coast.
- Foundation of the Azov Shipping Company.

1967–1968
- Oceanographic work was carried out on an experimental pier on Donuzlav Spit near the port of Mirnyi.

1969
- Research work of the underwater laboratory of the Leningrad Hydrometeorological Institute "Sadko-III" at the depth of 25 m near Sukhumi.
- A comprehensive 7-week voyage on the research vessel "Atlantis II" of the Woods Hole Oceanographic Institution (USA) on the Black Sea.
- The "Atlas of waves and wind of the Black Sea".

1970
- Foundation of the Romanian Marine Research Institute in Constanta.

1971
- In connection with the 100th anniversary of the Sevastopol Biological Station (SBS) and for the achievements in the development of the biological science the A.O. Kovalevskiy Institute of Biology of the Southern Seas of the Academy of Sciences of the Ukrainian Soviet Socialist Republic was awarded the Order of the Red Banner of Labour. In memory of the outstanding scientists there were put portrait sculptures of N.N. Miklukho-Maklay and A.O. Kovalevskiy in front of the Institute.
- Foundation of the Southern Scientific Center of the Academy of Sciences of the Ukrainian Soviet Socialist Republic in Odessa.

1972
- Publication of the first issue of the "Water Resources" journal of the Academy of Sciences of the USSR, which later paid attention to the problems of the Black Sea.
- On the basis of the part of the forces of the Soviet Black Sea Fleet, the Navy of Bulgaria and Romania there was created the Joint Black Sea Fleet.
- The Soviet Black Sea Fleet gave six torpedo boats to the Navy of Bulgaria.

1973
- Signing of the International Convention for the Prevention of Pollution from Ships (amended in 1978)—the main international convention regulating protection of the marine environment from pollution from ships (MARPOL—78).
- The first Bulgarian—Soviet experiment in the manned Soviet underwater laboratory "Chernomor", entitled "Shelf—Chernomor—73".
- Foundation of the Institute for Marine Research and Oceanology in Varna, Bulgaria.
- Kerch was awarded the title "Hero-City".
- Beginning of the construction of the sea commercial port "Yuzhnyi".
- Tuzla Island following the decision of the Krasnodar Krai Executive Committee was now under the administrative control of the Crimea.

– Creation of the Danube branch of the Black Sea State Reserve.
– Novorossiysk was awarded the title "Hero-City".

1974

– The second Bulgarian-Soviet experiment in the manned Soviet underwater laboratory "Chernomor", entitled "Chernomor—74".
– The port of Varna-West was built 30 km from the city of Varna—the most modern port of Bulgaria.
– The "Handbook on the Climate of the Black Sea" was compiled.

1975

– The special drilling vessel "Glomar Challenger" (USA) worked in the Black Sea.
– The research vessel "Chain" from the USA made a short voyage in the Black Sea.
– The Post Number One was opened in Novorossiysk on the 30th anniversary of the victory over Nazi Germany on the square, where the graves of the Heroes of the Soviet Union Ts.L. Kunikov and N.I. Sipyagin are situated.
– The territory of the Danube branch of the Black Sea State Reserve was identified as wetlands of international importance.
– Signing of the agreement between the USSR and Bulgaria on organization of the ferry service between Ilyichevsk and Varna.

1976

– Adoption of the Resolution of the Central Committee of the Communist Party of the Soviet Union (CPSU) and the Council of Ministers of the USSR "On measures to prevent pollution of the Black and Azov Sea basins".
– The seaport on the Dniester estuary was built in the city of Belgorod-Dnestrovskiy.
– Foundation of the Bulgarian Ship Hydrodynamics Centre.

1976–1978

– Implementation of the joint program of the Complex Research of the Black Sea (SKOICh) involving major organizations of the Black Sea.

1977

– The polygon Kamchiya of the Institute of Oceanology of the Academy of Sciences of Bulgaria came into service.
– The port of Ust-Dunaisk was founded in the Zhebriyanskaya Bay of the Danube Delta (Odessa region).

1977–1979

– International experiment at the polygon "Kamchiya" in Bulgaria under the program of the CMEA member countries.

1978
- Jacques-Yves Cousteau visited the Black Sea.
- Publication of the compilation book "Cherno more" ("The Black Sea") in Bulgaria, later (in 1983) it was published in Russian as "Chernoe more" ("The Black Sea").
- Italy joined the Montreux Convention.
- Publication of the works of Ovid "Tristia. Epistulae ex Ponto" ("Sorrows. Letters from the Black Sea").
- Publication of the paper by A.V. Fedorov "Ponto-Caspian Pleistocene".
- Foundation of the sea commercial port "Yuzhnyi" 20 km from Odessa in the Grigoryevskiy mouth. One of the three largest ports of Ukraine.

1980
- The nudibranch mollusk Doridella was for the first time detected in the north-western part of the Black Sea.
- Beginning of the formation of joint squadrons of the ships of the Soviet Black Sea Fleet and the Navy of the People's Republic of Bulgaria in the Black Sea.
- The canal "Danube—Sasyk" was built.

1981
- Tuapse for the courage and fortitude shown on a massive scale by its citizens during the Great Patriotic War was awarded the Order of the Patriotic War.
- Storm on the Black Sea near Cape Tarkhankut; waves washed away the floor of the oil platform at the height of 14 m above sea level.
- The scientific-production association "Gruzberegozaschita" was created on the basis of the laboratory of the sea coastal zone of the Institute of Geography of the Academy of Sciences of Georgia.
- Foundation of the State Reserve of the Academy of Sciences of Ukraine "Danube flooded lands".
- Publication of the book by V.N. Stepanov, V.N. Andreev "The Black Sea, resources and problems".

1982
- Signing of the UN Convention on the Law of the Sea.
- Turkey introduced new rules of navigation on the Bosphorus.
- The Ctenophore Mnemiopsis leidyi was detected in the northern part of the Black Sea.

1983
- Creation of the Black Sea State Biosphere Reserve.

1984
- Irish traveler and writer Tim Severin on the sailing oared ship "Argo" repeated the trip of the Argonauts.

– The ancient city of Acra founded by the Hellenes in the sixh century B.C. was found at the bottom of the Gulf of Kerch.

1985

– The Institute for Marine Research and Oceanology in Varna, Bulgaria, was transformed into the Institute of Oceanology.

1986

– Collision of the bulk carrier "Petr Vasev" with the passenger liner "Admiral Nakhimov" near Novorossiysk. The latter sank claiming 423 lives.
– The cruiser "Yorktown" and the destroyer "Caron" of the US Navy appeared at the coast of the Crimea near Feodosiya.

1987

– Publication of the book by M.V. Agbunov "Antique pilot of the Black Sea".
– Creation of the State Commission on the Crimean Tatars (Chairman—A.A. Gromyko).

1988

– International geological expedition to the Black Sea aboard the "Knorr", USA, that consisted of five trips.
– The Azov and Black Sea Research Institute for Fishery and Oceanography got a new name—YugNIRO.
– The cruiser "Yorktown" and the destroyer "Caron" of the US Navy appeared again in the Black Sea. They were going along the Soviet coast from the side of Sevastopol, violating the state border. The patrol ship "Bezzavetnyi" ("Selfless") twice rammed "Yorktown", after which the American ships came out of the Soviet territorial waters.

1989

– An uprising in Romania resulted in the overthrowing of the government, execution of Ceausescu, change of the country's name into Romania.

1990

– Adoption of the Declaration of the National Sovereignty of Ukraine.
– Publication of the Bulgarian-Soviet monograph "Practical Ecology of Marine Regions of the Black Sea".

1991

– Establishment of the program of joint scientific research of the Black Sea (CoMSBlack).
– Publication of the volume "The Black Sea" in the framework of the project "Sea" of the scientific—technical program "The World Ocean" of the USSR State Committee for Science and Technology.
– The Danube Delta was added to the UNESCO World Heritage List.

- The Supreme Council of the Ukrainian Soviet Socialist Republic declared the independence of Ukraine.

1992
- Creation of the Black Sea Fleet of Ukraine.
- Signing of the "Declaration on the Black Sea Economic Cooperation—BSEC" by the Heads of State and Government of Albania, Armenia, Azerbaijan, Bulgaria, Georgia, Greece, Moldova, Romania, Russia, Turkey, Ukraine in Istanbul.
- The Black Sea countries signed the Bucharest Convention on the Protection of the Black Sea Against Pollution.
- Creation of the JSC "Novorossiysk Commercial Sea Port".
- An extraordinary storm on the Black Sea. Large waves eroded beaches of the Crimea and flooded many ships in the port of Yalta. Losses exceeded 2.5 billion Rubles (at 1992 values).
- The Danube Delta became a biosphere reserve under the umbrella of UNESCO.
- Publication of the book "Peniteli Ponta" by E.V. Venikeev, L.T. Artemenko in Simferopol.
- From Poti to Novorossiysk there were drawn out 18 combat ships and boats; 3 marine vessels for various purposes; 17 roadsters and boats. Navy ships and vessels brought about five thousand military servants, members of their families, etc.
- Publication of the paper "The Black Sea ecosystem" by M.E. Vinogradov et al.

1993
- Creation of the Black Sea Environmental Programme (BSEP).
- Creation of the Black Sea Environmental Series for publication of works performed under the Black Sea Environmental Programme.
- The Odessa Declaration on the Black Sea protection by the Ministers of Environment.
- The Turkish government announced a review of the ships passage through the Bosporus and Dardanelles Straits due to "environmental security".
- Presidents of Ukraine and Russia signed an agreement on urgent measures for the formation of the Naval Forces of Ukraine and the Naval Fleet of Russia on the basis of the Black Sea Fleet.
- A large landing ship (BDK-69) and the "Konstantin Olshanskiy" ship carried out the delivery of humanitarian aid to the area of the Georgian-Abkhaz conflict and took refugees from the port of Sukhumi.
- At the request of the Government of Georgia a tactical operations group under the flag of Black Sea Fleet Commander Admiral E.D. Baltin was sent to the area of Poti to participate in restoring the constitutional order in Poti.

1994

- Creation of the Black Sea Regional Activity Centre for Environmental Aspects of Fisheries and other Marine Living Resources Management (Constanta, Romania).
- Turkey has introduced navigation regulations in the area of the Straits and the Sea of Marmara.
- An extraordinary storm at the Black Sea disrupted the planned NATO maneuvers.
- A collision of the crude oil tanker "Nassia" with the bulk carrier "Shipbroker" in the Bosporus. 34 people died. 9,000 tons of oil spilled and burnt. The strait was closed for several days.
- Foundation of the Black Sea Trade and Development Bank.
- The Maritime Safety Committee of the UN International Maritime Organization adopted a set of international rules for the passage through the Straits.
- Publication of the book "Accidents in the Black Sea" by E.F. Shnyukov, L.I. Mitin, V.P. Tsemke in Kiev.
- Creation of the Inter-departmental Commission on the Black Sea for implementation of the "Convention on the Protection of the Black Sea Against Pollution".

1995

- Publication of the "Black Sea Bibliography" (1974–1994) in the United States.
- The Turkish ferry "Avrasiya" was captured by Chechens in the Black Sea port of Trabzon.
- The workshop NATO-ARW «Sensitivity to Change: Black Sea, Baltic Sea and North Sea".
- Publication of the book "Pirates of the Black Sea" by E.F. Shnyukov in Kiev.

1996

- The Strategic Action Plan for the Rehabilitation and Protection of the Black Sea was approved in Istanbul—the Black Sea Day.
- The Moscow meeting of the Heads of State and Government of the participating countries of the BSEC. There was adopted a strategic political decision to transform the BSEC into an international organization of regional cooperation.
- Creation of the Black Sea Regional Commission (BSRC) within the Intergovernmental Oceanographic Commission.
- Creation of the Black Sea Regional Programme IOC/UNESCO.
- The Transboundary Diagnostic Analysis of the Black Sea was compiled.
- Creation of the independent group of observers of problems of the straits in Turkey.
- The first in the history of the Ukrainian state large-scale exercises of the Ukrainian Navy "More-96" ("Sea-96").
- Due to the worsening of the Georgian-Abkhaz conflict the Black Sea Fleet allocated a tactical group with a reinforced squadron of Marine Corps to assist ships of certain brigades of border patrol ships of the Russian Federal Border Service in the blockade of the Abkhazian coast.

– The Coordination Center of the BSEP in Miskhor held the first international conference on sustainable tourism development in the Black Sea region.

1997
– A meeting of the BSEC Foreign Ministers with the participation of Ministers of Economy in Istanbul. There was signed a declaration of intentions to create a Black Sea free trade zone and there was adopted an Action Plan for the implementation of provisions of the Moscow summit.
– The presidents of Russia and Ukraine signed the Treaty of Friendship, Cooperation and Partnership between Ukraine and the Russian Federation.
– There was signed an agreement on the status and conditions of the stay of the Black Sea Fleet of the Russian Federation on the territory of Ukraine.

1998
– The largest protest of representatives of 300 NGOs took place in the Bosporus against Caspian oil transportation through the Strait.
– Creation of the Danube Biosphere Reserve.

1999
– Creation of the Oceanological Center of the National Academy of Sciences of Ukraine, which united the Marine Hydrophysical Institute and its departments in Odessa and in the village of Katsiveli (Crimea), Institute of Biology of the Southern Seas of the National Academy of Sciences of Ukraine, Odessa branch of the Institute of Biology of the Southern Seas.
– The Russian tanker "Volganeft-248" crashed in the Bosporus with 43,000 tons of fuel oil. Almost 1,500 tons of oil spilled to the sea.
– Publication of the book "Environmental Degradation of the Black Sea: Challenges and Remedies" in the NATO Science Series.
– The International Coordinating Committee of the UNESCO Programme "Man and Biosphere" decided to recognize the estuarine territories of the Danube as the unified Romanian- Croatian biosphere reserve.
– Publication of the book "Russia and the Black Sea Straits" (eighteenth to twentieth centuries) in Moscow.
– Foundation of the Black Sea branch of the M.V. Lomonosov Moscow State University in Sevastopol.

2000
– The Verkhovna Rada of Ukraine adopted the law "On special (free) economic zone on the territory of the Odessa Sea Commercial Port".
– Creation of the Southern Federal District (originally North Caucasian District).
– Creation of the "Gosgidrografiya" institution ("State Hydrography") by order of the Ministry of Transport of Ukraine.

Twenty-First Century

2001
- President of the Russian Federation approved the Maritime Doctrine of the Russian Federation.
- The "National Programme for the protection and restoration of the environment of the Azov and Black Seas" was adopted and approved in Ukraine.

2002
- Completion of the construction of the world deepest gas pipeline along the Black Sea bottom "Blue Stream" (from Dzhubga, Russia to Samsun, Turkey).
- Tornado on the Russian coast of the Black Sea, which claimed hundreds of people.
- Publication of the book "The Black Sea ecology and oceanography" by Yu. Sorokin in the Netherlands.
- The French oil and gas company "Total" is involved in the exploration of hydrocarbons in the area of the Val Shatskiy oil field in the deep part of the Black Sea.
- Publication of the collection book "Comprehensive study of the northeastern part of the Black Sea".

2002–2003
- Negotiations between representatives of Russia and Ukraine on the legal status of the Sea of Azov and the Kerch Strait and delimitation of the Black Sea.

2003
- The Agreement on the Ukrainian-Russian state border was signed between the Russian Federation and Ukraine.
- Constanta became a free port.
- Construction of a dam in the Kerch Strait between the Island of Tuzla and land o fRussia.
- Beginning of territorial disputes on the Kerch Strait between Russia and Ukraine.
- The French oil and gas company "Total" signed an agreement with "Rosneft" on joint development and exploration of fields in the Tuapse Trough area of the Black Sea.
- Foundation of the Southern Scientific Center of the Russian Academy of Sciences in Rostov, which conducts oceanographic research expeditions in the Black and Azov Seas.
- A conference on the development of the port of Batumi was held in the Paris office of the World Bank.
- The Cabinet of Ministers of Ukraine approved the feasibility study for construction of the Danube-Black Sea deep-water fairway.

2004
– Romania joined NATO.

2005
– Russian President's visit to Samsun (Turkey) for the official opening of the "Blue Stream" gas pipeline.
– The international scientific conference "Modern state of ecosystems of the Black and Azov Seas", Donuzlav, Crimea.
– The traffic was blocked in the Bosporus for a few days because a bulk carrier with tanks of liquefied gas had sunk during a storm.

2006
– Large-scale maneuvers of the Russian Black Sea Fleet.
– The summit of the Black Sea Forum for Dialogue and Partnership was held in Bucharest.
– Russia and Italy announced drafting of the South European Gas Pipeline ("Southern Stream"), which is to be put at the bottom of the Black Sea.
– Ukraine resumed working on the Danube-Black Sea deep-water fairway.
– Publication of the book "Black Sea Encyclopedia" by S.R. Grinevetsky, I.S. Zonn, S.S. Zhiltsov by the "International Relations" publishing house in Moscow.
– The gas pipeline project Baku-Tbilisi-Erzurum was implemented.
– The EU has put forward a regional cooperation initiative for the Black Sea region—"Black Sea Synergy".
– Publication of the "Millennium around the Black Sea" by D.M. Abramov in Moscow.

2008
– Ukrainian President Viktor Yushchenko ordered the government to prepare a draft law on the termination of intergovernmental agreements on the temporary stay of the Russian Fleet on the territory of Ukraine starting from 2017.
– Turkey launched an initiative to create the Caucasus Stability and Cooperation Platform.
– The Charter on strategic partnership between Ukraine and the United States was signed in Washington.
– Publication of the book "The Black Sea Environment" by Kostianoy A.G. and Kosarev A.N. (Eds.) in Springer, Germany.

2009
– The UN International Court of Justice in The Hague defined the demarcation line between the continental shelf and the exclusive economic zone of Ukraine and Romania in the northwestern part of the Black Sea.
– Russia and Turkey agreed to establish a working group to develop a pipeline project called "Blue Stream-2" with the pipeline to be put along the bottom of the Black Sea.

- The international energy summit in Sofia adopted the Declaration on support of gas pipeline projects in the direction of Southern Europe. It emphasized the geopolitical importance of the Black Sea region.
- An intergovernmental agreement between Turkey, Hungary, Romania, Bulgaria and Austria to support the "Nabucco" project was signed in Ankara (Turkey).

2010
- Publication of the book "The Black Sea Region: a new shape" by I.S. Zonn and S.S. Zhiltsov in Moscow.
- Russia and Ukraine signed an agreement on the extension of the stay of the Black Sea Fleet of the Russian Federation in Sevastopol (Crimea, Ukraine) until 2042.
- The US company "Chevron" signed an agreement with the Russian company "Rosneft" to develop the Black Sea shelf (the Val Shatskiy oil field).
- The Second Black Sea Energy and Economic Forum was held in Istanbul (Turkey).

2012
- The traditional international naval exercises of the "Black Sea Naval Cooperation Task Group—BLACKSEAFOR" took place at the Black Sea.
- The Russian company "Rosneft" and the Italian company "Eni" in the framework of the previous agreements signed a financial agreement on development of the Black Sea areas.

2013
- President of the Russian Federation approved the concept of the Russian foreign policy, which reflects the importance of the Black Sea region.
- The Black Sea Oil and Gas conference took place in Istanbul (Turkey).

2014
- Dramatic political events in the Ukraine, the power overturn, and destabilization of the internal situation in the country in February and March 2014 led to a change of political map of the Black Sea Region. On 11 March 2014 Parliament of Autonomous Republic of Crimea adopted a Declaration of independence. On 16 March a referendum on the status of the Crimea took place. The official result from the Autonomous Republic of Crimea was a 96.77 % vote for integration of the region into the Russian Federation with an 83.1 % voter turnout. Based on these results, on 18 March 2014 the Crimea became a part of the Russian Federation, including the appropriate sea waters in the Black and Azov seas.
- On 12 May 2014 Donetsk and Lugansk People's Republics (former regions of the Ukraine) declared independence from the Ukraine.
- Civil war in the Ukraine.

List of Abbreviations

abs.	Absolute height over the World Ocean level (in meters)
A.D.	Anno Domini
ADB	Asian Development Bank
AS	Academy of Sciences
ASSR	Autonomous Soviet Socialist Republic
B.C.	Before Christ
C.A.	Central Asia
CC (CPSU)	Central Committee of the Communist Party of the Soviet Union
CIS	Commonwealth of Independent States
CPSU	Communist Party of the Soviet Union
cu.	cubic
EBRD	European Bank for Reconstruction and Development
ECO	Economic Cooperation Organization
ESCAP	The United Nations Economic and Social Commission for Asia and the Pacific
EU	European Union
FAO	Food and Agriculture Organization of the United Nations
GDP	Gross Development Product
GEF	Global Environment Facility
GGI	State Hydrological Institute (Russia)
GIS	Geographic Information System
GKNT (SCST)	The USSR State Committee for Science and Techniques
GNP	Gross National Product
GOIN	State Oceanographic Institute (Russia)
ha	hectare
IDB	Islamic Development Bank
HMS	Hydro-meteo station
HPS	Hydroelectric power station
IBRD	International Bank for Reconstruction and Development
ICID	International Commission on Irrigation and Drainage
ICSD	Interstate Commission for Sustainable Development
ICWC	Interstate Coordination Water (management) Commission

(continued)

S.R. Grinevetsky et al., *The Black Sea Encyclopedia*,
DOI 10.1007/978-3-642-55227-4, © Springer-Verlag Berlin Heidelberg 2015

INCO-Copernicus	The INCO-COPERNICUS programme promotes scientific and technological cooperation with the countries of Central Europe and the new independent states of the former Soviet Union
INTAS	International Association for the promotion of co-operation with scientists from the New Independent States of the former Soviet Union
IO RAS	P.P. Shirshov Institute of Oceanology, Russian Academy of Sciences
JSC	Joint stock company
m abs. elev.	Absolute height over the World Ocean level (in meters)
MSU	Moscow State University (Russia)
NATO CLG	NATO Cooperative Linkage Grant
NGO	A Non-governmental organization
OECD	Organisation for Economic Co-operation and Development
OIC	Organisation of the Islamic conference
OPEC	Organization of the Petroleum Exporting Countries
OSCE	Organization for Security and Co-operation in Europe
RAS	Russian Academy of Sciences
RF	Russian Federation
RFBR	Russian Foundation for Basic Research
RGS	Russian Geographical Society
RSFSR	Russian Soviet Federal Socialist Republic
SCO	The Shanghai Cooperation Organization
SCP	State Committee for Planning (USSR)
SCST (GKNT)	The USSR State Committee for Science and Technology
SNK	Council of People's Commissars (USSR, 1923–1946)
SOPS	Council for Investigation of Production Forces
sq.	square
SSR	Soviet Socialist Republic
TACIS	Technical Assistance for the Commonwealth of Independent States
thou	thousand
TSFSR	Transcaucasian Socialist Federative Soviet Republic
UN	United Nations
UNDP	The United Nations Development Programme
UNEP	The United Nations Environment Programme
UNESCO	United Nations Educational, Scientific and Cultural Organization
UNIDO	United Nations Industrial Development Organization
USAID	United States Agency for International Development
USSR	Union of Soviet Socialist Republics
USSR AS	The USSR Academy of Sciences
VKP(b)	All-Union Communist Party of bolsheviks (USSR, 1925–1952)
VNIIMORGEO	All-Union Scientific Research Institute for Marine Geology and Geophysics (Russia)
VNIRO	All-Russia Scientific Research Institute for Fishery and Oceanography
VSNKh	Supreme Council of the Russian (1917–1932) and the Soviet Union (1963–1965) National Economy
WMO	World Meteorological Organization

Bibliography

Altman, E. N., Gertman, I. F., & Golubeva, Z. A. (1987). *Climatic fields of the Black Sea water salinity and temperature*. Sevastopol: Gosudarstvennyi Okeanograficheskii Institut, Sevastopolskoye Otdeleniye (in Russian).

Babii, M. V., Bukatov, A. E., & Stanichny, S. V. (2005). *Atlas of the Black Sea surface temperature based on satellite data (1986–2002)*. Sevastopol: Marine Hydrophysical Institute (in Russian).

Besiktepe, S. T., Ünlüata, Ü., & Bologa, A. S. (Eds.). (1999). Environmental degradation of the Black Sea: Challenges and remedies. In *Proceedings of the NATO advanced research workshop* (NATO science partnership sub-series: 2, Vol. 56, 408 p), Constanta-Mamaia, Romania, October 6–10, 1997, Springer.

Bezborodov, A. A., & Eremeev, V. N. (1993). *Interaction zone between oxic and anoxic waters*. Sevastopol: Marine Hydrophysical Institute (in Russian).

Blatov, A. S., Bulgakov, N. P., Ivanov, V. A., Kosarev, A. N., & Tuzhilkin, V. S. (1984). *Variability of hydrophysical fileds in the Black Sea*. Leningrad: Gidrometeoizdat (in Russian).

Bogdanova, A. K., Dobrzhanskaya, M. A., & Lebedeva, M. N. (1969). *Water exchange through the bosporus strait and its influence on hydrology and biology of the Black Sea*. Kiev: Naukova Dumka (in Russian).

Degens, E. T., & Ross, D. A. (Eds.). (1974). *The Black Sea: Its geology, chemistry and biology*. American Association of Petroleum Geologists Memorial 20, Tulsa, OK.

Dumont, H. J., Shiganova, T. A., & Niermann, U. (Eds.). (2004). *Aquatic invasions in the Black, Caspian, and Mediterranean Seas* (NATO science series: IV earth and environmental sciences, Vol. 35, 313 p). Springer.

Eremeev, V. N. (1996). Hydrochemistry and dynamics of the hydrogene-sulphide zone in the Black Sea. In P.C. Griffiths (Ed.), UNESCO Rep Mar Sci 69, UNESCO.

Eremeev, V. N., Chudinovskikh, T. V., & Batrakov, G. F. (1993). Artificial radioactivity of the Black Sea. UNESCO Rep Mar Sci 59, UNESCO.

Fet, V., & Popov, A. (Eds.). (2007). *Biogeography and ecology of Bulgaria* (Monographiae Biologicae, Vol. 82, 690 p). Springer

Filippov, D. M. (1968). *Water circulation and structure of the Black Sea*. Moscow: Nauka (in Russian).

Goncharov, V. P., Neprochnov, Y. P., & Neprochnova, A. F. (1972). *Topography and deep structure of the Black Sea Basin*. Moscow: Nauka (in Russian).

Goptarev, N. P., Simonov, A. I., Zatuchnaya, B. M., & Gershanovich, D. E. (Eds.). (1991). *Hydrology and hydrochemistry of the seas*. Vol. V, the Azov Sea. St. Petersburg: Gidrometeoizdat (in Russian).

Goryachkin, Y. N., & Ivanov, V. A. (2006). *The Black Sea level: Past, present and future*. Sevastopol: MHI NASU (in Russian).

Greze, V. N. (1979). *Bases of biological productivity of the Black Sea*. Kiev: Naukova Dumka (in Russian).

Grinevetsky, S. F., Zonn, I. S., & Zhiltsov, S. S. (2006). *The Black Sea Encyclopedia* (p. 660). Moscow: Mezhdunarodnye Otnosheniya (in Russian).

Ismail-Zadeh, A. (Ed.). (2006). Recent geodynamics, georisk and sustainabe development in the Black Sea to Caspian Sea Region. In *Proceedings of the international workshop on recent geodynamics, georisk and sustainable development in the Black Sea to Caspian Sea Region* (AIP conference proceedings, Vol. 825, 161 p), Baku, Azerbaijan, July 3–6, 2005. Springer.

Ivanov, L., & Oguz, T. (Eds.). (1998). Ecosystem modelling as a management tool for the Black Sea. In *Proceedings of the NATO TU Black Sea project* (NATO science partnership sub-series: 2, Vol. 47, 774 p), Zori Rossii, Ukraine, June 15–19, 1997. Springer.

Izdar, E., & Murray, J. W. (Eds.). (1992). Black Sea oceanography. In *Proceedings of the NATO Advanced Research Workshop* (NATO science series C, Vol. 351, 508 p), Çesme, Izmir, Turkey, October 23–27, 1989. Springer.

Jaoshvili, Sh. (2002). *The rivers of the Black Sea* (Technical Report No. 71). European Environmental Agency.

Keondjyan, V. P., Kudin, A. M., & Terekhin, Y. V. (Eds.). (1990). *Practical ecology of marine regions. The Black Sea*. Kiev: Naukova Dumka (in Russian).

Kosarev, A. N., Tuzhilkin, V. S., Daniyalova, Z. H., & Arkhipkin, V. S. (2004). Hydrology and ecology of the Black and Caspian Seas. In *Geography, society and environment* (Vol. VI, Dynamics and interaction of the atmosphere and the hydrosphere). Moscow: Gorodetz (in Russian).

Kostianoy, A. G., & Kosarev, A. N. (Eds.). (2008). *The Black Sea environment: The handbook of environmental chemistry*. Berlin: Springer. 457 p.

Kotlyakov, V., Uppenbrink, M., & Metreveli, V. (Eds.). (1998). Conservation of the biological diversity as a prerequisite for sustainable development in the Black Sea Region. In *Proceedings of the NATO advanced research workshop* (NATO science partnership sub-series: 2, Vol. 46, 528 p), Batumi, Republic of Georgia, October 5–12, 1996. Springer.

Kovalev, A. V., & Finenko, Z. Z. (Eds.). (1993). *Plankton of the Black Sea*. Kiev: Naukova Dumka (in Russian).

Lagutov, V. (Ed.). (2012). *Environmental security in watersheds: The Sea of Azov* (NATO science for peace and security series C: Environmental security, Vol. VIII, 468 p). Springer.

Matishov, G., Gargopa, Y., Berdnikov, S., & Dzhenyuk, S. (2006a). *The regularities of ecosystem processes in the Sea of Azov*. Moscow: Nauka (in Russian).

Matishov, G., Matishov, D., Gargopa, G., Dashkevich, L., Berdnikov, S., Baranova, O., et al. (2006). Climatic Atlas of the Sea of Azov 2006. In G. Matishov & S. Levitus (Eds.), *NOAA Atlas NESDIS 59*. Washington, DC: U.S. Government Printing Office. http://www.nodc.noaa.gov/OC5/AZOV2006/start.html

Mikhailov, V. N. (Ed.). (2004). *Hydrology of the Danube Delta*. Moscow: GEOS (in Russian).

Mitin, L. I. (Ed.). (2006). *Atlas of the Black Sea and Sea of Azov nature protection*. St. Petersburg: GUNIO MO RF (in Russian).

Mordukhai-Boltovskoi, F. D. (1972). *Guide of the Black and Azov Seas Fauna*. Kiev: Naukova Dumka (in Russian).

Nath, B. (Ed.). (2001). Sustainable solid waste management in the Southern Black Sea Region. In *Proceedings of the NATO advanced research workshop* (NATO science partnership sub-series: 2, Vol. 75, 336 p), Sofia, Bulgaria, September 27–October 1, 1999. Springer.

Neretin, L. N. (Ed.). (2006). *Past and present water column anoxia* (NATO sciences series). Dordrecht: Springer.

Özsoy, E., & Mikaelyan, A. (Eds.). (1997). Sensitivity to change: Black Sea, Baltic Sea and North Sea. In *Proceedings of the NATO advanced research workshop on sensitivity of North Sea, Baltic Sea and Black Sea to Anthropogenic and Climatic Changes* (NATO science partnership sub-series: 2, Vol. 27, 536 p), Varna, Bulgaria, November 14–18, 1995. Springer.

Ryabinin, A. I., & Kravets, V. N. (1989). *Present state of the hydrogene sulphide zone in the Black Sea (1960–1986)*. Moscow: Gidrometeoizdat (in Russian).

Simonov, A. I., & Altman, E. N. (Eds.). (1991). *Hydrometeorology and hydrochemistry of the USSR Seas* (The Black Sea. Issue 1: Hydrometeorological conditions, Vol. IV). St. Petersburg: Gidrometeoizdat (in Russian).

Simonov, A. I., & Ryabinin, A. I. (Eds.). (1996). *Hydrometeorology and hydrochemistry of the Seas* (The Black Sea. Issue 3: Modern state of the Black Sea pollution, Vol. IV). Sevastopol: ECOSEA-Gidrofizika (in Russian).

Simonov, A. I., Ryabinin, A. I., & Gershanovich. D. E. (Eds.). (1992). *Hydrometeorology and hydrochemistry of the USSR Seas* (The Black Sea. Issue 2: Hydrochemical conditions and oceanographic bases of formation of biological productivity, Vol. IV). St. Petersburg: Gidrometeoizdat (in Russian).

Skopintsev, B. A. (1975). *Formation of present chemical composition of the Black Sea waters.* Leningrad: Gidrometeoizdat (in Russian).

Sorokin, Y. I. (1982). *The Black Sea: Nature, resources.* Moscow: Nauka (in Russian).

Sorokin, Y. (2002). *The Black Sea ecology and oceanography.* Leiden: Backhuys Publishers.

Vinogradov, M. E. (Ed.). (1991). *Variability of the Black Sea ecosystem: Natural and anthropogenic factors.* Moscow: Nauka (in Russian).

Vinogradov, M. E., Sapozhnikov, V. V., & Shushkina, E. A. (1992). *The Black Sea ecosystem.* Moscow: Nauka (in Russian).

Yanko-Hombach, V., Gilbert, A. S., Panin, N., & Dolukhanov, P. M. (Eds.). (2007). *The Black Sea flood question: Changes in coastline, climate and human settlement* (Vol. XXVIII). Dordrecht: Springer. 971 p.

Yilmaz, A. (Ed.). (2003). *Oceanography of the Eastern Mediterranean and Black Sea.* Ankara: Tubitak Publishers.

Zaitsev, Y. P. (2006). *An introduction into the Black Sea ecology.* Odessa: Aven (in Russian).

Zaitsev, Y. P., Aleksandrov, B. G., & Minicheva, G. G. (Eds.). (2006). *North-western part of the Black Sea: Biology and ecology.* Kiev: Naukova Dumka (in Russian).

Zaitsev, Y., & Mamaev, V. (1997). *Biological diversity in the Black Sea. A study of change and decline.* New York: UN Publications.

Zats, V. I., & Finenko, Z. Z. (Eds.). (1988). *Dynamics of waters and productivity of the Black Sea.* Moscow: Nauka (in Russian).

Zatsepin, A. G., & Flint, M. V. (Eds.). (2002). *Multi-disciplinary investigations of the North-East part of the Black Sea.* Moscow: Nauka (in Russian).